In Situ and On-Site Bioremediation: Volume 4

BIOREMEDIATION

The *Bioremediation* series contains collections of articles derived from many of the presentations made at the following technical meetings.

First International In Situ and On-Site Bioreclamation Symposium (April 1991, San Diego, California)

1(1) *On-Site Bioreclamation: Processes for Xenobiotic and Hydrocarbon Treatment*
1(2) *In Situ Bioreclamation: Applications and Investigations for Hydrocarbon and Contaminated Site Remediation*

Second International In Situ and On-Site Bioreclamation Symposium (April 1993, San Diego, California)

2(1) *Bioremediation of Chlorinated and Polycyclic Aromatic Hydrocarbon Compounds*
2(2) *Hydrocarbon Bioremediation*
2(3) *Applied Biotechnology for Site Remediation*
2(4) *Emerging Technology for Bioremediation of Metals*
2(5) *Air Sparging for Site Bioremediation*

Third International In Situ and On-Site Bioreclamation Symposium (April 1995, San Diego, California)

3(1) *Intrinsic Bioremediation*
3(2) *In Situ Aeration: Air Sparging, Bioventing, and Related Remediation Processes*
3(3) *Bioaugmentation for Site Remediation*
3(4) *Bioremediation of Chlorinated Solvents*
3(5) *Monitoring and Verification of Bioremediation*
3(6) *Applied Bioremediation of Petroleum Hydrocarbons*
3(7) *Bioremediation of Recalcitrant Organics*
3(8) *Microbial Processes for Bioremediation*
3(9) *Biological Unit Processes for Hazardous Waste Treatment*
3(10) *Bioremediation of Inorganics*

Bioremediation Series Cumulative Indices: 1991-1995

Fourth International In Situ and On-Site Bioremediation Symposium (April 1997, New Orleans, Louisiana)

4(1) *In Situ and On-Site Bioremediation: Volume 1*
4(2) *In Situ and On-Site Bioremediation: Volume 2*
4(3) *In Situ and On-Site Bioremediation: Volume 3*
4(4) *In Situ and On-Site Bioremediation: Volume 4*
4(5) *In Situ and On-Site Bioremediation: Volume 5*

For information about ordering books in the *Bioremediation* series, contact Battelle Press. Telephone: 800-451-3543 or 614-424-6393. Fax: 614-424-3819. Internet: press@battelle.org.

In Situ and On-Site Bioremediation: Volume 4

Toxicity and Geochemical Considerations
Genetically Engineered Microorganisms
Microbial Considerations
Emerging Technologies
In Situ Biobarriers
Nutrient Delivery and Transport in Porous Media
Regulatory, Economic, and Public Perception Issues
Crude Oil Bioremediation and Marine Environmental Considerations
Chemical and Physical Processes in Support of Bioremediation
Bioaugmentation
Microbial Transport in Porous Media

Papers from the Fourth International In Situ and On-Site Bioremediation Symposium
New Orleans, April 28 —May 1, 1997

Symposium Chairs
Bruce C. Alleman and Andrea Leeson
Battelle, Columbus, Ohio

BATTELLE PRESS
Columbus • Richland

Library of Congress Cataloging-in-Publication Data

International In Situ and On-Site Bioremediation Symposium (4th : 1997 :
New Orleans, La.)
In Situ and On-Site Bioremediation : Papers from the Fourth International In Situ and On-Site Bioremediation Symposium. New Orleans, April 28–May 1, 1997 / symposium chairs, Bruce C. Alleman and Andrea Leeson.
p. cm.
Includes bibliographical references and index.
ISBN 1-57477-031-4 (set : alk. paper)
1. Bioremediation--Congresses. 2. In situ remediation--Congresses.
I. Alleman, Bruce C., 1957- . II. Leeson, Andrea, 1962- .
TD192.5.I56 1997
628.5--DC21 97-7991
CIP

Five volume set: ISBN 1-57477-031-4
Volume 1: ISBN 1-57477-026-8
Volume 2: ISBN 1-57477-027-6
Volume 3: ISBN 1-57477-028-4
Volume 4: ISBN 1-57477-029-2
Volume 5: ISBN 1-57477-030-6

Printed in the United States of America

For content and ordering information on books in the Bioremediation series, see "Books from Battelle Press" at www.battelle.org or contact:
Battelle Press
505 King Avenue
Columbus, Ohio 43201, USA
1-614-424-6393 or 1-800-451-3543
Fax: 1-614-424-3819
Internet: press@battelle.org

— CONTENTS —

Foreword xv

Toxicity and Geochemical Considerations
Alison Thomas & Paul E. Flathman, Chairs

Application of Algal Tests on Soil Suspensions for Assessing Potential Bioavailable Toxicity. *A. Baun, N. Nyholm, J.V. Stoyanov, and K.O. Kusk* 1

Toxicity During the Aerobic Biodegradation of Naphthalene and Benzothiophene. *S. Dyreborg, E. Arvin, and H.H. Hansen* 7

Effect of Trichloroethylene Concentration on Oxygen Uptake Rates During Hydrocarbon Biodegradation. *B.A. Adams, R.R. Dupont, W.J. Doucette, and D.L. Sorensen* 9

TCE Product Toxicity and Internal Energy-Storage Effects on Methanotrophs. *K.-H. Chu and L. Alvarez-Cohen* 15

Ecological and Human Toxicology Studies on Weathered Petroleum-Contaminated Soils. *F.C. Hsu, M. Singh, and S. Cunningham* 17

Changes in Toxicity of Fuel-Contaminated Sediments Following Nitrate-Based Bioremediation: Column Study. *S.R. Hutchins, J.A. Bantle, and E.J. Schrock* 19

Survey of the Inherent Remediation and Attenuation Capacity of Phenanthrene in Diesel Fuel Added to Soil. *R.D. Maurice and D.L. Burton* 21

Descriptive Analysis of Iron Fouling in Groundwater Treatment Systems. *M.J. Tinholt and G.R. Wendling* 29

Importance of Soil-Water Relationships in Assessing End Points in Bioremediated Soils. *N. Sawatsky and X. Li* 35

Humic Substance and Iron Mineral Effects on Reductive Dechlorination Reactions. *R. Collins, S. Kim, and F. Picardal* 41

Enzymatic Activity of Soil Polluted by Copper Compounds. *A. Malachowska-Jutsz, K. Miksch, J. Pacha, and J. Surmacz-Górska* 43

Genetically Engineered Microorganisms
Gary S. Sayler, Chair

Recombinant Strain Applications for Bioremediation of Hydrophobic Organics in Soils. *G.S. Sayler* 49

Prospects for Use and Regulation of Transgenic Plants in Phytoremediation. *D.J. Glass* 51

Biodegradation of Aromatic Hydrocarbon Mixtures in Heavily Contaminated Streams by Immobilized Recombinant Microorganisms. *V.A.P. Martins dos Santos and J.L. Ramos* 57

Degradation of Trichloroethylene by Bacterial Mono- and Dioxygenase. *W. Takami, T. Shimizu, S. Imamura, H. Nojiri, H. Yamane, and T. Omori* 59

Transfer of Plasmid RP4:TOL to Legume Microsymbionts for Soil/Rhizosphere Bioremediation. *L.D. Kuykendall, W.K. Gillette, G.P. Hollowell, L.-H. Hou, H.E. Tatem, and S.K. Dutta* 65

Enhancement of Hydrocarbon Degradation and Catabolism in Bacilli by Transformation and Gene Fusion. *W.-H. Yang and J.-R. Yang* 71

Field Application of Genetically Engineered Microorganisms for PAH Bioremediation: Lessons Learned. *R.S. Burlage, A. Palumbo, U. Matrubutham, C. Steward, and G.S. Sayler* 77

Microbial Considerations

Laurie Lapat-Polasko, Chair

Biosurfactant Characterization and Production by *Burkholderia solanacearum* FH2. *P. Morris, S. Rawlin, H. Rogers, and R. Frontera-Suau* 79

Microbial Population Dynamics and Biodegradation Kinetics of Aromatic Hydrocarbon Mixtures. *J.D. Rogers, K.F. Reardon, and N.M. DuTeau* 81

The Mineralization of BTEX Mixtures by Environmental Cultures. *R.A. Deeb and L. Alvarez-Cohen* 83

Organic Matter in Aquifers as Carbon Source for Microorganisms. *C. Grøn, M. Krog, and R. Jakobsen* 85

Production of Extracellular Surface-Active Compounds by Microorganisms Grown on Hydrocarbons. *D.F. Carvalho, D.D. Marchi, and L.R. Durrant* 91

Biofilm Decay After In Situ Bioremediation: Post Treatment Water Quality. *V. Sudirgio, D. Johnstone, D. Yonge, and J. Petersen* 97

Biosurfactant Production by Organisms from an Ecology Study of a JP-4 Jet Fuel-Contaminated Site. *C.L. Bruce* 99

Spatial Distribution of Plants and Soil Microbial Communities in Soil Contaminated with TNT. *G.L. Horst, G. Krishnan, R.A. Drijber, and T.E. Elthon* 101

Microbial Population Dynamics During Hydrocarbon Biodegradation Under Variable Nutrient Conditions. *D.D. Cleland, V.H. Smith, and D.W. Graham* 105

Microbiological Monitoring of Contaminants in a Fractured Basalt Aquifer. *S.P. O'Connell, R.M. Lehman, F.S. Colwell, and M.E. Watwood* 111

Bioremediation Processes Demonstrated at a Controlled Gasoline Release Site. *S.M. Pfiffner, R. Siegrist, D.B. Ringelberg, and A.V. Palumbo* 117

Ecological Impacts of Bioremediation Field Study Experiments. *T.M. Wood, R.L. Lehman, and J. Bonner* 119

Bacterial Productivity in BTEX- and PAH-Contaminated Aquifers. *M.T. Montgomery, T.J. Boyd, B.J. Spargo, R.B. Coffin, and J.G. Mueller* 125

Microbial Nitrification Evidence in Bioremediation Experiments at 13°C. *P. Sacceddu, A. Robertiello, and P. Carrera* 131

Emerging Technologies

Carol D. Litchfield, Chair

Emerging Technologies: State and Future. *C. Litchfield* (Abstract not provided)

Bioremediation of Aqueous Pollutants Using Biomass Embedded in Hydrophilic Foam. *E.W. Wilde, J.C. Radway, J. Santo Domingo, M.J. Whitaker, P. Hermann, and R.G. Zingmark* 133

Thermally Enhanced In Situ Bioremediation of DNAPLs. *J.M. Kosegi, B.S. Minsker, and D.E. Dougherty* 135

Field Demonstration of Oxygen Microbubbles for In Situ Bioremediation. *P.M. Woodhull, D.E. Jerger, D.P. Leigh, R.F. Lewis, and E.S. Becvar* 141

Enhanced Biodegradation of MTBE and BTEX Using Pure Oxygen Injection. *S.R. Carter, J. M. Bullock, and W.R. Morse* 147

In Situ Management System Employing Integrated Funnel-and-Gate/GZB Containment and Recovery Technologies. *M.D. Brourman, M.D. Tischuk, E.J. Klingel, M. Sick, D.J.A. Smyth, E.A. Sudicky, S.G. Shikaze, S.M. Borchert, and J.G. Mueller* 149

Biodegradation and Impact of Phthalate Plasticizers in Soil. *C.D. Cartwright, I.P. Thompson, and R.G. Burns* 155

Biodegradation of Trichloroethane Under Denitrifying Conditions. *J.L. Sherwood, J.N. Petersen, and R.S. Skeen* 157

Bioremediation of Mineral Oil, PAH, and PCB in Dry Solid Reactors. *J. Gemoets, L. Bastiaens, D. Van Houtven, D. Springael, L. Hooyberghs, and L. Diels* 159

Feasibility and Compatibility of Groundwater Circulation Flows. *J. Stamm* 165

A Novel Liquid Foam Carrier for Use in Bioremediation. *A.J. Wilson and W.B.Betts* 167

In Situ Biobarriers

Gregory D. Sayles & Bruce Alleman, Chairs

Natural Attenuation and a Microbial Fence: Re-Engineering Corrective Action. *J.A.M. Abou-Rizk and M.E. Leavitt* 169

Migration Barriers: Sparging and Bioremediation in Trenches. *R.D. Norris and D.J. Wilson* 175

Passive and Semipassive Techniques for Groundwater Remediation. *M.J. Brown, M.L. McMaster, J.F. Barker, J.F. Devlin, D.J. Katic, and S.M. Froud* 181

Arrays of Unpumped Wells for Plume Migration Control or Enhanced Intrinsic Remediation. *R.D. Wilson and D.M. Mackay* 187

Biological Reactive Wall/Enhancement of Instrinsic Conditions. *S.A. Fam, M.F.Messmer, A. Lunt, and K. Marcott* 193

Reduction and Immobilization of Molybdenum by *Desulfovibrio desulfuricans*. *M.D. Tucker, L.L. Barton, and B.M. Thomson* 195

Gasoline Biodegradation in a Permeable Bioreactive Barrier. *L. Yerushalmi, M.F. Manuel, and S.R. Guiot* 197

In Situ Bioscreens. *H.H.M. Rijnaarts, A. Brunia, and M. van Aalst* 203

Semipassive Oxygen Release Barrier for Enhancement of Intrinsic Bioremediation. *S.W. Chapman, B.T. Byerly, D.J. Smyth, R.D. Wilson, and D.M. Mackay* 209

Management of a Hydrocarbon Plume Using a Permeable ORC® Barrier. *J.G. Johnson and J.E. Odencrantz* 215

Characterization of a New Support Media for In Situ Biofiltration of BTEX-Contaminated Groundwater. *D. Forget, L. Deschênes, D. Karamanev, and R. Samson* 221

Physico-Chemical Optimization of Biofilm Development in Fractured Rock Aquifer Conditions. *N. Ross, L. Deschênes, B. Clément, and R. Samson* 227

Active Biofilm Barriers for Waste Containment and Bioremediation: Laboratory Assessment. *M.J. Brough, A. Al-Tabbaa, and R.J. Martin* 233

Bioremediation of Pentachlorophenol-Contaminated Groundwater: A Permeable Barrier Technology. *J.D. Cole, S.L.Woods, P.J. Kaslik, K.J. Williamson, and D.B. Roberts* 239

Studies of Bioclogging for Containment and Remediation of Organic Contaminants. *C.D. Johnston, J.L. Rayner, D.S. De Zoysa, S.R. Ragusa, M.G. Trefry, and G.B. Davis* 241

The Use of Oxygen Release Compound (ORC®) in Bioremediation. *S. Koenigsberg, C. Sandefur, and W. Cox* 247

Sequential Treatment Using Abiotic Reductive Dechlorination and Enhanced Bioremediation. *S. Froud, R.W. Gillham, J.F. Barker, J.F Devlin, M.J. Brown, and M.L. McMaster* 249

Field Trial of an In Situ Anaerobic/Aerobic Bioremediation Sequence. *D.J. Katic, J.F. Devlin, J.F. Barker, M.L. McMaster, and M.J. Brown* 255

Nutrient Delivery and Transport in Porous Media

Rodney S. Skeen & James N. Petersen, Chairs

Site Characterization Methods for the Design of In Situ Electron Donor Delivery Systems. *S.D. Acree, M. Hightower, R.R. Ross, G.W. Sewell, and B. Weesner* 261

Nutrient Transport During Bioremediation of Crude Oil-Contaminated Beaches. *B.A. Wrenn, M.C. Boufadel, M.T. Suidan, and A.D. Venosa* 267

Pulsed Nutrient Injection for Improved Biomass Distribution. *B.M. Peyton, B.S. Hooker, M.J. Truex, and M.G. Butcher* 273

Laboratory- and Field-Scale TCE Biodegradation in Groundwater Under Aerobic Conditions. *L.T. LaPat-Polasko and N.R. Chrisman Lazarr* 275

Feasibility Screening Study for In Situ Bioremediation of Heavy Gas Oil in Refinery Soils. *D. Meo, III; W.T. Frankenberger, Jr.; M. Hillyer; and B.L. McFarland* 277

The Role of Soil Nitrogen Concentration in Bioremediation. *J.L. Walworth, C.R. Woolard, J.F. Braddock, and C.M. Reynolds* 283

Remedial Strategy for Petroleum Hydrocarbons: Enhanced Intrinsic Bioremediation. *C.H. Nelson, C.S. Wright, J.E. Goetz, and K. Van Rijn* 289

Case Study of Bioventing Including Nutrient Addition at Kincheloe AFB. *M.K. O'Mara* 291

Assessment of Nutrient-Contaminant Carbon Ratios for Enhancing In Situ Bioremediation. *R.B. Coffin, M.T. Montgomery, C.A. Kelley, and L.A. Cifuentes* 297

Enhanced Bioremediation of Soil and Groundwater Using Nutrient Injection. *R.S. Porter; M. Ghetti; W.S. Anderson, III; and S. Johnson* 299

Regulatory, Economic, and Public Perception Issues

Karl Nehring

Market Overview of the Bioremediation Market. *O.R. Jennings* 305

International Bioremediation: Recent Developments in Established and Emerging Markets. *D.J. Glass, T. Raphael, and J. Benoit* 307

Cost Evaluation of Anaerobic Bioremediation vs. Other In Situ Technologies. *G.E. Quinton, R.J. Buchanan, D.E. Ellis, and S.H. Shoemaker* 315

Cost Analysis of Risk-Based Corrective Action. *K.L. Davis, C. Kiernan, and G.D. Reed* 317

New Approaches Towards Promoting the Application of Innovative Bioremediation Technologies. *G.G. Broetzman, M.J. Chacón, P.W. Hadley, D.S. Kaback, R.W. Kennett, and N.J. Rosenthal* 323

Public Perception of Environmental Biotechnology: The Canadian Perspective. *K. Devine and T. McIntyre* 329

Practical Experience in Landfarming of Gasoline-Contaminated Soils. *D.P. Dunn, L.L. Schneider, and D.D. Dubrock* 335

Growing Markets of Environmental Protection in China. *Y.-Y. Zheng and L.-K. Wen* 337

Cost Effectiveness of Selected Remediation Technologies and Design Protocols. *J.C. Parker and M. Islam* 341

Interstate Acceptance of In Situ Bioremediation Technologies. *P.W. Hadley, L.C. Rogers, and S.R. Hill* 347

Risk-Based In Situ Bioremediation Design. *J. B. Smalley and B.S. Minsker* 353

Crude Oil Bioremediation and Marine Environmental Considerations

Roger Prince & Kenneth Lee, Chairs

Rates of Hydrocarbon Biodegradation in the Field Compared to the Laboratory. *A.D. Venosa, J.R. Haines, and E.L. Holder* 359

Marine Oil Spills: Enhanced Biodegradation with Mineral Fine Interaction. *K. Lee, A.M. Weise, and T. Lunel* 365

Application of Wastewater Sludge to Microbial Degradation of Crude Oil. *H. Maki, M. Ishihara, and S. Harayama* 371

Transformations of Polyaromatic Hydrocarbons in Sulfate- and Nitrate-Reducing Enrichments. *K.J. Rockne, H.D. Stensel, and S.E. Strand* 377

Effects of Crude Oil Contamination and Bioremediation in a Soil Ecosystem. *K. Lawlor, K. Sublette, K. Duncan, E. Levetin, P. Buck, H. Wells, E. Jennings, S. Hettenbach, S. Bailey, J.B. Fisher, and T. Todd* 383

Crude Oil Biodegradation in Saturated and Vadose Soil: Laboratory Simulation. *P. Carrera, P. Sacceddu, and A. Robertiello* 385

Biological Treatment of Highly Weathered Crude Oil-Affected Soils. *B.M. Haikola, M.Q. Henley, and R.M. Kabrick* 391

Demonstrated Cost Effectiveness of Bioventing at a Large Crude Oil-Impacted Site. *H.J. Reisinger, S.A. Mountain, V. Owens, J. Godfrey, D. Arlotti, G. Andreotti, and G. Di Luise* 393

Laboratory Simulation of the Bioremediation of Oil-Polluted Sand Beaches. *C. Dalmazzone and D. Ballerini* 395

Field Evaluation of Bioremediation to Treat Crude Oil on a Mudflat. *R.P.J. Swannell, D.J. Mitchell, D. M. Jones, A. Willis, K. Lee, and J. E. Lepo* 401

On-Site Bioremediation of Oil Sludge/Crude Oil-Contaminated Soil. *B. Lal and S. Khanna* 407

In Situ Bioremediation of Oil Sludges. *C. Infante, M. VialeRigo, M. Salcedo, J. Rodríguez, A. Melchor, E. Bilbao, and R. Arias* 409

BIOREN: Recent Experiment on Oil-Polluted Shoreline in Temperate Climate. *S. Lefloch, M. Guillerme, P. Tozzolino, D. Ballerini, C. Dalmazzone, and T. Lundh* 411

Assessment of Mixed Cultures for Bioremediation Product Testing. *J.R. Haines, E.L. Holder, and A.D. Venosa* 419

Sorption and Biodegradation Interactions During Phenanthrene Removal in Marine Sediment. *H.D. Stensel, T. Poeton, and S. Strand* 425

An Assessment of Organic Contaminant Biodegradation Rates in Marine Environments. *B. Krieger-Brockett, J.W. Deming, and R.P. Herwig* 427

In Situ Bioremediation Under Saline Conditions. *J. Raumin, B. Bosshard, and M. Radecki* 433

Biodegradation and Volatilization of Saudi Arabian Crude Oil. *S.E. Whiteside and S.K. Bhattacharya* 439

Mesocosm Assays of Oil Spill Bioremediation. *R. Santas, A. Korda, A. Tenente, E. Gidarakos, K. Buchholz, and P. Santas* 445

Biological Remediation of Oil Spills of the Saronikos Gulf, Greece. *A. Korda, A. Tenente, P. Santas, E. Gidarakos, M. Guillerme, and R. Santas* 451

Chemical and Physical Processes in Support of Bioremediation

Soon Haing Cho, Chair

Combining Oxidation and Bioremediation for the Treatment of Recalcitrant Organics. *R.A. Brown, C. Nelson, and M. Leahy* 457

Dechlorination by Methanogens or Co-Factors in the Presence of Iron. *P.J. Novak and G.F. Parkin* 463

Enhanced Biotransformation of Carbon Tetrachloride by an Anaerobic Enrichment Culture. *S.A. Hashsham and D.L. Freedman* 465

Biocompatibility of the Vitamin B_{12}-Catalyzed Reductive Dechlorination of Tetrachloroethylene. *K. Millar and S. Lesage* 471

Application of Chemical Processes for the Treatment of Leachate from Solid Waste Landfill. *S.H. Cho, Y.S. Choi, J.Y. Yoon, H.C. Yoo, and E.S. Lee* 477

Catalytic Stimulation of Perchloroethylene Reductive Dechlorination. *K.D. Schaller, B.D. Lee, W.A. Apel, and M.E. Watwood* 483

Applications of Fenton Oxidation Technologies to Remediate TNT- and RDX-Contaminated Water and Soil. *S.D. Comfort, Z. Li, M. Arienzo, E. Bier, and P.J. Shea* 485

The Role of Fenton's Reagent in Soil Bioremediation. *K.E. Stokley, E.N. Drake, R.C. Prince, and G.S. Douglas* 487

Bioaugmentation

James K. Fredrickson & Mary F. Deflaun, Chairs

Field Pilot Study of Bioaugmentation for Remediation of TCE Contamination in Fractured Bedrock. *M. Walsh, R.J. Steffan, and M.F. DeFlaun* 493

Inducer-Free Microbe for TCE Degradation and Feasibility Study in Bioaugmentation. *T. Imamura, S. Kozaki, A. Kuriyama, M. Kawaguchi, Y. Touge, T. Yano, E. Sugawa, Y. Kawabata, H. Iwasa, A. Watanabe, M. Iio, and Y. Senshu* 495

Bioaugmentation with *Burkholderia cepacia*: Trichloroethylene Cometabolism vs. Colonization. *J. Munakata-Marr, V.G. Matheson, L.J. Forney, J.M. Tiedje, and P.L. McCarty* 501

Evaluation of Bioaugmentation to Remediate an Aquifer Contaminated with Carbon Tetrachloride. *M.J. Dybas, S. Bezborodinikov, T. Voice, D.C. Wiggert, S. Davies, J. Tiedje, C.S. Criddle, O. Kawka, M. Barcelona, and T. Mayotte* 507

Aerobic Bioremediation of TCE-Contaminated Groundwater: Bioaugmentation with *Burkholderia cepacia* $PR1_{301}$. *A.W. Bourquin, D.C. Mosteller, R.L. Olsen, M.J. Smith, and K.F. Reardon* 513

Biodegradation of Dioxins in Soil. *R.U. Halden, B.G. Halden, and D.F. Dwyer* 519

Environmental Restoration Through Introduction of Microbial Consortia. *J. Yoshitani* 521

Approaches to Creation of Bacterial Consortium for Efficient Bioremediation of Oil-Contaminated Soil. *M.U. Arinbasarov, A.V. Karpov, S.G. Seleznev, V.G. Grishchenkov, and A.M. Boronin* 523

Pilot Cleanup of Chernaya River (Moscow Region) from Oil Pollution. *V.P. Murygina, M.U. Arinbasarov, E.V. Korotayeva, A.V. Stolyarova, and L.R. Peterson* 529

Remediation of Pentachlorophenol-Contaminated Soil by Bioaugmentation Using Soil Biomass-Activated in a Bioreactor. *C. Barbeau, L. Deschênes, Y. Comeau, and R. Samson* 535

In Situ Deep Soil Bioremediation of Petroleum Hydrocarbons. *V.H. Bess and R.P. Murray* 541

Low-Intervention Soil Remediation Approaches for Natural Gas Pipeline Facilities. *R.J. Portier and S.R. Chitla* 543

Evaluation of Commercial Products Used to Bioremediate Railroad Ballast. *N.R. Chrisman Lazarr, L.T. LaPat-Polasko, J. Heinicke, E.H. Honig, and B.D. Stewart* 549

Biostimulation and Bioaugmentation of Anaerobic Pentachlorophenol Degradation. *S. Zou, K.M. Anders, and J.F. Ferguson* 551

Microbial Transport in Porous Media

Brent M. Peyton, Chair

Rates and Mechanisms for Modeling Microbial Transport for Bioremediation Systems. *H.S. Fogler, D. Maurer, and D.-S. Kim* 553

Microbial Transport in a Pilot-Scale Biological Treatment Zone. *M.A. Malusis, D.J. Adams, K.F. Reardon, C.D. Shackelford, D.C. Mosteller, and A.W. Bourquin* 559

Bacterial Attachment and Transport Through Porous Media: The Effects of Bacterial Cell Characteristics. *K.L. Duston, M.R. Wiesner, and C.H. Ward* 565

Enhancing Biocolloid Transport to Improve Subsurface Remediation. *B.E. Logan, T. Camesano, B. Rogers, and Y. Fang* 567

Quantitative Characterization of Bacterial Migration Through Porous Media. *R.M. Ford, M. Jin, P.T. Cummings, and K.C. Chen* 573

Bioaugmentation and Numerical Simulation of Carbon Tetrachloride Transformation in Groundwater. *M.E. Witt, D.C. Wiggert, M.J. Dybas, K.C. Kelly, and C.S. Criddle* 575

Author Index **581**

FOREWORD

This is one of five volumes published in conjunction with the Fourth International In Situ and On-Site Bioremediation Symposium, held in New Orleans, Louisiana in April 1997. The volumes contain technical notes or abstracts, as received from the authors, for 100% of the 653 platform and poster presentations scheduled for the Symposium program as of February 28, 1997, when the volumes were submitted to the publisher.

The growth of the bioremediation field is reflected in the growth in numbers of abstracts submitted for the program. We received approximately 800 abstracts to be considered for the 1997 Symposium. This was a 20% increase over the number submitted for the 1995 Symposium and a 500% increase over the first Symposium, held in 1991. During these six years, we have seen bioremediation progress from an innovative to an established field, with bioremediation technologies now in common use as remedial alternatives for site closure. Innovative bioremediation technologies are being developed for recalcitrant compounds previously thought to be nearly impossible to bioremediate, while technologies for the more readily biodegradable compounds are being optimized. This Symposium provides a comprehensive forum for research being conducted in all areas of bioremediation, encompassing laboratory, bench-scale, and full-scale field studies for a variety of technologies and contaminants.

Despite the increased number of abstracts submitted for this year's Symposium, we determined that it was not feasible to increase the size of the Symposium program. Therefore, for the first time, it was necessary to limit the number of presentations accepted for the program. This proved to be difficult—good abstracts were submitted for which we simply did not have space. A Technical Review Committee, made up of experienced bioremediation researchers, assisted us in the selection process. Each of the 800 abstracts submitted to the Symposium was reviewed by at least four members of the Technical Review Committee and evaluated on technical merit and relevance to the scope of the Symposium. Once the technical review had reduced the number being considered, abstracts were tentatively assigned to sessions and then sent to the chairs and co-chairs of those sessions for final recommendations on placement. We conducted a final review of all decisions. We are indebted to all who donated time to assist in continuing the high quality of the Symposium. Their input was crucial in selecting the abstracts to be presented. The Technical Review Committee was composed as follows:

C. Marjorie Aelion *(University of South Carolina)*
Robert G. Arnold *(University of Arizona)*
Bruce Bauman *(American Petroleum Institute)*
Janet E. Bishop *(URS Consultants)*
Charles A. Bleckmann *(Air Force Institute of Technology)*
Christian F. Bocard *(Institut Français du Pétrole)*
Robert Booth *(Water Technology International Corporation)*

Robert C. Borden *(North Carolina State University)*
Fred Brockman *(Battelle)*
Bruce Buxton *(Battelle)*
Abraham Chen *(Battelle)*
Soon Haing Cho *(Ajou University)*
Gordon Cobb *(Environ)*
Stuart Coffa *(Geo & Hydro Milieu)*
Margaret Findlay *(Bioremediation Consulting, Inc.)*
Christopher D. Finton *(Battelle)*
Paul E. Flathman *(OHM Remediation Services Corporation)*
Eric A. Foote *(Battelle)*
Arun R. Gavaskar *(Battelle)*
James T. Gibbs *(Battelle)*
John A. Glaser *(U.S. Environmental Protection Agency)*
William J. Guarini *(Envirogen, Inc.)*
Joop Harmsen *(DLO Winand Staring Centre)*
Ed Heyse *(Air Force Institute of Technology)*
Pat Hicks *(Fluor Daniel GTI)*
Robert E. Hinchee *(Parsons Engineering Science)*
Daniel Janke *(Battelle)*
Richard Johnson *(Oregon Graduate Institute)*
Ben Keet *(Geo & Hydro Milieu)*
Richard Lamar *(U.S. Dept. of Agriculture)*
Kenneth Lee *(Fisheries & Oceans Canada)*
Dennis J. Lew *(General Physics Corp.)*
Carol D. Litchfield *(George Mason University)*
Bruce E. Logan *(The University of Arizona)*
Victor S. Magar *(Battelle)*
David McWhorter *(Colorado State University)*
Jeffrey Means *(Battelle)*
Ralph Moon *(HSA Environmental, Inc.)*
Thomas G. Naymik *(Battelle)*
Karl Nehring *(Battelle)*
Robert D. Norris *(Eckenfelder, Inc.)*
Robert Olfenbuttel *(Battelle)*
Say Kee Ong *(Iowa State University)*
James N. Petersen *(Washington State University)*
Brent M. Peyton *(Battelle)*
Flynn W. Picardal *(Indiana University)*
Albert J. Pollack *(Battelle)*
Roger Prince *(Exxon Research & Engineering)*
P. H. (Hap) Pritchard *(U.S. Naval Research Laboratory)*
Michele Puchalski *(Minnesota Dept. of Agriculture)*
H. James Reisinger *(Integrated Science & Technology, Inc.)*
Bruce Sass *(Battelle)*
Nagappa Sathish *(CH2M Hill)*
Gregory D. Sayles *(U.S. Environmental Protection Agency)*
Ronald C. Sims *(Utah State University)*
Rodney S. Skeen *(Battelle)*
Larry Smith *(Battelle)*
Alison Thomas *(U.S. Air Force)*
Catherine Vogel *(U.S. Air Force)*
F. Michael von Fahnestock *(Battelle)*
Peter Werner *(Technical University of Dresden)*
John T. Wilson *(U.S. Environmental Protection Agency)*
Patrick M. Woodhull *(OHM Remediation Services Corp.)*
Robert Wyza *(Battelle)*
George Yu *(Battelle)*
Thomas C. Zwick *(Battelle)*

The author of each presentation accepted for the Symposium was invited to prepare a technical note, to be no longer than six pages and submitted in final copy, formatted according to specifications provided by the Symposium Organizing Committee. Every technical note received by the Committee appears in these volumes as submitted by the author—no peer review, copy editing, or typesetting was performed. Because the technical notes were published as received, differences in national convention and personal style lead to variations in such matters as word usage, spelling, abbreviation, manner in which numbers and measurements are presented, and type style and size. Presentations for which neither technical notes nor revised abstracts were received are represented by the abstracts the authors submitted in summer 1996 to be considered for placement in the program.

The technical notes and abstracts are grouped in the five volumes according to the sessions in which the corresponding presentations will be made at the Symposium. For convenience, the table of contents in each volume matches the sequence of the program schedule, with the platform papers listed first in each session, followed by the poster presentations. The session titles are listed below according to volume. The chairs and co-chairs for each session are listed in the table of contents for each volume.

VOLUME 1—Sessions A1 through A8

Natural Attenuation of Petroleum Hydrocarbons
Air Sparging and Related Technologies
Process Monitoring of Petroleum Biodegradation in Soil
Bioslurping
Cold Region Applications
Bioventing Applications and Extensions
Integrated Approaches to Bioremediation
Biopiles

VOLUME 2—Sessions B1 through B10

Explosives and Nitroaromatics
Composting of Contaminated Soils and Sludges
Natural Attenuation of Recalcitrant Compounds
Landfarming
Polycyclic Aromatic Hydrocarbons
Pesticides and Herbicides
Field Methods and Process Monitoring
PCBs and Chlorinated Aromatics
Fungal Technologies
Surfactant-Aided Bioremediation

VOLUME 3—Sessions C1 through C8

Anaerobic Degradation of Chlorinated Solvents
Aerobic Degradation of Chlorinated Solvents
Natural Attenuation of Chlorinated Solvents
Anaerobic/Aerobic Processes for Chlorinated Solvent Degradation
Remediation Technologies Development Forum
Phytoremediation
Inorganics
Manufactured Gas Plant Site Remediation

VOLUME 4—Sessions D1 through D11

Toxicity and Geochemical Considerations
Genetically Engineered Microorganisms
Microbial Considerations
Emerging Technologies
In Situ Biobarriers
Nutrient Delivery and Transport in Porous Media
Bioaugmentation
Regulatory, Economic, and Public Perception Issues
Crude Oil Bioremediation and Marine Environmental Considerations
Chemical and Physical Processes in Support of Bioremediation
Microbial Transport in Porous Media

VOLUME 5—Sessions E1 through E13

Alternative Electron Acceptors
Bioreactors
Bioslurry Reactors
Vapor-Phase Bioreactors
Biological Treatment of Wastewater
Biological Treatment of Landfill Leachate
Molecular Monitoring of Bioremediation Potential
Environmental Statistics for Site Remediation
Bench-Scale to Full-Scale Interpretations
In Situ and On-Site Case Histories
Low-Permeability Soils Applications
Modeling
Bioavailability Considerations

We would like to thank the Battelle staff who assembled this book and its companion volumes and prepared them for printing. Carol Young, Timothy Lundgren, Gina Melaragno, Lynn Copley-Graves, and Loretta Bahn spent many long hours on production tasks—developing the detailed format specifications sent to each author; examining each technical note and abstract to ensure that it met basic page layout requirements and making adjustments when necessary; assembling the volumes; applying headers and page numbers; compiling the tables of contents and author index; and performing a final check of the pages before submitting them to the publisher. Joseph Sheldrick, manager of Battelle Press, provided valuable advice during production planning and coordinated with the printer; he and Daniel Muko designed the volume covers.

The Symposium was sponsored by Battelle Memorial Institute with support from the following co-sponsors and participating organizations.

Ajou University, College of Engineering
American Petroleum Institute
Asian Institute of Technology
Biotreatment News
Fluor Daniel GTI
Gas Research Institute (GRI)
Institut Français du Pétrole
The Japan Research Institute, Limited
Mitsubishi Corporation
National Center for Integrated Bioremediation Research and Development
OHM Remediation Services Corporation
Parsons Engineering Science, Inc.
Umweltbundesamt (Germany)
U.S. Air Force Armstrong Laboratory Environics Directorate
U.S. Air Force Center for Environmental Excellence
U.S. Environmental Protection Agency
U.S. Naval Facilities Engineering Service Center
Waste Management, Inc.
Water Technology International Corporation
Western Region Hazardous Substance Research Center

Neither Battelle nor the co-sponsors or supporting organizations reviewed the materials published in these volumes, and their support for the Symposium should not be construed as an endorsement of the content.

Andrea Leeson & Bruce Alleman
February 1997

APPLICATION OF ALGAL TESTS ON SOIL SUSPENSIONS FOR ASSESSING POTENTIAL BIOAVAILABLE TOXICITY

Anders Baun, Niels Nyholm, Jivko V. Stoyanov, and K. Ole Kusk
(Groundwater Research Center/Department of Environmental Science and Engineering, Technical University of Denmark, Denmark)

ABSTRACT: An algal test procedure with soil suspensions is proposed and evaluated. The freshwater green alga *Selenastrum capricornutum* was used as test organism. Five unpolluted Danish soils were tested and no growth reduction was detected for soil contents as high as 20 g/L. Comparisons between toxicities of soil elutriates and soil suspensions were made. The direct application of PAH-contaminated soil in the tests showed toxic responses 1-2 orders of magnitude higher than the elutriates of the same soils. Algal tests on unpolluted soil spiked with the compounds: Atrazine, 3,4-dichloroaniline, pentachlorophenol, and phenanthrene yielded toxicities slightly less than those obtained with aqueous solutions of the pure compounds.

INTRODUCTION

In comparison to sophisticated chemical-analytical procedures, simple bioassays provide a more direct measure of environmental relevant toxicity of contaminated soils. Ecotoxicological tests performed directly on soil give information on the total toxicity of samples and may facilitate evaluation of potential bioavailability of pollutants. Aqueous elutriates of soils and sediments are commonly used for toxicity testing, even though elutriates in many cases may underestimate the bioavailable fraction of sorbed compounds. On the other hand, extractions with organic solvents possibly overestimate bioavailability. Traditionally, toxicity testing directly on soil is carried out with higher plants or terrestical organisms (for instance, earthworms and springtails), but also some screening methods for direct solid phase toxicity testing using microorganisms are available (e.g. *Photobacterium phosphoreum* and *Escherichia coli*). Algal tests on suspensions obtained after a standard aqueous extraction could be better suited for an initial toxicity assessment than most of the existing tests, because of likely high sensitivity, potential cost-effectiveness, and easy quantification. The study of Miller *et al.* (1985) showed that algal tests were the most sensitive to the constituents in soil elutriates of five short term bioassays studied. In a recent review, Keddy *et al.* (1995) recommend to include the freshwater alga *Selenastrum capricornutum* as test organism in biotest batteries for screening and definitive testing of soil elutriates in relation to soil quality assessment.

The idea behind the present study is, that algal test on soil suspensions may be used for assessing the potential bioavailable toxicity, because a very large biological surface is in direct contact with contaminated soil particles. The suggested practical procedure is to prepare an aqueous elutriate and a concentrated suspension at the same time. Instead of proceeding only with the elutriate, also the suspension is tested. The feasibility of this procedure depends on, how addition of uncontaminated soil to the aqueous medium will affect algal growth and whether suspensions of contaminated soil will inhibit algal growth rate.

MATERIALS AND METHODS

Tests were carried out with 7 different soils with the characteristics listed in table 1. Prior to testing, the soils were sieved through a 2 mm mesh. To prevent loss of volatile compounds, the soils were not dried before application in tests. The Gladsaxe soil originates from a Danish gaswork and is contaminated with creosote and diesel oil. The source of the Allerød soil is a former gas station, thus contamination is mainly gasoline and diesel oil. Additional chemical analysis (water suspensions extracted with dichloromethane and analyzed by GC-FID) were carried out to assess the total concentration of alifatic and mono-aromatic hydrocarbons (ΣHC) and selected polyaromatic hydrocarbons (ΣPAH: Naphthalene, Phenanthrene, Flourathene, Pyrene, Benz(a)anthracene, Crysene, and Benz(a)pyrene). Chemical analysis were carried out by accredited Danish analytical laboratories.

TABLE 1. Soil characteristics for the soils applied in algal tests.

	% Clay (<2 μm)	% Silt (2-20 μm)	% Fine sand (20-200 μm)	% Coarse sand (200-2000 μm)	% Humus
Jyndevad	3	4	19	72	1-7
Roskilde	10	17	49	21	2.5
Flakkebjerg	16	20	39	21	3.2
Svanholm	8	14	47	29	2
DTU	6	7	68	17	2
Gladsaxe	15	10	43	30	2
Allerød	9	1	60	28	2

	pH (H_20)	pH (0.01M $CaCl_2$)	% Water	% Loss of ignition	ΣPAH (mg/kg)	ΣHC (mg/kg)
Jyndevad	7.0	5.9	1.9	3.7	*	*
Roskilde	7.2	6.4	3.5	3.4	*	*
Flakkebjerg	6.5	6.1	4.1	4.7	*	*
Svanholm	6.9	6.8	2.0	8.4	*	*
DTU	7.0	6.6	32	3.5	*	*
Gladsaxe	7.2	7.1	21	6.2	2690	16700
Allerød	7.1	6.9	22	2.3	n.a.	410-2000

*: uncontaminated reference soils
n.a.: not analyzed

Soil elutriates and suspensions for algal tests were prepared, placing 200 g of soil in 800 ml of ISO-medium (ISO, 1989). The slurry was treated with ultra-sound (47 kHz for 5 min.) and placed on a shaking table (100 rpm) at 15 °C for 24 hours. The pH of the suspension was adjusted to pH 7.0. A part of the slurry was centrifuged at 5000 rpm for 10 min. Both for this elutriate and the soil suspension, ISO-medium was used to make dilution series. Test flasks were inoculated with an exponentially growing pre-culture of *Selenastrum capricornutum* to 10^4 cells per mL. Lag-phases during the first day of incubation were prevented by applying 2 g/L of uncontaminated soil (DTU soil) to the pre-culture. Algal growth inhibition tests were conducted according to ISO (1989) either as open mini-scale tests (Arensberg *et al.*, 1995) or as closed tests with the modifications described by Halling-Sørensen *et al.* (1996). Using the latter method, testing of volatile pure compounds and soils contaminated with gasoline and diesel oil was possible.

For all tests, five to seven concentrations of elutriate or soil in three replicates were used. Three controls containing only algae and ISO-medium were used. The test flasks were placed in continous light (60-70 $\mu E/m^2/s$) in a temperature controlled chamber (22 ± 1 °C). From each flask, samples were withdrawn at 0, 24, and 48 hours. The samples were extracted in acetone/DMSO (1:1 v/v) and measured on a Hitachi F-2000 Fluorescence Spectrophotometer (Exitation: 420 nm, Emission: 671 nm). Background fluorescence in non-inoculated blanks was substracted from the fluorescence in test flasks. Growth rate inhibition was used as endpoint and concentration-response curves were described by the Weibull equation, which was fitted to data using weighted non-linear regression applying a computer program developed by Andersen (1994).

Additional algal growth inhibition tests were carried out on four suspensions of DTU soil each spiked with one of the compounds: Atrazine, 3,4 Dichloroaniline, Pentachlorophenol (PCP) or Phenanthrene. All chemicals used were analytical grade. To obtain equlibrium, the spiked suspension (10-20 g soil/L) was shaken for 24 hours at 100 rpm (15 °C). In order to assess the phase distribution, Atrazine, PCP and Phenanthrene were also added in ^{14}C-labelled form. Samples containing ^{14}C-labelled compounds were filtered through Whatman GF/C filters (approx. 1.2 µm). Filters were placed in 20 ml scintillation vials with 5 ml Ecoscint A (*National Diagnostics*). The filtrate (2-5 ml) was collected and mixed with 10 ml Optiphase HiSafe (*Wallac Scintillation Products*). Samples were counted on a Tri-Carb 2000 Liquid Scintillation Counter (*United Technologies Packard*).

RESULTS AND DISCUSSION

The influence of uncontaminated soil on algal growth was studied on the reference soils listed in table 1. Suspensions of up to 40 g/L were examined. Figure 1 shows that growth was not inhibited by soil contents up to 20 g/L of the DTU soil. In fact, the presence of soil slightly stimulated growth. For all reference soils similar results were obtained. Above 20 g/L the suspensions were so dense that an even distribution of light in the test medium was unlikely. From the results it was not possible to conclude, if soil contents higher than 20 g/L was the sole source of the reduced growth rates. At these high soil contents the background fluorescence was very high, which made proper determinations of algal biomass uncertain. Therefore, a practical upper limit for soil suspensions was set at 20 g/L and almost constant growth rates were obtained using uncontaminated soil.

Figure 2a shows the results obtained testing the contaminated Gladsaxe soil as aqoeous elutriates compared to results obtained with unpolluted DTU soil elutriates. Results are expressed as the corresponding amount of soil used for

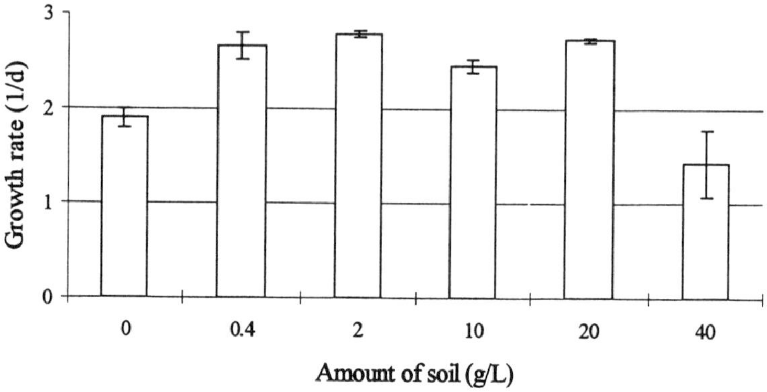

FIGURE 1. Effect of uncontaminated soil suspensions (DTU soil) on growth rate of *Selenastrum capricornutum*. Bars indicate standard deviation on triplicate determinations.

preparation of the elutriate. Exposed to more than to 12.5 g soil/L of the elutriates from the Gladsaxe soil, the algal growth was totally inhibited. Figure 2b shows test results by direct application of the Gladsaxe and DTU soils in test flasks. It was observed that 0.16 g/L of the contaminated soil inhibited algal growth completely. The tests on the highly contaminated Gladsaxe soil, showed that tests of soil suspensions were more than 78 times more sensitive than tests on the respective elutriates. No final explanations for higher toxicities of soil suspensions as compared to elutriates can be given by this study, but have to be further investigated. The mechanisms involved in revealing particulate bound toxicity in an algal test on suspensions could be 1) solubilization of toxicants mediated by the algae (via algal/particular contact or resolubilization from algal excretion products), or 2) direct interactions between particulate bound toxicants and the algal surface.

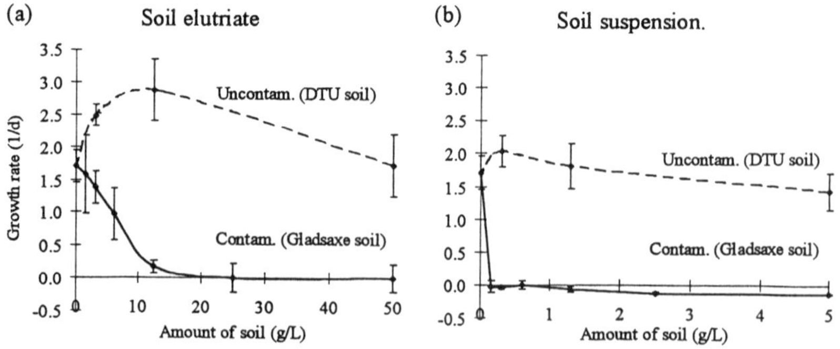

FIGURE 2. Toxicity of contaminated and uncontaminated soil elutriate (a) and the respective soil suspensions (b) in 48 hours algal test. Note: Different scale on the primary axis.

Algal tests on Allerød soil suspensions showed no toxicity when up to 20 g/L soil was tested. Testing the elutriate of the Allerød soil, 35% growth rate inhibitions were seen at a concentartion corresponding to 200 g/L. In this case the higher exposure concentrations possible in elutriate tests enabled detection of toxicity. On both contaminated soils investigated in the present study, significant toxicities were detected in the aqueous elutriates, therefore the proposed method should be used as an addition to "traditional" elutriate testing, not as an alternative.

The toxicity modifying effect of soil in algal tests were investigated using the DTU reference soil and four pure chemicals. Table 2 shows the EC_{50} of tests carried out on four different suspensions of the DTU soil each spiked with one of the pure compounds: Atrazine, 3,4 Dichloroaniline, PCP, and Phenanthrene. For PCP the presence of soil did not reduce toxicity, whereas 25-40% reductions were observed for 3,4 Dichloroaniline, Atrazine and Phenanthrene. The EC_{50}'s on soil suspensions were in all cases in the same order of magnitude as the respective EC_{50} from test on the pure compound.

TABLE 2: EC_{50}-values and 95% confidence interval for pure compounds and suspensions of DTU soil spiked with the pure compounds in the 48 hours *Selenastrum capricornutum* growth inhibition test at pH 7.0.

	Pure compound	Soil suspension
Atrazine	105 µg/L $[93;117]_{95\%}$	159 µg/L $[145;173]_{95\%}$
3,4 Dichloroaniline	1340 µg/L $[1270;1410]_{95\%}$	1900-2000 µg/L $[1800;2100]_{95\%}$ [1]
Pentachlorophenol	22 µg/L $[20;25]_{95\%}$	18 µg/L $[13;49]_{95\%}$
Phenanthrene	302 µg/L $[272;334]_{95\%}$ [2]	498 µg/L $[462;537]_{95\%}$

[1]Roskilde, Flakkebjerg, and Jyndevad soils
[2]ref. Halling-Sørensen *et al.*, 1996.

Related to dissolved concentration of Phenanthrene in the soil suspension, as estimated from radiolabelled counting, the EC_{50} is 244 µg/L with $[185;321]_{95\%}$, which is not significant different from the test results obtained with Phenanthrene in aqueous solution (EC_{50}=302 µg/L $[272;334]_{95\%}$). This indicates that the lower toxicity of Phenanthrene in the soil suspension can be explained by the phase distribution of the compound, i.e. sorbed Phenanthrene had less toxic effect than dissolved Phenanthrene. There was no significant difference between the EC_{50}'s obtained for PCP with and without soil added, therefore it is concluded that the PCP sorbed on soil may also exibit toxicity and that this toxicity is in the same order of magnitude as the PCP in aqueous solution.

CONCLUSIONS

The present study shows, that algal test can be carried out on soil suspensions with a practial upper limit of 20 g/L under the test conditions used.

For the tested unpolluted Danish soils, no growth inhibition was detected at soil contents up to 20 g/L. Using algal tests on soil suspensions, the presence of bioavailable toxic contaminants in the solid phase could be detected. Contaminated Danish soils were very toxic towards algae and tests on soil suspensions were 1-2 orders of magnitude more sensitive than tests on aqueous elutriates. This difference indicate that algal tests on soil suspensions may provide information on the potential bioavailable toxic fraction of soil contaminants. Algal tests on unpolluted soil spiked with the compounds: Atrazine, 3,4-dichloroaniline, pentachlorophenol, and phenanthrene yielded toxicities slightly less than or equal to those obtained with aqueous solutions of the pure compounds. Used a screening method we recommend algal testing on soil suspensions as a supplement to elutriate testing due to its high sensitivity and reproducebility. If further developed and automated algal testing could be a cost-effective screening tool for contaminated soil, supporting and guiding chemical analysis and contributing to remediation decisions.

ACKNOWLEDGEMENTS

Part of this work was supported by a grant from the Technical Research Council of Denmark. Vicky Hemmingsen and Pia Arensberg are acknowledged for initial investigations on application of soil in algal test systems. Lene Kløft and Birte Ebert are acknowledged for their technical assistance.

REFERENCES

Andersen, H (1994). "Statistical Methods for the Assessment of the Toxicity of Wastewater" (in Danish). M.S. Thesis 7/94. Institute of Mathematical Modelling, Technical University of Denmark, Lyngby, Denmark.

Arensberg, P., V. Hemmingsen, and N. Nyholm (1995). "A Miniscale Algal Toxicity Test." *Chemosphere*, 30, 2103-2115.

Halling-Sørensen, B., N. Nyholm, and A. Baun (1996). "Algal Toxicity Tests with Volatile and Hazardous Compounds in Air-tight Test Flasks with CO_2 Enriched Headspace." *Chemosphere*, 32, 1513-1526.

International Organization for Standardization, ISO (1989). *Water quality - Fresh Water Algal Growth Inhibition Test with Scenedesmus subspicatus and Selenastrum capricornutum.* ISO 8692:1989. Geneve.

Keddy, C. J., J. C. Greene, and M. A. Bonnell (1995). "Review of Whole-Organism Bioassays: Soil, Freshwater Sediment, and Freshwater Assessment in Canada." *Ecotoxicol. Environ. Safety,* 30, 221-251.

Miller, W.E., S.A. Peterson, J.C. Greene, and C.A. Callahan (1985). Comparative Toxicology of Laboratory Organisms for Assessing Hazardous Waste Sites." *J. Environ. Qual.*, 14, 569-574.

TOXICITY DURING THE AEROBIC BIODEGRADATION OF NAPHTHALENE AND BENZOTHIOPHENE

Søren Dyreborg, Erik Arvin, and Hannah Houmann Hansen
Groundwater Research Centre, Department of Environmental Science and Engineering
Technical University of Denmark, Building 115, DK-2800 Lyngby, Denmark

ABSTRACT: Naphthalene and benzothiophene are commonly detected in soil and groundwater at oil- and creosote-contaminated sites. The purpose of the study was to investigate the toxicity toward nitrifying bacteria (*nitrosomonas)* and algae (*Selenastrum capricornutum*) during the aerobic biodegradation of naphthalene and benzothiophene. It was the intention to investigate if metabolites formed during the biodegradation could be detected by normal chemical analysis and if any changes in the toxic effect on nitrification or on the growth of algae could be observed during the degradation. The chemical analysis was carried out on a gas chromatograph with a flame ionization detector. Toxic effects on nitrification were tested in a short-time (two hours) toxicity test modified from ISO 9509. Toxic effects on algae growth were tested in a 48-hour test modified from ISO 8692.

Three microcosms with water and media were inoculated with groundwater microorganisms. Naphthalene, benzothiophene, and both naphthalene and benzothiophene, respectively, were added to the three different microcosms in an initial concentration of 4-6 mg/L. Water samples for chemical analysis and for determination of toxicity toward nitrifying bacteria and algae were collected during 24 days of incubation.

Naphthalene was rapidly biodegraded after a lagphase of one day in the microcosm where it was present as a sole source of carbon and energy. No metabolites were detected by the GC-analysis during the biodegradation of naphthalene, and after seven days naphthalene was undetectable too. Naphthalene was not toxic to algae and nitrifying bacteria at the initial concentration and no significant toxicity to algae or nitrifying bacteria could be detected during the biodegradation of naphthalene. An obvious reason for the lack of toxicity was that no metabolites were formed during the degradation. However, it may also be explained by the metabolites formed during the biodegradation were nontoxic to the two test-organisms or they were formed in nontoxic concentration levels. Dose-response experiments with two possible metabolites (1-naphthol and 2-naphthol) showed that the two metabolites were more toxic to nitrifying bacteria than naphthalene (the 50% effect-concentration, EC_{50}, was 1.8, 0.8, and 8.9 mg/L, respectively) whereas similar toxicity data were obtained using the algae-toxicity test (EC_{50} was 6.9, 3.8, and 6.0). However, a small inhibiting effect on the nitrifying bacteria could be detected at low concentrations (0.1 mg/L) of the metabolites. This suggests that no toxic metabolites were formed during the biodegradation of naphthalene.

Benzothiophene was not degraded during 24 days of incubation in the microcosm where it was present without an extra carbon source. These results showed that benzothiophene could not be biodegraded as a sole source of carbon and energy. The toxicity toward nitrifying bacteria and algae was constant during the incubation.

A lagphase of eight days before the biodegradation of naphthalene and benzothiophene was initiated was observed in the microcosm where the two compounds were present together. The biodegradation of the two compounds occurred concomitantly and stopped after 19 days resulting in a residual concentration of 0.04 mg/l and 0.02 mg/L of naphthalene and benzothiophene, respectively. No other compounds could be detected by the chemical analysis. The two tests showed different patterns of toxicity. The toxicity to nitrifying bacteria decreased during the biodegradation of naphthalene and benzothiophene and reached a nontoxic level at the end of the biodegradation. In the algae toxicity test, a decrease in the toxicity was observed during the biodegradation, but a high residual toxicity could be observed at the end of the biodegradation. This latter phenomenon was likely due to the formation of persistent metabolites. A dose-response experiment with a potential benzothiophene metabolite (benzothiophene-sulfone) showed that the metabolite was more toxic to algae than benzothiophene (EC_{50} was 3.6 mg/L and 5.6 mg/L, respectively), whereas it was nontoxic to nitrifying bacteria (EC_{50} = 3.8 g/L). This supports the hypothesis that toxic persistent metabolites were formed during the biodegradation of naphthalene and benzothiophene, and it is likely that the metabolites are formed from the biodegradation of benzothiophene.

The biodegradation experiment showed that benzothiophene could not be degraded as a sole source of carbon and energy but only when a concomitant degradation of naphthalene took place. GC-analysis of samples withdrawn from the microcosms during the biodegradation showed no evidence of any metabolites. However, the toxicity-tests indicated that some metabolites were formed, determined as a residual toxicity in samples where the parent compounds were in nontoxic concentration levels. The experiments showed that metabolites formed during the biodegradation-process can be more toxic than the parent-compounds, but also that organic compounds can have different effects in different toxicity tests. These observations suggest that toxicity-tests should be conducted concomitantly with normal chemical analysis when evaluating the pollution from oil- and creosote-contaminated sites, but also that a battery of different tests are preferable.

EFFECT OF TRICHLOROETHYLENE CONCENTRATION ON OXYGEN UPTAKE RATES DURING HYDROCARBON BIODEGRADATION

Brent A. Adams, ***R. Ryan Dupont***, William J. Doucette, and Darwin L. Sorensen (Utah Water Research Lab, Utah State University, Logan, Utah)

ABSTRACT: Laboratory batch reactors spiked with fuel contaminant and trichloroethylene (TCE) were used to examine the effect of TCE concentration on oxygen consumption during hydrocarbon biodegradation. Soil reactors were spiked with a mixture of weathered JP-4 and diesel fuel #2 at a concentration of 1,000 mg/kg and TCE at concentrations of 0, 1, 10, and 100 mg/kg. Oxygen consumption was represented using the biochemical oxygen demand (BOD) model with a lag phase. Results indicate that the presence of TCE inhibited the initial rate of oxygen consumption during hydrocarbon degradation by an average of 25% compared to the treatment without TCE. The highest concentration of TCE (100 mg/kg) also resulted in a significantly longer lag phase than the other treatments. Ultimate oxygen consumption values were significantly larger in the treatments containing TCE than in the treatment without TCE, possibly because of variations in incubation time affecting total oxygen consumption. Based on the results of this study, it is possible to bioremediate soils contaminated with fuels and up to at least 100 mg/kg TCE, although remediation times may be longer than if TCE was not present as a co-contaminant.

INTRODUCTION

It is not uncommon for contaminated sites to have a variety of contaminants present. U.S. Air Force bases, in particular, have sites with both fuels and TCE present as co-contaminants. It has been shown at many field sites that bioventing is an effective in situ remediation technique for soils contaminated with fuels, particularly JP-4 jet fuel and diesel fuel. The effect of co-contaminants, such as TCE, on bioventing effectiveness, however, has not been studied extensively.

Laboratory studies examining cooxidation of TCE in the presence of methane or toluene have shown TCE to be toxic at aqueous concentrations ranging from 7.8 to 42 mg/L (Wackett and Gibson, 1988; Oldenhuis et al., 1989; Strand et al., 1990; Broholm et al., 1990; Oldenhuis et al., 1991). TCE has also been shown to affect biodegradation rates of JP-4 vapor in soils (Kampbell and Wilson, 1994). TCE at a concentration of 1.2 mg/kg resulted in minimal decrease in degradation rates, whereas TCE concentrations in the range of 10 mg/kg decreased JP-4 vapor degradation rates by approximately 25%. TCE concentrations of 4,500 mg/kg were very inhibitory, resulting in a 99.7% decrease in JP-4 vapor degradation rate.

A laboratory treatability study was initiated to determine the effect of varying TCE concentrations on biodegradation of fuel contaminants in soil. Oxygen consumption results from this study are presented.

Objective. The objective of this study was to determine the effect of TCE over a range of concentrations on the rate and pattern of oxygen consumption during biodegradation of fuel contaminants.

MATERIALS AND METHODS

Batch microcosms consisted of 120 mL glass vials stoppered with Mininert® valves containing 40 g dry weight soil. The soil moisture was adjusted to 75% of field capacity and the microcosms were maintained at 11°C for 23 to 30 days of incubation. Unspiked controls were included to determine background respiration rates and control reactors poisoned with mercuric chloride were monitored to examine abiotic processes. All reactors were inverted in water to minimize exchange of gases with the atmosphere.

Headspace oxygen concentrations were periodically sampled and analyzed using a gas chromatograph with a thermal conductivity detector. An equal volume of air was injected into the reactor to replace the sample volume to avoid formation of a partial vacuum over time. When headspace oxygen concentrations reached approximately 5 vol%, the reactor headspace was purged with approximately three purge volumes of compressed air. Triplicate reactors were sacrificed initially, at the first purge event, and after the second oxygen depletion cycle for hydrocarbon analyses (data not shown).

Data reduction. Oxygen consumption rates were determined using the BOD model. This model states that the oxygen consumption rate is first order with respect to the amount of oxygen demanding material remaining in the system, as shown in Equation 1:

$$\frac{dL}{dt} = -k\,L \tag{1}$$

where L = the oxygen demand remaining, mg; t = time, hr; and k = the first order reaction rate constant, 1/hr. Equation 1 can be integrated and modified to model the oxygen consumed and include a lag phase:

$$y = L_0\left(1 - e^{-k(t - t_{lag})}\right) \tag{2}$$

where y = the oxygen consumed to time t, mg; L_0 = the oxygen demand at time 0, mg; t = elapsed time from start of batch, hr; and t_{lag} = length of lag phase, hr.

The actual oxygen consumption rate, k_{O2}, is the first derivative with respect to time of Equation 2. The oxygen consumption rate decreases with time according to the BOD model and initial oxygen consumption rates, $k_{O2,i}$, were selected as an additional means of examining the effect of treatments on oxygen consumption rate. The initial oxygen consumption rate can be obtained by differentiating Equation 2 and letting $t = t_{lag}$, producing Equation 3:

$$k_{O2,i} = \left.\frac{dy}{dt}\right|_{t=t_{lag}} = k\,L_0 \qquad (3)$$

Oxygen consumption determined from changes in headspace oxygen concentrations over time were corrected for addition of oxygen during sample volume replacement. These oxygen consumption volumes were converted to mass values and accumulated on a reactor-specific basis over the period of incubation. To determine the amount of oxygen consumed from the biodegradation of spiked contaminants the cumulative oxygen consumption data were corrected for background respiration using oxygen consumption results from unspiked reactors.

The net oxygen consumption data for all treatment reactors for each batch were then combined for the determination of oxygen consumption model parameters. Apparent outlier values, caused by sampling error such as a plugged syringe needle or leaking syringe, or a reactor septa leak during sampling, were not included in the determination of model parameter estimates. The cumulative oxygen consumption data were analyzed by nonlinear regression of cumulative oxygen consumed versus time, resulting in estimates of L_0, k, and t_{lag}. F-values were calculated using regression output and p-values were determined to test the significance of these regressions. The coefficient of determination, r^2, was also calculated as a measure of the fit of the regression. In addition, the initial oxygen consumption rate, $k_{O2,i}$, was calculated for each batch, with the standard error determined using the method described in Meyer (1975). Differences in model parameter estimates and $k_{O2,i}$ among treatments were determined using Duncan's New Multiple Range Test procedure as described in Dowdy and Wearden (1983).

RESULTS AND DISCUSSION

Figure 1 shows the fit of the oxygen consumption model to the 0 and 100 mg/kg TCE treatments. All regression results were significant at the 95% confidence level and the BOD equation modelled the data very well, with r^2 values greater than 0.98 for all regressions.

Table 1 shows the BOD parameter estimates with standard errors and results of the statistical comparisons of parameters among treatments. Statistical analyses of the k and $k_{O2,i}$ estimates indicate that the presence of TCE decreases both of these rate constants. The estimates of k and $k_{O2,i}$ for the treatment without added TCE were significantly greater than the estimates for the treatments containing TCE, with no significant difference among estimates for treatments containing TCE. The initial oxygen consumption rates of the reactors containing TCE were an average of 25% less than the initial rate of the reactors without TCE. This decrease is similar to that observed by Kampbell and Wilson (1994) for moderate (10 mg/kg) TCE concentrations. The impact of TCE concentration

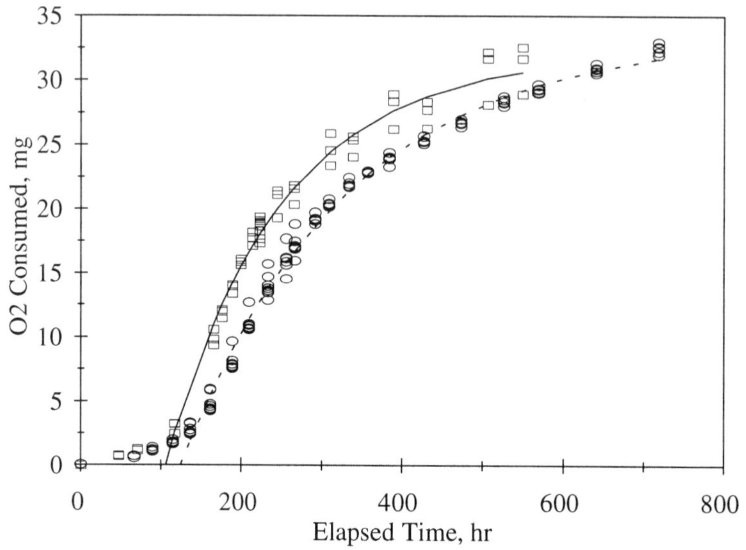

FIGURE 1. Oxygen consumption data from 0 mg/kg TCE (□) and 100 mg/kg TCE (○) treatments with oxygen consumption model fits shown as solid and dashed lines, respectively.

on biodegradation rates observed by Kampbell and Wilson (1994) was not observed at the TCE concentrations examined in this study.

The results from the comparison of lag times indicate an inhibitory effect due to TCE only at high concentrations. No significant difference in lag time estimates were apparent except for the 100 mg/kg TCE treatment, which had a significantly longer lag time than the other treatments.

An unexpected result of this study was the significantly lower ultimate oxygen consumption value for the treatment with no TCE added compared to treatments containing TCE (all of which showed no significant difference in ultimate oxygen consumption). These results could be related to the total amount of oxygen consumed in each treatment as the estimated magnitude of L_0 is closely related to the actual amount of oxygen consumed. The treatments containing TCE were incubated over a longer time, resulting in a slightly greater net oxygen consumption than the treatment without added TCE.

These results show that the presence of TCE inhibits oxygen consumption through decreased oxygen consumption rate constants and initial oxygen consumption rates at all TCE concentrations, and increased lag phase times at 100 mg/kg TCE. Although TCE can have a negative impact on oxygen consumption rates even at relatively low concentrations, and can lengthen lag times at higher concentrations, considerable respiration and biodegradation of compounds can still occur at the range of concentrations examined. The increase in lag time of 125 hours at a TCE concentration of 100 mg/kg would not seriously impact

TABLE 1. Oxygen consumption model parameter estimates and statistical test results for effect of TCE concentration on oxygen consumption.

Parameter	TCE Concentration, mg/kg			
	0	1	10	100
Incubation period, hr	549	691	688	718
Net O_2 consumed, mg	31.1	33.5	32.9	32.4
L_0, mg O_2	32.1 ± 0.600	34.6 ± 0.345	33.7 ± 0.290	33.6 ± 0.330
L_0 statistical comparison	a	b	b	b
k, 1/hr	0.00696 ± 0.000390	0.00495 ± 0.000129	0.00497 ± 0.000112	0.00484 ± 0.000125
k statistical comparison	a	b	b	b
$k_{O2,i}$, mg/hr	0.223 ± 0.00912	0.171 ± 0.00302	0.168 ± 0.00257	0.162 ± 0.00283
$k_{O2,i}$ statistical comparison	a	b	b	b
t_{lag}, hr	105 ± 3.10	106 ± 1.39	105 ± 1.22	125 ± 1.44
t_{lag} statistical comparison	a	a	a	b

± values are standard errors of parameter estimates

Statistical comparison results with identical letters can not be distinguished at the 95% confidence level.

remediation times, given typical bioventing timeframes. Bioremediation of soils with TCE concentration ranges similar to those examined in this study would therefore be possible, although remediation times may be longer than if TCE was not a co-contaminant with aerobically degradable waste components.

REFERENCES

Broholm, K., B.K. Jensen, T.H. Christensen and L. Olsen. 1990. "Toxicity of 1,1,1-trichloroethane and trichloroethene to a mixed culture of methane-oxidizing bacteria." *Applied and Environmental Microbiology*. 56():2488-2493.

Dowdy, S. and S. Wearden. 1983. *Statistics for Research*. John Wiley and Sons, New York, NY.

Kampbell, D.H. and B.H. Wilson. 1994. "Bioremediation of Chlorinated Solvents in the Vadose Zone." In R.E. Hinchee, A. Leeson, L. Semprini, and S.K. Ong (Eds.), *Bioremediation of Chlorinated and Polycyclic Aromatic Hydrocarbon Compounds*, pp. 255-258. Lewis Publishers, Boca Raton, FL.

Meyer, S.L. 1975. *Data Analysis for Scientists and Engineers*. John Wiley and Sons, New York, NY.

Oldenhuis, R., R.L.J.M. Vink, D.B. Janssen, and B. Witholt. 1989. "Degradation of chlorinated aliphatic hydrocarbons by *Methylosinus trichosporium* OB3b expressing soluble methane monooxygenase." *Applied and Environmental Microbiology*. 55(11):2819-2826.

Oldenhuis, R., J.Y. Oedzes, J.J. van der Waarde, and D.B. Janssen. 1991. "Kinetics of chlorinated hydrocarbons degradation by *Methylosinus trichosporium* OB3b and toxicity of trichloroethylene." *Applied and Environmental Microbiology*. 57(1):7-14.

Strand, S.E., M.D. Bjelland, and H.D. Stensel. 1990. "Kinetics of chlorinated hydrocarbon degradation by suspended cultures of methane-oxidizing bacteria." *Research Journal of the Water Pollution Control Federation*. 62(2):124-129.

Wackett, L.P. and D.T. Gibson. 1988. "Degradation of trichloroethylene by toluene dioxygenase in whole-cell studies with *Pseudomonas putida* F1." *Applied and Environmental Microbiology*. 54(7):1703-1708.

TCE PRODUCT TOXICITY AND INTERNAL ENERGY-STORAGE EFFECTS ON METHANOTROPHS

Kung-Hui Chu and ***Lisa Alvarez-Cohen***
(University of California, Berkeley)

ABSTRACT: Trichloroethylene (TCE) product toxicity and the presence of internal energy storage products, such as poly-β-hydroxybutyrate (PHB), are important factors for predicting the success of chlorinated solvent bioremediation by methane-oxidizing cultures in both in situ and reactor applications. Mixed cultures of nitrogen-fixing methane oxidizers have been demonstrated to have less significant product toxicity and higher cellular PHB than nitrate- or ammonia-supplied methane-oxidizing mixed cultures. However, the fundamental basis of this phenomenon has not yet been evaluated. This study addresses our understanding of the effects of nitrogen sources on cellular growth, internal energy-storage, and TCE degradation by evaluating the nature and extent of TCE product toxicity using a variety of pure and mixed methane oxidizing cultures.

A series of experiments were conducted to evaluate the growth and TCE degradation of pure and mixed microbial cultures provided with either nitrate, ammonia, or molecular nitrogen as nitrogen source. The nature of the toxic effects of TCE transformation on cells was examined through a series of bioassays, including the naphthalene assay for measurement of soluble methane monooxygenase, a combination of epifluorescent cellular stains for measurements of viable cell numbers as a fraction of total cell counts, and measurements of protein content and methane oxidation rates following TCE oxidation. The activity of soluble methane monooxygenase and methane oxidation rates of cells were used to represent toxicity directly affecting the enzyme responsible for TCE degradation. Non-specific cellular toxicity was evaluated by measuring the change of the respiratory activity of methane oxidizers following oxidation of TCE.

Preliminary results indicate that initial specific TCE degradation rates of pure cultures were unaffected by the provided nitrogen source. Soluble methane monooxygenase activity was highest for nitrogen-fixing cells and was significantly decreased following TCE oxidation. TCE product toxicity damaged both the soluble methane monooxygenase enzyme and general cellular respiration, suggesting both localized and non-specific toxicity.

ECOLOGICAL AND HUMAN TOXICOLOGY STUDIES ON WEATHERED PETROLEUM-CONTAMINATED SOILS

Francis C. Hsu, Manjari Singh, and Scott Cunningham
(DuPont, Newark, DE)

ABSTRACT: The process of oil and gas exploration, production, tankage, transportation, refining, and marketing has resulted in the inadvertent spillage and leakage of petroleum hydrocarbon materials to soil. Although much of these materials are lost over time through volatilization and physical and biological degradation processes, some portion can persist, being only slowly available for interactions with living organisms. Because of soil sequestration mechanisms, residual hydrocarbons often have very limited availability, mobility, and toxicity. This report summarizes the findings in microbial, plant, ingestion and dermal toxicities of ten weathered petroleum-contaminated soils. The ten soils are from North America petroleum production, refinery, terminal and manufacturing gas plant sites representing contaminations by diverse types of hydrocarbons with varying degrees of weathering. For all tests, closely matched uncontaminated background soils were used as controls. Phytotoxicity was measured as inhibition of seed germination and seedling growth of 10 plant species in different soils against the uncontaminated control soil. Microbial toxicity was measured for various soil extracts in the *Photobacterium phosphoreum* Microtox test and compared to their counterparts from the uncontaminated control soils. Simulated dermal and intestinal permeation tests of C_{14}-labeled hydrocarbons were developed using a polymer membrane and cultured Caco-2 cell layer respectively. The apparent permeability coefficients (P*app*) for compounds were calculated from the permeation data and plotted against compounds' log K*ow* values. This P*app* vs. log K*ow* relationship will be used for estimating unknown compounds' skin and intestinal uptake into the bloodstream. Various soil extracts tested in these tests can be used to estimate the amount of bloodstream uptake of hydrocarbons under different scenarios.

CHANGES IN TOXICITY OF FUEL-CONTAMINATED SEDIMENTS FOLLOWING NITRATE-BASED BIOREMEDIATION: COLUMN STUDY

S.R. Hutchins (Robert S. Kerr Environmental Research Lab, U.S. EPA)
J.A. Bantle, E.J. Schrock (Oklahoma State University)

ABSTRACT: A laboratory column study was set up to evaluate changes in sediment toxicity following nitrate-based bioremediation, and to correlate toxicity reduction with loss of fuel components. Glass columns, 23 cm X 3.8 cm ID, were packed with sediment from an aquifer at Eglin AFB, FL, which had been contaminated with JP-4 jet fuel. The columns were packed inside an anaerobic chamber to preclude oxygen intrusion and were operated outside of the glovebox under anaerobic conditions at 20°C. Feed solution was made with spring water and amended with 10 mg/L sulfate to simulate the ground water at Eglin AFB. Nitrate was added as an alternative electron acceptor at the design concentration of 20 mg/L. The feed solution was purged with helium and then pumped at 0.05 mL/min through a degasser which contained permeable tubing exposed to a helium atmosphere to remove oxygen. Column influents and effluents were monitored for fuel aromatic hydrocarbons (BTEXTMB), electron acceptors, nutrients, and dissolved gases. Duplicate columns were sacrificed after one, six, and twelve months, and core material was analyzed for chemical constituents. Core material was evaluated for toxicity using FETAX, an established developmental toxicity test employing frog embryos.

Data from the first pair of sacrificed columns have been obtained and analyzed. Denitrification has been established in all columns, with nitrate reduction and transient production of nitrite and nitrous oxide. Aqueous total organic carbon levels are dropping, but BTEXTMB continues to leach from the contaminated sediments. After one month of operation, BTEXTMB levels in the core have dropped from 167 ± 25 mg/kg to 86.0 ± 9.8 mg/kg, and JP-4 levels have dropped from 5400 ± 620 mg/kg to 4970 ± 42 mg/kg. In contrast, mortality has dropped from 93.3% to 11.7% in one column, and remained unchanged in the second. Similarly, malformation has dropped from 100% to 15.1% in that one column, and has dropped to only 75.0% in the second. Toxicity reduction does not appear to be directly correlated with either BTEXTMB or JP-4 levels for this first column pair, and this may be due to variable production of more toxic water-soluble intermediates. The next two column pairs will be analyzed similarly to provide quantitative data on the correlation of toxity reduction with nitrate consumption.

Survey of the Inherent Remediation and Attenuation Capacity of Phenanthrene in Diesel Fuel added to Soil

R. D. Maurice and D. L. Burton
Department of Soil Science, University of Manitoba, Winnipeg, MB, Canada

ABSTRACT: In order to interpret the relationship between the physical, chemical, and biological properties of soils and the rate of biodegradation of diesel fuel components, a survey of the bioremediation potential of a range of Manitoba soils was undertaken. The organic matter content, texture and climate of the soils varied dramatically. Samples were collected from the 0-10 cm and 90-100 cm depth in order to compare surface and subsurface rates of degradation. Determination of degradation rates were conducted in microcosms incubated with field moist soil from each site. Radiolabeled phenanthrene was added along with 5000 µg of diesel fuel g^{-1} soil in order to assess the degradation potential of that compound in diesel fuel. Volatilization was also determined in the experiment for each soil type.

INTRODUCTION

Diesel fuel spills are a concern to many industrial agencies due to toxic effects on the environment. When they do occur two remedial options may be used either excavation followed by treatment or treatment *in situ*. The relative costs and degree of site disturbance of these two approaches vary dramatically. The decision as to the most appropriate method depends upon an understanding of the potential threat to surrounding environments. The potential for transport of the contaminant off site or into more sensitive components of the ecosystem may limit the ability to treat the contaminant *in situ*. The ability to predict the fate of diesel fuel, based on fundamental soil properties, would allow estimation of potential risk of a spill prior to site investigation. Knowledge of the attenuation and remediation capacity of the soil may support the recommendation of intrinsic remediation (Alexander 1994).

Objective. The objective of this study is to determine the degree of mineralization of phenanthrene, as a component of diesel fuel #2, over a range of Manitoba soils.

Soils. The soils selected for the study range dramatically in organic matter, texture and climatic conditions (Table 2). Soil microcosms were implemented to monitor volatility, and mineralization at 20°C. Sites # 8 and #10 had prior hydrocarbon exposure in the last 5 years as a result of diesel fuel and crude oil spills respectfully. Sampling of the soil was conducted in late July using an auger. Soil was collected at intervals of 0-10 cm and 90-100 cm. Reps were completed in quadruplicate and stored in air tight plastic bags in the dark at 4°C until the beginning of the experiment. The field moist soil equivalent to 40 g oven dry mass was added to 50 ml glass beakers then brought to field capacity. The beaker was inserted into a microcosm and incubated at 20 $\pm$ 2°C for one week to allow equilibration of the microbial biomass.

TABLE 1. Site Classifications for Remediation Study

Site	Classification	Site Location
Site #1	Calcareous Black (Calcic Cryoboroll)	South facing catena sampled at mid-slope
Site #2	Orthic Dark Grey (Boralfic Cryoboroll)	South facing catena sampled at mid-slope
Site #3	Orthic Black (Typic Cryoboroll)	South facing catena sampled at mid-slope
Site #4	Gleyed Rego Black (Aquic Cryoboroll)	Sampled on moderate incline slope
Site #5	Orthic Black (Typic Cryoboroll)	South facing catena sampled at mid-slope
Site #6	Humic Luvic Gleysol (Argiaquic Cryoboroll)	Sampled in a depression area
Site #7	Orthic Grey Luvisol (Typic Cryoboralf)	South facing catena sampled at mid-slope
Site #8	Gravel Fill	Diesel storage tank leak
Site #9	Orthic Black (Typic Cryoboroll)	Control for Site #10
Site #10	Gleyed Rego Black (Aquic Cryoboroll)	Crude oil spill in 1994 sampled in a depression area

MATERIALS AND METHODS

Microcosms. Microcosms were purchased from Richards Packaging (Winnipeg, MB) as 500 ml glass jars with sealed metal lids. To the microcosm a 7 ml scintillation vial for $^{14}CO_2$ trapping, a polyurethane foam plug (PUF) to trap any volatile phenanthrene and a 20 ml scintillation vial containing 10 ml water (pH 3) was included to humidify the air.

^{14}C Phenanthrene and Diesel Fuel. ^{14}C phenanthrene was purchased from the Sigma Chemical Co. (St. Louis, MO) as phenanthrene-9-^{14}C (13.3 mCi/mmol, >98% purity). Stock solutions were first made up in Hexane then transferred to diesel fuel #2 for addition to soil. At the commencement of each experiment 5000 µg/g soil of diesel-^{14}C phenanthrene mixture was added with approximately 0.05 µCi of activity. Stock diesel fuel added contained about 0.7059µg unlabelled and 0.0586µg of labeled phenanthrene per gram of soil.

TABLE 2. Physical Properties of Sites Selected For Remediation Survey

Site	% Sand	% Silt	% Clay	Class	% Organic Carbon	Field Capacity (%)
Site #1 0-10 cm	87	4	9	S		25.3
90-100 cm	91	2.5	6.5	S		11.3
Site #2 0-10 cm	43.5	29	27.5	CL	2.6 ± 0.4	26.8
90-100 cm	72	14.5	13.5	SL		18.6
Site #3 0-10 cm	45	28.5	26.5	L	3.0 ± 0.1	28.5
90-100 cm	39	36.3	24.7	L		24.8
Site #4 0-10 cm	4	19	77	HC	2.9 ± 0.1	42.2
90-100 cm	1	16	83	HC		38.3
Site #5 0-10 cm	28.5	33.5	38	CL	1.2 ± 0.2	34.4
90-100 cm	32.5	34.5	33	CL		21.9
Site #6 0-10 cm	31.5	35.5	33	CL		30.7
90-100 cm	34.5	30.5	35	CL		29.4
Site #7 0-10 cm	34.5	33.5	32	CL	3.5 ± 0.2	27.5
90-100 cm	33	35.5	31.5	CL		22.8
Site #8 0-1 m *	91	8	1	S		4.4
1-2 m *	95	4.5	0.5	S		27.2
Site #9 0-10 cm	13	44	43	SiC		34.0
90-100 cm	47.5	32.5	20	L		24.7
Site #10 0-10 cm *	16.3	44	39.7	SiC		42.5
90-100 cm *	32.5	42	25.5	L		29.8

* Prior sensitization of hydrocarbons on site

$^{14}CO_2$ Traps. 0.5 ml (+/-)-α-phenylethylamine and 0.5 ml methanol were added to a 7 ml scintillation vial in order to determine the amounts of $^{14}CO_2$ evolved from the soil (the methanol was added to prevent crystallization). During the experiment traps were changed at weekly intervals until there was no longer any significant radioactivity recovered.

Volatilization. When the experiment was completed, the polyurethane foam plugs were removed from the microcosms and placed in 100 ml French Square Bottles (EPA §796.3400). To these bottles 20 ml Methanol was added for total extraction of the phenanthrene. Bottles were then stoppered and shaken on a lateral shaker for 2 minutes. The foam plugs were then removed and placed in a 50 ml syringe to extract all of the Methanol. From the extract, a 1 ml subsample was collected and counted by the scintillation counter.

Liquid Scintillation Counting. Each 1 ml CO_2 trap removed from the microcosm was combined with 5 ml Ecolite (+) Liquid Scintillation Fluid (ICN Biochemicals Inc. Aurora, OH). A Beckman LS 7500 scintillation counter was implemented to give final results of disintegrations per minute (DPM). Final DPMs were corrected for background and blanks then related to the original radioactivity added to each microcosm to give the percent mineralization of phenanthrene in diesel.

Kinetic Analysis of $^{14}CO_2$ Evolution. The first order rate model was used to determine the rate of mineralization of Phenanthrene in diesel #2. The calculation describes the reaction over time as a single first order component:

$$P = A[1-e^{(-kt)}] \tag{1}$$

P is the percent phenanthrene mineralized at time t, A is the percent of compound evolved as CO_2 at $t = \infty$, and k is the rate constant for $^{14}CO_2$ evolution (day^{-1}). If a lag in mineralization was seen, the lag time was noted and kinetic constants calculated from the post-lag CO_2 flux. Quadruplicate reps were analyzed, then averaged including a standard deviation.

Adsorption and Biomass. Adsorption of phenanthrene by the soil and incorporation in the microbial biomass are being determined but are unavailable at this time.

RESULTS AND DISCUSSION

Volatilization. Based on the low vapour pressure (0.113 Pa at 25°C) and high log K_{ow} (4.53 at 26°C) (Piatt et al., 1996), the volatility of phenanthrene in soil was considered to be low. The total percent phenanthrene evolved after 91 days from each site is listed in Table 3. Little of the added phenanthrene was captured in the volatile traps which is consistent with a low vapour pressure of this compound. There was not a significant correlation between volatilization and soil texture for the amount volatilized at each site.

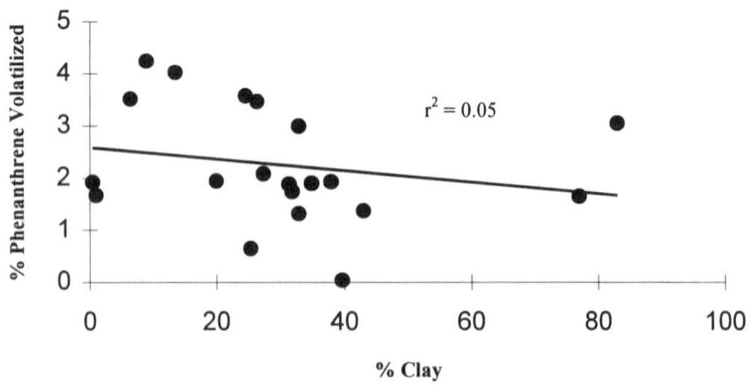

FIGURE 1. Correlation between percent phenanthrene volatilization and clay content for 10 sites examined

Phenanthrene Mineralization. The $^{14}CO_2$ production did not vary greatly between the soils sampled in the province (Table 3).

TABLE 3. Volatilization, Mineralization and Mineralization Half-life for Soils Examined in Survey

Site	% Volatilized	% Mineralized	Mineralization Half-life (days)
Site #1 0-10 cm	4.3 ± 0.9	3.8 ± 0.9	26
90-100 cm	3.5 ± 0.8	1.8 ± 0.6 *	27
Site #2 0-10 cm	2.1 ± 0.7	1.5 ± 0.2	33
90-100 cm	4.0 ± 0.7	2.2 ± 0.8 *	34
Site #3 0-10 cm	3.5 ± 0.7	3.8 ± 0.9	52
90-100 cm	3.6 ± 1.1	2.2 ± 0.6 *	25
Site #4 0-10 cm	1.7 ± 0.7	18.0 ± 17.1	29
90-100 cm	3.1 ± 0.5	1.9 ± 0.8 *	23
Site #5 0-10 cm	1.9 ± 0.7	2.3 ± 1.4	51
90-100 cm	3.0 ± 1.1	2.0 ± 0.5 *	34
Site #6 0-10 cm	1.3 ± 0.2	1.7 ± 0.8 [t]	62
90-100 cm	1.9 ± 0.2	1.1 ± 0.4 *	31
Site #7 0-10 cm	1.8 ± 0.5	2.6 ± 0.3	51
90-100 cm	1.9 ± 0.3	1.7 ± 0.7 *	43
Site #8 0-1 m	1.7 ± 0.2	1.0 ± 0.3 *	19
1-2 m	1.9 ± 0.4	4.8 ± 4.2	43
Site #9 0-10 cm	1.4 ± 0.6	8.9 ± 7.1	15
90-100 cm	2.0 ± 0.4	3.2 ± 1.7	168
Site #10 0-10 cm	0.1 ± 0.1	55.1 ± 12.6	2
90-100 cm	0.7 ± 0.4	43.5 ± 8.9	7

* Initial lag of 21 days prior to CO_2 production
[t] Initial lag of 14 days prior to CO_2 production

The greatest degradation resulted in the site where prior hydrocarbon sensitization has occurred (Site #10) at the 0-10 cm depth. The least amount of degradation was found in the surface (0-1 m) of Site #8. When considering the sites without prior sensitization of hydrocarbons, Site #4 had the greatest total mineralization of phenanthrene in diesel fuel #2. The general trend to be noted is that surface soils (0-10 cm) had greater mineralization compared to sub-surface soils, probably due to increased microbial activity and organic matter. In most cases the rate of mineralization was less that 5% of the added phenanthrene in diesel fuel. Where large amounts of degradation was observed, the results were often inconsistent having standard deviations as large as the mean (Site #4, 8 and 9). The exception being Site #10 where prior exposure has resulted in a more consistent increase in rates of mineralization. The inconsistency of elevated rates of inherent mineralization in sites without prior exposure suggest this may be a relative sporadic occurrence in these soils.

Clay content is an important characteristic of the degradative environment providing extensive surfaces for the stabilization or reaction of organic compounds. In this study particle size did not have a consistent influence on the mineralization of phenanthrene in diesel. Sites #4 had the greatest degradation of the non sensitized sites and had the greatest clay content. The sub-surface of Site #4 had a relatively low rate of degradation despite having a clay content greater than 80%. Clay content alone cannot account for differences in the degradation ratio. Figure 2 indicates the distant relationship between texture and mineralization in the soils that have not had prior hydrocarbon stimulation.

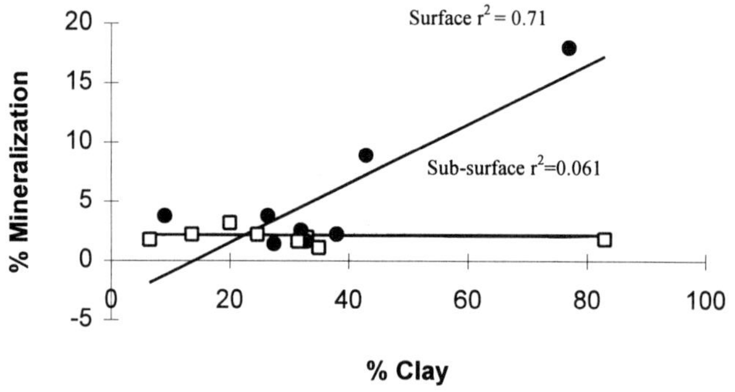

FIGURE 2. Correlation between percent mineralization and percent clay in the Surface(●) and Sub-surface(□) of 8 sites investigated without prior hydrocarbon exposure.

Prior sensitization of hydrocarbons on the soil decreased the half life and acclimation period of phenanthrene degradation in diesel fuel due to prior enzyme stimulation. The same trend was also seen by Alexander (1994). This was evident in both the surface and subsurface of Site #10 where a large amount of degradation was seen and a lag phase was not present. When comparing literature values of degradation half lives to our own, the aerobic half life can range from 16-200 days (Howard et al., 1991) depending on the soil type involved therefore the varying results in this study are reasonable.

CONCLUSIONS

Four conclusions can be drawn. The first was that texture (clay content) had little effect on the biodegradation of this compound. The second conclusion drawn from the data was prior hydrocarbon exposure to the soil greatly decreased the half life and increase the total amount of $^{14}CO_2$ evolved. This indicates the presence of microbial populations with the capability of enhanced hydrocarbon breakdown. These populations would have been selected in the initial spill and further enhanced in this experiment. The third conclusion centers around the fact that phenanthrene degradation in diesel fuel was consistently low (<5%) for most of the soils examined signifying poor, if any inherent degradation. A fourth conclusion was that the mineralization of phenanthrene in diesel fuel was found to be dependent on depth from surface as the sub-surface usually had less mineralization.

ACKNOWLEDGMENTS

The research was funded by Manitoba Hydro.

REFERENCES

Alexander, M. 1994. *Biodegradation and Bioremediation.* Academic Press, Inc., San Diego, CA.

EPA §796.3400. 1992. *Code of Federal Regulations: Inherent biodegradability in soil.* CFR §796.3400 US Government Printing Office, Washington, DC.

Howard, P. H., R. S. Boethling, W. F. Jarvis, W. M. Meylan, E. M. Michalenko, and H. T. Printup (Eds.). 1991. *Environmental Degradation Rates.* Lewis Publishers, Inc., Chelsea, MI.

Piatt, J. J., D. A. Backhus, P. D. Capel, and S. J. Eisenreich. 1996. "Temperature-Dependent Sorption of Naphthalene, Phenanthrene, and Pyrene to Low Organic Carbon Aquifer Sediments." *Environ. Sci. Technol.* 30(3): 751-760.

Descriptive Analysis of Iron Fouling in Groundwater Treatment Systems

Mark J. Tinholt, B.A.Sc.
Gilles R. Wendling, Ph.D.
Morrow Environmental Consultants Inc., Burnaby, British Columbia,Canada

ABSTRACT: Hydrocarbon contaminated groundwater treatment systems commonly experience iron related problems consisting of elevated dissolved iron levels present in the groundwater and the formation of iron hydroxide precipitate and iron bacteria biofilms. The precipitate/biofilms can jeopardize the proper operation of treatment systems and are associated to high maintenance cost. A study was completed by compiling geochemical data at several sites contaminated by hydrocarbon products with active groundwater treatment systems. Subsurface aerobic and anaerobic conditions were determined through the analysis of electron acceptors (dissolved oxygen, nitrate, iron, manganese, sulphate). The hydrocarbon contamination was characterized through benzene, ethyl-benzene, toluene and xylenes (BETX) analyses. The data indicates that in saturated zones where oxygen and nitrate had been depleted, dissolved iron and manganese concentrations increased according to an inverse exponential function of the BETX concentrations. In addition, water samples collected from the influent and effluent of groundwater treatment systems were analyzed to determine how changes in chemical conditions through the treatment system promote iron precipitation and bacterial growth. The results indicated that within the treatment system, an increase in pH and redox potential promotes rapid iron oxidation and precipitation and conditions are favourable for the growth of iron oxidizing bacteria. Based on data collected from the influent and effluent of groundwater treatment systems during pilot testing, the final design can be adjusted to reduce the iron related problems.

INTRODUCTION

Air strippers and oil water separators are commonly used for the treatment of hydrocarbon contaminated groundwater. Iron related problems experienced with these types of systems are due to 1)elevated dissolved iron concentrations present in the groundwater on the order of 50-80 mg/l (normal background levels: 0.01 - 10 mg/l).and 2)prolific precipitation of iron hydroxides and the development of aerobic iron bacteria biofilms within the treatment systems. The generated colloids, sludge or scaling in the treatment train can jeopardize the proper operation of the system and are associated with high maintenance costs.

MATERIAL AND METHODS

Groundwater data from several hydrocarbon contaminated sites were compiled to characterize the aerobic and anaerobic conditions in the subsurface. Benzene, ethylbenzene, toluene, and xylenes (BETX) concentrations were used to characterize the level of hydrocarbon contamination, and dissolved oxygen, nitrate, dissolved iron and manganese, and sulphate were used to assess the role played by the main electron acceptors in the bioremediation process. In saturated zones where oxygen and nitrate had been depleted, iron and manganese concentrations were compared to BETX concentrations.

Water samples collected from both the influent and effluent of groundwater treatment systems were analyzed to determine how changes in chemical conditions through the treatment system promoted iron precipitation and bacterial growth. The data was plotted in an iron stability diagram to assess the effect of the treatment system conditions on iron solubility. The speed and ease of iron oxidation within the treatment system was determined through oxygenation kinetics. The Eh-pH-Fe regime within the treatment systems were examined to determine whether environmental conditions are favourable for the growth of iron oxidizing bacteria. Microscopic analysis and "biological activity and reactivity tests" (BART) were performed on biofilm samples to assess the presence and role of iron oxidizing bacteria and samples of the precipitate sludge were laboratory analyzed to determine the chemical make-up of the sludge.

RESULTS AND DISCUSSION

The nitrate and dissolved oxygen concentrations detected in all of the water samples were less than 0.5 mg/l indicating that these preferred electron acceptor reserves had been depleted. The dissolved iron and manganese concentrations (reduced form) increased according to an inverse exponential function of the BETX concentrations as shown on Figure 1. The results are consistent with anaerobic bacterial reduction of iron hydroxides in the sediments coupled with the oxidation of hydrocarbons.

The following results were observed through the treatment systems:

- In general, the influent is slightly acidic with pH ranging from 6.1 to 6.8, while the effluent increased to almost neutral pH at 6.8 to 7.7. The dissolved oxygen and oxidation redox potential increased significantly through the treatment system. The Eh-pH diagram in Figure 2 shows that the influent conditions are in the dissolved iron stability zone, while the effluent conditions are in the ferric hydroxide stability zone.
- Concentrations of total iron entering the systems ranged from 10 to 90 mg/l of which 40 to 85% was dissolved. Since the water is from a properly screened and well developed groundwater pumping well, it is reasonable to assume that most of the "total" metals were in solution at point of entry into the well and some oxidation and precipitation of iron occurred during delivery of the water to the treatment system. 50 to 100 % of the dissolved iron precipitated as amorphous iron hydroxide through the

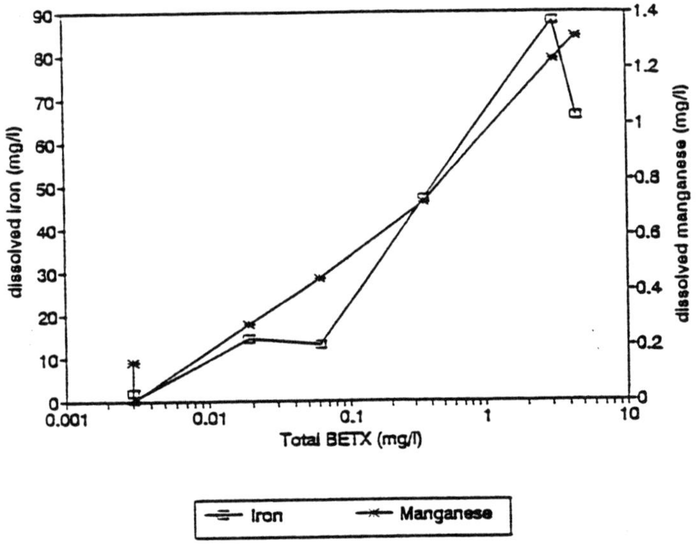

Figure 1. The effect of hydrocarbons (BETX) on dissolved Iron and Manganese concentrations.

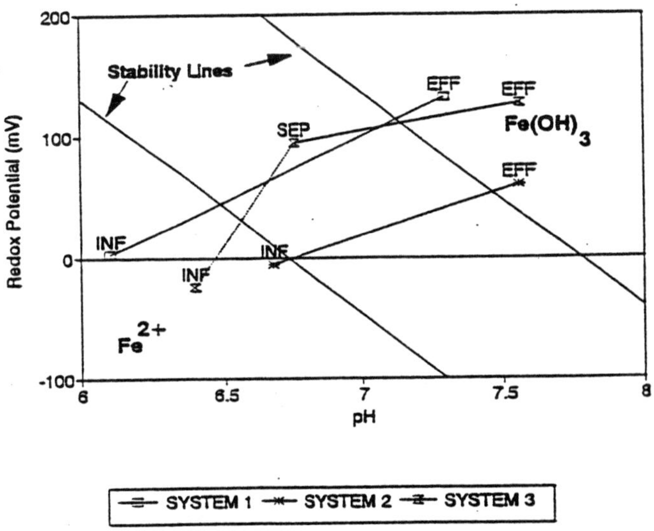

Figure 2. Iron Stability Diagram showing Treatment System Influent and Effluent Conditions. adapted from Garrels, R.M and Christ, C.L. 1960. *Solutions, Minerals and Equilibria*. Harper and Row, New York.

treatment system. Based on reaction kinetics, 80 % of the precipitation occurred within 5 minutes of aeration.

- The precipitate sludge contained high concentrations of aluminium, barium, calcium, copper, iron, lead, magnesium, manganese and zinc. Iron was present at a concentration much higher then the other analytes (88.8 %) and formed 14.7% of the total sludge mass.
- Conditions were favourable for the growth of aggressive iron oxidizing bacteria in specific zones of the treatment system. The iron bacteria "compete" with chemical oxidation of iron and are most active within a certain Eh - pH range, as represented by the triangular area on Figure 3. The BART tests indicated large and aggressive bacterial communities present in all of the samples analyzed. Large and aggressive iron related bacteria populations were detected in samples collected from the oil water separators.

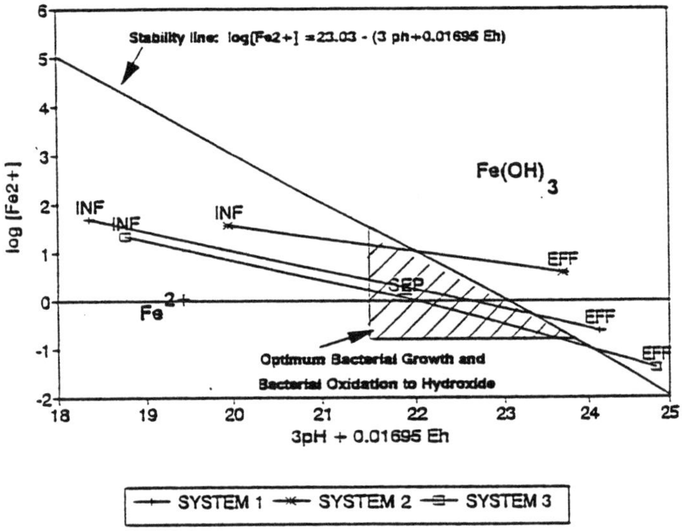

Figure 3. Modified Iron Stability Diagram showing Zones of Optimum Bacterial Growth. adapted from Lindblad-Passe, 1988, "Clogging Problems in Groundwater Heat Pump Systems in Sweden" *Water Science and Technology.* 20(3): 133-140.

CONCLUSIONS

Data collected on the influent and effluent of groundwater treatment systems during pilot testing can be plotted on a the stability diagram (Figure 3). The curve representing the variation of iron concentrations, pH and redox conditions through the treatment process will show if conditions pass through the zone in the diagram which represents where conditions are associated to optimum iron related bacteria growth and bacterial oxidation of dissolved iron to iron hydroxide. The final design of the treatment system may then be adjusted to avoid the critical zone and reduce the iron related problems.

REFERENCES

Appelo, C.A.J. and Postma, D. 1993. *Geochemistry, Groundwater and Pollution.* A.A. Balkema, Rotterdam , Netherlands.

Barbic, F. et al. 1986. "Ecology of Iron and Manganese Bacteria in Underground Water". *Proceedings of the International Symposium on Biofouled Aquifers: Prevention and Restoration.* American Water Resources Association, Maryland US.

Borden, R.C. and Gomez, C.A. and Becker, M.T. 1995. "Geochemical Indicators of Intrinsic Bioremediaton" *Groundwater.* Vol. 33 No.2

Cullimore, D.R. 1986. *Practical Manual of Groundwater Microbiology.* Droycon Bioconcepts Inc. Regina, Saskatchewan.

Cullimore, D.R. 1986. "Physio-Chemical Factors in Influencing the Biofouling of Groundwater" *Proceedings of the International Symposium on Biofouled Aquifers: Prevention and Restoration.* American Water Resources Association, Maryland US.

Garrels, R.M and Christ, C.L. 1960. *Solutions,Minerals and Equilibria.* Harper and Row, New York.

Ghiorse, W.C. 1986. "Biology of Leptothrix, Gallionella and Crenothrix; Relationship to Plugging" *Proceedings of the International Symposium on Biofouled Aquifers: Prevention and Restoration.* American Water Resources Association, Maryland US.

Hackett, G. and Lehr, J.H. 1985. *Iron Bacteria Occurrence, Problems & Control Methods in Water Wells.* USACE Report, Sep 30, fed govt report

Hallberg, R.O. and Nalser, C. 1986. "Oxidation Processes of Iron in Groundwater - Causes and Measures" *Proceedings of the International Symposium on Biofouled Aquifers: Prevention and Restoration.* American Water Resources Association, Maryland US.

Lindblad-Passe, 1988, "Clogging Problems in Groundwater Heat Pump Systems in Sweden" *Water Science and Technology.* Vol 20 No. 3 pp 133-140. Great Britain.

Lovley, D.R. et al. 1989. "Oxidation of Aromatic Contaminants Coupled to Microbial Iron Reduction" *Nature*, May 25, v339, n6222, p297(4) research article

Manahan, S.E. 1994. *Environmental Chemistry.* CRC Press Inc. Florida, USA.

Smith, S.A. and Tuovinen, O.H. 1985. "Environmental Analysis of Iron-Precipitating Bacteria in Ground Water and Wells". *Ground Water Monitor Rev*, v5, n4, p45(8) journal article

Smith, S.A. 1995. *Monitoring and Remediation Wells: Problem Prevention, Maintenance and Rehabilitation* CRC Press USA

Tuhela, L. and Smith, S.A. and Tuovinen, O. 1993. "Microbial Analysis of Iron-Related Biofouling in Water Wells and a Flow-Cell Apparatus for Field and Laboratory Investigations" *Groundwater*, v31, n6, p982(7).

Water-Science-and-Technology. 20. (3). p. 249. International Association on

Wojcik, W. 1986. "Monitoring Biofouling" *Proceedings of the International Symposium on Biofouled Aquifers: Prevention and Restoration.* American Water Resources Association, Maryland US.

IMPORTANCE OF SOIL-WATER RELATIONSHIPS IN ASSESSING ENDPOINTS IN BIOREMEDIATED SOILS

N. Sawatsky (Alberta Research Council, Alberta, Canada)
X. Li (Alberta Research Council, Alberta, Canada)

ABSTRACT: Soil physical properties and plant growth were measured in hydrocarbon contaminated soils. Studies were conducted on 3 materials, a flare pit sludge, an agricultural top soil contaminated with crude oil, and a diesel invert mud residue. All three materials were treated in a bioreactor for 1 year. For the agricultural soil, it was found that barley growth and yield was significantly reduced by oil contamination. Bioremediation did not improve yield of barley. The lack of effect from bioremediation was attributed to poor soil water absorption by the contaminated material. Water sorption seemed to be highly dependant on soil water potential and became critical as the water potential dropped below -5 to -10 bars. Results suggest that water availability in contaminated soils will be highly dependent on soil water properties as water potential approaches the permanent wilting point (-15 bars). For the field site, water preferentially moved through the contaminant layer without being absorbed by the soil. However, water movement in the surrounding uncontaminated soil layers behaved similarly to that observed in the control material.

INTRODUCTION

Hydrocarbon contaminated soils often are severely water repellent. This can have significant impacts on growth of plants. Clean-up of hydrocarbon contaminants is necessary to restore soil-water conditions on these soils. One potentially efficient clean-up methods is bioremediation. However, bioremediation often leaves a significant amount of hydrocarbon residues in the soil. After the initial, quick degradation of the labile fraction, further biodegradation of the residual contaminants can be very slow, mainly because of the low bioavailability of residual contaminants to soil microorganisms (Blackburn and Halker 1993; Mihelcic et al. 1994). Although the residual hydrocarbon pool can be as high as 40 to 50% of the initial total extractable hydrocarbon (TEH) (Johnson and Danielson 1994), this pool is not readily available to the organisms and shows extremely low rates of biodegradation.

Clean-up criteria for TEH are generally very low (often < 1000 mg kg^{-1}). With bioremediation, such levels are nearly impossible to reach when starting TEH concentrations are high or a significant portion of the hydrocarbon contaminant is represented by high molecular weight, low water soluble hydrocarbons. Ecotoxicity tests have shown that the materials with residual TEH as high as 20,000 mg kg^{-1}, have no effect on earthworm survival, seed germination, and root elongation (Visser and Danielson 1995). However, such toxicity tests do not assess the capability of the remediated soil to support plant growth, which is the intended end-use for the soil.

Commonly, ecotoxicity tests are conducted under optimal soil environmental conditions, including a soil water content of at least 70% field capacity and optimal nutrient contents (Wang 1991; Shepard et al. 1993; Visser and Danielson 1995). However, soil moisture conditions can be highly variable in the field. Soils contaminated with hydrophobic materials, such as organic hydrocarbons have the potential to develop water repellency. If the soil has potential to become water repellent, the degree of water repellency will be dependent on the soil water content (King 1981). Thus, when the soil is maintained near field capacity, soil water repellency can be almost undetectable but as the soil dries, water repellency can be severe. The question that needs to be answered is: can soils with relatively high residual hydrocarbon contamination but showing no ecotoxcity support adequate plant growth?

MATERIALS AND METHODS

Soil Treatments. Three wastes common to the oil and gas industry in Alberta were used for this study, a crude oil and brine contaminated agricultural topsoil (waste 1), a diesel invert mud residue (waste 2), and a flare pit sludge (waste 3). Waste 1 contained 4% TEH, waste 2 contained 10% TEH and waste 3 contained 9% TEH. Waste materials were bioremediated for at least 1 year and leached to reduce salt contamination. All waste materials, however, still had TEH concentrations exceeding Alberta Tier 1 criteria.

Laboratory Studies. Laboratory studies measured barley growth and soil water properties in waste 1, bioremediated waste 1 and the control soil. For the plant growth studies, the water content of the soil was allowed to fluctuate during the growing period. Enough water was added to bring the control soil to field capacity when the water content dropped below 20% (v/v). The same amount of water was applied to all treatments and the water drained from each treatment was monitored. Seed germination, growth rate, shoot and root dry mass, and yield were measured.

To measure water absorption by the soil, soil samples were air dried, ground to pass through a 2 mm sieve and packed into 1.5 cm soil columns. The soil column was placed in a water bath and water was allowed to infiltrate by capillary movement from the base of the column. Water uptake was monitored for periods between 30 seconds and 8 hours.

A breakthrough time was defined as the time for water to infiltrate to the surface of a 1.5 cm soil column. A series of soil samples were prepared with varying moisture contents from air dried (2% w/w) to 37% w/w and the water breakthrough time was monitored as a function of water content.

Field Studies. The field site is shown in figure 1. Each treatment consists of:

1. a replicated 2x4 m field site and,
2. a replicated lysimeter site consisting of 1 m deep and 0.8 m diameter lysimeters.

Prior to use, the entire site was excavated to 60 cm. Excavation was done incrementally to preserve the A, B and BC horizons in the soil. A plastic liner was placed along the circumfrance of the plot to prevent water movement between different soil treatments.

For wastes 2 and 3, the material was buried from 20 to 60 cm below the soil surface. The remaining B and A horizons were then reconstructed on the waste material.

For the control and waste 1, the BC and B horizons were reconstructed between 60 and 15 cm. This was followed by the addition of either a 15 cm layer A material (contol treatment) or 15 cm of waste 1.

Prior to excavation, soil bulk density was measured at the field site. During reconstruction, soil materials were compacted so that the bulk density of the final soil profile would be near that of the original site.

After reconstruction of the site, the entire site was seeded to hay (brome grass, alfalfa). TDR probes and water sampling ports were installed in each lysimeter to measure water and solute movement.

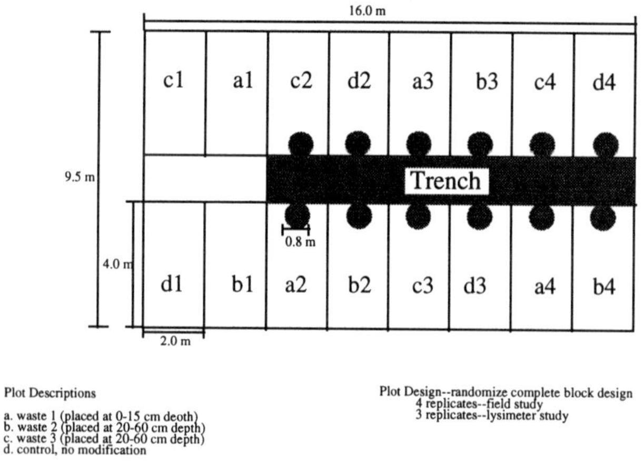

FIGURE 1. Plot diagram for the field site.

RESULTS AND DISCUSSION

Barley growth was significantly reduced by oil-contamination of agricultural topsoil. Bioremediation did not improve the yield of barley. The lack of effect of bioremediation on plant growth may be attributed to the poor ability of the contaminated and bioremediated soils to absorb water due to water repellence.

Thus, even though chemical analysis and toxicity tests may show remarkable improvement in contaminated soil after bioremediation, the residual compounds can still alter the soil physical properties, reducing plant growth.

Figure 2 shows the rate of barley growth for the three soils. This is compared with the average water content in the soils (Figure 3). As water demand increased during stem elongation, water content in the bioremediated and contaminated soils approached the permanent wilting point. Plants grown on these soils became water stressed despite the fact that water additions should have been able to maintain all soil treatments well above the permanent wilting point.

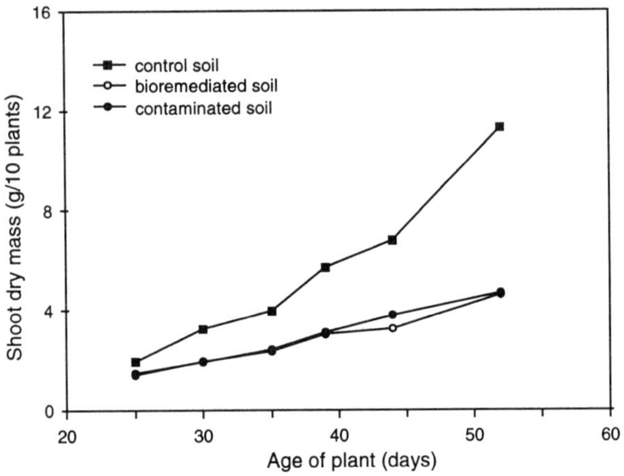

FIGURE 2. Barley shoot growth during the stem extension stage.

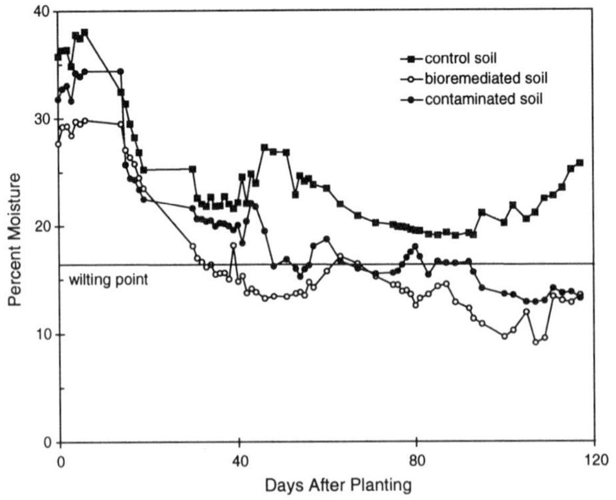

FIGURE 3. Soil water content measured by time domain reflectometer (TDR) during the plant growing season.

The rate of water absorption by the soil decreased with bioremediation. Water repellency is also highly dependent on soil water content (Figure 3). At higher water contents, the time required for water adsorption in the 1.5 cm soil column approached that of the uncontaminated soil. As water content decreased, the time required to wet the soil column increased by up to three orders of magnitude compared with the uncontaminated soil. For the plant growth studies, as water demand on the contaminated soils increased, water adsorption by the soil decreased, and plants became water stressed.

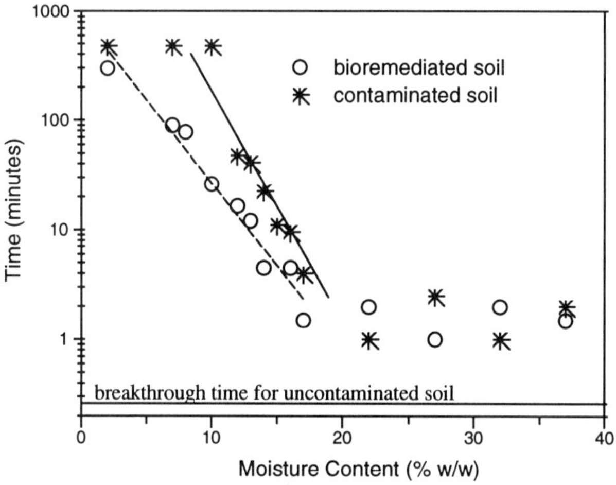

FIGURE 3. Time to water breakthrough from a 1.5 cm. soil column under varying initial soil moisture conditions.

The critical soil water content seemed to be more closely related to water potential of the soil rather than water content. Between moisture potentials of 0 to -10 bars, there seemed to be a reasonable relation to water potential and water infiltration rate. Water absorption by the soil only becomes critical at water contents below about 5 to 10 bars, which would not be accounted for in normal toxicity tests where optimum water conditions are maintained.

Similarly for the field site, water sorption by the contaminated soil horizon became problematic with the natural fluctuation in water content. The development of water repellency caused preferential flow of water through the contaminant layer and drought conditions within the contaminant layer. This did seem to affect germination and early plant growth on contaminated sites. However, preferential flow did not affect the surrounding soil horizons. Water content in the surrounding horizons remained at or above that observed in the control soil.

Based on this study, it is suggested that:

- Potential for water repellency should be included in assessing hydrocarbon contaminated soils.
- Water repellency in these soils seems to be dependent on soil water properties at low matrix potentials (near about -1.5 MPa). Further research

on the nature of water repellent soils should emphasize the relationship between the hydrocarbon contaminant and water properties around this soil water potential.

- It is still unclear what controls the critical soil water content. Although the critical soil water content can vary widely between different soils, there seems to be a relatively consistent water potential at the critical soil water content. Thus, the relation between water potential and water adsorption may be more important in studying water repellency than the relation between water content and water adsorption rates.
- If water potential is important in the sorption of water in repellent soils, sorption may also be different for the wetting and drying curve in the soil (i.e. it may be subject to hysterisis). The nature of hysterisis in these soils, therefore, may be important in understanding water absorption.
- Wetting front instability (development of fingered flow patterns) will be common in all repellent soil materials. This instability will be one of the most important parameters controlling water infiltration, surface runoff and solute transport in severely repellent soils.

REFERENCES

Blackburn, J. and Halker, W.R. 1993. "The impact of biochemistry, bioavailability and bioactivity on the selection of bioremediation techniques." *Bioremed.* 11:328-333.

Johnson, R.L. and R.M. Danielson. 1994. "The bioremediation of salt and hydrocarbon contaminated topsoil." The bioreactor project: summary for 1992-93. CAPP report.

King, P.M. 1981. "Comparison of methods for measuring severity of water repellence of sandy soils and assessment of some factors that affect its measurement." *Aust. J. Soil Res.* 9:275-285.

Mihelcic, J.R., D.R Lueking, R.J. Mitzell, and J.M. Stapleton. 1993. "Bioavailability of sorbed- and separated-phase chemicals." *Biodeg.* 4:141-153.

Shepard, S.C., Evenden, W.G., Abboud, S.A. and Stephenson, M., 1993. "A plant life-cycle bioassay for contaminated soil, with comparison to other bioassays: mercury and zinc." *Arch. Environ. Contam. Toxicol.* 25:27-35.

Visser, S and Danielson, R.M., 1995. "Bioreactor project: Ecotoxicological testing of hydrocarbon and salt-contaminated wastes following bioremediation in a bioreactor." CAPP Pub. 1995-0004.

Wang , W. 1991. "Literature review on higher plants for toxicity testing." *Water, Air and Soil Pollution.* 59:381-400.

HUMIC SUBSTANCE AND IRON MINERAL EFFECTS ON REDUCTIVE DECHLORINATION REACTIONS

Richard Collins, Sanggoo Kim, and ***Flynn Picardal***
(Indiana University, Bloomington, IN)

ABSTRACT: The effects of (i) natural organic matter and (ii) iron minerals on rates and extent of reductive dehalogenation were studied. Our experimental system utilized the facultative anaerobe, *Shewanella putrefaciens* 200, and tetrachloromethane (CT) as the model bacterium and chloroorganic compound. Biomimetic studies were also done in which the bacterium was replaced by 20 *mM* dithiothreitol.

The rate and extent of anaerobic CT dechlorination was enhanced by the presence of a high-organic content ($f_{oc} = 0.13$) soil. Oxidation of organic matter in the soil by hydrogen peroxide eliminated most of the catalytic activity of the soil. When the soil was separated into various particle sizes, the fractions with the smallest size were generally most effective in mediating CT transformation. Humic substances extracted from the soil also showed the ability to increase CT dechlorination. Although enhanced CT degradation was observed in the presence of extracted humin, humic acid, and fulvic acid, the humic acid and humin fractions produced the greatest effects. The degree of enhancement was also found to be a function of pH with reduced effects at low pH. Based on an analysis of fulvic and humic acid functional groups, no clear correlation could be established between catalytic ability and (i) total acidity, (ii) carboxyl group, (iii) total carbonyl group, or (iv) quinone content.

Rates of CT transformation by both *S. putrefaciens* and dithiothreitol were also increased in the presence of Fe(III)-containing minerals. Measurement of soluble and acid-extractable ferrous iron showed that increased CT transformation rates were proportional to microbially reduced, surface-bound Fe(II) rather than soluble Fe(II). The ability of the iron minerals tested to enhance anaerobic CT transformation rates was in the order: high-surface-area goethite > amorphous Fe(III) hydroxide > medium- or low-surface-area goethite > magnetite > hematite. These experiments extend our knowledge of the interaction of abiotic and biological processes in the transformation of pollutants in environmental matrices.

ENZYMATIC ACTIVITY OF SOIL POLLUTED BY COPPER COMPOUNDS

Anna Małachowska-Jutsz, ***Korneliusz Miksch,*** Jerzy Pacha, Joanna Surmacz-Górska (Silesian Technical University, Poland)

ABSTRACT: Research carried out had in view the examination of copper salts (acetate and sulphate) influence on enzymatic activity of soil. The concentrations of salts were changed in range of 100-10 000 ppm. The salt influence was estimated by measurements of selected enzymes activity. The effects of measurements were compared with the blank control was made the soil not treated with the copper salts. Moreover the soil polluted with 5 000 ppm was remedied by addition of agents modifying the soil properties. The modification of biological properties was started four months after the soil pollution with the copper salts. the contents of copper in organic matter was measured in 11th month of experiment. The results showed that the copper salts inhibited the amylolitic activity in soil in a strongest way but the cellulase activity C_x was inhibited the most weakly. The toxicity of examined copper salts increased with time. In case of amylases, cellulase C_x and proteases, copper acetate inhibited their activity stronger than the copper sulfate. The addition of water extract of pine needles and glucose as well as the liming of soil polluted by copper salts provoked the increasing of enzymatic activity of examined soil.

INTRODUCTION

Research of recent years introduced a lot of new information on effect of heavy metals on soil as an environment of micro-organisms existence. However, experiments carried out frequently to determine an influence of heavy metals on micro-organisms are performed „in vitro" using pure bacterial culture. There is only few references on direct effect of heavy metals on natural environment , including an effect on soil. Some authors prove the necessity to carry out ecotoxicologic research, maintaining, however, environment reality. This means that examinations of influence of given xenobiotic on environment should be performed directly in environment, considering it as a whole and not separate its biotic part from the abiotic one.

An additional problem is how to determine the period of time, within which, the influence of xenobiotics on environment is examined. Thus, it seems very essential that examinations would include a period of at least several months, because the present physiologic condition of micro-organisms is an result of influence of heavy metals on microbes existing in soil, its structure as well as type of assimilated substrate. Chemical compound included in substrate subject to decomposition in determined sequence, accompanied usually by characteristic succession of micro-organisms.

This paper presents an influence of copper salts on biological activity of soil. This influence was determined by means of measurements of selected enzymes activity, i.e. proteases, amylases and cellulases Cx. Besides, an influence of modifying agents was examined, such as calcium carbonate, an extract of pine needles or glucose on activity of selected enzymes in soil polluted with 5000 ppm of sulfate or copper acetate.

MATERIALS AND METHODS

Soil of sanctuary forest was used for examinations. 1800g samples of soil under examination were preliminary air dried, than sieved by screen of 1 mm sieve mesh and than put into buckets. 200g of pine needles were added to each sample, than they were mixed and soaked to obtain moisture by weight of approx. 26% (moisture was kept on this level within the whole period of time). Two series of examinations were carried out. In the first series, copper occurred as $CuSO_4$, and in the second one as $Cu(CH_3COO)_2$. The consecutive samples of soil in each series of examinations were polluted with doses of 100, 500, 1000, 2500, 5000 and 10000 ppm of copper. Additionally, in the third series of examinations, three soil samples polluted with 5000 ppm of Cu and $CuSO_4$ and three samples polluted with 5000 ppm of Cu and $Cu(CH_3COO)_2$ were subjected to reaction of modifying agents, i.e. an extract of pine needles, 1% of glucose solution and $CaCO_3$.

Activity of separate enzymes in soil polluted with copper salts are presented as percentage of activity of these enzymes in control soil (Russel, 1972). Modifying compounds were added to separate samples after 4 months from the moment of pollution of soil with copper compounds. Content of copper in needles was determined in eleventh month of experiment using AAS method.

RESULTS

Activity of Cx cellulase in soil polluted with copper sulfate maintained at the same level during the whole period of examinations, independently from dose of xenobiotic and fluctuated within the limits of 105% of activity of test sample (Figure 1). However, in case of copper acetate the effect of toxic reaction of this compound was present. As metal dose increased, activity of cellulase Cx decreased beginning from value of 100% to 60% in eleventh month of examination.

After three month of experiment, the very high decrease of amylolytic activity under influence of copper sulfate was observed. Already at 100 ppm dose of Cu, amylolytic activity decreased up to 25% of activity of test sample and as dose of metal increased, it still decreased (Figure 2). After 7 months of examinations, an increase of amylolytic activity in soil was found up to the value of 40 to 85% depending on dose of copper. After 11 months, activity of amylases was constant independently from dose of copper sulfate and was approximately 30 - 40% of control activity. In case of copper acetate, decrease of amylolytic activity from 65 to 15% was found after the third month of experiment together with increase of dose of acetate. After 7 and 11 months of examinations activity of amylases did not depend on dose of copper and fluctuated within the limits of 30 - 40% of control activity.

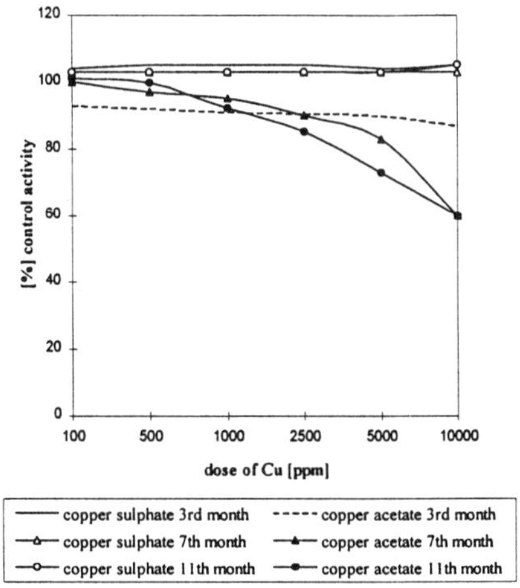

Figure 1 Changes of cellulase Cx activity provoked by copper compounds.

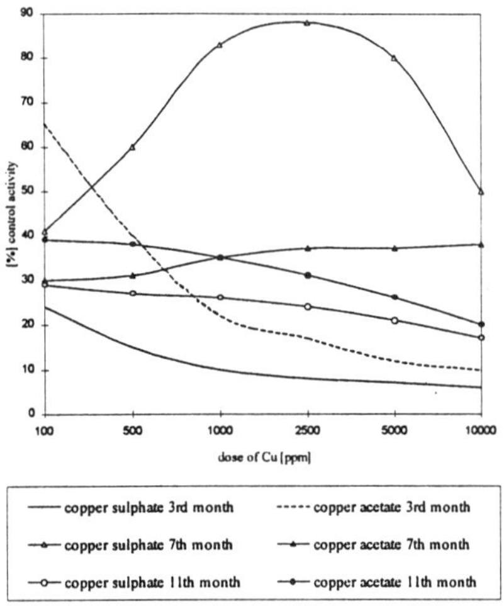

Figure 2 Changes of amylase activity provoked by copper compounds.

Dose of 100 ppm of Cu in copper sulfate did not cause decrease of proteolytic activity. As dose increased, slight decrease of activity of up to 70% for 10 000 ppm of Cu was observed (Figure 3). In seventh month, activity curve was similar, whereas in eleventh month - proteolytic activity decrease was found only above 1000 ppm of Cu. For dose of 10 000 ppm of Cu the value of activity was 50% of control activity. Proteolytic activity decreased rapidly in the presence of copper acetate after the first three months of examination. From the initial value of 97% of activity for the dose of 100 ppm Cu, proteolytic activity reached zero for the dose of 10 000 pp Cu. As time went by, the influence of copper acetate on products activity was decreasing and in the eleventh month had been almost entirely reduced. Decrease of activity for the greatest dose of copper was only 10% of proteolytic activity in test sample.

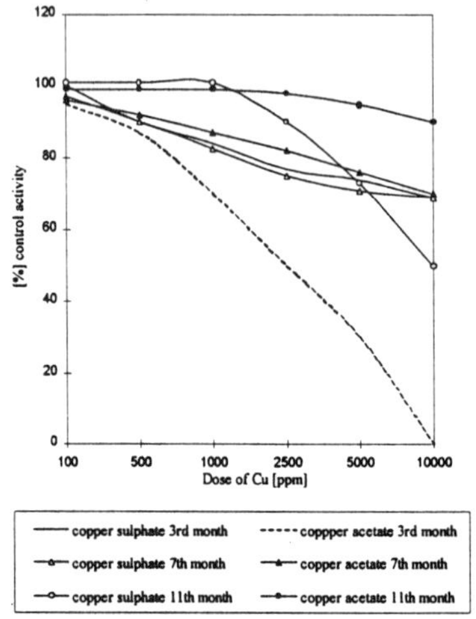

Figure 3 Changes of protease activity provoked by copper compounds.

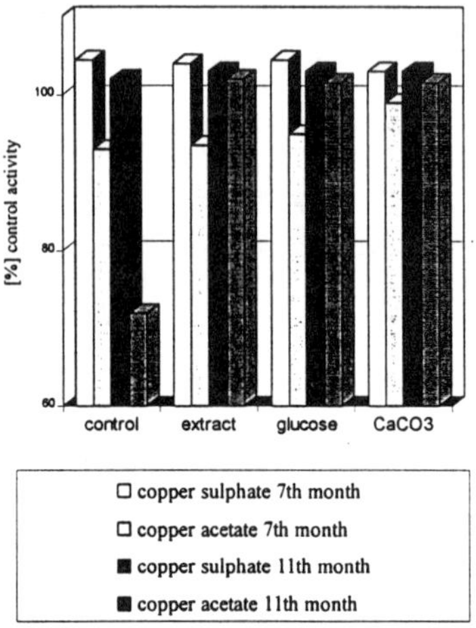

Figure 4 Cellulase activity in soil enriched in modification compounds polluted by 5 000 ppm Cu.

Compounds modifying the biological properties of soil, i.e. extract of pine needles, glucose and calcium carbonate had no significant influence on change of proteolytic activity , both in soil polluted with sulfate and copper acetate. In soil subjected to reaction of copper acetate, an addition of modifying agent caused en distinct increase of activity of cellulase Cx, reducing entirely the toxic activity of this compound in eleventh month of experiment (Figure 4).

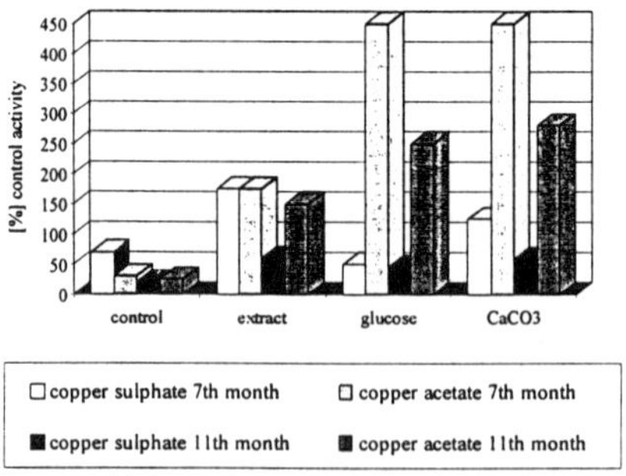

Figure 5 Amylase activity in soil enriched in modification compounds polluted by 5 000 ppm Cu

Amylolytic activity of soil subjected to reaction with copper acetate increased significantly after adding calcium carbonate and glucose, up to 450% of control activity (Figure 5). In soil polluted with copper sulfate, extract of pine needles and calcium, both in seventh and eleventh month, caused increase of amylolytic activity cancelling entirely toxic activity of this salt (Figure 5).

In soil subjected to reaction of copper acetate, accumulation of copper in needles was lower than in needles of soil subjected to reaction with copper sulfate for the same dosages of metal (Figure 6 and 7). Agents modifying biological properties of soil had not distinct effect on quantity of copper accumulated in needles.

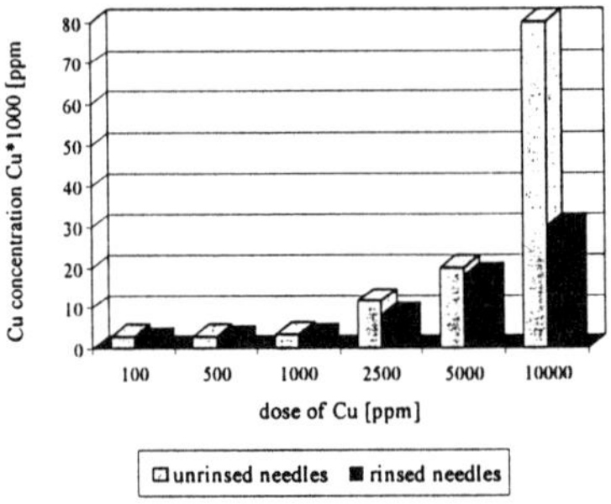

Figure 6 Copper concentration in pine needles after 11- month- experiment with copper acetate.

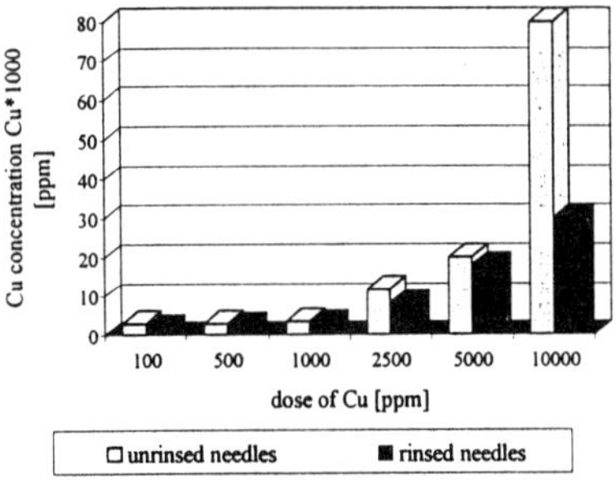

Figure 7 Copper concentration in pine needles after 11- month experiment with copper sulphate.

DISCUSSION

Examinations carried out proved that enzymatic activity of soil micro-organisms depends on a dose as well as on form in which copper was introduced into environment. Copper acetate as salt of weak acid may subject to hydrolysis and stable immobilisation in soil. However, copper sulfate is salt of strong acid and is immobilised more weakly and therefore is more reactive one.

Toxicity of copper compounds could partly result from low reaction of soil solution. An environment of low reaction is disadvantageous for growth of micro-organisms, especially of bacteria. At low reaction, accessibility of necessary elements such as carbon, nitrogen and phosphorus is changed (Aleksander, 1980). Decrease of the reaction has disadvantageous effect on catalytic activity of soil enzymes, because of occurrence of denaturation changes in enzymatic proteins structure.

In soil subjected to reaction with copper acetate, within the scope of dosages 100 - 10 000 ppm Cu, accumulation of copper in pine needles was lower than in needles of soil polluted with copper sulfate by organic compounds existing in pine needles, because as salt of weak acid and weak alkali is subjected to hydrolysis and stable immobilisation on soil colloids.

Amylolytic activity of soil micro-organisms was mostly inhibited by copper salts. Supplying to soil subjected to reaction with copper salts of easily assimilable nutrition substances in form of extract of pine needles caused increase of amylolytic activity as compared with soil with addition of only salts.

Increase of physiological activity of micro-organisms could be a result of supplying to soil, together with extract of needles, not only nutrition substances but also growth stimulators. One of them is shikimic acid occurring in needles extract.

The best effect of action of modifying agents were found in case of soil, where calcium carbonate was added, however water addition of pine needles extract had the least influence.

Addition of calcium carbonate to soil caused increase of pH value, and at the same time, increase of enzymatic activity of soil (Figure 4 and 5). Whereas addition of easily assimilable carbon source in form of glucose caused increase of amylolytic and cellulolytic activity of soil micro-organisms (Figure 4 and 5), destroying entirely toxic action of this metal.

Presentation of results obtained in form of complete hypothesis is a quite difficult problem. Decrease of enzymatic activity influenced by added heavy metal could be caused both by direct and indirect influence on properties of soil environment.

REFERENCES

Aleksander, M. 1980. "Effects of acidity on microorganisms and microbial processes in soil". *Effects of Acid Precipitation on Terrestrial Ecosystems.* T.C. Hutchinson, M. Havas, New York, Plenum Publishing Corp. 363-374.

Russel, S. 1972. *Methods of soil enzymes determination.* (Polish). Polish Society of Pedology. Warsaw.

RECOMBINANT STRAIN APPLICATIONS FOR BIOREMEDIATION OF HYDROPHOBIC ORGANICS IN SOILS

Gary S. Sayler (The University of Tennessee, Knoxville, Tennessee)

ABSTRACT: Premanufacturing Notifications (PMNs) to the U.S. EPA have been made for the release of genetically modified bacteria in bioremediation process research and development. Strains of *Pseudomonas fluorescens* and *Alcaligenes eutrophus* have been developed as field application vectors (FAV) for use in PCB-contaminated soil bioremediation. The PMN filed on these organisms seeks to use these strains in conjunction with surfactant soil flushing and bioremediation. The FAV strains were selected for growth on commercial surfactants and were modified by transposon introduction of a biphenyl degradation operon, to permit cometabolic PCB transformation at the expense of surfactant as a growth substrate. Field tests with these strains in bioremediation of electrical substation soil are planned following regulatory approval.

On October 30, 1996 *Pseudomonas fluorescens* strain HK44 was released in large-scale soil lysimeter test facilities at Oak Ridge National Laboratory. Strain HK44 maintains an introduced NAH7-like naphthalene degradative plasmid with a *lux* CDABE bioluminescent gene cassette, introduced via transposon TN4431, in the *nah*G gene (salicylate hydroxylase). Approximately 10^{14} total cells were released to five lysimeters by spray application of the organisms to layered (4") soil lifts to build the three-foot treatment bed of the lysimeter. The objective of the ongoing study with this genetically modified strain is to explore the use of on-line bioluminescent sensing in bioremediation process monitoring. A number of factors influencing the ongoing field use of the recombinant strains will be discussed.

PROSPECTS FOR USE AND REGULATION OF TRANSGENIC PLANTS IN PHYTOREMEDIATION

David J. Glass (D. Glass Associates, Inc., Needham, Massachusetts)

ABSTRACT: The plant species currently being developed for phytoremediation seem capable of effective bioaccumulation of targeted contaminants, but efficiency might be improved through the use of transgenic (genetically engineered) plants. Transgenic plants were first field tested in the United States in 1986, and thousands of research field tests have since taken place in the U.S., Canada and Europe under reasonable regulatory regimes. Many specific transgenic varieties have been exempted from regulation based upon a record of safe research use, and many novel crop varieties are being sold and used commercially. It should be possible to routinely obtain government approvals for field testing and ultimate commercial use of transgenic plants in phytoremediation.

PROSPECTS FOR GENETIC ENGINEERING

All commercial and research activity to date in phytoremediation has used naturally occurring plant species. However, many of these are species that can be genetically engineered, including *Brassica juncea,* which is being investigated for phytoremediation of heavy metals from soils (Dushenkov et al., 1995), sunflower, *Helianthus annuus*, being tested for rhizofiltration of uranium (Dushenkov et al., 1995) and poplar trees (*Populus deltoides nigra*), being investigated for the accumulation of nitrates and other organic chemicals from soil (Schnoor et al., 1995). In general, any dicotyledonous plant species can be genetically engineered using the *Agrobacterium* vector system, while most monocotyledonous plants can be transformed using particle gun or electroporation techniques.

Genetic engineering might be used to improve heavy metal phytoremediation by introducing biochemical traits that enhance hyperaccumulation. Examples might include genes controlling the synthesis of peptides that sequester metals, like phytochelatins (e.g., the Arabidopsis *cad*1 gene of Howden et al. 1995), genes encoding transport proteins, such as the Arabidopsis *IRT*1 gene that encodes a protein that regulates the uptake of iron and other metals (Eide et al. 1996) or genes encoding enzymes that change the oxidation state of heavy metals, like the bacterial *mer*A gene encoding mercuric oxide reductase (Rugh et al. 1996).

REGULATION OF TRANSGENIC PLANTS

United States. Genetically engineered plants are regulated in the United States by the U.S. Department of Agriculture (USDA) under regulations first promulgated in 1987 (52 Federal Register 22892-22915). Although these regulations arose from the debates over "deliberate releases" of genetically engineered organisms in the mid 1980s, field tests of plants have never been unusually controversial (see Glass 1991 for a historical review). Today these rules present only a minimal barrier against research field tests, and also allow commercial use of transgenic plants under a reasonable regulatory regime.

Under these regulations, USDA's Animal and Plant Health Inspection Service (APHIS) uses the Federal Plant Pest Act to regulate outdoor uses of transgenic plants. Originally, permits were required for most field tests of genetically engineered plants. Permit applications must include a description of the modifications made to the plant, data characterizing the stability of these changes, and a description of the proposed field test and the procedures to be used to confine the plants in the test plot. Submitters must also assess potential environmental effects, such as those shown in Table 1. APHIS review of these field test proposals has usually been completed well within the 120 days allowed by the regulations.

Table 1. Key Scientific Concerns: Field Testing Transgenic Plants.

Introduced DNA
- Characterization of vector system, its stability and expression in the plant cell.
- Avoidance of infectious, pathogenic, toxic or deleterious functions encoded by introduced DNA.

Host Plant
- Reproduction and pollen/seed dispersal mechanisms.
- Ability to outcross with related species (particularly wild relatives).
- Status as a weed, characteristics involved in weediness (ability to compete, survive and spread in the environment).

Environmental Impacts
- Comparison of transgenic to wild type (e.g., competitiveness).
- Unintended effects (e.g. effect on birds, insects, etc.).
- Possibility of gene transfer to other plants.

Test Conditions
- Initial field trials may require monitoring, confinement procedures, pollination controls.

These regulations were substantially relaxed in 1993 (58 Federal Register 17044-17059) to create two procedures to exempt specific plants. Under the first, transgenic plants of six specific crops (corn, soybean, tomato, tobacco, cotton and potato) can be field tested merely upon notifying the agency 30 days in advance, provided the plants did not contain any potentially harmful genetic sequences and the applicant provided certain information and submitted annual reports of test results. The second procedure allowed applicants to petition that specific transgenic plant varieties be "delisted" following several years of safe field tests, to proceed to commercial use and sale without the need for yearly permits. In August 1995, APHIS proposed further revisions under which most transgenic plants could be field tested under a notification rather than a permit, and where annual reports would only be required if the tests showed unexpected adverse effects (60 Federal Register 43567-43573). At this writing, this regulation has not yet been adopted.

The USDA regulations have allowed a large number of field tests to be carried out with moderate levels of government oversight: through October 1995, APHIS had received about 1,800 permits or notifications for field tests of at least 17 different plant species, representing several thousand discrete experiments (USDA 1996, see Figure 1).

Through the end of 1996, 27 different transgenic varieties had been delisted for commercial use (USDA 1996). Although other government regulation may apply for plants used for traditional agricultural purposes (for example, Food and Drug Administration review of food crops), transgenic plants for phytoremediation could be commercialized upon USDA delisting, subject to regulation under the applicable hazardous waste laws.

Regulation in Other Countries. Under a regulatory scheme similar to that of the U.S., Agriculture Canada has reviewed over 600 submissions since 1988 requesting permission for over 3,000 research field tests, about half of which involved canola (Agriculture Canada 1996). As of July 1, 1996, the agency had granted commercial approval for 15 different plant varieties (Yanchinski 1996).

In the European Union (EU), outdoor use of genetically engineered plants is governed by national laws adopted to implement an April 1990 directive (90/220/EEC). Member states can approve R&D uses within their jurisdiction, after notification to the European Commission (EC) and a 90 day review period. Applications for commercial use are to be made to a single member state, which would forward its recommendation to the EC within 90 days. A product approved by one nation could be marketed across the entire EU in the absence of objections from other states within 60 days. Individual national laws vary widely, with some countries (e.g. France) offering quick, inexpensive reviews and others like the U.K. and Germany requiring six month review periods (Lloyd-Evans and Barfoot 1996). Although environmental activism is still a problem in some EU nations, most has been directed against food uses. Nevertheless, there have been over 500 approved field releases in the EU between 1991 and 1995 (Lloyd-Evans and Barfoot 1996).

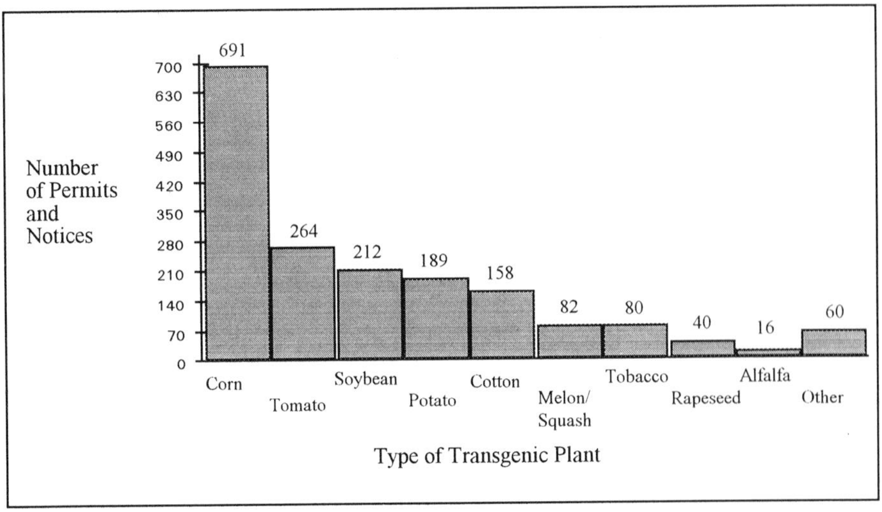

FIGURE 1. U.S. Field Tests of Transgenic Plants, 1987-1995, by type of plant (Source: USDA 1996).

In Japan, which has historically been very conservative about allowing open-field experiments with engineered organisms, there have now been five field tests of transgenic plants (OECD 1996).

USE OF TRANSGENIC PLANTS IN PHYTOREMEDIATION

There has not yet been a field test of a transgenic plant engineered for enhanced phytoremediation, although there have been field tests of tobacco plants engineered to express the metallothionein gene in their roots to avoid accumulation of metals in the leaves (USDA 1996). Proposals for research field trials of genetically engineered plants for phytoremediation should be handled routinely by the appropriate agency, and requests for commercial use could be expected to be straightforward once initial trials were conducted safely. The following are some considerations that applicants might keep in mind in preparing for such field tests.

Host Plant Biology. The two most important environmental issues relate to possible enhancement of the weediness of the plant, and its potential to outcross to related species. These will be discussed using examples of plant species likely to be used in phytoremediation: *B. juncea* and *H. annuus*.

Weediness. Single gene changes can enhance weediness, although more often multiple changes are needed (Keeler 1989, NRC 1989), but crops that have been subject to extensive agricultural breeding are less likely to revert to a weedy phenotype by simple genetic changes (NRC 1989). However, many plant species used in phytoremediation are not as well-characterized or as long-cultivated as agricultural crop species, and some may be, or be related to, weeds. For example, certain varieties of *B. juncea* are known to be weeds in the U.S., as are several related species like *B. nigra* and other species within the *Brassica* genus (USDA 1994). The genus *Helianthus* is also known to include a number of wild and weedy species (USDA 1991). It might be necessary to consider whether genes encoding an enhanced hyperaccumulation phenotype would confer any growth advantage or enhance weediness, either through literature evidence or laboratory experiments on the growth rates of the transgenic plants.

Cross-pollination to weedy relatives. A transgene introduced via cross-pollination into a weedy relative might give the recipient plant a competitive advantage or confer a weedy trait. Almost all plants have wild relatives (NRC 1989), so every plant species of commercial utility would have some potential to interbreed with wild, perhaps weedy, species. Commercially-cultivated plants are more likely to cross-breed when grown in the region of the world from which they originated, which is where close genetic relatives would most likely be found.

Different *Brassica* species can interbreed: for example, it is known that *B. juncea* and the oilseed crop *B. napus* can cross, at least where *napus* is the pollinator, and genetic material can be transferred using an intermediate species to bridge the gap between otherwise incompatible species (USDA 1994). *B. juncea* is believed to have originated in China, and other important *Brassica* species either in China, temperate Asia or Europe (NRC 1989), lessening the likelihood of outcrossing to wild relatives for uses in North America and regions of Europe. *Brassica* are generally pollinated by honeybees, but years of cultivation of rapeseed and other crops have led to a knowledge of the required isolation distances needed to prevent accidental outcrossing from one species to another (USDA 1994).

Cultivated sunflower is known to be capable of crossing to wild and weedy forms of *H. annuus*. Honeybees are the primary pollinator, and sunflower pollen can be transported very long distances, requiring long isolation distances for commercial breeding (USDA 1991). *Helianthus* is believed to have arisen in North America, increasing the chance of outcrossing to a wild relative (NRC 1989).

In small-scale agricultural field trials, the possibility of cross-pollination has generally been mitigated by preventing pollination, for example, by bagging or removing the pollen-producing organs or harvesting biomass before flowering. The need for pollen control has generally been relaxed for any given transgenic variety, as more is known about its safety. However, phytoremediation experiments are likely to take place in urban or industrial areas and are likely to be far from related plants. For those uses that happen to be near agricultural areas, the recognized procedures for pollen control can be utilized, particularly if the plants need to be harvested frequently to remove accumulated contaminants.

Cross-pollination to food crops. For phytoremediation, one must also be concerned over transfer of a hyperaccumulation phenotype into crop plants, causing contaminants to enter the food chain. Among the many *Brassica* species are a number that are used as foods or condiments (e.g., cabbage, broccoli, mustard) and *Helianthus tuberosus* is the Jerusalem artichoke. However, the proximity of such food crops to a hazardous waste site would itself be more immediately troubling than the remote chance of gene transfer from a transgenic.

Other Issues. For all proposed field tests, regulatory agencies would want to be certain that the products of the introduced genes are not toxic or pathogenic. One concern unique to phytoremediation might be the potential risks to birds and insects who might feed on plant biomass containing high concentrations of hazardous substances, particularly metals. Questions relating to the proper disposal of plants after use would also arise, and commercial approvals may require restrictions on the use of the harvested plant biomass for human or animal food. However, such concerns would be common to all uses of phytoremediation.

IMPLEMENTATION PLAN

Government regulation should not be a significant obstacle to the use of transgenic plants in phytoremediation. Applicants should consult with the appropriate regulatory agency in advance of any actual permit request, to ascertain the data the agency expects to see, which should not take longer than 2 to 3 months to obtain. Permit applications should be filed approximately 4 months before the planned starting date, but in most jurisdictions, approval for research field tests should be routine. As commercial use of a given variety approaches, close consultation with the regulators is also recommended. One can expect the agricultural review agency to consult with any applicable hazardous waste regulators, to ensure proper communication.

REFERENCES

Agriculture Canada. 1996. *Field Trials and Registered Products of Biotechnology Under Agriculture and Agri-Food Canada's Acts*, available at http://aceis.agr.ca/./fpi/agbiotec/fieldt96.html.

Glass, D. J. 1991. "Impact of Government Regulation on Commercial Biotechnology." In R. D. Ono, (Ed.), *The Business of Biotechnology: From the Bench to the Street*, pp. 169-198. Butterworth-Heinemann, Stoneham, MA.

Dushenkov, V., P. B. A. Nanda Kumar, H. Motto, and I. Raskin. 1995. "Rhizofiltration: The Use of Plants to Remove Heavy Metals from Aqueous Streams." *Env. Sci. Tech.* 29(5): 1239-1245.

Eide, D., M. Broderius, J. Fett, and M. L. Guerinot. 1996. "A Novel Iron-Regulated Metal Transporter from Plants Identified by Functional Expression in Yeast." *Proc. Natl. Acad. Sci.* 93(11):5624-5628.

Howden, R., P. B. Goldsborough, C. R. Anderson, and C. S. Cobbett. 1995. "Cadmium-Sensitive, *cad*1 Mutants of *Arabidopsis thaliana* are Phytochelatin Deficient." *Plant Physiol.* 107:1059-1066.

Keeler, K. H. 1989. "Can Genetically Engineered Crops Become Weeds?" *Bio/Technology*. 7(11): 1134-1139.

Lloyd-Evans, L. P. M., and P. Barfoot. 1996. "EU Boasts Good Science Base and Economic Prospects for Crop Biotechnology." *Gen. Eng. News.* 16(13): 16.

National Research Council. 1989. *Field Testing Genetically Engineered Organisms: Framework for Decisions*. National Academy Press, Washington, DC.

Organization for Economic Cooperation and Development. 1996. *OECD's Database on Field Trials*, available at http://www.oecd.org/ehs/biotrack.

Rugh, C. L., H. D. Wilde, N. M. Stack, D. M. Thompson, A. O. Summers, and R. B. Meagher. 1996. "Mercuric Ion Reduction and Resistance in Transgenic Arabidopsis Thaliana Plants Expressing a Modified Bacterial MerA Gene." *Proc. Natl. Acad. Sci.* 93(8): 3182-3187.

Schnoor, J. L., L. A. Licht, S. C. McCutcheon, N. L. Wolfe, and L. H. Carreira. 1995. "Phytoremediation of Organic and Nutrient Contaminants." *Env. Sci. Tech.* 29(7): 318A-323A.

USDA. 1996. *Biotechnology Permits Homepage*, available at http://www.aphis.usda.gov/bbep/bp.

USDA. 1991. *Environmental Assessment and Finding of No Significant Impact, Permit 91-067-01 to Pioneer Hi-Bred*, available at USDA (1996).

USDA. 1994. *Response To Calgene Petition 94-090-01p For Determination Of Nonregulated Status For Laurate Canola Lines*, available at USDA (1996).

Yanchinski, S. 1996. "Canada Watch." *Gen. Eng. News.* 16(18):11, 31.

BIODEGRADATION OF AROMATIC HYDROCARBON MIXTURES IN HEAVILY CONTAMINATED STREAMS BY IMMOBILIZED RECOMBINANT MICROORGANISMS

Vitor A. P. Martin dos Santos and Juan Luis Ramos (Consejo Superior de Investigaciones Científicas, Estación Experimental del Zaidín, Granada, Spain

ABSTRACT:

Genetically Engineered Microorganisms. The use of genetically engineered microorganisms (GEMs) has been often acknowledged as having an enormous potential for the treatment of toxic wastes in general and xenobiotic compounds in particular. Microorganisms with specific metabolic capabilities for biodegradation are specially indicated in the frequent situations where, for instance, indigenous microorganisms are unable to degrade the target pollutants, when degradation should occur rapidly or when the unwanted chemical is present at inhibiting concentrations for the native populations. However, despite the promising possibilities, the actual use of recombinant microorganisms in the treatment of soils, groundwaters and waste streams is still very limited. The additional complexity of the microorganisms, their limited survival and genetic (in)stability in natural and engineered systems, and ecological concerns about the release of GEMs in the environment often hamper their use in bioremediation. Moreover, that specialized microorganisms (such as GEMs) possess better or unique intrinsic biodegradation abilities does not guarantee success upon their introduction in contaminated ecosystems or in treatment units. These microorganisms should also be able to cope with the stresses in the environments where they are introduced.

Thus, there is a strong need to develop appropriate tools and to extend the knowledge of the factors that affect the behavior of GEMs in natural and engineered environments for their effective use in biodegradation and to develop ways to "preserve" the microorganisms in the chosen target ecosystem.

Cell Immobilization. Immobilization of recombinant cells in solid supports can largely reduce some of the above-mentioned problems. Encapsulated cells are shielded from the surroundings while the target pollutants still can flow into the supports and be metabolized there. Immobilization can be itself a form of biocontainment since it provides a way to control, to a certain extent, the spreading of recombinant cells in the environment. Furthermore, immobilization has been reported to increase the tolerance of cells to organic solvents and to increase genetic stability of recombinant microorganisms. Additionally, the immobilization of high cell densities in compact reactors results in enhanced biodegradation rates when compared to conventional systems. Since such a system is much less dependent of the growth rates of the microorganisms involved, short retention times can be applied and thus high rates attained. These characteristics make the use of immobilized cells particularly attractive for the treatment of groundwaters and aquifers heavily contaminated with toxic, relatively soluble pollutants such as BTEX compounds. Most of these pollutants can be degraded at low concentrations under aerobic or anaerobic conditions (or in aerobic/anaerobic sequences). However, at high concentrations they remain mostly unattacked or require extremely long adaptation periods because most microorganisms lack adequate tolerance mechanisms. We have recently isolated a *Pseudomonas putida* strain (DOT-T1-5) able to grow in media containing up to 90% (v/v) toluene. The catabolic potential of the isolate was expanded to include the degradation of m- and p-xylene and related hydrocarbons by transfer of the TOL plasmid pWW0-Km. These characteristics make this strain particularly suitable for the treatment of streams heavily contaminated with mixtures of alkylsubstituted benzenes.

In this work we describe, characterize and evaluate a system with immobilized recombinant cells for in-situ bioremediation of heavily contaminated groundwater using above-ground air-driven reactors. Degradation of (mixtures of) alkylsubstituted benzenes at high concentrations was used as a model to assess the viability of this system. The behaviour of free and immobilized *Pseudomonas putida* DOT-T1-5 TOL plasmid pWW0-Km was studied with respect to its biodegradation activity, genetic stability, tolerance to high concentrations of mixtures of substrates, and survival capacity in a continuous reactor.

DEGRADATION OF TRICHLOROETHYLENE BY BACTERIAL MONO- AND DIOXYGENASE

Wako Takami, Toshio Shimizu and Shigeyuki Imamura (Asahi Chemical Industry Co.,Ltd., Fuji city, Japan)
Hideaki Nojiri, Hisakazu Yamane and ***TOSHIO OMORI*** (Biotechnology Research Center, The University of Tokyo, Japan)

ABSTRACT: It was demonstrated that the *Acinetobacter* sp. 20B growing on dimethyl sulfide (DMS) and *Pseudomonas fluorescens* IP01 growing on isopropylbenzene (cumene) were capable of degrading trichloroethylene (TCE). *Escherichia coli* MV1184 was transformed with DMS monooxygenase genes (*dsoABCDEF*) from strain 20B, and/or cumene dioxygenase genes (*cumA1A2A3A4*) from strain IP01. The TCE degradation abilities of the recombinants were examined. The recombinant *E. coli* harboring cumene dioxygenase genes degraded 10 ppm (v/v) TCE (73 μg/vial) for 3 hours completely and the recombinant *E. coli* harboring DMS monooxygenase genes degraded 10 ppm (v/v) TCE for 5 hours completely. The recombinant *E. coli* harboring genes both cumene dioxygenase and DMS monooxygenase genes degraded 10 ppm (v/v) TCE for an hour completely.

INTRODUCTION

Trichloroethylene (TCE), which is widely used as an organic solvent and degreasing agent, has been known to contaminate ground water and soil. It is necessary to remove TCE from polluted sites so as to protect drinking water supplies from contamination, because TCE is known to persist for a long period of time in the environment and to be toxic and carcinogenic. There have been several reports on TCE degradation by various types of aerobic bacteria including methane- (Litlle et al., 1988) and aromatic compounds-oxidizing bacteria (Nelson et al., 1988).

Biodegradation of TCE by these wild strains is efficient at a site contaminated with low concentrations of TCE, however it is difficult to apply these bacteria to bioremediation of a site contaminated with high concentrations of TCE. In this study, we found that the use of the recombinants containing various oxygenase genes in the closed bioremediation system is effective to degrade high concentrations of TCE.

This paper describes TCE degrading activities of *P. fluorescens* IP01, *Acinetobacter* sp. 20B and recombinant *E. coli* harboring the genes responsible for TCE degradation in strains IP01 and 20B.

MATERIALS AND METHODS

Bacterial strains and plasmids. *P. fluorescens* IP01 (Aoki et al., 1996) and *Acinetobacter* sp. 20B (Horinouchi et al., unpublished) were isolated from soil samples by enrichment culture technique. *E. coli* MV1184 was utilized as the host strain for all plasmids. The pUC119 and pUC118 vectors [*lacZ'*, ampicillin (AP) resistant] were used as cloning vectors. All plasmids were constructed by conventional recombinant techniques, and the host strains were transformed with the recombinant plasmids.

TCE Degradation. *P. fluorescens* IP01 was grown on a carbon-free minimal medium (CFM) (1 *l*) containing 2.2 g of Na_2HPO_4, 0.8 g of KH_2PO_4, 3.0 g of NH_4NO_3, 10 mg of $FeSO_4{\cdot}7H_2O$, 10 mg of $CaSO_4{\cdot}2H_2O$, 10 mg of $MgSO_4{\cdot}7H_2O$, and 50 mg of yeast extract with 0.1 % (v/v) cumene at 30 °C. *Acinetobacter* sp. 20B was grown on a sulfur-free minimal medium (SFM) (1 *l*) containing 10 g of disodium succinate, 2.2 g of Na_2HPO_4, 0.8g of KH_2PO_4, 3.0 g of NH_4NO_3, 10 mg of $FeCl_2{\cdot}2H_2O$, 10 mg of $CaCl_2{\cdot}2H_2O$, 10 mg of $MgCl_2{\cdot}6H_2O$, and 20 ml of a modified Hunt's vitamin-free mineral base with 0.1 % (v/v) of DMS at 30°C. *E. coli* was cultured in 2YT medium (Trypton, 16g; yeast extract, 10 g; NaCl, 5 g/L) at 37 °C. To induce the oxygenases, cumene, DMS or IPTG was used. Wild strains were cultured in several medium (CFM with cumene, SFM with DMS) at 30 °C on a rotary shaker at 120 rpm. When the optical density of the culture attained to 1.0 at 600 nm, the cells were harvested by centrifugation at 4,000 x g for 10 min at 4 °C. *E. coli* cells carrying recombinant plasmids were cultured in 2YT with 50 mg/l of ampicillin at 37 °C on a rotary shaker at 120 rpm. When the optical density of the culture attained to 0.5 at 600 nm, IPTG was added to a final concentration of 1 mM and the cell growth was continued. When the optical density of the culture attained to 2.0 at 600 nm, the cells were harvested by centrifugation in the same manner as the harvest of the wild strains. The harvested cells were washed in 50 mM phosphate buffer (pH, 7.0) at three times, and were resuspended. A 5 ml aliquot of the cell suspension was added to a 14 ml crimp-seal glass vial, which was sealed with a Teflon-coated rubber septum and an aluminium crimp seal. TCE or chlorinated compound (1000 ppm stock solution) was added to the glass vial through the Teflon-coated rubber septum. The vials were incubated at 30 °C on a rotary shaker at 300 rpm. A 100-μl portion of the gas phase in each sealed glass vial was taken by gas tight syringe periodically and injected into a gas chromatograph equipped with a flame ionization detector.

Analysis of Chloride Ion. Chloride ion released from TCE during the degradation test was quantified by ion chromatography. The test sample was centrifuged (10,000 x g, 10 min at 4 °C) and the supernatant was used for the analyses.

RESULTS AND DISCUSSION

TCE Degradation by wild strains. *P. fluorescens* IP01 cultivated on cumene and *Acinetobacter* sp. 20B cultivated on DMS were able to degrade TCE (Fig.1). *P. fluorescens* IP01 degraded about 90% of added TCE (7.3 μg/vial) and *Acinetobacter* sp. 20B degraded 24% of added TCE (7.3 μg/vial) after 24 hours. However, both wild strains degraded neither tetrachloroethylene nor 1,1,1-trichloroetane (data not shown).

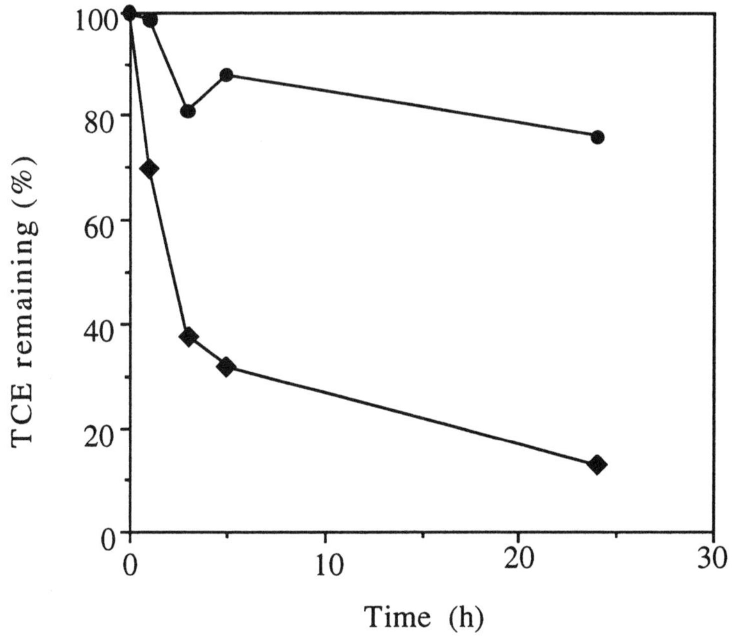

FIGURE 1. TCE degradation by *P. fluorescens* IP01(◆)and *Acinetobacter* sp. 20B (•). TCE was added into the buffer at an initial concentration of 1ppm (7.3 μg/vial).

Recombinant Plasmids and *E.coli* harboring the Recombinant Plasmids. DMS monooxygenase genes (*dsoABCDEF*) in strain 20B and cumene dioxygenase genes tandemly linked with cumene dihydrodiol dehydrogenase genes (*cumA1A2A3A4B*) were ligated to the plasmids pUC118 (pAU96) and pUC119 (pIP107) respectivily, *cumA1A2A3A4B* and *dsoABCDEF* were also tandemly linked to the plasmid pUC119 (pIO720). *E. coli* MV1184 was transformed with these recombinant plasmids. The recombinant *E. coli* MV1184 (pAU96) generated dimethyl sulfoxide when incubated with DMS, and the recombinant *E. coli* MV1184 (pIP107) generated 3-isopropylcatechol when incubated with cumene (Fig.2).

TCE Degradation by *E.coli* harboring the Recombinant Plasmids. TCE degradation by *E.coli* harboring the recombinant plasmids is shown in Fig.3. The recombinant *E. coli* harboring cumene dioxygenase genes (pIP107) degraded up to 75% of the added TCE (365 μg/vial) after 24 hours and the recombinant *E. coli* harboring DMS monooxygenase genes (pAU96) degraded up to 50% of the added TCE (365 μg/vial) after 24 hours. The recombinant *E. coli* harboring both cumene dioxygenase and DMS monooxygenase genes (pIO720) degraded up to about 90% of the added TCE (365 μg/vial) after 24 hours. These recombinants released chloride ion from TCE stoichiometrically (Table 1). From 1 mol of TCE, 2.1 to 2.6 mol of chloride ions were released. This suggests that almost complete dechlorination occurred in degradation of TCE by the recombinants. These recombinants were able to degrade 1,2-*cis*-dichloroethylene (1,2-*cis*-DCE) (320 μg/vial) completely after 24 hours. However, degradation and dechlorination activity of 1,1-DCE by the recombinant *E. coli* MV1184 (pIP107) was lower than the recombinant *E. coli* MV1184 (pAU96) (data not shown). These results indicate that monooxygenation of 1,1-DCE occurs easier than dioxygenation of 1,1-DCE. The recombinant *E. coli* MV1184 (pIO720) was able to dechlorinate 1,1-DCE nearly completely (date not shown). In conclusion, this study indicates that the recombinant *E. coli* harboring cumene dioxygenase and/or DMS monooxygenas genes are/is quite useful tool for bioremediation of TCE in the closed system.

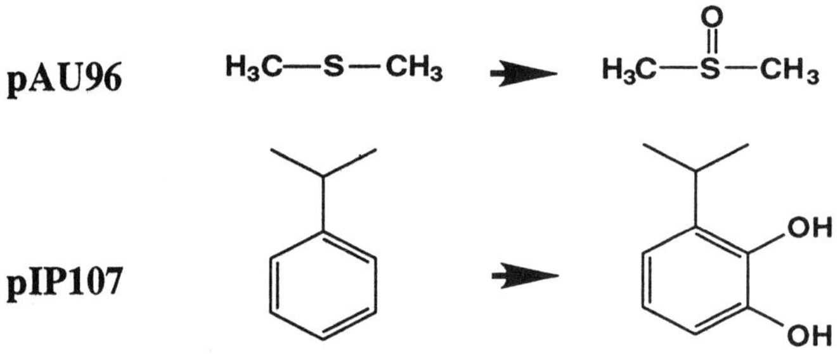

FIGURE 2. Products from DMS and cumene by *E.coli* recombinants with pAU96 and pIP107.

TABLE 1. Generation of chloride ion during TCE degradation by *E. coli* recombinants.

Strain	Initial TCE (μmol/vial)	TCE degraded (μmol/vial)	Chloride ion generated (μmol/vial)	Ratio*
E.coli MV1184				
pIO720	2.786	2.459	6.168	2.5
pIP107	2.786	2.183	5.730	2.6
pAU96	2.786	1.658	3.493	2.1

* Ratio of moles of chloride ion generated to moles of TCE degraded.

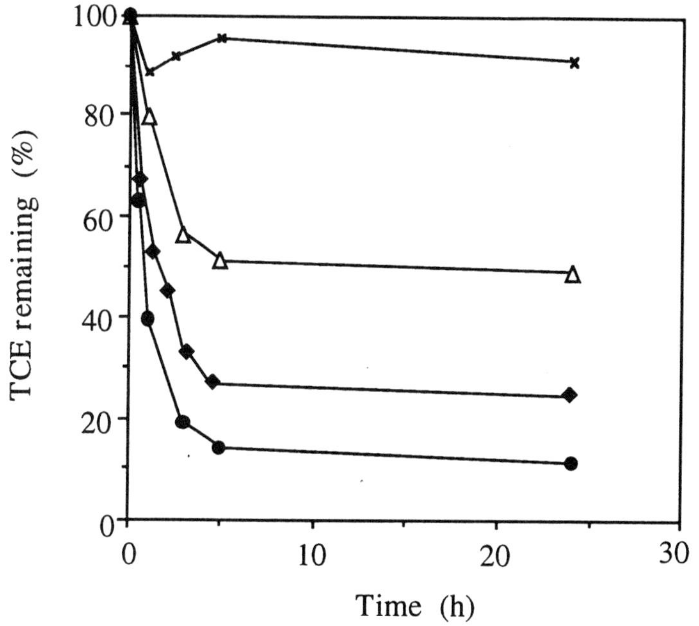

FIGURE 3. TCE degradation by the recombinant *E. coli* MV1184 (pAU96) (Δ), *E. coli* MV1184 (pIP107) (◆), *E. coli* MV1184 (pIO720) (•), and *E. coli* MV1184 (X). TCE was added into the buffer at an initial concentration of 50 ppm (365 μg/vial).

REFERENCES

Aoki, H., K. Toshiaki, H. Hiroshi, Y. Hisakazu, K. Tohru, and O. Toshio. 1996. "Cloning, Nucleotide Sequence, and Characterization of the Genes Encoding Enzymes Involved in the Degradation of Cumene to 2-Hydroxy-6-Oxo-7-Methylocta-2,4-Dienoic Acid in *Pseudomonas fuluorescens* IP01" *J. Ferment. Bioeng.* 81(3): 187-196.

Little, C. D., A. V. Palumbo, S. E. Herbes, M. E. Lindstrom, R. L. Tyndall, and P. J. Glimer. 1988. "Trichloroethylene Biodegradation by a Methane-oxidizing Bacterium" *Appl. Environ. Microbiol.* 54: 951-956.

Nelson, M. J. K., S. O. Montgomery, and P. H.Pritchard. 1988. "Trichloroethylene Metabolism by Microorganisms that degrade Aromatic Compounds" *Appl. Environ. Microbiol.* 54: 604-606.

TRANSFER OF PLASMID RP4: TOL TO LEGUME MICROSYMBIONTS FOR SOIL/RHIZOSPHERE BIOREMEDIATION

L.D. Kuykendall, W.K. Gillette, G.P. Hollowell (USDA, Beltsville, MD),
L.-H. Hou (Howard Univ., Washington, DC),
H.E. Tatem (Army Environmental Res., WES, Vicksburg, MS)
and S.K. Dutta (Howard Univ., Washington, DC)

ABSTRACT: Legume microsymbionts are good candidates for genetic modification to enhance soil/rhizosphere bioremediation since much is known of their genetics and their growth and metabolism in soil are both energetically and specifically stimulated by their host plants. Furthermore, together with their eukaryotic hosts, they fix nitrogen, permitting high, sustainable biomass yields that are potentially useful as energy fuel. As a first step, the broad-host-range plasmid RP4::TOL that carries the benzene, toluene, and xylene (BTX) degradative gene cluster was transferred in crosses between *Escherichia coli* C600 and symbiotically elite strains of alfalfa-nodulating *Rhizobium meliloti* and of soybean-nodulating *Bradyrhizobium japonicum* and *B. elkanii*. Reciprocal crosses with *E. coli* RR28 verified the plasmid-bearing *R. meliloti* exconjugants since they donated the original RP4::TOL plasmid, but only derivative plasmids with a substantial deletion of part of the TOL region were found in *E. coli* plasmid-bearing exconjugants from matings with *B. japonicum* and *B. elkanii* donors. *R. meliloti* (RP4::TOL) exconjugants completely metabolized toluate and transformed trinitrotoluene to a much greater extent than did the parental strain.

INTRODUCTION

Bioremediation research on soil contaminated with toxic chemicals is concerned with the following: (i) removal of or reduction in the amount of toxic substances so that the soil is no longer harmful to humans, animals, or plants, and (ii) prevention of the spread of toxic substances beyond the original localized area, avoiding contamination of other natural resources such as ground water and the atmosphere. Toxic aromatic hydrocarbons, such as benzene, toluene, ethylbenzene and xylene (BTEX), are widespread environmental pollutants. These chemicals are used as gasoline additives, in the production of plastics, dyes, solvents, and as components to make explosives, such as TNT, that also contaminate some soils. In soil, BTEX exist in an absorbed state but still may participate in chemical reactions or be subject to bioremediation. Since the general feasibility of rhizosphere approaches to remediation has been demonstrated (Anderson and Coats, 1994), our goal is to use genetically modified legume microsymbionts and their host plants for rhizosphere bioremediation of TNT and related compounds.

Bioremediation of TNT in soil has been studied. Medary (1992) developed a rapid field test for detecting TNT contamination in soil. Pennington and Patrick (1990) and Comfort et al. (1995) studied the fate of TNT in soil. Caton et al. (1994) characterized insoluble fractions of TNT transformed by composting. Bradley and

Chapelle (1995) studied factors affecting TNT mineralization in contaminated soil. Duque et al. (1993) developed a *Pseudomonas* strain that mineralized 2,4,6-trinitrotoluene (TNT). This construction partially relied on the introduction of pTOL from *P. putida* to isolates deaminating TNT. This research suggested to us that the introduction of the RP4::TOL plasmid could be used to enhance remediation by *Rhizobium*.

Whereas Williams and Murray (1974) originally described the TOL plasmid of *P. putida*, Nakazawa et al. (1980) detailed the molecular biology of RP4-TOL plasmids derived from recombination of RP4 and the TOL plasmid of *Pseudomonas putida* strain mt-2. For a review of the literature, see Burlage et al. (1989). Jeenes and Williams (1982) isolated *Pseudomonas* strain WR211 that had undergone a 39-kb deletion of pTOL concomitant with the integration of at least 17 kb of pTOL into the chromosome. Jeenes et al. (1982) also found a spontaneous derivative of strain WR211, WR216, which was unable to grow on *m*-xylene, but grew on *m*-toluate. This strain had a plasmid with a smaller deletion of 19 kb instead of 39 kb, and the plasmid had two 3-kb insertions, of undefined origin, causing an interruption of *xylE*, the structural gene for catechol 2,3-oxygenase. Love and Grady (1995) demonstrated the generation of a viable non-culturable state of *P. putida* during continuous culture on *m*-toluate (or benzoate). Thus genetic rearrangements may result from a very strong positive selection for the loss of certain functions encoded by TOL (Bayley et al., 1977).

Objective. Our objective was the intergeneric transfer of the RP4::TOL plasmid, originally from *Pseudomonas*, carrying genes that specify enzymes for toluene utilization to the alfalfa microsymbiont *Rhizobium meliloti* and the soybean microsymbionts *Bradyrhizobium japonicum* and *B. elkanii*.

MATERIALS AND METHODS

Escherichia coli carrying the RP4::TOL plasmid was obtained from Dr. Jeffrey S. Karns, Soil Microbial Systems Laboratory, Natural Resources Institute, Beltsville Agricultural Research Center, ARS, USDA, Beltsville, MD 20705. Intergeneric matings were by patching freshly grown, dense bacterial cultures onto KO, i.e., Kuykendall-O'Neill, nonselective growth medium (Gillette and Elkan, 1996), previously known as A1EG. Control plates and mixtures were incubated at 30°C for 16h in the case of fast-growing *Rhizobium*, and for 48h in cases where the recipients were *Bradyrhizobium* (Kuykendall, 1979). In intergeneric matings where *E. coli* was the recipient, LB agar was used and growth was at 37°C.

Verification of Exconjugants. Carbenicillin resistance (Cb^R), an unselected R factor marker, was useful in verifying the transconjugants whereas tetracycline resistance (Tc^R) was not (Table 1). By screening for Cb^R, transconjugants were distinguished from background kanamycin-resistant (Kn^R) mutants and thus verified as recombinants. Tc^R screening was not useful in verifying recombinants presumably since the TOL transposon is within this locus, rendering the gene inactive. The verified exconjugants were unambiguously confirmed by testing for their ability to serve as donors of RP4::TOL in crosses with *E. coli* RR28.

Collection of metabolites. Cells, grown aerobically at 30°C, were removed by centrifugation at 5000 × *g* for 10 min. Ten mL of supernatant, or culture fluid, was diluted with an equal volume of water then extracted twice with three volumes of ethyl acetate or methylene chloride. The extracts were dried over anhydrous sodium sulfate and excess solvent was removed by evaporation under reduced pressure at 42°C. The metabolites were then dissolved in one mL of water or methanol.

Analytical methods. Five-μL samples were analyzed with high-performance liquid chromatography (HPLC) by using a SUPELCOSIL™ LC-18, 150 × 4.6 mm column with water-methanol as the mobile phase with a flow rate of one mL min^{-1}. Compounds were detected at 254 nm with a Hewlett-Packard UV-Visible detector.

RESULTS

Transfer of the broad-host-range RP4::TOL degradative plasmid from *E. coli* strain C600 (RP4::TOL) to *Rhizobium meliloti* USDA 1936 Rif ^{R} was successful (Table 1). On the other hand, the results obtained when using either *B. japonicum* or *B. elkanii* first as recipients of this plasmid and then as donors clearly indicated that partial deletion or rearrangement of the TOL region occurred (Table 1).

TABLE 1. Intergeneric transfer of the broad-host-range plasmid RP4::TOL to *Rhizobium* and *Bradyrhizobium*.

Donor	Recipient	Transconjugant
E. coli C600 (RP4::TOL) Kn^R, kanamycin resistance, & Cb^R, carbenicillin resistance, are plasmid markers	*R. meliloti* USDA 1936 Rif^R, rifampicin resistance, a chromosomal marker	*R. meliloti* USDA 1936 Rif^R (RP4::TOL), Kn^R, Cb^R, donates plasmid in reciprocal crosses
E. coli C600 (RP4::TOL) Kn^R & Cb^R	*B. japonicum* USDA I-110 Rif^R	*B. japonicum* USDA I-110 Rif^R (R derivative), Kn^R & Cb^R, donor of plasmids differing from RP4::TOL in DNA restriction pattern
E. coli C600 (RP4::TOL) Kn^R & Cb^R	*B. elkanii* USDA 61 Nal^R, nalidixic acid resistance	*B. elkanii* USDA 61 Nal^R (R derivative), Kn^R & Cb^R, donor of plasmids missing 7.2-kb and 7.8-kb *Eco*RI fragments

Intergeneric Plasmid Transfer Back to *E. coli* for DNA Characterization. *E. coli* C600 (RP4::TOL) and the *E. coli* transconjugants carrying the RP4::TOL plasmid from *R. meliloti* strain USDA 1936 had plasmid DNAs with an identical *Eco*RI restriction pattern. However, *E. coli* exconjugants from matings with *Bradyrhizobium* plasmid donors had plasmids lacking both a 7.2-kb and a 7.8-kb *Eco*RI fragment, showing deletion of a minimum of 15 kb.

Utilization of Toluate and TNT Transformation. Growth stimulation of all exconjugants on minimal medium by 5mM *p*- and *m*-toluates was observed as was the accumulation of a brownish-red pigment in toluene-containing medium, rich or minimal. The *R. meliloti* USDA 1936 Rif^R parent strain and *R. meliloti* USDA 1936 Rif^R exconjugant carrying RP4::TOL were tested for utilization of *m*-toluate and transformation of TNT. There was 52% utilization of *m*-toluate by the parent strain, however, the exconjugant showed 100% utilization of this compound. HPLC analysis indicated about 18% TNT transformation by the parent strain, whereas the RP4::TOL-carrying exconjugant strain evidently transformed about 60% of the TNT initially present (Table 2).

TABLE 2. Decrease in concentration of toluate or TNT in 7-day-old cultures of parental and RP4::TOL-bearing exconjugant strains.*

Treatment	*m*-Toluate (mM)	TNT ($\mu g/\mu L$)
None (=Initial Concentration)	5.0	25.0
R. meliloti USDA 1936	2.4	20.6
R. meliloti (RP4::TOL)	0.0	10.3

*Two experiments.

DISCUSSION

Intergeneric transfer of plasmid RP4::TOL to *R. meliloti* was accomplished, but instability of the TOL region in *Bradyrhizobium* was evident. A deletion may have occurred due to one of the mechanisms described in the literature for TOL instability or a genetic rearrangement may have occurred due to a chromosomal recombination event as described by Berry and Atherly (1984).

We are presently evaluating the ability of *R. meliloti* (RP4::TOL) strains to carry out TNT degradation in contaminated soil via alfalfa rhizosphere-accelerated metabolism. Data currently available suggest mineralization of TNT, and radioactive tracer studies are planned.

ACKNOWLEDGMENTS

Supported in part by U.S. Army Research Grants # DACA39-95-K-0091 and 34509-RT-AAS. The authors gratefully acknowledge Michael A. Behler for helping to prepare the manuscript.

REFERENCES

Anderson, T.A., and J.R. Coats (Eds.). 1994. *Bioremediation Through Rhizosphere Technology*. American Chemical Society, Washington, DC.

Bayley, S.A., C.J. Duggleby, M.J. Worsey, P.A. Williams, K.G. Hardy, and P. Broda. 1977. "Two Modes of Loss of the Tol Function from *Pseudomonas putida* mt-2." *Mol. Gen. Genet. 154*:203-204.

Berry, J.O., and A.G. Atherly. 1984. "Induced Plasmid-Genome Rearrangements in *Rhizobium japonicum*." *J. Bacteriol. 157*:218-224.

Bradley, P.M., and F.H. Chapelle. 1995. "Factors Affecting Microbial 2,4,6-Trinitrotoluene Mineralization in Contaminated Soil." *Environ. Sci. Technol. 29*:802-806.

Burlage, R.S., S.W. Hooper, and G.S. Sayler. 1989. "The TOL (pWW0) Catabolic Plasmid." *Appl. Environ. Microbiol. 55*:1323-1328.

Caton, J.E., C.-h. Ho, R.T. Williams, and W.H. Griest. 1994. "Characterization of Insoluble Fractions of TNT Transformed by Composting." *J. Environ. Sci. Health 29(4)*:659-670.

Comfort, S.D., P.J. Shea, L.S. Hundal, Z. Li, B.L. Woodbury, J.L. Martin, and W.L. Powers. 1995. "TNT Transport and Fate in Contaminated Soil." *J. Environ. Qual. 24*:1174-1182.

Duque, E., A. Haidour, F. Godoy, and J.L. Ramos. 1993. "Construction of a *Pseudomonas* Hybrid Strain That Mineralizes 2,4,6-Trinitrotoluene." *J. Bacteriol. 175*:2278-2283.

Gillette, W.K., and G.H. Elkan. 1996. "*Bradyrhizobium* sp. (*Arachis*) Strain NC92 Contains Two *nod*D Genes Involved in the Repression of *nod*A Gene Required for the Efficient Nodulation of Host Plants." *J. Bacteriol. 178*:2757-2766.

Jeenes, D.J., W. Reineke, H.-J. Knackmuss, and P.A. Williams. 1982. "TOL Plasmid pWW0 in Constructed Halobenzoate-Degrading *Pseudomonas* Strains: Enzyme Regulation and DNA Structure." *J. Bacteriol. 150*:180-187.

Jeenes, D.J., and P.A. Williams. 1982. "Excision and Integration of Degradative Pathway Genes from TOL Plasmid pWW0." *J. Bacteriol. 150*:188-194.

Kuykendall, L.D. 1979. "Transfer of R Factors to and Between Genetically Marked Sublines of *Rhizobium japonicum*." *Appl. Environ. Microbiol. 37*:862-866.

Love, N.G., and C.P.L. Grady, Jr. 1995. "Impact of Growth in Benzoate and *m*-Toluate Liquid Media on Culturability of *Pseudomonas putida* on Benzoate and *m*-Toluate Plates." *Appl. Environ. Microbiol. 61*:3142-3144.

Medary, R.T. 1992. "Inexpensive, Rapid Field Screening Test for 2,4,6-Trinitrotoluene in Soil." *Analytica Chimica Acta 258*:341-346.

Nakazawa, T., S. Inouye, and A. Nakazawa. 1980. "Physical and Functional Mapping of RP4-TOL Plasmid Recombinants: Analysis of Insertion and Deletion Mutants." *J. Bacteriol. 144*:222-231.

Pennington, J.C., and W.H. Patrick, Jr. 1990. "Adsorption and Desorption of 2,4,6-Trinitrotoluene by Soils." *J. Environ. Qual. 19*:559-567.

Williams, P.A., and K. Murray. 1974. Metabolism of Benzoate and the Methylbenzoates by *Pseudomonas putida* (*arvilla*) mt-2: Evidence for the Existence of a TOL Plasmid." *J. Bacteriol. 120*:416-423.

ENHANCEMENT OF HYDROCARBON DEGRADATION AND CATABOLISM IN BACILLI BY TRANSFORMATION AND GENE-FUSION

Wen-Hsun Yang **and Jen-Rong Yang**
Bioremediation Education, Science and Technology Center and Biology Department, School of Science and Technology
Jackson State University, Jackson, Mississippi

Abstract: Various strains of wild type selenite-hyperresistant bacilli, have been isolated from the Mississippi River with strong performance in bioremediation of selenite and several other Group V and VI elements. None of them was capable of catabolizing either aliphatic or aromatic hydrocarbons as a single source of carbon in the test using basic medium (BM) containing 79.8 mM K_2HPO_4, 44.1 mM KH_2PO_4, 15.1 mM $(NH_4)_2SO_4$, 6.5 mM Sodium Citrate, 0.8 mM $MgSO_4$, 0.1 mM $Ca(NO_3)_2$, 0.1 mM $MnCl_2$, 0.001 mM $FeSO_4$, and 2% agar with the addition of 2% (v/v) hydrocarbons. However, following transformation and gene-fusion of Bacillus mycoides with a plasmid, pTV1T, nearly 20% of the bacilli were able to catabolize either diesel oil, gasoline or benzene as the only source of energy and carbon in the BM medium. Further transformation and gene-fusion of the pTV1Ts-fused bacilli with plasmid, $pLTV_1$ enhance the capability of the gene-fused bacilli to catabolize hydrocarbons more.

Introduction

Acinetobacter calcoaceticus RAG-1 (Shabtai et al 1994) and several other Pseudomonas (Harvey et al) were previously identified as aliphatic hydrocarbon-degradation microorganism in earlier studies and actively applied for bioremediation of the oil-contaminated beaches such as Exxon Valdez oil spill in 1989. Various other bacteria capable for degradation of aromatic hydrocarbons, were also identified (Gibson et al 1984). The bacteria of the genus Bacillus were, in general, inhibitory for their growth in the presence of hydrocarbons, and inadequate for bioremediation of either aliphatic or aromatic hydrocarbons. Nevertheless, Bacilli were well-known spore-forming soil bacteria, non-pathogenetic, convenient, and effective for bioremediation of many inorganic pollutants. The addition of hydrocarbon-degradation capability to the bacillus will increase greatly the success performance of bacilli for bioremediation of mix-contaminated site with both organic and inorganic pollutants

Objective. The objective of this study is to investigate the feasibility of transforming the wild type bacilli into hydrocarbon-resistant bacilli for bioremediation of either aliphatic or aromatic hydrocarbons.

MATERIALS AND METHODS

Bacterial strains and plasmids. A bacterial strain, PY313, containing plasmid, pTV1Ts ($Tn917Erm^{r}Cm^{r}Ts^{rep}$) [Youngman et al. 1989] was obtained from Dr. Youngman. The host cell for the plasmid were originated from BD170 (trpC2 thr5), one of **Bacillus subtilis** 168 derivative [Dubnau et al. 1969]. The recipient of the plasmid in the current transformation experiment was a strain of wild type Bacillus mycoides (MR-1) which was isolated from Mississippi River. The wild type selenite hyperresistant spore-forming bacteria was classified into the **B. mycoides** according to the difference in carbon source utilizations by microplate incubation method of Biolog Inc. [Miller et al, 1991]

Culture media and reagents. Luria-Bertani broth (LB) containing 0.8% tryptone, 0.3% yeast extract was used for regular liquid culture. Tryptose Blood Agar Base (TBAB) medium containing 0.5% NaCl, 1.0% tryptose, 0.3% beef extract and 1.5% agar was used for regular agar plate culture. Hydrocarbon catabolism was tested by measuring the growth of bacteria in the Basic Medium (BM) containing 79.8 mM K_2HPO_4, 44.1 mM KH_2PO_4, 15.1 mM $(NH_4)_2SO_4$, 6.5 mM Sodium Citrate, 0.8 mM $MgSO_4$, 0.1 mM $Ca(NO_3)_2$, 0.1 mM $MnCl_2$, 0.001 mM $FeSO_4$, and 2% agar with the addition of 2%(v/v) diesel oil (DO), gasoline oil (GO) or benzene(BZ).

Preparation of plasmids. For isolation of $pTV1T_s$, a single colony of PY313 was cultured in LB containing erythromycin (2 ug/ml) and chloramphenicol (12.5 ug/ml)[LB+EC] at 30°C with 250 rpm for 24 hours in a shaker incubator. After centrifugation at 3000 rpm for 20 minutes, bacterial pellet was obtained for preparation of plasmid according to a modified method using alkasline lysis of host bacteria [Schuldiner et al. 1992]. The purified plasmid DNA was precipitated and kept in 70% ethanol until the time of further experiment. Similar method has been employed for isolation of $pLTV_1$ from a strain of host bacteria PY1177 containing the plasmid.

Transformation of B. subtilis prototrophic or wild type selenite-hyperresistant bacilli (MR1) with pTV1Ts. After overnight culture of a single colony of **B. mycoides** (MR1) in 12 ml LB at 30°C for 16 hours, a stationary phase of bacterial growth was obtained. After centrifugation at 3000 rpm for 20 minutes, the cell pellet was isolated. One ml of supernatant was used to resuspend the cell pellet, another ml was used to dissolve plasmid DNA following removal of 70% ethanol from the plasmid storage tube. Following the combination of cell suspension with the plasmid preparation, shaking culture was continued for another 6 hours to 12 hours in a smaller tube (13 x 100 mm) for cellular uptake of plasmid DNA. Thereafter, 200 ul of the mixture was spread cultured on MG plate containing erythromycin (2 ug/ml) and chloramphenicol (12.5 ug/ml) to express the transformed cells. For control, 200 ul of the sequentially diluted culture was spread cultured on MG plate for both transformed and untransformed cells to grow. Similar procedures were used for transformation of pTV_1Ts-fused **B. mycoides** with $pLTV_1$,

Mutagenetic treatment of transformed cell. A strain of pTV1Ts-transformed bacteria from **B. mycoids** (WH-1) was cultured in LB broth containing erythromycin(2 ug/ml) and chloramphenicol (12.5 ug/ml) for 24 hours with a temperature below 30^0 C at 250 rpm for transformed bacteria to grow. After centrifugation at 3,000 rpm for 20 min, the pellet was resuspended into similar LB broth with antibiotics for further culture at 40^0C for 24 hours. Following the second procedure for elevation of culture temperature well above the restriction temperature, the plasmid replication was suppressed and host cell without insertion of plasmid was suppressed to grow, only the host cell with inserted plasmid continue to grow in the medium containing antibiotics at 40°C. [Youngman et al. 1989]. The heat-treated plasmid-inserted cells were collected and resuspended in 5 ml of LB containing 15% glycerol for frozen storage at -80°C. After thawing these cells they were spread cultured on TBAB plate containing 10 mM selenite for selection of selenite resistant cell. Those selenite resistant cells were and then patch-culture with toothpick to plate containing BM with the addition of 2% diesel oil, gasoline oil, or benzene for monitoring of colony growth. Similar procedures as above were taken for fusion of pTV_1Ts-fused **B. mycoides** with $pLTV_1$ also.

Assessment and Monitoring of Bacterial Growth on Agar Plate. A single colony of the bacterium was isolated and patched on the center of each plate in a set of 4 to 6 plates containing TBAB or BM with the addition of various amounts of hydrocarbons or selenite for assessment of their effect on growth or metabolism. Colony size was daily monitored by photography of the plate for record of the colony image. The colony size was thereafter calculated by the number of grids(1.25 x1.25 mm) filled by images in a transparent section paper.

RESULTS AND DISCUSSION

As demonstrated in Fig. 1, B. mycoides, grew rapidly

to the colony size of 721 aq. mm on TBAB agar plate during 4 days of incubation at 30°C following patching transfer of a fresh colony to the plate with a toothstick. Similar transfer from the same colony to the test plate containing TBAB with the addition of 2% diesel oil resulted in 23% reduction in the growth of the colony. Transformation of the bacillus with the plasmid ,pTV_1Ts, however, resulted in 140% increase in the colony size by the additional nutrition of 2% diesel oil in TBAB agar plate as compared with the control plate using TBAB alone for the growth of transformed cell. Nevertheless, transformation alone, was unable for the bacillus to obtain the capability to catabolize hydrocarbons as the only carbon source for growth of body mass. Both intact or transformed bacillus were incapable to grow in the basic medium containing 2% diesel oil, 2% gasoline, or 2% benzene as the only carbon sources (Fig. 1). Further treatment of the transformed cell with increased temperature to 50°C for 24 hours resulted in subsequent mutation of the bacillus with the addition of the plasmid to the chromosomal DNA. About 20% of such mutants were capable of catabolizing either 2% diesel oil, 2% gasoline or 2% benzene in the BM as the only source of carbon for increase of biomass (Fig. 2). The independent hydrocarbon-catabolizing capability of the mutant (strains 1 and 2) could be enhanced further by transformation of the pTV_1Ts-fused cell with the second plasmid, $pLTV_1$ (strains 3 and 4) or further more by fusion of the the second plasmid,$pLTV_1$ to the host chromosome DNA which had alredy fused with the first plasmid, pTV_1Ts (Strain 5)(Fig. 3). Since both plasmids, pTV_1Ts and $pLTV_1$ have been constructed from the same origin sharing the fragment containing Tn918 and erythromycin. It is possible that, this fragment of the plasmids may contain gene(s) which are responsible for the production of protein enzyme(s) for catabolism of hydrocarbons.

REFERENCES

Dubnau, D., Davidoff-Abelson, R., and Smith, I. 1969. *"Transformation and transduction in Bacillus subtilis: evidence for separate modes of recombinant formation." J. Mol. Biol.* 45:155

Gibson, D. T., and V. Subramanian, 1984. *"Microbial Degradation of Aromatic Hydrocarbons, in Microbial Degradation of Organic Compounds,"* D. T. Gibson, (ed.) Marcel Dekker, Inc., New York, pp. 182-252.

Harvey, S. I., Elashvili, J. J. Valdes, D. Kamely, and A. M. Chakrabarty. 1990. *"Enhanced removal of Exxon Valdez spilled oil from Alaskan Gravel by a microbial surfactant." 8: 228-230.*

Miller, J. M. and D. L. Rhoden. *"Preliminary evaluation of Biolog, a carbon source utilization methods for bacterial identification." J. of Clinical Microbiology.* 29: 1143-1147.

Schuldiner, I.R. and Tanner, K.. 1992. *"A simple procedure for maximum yield of plasmid DNA." Biotechnique* 9: 676-679.

Shabtai, Y. and D. L. Gutnick. 1985. *"Extracellular esterase and emulsan release from the cell surface of Acinetobacter calcoaceticus." J. Bacteriol. 161: 1176-1181.*

Youngman, P., Poth, H., Green, B., York, K., Olmedo, G., and Smith, K. 1989. *"Method for Genetic Manipulation, Cloning, and Functional Analysis of Sporulation Gene in Bacillus subtilis." In Regulation of Procaryotic Development,* eds. Smith, I., R. A. Slepecky, and P, Setlow. p. 65-87. Amer. Soc. for Microbiology, Washington, D.C.

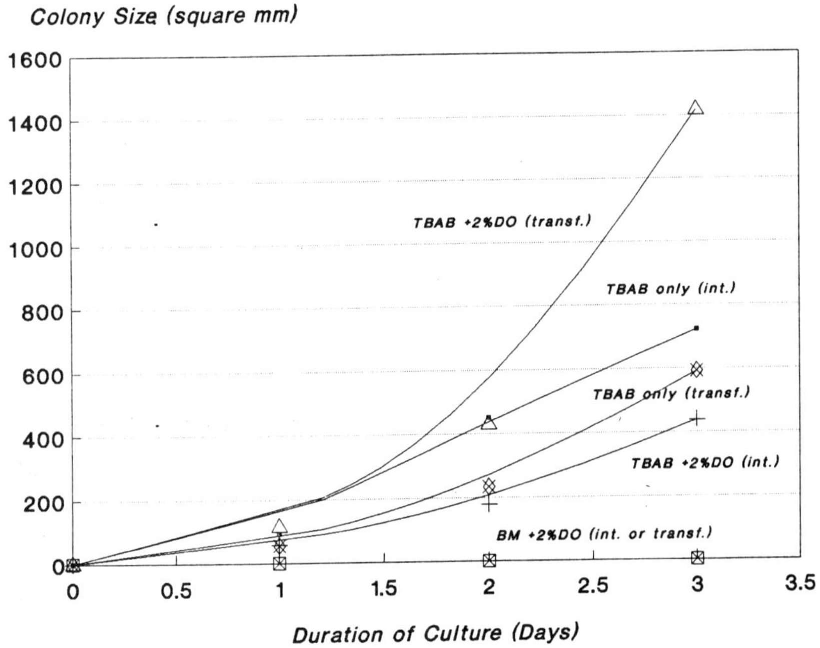

FIGIRE 1. Growth of <u>B</u>. <u>mycoides</u> or transformed bacillus in the TBAB or BM agar plate with the addition of 2% diesel oil (DO), gasoline (GO), or benzene (BZ).

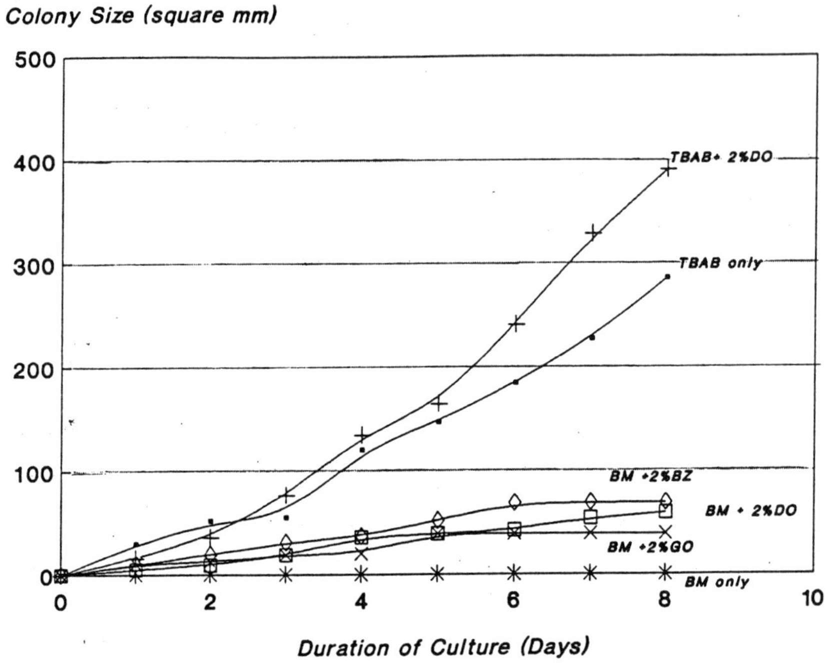

FIGURE 2. Growth of a pTV_1Ts-fused B. mycoides in TBAB or BM with the addition of 2% diesel oil (DO), gasoline (GO), or benzene (BZ).

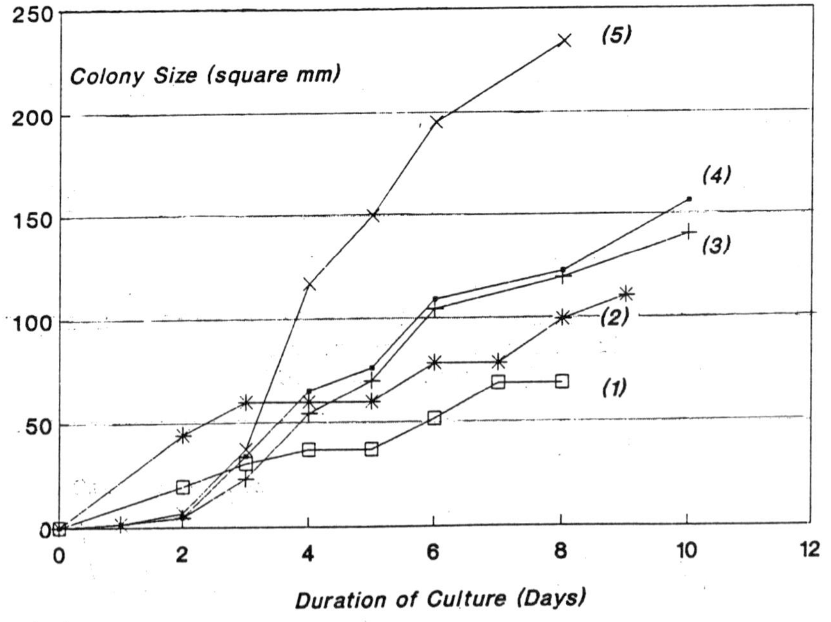

FIGURE 3. Growth of pTV_1Ts-fused strains (1,2), pTV_1Ts-fused and $pLTV_1$-transformed strains (3,4), and pTV_1Ts- and $pLTV_1$- fused strain (5) in BM with the addition of 2% benzene (BZ).

FIELD APPLICATION OF GENETICALLY ENGINEERED MICROORGANISMS FOR PAH BIOREMEDIATION: LESSONS LEARNED

Robert S. Burlage and Anthony Palumbo (Oak Ridge National Laboratory[1], Oak Ridge, TN); Uday Matrubutham, Charles Steward, and Gary S. Sayler (University of Tennessee, Knoxville, TN)

ABSTRACT: During the summer of 1996 a field-scale experiment involving the application of a genetically engineered microorganism (GEM) was initiated at a site on the Oak Ridge Reservation. The genetically engineered strain is a *Pseudomonas fluorescens* that produces a bioluminescent response when naphthalene is bioavailable and undergoing biodegradation. The release experiment was designed for large (800 cubic ft) lysimeters that were heavily instrumented to monitor physiological processes and environmental conditions. During the preparation for this experiment, certain valuable insights relating to the initiation of such an experiment have become apparent. Regulatory approval for field use of the GEM in *in situ* bioremediation process monitoring and control was received from the U. S. EPA, as administered through its TSCA statute. This is an arduous but manageable process, and ultimately resulted in a more clearly defined and characterized experimental protocol. Absolute requirements for TSCA approval were identified, and will be of interest to anyone proposing a release experiment. Adherence to other regulatory statutes (RCRA, NEPA, OSHA) required project-specific analysis. Engineering modifications were made to permit unique experimental changes throughout the experiment. The planning and logistics necessary for successful field experimentation demonstrate the complexity of field application experiments, and provide a useful checklist for other researchers.

[1] Research sponsored by the Of fice of Health and Environmental Research, U.S. Department of Energy. Oak Ridge National Laboratory is managed by Lockheed Martin Energy Research Corp. for the U.S. Department of Energy under contract number DE-AC05-96OR22464.

BIOSURFACTANT CHARACTERIZATION AND PRODUCTION BY *BURKHOLDERIA SOLANACEARUM* FH2

P. Morris, S. Rawlin, H. Rogers, and R. Frontera-Suau

ABSTRACT: A non-fermenting, Gram-negative bacilli, designated FH2, was isolated from a crude oil-degrading enrichment culture. Isolate FH2 was shown to reduce the surface tension of a defined medium to the same extent as *Pseudomonas aeruginosa* ATCC 9027. This microorganism has been identified by fatty acid analysis (Microbial Identification Systems, Inc.) as *Burkholderia solanacearum*, and confirmatory biochemical tests in the laboratory have substantiated this identification. The effect of biosurfactant production on cell surface hydrophobicity for isolate FH2 and *Pseudomonas aeruginosa* has been compared. Both microorganisms were able to lower the surface tension of a proteose-peptone ammonia salts medium containing glucose (PPGAS) from 68 to 29 dynes/cm. When these microorganisms were grown on PPGAS and harvested during stationary phase, adherence to *n*-hexadecane was determined by the bacterial adhesion to hydrocarbons (BATH) assay. With this assay, *Burkholderia solanacearum* FH2 exhibited a 6.6% adhesion to *n*-hexadecane compared to 25% adhesion for *Pseudomonas aeruginosa*. Removal of the biosurfactant prior to the assay resulted in reduction of adherence to 45% and 70% for *Burkholderia solanacearum* FH2 and *Pseudomonas aeruginosa*, respectively. Hydrophobic interaction chromatography (HIC) was conducted, and the retention of *Burkholderia solanacearum* FH2 and *Pseudomonas aeruginosa* to a hydrophobic gel (octyl sepharose) was 94% and 25%, respectively. However, when biosurfactant was removed prior to the assay, retention was reduced to 44% for *Burkholderia solanacearum* FH2 and *Pseudomonas aeruginosa*.

Preliminary studies in the laboratory have shown that *Burkholderia solanacearum* FH2 is more effective in hydrocarbon degradation when grown on a medium (PPGAS without the glucose addition) that stimulates biosurfactant production or if exogenous biosurfactant is added. Chemical characterization of the biosurfactant produced by *Burkholderia solanacearum* FH2 is being conducted using thin layer chromatography and spectroscopy.

MICROBIAL POPULATION DYNAMICS AND BIODEGRADATION KINETICS OF AROMATIC HYDROCARBON MIXTURES

J. D. Rogers, ***K. F. Reardon***, and N. M. DuTeau
(Colorado State University, Fort Collins, CO)

ABSTRACT: Contamination of groundwater and soil is most often a mixture of ollutants, and these mixtures are transformed by mixtures of microorganisms. However, ew quantitative studies of biodegradation kinetics have addressed these complex mixture spects. The goal of this work is to measure changes in both the chemical and microbial oncentrations during mixed culture-mixed pollutant biodegradation.

A simple two-species microbial system was chosen as the first test of our ethods. *Pseudomonas putida* F1 and *P.* sp. JS150 are capable of growth on various romatic hydrocarbons; however, *P.* sp. JS150 has a wider substrate range that includes hlorobenzene. Binary mixtures of toluene and phenol were used with the two-species icrobial system in biodegradation experiments. In situ hybridization using species-pecific fluorescently labeled oligonucleotide probes was used for quantitation of each pecies during biodegradation experiments. Mathematical modeling was used to etermine the relationships between the composition of the microbial and pollutant ixtures and to describe the interactions between the two species. Kinetic constants were und for each species degrading a single pollutant, and these parameters were used in the odel for microbial growth and utilization of binary substrate mixtures.

The population dynamics of the mixed microbial population was found to be lated to the changes in the composition of the pollutant mixture, and interaction tween the two species was apparent. These results suggest that single species/single llutant biodegradation kinetics are not readily extended to more complex systems.

is work was supported by grant number 5 P42 ES05949-05 from the National Institute Environmental Health Sciences, NIH.

THE MINERALIZATION OF BTEX MIXTURES BY ENVIRONMENTAL CULTURES

Rula Anselmo Deeb and ***Lisa Alvarez-Cohen***

ABSTRACT: Benzene, toluene, ethylbenzene and xylenes (BTEX compounds) are priority environmental pollutants commonly found in complex mixtures at gasoline-contaminated subsurface sites. This study involves experiments aimed at identifying the key inhibitory and stimulatory processes governing the aerobic biotransformation of BTEX mixtures.

An enriched culture obtained from a gasoline-contaminated aquifer has been shown to mineralize each BTEX compound individually at high substrate concentrations (>80 mg/L). Studies with BTEX mixtures have also shown complete biotransformation of all BTEX compounds with corresponding production of CO_2; however, it is unclear whether mineralization of each of the BTEX components within the mixture took place. Studies are underway to trace compound mineralization using radiolabeled BTEX compounds. Experimental studies are also aimed at characterizing the effect of substrate interactions on the mineralization of BTEX mixtures. Preliminary results have shown ethylbenzene, a compound rarely considered in reported laboratory studies, to be a potent inhibitor of the biodegradation of benzene, toluene and xylenes. These observed inhibition effects are being quantified.

Pure cultures have been isolated from the mixed microbial consortium and are being tested for mineralization of BTEX mixtures. Preliminary results suggest that the pure cultures exhibit a variety of biodegradation properties. It is our hypothesis that although none of the pure cultures alone may be capable of completely mineralizing BTEX mixtures, the parent mixed microbial consortium exhibits a combination of the degradation properties of the pure cultures and thus the ability to completely mineralize BTEX mixtures, a capability that has not previously been reported in the literature.

Characterizing the influence of substrate interactions (inhibitory and stimulatory) on the biodegradation of BTEX mixtures will contribute to a better understanding of the fate of these hazardous organics in the environment and will provide useful insights leading to new approaches to in situ bioremediation.

ORGANIC MATTER IN AQUIFERS AS CARBON SOURCE FOR MICROORGANISMS

Christian Grøn (Risø National Laboratory, Denmark)
Marianne Krog (Aarhus University, Denmark)
Rasmus Jakobsen (Technical University of Denmark)

ABSTRACT: Studies of organic matter contents and aquifer microbial activity, measured as sulfate reduction rates, at two Danish sites demonstrated that bulk contents of dissolved and sedimentary organic carbon (DOC and TOC) were not determining the microbial activity. Conversely, high contents of biomolecule residues in groundwater humic substances and large fractions of acid desorbable organic matter in the sediments were related to high microbial activity. The relative substrate value of aquifer organic matter for microbial processes may thus be assessed.

INTRODUCTION

The microorganisms responsible for biodegradation of organic contaminants in aquifers depend, in many cases, upon the availability of degradable organic matter as primary carbon source. High contents of organic matter are observed in most soils (Tate, 1987), and even in the subsoils and in aquifer sediments, measurable contents of organic carbon are found (Ball et al, 1990). Dissolved organic matter in groundwaters is mostly low, below 2 mg/L (Thurman, 1985). The structure of sedimentary organic matter varies from the soil organic matter with high contents of residual biomolecules, such as carbohydrates and amino acids (Tate, 1987), over the humified organic matter adsorbed to mineral particles in the subsoil (Stevenson, 1985), to the coalified lignite particles found in some aquifers (Filip and Smed-Hildmann, 1992). The organic matter dissolved in groundwaters exhibits major structural differences as well (Grøn et al, 1996). The utility or substrate value of the organic matter as carbon source for microorganisms is expected to vary immensely with different degrees of degradation/diagenesis in aquifers, as already found for marine sediments (*e.g.:* Boudreau and Ruddick, 1991). Still, the organic matter contents of aquifer systems are mostly quantified only as bulk dissolved organic carbon (DOC) of groundwaters or total organic carbon (TOC) of sediment.

MATERIALS AND METHODS

Study sites. Groundwater and sediment samples were taken from two Danish field sites at Tuse and Rømø. The Tuse aquifer is in ≈20 m of glacial, calcareous sand deposited during the Weichsel glaciation (estimated 70 000 - 14 000 years B.P. in this region) underlain first by glacial clay till and then by marine clay, merl and limestone from the Selandien period. The aquifer cover is a ≈20 m clay, sand and gravel sequence, where the top 5-10 m are glacial till, followed by what have just recently been suggested to be a 9-12 m marine clay sequence of interstadial origin. Redeposited lignite fragments are found in the deeper parts of the sandy aquifer. The water table is in ≈15 m.b.s. (m below the surface). The studied site on Rømø features a shallow, phreatic Holocene postglacial and eolian sand aquifer extending approximately 25 m.b.s. The homogenous, upper 5-6 m of sediment studied are eolian sands deposited in historic time. The water table is found 0.5 -1.5 m.b.s and fluctuates up to 1 m per year.

Organic carbon. Dissolved organic carbon (DOC) was determined after 0.1 μm membranefiltration, acidification and purging of inorganic and volatile organic carbon by UV/peroxodisulfate oxidation and infrared detection on a Dohrmann DC 180 TOC Analyzer. Sedimentary organic carbon was determined separately (modified after Roberts et al, 1973) as acid desorbable (ADOC) and non-acid desorbable organic carbon (NADOC), with total organic carbon (TOC) = ADOC + NADOC. Inorganic carbon was removed from dried, ground sediment samples by successive washes with 2 M nitric acid and ADOC was determined as above by organic carbon analysis of the supernatant after centrifugation. The organic carbon fraction, *i.e.:* NADOC, not desorbable with 2 M nitric acid was determined by carbon analysis of the acid washed, centrifuged pellet using a Leco CS-225 Carbon-Sulfur Determinator.

Sulfate reduction. In situ rates of sulfate reduction were made with S-35 sulfate as a tracer (Jakobsen and Postma 1994). The tracer was injected into freshly taken sediment cores, where it was reduced along with the natural sulfate in the sediment during the 22-40 h incubation. Dissolved sulfate was determined with ion chromatography.

Dissolved hydrogen. Measurements of dissolved hydrogen were done on-site after bubble stripping using a molecular sieve column for separating gases and a TraceAnalytical RGD2 Reduced Gas Detector (Chapelle and McMahon, 1991).

Humic substances. The humic substances dissolved in groundwater were isolated according to a modified XAD-8 method (Krog and Grøn, 1995). Total amino acids and carbohydrates were hydrolysed off the humic substances and determined using ion exchange chromatography as described previously (Grøn et al, 1996).

RESULTS AND DISCUSSION

Comparing two selected sampling positions at Tuse and Rømø, we could observe higher concentrations of sulfate and of dissolved and sedimentary organic carbon at Tuse (Table 1), but a 10 fold lower sulfate reduction rate. At both sampling positions, the concentrations of dissolved hydrogen indicated sulfate reduction as the most likely microbial electron-accepting process during organic matter degradation (Lovley and Goodwin, 1988). The lower sulfate reduction rate found with higher substrate concentrations (sulfate and organic matter) and suitable redox conditions demonstrates that the microbial consortia are much less efficient or, as suggested by Jakobsen and Postma (1994), that the organic substrate is of a poorer quality.

TABLE 1. Concentrations of organic matter, sulfate and hydrogen, and sulfate reduction rates at two selected sampling positions at the Tuse and Rømø sites, Denmark.

	Groundwater			Sediments	
	Tuse	Rømø		Tuse	Rømø
DOC (mg C/L)	29	19	TOC (%C)	0.07	0.01
Sulfate (mg/L)	115	65	ADOC/NADOC	1.1	1.7
Hydrogen (nmol/L)	2.7	1.1	Sulfate reduction rate (mg sulfate/L/year)	6.9	79

Organic matter dissolved in groundwater, DOC. Through 6 detailed profiles at Rømø, we observed (data not shown) that high DOC concentrations in the groundwater did not increase the sulfate reduction rates. Comparable observations were done at the Tuse site but with a more limited data set.

A much higher content of hydrolyzable carbohydrates and slightly higher content of hydrolyzable amino acids in fulvic acids at Rømø than at Tuse (Table 2) point to a higher substrate value of the Rømø DOC. Other structural differences could be observed for the fulvic acids from the two sites, but none that could be interpreted directly in terms of different substrate value. The fulvic acids are a fraction of the humic substances found dissolved in groundwater and constituted 60% and 58% of the DOC at the Rømø and Tuse sites, respectively. Studies of the aquatic diagenesis in aquifer humic substances have shown, that the biomolecules residues are degraded faster than other structural fragments of the humic substances (Grøn et al, 1996). Still, the hydrolyzable biomolecule residues in groundwater fulvic acids amounted to less than 1% of the total, dissolved organic matter and can not be the only organic substrate for the observed microbial sulfate reduction activity.

Whereas the DOC of groundwater did not determine the microbial activty measured as sulfate reduction rates in the two aquifers, more reactive fragments, *e.g.:* biomolecule residues, appeared of greater importance. Still, more data are needed to establish a clear relationship between microbial reactivity and DOC properties.

Sedimentary organic carbon, TOC. For the sedimentary organic carbon, a larger fraction of the TOC was acid desorbable (ADOC), *e.g.:* primarily bound to iron oxides, clay minerals or calcite, in the aquifer with higher sulfate reduction activity at Rømø (Table 1).

TABLE 2. Hydrolyzable biomolecules in humic substances (fulvic acids) from the Tuse and Rømø aquifers, Denmark.

	Tuse	Rømø
Total hydrolyzable amino acids (nmol/mg)	27	64
Total hydrolyzable carbohydrates (mg/g)	0.33	3.0

The relation between high sulfate reduction rates and a high content of ADOC was also observed in a profile through the upper 5-6 m of the Rømø aquifer (Figure 1). In the deepest part of the profile, the ADOC increased without a concomitant increase in sulfate reduction rate but in this section, the sulfate concentration was rapidly declining and may thus have limited the reduction rate. The fraction of sedimentary organic carbon that is not acid desorbable (NADOC) did not exhibit any discernible relation to the observed sulfate reduction rates.

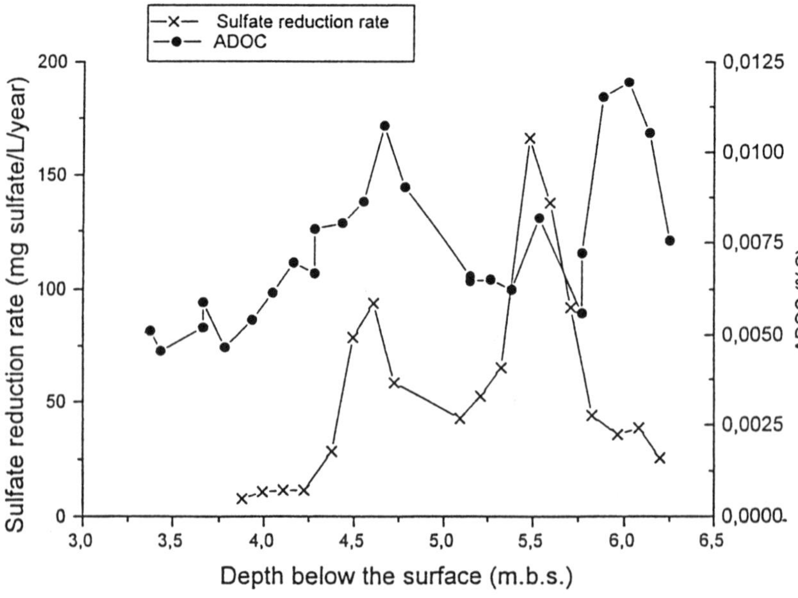

FIGURE 1. Acid desorbable, sedimentary organic carbon and sulfate reduction rates through an aquifer profile at Rømø, Denmark.

Figure 2 gives the variations of TOC and the ratio ADOC/NADOC of acid desorbable to non acid desorbable organic carbon for a profile through the Tuse aquifer. The microbial sulfate reduction rate was high (28±8.0 mg sulfate/L/year, 7 determinations over a 0.8 m core) at 22 m.b.s. with low, but comparatively desorbable TOC, and lower (6.7±0.4 mg sulfate/L/year, 3 determinations over a 0.2 m core) at 35 m.b.s. with much higher, but less desorbable TOC.

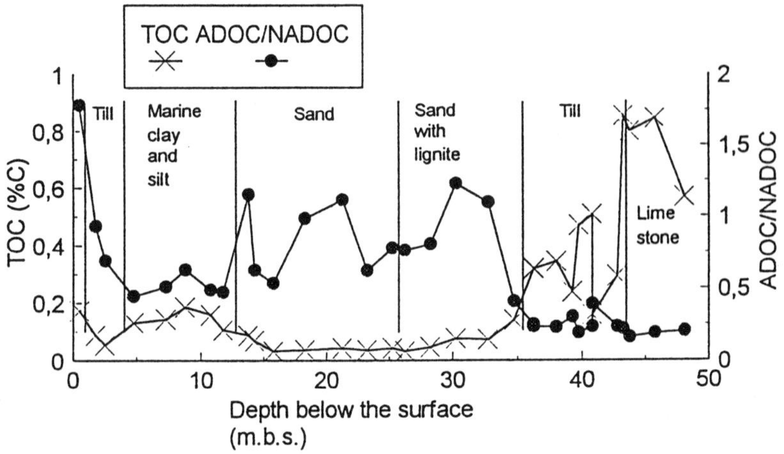

FIGURE 2. Total organic carbon and ADOC/NADOC in sediments through an aquifer profile at Tuse, Denmark.

The microbial activity measured as the sulfate reduction rate is clearly related to the size of the acid desorbable fraction of sedimentary organic carbon at both sites, whereas no relation could be observed with the non desorbable fraction. Still, we can not, with the data available, exclude the possibility that the relation between ADOC and sulfate reduction rates is the result of the microbial activity, rather than its cause.

CONCLUSIONS

Data on microbial activity in aquifers, measured as sulfate reduction rates, and on groundwater and sedimentary organic matter contents suggest, that bulk measurements of total organic carbon are insufficient to assess the substrate value of the organic matter in an aquifer for microbial action.

Biomolecule residues in groundwater humic substances and acid desorbable organic carbon in sediments are potential indicators of organic matter substrate value.

REFERENCES

Ball, W. P., Buehler, C. H., Harmon, T. C., MacKay, D. M. and P. V. Roberts, 1990. "Characterization of a Sandy Aquifer Material at the Grain Scale". *J. Contam. Hydrol. 5:* 253-295.

Boudreau, B. P. and B. R. Ruddick, 1991. "On a Reactive Continuum Representation of Organic Matter Diagenesis." *Am. J. Sci. 291*: 507-538.

Chapelle, F. H. and P. B. McMahon, 1991. "Geochemistry of Dissolved Inorganic Carbon in a Coastal Plain Aquifer. 1. Sulfate from Confining Beds as an Oxidant in Microbial CO_2 Production." *J. Hydrol. 127*: 85-108.

Filip, Z. and R. Smed-Hildmann, 1992. "Does Fossil Plant Material Release Humic Substances into Groundwater?" *Sci. Tot. Environ. 117/118*: 313-324.

Grøn, C., Wassenaar, L. and M. Krog, 1996. "Origin and Structures of Groundwater Humic Substances from Three Danish Aquifers." *Environ. Int. 22*: 519-534.

Jakobsen, R. and D. Postma, 1994. "In Situ Rates of Sulfate Reduction in an Aquifer (Rømø, Denmark) and Implications for the Reactivity of Organic Matter." *Geology 22*: 1103-1106.

Krog, M. and C. Grøn, 1995. "Isolation of Haloorganic Groundwater Humic Substances." *Sci. Tot. Environ. 172*: 159-162.

Lovley, D. R. and S. Goodwin, 1988. "Hydrogen Concentrations as an Indicator of the Predominant Electron-Accepting Reactions in Aquatic Sediments of the Atlantic Coastal Plain." *Geochim. Cosmochim. Acta 52*: 2993-3003.

Roberts, A. A., Palacas, J. G. and I. C. Frost, 1973. "Determination of Organic Carbon in Modern Carbonate Sediments." *J. Sediment. Petrol. 43*: 1157-1159.

Stevenson, F. J., 1985. "Geochemistry of Soil Humic Substances." In G. R. Aiken, D. M. McKnight, R. L. Wershaw and P. MacCarthy (Eds.), *Humic Substances in Soil, Sediment, and Water. Geochemistry, Isolation, and Characterization,* pp. 13-52. Wiley & Sons, New York.

Tate, R. L., 1987. *Soil Organic Matter. Biological and Ecological Effects*. Wiley & Sons, New York.

Thurman, E. M., 1985. *Organic Geochemistry of Natural Waters*. Kluwer Academic Publishers, Dordrecht, The Netherlands.

PRODUCTION OF EXTRACELLULAR SURFACE-ACTIVE COMPOUNDS BY MICROORGANISMS GROWN ON HYDROCARBONS

Daniela F. Carvalho, Daniela D. Marchi and **Lucia R. Durrant**
DCA/FEA - UNICAMP (Campinas State University), Campinas-SP, Brazil.

ABSTRACT: The isolation of microorganisms involved in the biodeterioration of diesel oil, from soil samples collected near Paulinia's oil refinery near Campinas, SP, Brazil, was carried out. Thirty-one microbial strains were selected and the production of biosurfactants in media containing 1.5% of either kerosene, toluene, vaseline or olive oil as the sole carbon sources was determined through the measurement of their emulsification activities, using various hydrocarbons and also olive oil as substrates. The increases in absorbance of the oil-in-water emulsions were measured at 610 nm, whereas the water-in-oil emulsions were expressed as the height (cm) of the emulsion layers formed. Highest emulsification activities were obtained after growth of the strains in kerosene, vaseline and olive oil. Toluene was the best emulsified hydrocarbon, followed by xylene. A bacterium identified as *Planococcus citreus* and a *Bacillus sp* presented the highest bioemulsification of toluene after 48 hours of growth in toluene and vaseline respectively. Two yeast strains (L7A and L5A), showed highest activities towards toluene and xylene after growth for 48 hours in vaseline and kerosene, respectively. Two filamentous fungi were only able to emulsify olive oil, regardless of the growth substrate or the cultivation time.

INTRODUCTION

Surface-active compounds of biological origin (biosurfactants) have received increasing attention in recent years because of their role in the growth of microorganisms on water-insoluble, hydrophobic materials such as hydrocarbons or other lipophilic substrates. They are a diverse group of bio-molecules which share the same properties as synthetic surfactants, and in some cases, are superior in creating water-in-oil or oil-in-water microemulsions (Ashtaputre et al, 1995). Owing to their biodegradability and production from renewable resources, biosurfactants have been gaining prominence; their successful applications in the cleaning-up of oil spills from the environment, remediation of metal-containing soils or waste streams, and enhanced oil recovery have been reported. They have the properties of reducing surface tension, critical micelle concentration and interfacial tension, and of stabilizing emulsions in both aqueous solutions and hydrocarbon mixtures. Their appearance in the culture medium is often regarded as a prerequisite for interaction of hydrocarbons with microbial cells (Hommel, 1990), which improves the accessibility of these substrates to microorganisms. Many hydrocarbon-degrading microorganisms have highly hydrophobic cell surfaces, allowing microbial cells to grow on hydrocarbons through adhesion to the growth substrates.

When properly stimulated, biosurfactant-producing micro-organisms can aid in the bioremediation of oil-contaminated soil, of hydrocarbon contaminants in the environment, and can stimulate the biodegradation of fats and oils present in wastewater produced by various food industries (Carrillo et al, 1996).

MATERIALS AND METHODS

Isolation of the microbial strains. Samples from soil contaminated with diesel oil were collected near Paulinia's petroleum refinery in Campinas (SP, Brazil), and used for the isolation of microorganisms as follow: homogenization of soil samples, dilution of samples (1:100 wt/vol), surface plating, incubation at 30 ^{0}C and isolation of pure colonies.

Identification of the bacterial strains. For the tentative identification of the bacteria, various biochemical tests were undertaken and observation was made of all morphological characteristics.

Culture conditions. The microbial strains were cultivated in a medium containing 1.5 % (vol/vol), of either kerosene, or vaseline or toluene, or xylene or olive oil as the sole carbon source plus 0,5 g $MgSO_4$, 3 g $NaNO_3$, 1 g KH_2PO_4, 1 g yeast extract and 0,3 g peptone per liter of medium. Cultures were grown in 100 mL Erlenmeyer flasks with 75 mL of medium and incubated with shaking at 30 ^{0}C for 120 hours. Duplicate flasks were collected every 24 hours and the emulsification activities were determined as described below.

Emulsification activity measurements. Culture broth was made cell free by centrifugation. 3,5 ml of the cell free broth was vigorously shaken with 2 ml of hydrocarbon on a vortex shaker and left undisturbed. After one hour, optical density of the oil-in-water emulsion phase was measured at 610 nm. The O.D. was reported as emulsification activity (Johnson et al, 1992). After 24 hours the height of the emulsion layer (water-in-oil) was measured and emulsification activity was expressed in cm (Cooper, 1987).

Determination of the cells surface hydrophobicity. The selected bacteria and yeasts were grown at 30 ^{0}C with shaking in nutrient broth. After 24 hour incubations, cells were harversted, washed twice and ressuspended in PUM buffer (Rosenberg et al., 1980). The surface hydrophobicities of the cells were then measured as decribed by Rosenberg et al., 1980. The hydrocarbons tested were: kerosene, toluene, vaseline, xylene, hexadecane and octane.

RESULTS AND DISCUSSION

Eighty microbial strains (42 bacteria, 27 fungi and 11 yeasts), were isolated. All the bacterial and yeasts strains but only three fungi were tested for growth on hydrocarbons and olive oil and for the production of emulsification activity by the cell-free supernatants. Four bacteria and three yeasts were selected and the results obtained are shown below. According to the results obtained after the biochemical tests plus observation of all morphological characteristics the strains were identified as follows:

4	*Chromobacterium sp.*	9	*Planococcus citreus*	B1A	*Enterobacter sp.*
20	Bacillus sp.	L5A	yeast	L7A	yeast	L9A	yeast

For all strains tested, best results for emulsification activities were obtained after growth in kerosene, vaseline, toluene or olive oil as the carbon sources and the best emulsified substrates were toluene, xylene and olive oil. Therefore, only the results obtained under the abovementioned conditions will be presented bellow (Table 1).

TABLE 1. Emusification activities present in the cell-free cultures supernatants.

	Emulsification Measurement (cm)														
	Growth time (h)														
	24			48			72			96			120		
oil tested (a)	Oo	T	X	Oo	T	X	Oo	T	X	Oo	T	X	Oo	T	X
strain/growth substrate (b)															
4/Oo	1,5	1,2	1,8	ND	0,6	0,4	ND	1,8	1,3	ND	0,6	0,7	ND	1,5	1,6
4/K	1,8	2,1	2,5	ND	1,8	1,9	ND	1,8	1,4	ND	1,8	2,4	ND	1,5	1,7
4/V	1,6	2,3	2,5	ND	0,6	1,5	ND	0,2	0,2	ND	2,0	1,8	ND	1,1	1,5
B1A/Oo	2,5	0	0,7	2,2	0,7	1,3	1,8	1,0	0	1,7	1,2	0	2,2	1,2	0,4
B1A/K	1,7	2,1	2,2	2,0	2,0	2,5	2,0	2,2	2,5	1,8	2,1	2,0	1,6	2,2	2,5
B1A/T	1,9	2,2	1,7	2,0	2,3	2,5	1,6	2,2	2,5	2,0	1,7	1,7	2,0	2,0	2,0
B1A/V	2,0	2,1	2,5	2,0	2,2	2,5	1,8	2,1	2,5	1,8	2,1	2,1	1,8	2,1	2,5
L5A/Oo	1,6	0	0,2	1,5	0	0	1,6	0,2	0,4	1,8	0,7	1,2	1,8	0,2	0,4
L5A/K	2,0	1,8	2,0	2,0	2,1	2,1	2,2	2,2	2,4	2,1	2,0	2,1	2,2	1,8	2,0
L5A/T	1,7	0	0,2	1,8	1,8	2,1	1,5	0	0	1,8	1,3	1,0	1,5	0,3	0,2
L5A/V	2,0	1,8	2,2	1,8	0,6	0,9	1,5	0,1	0,2	1,8	0,2	0,3	1,8	0,4	0,6
L9A/Oo	1,1	0	0,1	2,0	0,1	0,4	1,5	0,2	0,6	1,5	1,7	2,0	1,8	1,3	1,6
L9A/K	1,9	1,4	1,8	1,9	0	0,5	1,8	1,8	2,1	1,8	2,1	2,1	1,5	2,3	2,2
L9A/T	1,5	0	0,1	1,7	0,2	0,3	1,6	0,2	0,2	1,5	0,1	0,1	1,8	1,6	1,9
L9A/V	1,8	1,8	2,1	1,8	0,3	0,5	1,6	0,3	0,2	1,7	0,4	0,4	1,8	1,7	2,0

(a) Oo: olive oil; T: toluene; X: xylene (b) Oo: olive oil; K: kerosene; T: toluene; V: vaseline
ND: not determined

TABLE 2. Adherence (% Absorbance Reduction) of bacterial cells to various hydrocarbons.

	Absorbance Reduction (%)					
oil tested (a)	K	T	V	X	H	O
strain						
4	28,57	0	50,00	4,08	16,32	20,40
9	88,23	74,28	86,11	80,55	88,57	83,33
20	78,57	61,90	71,42	69,04	76,19	78,57
B1A	23,94	33,09	20,54	42,95	18,30	12,67
L5A	49,94	54,11	48,90	53,07	58,28	48,90
L7A	75,00	87,50	9,50	85,70	71,40	71,40
L9A	92,17	88,00	91,65	89,05	93,22	90,61

(a) K: kerosene; T: toluene; V:vaseline; X: xylene; H: hexadecane; O: octane

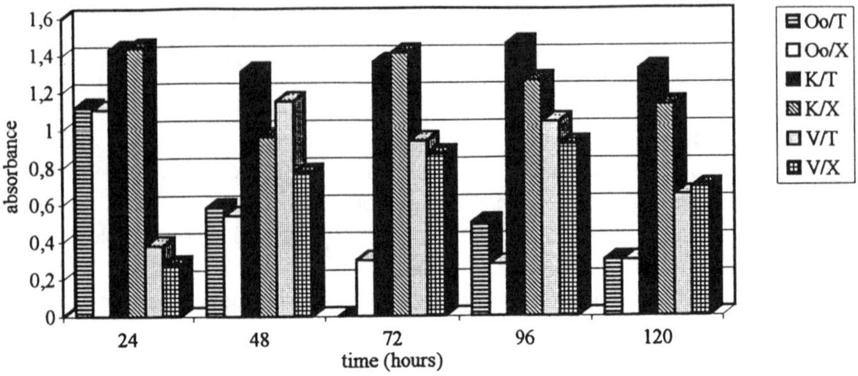

FIGURE 1. Emulsifying activity in cell-free culture broths of strain 20 after growth in olive oil (Oo), kerosene (K), or vaseline (V), and tested with toluene (T), and xylene (X).

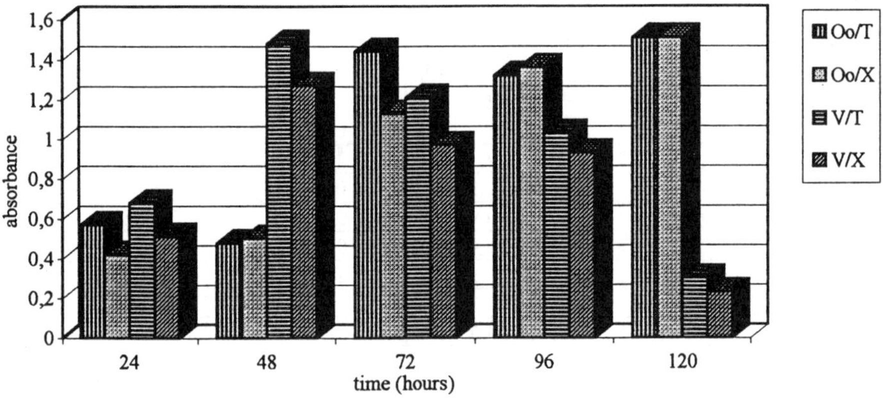

FIGURE 2. Emulsifying activity in cell-free culture broths of strain L7A after growth in olive oil (Oo), or vaseline (V), and tested with toluene (T), and xylene (X).

As shown in Table 1, strain B1A was one of the best strains for producing water-in-oil emulsification activity. Its activities after growth in toluene were the highest produced among the selected microbial strains, for this carbon source, and all the three compounds tested served as substrates for emulsification. Considering the four organisms presented in Table 1, the best emulsifying activities were produced after growth in kerosene or vaseline.

Figures 1 and 2 show the oil-in-water emulsions produced by the cell-free culture fluids of strains 20 and L7A, respectively, obtained after their growth in olive oil, kerosene or vaseline. Kerosene was the best carbon source for strain 20, followed by vaseline. For the yeast strain L7A, vaseline was the best.

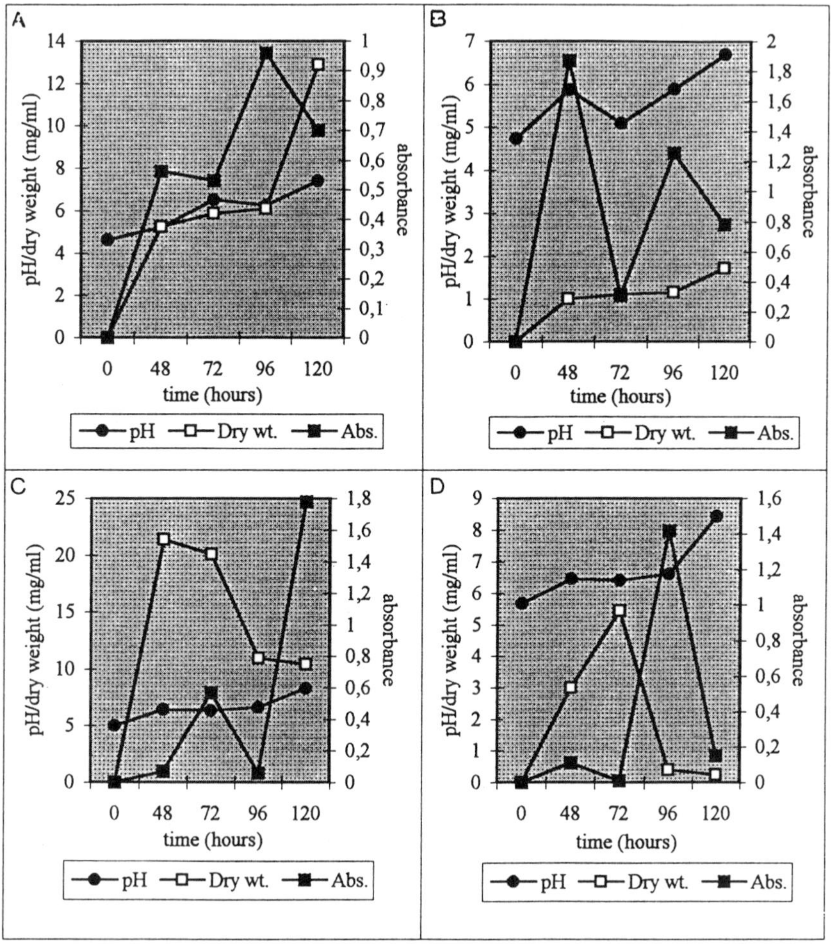

FIGURE 3 Growth and emulsification activity towards toluene produced by strain 9. A: grown in olive oil. B: grown in kerosene. C: grown in vaseline. D: grown in xylene.

Growth, pH variation and emulsification activities towards toluene produced by strain 9 during growth in various carbon sources are presented in Figure 3. For all the carbon sources used, a slight increase in pH was observed towards the end of the cultivation period.

The best substrate for biomass production was vaseline. Interestingly, best emulsification activity was obtained following growth in kerosene, which was the carbon source producing lowest growth. The appearance of emulsification activity varied with the carbon source and the time of growth. Peaks of activity were observed either at an early or in the late logarithimic and the stationary phases of growth. It should be noted here that this behavior was also observed with the other six strains. Many hydrocarbon-degrading microorganisms produce extracellular emulsifying agents, which are in some cases induced by their growth in these lipophilic compounds. However, biosurfactants should not be regarded only as a prerequisites of hydrocarbon uptake, but also as secondary metabolic products, since emulsification may be a by-product of a cell/hydrocarbon detachment process.

As shown in Table 2, most of the strains selected were able to adhere to all the hydrocarbons tested, indicating that the ability of these microbial cells to grow on hydrocarbons may be initially through direct contact with the immiscible substrate.

The growth of microorganims on lipophilic compounds involve some of the following steps: a) adherence to the surfaces of oil droplets because of the hydrophobic oil surface and the cell hydrophobicity, b) growth of the adhering cells at the oil-water interface of the droplet, c) production of surfactants and emulsification of the growth substrate, d) mechanisms for desorption from the lipophilic substrate. Our results suggest that some of the strains selected follow the abovementioned steps, for instance, strains 9, 20 and L9A both which exhibited high cell surface hydrophobicities and emulsification activities at the early and late phases of growth.

ACKNOWLEDGEMENTS

We thank FAPESP and CNPq for financial support.

REFERENCES

Ashtaputre, A. A. and Shah A. K. 1995. "Emulsifying Property of a Viscous Exopolysaccharide from *Sphingomonas paucimobilis.*" *World Journal of Microbiology and Biotechnology*. 11: 219-222.

Carrillo, P. G., Mardaraz, C., Pitta-Alvarez, S. I. and Giulietti, A. M. 1996. "Isolation and Selection of Biosurfactant-Producing Bacteria." *World Journal of Microbiology & Biotechnology*. 12: 82-84.

Cooper, D. and Goldenberg, B. G. 1987. "Surface Active Agents from Two Bacillus species." *Applied and Environmental Microbiology*. 53: 224-229.

Hommel, R. K. 1990. "Biosurfactants in Hydrocarbon Utilization." *Biodegradation*. 1: 107-119.

Johnson, V., Singh, M., Saini, V. S., Adhikari, D. K., Sista, V. and Yadav, N. K. 1992. "Bioemulsifier Prodution by an Oleaginous Yeast *Rhodotorula glutinis* IIP-30." *Biotechnology Letters*. 6: 487-490.

Rosemberg, M., Gutnick, D. and Rosemberg, E. 1980. "Adherence of Bacteria to Hydrocarbons: A Simple Method for Measuring Cell-Surface Hydrophobicity." *FEMS Microbiology Letters*. 9: 29-33.

BIOFILM DECAY AFTER IN SITU BIOREMEDIATION–POST TREATMENT WATER QUALITY

Vivien Sudirgio, ***Donald Johnstone***, David Yonge, and James Petersen

ABSTRACT: A thorough understanding of general groundwater quality and microbial succession is crucial to assessing the long-term effects following in situ bioremediation. At the termination of each or final treatment phase, normal groundwater flows are generally reestablished. The primary objective of this research was to ascertain the degree to which water quality returns to upgradient quality following establishment of biomass in porous media.

A bench-scale continuous flow groundwater system was established using a porous media biofilm reactor operated under constant substrate loading conditions (0.014 mg min^{-1} cm^{-2} nitrate and 0.016 mg min^{-1} cm^{-2} acetate). A consortium of denitrifying bacteria (carbon tetrachloride degrading, Hanford, WA) was inoculated into the porous media and stimulated to grow by feeding acetate and nitrate in simulated groundwater medium (SGM). When cell growth stabilized, acetate and nitrate were eliminated from the SGM flow. On the effluent side of the reactor the following were monitored: biofilm decay, suspended cells, organic acids, and non-biological water quality parameters. Residence time and pressure drop were also monitored. All parameters were analyzed to determine the effects of nutrient shut-down on water quality.

The concentration of effluent biomass showed a significant decrease during the nutrient-depleted conditions once the nutrients were stopped. Electron microscopy showed that the biofilm decreased in mass, yet a significant film remained in the porous media with most free cells washing out of the reactor. The change in the density of the biofilm directly affected the column pressure drop and hydraulic residence time. Hence, these two parameters were used to determine when to terminate the experiment. The general water quality parameters indicated a return close to background within 15 days after nutrient shut-down. Pressure drop appears to be the best indicator of biofilm decline and commencement of permeability recovery.

BIOSURFACTANT PRODUCTION BY ORGANISMS FROM AN ECOLOGY STUDY OF A JP-4 JET FUEL-CONTAMINATED SITE

Cristin Lee Bruce

ABSTRACT: The ability of subsurface microbial populations to degrade organic contaminants is affected by nutrient availability. A study was performed to measure the change in microbial ecology of a weathered JP-4 jet fuel contaminated site (Eglin Air Force Base, Florida) over the course of a nitrate-based treatment in terms of heterotrophs, JP-4 degraders, oligotrophs, biosurfactant producers, bioemulsifier producers, and predators.

Microbial numbers increased and species diversity decreased in the nitrate-amended site significantly more than in background areas. The surfactant-producing potential of native microbial communities decreased significantly with nitrate addition. The bioemulsification capacity of these communities was slightly lowered with addition of nitrate. Aerobic predators decreased over the course of the experiment, while microaerophilic predators increased significantly.

Due to fluctuations in measured JP-4, no conclusions could be drawn about the success of nitrate addition as a remedial enhancement technique at this site.

SPATIAL DISTRIBUTION OF PLANTS AND SOIL MICROBIAL COMMUNITIES IN SOIL CONTAMINATED WITH TNT.

Garald L. Horst, Gopal Krishnan, Rhae A. Drijber, and Tom E. Elthon
(University of Nebraska-Lincoln)

ABSTRACT: Former munition operations at Ordnance Plant near Mead, Nebraska have resulted in environmental and public health concerns. A site survey was conducted in order to assess plant distribution and species composition in response to soil levels of TNT (2,4,6 trinitrotoluene). Soil cores were taken along transects at 3 depths parallel to the ditch through which TNT-laden waste waters were discharged from the munitions plant, and then composited. Soil samples were also taken from several patches bare of vegetation. Soils were analyzed for water-extractable and total TNT, pH, electrical conductivity and mineral nitrogen. Microbial community structure was assessed using fingerprints of fatty acid methyl esters (FAMEs) derived from the soil microbial cells. Presence of plants decreased TNT concentrations within the top layer of soil suggesting rhizosphere based microbial degradation of TNT. Microbial community structure of TNT contaminated soils was clearly influenced by the presence of plants based on cluster analysis of FAME fingerprints.

INTRODUCTION:

Current remediation technologies for munitions contaminated soils include incineration, composting , and soil slurry treatments. Incineration is commonly used which could be expensive and results in a product far removed from the original soil. Phytoremediation can be a cost-effective alternative for remediating contaminated soils. Plants can prevent further movement of contaminants to groundwater, improve uptake and transformation of the pollutant, and can also stimulate indigenous rhizosphere microorganisms capable of metabolizing contaminants by providing carbon substrates (Anderson et al., 1994).

Different plant species found at the site include bromegrass, tall fescue, annual ragweed, barnyardgrass, kentucky bluegrass, sunflower, and crown vetch among others (Peterson, 1996). Bromegrass was predominant since it was planted around the site. Plants were growing both in contaminated and uncontaminated soils. Since TNT is phytotoxic, the question arises as to the role of rhizosphere community associated with the plants in protecting the plant from TNT. A first attempt to address this question was made by characterizing in situ soil microbial community through analysis of FAMEs extracted from microbial cells within the soils.

Objective. Understand shifts in microbial communites due to presence or absence of TNT and plants. Presence of TNT can change the composition of the microbial community and its physiological status which results in the shift of its FAME 'fingerprint'.

Site Description. The site has a ditch through which washings of shells from the plant was drained. It is highly contaminated with TNT, and its degradative products. Plant species mentioned above were growing in this ditch. There were also some bare patches away from the ditch which was also contaminated with TNT and grasses were seen growing around these patches. Initial contamination with respect to depth of soil sampling, plant distribution is unknown since the site was contaminated for over 45 years.

TABLE 1. Characteristics of soil sample from patches with and without plants.

	Acetonitrile extractable TNT (mg L^{-1})	Water extractable TNT (mg L^{-1})	EC (mmhos)	pH
Vegetated				
0-7.5 cm	27	2	0.39	5.50
7.5-15 cm	725	109	0.14	6.12
15-30 cm	53	1	0.10	6.34
Unvegetated				
0-7.5 cm	1609	101	0.18	6.18
7.5-15 cm	85	34	0.15	6.00
15-30 cm	15	6	0.14	5.57

MATERIALS AND METHODS

Transects separated by 2 m intervals were marked parallel to the ditch. At 8 locations along the transects soil cores were taken at three depths: 0-7.5 cm, 7.5-15 cm, and 15-30 cm, and composited. Similarly soils cores were taken from within patches bare of vegetation (unvegetated), and from the vegetated edge (vegetated). Soil samples were maintained at field moist condition prior to passing through a 2 mm sieve to remove stones and roots, and to allow uniform mixing of the composite soil samples. Composite samples were used to get water extractable and acetonitrile extractable TNT, pH, electrical conductivity (EC), mineral nitrogen (NO_3 and NH_4), and bulk density.

Microbial community structure was based on extraction of total FAME's from the soil community. Samples were first hydrolyzed using freshly prepared methanol-potassium hydroxide, and the resulting FAMEs were partitioned into hexane. Nonahexadecanoic acid (C19:0) was added as an internal standard. FAMEs were quantified using gas chromatography with FID detector and GC-MS confirmation of peak identity. Amounts of FAMEs containing 14-20 carbons were normalized to hexadecanoic acid methyl ester (C16:0), a FAME correlated with total microbial biomass, before analysis by STATISCA®. Single linkage method was used for cluster analysis.

RESULTS AND DISCUSSION

Soil samples from the transect parallel to the ditch did not show differences in all quantities measured. Our initial hypothesis was that transects located further from the ditch would be lower in TNT concentration. Transects located 1 m from the ditch; however, were not contaminated with TNT, indicating that TNT contamination was largely confined to the ditch. Nevertheless, TNT was found in several patches located distant from the ditch. In vegetated patches, TNT concentration was lower at 0-7.5 cm depth, increased concentration at 7.5-15 cm depth, and decreased concentration at 15-30 cm depth (Table 1). In unvegetated patches; however, TNT concentration was higher at 0-7.5 cm layer and decreased with increasing depth. This was seen in most of the samples. Plants may support microbial communities in the rhizosphere which may help them to establish and grow in presence of TNT. Another possibility could be that microorganisms protected the plants from exposure to TNT (Pfender, 1996). A third possibility, is over the years, formation of plant residue could have bound TNT, thus lowering the concentration.

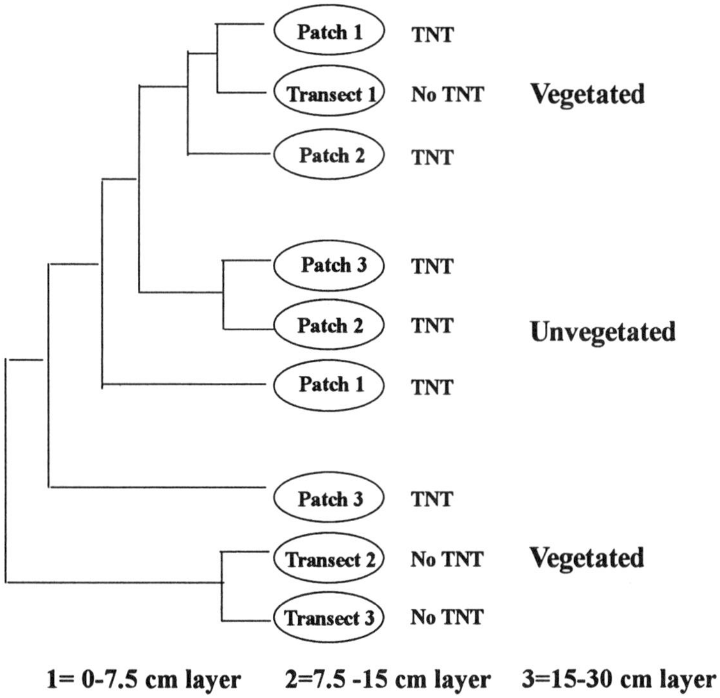

FIGURE. 1. Cluster analysis of FAME profiles by single linkage method.

FAME analysis. In vegetated soil samples at 0-7.5 cm layer FAME profiles clustered separately with regard to TNT concentration, but closely together compared to other clusters (Fig 1). Rhizosphere may be stimulating similar communities regardless of TNT concentration. FAMEs from deeper layers (7.5-15 cm and 15-30 cm) clustered separately and further apart from surface layer, thus having different communities.

FAMEs from unvegetated soil samples clustered separately according to depth of soil samples, but were closely together. This may be due to influence of TNT on microbial communities or possible differences in soil properties with depth. Deeper layers were clustered closer to surface layer of vegetated samples, which suggests low TNT concentration with similar microbial communities. This is also supported by 0-7.5 cm layer with high TNT concentration, which was clustered together and away from deep layers. Statistical analysis like principal component analysis will help to better understand these FAME profiles.

REFERENCES

Anderson, T.A., E.L. Kruger, and J.R. Coats. 1994. "Biological Degradation of Pesticide Wastes in the Root Zone of Soils Collected at an Agrochemical Dealership." In T.A. Anderson and J.R. Coats (Eds.), *Bioremediation through rhizosphere technology*, p. 199-209. Am. Chem. Soc. Symp. Ser. 0097-6156. Am. Chem. Soc., Washington, DC.

Peterson, M.M. 1996. "Germination and early seedling development of tall fescue, switchgrass, and smooth bromegrass as influenced by 2,4,6- trinitrotoluene and 4-amino-2,6-dinitrotoluene". Masters Thesis. University of Nebraska-Lincoln, Lincoln, NE.

William F. Pfender. 1996. "Bioremediation bacteria to protect plants in pentachlorophenol-contaminated soil." *J. Environ. Qual.* *25*:1256-1260.

MICROBIAL POPULATION DYNAMICS DURING HYDROCARBON BIODEGRADATION UNDER VARIABLE NUTRIENT CONDITIONS

D. D. Cleland (Department of Geography and Environmental Engineering, United States Military Academy, West Point, NY)
V. H. Smith (Department of Systematics and Ecology, University of Kansas, Lawrence, KS)
D. W. Graham (Department of Civil and Environmental Engineering, University of Kansas, Lawrence, KS)

ABSTRACT: Natural soils contain a wide variety of microbial species capable of mineralizing both long-chain alkanes and polycyclic hydrocarbon (PAH) compounds. Using a batch microcosm-based screening procedure, a range of nitrogen (N) and phosphorus (P) external supply conditions, and phenanthrene as a hydrocarbon source, we determined that N and P not only influenced biodegradation rates but also caused population shifts to occur amongst the microfauna in a soil system. Results indicated that both the absolute amount and relative proportion of N and P impacted the size and diversity of the biodegrading microbial populations. In addition to effecting microorganism growth-rates, nutrient supply conditions influenced the individual species dominance. When examined in conjunction with bioavailable N and P, the dominant species appear to fulfill consistent roles within each consortium.

INTRODUCTION: Applied scientists have long been interested in methods to improve hydrocarbon biodegradation. Microorganisms capable of degrading most hydrocarbons can be found in soil (Atlas & Bartha, 1972; Kästner et al., 1994). One factor influencing biodegradation rates is the availability of the key nutrients required by the microorganisms already present in the soil, specifically nitrogen (N) and phosphorus (P). Amendment of N and P has been shown to have a positive effect on degradation rates (Atlas & Bartha, 1972, 1973; Dibble & Bartha, 1979; Lindstrom et al., 1991; Bragg et al., 1994), although the form and amounts have not been well defined. Graham et al. (1995, 1997) defined both the ideal ratio of nitrogen and phosphorus for degradation of selected hydrocarbons and examined the forms of each which were bioavailable for short-term utilization. The purpose of the preliminary experiments described here was to determine if the nutrient amendments which increased degradation rates also effected the distribution of hydrocarbon-degrading microbial species in the degradation environment.

MATERIALS AND METHODS

The soil used throughout this experiment was an unacclimated sandy silt fill from near a crude oil holding tank (see Graham et. al., 1997). Total organic carbon and total soil nitrogen, as determined by Carlo-Erba C-N analyzer, averaged 13100 mg-C/kg and 960 mg-N/kg. Total soil phosphorus was determined by digestive colormeteric method and found to be 90 mg-P/kg. Further analysis of the soil nitrogen after Recous et. al. (1990) indicated nitrogen subfractions of: NH_3-N, 2.2 mg/kg; NO_3-N, 5.8 mg/kg; NO_2-N, 26.0 mg/kg; organic-N, 926 mg/kg. Phosphorus subfractions were: inorganic-P,PO_4, 24.3 mg/kg; extractable organic-P, 39.6 mg/kg; and nonextractable organic-P, 26.1 mg/kg (Graham at. al., 1997).

Microcosms were prepared after Graham et. al. (1995), with variable N and P amendments and 15 mmol phenanthrene (per microcosm). The four amendment conditions corresponded to a N:P ratio of 15:1 with a "high" N and P supply (6.42 mM N), a "very low" N and P supply (1.24 mM N), a P amendment without N (0.43mM P) and a N amendment without P (6.42 mM N). Blank and killed controls were prepared for each nutrient condition by omitting the hydrocarbon or adding sodium azide respectively.

Plates for enumeration were made using NSM media with 6.42 mM N, 0.43 mM P in 15 g/L noble agar. Plates were coated with phenanthrene after Kästner et al.(1994) and Kiyohara et al. (1982). The hydrocarbons was dissolved in acetone (2 mg per 4 ml acetone) and then 1 ml was spread on each plate.

Carbon dioxide (CO_2) production in microcosm headspace was measured using gas chromatographic (GC) analysis and used as an indicator of bioactivity. Because the carbon source added to the soil was at least 60-fold larger than the soil carbon this is an accurate measure of hydrocarbon utilization (Graham at. al., 1997). The GC used was a Carle 311 Analytical Gas Chromatograph with a Poropack Q 80/100 capillary column, thermal conductivity detector (TCD), and Hewlett-Packard HP3392A Integrator.

The microcosms were subjected to destructive sampling at 0, 14 (when microbial activity was highest for most microcosms), and 42 days for enumeration. Plates were inoculated with 100 μl serially diluted soil slurry from each nutrient amendment condition, spread and incubated at 23° C in the dark for approximately 2 weeks. Colonies which formed a zone of clearance in the hydrocarbon around themselves and then were counted under 100x magnification. Microorganisms were labeled G to K according to colony morphology, color and other distinguishing characteristics. Microorganisms of interest were plated on BIOLOG BUGUM agar and identified with BIOLOG II gram negative identification kits.

RESULTS

CO_2 Evolution Under Varied Nutrient Conditions: Figure 1 shows cumulative CO_2 production for phenanthrene degradation under five different external nutrient supply conditions over the 42-day study period. The High nutrient amendment had the highest CO_2 production, followed by the Very Low and N-Only conditions. The Very Low and N-Only conditions produced approximately 20% less net CO_2 as compared with the High condition. It should be noted that the Natural and P-Only conditions produced virtually no CO_2 over 42 days.

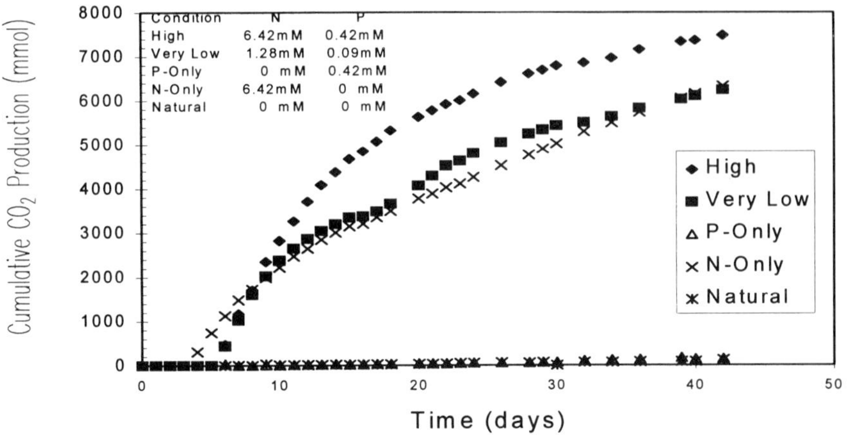

FIGURE 1 - Effects of Nutrient Supply Conditions on CO_2 Production Under Phenanthrene Biodegradation

Distribution and Enumeration of Hydrocarbon-degrading Microorganisms: Figure 2 shows the 14-day results of species distribution for phenanthrene degradation, using the colony labels to depict shifts in predominance by nutrient condition. Figure 2a represents the percent of the population each labeled colony represented, while 2b represents the estimated number of those same species, projected on a logarithmic scale. Figure 3 shows an isocline of the two most dominant species for each nutrient supply condition plotted against the external nitrogen and phosphorus supply levels utilized. Using BIOLOG II identifiaction, microorganism H was tentatively identified as a *Micrococcus* species, J as an *Actinomycetes* species, and K as a *Corynebacterium* species. Microorganisms G and I were not identified.

DISCUSSION

Engineers have long been interested in the means to improve biodegradation rates, and over the past twenty years, the effect of the key nutrients nitrogen and phosphorus has been explored in that light. The results shown in Figure 1 confirm the positive effects of such nutrient amendments in other work. Optimal phenanthrene degradation resulted under the highest external N and P supply conditions (15:1 N:P, High amendment). Interestingly, in this soil, external N supply was much more critical to effective biodegradation than the P supply.

Figure 2 demonstrates a clear effect of nutrient amendment on species distribution. As the bioavailable levels of N and P were altered, the distribution of the culturable species varied. These results suggest that the biodegradation of hydrocarbons such as phenanthrene is a function of both the nutrients available for microbial use and the resulting species present. The figure also suggests the species dominance in the degradation environment is largely a function of the nutrient supply condition provided.

Figure 3 presents the most interesting results of these experiments. When

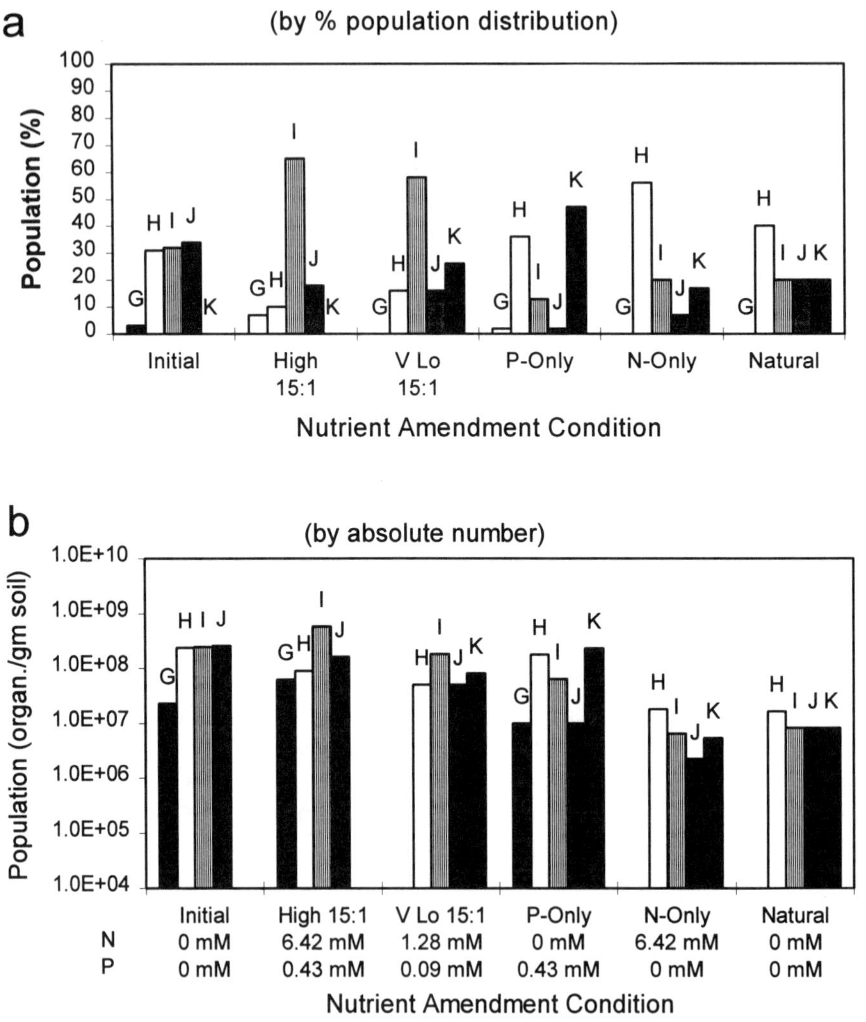

FIGURE 2 - Change in the Population Distribution at Day 14 of Phenanthrene-Degrading Species as a Function of Nutrient Supply Condition

plotted against the N and P external supply conditions, predominant microorganisms exhibit consistent nutritional growth-strategy characteristics. For example, the microorganism labeled H appears to predominate under both N and P nutrient stress, possibly scavenging under N- and/or P-limitation. Microorganism I appears to be an ecological opportunist, becoming dominant when N is in excess. Microorganism K appears to have a niche when there is low N, suggesting that it might have the capability to conserve N, fix N, or use less bioavailable forms of soil N. J appears to be an organism which thrives only under excess N and P. While these suggested roles are only speculation, the possible niche patterns for the selected species are surprisingly consistent given the various microcosms evaluated.

Under the conditions prescribed in these preliminary experiments, the

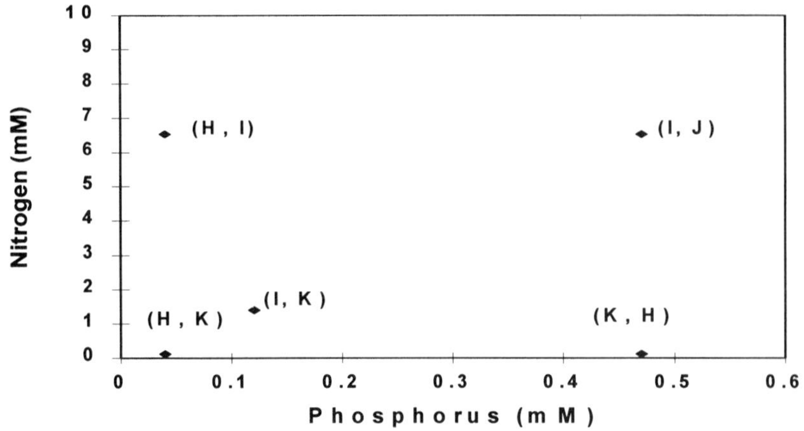

FIGURE 3 - Dominant Microbial Species Under Varied N and P Supply Conditions for Phenanthrene Degradation (at 14 days)

results show that nutrient conditions have an effect on both bioactivity as measured by CO_2 production, and on the microbial species distribution. Plots of the dominant microbial species against the different nutrient supply conditions used here exhibit clear nutritional regions where each species dominates and suggests various species' ecological roles. Further work along these lines would define the effects of subtle changes in nutrient conditions, especially those mimicking natural nutritional shifts.

REFERENCES

Atlas, R. M., and R. Bartha. 1973. " Abundance, Distribution, and Oil-Biodegrading Potential of Microorganisms in Rartian Bay." ***Environmental Pollution, 4:*** 291-300.

Atlas, R. M., and R. Bartha. 1972. " Degradation and Mineralization of Petroleum in Sea Water: Limitation by Nitrogen and Phosphorous." ***Biotechnology and Bioengineering, 14:*** 309-318.

Bragg, James R., Roger C. Prince, E. James Harner, and Ronald M. Atlas. 1994. "Effectiveness of Bioremediation for the *Exxon Valdez* Oil Spill." ***Nature, 368(6470):*** 413-418.

Dibble, J. T., and R. Bartha. 1979. " Effect of Environmental Parameters on the Biodegradation of Oil Sludge." ***Applied and Environmental Microbiology, 37(4):*** 729-739.

Graham, D. W., Val H. Smith, and Kam P. Law. 1995. " The Application of Variable Nutrient Supplies to Optimize Hydrocarbon Biodegradation." In R. Hinchee, F. Brockman, and C. Vogel (eds.), ***Proceedings of the International***

In Situ and On-Site Bioreclaimation Symposium; Bioremediation of Recalcitrant Organics, 331-340. Batelle Press, Columbus, OH.
Bragg, James R., Roger C. Prince, E. James Harner, and Ronald M. Atlas. 1994. "Effectiveness of Bioremediation for the *Exxon Valdez* Oil Spill." ***Nature, 368(6470):*** 413-418.

Dibble, J. T., and R. Bartha. 1979. " Effect of Environmental Parameters on the Biodegradation of Oil Sludge." ***Applied and Environmental Microbiology, 37(4):*** 729-739.

Graham, D. W., Val H. Smith, and Kam P. Law. 1995. " The Application of Variable Nutrient Supplies to Optimize Hydrocarbon Biodegradation." In R. Hinchee, F. Brockman, and C. Vogel (eds.), ***Proceedings of the International In Situ and On-Site Bioreclaimation Symposium; Bioremediation of Recalcitrant Organics,*** 331-340. Batelle Press, Columbus, OH.

Graham, D. W., Val H. Smith, Dale Cleland, and Kam P. Law. 1997. "Effects of Nitrogen and Phosphorous Supply Conditions on Nutrient Limitation for Biodegradation in Soil Systems." *Soil, Water, and Air Pollution*, Submitted.

Kästner, M., M. Breuer-Jammali, and B. Mahro. 1994. " Enumeration and Characterization of the Soil Microflora from Hydrocarbon-Contaminated Soil Sites Able to Mineralize Polycyclic Aromatic Hydrocarbons (PAH)." ***Applied Microbiology and Biotechnology***, ***41(2):*** 267-273.

Kiyohara, Hohzoh, Kazutaka Nagao, and Keiji Yana. 1982. " Rapid Screen for Bacteria Degrading Water-Insoluble, Solid Hydrocarbons on Agar Plates." ***Applied and Environmental Microbiology***, ***43(2):*** 454-457.

Lindstrom, Jon E., Roger C. Price, James C. Clark, Matthew J. Grossman, Thomas R. Yeager, Joan F. Braddock, and Edward J. Brown. 1991. "Microbial Populations and Hydrocarbon Biodegradation Potentials in Fertilized Shoreline Sediments Affected by the T/V *Exxon Valdez* Oil Spill." ***Applied and Environmental Microbiology***, ***57(9):*** 2514-2522.

Recous, S., B. Mary, and G. Faurie. 1990. " Soil Inorganic N Availability: Effect on Maize Residue Decomposition." ***Soil Biology and Biochemistry***, ***7:*** 913-922.

MICROBIOLOGICAL MONITORING OF CONTAMINANTS IN A FRACTURED BASALT AQUIFER

S.P. O'Connell (Idaho National Engineering Laboratory and Idaho State University), R.M. Lehman, F.S. Colwell (Idaho National Engineering Laboratory), and M.E. Watwood (Idaho State University)

ABSTRACT: Indigenous groundwater microbial communities were evaluated for their usefulness in reflecting the presence of contaminants including trichloroethylene (TCE). Groundwater was collected from wells in and around a contaminant plume in the Snake River Plain Aquifer and inoculated directly into Biolog GN microplates to produce a community-level physiological profile (CLPP) based on the respiration of 95 sole carbon sources. Principal components analyses of the carbon source utilization pattern (CSUP) produced by the CLPP method allow rapid, quantitative comparison of microbial communities. Preliminary results demonstrated differences in CSUP from wells with dissimilar organic carbon concentration, lithology, and screened-interval depth. Groundwater from wells with similar screened-interval lithology, depth, and magnitude of gross CLPP response but different TCE concentrations exhibited significantly different CSUP. Trends in community response from TCE contaminated wells followed a pattern from most impacted (6,200 and 3,900 μg/L) to lesser impacted (0.9 to 56 μg/L) groundwaters. Even in a geologically heterogeneous site, CLPP of indigenous microorganisms can be correlated with contaminant concentration and may be useful for ecological risk assessment and the determination of relevant remediation endpoints.

INTRODUCTION

Traditional water chemistry data and toxicity testing serve useful functions in describing the extent of pollution by single or multiple contaminants, but these measurements do not necessarily reflect *in situ* effects on indigenous organisms (Tay et al., 1992). In most groundwater studies, the interactive effects of contaminants and the role of physical and chemical parameters on the indigenous biota need to be better understood. This is especially true for proposed *in situ* remediation of aquifers where access to the pollutants is limited. Because of their intimate interaction with groundwater and unique ability to degrade toxic organic compounds, native microorganisms have great potential in assisting remediation efforts and in acting as environmental monitors (Gounot 1994).

In this study a community-level physiological profile (CLPP) was used to assess microbial community differences due to contaminated groundwater. The CLPP was first described by Garland and Mills (1991) as a rapid means of characterizing microbial communities based on their ability to use 95 different carbon sources in Biolog GN microplates. Use of each of the 95 sources is indicated by color produced via reduction of a tetrazolium dye. The resulting carbon source utilization pattern (CSUP), sometimes likened to a "microbial community fingerprint", is used to compare environmental samples. The CLPP is a sensitive and reproducible measure for characterizing microbial communities in environmental samples (Garland and Mills 1991; Bossio and Scow 1995; Colwell and Lehman 1997; Haack et al., 1995).

The Snake River Plain Aquifer (SRPA) at Test Area North (TAN) of the Idaho National Engineering Laboratory was chosen as a study site because a contaminant plume originating there has been the subject of ongoing and multidisciplinary work that provides data from geochemical, hydrological, and biochemical studies used in conjunction with CLPP assessment. Chlorinated solvents, radionuclides, and sewage were disposed of into the eastern SRPA via a waste injection well (TSF-05) until twenty years ago. Since then, trichloroethylene (TCE) has migrated 3000 m downgradient (Figure 1) while the sewage constituents and radionuclides have been limited to a few hundred meters migration. Concern has been raised over the quality of the SRPA groundwater, which is used for irrigation and municipal purposes (Sorenson et al., 1996).

The purpose of this study was to investigate the potential of the CLPP to serve as an indicator of groundwater quality and as a marker for remediation progress that relies on indigenous microbial communities.

METHODS AND MATERIALS

The SRPA near TAN is comprised of layered basalt flows, basaltic rubble, and sediment with a depth to water at 60-70 m below ground level. The aquifer is unconfined and highly heterogeneous with hydraulic conductivity varying over five orders of magnitude (Sorenson et al., 1996).

Triplicate water samples were collected in 1996 from twenty-three wells near TAN in 50-mL sterile centrifuge tubes and stored on ice for transport to the lab. Wells were routinely sampled after three well volumes had been purged and pH and conductivity had stabilized. Triplicate groundwater samples were inoculated directly into the Biolog GN microplates on the day of collection. Color development produced by reduction of the tetrazolium dye in the plates was assessed every two hours over a week of incubation. An analysis threshold of average absorption (average well color development or AWCD) of 0.5 was chosen for each replicate so that the data were normalized to color development rather than time of incubation (Garland 1996). The resulting carbon source utilization patterns (CSUP's) were compared using the first two factor scores from principal components analysis (PCA) as per Garland and Mills (1991; see for more detailed explanation of the CLPP). The well with the highest TCE concentration (TSF-05) and the well with the lowest concentration (TAN24A) were selected as endmembers to determine which carbon substrates were most instrumental in defining differences between groundwater communities exposed to TCE. From

Factor 1 of the PCA of those endmember wells, the highest loading carbon sources (those at or over a threshold of 0.899) were used for further analyses of a selected set of additional wells that were similar in borehole lithology (basalt with no fractures or sediment), depth of pump (70-90 m), and number of Biolog carbon sources used (50-85). The selected set of wells was broken into two treatment groups: high (>3900 µg TCE/L) and low (<100 µg TCE/L) for multivariate analysis of variance (MANOVA) testing of significance ($p < 0.5$) due to presence of TCE. PCA analyses for wells that differed in other factors (low Biolog response, high dissolved organics, fractured basalt lithology) were also performed to explore the effect of factors other than TCE on CSUP.

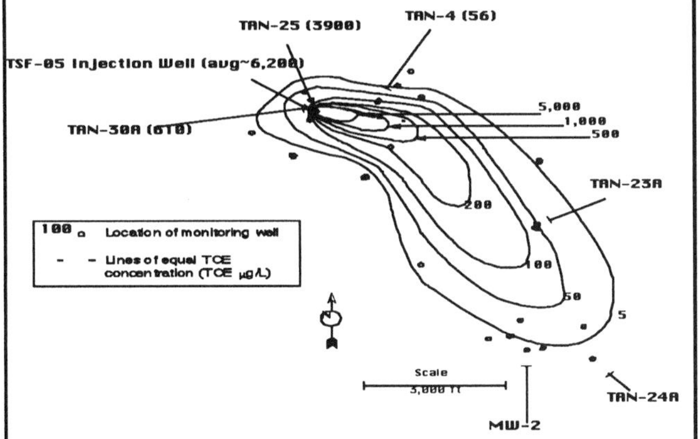

Figure 1 . Diagram of Test Area North showing injection well (TSF-05) as source of trichloroethylene plume and monitoring well locations with TCE concentration.

RESULTS

Bacterial communities from the highly impacted injection well (TSF-05) and a well from outside of the contaminant plume (TAN24A) at TAN were distinguished by their CSUP (Figure 2; see Figure 1 for well locations and TCE concentrations). PCA Factor 1 loadings for the following thirteen carbon sources were all at or above 0.899; alpha-cyclodextrin (–0.962), dextrin (–0.983), glycogen(–0.966), cellobiose (–0.960) maltose (–0.987), D-galactonic acid lactone (0.938), p-hydroxyphenylacetic acid (0.940), sebacic acid (0.899), glycyl-L-glutamic acid (–0.983), L-proline (–0.938), gamma-amino butyric acid (0.921), inosine (0.937), and thymidine (0.933). The communities from a larger set of wells selected primarily to differ in TCE concentration grouped according to low TCE values (wells TAN4, TAN23A, and TAN24A) and high TCE values (wells TSF-05 and TAN25; Figure 3). MANOVA of the two groups of wells (high and low TCE) showed a significant difference ($p < 0.05$, Wilks' Lambda statistic) based on the utilization of the thirteen carbon sources which had distinguished between endmember wells.

Wells in which groundwater was collected from screened intervals adjacent to fractured basalt (MW2 and TAN30A) versus intact basalt (TAN23A and TAN24A) were distinct (Figure 4). TCE concentration (µg/L) was 0.3 for MW2, 610 for TAN30A, 77 for 23A, and 0.9 for 24A.

DISCUSSION

Microbial communities have been shown to be distinct in aquifers based largely upon sediment mineralogy and grain size (Gounot 1994) and less dependent on hydraulic conductivity [Colwell and Lehman 1997 (in press)]. Groundwater contains significant microbial populations whose community structure and function are greatly affected by dissolved substances rather than the massive rock they are flowing through (Gounot 1994). The SRPA presents difficulties in interpretation of contaminant effects on the biota due to the heterogeneous fractured basalt with varied local lithologies and groundwater chemistries. TAN well completion logs show a high variability in well design that, together with biogeochemical factors, interfere with the interpretation of effects of TCE and other contaminants on the aqueous microbial communities. These well construction differences have been shown to be significant in microbial community structure and function in past studies (Hirsch and Rades-Rohkohl 1988).

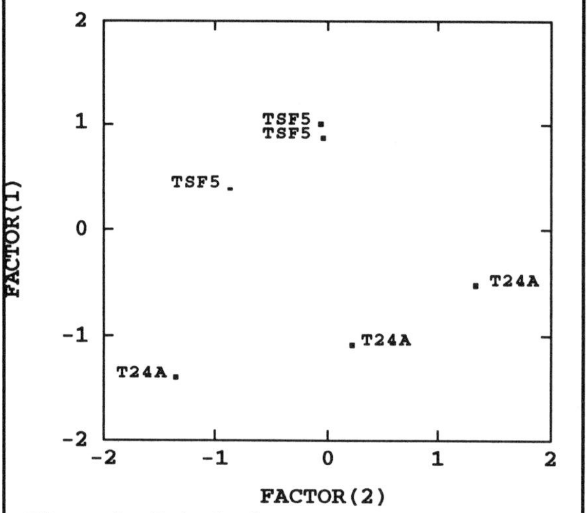

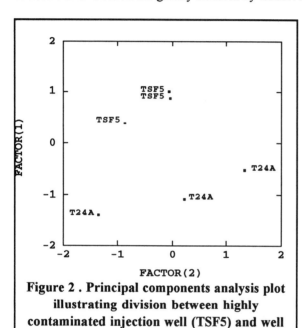

Figure 2. Principal components analysis plot illustrating division between highly contaminated injection well (TSF5) and well from outside TCE plume at TAN (T24A).

The highly contaminated well, TSF-05, produces groundwater communities distinct from TAN24A, a well presumed to be free of TCE (Figure 2). Performance of PCA on additional wells using only the limited number of carbon sources responsible for distinguishing communities from endmember wells allows reduction of the variability due to factors other than the presence of contaminants. Additionally, use of a smaller set of variables allows testing by MANOVA between groups of wells thought to differ by a given factor, e.g., TCE concentration. In order to detect the influences of one factor (e.g., contaminants) on groundwater microbial communities, wells were selected that were similar in all other aspects. Using a selected set of wells and a reduced number of variables, TCE did have a significant effect ($p<0.05$) on the CSUP of groundwater communities in the SRPA (Figure 3).

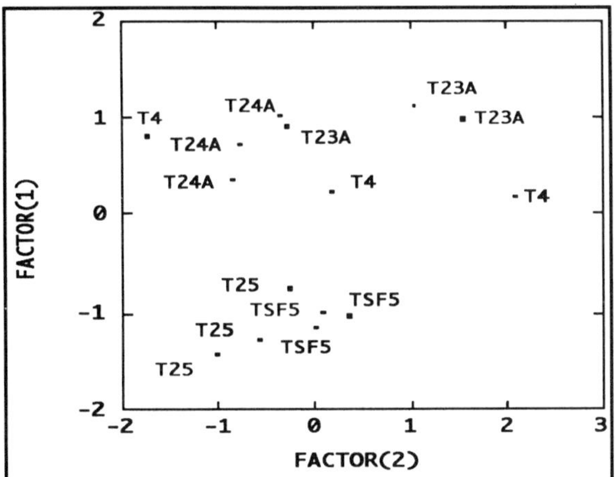

Figure 3 . Principal components analysis of 13 carbon variables that distinguish between highly impacted groundwater communities (TSF5 and TAN25) and communities from less impacted well locations.

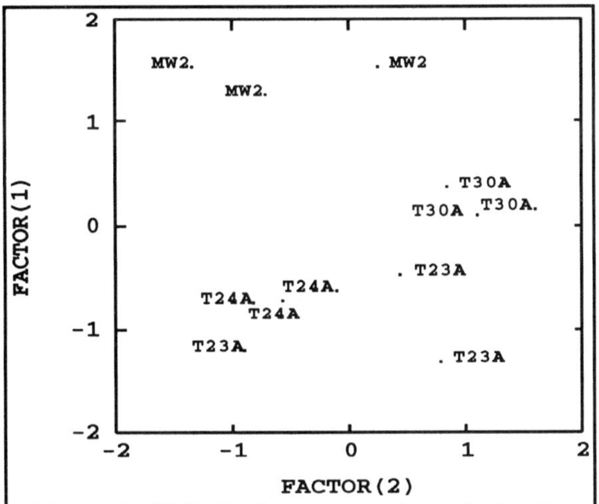

Figure 4 . Principal components analysis plot discriminating between TAN groundwater communities sampled from fractured basalt (MW2, T30A) and intact basalt (T24A, T23A) at screened-level of pump.

Even when analyzing wells that vary in three orders of magnitude of TCE concentration, factors such as the hydrogeology of the screened interval may interfere with interpretation of the TCE contamination (Figure 4). The wells MW2 and TAN24A are within 300 m of one another, 2800 m from the injection well TSF-05, and have identical TCE concentrations while TAN30A and TAN23A have higher TCE concentrations with TAN30A about 80 m from TSF-05 and TAN23A about 1800 m from the TCE source well. Yet MW2 and TAN30A are segregated from the other two wells, despite the TCE differences. This finding is possibly due to fracturing across the screened intervals of MW2 and TAN30A. This result shows the inherent complications in analyzing data from a heterogeneous environment.

This study has demonstrated that groundwater microbial communities are highly sensitive to multiple chemical and physical factors in a heterogenous aquifer. When datasets were composed of samples from wells with similar lithology, depth of sampling, etc., CLPP distinguished between indigenous microbial communities from wells contaminated with TCE and from wells not impacted. This technique has the potential to be used in remediation activities where progress must be rapidly and economically evaluated and an ecologically relevant endpoint is desired, e.g., an indigenous community resembling that of an unperturbed site.

ACKNOWLEDGMENTS

Research was supported by US Department of Energy Environmental Management Proof-of-Concept funding and by the Subsurface Science Program of the US Department of Energy, Office of Health and Environmental Research (Contract DE-AC07-76ID01570). Don Maiers, Joe Lord, and Steve Ugaki greatly assisted in sample acquisition.

REFERENCES

Bossio DA and KM Scow. 1995. "Impact of Carbon and Flooding on the Metabolic Diversity of Microbial Communities in Soils." *Appl.Environ. Microbiol.* 61: 4043-4050.

Colwell FS and RM Lehman. 1997. "Microbial Communities from Hydrologically Distinct Zones of a Basalt Aquifer." *Microb. Ecol.* (in press).

Garland JL and AL Mills. 1991. "Classification and Characterization of Heterotrophic Microbial Communities on the Basis of Patterns of Community-Level Sole-Carbon-Source Utilization." *Appl. Environ. Microbiol.* 57: 2351-2359.

Garland JL. 1996. "Analytical Approaches to the Characterization of Samples Using Patterns of Potential C Source Utilization." *Soil Biol. Biochem*. 28: 213-221.

Gounot AM. 1994. "Microbial Ecology of Groundwaters." In: Groundwater Ecology (Gibert J, DL Danielopol, and JA Sanford, Eds.), pp. 189-215. Academic Press, San Diego.

Haack SK, H Garchow, MJ Klug, and LJ Forney. 1995. "Analysis of Factors Affecting the Accuracy, Reproducibility, and Interpretation of Microbial Community Carbon Source Utilization Patterns." *Appl. Environ. Microbiol*. 61: 1458-1468.

Hirsch P and E Rades-Rohkohl. 1988. "Some Special Problems in the Determination of Viable Counts of Groundwater Microorganisms." *Microb. Ecol*. 16: 99-113.

Sorenson, KS, AH Wylie, and TR Wood. 1996. *Test Area North Hydrogeologic Studies Test Plan; Operable Unit 1-07B*. Idaho National Engineering Laboratory. INEL-96/0105.

Tay K-L, KG Doe, SJ Wade, DA Vaughan, RE Berrigan, and MJ Moore. 1992. "Sediment Bioassessment in Halifax Harbour." *Environ. Toxicol. Chem*. 11: 1567-1581.

BIOREMEDIATION PROCESSES DEMONSTRATED AT A CONTROLLED GASOLINE RELEASE SITE

S.M. Pfiffner (Oak Ridge Institute for Science and Education), R. Siegrist (Colorado School of Mines), D. B. Ringelberg (University of Tennessee), and A.V. Palumbo (Oak Ridge National Laboratory[1]).

ABSTRACT: After a controlled release of a synthetic gasoline-like mixture of petroleum hydrocarbons at a site consisting of highly fractured clay-rich sediments, various physical treatment methods were applied. Fracturing and air sparging removed a measured portion of the contamination and an inventory of the contaminants remaining in the sediments suggested either a poor recovery of the contaminants from the sediments or an unmeasured loss of the contaminants. Bioremediation, stimulated by the fracturing and air sparging, was considered as a potential unmeasured sink of the contaminants. The potential effects of bioremediation on the mass balance of contaminants was evaluated. Microbial characterization involved 1) MPN enumeration of various physiological types of microorganisms, 2) measurement of toluene mineralization rates, and acetate incorporation rates, and 3) assessment of community structure and nutritional status analysis by phospholipid fatty acids techniques. Sediments containing high levels of contaminants showed low microbial abundance and viable biomass. Also, those sediments had higher indicators of stress (e.g., lipid biomarkers and evidence for stimulatory effects in the acetate incorporation studies). Although the sediments with high contamination showed lower total microbial biomass, toluene mineralization rates were higher in these samples than in less contaminated and uncontaminated samples. There are two possible explanations for these observations and these will be evaluated. First, although the total population was lower, the population of BTEX degraders may have been higher in the contaminated samples. Second, the bacteria present may have had a higher per cell activity. The effect of the air sparging and fracturing may have been primarily on the degraders since the contaminants may have been the most available carbon in the system. The evaluation of the potential for microbial degradation in these sediments indicates that the unexplained loss of contaminants could have been due to biodegradation.

---

[1]Oak Ridge National Laboratory is managed by Lockheed Martin Energy Research, for the U.S. Department of Energy under contract DE-AC05-96OR22464.

ECOLOGICAL IMPACTS OF BIOREMEDIATION FIELD STUDY EXPERIMENTS

Terri M. Wood (Center for Coastal Studies, Texas A&M University-Corpus Christi, Corpus Christi, Texas)
Roy L. Lehman (Center for Coastal Studies, Texas A&M University-Corpus Christi, Corpus Christi, Texas)
James Bonner (Texas A&M University, College Station, Texas)

ABSTRACT: The assurance that further damage is not inflicted upon an already impacted biotic community should be included in determining the effectiveness of bioremediation. The Texas General Land Office proposed utilizing a wetland area impacted by the October 1994 San Jacinto River flood and oil spill as an experimental bioremediation field study site, with ecological monitoring included as a component of the study. Baseline sampling was conducted in July 1995, and yielded 27 benthic species, representing five phyla, with 1196 total specimens collected. Baseline study results indicated low species richness and abundance. Benthic samples were collected monthly for one year, from November 1995 until October 1996 with species abundance and richness increasing in comparison to the July 1994 baseline results. Bioremediation treatments began in February 1996, and monthly sampling monitored changes in the wetland ecological community structure. The dominant benthic organism in both the baseline and year long study was the polychaete *Laeonereis culveri*, accounting for 73% of the baseline benthic organisms and 59.8% of the year long study benthic organisms. The analysis of this preliminary data indicates bioremediation treatments utilized in this study did not significantly affect benthic communities.

INTRODUCTION

Coastal wetlands are ecologically critical, functioning as nursery and spawning grounds for estuarine organisms, providing food, habitat and protection for many species. Oil spills have serious consequences on estuarine communities, depending on organism seasonal activity, life cycle, duration of oil exposure, amount and type of spilled petroleum, as well as sediment composition (Mielke, 1990). Permeating into sediments up to 70 cm, traces of oil have been detected in organisms at multiple trophic levels for as long as one year after spills occur (Burns and Teal, 1971 *in* N.A.S., 1975).

Studies investigating the effects of oil cleanup methods indicate that some procedures, such as hot-water high-pressure washing, considerably altered communities of invertebrates which were unaffected by the oil spill (Houghton et al., 1991). Other studies have concluded that detergent emulsification of oil, removal of oiled biota, and burning of oiled areas are less effective than leaving the oiled areas untreated, although not all researchers are in agreement (Freedman, 1989; Tunnell et al., 1994).

Cleanup methods which inflict minimal ecological impacts are essential, and bioremediation shows promise in effective restoration of contaminated wetlands. Yet, many bioremediation field studies lack sufficient monitoring of the effects of treatments on ecological communities (Mearns et al., 1993). Although nutrient addition increases rates of microbial degradation of hydrocarbons, previous bioremediation studies have called for further monitoring of biological effects of treatments due to potential toxicity of micronutrients to organisms (Bragg et al., 1992; Mearns et al., 1993). Sensitivity to environmental disturbances, restricted mobility, and high population growth rates of benthic organisms support their use as indicators of environmental stress and ecological community health (Longley, 1994).

Inundated by rainfall in mid-October 1994, the San Jacinto River flooded, and flood debris ruptured underwater pipelines, spilling over 2 million gallons petroleum products down river. A baseline study was conducted to permit limited analysis of community structure before beginning a year-long ecological study. The purpose of the ecological study is to monitor the effects and impacts of bioremediation treatments on ecological community structure. Analysis of benthic organism abundance and species richness will aid in determining the ecological community health throughout the bioremediation experiments.

Study Site. The study site, Parker's Cove, is approximately 300 m long by 100 m wide and is located on the San Jacinto River in Harris County, Texas, approximately 24 km north of the San Jacinto River and Houston Ship Channel Gulf entrance (29°47′.9″N, 95°04′.2″W). The 1994 flooding and spill rendered the site ideal for experimental bioremediation field studies. Less affected by the spill, a control site was established across the river and two additional control sites were established within the cove. Sampling design for Parker's Cove was established by the TAMU environmental engineering research team, and consists of six blocks containing three plots each. Each plot within a block underwent a different bioremediation treatment during testing, allowing for six replicate samples of each treatment type. The Parker's Cove control sites (plots 32 and 33) are located along the cove east shoreline, Block 3 (plots 8, 10, and 12) is located along the west shoreline, and the original control (OC) is located across the river.

METHODS AND MATERIALS

Hydrological parameters of water temperature, salinity and pH were recorded during each monthly sampling event. Benthic samples were collected using a 10-cm diameter PVC core, inserted 10 cm into the substrate. Two replicate core samples were taken monthly at each location. Samples were fixed in a 10% formalin/rose bengal solution. Laboratory analysis consisted of washing samples through nested sieves (minimum mesh = 0.5 mm), sorting, identifying, and counting organisms before drying for biomass amounts at 90° celsius for two days.

Data presented in this paper consists of monthly benthic samples collected from Block 3 (plots 8, 10, and 12), the cove control sites (PCC = plots 32 and 33) and the original control site (OC) for November 1995 through July 1996.

Application of oil and bioremediation treatments began in March 1996, but were not applied to the original control site or to Parker's Cove controls. Oil was applied to all plots in Block 3 with Plot 12 receiving no nutrients or oxidants, Plot 8 receiving both nutrients and oxidants, and Plot 10 receiving nutrients only. Throughout the experiment, acceptable levels of treatments were maintained by reapplication of nutrients and oxidants.

RESULTS AND DISCUSSION

Water temperature, salinity, pH, and dissolved oxygen values were recorded from both Parker's Cove and the original control site on sampling days. Water temperatures ranged from a low of 9.7°C in February 1996 to a high of 29.2°C in July 1996. Due to drought conditions in Texas, limited amounts of freshwater were released from the dam upriver and salinities varied monthly, generally increasing from 8.6 ppt in November 1995 to a high of 17.3 ppt in April 1996. Following rainfall which alleviated area drought conditions in July 1996, salinities decreased to 7.8 ppt. Dissolved oxygen ranged from 5.9 to 11.7 mg l^{-1} over the course of the study while pH remained fairly stable, ranging from a high of 8.6 in March 1996 to a low of 7.4 in January 1996.

Monthly benthic core samples produced a total of 6630 individual organisms. A total of 35 species representing four phyla were identified; 17 species of annelids, primarily polychaetes, dominated core samples. Arthropods consisted of nine species of crustaceans and four species of insects, and molluscs yielded three bivalve and one gastropod species. Nemerteans represented the remaining phyla, with one unidentified species collected.

Comparison of species richness in Block 3 plots show Plot 10 (oil and nutrient application) yielded the highest species richness, with 22 species collected. Species richness was only slightly lower in plots 8 (oil, nutrient and oxidant) and 12 (oil only), with 20 species collected for each location. Plot 12 showed a gradual decline in species richness from a high December 1995 of 14 species, until February 1996 with 8 species. After bioremediation treatments were applied in March, Plot 12 experienced an increase in species richness. Species richness of PCC plots, which had no oil or bioremediation treatments, were comparable to Block 3 plots, with Plots 32 and 33, producing a total of 19 and 21 species, respectively. April mean monthly densities decreased at all sites, except for Plot 8, and generally decreased in monthly densities reaching lowest densities in July.

Comparison of monthly mean densities from November 1995 until March of 1996, show PCC to be greater than Block 3, and Block 3 greater than OC (Figure 1). While the mean densities and abundance of PCC and Block 3 rise each month through February, both locations declined slightly in March. OC mean monthly density for each month continued to rise and fall with each month from November 1995, declining in December, until July 1996. Although no significant differences existed between plot densities, all samples were dominated by large numbers of polychaetes being collected, and this taxa typically represented greater than 76.7% of all organisms occurring in the cores.

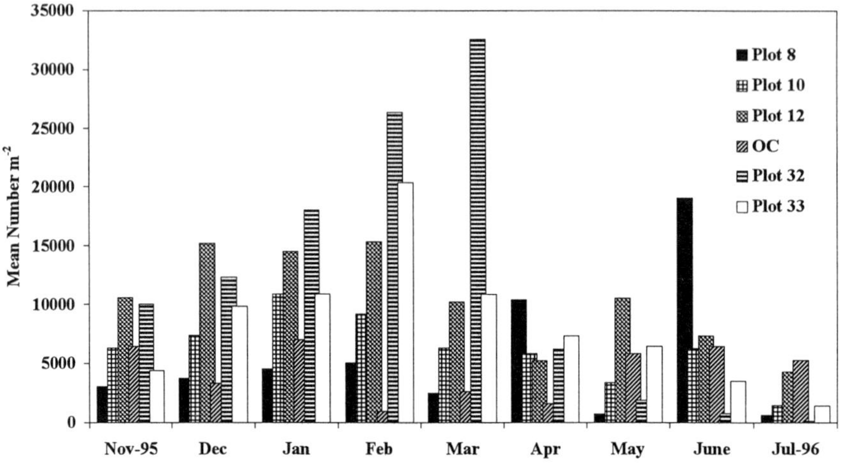

FIGURE 1. Mean monthly benthic density at Plots 8, 10, 12, Original Control (OC), Plot 32, and Plot 33, November 1995 through July 1996.

Monthly densities of plots in Block 3 decreased from February to March after application of oil and bioremediation treatments, but Plot 10 density began decreasing in January 1996, prior to treatments, and continued decreasing from February until March. However, after application of the oil and bioremediation treatments, there are no statistically significant differences between the three treated plots in Block 3 ($p=0.763$).

In both baseline and monthly sampling, the original control (OC) samples yield densities and species richness much lower than those samples from within Parker's Cove. One possibility is that other contamination may be impacting OC and not influencing Parker's Cove. In addition, this analysis was on a limited data set, and ongoing analysis of replicate samples may provide differing results.

July 1995 baseline data showed the cove to be a fairly depauperate area, exhibiting low species richness and abundance, most likely due to a composite of impacts. Manmade and natural perturbations, including flooding, oil spills, normal seasonal population variance, and general poor health of the Galveston Bay system should be considered. Additionally, continual stress such as fluctuating salinity and hydrological characteristics may be affecting the ecosystem, which lies midway between freshwater and saline environments.

Salinities reached a high in April of 1996. This high salinity possibly impacted April benthic densities, since samples from the experimental plots (plots 8, 10, 12) remained in the usual sequential order rank in March 1996 (12.4 ppt) following bioremediation treatments, with plot 8 density less than plot 10, and plot 10 less than plot 12. While densities again declined in April except for Plot 8, the usual ranked order of density changed (Plot 12<10<8). In July 1996, when salinity dropped to 7.8 ppt, the densities returned again to the usual order of Plot 8<10<12. Plot 8 elevation is slightly higher than that of Plot 10 and Plot 12, and so the benthic mean monthly density could have been less impacted by higher salinity.

There was a decrease in mean monthly densities in March following bioremediation treatments, however, the untreated control plots within Parker's Cove also declined. These results may indicate a strong possibility that salinity was affecting the entire cove. Along with the decline in mean monthly densities from March to April 1996, species richness also declined. The number of species per block had remained constant in benthic samples from February 1996 to March, furthering the possibility that salinity caused declines in densities.

Comparison of July 1996 benthic samples to the July 1995 baseline study (Wood et al., 1995) showed that the species richness and densities have improved. Results show the relative order of species richness and abundance to be the same in both July 1995 and July 1996 samples, indicating no differences between improvement of treated or control plots.

Laeonereis culveri was the dominant species from both July 1995 and July 1996. However, there were 147 individual *L. culveri* collected from 25 sample sites in 1995, accounting for 73 % of the total benthic organisms collected, and in July 1996 there were 100 individual *L. culveri* collected from the six sample sites included in this study, and accounting for only 48 % of the benthic organisms collected. The lower proportion of the dominant species indicates a slightly healthier ecosystem, just four months after the bioremediation test oiling and continual application of additional nutrients and oxidants.

Laeonereis culveri, is an extremely adaptable organism that survives well in environmental conditions other species cannot tolerate. *Hobsonia florida*, a polychaete representing the second most dominant species identified in this study, and *Rangia cuneata*, the dominant bivalve are less saline tolerant; and species distribution is generally restricted to a limited salinity range in Texas bay waters. *H. florida* increased in abundance from November 1995 until it reached a high of 190 in February 1996, declining to 1 collected in July 1996. Since *H. florida* is not tolerant of high salinities, further emphasis is placed on the possibility of lower benthic densities and species richness being affected by salinity rather than bioremediation treatments.

While there were no significant differences in benthic community structure between plots, there were indications of statistical differences between months. Comparisons between mean densities of plots does indicate some unusual patterns. When compared to the original control, all plots had higher mean species density, yet when compared to cove control site 32, all plots experienced lower mean species density. Therefore, although initial comparisons and analysis have offered other possible causes for reduced numbers, conclusions cannot be substantiated at this time due to the limited data set available. In addition, with only a preliminary data set completed, and with many variables affecting ecological community structure, conclusive assessment of bioremediation treatments effects upon an ecosystem is difficult to discern.

REFERENCES

Bragg, J. R., R. C. Prince, J. B. Wilkinson, and R. M. Atlas. 1992. *Bioremediation for Shoreline Cleanup Following the 1989 Alaskan Oil Spill*. Exxon Company, U.S.A., Houston, TX.

Burns, K., and J. Teal. 1971. *Hydrocarbon Incorporation Into the Salt Marsh Ecosystem from the West Falmouth Oil Spill.* Woods Hole Oceanographic Institution Technical Report 71-69. Woods Hole, MA.

Freedman, B. 1989. *Environmental Ecology*. Academic Press, San Diego, CA.

Houghton, J. P., D. C. Lees, W. B. Driskell, and A. J. Mearns. 1991. "Impacts of the *Exxon Valdez* Spill and Subsequent Cleanup on Intertidal Biota – 1 year later." In *Proceedings 1991 International Oil Spill Conference*, pp. 467-475. American Petroleum Institute, Washington, DC.

Longley, W. L. (Ed). 1994. *Freshwater Inflows to Texas Bays and Estuaries: Ecological Relationships and Methods for Determination of Needs*. Texas Water Development Board and Texas Parks and Wildlife Department, Austin, TX.

Mearns, A. J., P. Roques, C. B. Henry, Jr. 1993. "Measuring Efficacy of Bioremediation of Oil Spills: Monitoring, Observations, and Lessons From the *Apex* Oil Spill Experience." In *Proceedings of 1993 International Oil Spill Conference*, pp. 335-343. American Petroleum Institute, Washington, DC.

Mielke, J. E. 1990. *Oil In The Ocean: The Short- and Long-Term Impacts of a Spill*. CRS Report for Congress. Congressional Research Service. The Library of Congress. Washington, DC.

National Academy of Sciences (N.A.S.). 1975. *Petroleum in the Marine Environment.* National Research Council. Washington, DC.

Tunnell, J. W. Jr., D. W. Hicks, and B. Hardegree. 1994. *Environmental Impact and Recovery of the Exxon Pipeline Oil Spill and Burn Site, Upper Copano Bay, Texas: Year 1*. Center for Coastal Studies Report TAMU-CC-9402-CCS. Texas A&M University-Corpus Christi. Corpus Christi, TX.

Wood, T. M., B. A. Nicolau, E. H. Smith, and R. L. Lehman. 1995. *Baseline Ecological Investigation of the San Jacinto Bioremediation Field Study Site. Center For Coastal Studies*. Interim Report for Texas General Land Office.

BACTERIAL PRODUCTIVITY IN BTEX- AND PAH-CONTAMINATED AQUIFERS.

Michael T. Montgomery (Geo-Centers, Inc., Fort Washington, Maryland)
Thomas J. Boyd, Barry J. Spargo, and Richard B. Coffin
(Naval Research Laboratory, Washington, DC)
James G. Mueller (SBP Technologies, Inc., Pensacola, Florida)

ABSTRACT: Hydrocarbon degradation by natural bacterial assemblages can significantly lower the cost of *in situ* treatment strategies by reducing the amount of contaminant that has to be removed from the site by abiotic transfer to another medium (e.g. air stripping). *In situ* treatment strategies for these aquifers are often purported to stimulate bacterial activity, however, the evidence presented for contaminant biodegradation is often qualitative and rarely unequivocal. Typically, investigators conclude that bioremediation has occurred when an increase in bacterial abundance is coupled with a decrease in contaminant concentration. Increase in bacterial abundance in groundwater is often seen as the measure of enhanced activity even though changes in abundance are often independent of heterotrophic activity (as measured as other means). By using an approach common in marine studies, we were able to sensitively and quantitatively measure changes in bacterial production and relate these changes to increased carbon demand by indigenous heterotrophic bacteria. Bacterial production (cells ml^{-1} h^{-1}) was measured by rate of leucine incorporation into protein (protein synthesis) which can be related to carbon demand (mg C ml^{-1} h^{-1}) using conversion factors for amount of carbon per cell and metabolic efficiences of bacteria grown on BTEX or PAHs. As part of the SERDP CU-030 program, bacterial production was one of several measures of treatment efficacy used to evaluate the biological component of groundwater circulation well strategies at two sites. The BTEX-contaminated aquifer was at the National Hydrocarbon Test Site at Port Hueneme, California and the PAH-contaminated aquifer was at a wood treatment facility in central Florida. Over twelve months, bacterial productivity increased three to four orders of magnitude in groundwater influenced by circulation wells at both hydrocarbon-contaminated sites. By combining this information on bacterial production with $\delta^{13}C$ measurements of bacteria and CO_2, the net flow of contaminant carbon through the heterotrophic bacterial assemblage can be estimated.

INTRODUCTION

The basic concept behind most *in situ* bioremediation strategies is that they stimulate metabolism of carbon by the natural bacterial assemblage in groundwater and on soil surfaces. Increasing overall bacterial activity (or production) increases the carbon demand at the treated site and subsequently petroleum hydrocarbons are biodegraded as they are used as a carbon source to produce bacterial biomass. Accurate estimates of bacterial production in groundwater are important for determining efficacy of bioremediation treatments in the field.

In typical bioremediation and biodegradation field studies, bacterial abundance is measured during treatment. If the observed abundance of bacteria increases and with a concomittment decrease in contaminant concentration, it is often concluded that the contaminant has been used as a carbon source by the population of bacteria and has stimulated the increase in their abundance (see rev. by National Research Council, 1993). There is an assumed correlation between increased abundance of bacteria and increased productivity of the assemblage (National Research Council, 1993). In these studies, bacteria are often enumerated using heterotrophic plate counts or dilution tubes, though both techniques rely on culturing the strains in high nutrient conditions. Given that culturing methods are known to underestimate total abundance of viable bacteria by upwards of three orders of magnitude, this is a poor strategy for measuring overall bacterial activity and the capacity of the natural assemblage to metabolize a given contaminant. In this study, production of the bacterial assemblage was measured by rates of leucine incorporation into bacterial protein synthesis (Kirchman et al. 1985, Smith and Azam, 1992).

Here, we report on two field studies where groundwater circulation wells (GCWs) were used as an *in situ* treatment for a hydrocarbon-contaminated aquifer. GCWs may stimulate bacterial production by circulating limiting nutrients, enhancing bioavailability of hydrocarbons, or by removing inhibitory low molecular weight compounds through abiotic removal (e.g. air stripping). In one study, a GCW was used to treat groundwater contaminated with 11,000 gal of leaded and unleaded gasoline (BTEX) at the National Hydrocarbon Test Site in Port Hueneme, CA. In the second study, a GCW was used to treat groundwater chronically-contaminated with polycyclic aromatic hydrocarbons (PAHs) from creosote at a wood treatment facility in central Florida.

Objective. The objective of the study is to determine efficacy of *in situ* biological treatment of petroleum hydrocarbons-contaminated groundwater a using GCW strategy. This report focuses on using measurements of production to detect changes in bacterial activity effected by operation of the GCWs.

MATERIALS AND METHODS

GCWs were installed in aquifers at two sites; one contaminate with gasoline (BTEX), and the other with creosote (PAHs). The effect of the GCWs on bacterial production was measured in groundwater samples taken within the GCW and from twelve shallow (1 m below water level) and deep monitoring wells placed either within or outside the zone of influence. More detail on site description and sampling can be found in Mueller et al. (1996, 1996a).

Bacterial Productivity. The centrifugation method for measuring bacterial productivity was performed as described by Smith and Azam (1992) and Montgomery et al. (1996). Briefly, 5.1 ul of ^{3}H-Leu or ^{3}H-TdR (spec. act. = 1.54-1.58 Ci mmol^{-1}, Amersham) was added into sterile 2.0 ml capacity screw cap microfuge tubes with O-rings (Fisher) prior to the addition of 1.7 ml of groundwater.

TCA was added to blanks (final concn. = 5%) and to end incubations. Samples were incubated at 21°C for 1 to 2 h to allow incorporation of radiolabel into bacteria, and then incubated an additional 30 min after addition of TCA (at 21°C for ^{3}H-Leu) prior to centrifugation (16 000 x g, 10 min; Eppendorf, 5415C). Supernatant was removed by aspiration and the pellet was washed with 1.5 ml of TCA (5%), the samples respun and aspirated. The pellets were then washed with 0.5 ml of ethanol (80%), respun, and aspirated. Liquid scintillation cocktail (0.5 ml) was added prior to radioassay in a liquid scintillation counter (Beckman LS 6000TA). Because of difficulties determining empirical factors for converting ^{3}H-Leu uptake into bacterial biomass due to unknown losses from viruses, we used the theoretical conversion factor of Kirchman et al. (1985), 1.0 x 10^{17} cells mol^{-1} ^{3}H-Leu incorporated, and assumed a dilution factor of 2.

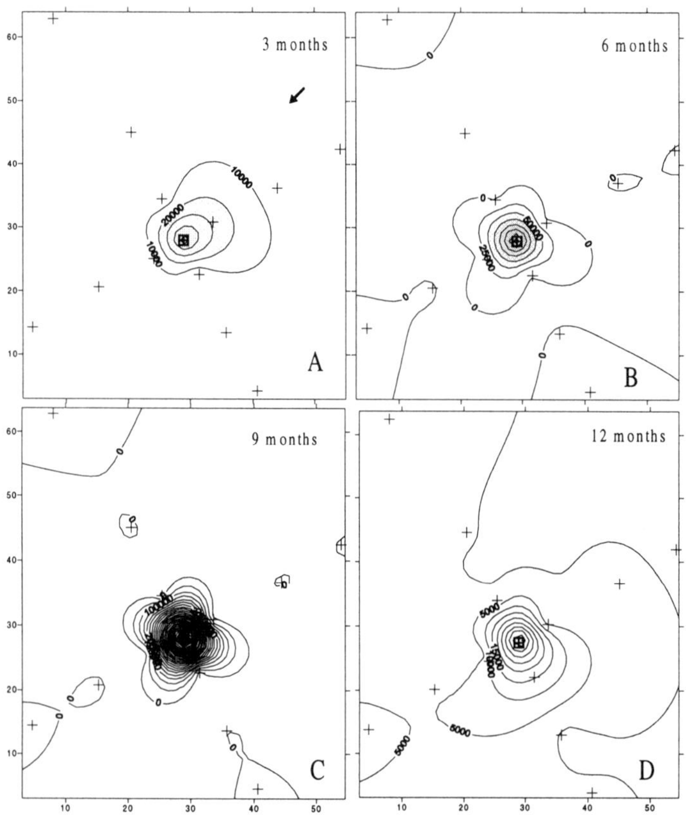

FIGURE 1. Bacterial production (cells ml^{-1} h^{-1}) in shallow wells (+) surrounding a GCW (■) in a BTEX contaminated aquifer. Sampling was performed at 3 months (A), 6 months (B), 9 months (C), and a year (D).

RESULTS AND DISCUSSION

BTEX Impacted Site. Bacterial productivity (^{3}H-Leu incorporation) was measured in and around the main GCW for shallow (Fig. 1) and groundwater samples in 1995 from April (Fig. 1A), August (Fig. 1B), November (Fig. 1C) and December (Fig. 1D). Production in all samples ranged from 880 to 7.7 x 10^4 cells ml^{-1} h^{-1} in the beginning of the experiment (Fig 1A) prior to increasing to 2.0 x 10^5 cells ml^{-1} h^{-1} in the GCW in August (Fig 1B), 7.8 x 10^5 cells ml^{-1} h^{-1} in November (Fig. 1C) and finally doubling again to 1.6 x 10^6 cells ml^{-1} h^{-1} in December (Fig. 1D). Bacterial turnover times for the free living assemblage decreased to ca. 1 h in December. Production increased in the deep samples of the GCW also, but not as dramatically (range: 1.0 x 10^4 to 6.2 x 10^4 cells ml^{-1} h^{-1}) (data not shown).

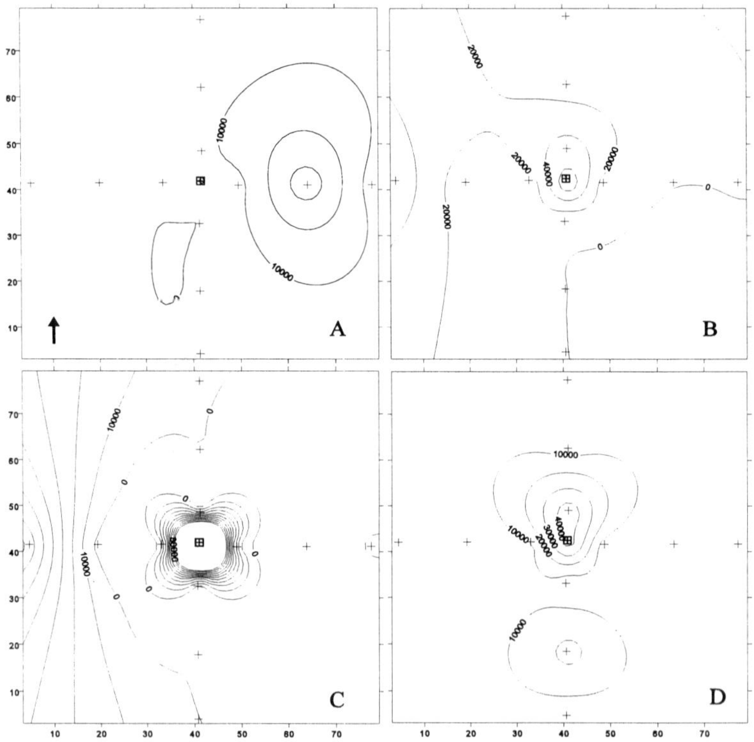

FIGURE 2. Bacterial production (cells ml^{-1} h^{-1}) in shallow wells (+) surrounding a GCW (▣) in a PAH contaminated aquifer. Sampling was performed at 0 months (A), 3 months (B), 6 months (C), and 9 months (D).

PAH Impacted Site: Prior to operation of the GCW, bacterial productivity ranged from undetectable to 1.1 x 10^4 cells ml^{-1} h^{-1} across the survey site (Fig. 2A). Productivity was measured in shallow and deep groundwater samples from 12 monitoring wells and the groundwater circulation well after 3 mos. (Fig. 2B), 6 mos.

(Fig. 2C), and 9 mos. (Fig. 2D) and increased across the zone of influence (ZOI) of the GCW from near detect (100 cells ml^{-1} h^{-1}) outside the ZOI to 4.8 x 10^5 cells ml^{-1} h^{-1} within the GCW.

Effect of GCWs on Production: GCWs may stimulate bacterial production by circulating limiting nutrients, enhancing bioavailability of hydrocarbons, or by removing inhibitory low molecular weight compounds through abiotic removal (air stripping). We found that bacterial production was stimulated across the ZOI of a GCW placed in an freshwater aquifer at a gasoline-contaminated site. Changes in bacterial abundance was not a useful parameter to evaluate stimulation of the natural assemblage resulting from the well operation, whereas productivity increased in response to a properly functioning wells and decreased when well circulation was poor. Thus, measuring instantaneous rates of heterotrophic production may provide a rapid assessment (< 1 d) of well performance. Bacterial productivity increased from ca. 10^3 to over 10^6 cells ml^{-1} h^{-1} across the ZOI of the groundwater circulation well suggesting that this treatment stimulates overall biodegradation of predominanct carbon sources present in groundwater (BTEX). Production continued to increase in the GCW over the 9 month experiments up to a rate of 1.6 cells ml^{-1} h^{-1}. This extremely rapid rate results in turnover time of the free living assemblage of ca. 1 h.

Likewise, at the PAH impacted site, we found that heterotrophic bacterial production increased dramatically in the GCW itself, and to a lesser extent in the nearest (ca. 15 ft. away) tier of monitoring wells. This is among the first evidence that operation of the GCW increases microbial activity *in situ* at a creosote-contaminated site. Increases in overall heterotrophic activity is not *de facto* evidence that petroleum hydrocarbons are being metabolized to account for this three orders of magnitude increase in carbon demand. However, the fact that inside the ZOI, there was an increase in the instantaneous mineralization rate for benzene and toluene at the BTEX site (Boyd et al. 1997), and phenanthrene and naphthalene at the PAH site (Boyd et al. 1996), suggests that petroleum hydrocarbon biodegradation is occurring among free living bacteria in response to operation of the system.

REFERENCES

Boyd, T. J., B. J. Spargo, and M. T. Montgomery. 1996. "Improved method for measuring biodegradation rates of hydrocarbons in natural water samples." *In Situ Bioremediation and Efficacy Monitoring.* B. J. Spargo (ed.) NRL/PU/6115-- 96-317.

Boyd, T. J., M. T. Montgomery, and B. J. Spargo. 1997. "Mineralization rates of benzene and toluene from a BTEX-contaminated aquifer." *Proceedings from the Fourth International Symposium on In Situ and On-Site Bioremediation.*

Kirchman, D. L., E. K'Nees, and R. Hodson. 1985. "Leucine incorporation and its potential as a measure of protein synthesis by bacteria in natural aquatic systems." *Appl. Environ. Microbiol.* 49:599-607.

Montgomery, M. T., Boyd, T. J., Spargo, B. J., Mueller, J. G., Coffin, R. B., and D. C. Smith. 1996. "Bacterial productivity in a BTEX-contaminated aquifer: a comparison of ^{3}H-leucine and ^{3}H-thymidine methods." *In Situ Bioremediation and Efficacy Monitoring.* B. J. Spargo (ed.) NRL/PU/6115--96-317.

Mueller, J. G., E. Klingel, M. D. Brourman, R. B. Coffin, L. A. Cifuentes, W. W. Schultz, and B. J. Spargo. 1996. "Field Demonstration of In Situ Chemical Containment Technologies Case Study: Naval Battalion Command, Port Hueneme, CA." *In Situ Bioremediation and Efficacy Monitoring.* B. J. Spargo (ed.) NRL/PU/6115--96-317.

Mueller, J. G., E. Klingel, M. D. Brourman, R. B. Coffin, L. A. Cifuentes, W. W. Schultz, and B. J. Spargo. 1996a. "Field Demonstration of In Situ Chemical Containment Technologies Case Study: Cabot Carbon/Kopper's Superfund Site, Gainesville, FL." *In Situ Bioremediation and Efficacy Monitoring.* B. J. Spargo (ed.) NRL/PU/6115--96-317.

National Research Council. 1993. *In situ bioremediation: when does it work?* National Academy Press, Washington, DC. 208 pp.

Smith, D. C., and F. Azam. 1992. "A simple, economical method for measuring bacterial protein synthesis rates in seawater using ^{3}H-leucine." *Mar. Microb. Food Webs* 6(2):107-114.

MICROBIAL NITRIFICATION EVIDENCE IN BIOREMEDIATION EXPERIMENTS AT 13°C

P. Sacceddu, A. Robertiello and *P. Carrera.*
Eniricerche S.p.A., Monterotondo, Rome,.Italy.

ABSTRACT: It is well-known from scientific literature, that the enhancement of nitrifying bacteria activity in soils greatly depends from temperature conditions (optimum values 25-35°C) and is related not only to bacterial number increase but dlso to a stimulation mechanism probably due to the presence of *Protozoa Infusoria.*

In the context of laboratory experiments simulating bioremediation, on crude oil contaminated soils, different mesocosms were set up, run at the constant temperature of 13°C, and fed with two different fertilizers solutions, soluble (ammonium nitrate) and slow-relase (ammonium magnesium phosphate).

After 5 weeks from fertilizers addition, chemical analysis of interstitial water showed, for all
mesocosms, the presence of nitrates and the corresponding decrease in ammonium ions.

Microbiological analysis indicated a conspicuous presence *of Protozoa Infusoria (Paramecium, Colpidium)* consequent to the development of a relevant hydrocarbon oxidizing bacterial charge. On the other side, a limited number of nitrifying bacteria has been monitored.

The interpretation of the microenvironment present in the soil under study, suggests that the presence of *Protozoa* plays an important and profitable role in the detoxification of a hydrocarbon contaminated biotope; thus contributing, in the global economy of the site itself, to improve the decontamination process taking part to make available nitrogen sources in all the forms necessary to the development of a rich hydrocarbon oxidizing bacterial flora. Besides, it is well known that nitrate ions are more mobile of ammonium ions and thus more available for the plants.

The present work reports experimental evidence of nitrifying phoenomena acting in a particular hydrocarbon oxidizing biosystem kept at 13°C, by comparing the variation of nitrogen chemical species present in the soil together with the presence of bacterial microflora and bacterial-eating microfauna.

BIOREMEDIATION OF AQUEOUS POLLUTANTS USING BIOMASS EMBEDDED IN HYDROPHILIC FOAM

E.W. Wilde, J.C. Radway, J. Santo Domingo, M.J. Whitaker (Savannah River Technology Center); P. Hermann (Frisby Technologies); and R.G. Zingmark (University of South Carolina)

ABSTRACT: This project involves development of a bioremediation process using hydrophilic polyurethane foam as an immobilization medium for algae, bacteria, and other types of biomass. The process has potential for cleaning up waters contaminated with heavy metals, radionuclides and toxic organic compounds. Initial investigations focused on the bioremoval of heavy metals from wastewaters at the Savannah River Site using immobilized algal biomass. This effort met with limited success for reasons which included interference in the binding of biomass and target metals by various non-target constituents in the wastewater and the unavailability of bioreactor systems capable of optimizing contact of target pollutants with sufficient biomass binding sites. Subsequent studies comparing algal, bacterial, fungal, and higher plant biomass demonstrated that other biomass sources were also ineffective for metal bioremoval under the limited test conditions. Radionuclide bioremoval using a Tc-99 source provided more promising results than the metal removal studies with the various types of biomass. The most promising results were obtained when we explored the use of hydrophilic foam to embed microbes for use in the metabolically mediated biodegradation of toxic organic compounds. Optimized conditions for foam/biomass production were developed to achieve a high degree of biomass retention within the matrix while allowing the embedded microorganisms to remain viable. The embedded microbes were shown to be capable, when properly induced, of producing enzymes capable of degrading target organic pollutants. Further process development work is in progress.

THERMALLY-ENHANCED IN SITU BIOREMEDIATION OF DNAPL'S

Jeremy M. Kosegi (University of Illinois, Urbana-Champaign, IL)
Barbara Spang Minsker (University of Illinois, Urbana-Champaign, IL)
David E. Dougherty (University of Vermont, Burlington, VT)

ABSTRACT: A new treatment technology is being developed to remediate sites in which dense nonaqueous phase liquids (DNAPLs) have become trapped in soil pores. It has been hypothesized that coupling thermal treatment with *in situ* bioremediation will enhance aquifer remediation by increasing both bioavailability and the rate of biodegradation. A mathematical model has been developed to describe the process and to test the efficacy of this scheme. The model allows for both equilibrium partitioning and first-order kinetic descriptions of dissolution and desorption as well as dual-Monod kinetics. Only isothermal cases are considered, which eliminates the need to model heat explicitly. Equilibrium and rate constants are calculated as functions of temperature using the Van't Hoff and Arrhenius equations respectively. With this model, the optimal temperature at which the increase in bioavailability equals the increase of biodegradation may be obtained and the efficacy of maintaining this temperature evaluated.

INTRODUCTION

Chlorinated ethenes are extremely difficult to remediate in groundwater because of their high density, low viscosity, and low solubility. These properties cause the contaminants to sink through the water table, leaving a trail of residual contaminants trapped in the soil matrix. The entrapped chemicals can serve as long-term sources of groundwater contamination. Conventional remediation technologies such as pump-and-treat have not been successful at removing chlorinated ethenes (National Research Council, 1994), but innovative technologies are under development which offer more promise. A United States Environmental Protection Agency (USEPA) report evaluating technologies for remediating DNAPLs (Grubb and Sitar, 1994) concluded that thermally-based technologies are among the most promising for highly contaminated areas, but may not be able to achieve the currently mandated cleanup standards. The report recommends that thermal technologies be combined with technologies such as *in situ* bioremediation, which hold the best promise for long-term plume management. This work is the first to examine the efficacy of combining these two technologies.

In situ bioremediation is an attractive choice for coupling with thermal treatment because it destroys the contaminants in place without generating large quantities of hazardous waste which must be transferred to other media. Two approaches to *in situ* bioremediation of chlorinated ethenes can be taken: aerobic or anaerobic. The anaerobic process is more promising than the aerobic process because it eliminates the potential problem of oxygen availability associated with the aerobic system. Anaerobic reductive dehalogenation of chlorinated ethenes is also an energy-yielding reaction which has shown to be coupled with microbial growth. Under aerobic conditions, the contaminants must be cometabolized using another growth substrate which may be preferentially degraded before the chlorinated ethenes. The model discussed in this paper has been developed to assess the feasibility of remediating a site contaminated with tetrachloroethylene (PCE) which is resistant to aerobic microbial degradation. Hence this new technology will focus on anaerobic biodegradation.

If *in situ* bioremediation were used as a "polishing" method for thermal technologies, considerable heat would remain in the subsurface for extended periods after the thermal technology is completed (Parr et al., 1983). Two approaches to thermal enhancement will be investigated using an isothermal assumption. The first would rely upon residual heat from thermal treatment of highly contaminated areas. In order to create an isothermal environment additional heat would have to be supplied to the system by injecting hot water at a constant temperature. The second approach would involve injecting hot water at a constant temperature to create and maintain an isothermal system.

The remainder of this paper is devoted to the development of a mathematical model that describes this thermally-enhanced *in situ* bioremediation process. The best data currently available will be used in this model for preliminary evaluation of the enhanced treatment. Laboratory studies are planned to generate data on the complex relationships between the physical, chemical, and biological processes at ambient and elevated temperatures but the data is unavailable at this time. Future work will also evaluate nonisothermal conditions which might result from gradual cooling or heating or from temperature pulses.

MODEL DEVELOPMENT

Advection-Dispersion. Based on the work of Chrysikopoulos et al. (1990), who showed that for certain aquifers flow may be modeled as 1-dimensional with uniform velocity when considering well-to-well recirculation, a 1-D, uniform flow field has been assumed. The equation governing the fate and transport of PCE takes the form shown in equation (1):

$$\frac{\partial(\phi\,\rho_{aq}S_{aq}W_{C,aq})}{\partial t} = -V\frac{\partial(\phi\,\rho_{aq}S_{aq}W_{C,aq})}{\partial x} + \frac{\partial}{\partial x}\left(\phi\,S_{aq}D\frac{\partial(\rho_{aq}W_{C,aq})}{\partial x}\right)$$

$$-R_{bio} + R_{naq\to aq} + R_{s\to aq} \qquad (1)$$

where ϕ is the porosity of the soil system (unitless); ρ_{aq} is the density of the aqueous phase (mass of the aqueous phase / volume of the aqueous phase); S_{aq} is the aqueous saturation (volume of the aqueous phase / pore volume); $W_{C,aq}$ is the mass fraction of contaminant in the aqueous phase (mass contaminant / mass aqueous phase); t is time; V is the average linear velocity of groundwater flow (length / time); x is the direction of groundwater flow; D is the hydrodynamic dispersion coefficient (length^2 / time); R_{bio} is the biological decay term; $R_{naq \to aq}$ is the dissolution term; and $R_{s \to aq}$ is the desorption term. Note that for the purposes of this model it has been assumed that all pooled PCE has been previously remediated using one of the techniques discussed by Grubb and Sitar (1994) and that residual DNAPL is all that remains. With this assumption the nonaqueous phase may be treated as immobile. Also, the residual DNAPL has been assumed to be trapped in the saturated zone of the aquifer, thereby eliminating the need for modeling the mobile gas phase DNAPL components. When heat is added to the system, however, a constant, immobile, temperature-dependent gas phase is included by reducing the aqueous saturation, S_{aq}. With these assumptions, equation (1) alone describes the fate and transport of the DNAPL in the aquifer.

Biodegradation. It has been assumed that the degradation occurs by anaerobic dehalogenation via the following pathway: PCE→TCE→DCE→VC→ETH, where TCE is trichloroethylene, DCE is dichloroethylene, VC is vinyl chloride and ETH is ethylene. As shown by equation (1), this model only tracks one species (C equals PCE); however, it is intended to simulate the transformation of PCE to ETH. Therefore, the biodegradation term uses the rate constants and coefficients for the transformation of VC to ETH, which has been shown under anaerobic conditions to be the rate-limiting step in the pathway shown above (Tandol et al., 1994). The biodegradation term uses the following form of Monod kinetics in which either the electron donor or acceptor may be limiting (equation (2)):

$$R_{bio} = -\mu_{max} \phi \, \rho_{aq} S_{aq} W_B \alpha \left(\frac{\phi \, \rho_{aq} S_{aq} W_{P,aq}}{K_S^P + \phi \, \rho_{aq} S_{aq} W_{P,aq}} \right) \left(\frac{\phi \, \rho_{aq} S_{aq} W_{C,aq}}{K_S^C + \phi \, \rho_{aq} S_{aq} W_{C,aq}} \right) \tag{2}$$

where μ_{max} is the maximum utilization rate of the contaminant (mass contaminant per volume per time); W_B is the mass fraction of attached biomass; α is the ratio of contaminant transformed to primary substrate utilized; $W_{P,aq}$ is the mass fraction of primary substrate (mass primary substrate per mass solution); K_S^P is the half-saturation constant for the primary substrate (mass per volume); and K_S^C is the half-saturation constant for the contaminant (mass per volume). Vogel and McCarty (1985) have shown that the primary substrate serves as the electron donor and the contaminant serves as the electron acceptor under anaerobic conditions.

In this model, the biomass is assumed to be attached and fully penetrated by both the primary substrate and the contaminant. Equation (3) describes the growth of the biomass. In equation (3) Y is the microbial yield coefficient (mass biomass / mass contaminant transformed); K_D is the first-order decay coefficient for the biomass (1 / time); K_{bc} is the first-order rate constant for the utilization of background carbon (1 / time); Y_{bc} is the microbial yield coefficient for background carbon (mass cells / mass background carbon); C_d is the concentration of background carbon in the aquifer (mass / volume).

$$\frac{\partial(\phi\ S_{aq}\ \rho_{aq} W_B)}{\partial t} = \mu_{max} \phi\ \rho_{aq} S_{aq} W_B Y \left(\frac{\phi\ \rho_{aq} S_{aq} W_{P,aq}}{K_S^P + \phi\ \rho_{aq} S_{aq} W_{P,aq}} \right) \left(\frac{\phi\ \rho_{aq} S_{aq} W_{C,aq}}{K_S^C + \phi\ \rho_{aq} S_{aq} W_{C,aq}} \right)$$

$$-K_D \phi\ S_{aq}\ \rho_{aq} W_B + K_{bc} Y_{bc} C_d \qquad (3)$$

Since it has been assumed that both the primary substrate and the contaminant may be limiting for the biodegradation process, the advection, dispersion and decay of the primary substrate must also be modeled. This is shown as equation (4). It has been assumed in this equation that the primary substrate does not adsorb or volatilize.

$$\frac{\partial(\phi\ \rho_{aq} S_{aq} W_{P,aq})}{\partial t} = -V \frac{\partial(\phi\ \rho_{aq} S_{aq} W_{P,aq})}{\partial x} + \frac{\partial}{\partial x} \left(\phi\ S_{aq} D \frac{\partial(\rho_{aq}\ W_{P,aq})}{\partial x} \right)$$

$$-\mu_{max} \phi\ \rho_{aq} S_{aq} W_B \left(\frac{\phi\ \rho_{aq} S_{aq} W_{P,aq}}{K_S^P + \phi\ \rho_{aq} S_{aq} W_{P,aq}} \right) \left(\frac{\phi\ \rho_{aq} S_{aq} W_{C,aq}}{K_S^C + \phi\ \rho_{aq} S_{aq} W_{C,aq}} \right) \qquad (4)$$

Dissolution. Describing the dissolution process of DNAPLs into the aqueous phase is not a trivial task. This term, however, must be included to determine when remediation is complete. A common assumption concerning this process is that it may adequately be described as either equilibrium partitioning or as a first-order kinetic process. Details on equilibrium partitioning are discussed in Abriola and Pinder (1985) and on first-order kinetic processes are given in Pankow and Cherry (1996) respectively. Both processes are included in our model through the $R_{naq \to aq}$ term shown in equation (1).

Sorption. One of the main factors that influences the aqueous concentration of DNAPLs is the process of sorption. This process is controlled by factors such as soil type, amount of organic carbon present in the soil, solvent characteristics, aqueous phase concentration, pH, and temperature. When modeling the sorption process, assumptions similar to those made for dissolution are common; that is, sorption is usually modeled using an equilibrium partition coefficient or as a first-order kinetic process. Equations that describe these two scenarios are discussed in Semprini and McCarty (1991). When dealing with hydrophobic contaminants such as PCE, the equilibrium partition coefficient, K_d, is often adjusted to account for the organic content of the soil, f_{oc} ($K_d = f_{oc} K_{oc}$ where K_{oc} is the organic carbon partition coefficient). K_{oc} is calculated using equation (5):

$$\log K_{oc} = a\ \log K_{ow} + b \qquad (5)$$

where K_{ow} is the octanol-water partition coefficient and a and b are known constants for a given class of compounds. Values for these parameters are shown in Pankow and Cherry (1996). Both equilibrium and first-order desorption are included in our model through the $R_{s \to aq}$ reaction term in equation (1).

HEAT EFFECTS

When heat is added to the system, the set of governing equations, (1), (3), and (4), must be modified. For the purposes of this model heat itself will not be explicitly modeled; rather, only the isothermal case will be investigated. Although it has been assumed that no gas phase is present in the system, gas bubbles will form when heat is applied to the system. It has been assumed that these bubbles will be surrounded by the aqueous phase only; therefore, volatilization from only the aqueous phase will be considered. The bubbles are assumed to remain in the aquifer in equilibrium with the aqueous phase. Henry's Law is used to represent this mass transfer. Values for Henry's Law constant and the saturated vapor pressure, $P_{sat,vp}$, are tabulated in Pankow and Cherry (1996). Depending on whether equilibrium partitioning or first-order reactions are used to describe dissolution and desorption determines how the equations are modeled as functions of temperature. If equilibrium is assumed, Van't Hoff's equation may be used to calculate the partition coefficients at the desired temperature. Likewise if first-order kinetics are assumed, then the Arrhenius equation may be utilized to determine the new rate constants. However, Ratkowsky et al. (1983) have shown that this relationship is not valid for the maximum utilization rate constant, μ_{max}, in the temperature range of interest, 0-40°C. They developed a new way to calculate μ_{max} at the desired temperature which is used in our model. This is shown as equation (6):

$$\sqrt{\mu_{max}} = b(T - T_{min})\{1 - \exp[c(T - T_{max})]\} \qquad (6)$$

where b and c are microorganism-dependent constants; T_{min} and T_{max} are the minimum and maximum temperatures at which the growth rate is zero; and T is the temperature of interest.

CONCLUSION

A mathematical model has been developed to describe a thermally-enhanced bioremediation process. Currently, much of the data needed for the model's equations is unavailable; therefore, "what if" scenarios will be used to explore the feasibility of this new remediation technology. Results of these hypothetical cases will be presented at the Bioremediation Symposium.

REFERENCES

Abriola, L. M., and G. F. Pinder. 1985. "A Multiphase Approach to the Modeling of Porous Media Contamination by Organic Compounds. 1. Equation Development." *Water Resour. Res. 21*(1): 11-18.

Chrysikopoulos, C. V., P. V. Roberts, and P. K. Kitanidis. 1990. "One-Dimensional Solute Transport in Porous Media With Partial Well-to-Well Recirculation: Application to Field Experiments." *Water Resour. Res. 26*(6): 1189-1195.

Grubb, D. G., and N. Sitar. 1994. *Evaluation of Technologies for In-Situ Cleanup of DNAPL Contaminated Sites.* USEPA Technical Report, EPA/600/R-94/120, R. S. Kerr Environmental Research Laboratory, Ada, OK.

National Research Council. 1994. *Alternatives for Groundwater Cleanup.* National Academy Press, Washington DC.

Pankow, J. F., and J. A. Cherry. 1996. *Dense Chlorinated Solvents and Other DNAPLs in Groundwater.* Waterloo Press, Ontario, Canada.

Parr, A. D., F. J. Molz, and J. G. Melville. 1983. "Field Determination of Aquifer Thermal Energy Storage Parameters." *Ground Water. 21*(1): 22-35.

Ratkowsky, D. A., R. K. Lowry, T. A. McMeekin, A. N. Stokes and R. E. Chandler. 1983. "Model for Bacterial Culture Growth Rate Throughout the Entire Biokinetic Temperature Range." *J. Bacteriol. 154*(3): 1222-1226.

Semprini, L., and P. L. McCarty. 1991. "Comparison Between Model Simulations and Field Results for In-Situ Biorestoration of Chlorinated Aliphatics: Part 1. Biostimulation of Methanotrophic Bacteria." *Ground Water.* 29(3): 365-373.

Tandol, V., T. D. DiStefano, P. A. Bowser, J. M. Gossett, and S. H. Zinder. 1994. "Reductive Dehalogenation of Chlorinated Ethenes and Halogenated Ethanes by a High-Rate Anaerobic Enrichment Culture." *Environ. Sci. Technol. 28*(5): 973-979.

Vogel, T. M., and P. L. McCarty. 1985. "Biotransformation of Tetrachloroethylene to Trichloroethylene, Dichloroethylene, Vinyl Chloride, and Carbon Dioxide Under Methanogenic Conditions." *Appl. Environ. Microbiol. 49*(5): 1080-1083.

FIELD DEMONSTRATION OF OXYGEN MICROBUBBLES FOR IN SITU BIOREMEDIATION

Patrick M. Woodhull, P.E. (OHM Remediation Services Corp., Findlay, OH)
Douglas E. Jerger, Ph.D. (OHM Remediation Services Corp., Findlay, OH)
Daniel P. Leigh, P.G. (OHM Remediation Services Corp., Findlay, OH)
Ronald F. Lewis, Ph.D. (USEPA, Cincinnati, OH)
Erica S. Becvar, M.S. (Armstrong Laboratory, Tyndall AFB, FL)

ABSTRACT: OHM Remediation Services Corp. (OHM) is participating in a USEPA SITE Emerging Technology Demonstration using microbubbles for delivery of oxygen to enhance *in situ* bioremediation. The process utilizes oxygen microbubbles to treat contaminated soils and groundwater in the saturated zone. The objective of the demonstration was to determine the subsurface oxygen transfer to the groundwater, retention and migration of the microbubbles in the soil matrix, biodegradation of the petroleum hydrocarbons, and the economic feasibility of the technology. Preliminary laboratory and field pilot tests were conducted to determine microbubble production rates, quality, and the feasibility of microbubble migration through the soil.

INTRODUCTION

The application of aerobic microbial degradation for petroleum hydrocarbon contaminated soils and groundwater has become a common and wide-spread practice. One of the main difficulties for *in situ* bioremediation is the delivery of oxygen to the subsurface, especially the saturated zones. Oxygen microbubble technology may be effective in overcoming this limitation. Microbubbles are generated by mixing water with a surfactant at a concentration of approximately 200 milligrams per liter (mg/l). The oxygen is added as a gas under pressure and the mixture pumped to a high-shear, continuous mixer. The resulting dispersion contains 60 to 80% oxygen by volume and the microbubbles ranging in size from 50 to 100 microns.

Laboratory Pilot Testing. Prior to performing field testing, a laboratory pilot test was performed to determine if microbubbles could be injected through saturated soils with similar properties expected in the field. A 25-centimeter (cm) diameter by 2 meter high vertical, steel cylinder was fabricated in OHM's Treatability laboratory. The column was packed with approximately 12 cm of pea gravel at the base and covered by approximately 1.5 meters of a fine- to medium-grained, washed sand. The column had four observation and sampling ports located around the perimeter every 25 cm up the column. A laboratory-scale spinning-disc microbubble generator (Sebba, 1985) was provided by Aphron Technologies, Inc. (Blacksburg, VA). The microbubbles were injected into the bottom of the sand column with a peristaltic pump. The movement of microbubbles was monitored by opening the sample ports and observing the presence of microbubbles. Two observations were made during the laboratory testing: 1) microbubbles could be injected through approximately 1.2 meters of sand, and, 2) the microbubbles were persistent in the column longer than air from a sparging test.

Field Pilot Testing. A field pilot test was performed to determine microbubble movement through an unconfined aquifer. The site selected was the former Fire Training Area No. 23 (FT-23) at Tyndall Air Force Base (AFB), Panama City, Florida. The pilot test was performed in an uncontaminated area upgradient from the contaminant plume at FT-23. The pilot test "cell" consisted of three monitoring wells and an AMS Retract-A-Tip probe utilized as a microbubble injection point. The test cell was constructed near an existing monitoring well (TY22-FTA).

The microbubbles were generated using equipment described in a following section. For the pilot test, the microbubbles were generated with air and included helium as a tracer gas. The movement of microbubbles in the subsurface was monitored through changes in the groundwater elevation, increases in groundwater dissolved oxygen (DO) concentrations, and by the presence of the tracer gas.

This pilot test verified that microbubbles could be injected at rates of up to 4 liters per minute (lpm) of foam into the shallow aquifer. The microbubbles were observed to be persistent in the aquifer for longer periods of time compared to conventional air sparging techniques. Based on the conclusions from this pilot testing, a more extensive field demonstration test was designed to evaluate the effects of microbubbles in saturated and unsaturated soils.

FIELD DEMONSTRATION TESTING

Site Description. The site selected for the demonstration phase of the project is the POL-B site at Tyndall AFB (Figure 1). This location was selected because of the numerous site characterization activities and previous technology demonstrations performed at the site. The shallow soils underlying the POL-B site consists predominantly of a fine- to medium-grained quartzose sand. The static depth to groundwater ranges from 1 to 2.5 meters below ground surface (bgs), indicating unconfined conditions. A extensive site description has been previously documented (GSI, 1995).

Microbubble Injection Points. The microbubbles were injected into the shallow aquifer at the site through six locations arranged in a hexagonal pattern. The injection points were installed by a Geoprobe® direct push method to a total depth of 6.1 meters bgs. Each injection point consisted of a 53-cm long stainless-steel probe connected by a Teflon® tube to the ground surface. The probe tip is a 145-micron mesh that allowed the microbubbles to be injected unimpeded while preventing silting of the point between injections.

Microbubble Generator System. The microbubble generator system used for the field demonstration consisted of a high-shear mixer, oxygen and helium storage cylinders, surfactant/water storage tank, and a flow control panel. A process schematic of the system is shown in Figure 2. Oxygen and helium gas were stored and supplied to the system from standard gas cylinders with pressure regulators.

The surfactant used in the field test was sodium dodecyl benzyl sulfonate (NaDBS; BioSoft D-40, Stepan Chemical Company, IL). This surfactant was selected based on its biodegradability, laboratory testing (OHM unpublished data), and published literature (Enzien, et.al., 1995). The target NaDBS concentration was 200 to 250 mg/l. This concentration was selected to minimize the oxygen demand for surfactant degradation while producing acceptable microbubbles. Prior to microbubble injection, the surfactant was approved by the Florida Department of Environmental Protection after laboratory analysis confirmed that primary and secondary drinking water standards would not be exceeded. The surfactant and water solution was mixed and stored in two 500-liter storage tanks.

A positive displacement pump (Moyno Industrial Products, Springfield, OH), capable of 3 to 5 lpm at 60 pounds per square inch guage (psig), pumped the water and surfactant solution to a control panel. To ensure oxygen compatibility, the panel was constructed of stainless-steel tubing, flowmeters, control and check valves, and pressure relief valves. A pressure regulator and solenoid valve automatically stopped oxygen flow to the generator in the event of loss of water flow. The microbubble flowrate and quality were controlled by independently adjusting the liquid and oxygen flow at the control panel.

The microbubble generator was a "Continuous Mixer/Foamer" manufactured by Goodway Industries, Inc. (Bohemia, NY). The original seals and gaskets were modified to either Teflon® or Viton® and the wetted-components (i.e., rotor housing, mixing head, shaft, etc.) were constructed of stainless steel. The mixing chamber consisted of a mixing head with a several rows of "teeth" that fit in close tolerance with the stationary head. The mixing head rotated at 800 to 850 rpm, which produced a high-shear environment in the mixing chamber. The high shear formed a microbubble solution with a typical bubble diameter of 50 to 100 microns. The microbubbles flowed from the mixing chamber to each of the injection points, as required, through 0.63-cm inside-diameter re-enforced tubing.

The quality and stability of the microbubbles produced were measured by collecting a sample in a 250-milliliter (ml) graduated cylinder. The foam quality was determined by weighing the foam in the cylinder and calculating the volume of gas contained in the foam. The result was expressed as a volume percent of oxygen. Stability of the foam was determined by allowing the foam to coalesce for 30 seconds in the graduated cylinder and measuring the amount of free liquid in milliliters. Based on the laboratory and field pilot testing, a target foam quality and stability was 65% to 75% and less than 15 ml, respectively. This range corresponds to published literature (Enzien, et.al., 1995; Longe, et.al., 1995) of microbubble properties for similar applications.

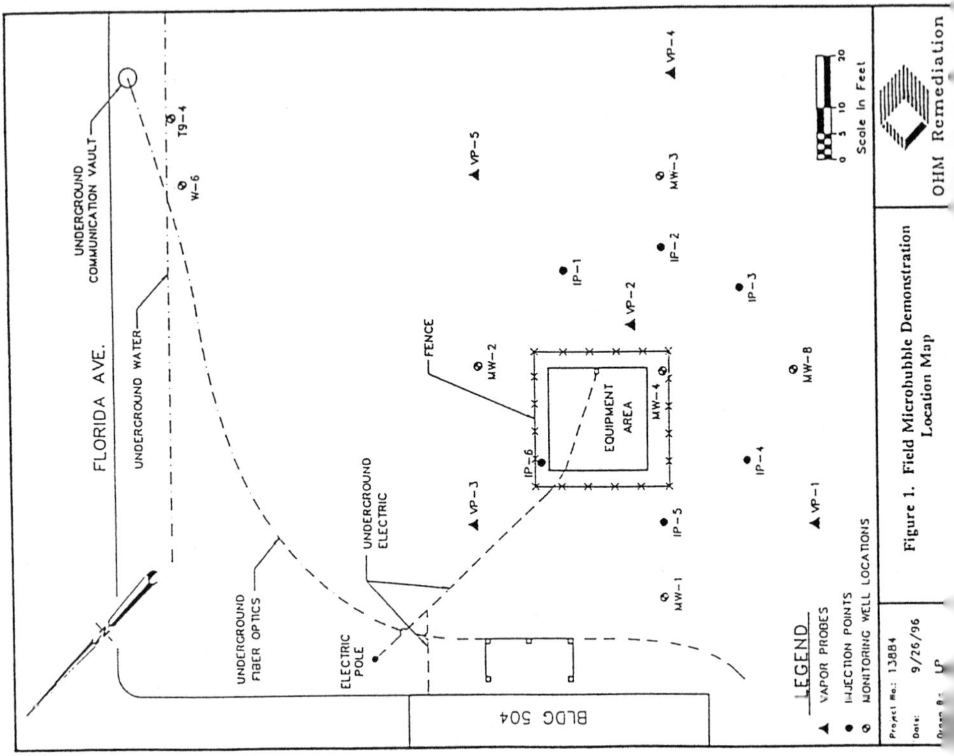

Figure 1. Field Microbubble Demonstration Location Map

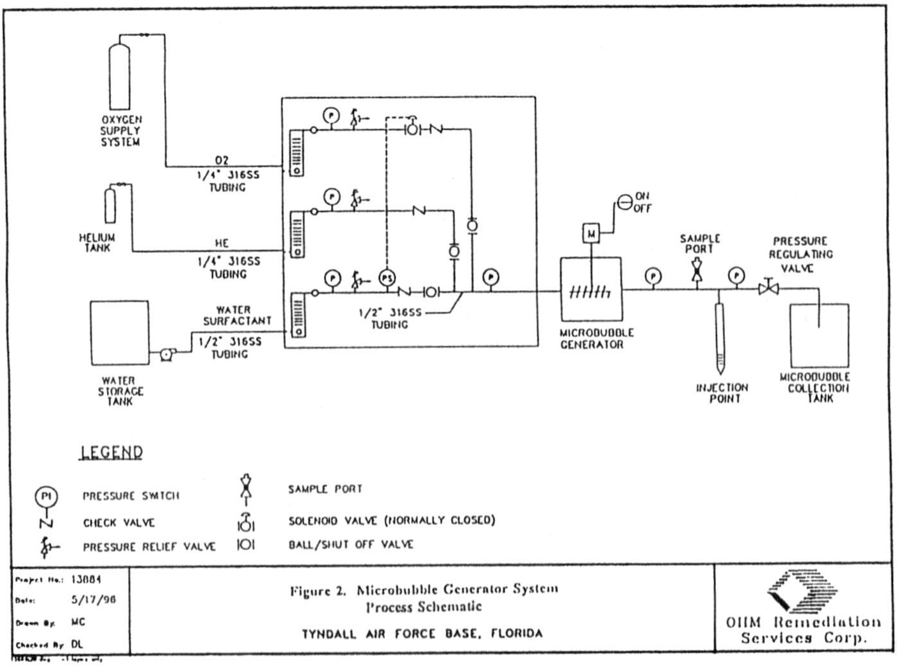

Figure 2. Microbubble Generator System Process Schematic

FIELD DEMONSTRATION RESULTS

Microbubble Production. The installation, setup, and startup of the microbubble system was performed in August and September 1996. The system was leak-tested by pumping microbubbles into a temporary storage tank. The microbubble stability and quality were measured and operation of the system adjusted until an acceptable foam was produced. Continuous microbubble injection into the subsurface at the POL-B site was initiated on September 24, 1996, and continued until November 18, 1996, when the system was temporary shutdown. The microbubbles were sampled and analyzed daily for quality and stability. The average foam quality was approximately 75% and the foam stability less than 10 ml during the entire injection period. The overall, average foam production rate from September through November was 3.0 to 3.5 lpm.

Microbubble Injection. Injection of the foam was rotated between the six injection points at varying periods of 1 hour to 4 hours at each injection point. Between October 30, and November 8, 1996, the microbubbles were injected continuously into a single injection point (IP-2) for approximately 6 to 8 hours per day. The objective was to evaluate the effect of injection at a single location.

Throughout the testing, injection pressure was monitored and recorded. The injection pressure remained relatively constant at 35 to 40 psig, both during the period of injection at each point and from day to day. No short-circuiting or fractures to the soil surface were observed during injection.

Daily, on-site measurements in the monitoring wells did not detect an increase in either DO or the redox potential of the groundwater during microbubble injection. During low-flow purging of the groundwater monitoring wells, the presence of small bubbles in the groundwater was observed, especially in the deep, down-gradient wells. However, the DO and redox were low in the groundwater and it was theorized that the bubbles were carbon dioxide.

Modified Microbubble Injection. In December 1996, additional monitoring wells were installed and a modified injection schedule was implemented based on the low groundwater DO measurements. This modification was based on two assumptions: 1) the oxygen was being utilized by the indigenous microflora prior to reaching the existing monitoring wells, and, 2) the vertical groundwater gradient was carrying the oxygen downward, away from the monitoring wells.

Six additional groundwater monitoring points were installed near IP-2 and ranged in depth from 3.6 to 7.6 meters bgs. The points were located laterally from IP-2 a distance of approximately 1 to 5 meters. The locations were selected to visually observe microbubble movement through the formation and to evaluate the potential for the vertical groundwater gradient to carry microbubbles downward.

During microbubble injection into IP-2, groundwater DO rapidly increased to >20 mg/l and microbubbles were readily present in the closest monitoring point, a distance of approximately 1 meter. However, due to freezing weather conditions affecting the monitoring equipment, the test was suspended without detecting elevated DO or microbubbles in additional monitoring points.

Temperature appeared to affect the microbubble quality and stability. The water/surfactant temperature during December was 5 to 10 °C, compared to 20 to 25 °C during September and October. The microbubble quality in December averaged 75 to 80% and the stability was significantly less than 10 mL In addition, based on visual observations, the microbubbles had an increased viscosity.

CONCLUSIONS

Based on the results of the field pilot test at FT-23 and the field demonstration test at POL-B, the following conclusions were made:

- microbubbles with acceptable quality and stability can be generated using commercially available equipment,
- microbubbles can be injected short distances through the site matrix,
- continued field demonstration testing is required to determine the migration distance in the soil, and,
- additional dissolved gas analyses of the groundwater is required to determine if the oxygen is rapidly utilized and converted to carbon dioxide.

ACKNOWLEDGMENTS

The authors would like to acknowledge Catherine Vogel of the USAF Armstrong Laboratory, Environics Directorate, and Mike Jones of the USAF 325th CES/CEV, Tyndall AFB for their valuable assistance on this project.

REFERENCES

Enzien, M.V., D.L. Michelsen, R.W. Peters, J.X. Bouillard, and J.R. Frank. 1995. "Enhanced In Situ Bioremediation Using Foams and Oil Aphrons." In R.E. Hinchee, R.N. Miller, and P.C, Johnson, (Eds.), *In Situ Aeration: Air Sparging, Bioventing, and Related Remediation Processes*, pp. 503-509. Battelle Press, Columbus, OH.

Groundwater Services, Inc. November 13, 1995. "Natural Attenuation Study/Risk Assessment: POL B Site, Tyndall AFB, Florida." Prepared for the Air Force Center for Environmental Excellence (AFCEE).

Longe, T.A., J.X.Bouillard, and D.L. Michelsen. 1995. "Use of Microbubble Dispersion for Soil Scouring." In R.E. Hinchee, R.N. Miller, and P.C, Johnson, (Eds.), *In Situ Aeration: Air Sparging, Bioventing, and Related Remediation Processes*, pp. 511-518. Battelle Press, Columbus, OH.

Sebba, F. 1985. "An Improved Generator for Micron-Sized Bubbles." *Chemistry & Industry Feb.* 4(3):91-92.

ENHANCED BIODEGRADATION OF MTBE AND BTEX USING PURE OXYGEN INJECTION

Sean R. Carter, J. Michael Bullock and William R. Morse

ABSTRACT: A novel approach to groundwater remediation at a petroleum release site is presented. The technique involves the injection of pure oxygen into groundwater via multiple injection points at flow rates substantially lower than traditional air sparging. The remediation system consists of an AirSep AS80 pressure swing adsorption oxygen generator which produces oxygen at a rate of 80 standard cubic feet per hour (scfh). The oxygen is stored in a 60-gallon receiver tank and pulse sparged to seven (7) injection points at a rate of approximately 7 scfh per point. The injection points are located directly downgradient of the source area and spaced to sufficiently increase dissolved oxygen levels throughout the site. The surficial geology consists of alluvial sands underlain by a low-permeability clay. Depth to groundwater is approximately 4 feet in the source area and drops to greater than 10 feet off site. Due to the steep groundwater gradient, oxygen transport was considered sufficient for off-site remediation.

After 6 weeks of operation, groundwater dissolved oxygen levels increased from a background level of 0 parts per million (ppm) to 6 to 10 ppm in both on-site and off-site monitoring wells. Dissolved oxygen remained at high levels as long as the system was operational. Concentrations of both MTBE and BTEX decreased with increasing dissolved oxygen levels. Four months after system startup, MTBE and BTEX had decreased by an order of magnitude in both on-site and off-site monitoring wells. A 48-hour endpoint assay using soils obtained from the groundwater interface zone confirmed the biodegradation of MTBE by *Pseudomonas fluorescens* type G.

The relatively low oxygen injection rate per point and high transfer efficiency into groundwater negates the need for vapor control via vadose zone extraction. This eliminates costly air treatment, provides a high degree of certainty for plume control, is suitable for shallow groundwater, and results in rapid biodegradation of MTBE and BTEX.

IN SITU MANAGEMENT SYSTEM EMPLOYING INTEGRATED FUNNEL-AND-GATE/GZB CONTAINMENT AND RECOVERY TECHNOLOGIES.

Mitchell D. Brourman and Michael D. Tischuk (Hanson Environmental & Legal Group, Inc., Pittsburgh, PA)
Eric J. Klingel and Marc Sick (IEG Technologies, Corp., Charlotte, NC)
David J.A. Smyth, Edward A. Sudicky and Steven G. Shikaze, (Waterloo Centre for Groundwater Research, University of Waterloo, Ontario, Canada)
Susanne M. Borchert and ***James G. Mueller*** (SBP Technologies, Inc., Pensacola, FL)

ABSTRACT: An *in situ* management system was designed to provide containment and recovery of creosote NAPL at a former wood treating facility in Nashua, New Hampshire where an existing pump-and-treat system has proved to be technically impracticable and economically inefficient in affecting source control. The *in situ* system was based on the integration of a patented technology for physical source containment and management (*i.e.*, funnel-and-gate barrier technology) with patented vertical groundwater circulation well technologies for *in situ* product recovery (*i.e.*, GZB™ technology) and/or *in situ* biodegradation (*i.e.*, UVB™ systems). Mathematical modeling of the combined technologies led to the selection of a Waterloo Barrier™ metal sheet pile wall along 650 feet of the river's shoreline. The purpose of the physical barrier is to prevent the continued migration of creosote NAPL into the Merrimack River by diverting groundwater and NAPL flow through engineered gate areas. Multiple GZB wells would be placed strategically within the system to enhance the recovery of contained creosote NAPL. This combination of technologies promises an effective, cost-efficient approach for long-term management of environmental concerns at Nashua, and at many other sites.

INTRODUCTION

Recognized limitations in the effectiveness and cost efficiency of conventional approaches for managing environments impacted by non-aqueous phase liquids (NAPL) mandate the application of alternative solutions. For several years, we have participated in the development of two *in situ* technologies aimed at providing innovative solutions to such problems: i) funnel-and-gate physical barriers, and ii) vertical groundwater circulation cell technology (i.e., GZB™) for enhanced product recovery.

The funnel-and-gate system (Smyth *et al.*,1995; Starr and Cherry, 1994) is an *in situ*, physical barrier technology that can be used, potentially, to manage and contain constituents of interest (COI) in groundwater, including NAPL. The system consists of low conductivity cutoff walls (funnels) with one or more openings (gates). The purpose of the funnel is to divert groundwater and/or NAPL flow through the gate area which contains a system for recovery and/or treatment (*i.e.*, Grundwasser Zirkulation Brunnen [GZB] technology). Several factors pertaining to hydraulic control and performance are important to consider in the site-specific design of funnel-and-gate systems. These include: i) accurate site characterization including spatial distribution of varying hydraulic conductivities, ii) changing hydraulic gradients that influence the pathways and velocities of fluid flow, iii) the hydraulic conductivity at the treatment gate, and iv) physical parameters governing COI transport, to list a few. While minimizing the number of gates has certain construction and economic advantages, system design must accommodate satisfactory capture and recovery of NAPL.

A number of design and construction methods exist that may be employed to generate the physical barrier (*i.e.*, funnels) (Smyth *et al.*, 1995). Conventional construction techniques include compacted clay barriers and slurry trenches, which typically incorporate soil bentonite, soil attapulgite and/or cement bentonite mixtures in the barrier. Other technologies for barrier construction include vibrated beam cutoff walls, deep soil mixing or auger cast walls, jet grouted walls, geomembrane barriers, and sealable joint steel sheet piling. Given this variety, it is reasonable to assume that there will be technical and cost advantages of using particular barrier walls in different situations.

The GZB technology creates a vertical groundwater circulation cell in the aquifer to effect *in situ* soil flushing (Figure 1). This is accomplished by moving water (and NAPL) from the lower portion of the aquifer into the GZB well through a lower screen section of the well. The water is then pumped vertically upward through the GZB well casing, allowing it to exit through an upper screen section of the same well: NAPL is separated from the circulating groundwater, collected in an internal NAPL sump, and recovered. Exiting water (NAPL-free) mounds in the area surrounding the upper screen section of the GZB well thus producing a positive hydraulic head. Natural equilibration of hydraulic potential (negative at the GZB well bottom, positive at the well top) results in circular ground water flow through the aquifer, with both a horizontal and vertical component. As such, the GZB well facilitates *in situ* soil flushing using circulating ground water as a carrier and encourages NAPL movement in the direction of the GZB well wherein it is collected and recovered for reuse.

Objectives. The relative inefficiency and high operating expense of an existing pump-and-treat approach to contain NAPL at the Nashua site highlighted the need for more cost efficient approaches for *in-situ* source management. In further consideration of challenges posed by similar problems at other sites nationwide, three main objectives for this project were identified: i) to prevent the continued seepage of NAPL to the Merrimack River in a scientifically valid, more cost-effective manner, ii) to reduce the life-cycle cost of managing the Nashua site, and iii) to demonstrate, under field conditions, the successful application of advanced methods to manage similar problems at other sites.

TECHNICAL APPROACH

Site Description: Wood treating operations at the Nashua, New Hampshire site were conducted between the years of 1923 until 1983. The site is approximately 97 acres in size, and is bounded by the Merrimack River on the eastern and northern sides. In 1981, oily seepage into the Merrimack River was reported. A series of site investigations concluded that the former Lagoon Area (where two wastewater impoundments were formerly located) was a source of oily, NAPL seepage into the Merrimack River. Creosote NAPL appeared to be a blend of creosote/PCP with refined petroleum hydrocarbons that is migrating west to east (into the Merrimack River) preferentially through the higher-permeable layer of fine-grained sand and silt overlying a discontinuous till layer above fractured bedrock.

In 1985, Beazer installed a series of product recovery wells, groundwater extraction wells, and a groundwater pre-treatment system in an effort to manage the migration of COI into the Merrimack River. Product recovery wells were installed in the former Lagoon Area, along with a series of interceptor wells between the Lagoon Area and the Merrimack River. The pump-and-treat system has been operating continuously since 1987. As of February 1995, the system had removed 186,000,000 gallons of groundwater and recovered approximately 103,000 gallons of NAPL. But the system remains only partially successful in controlling NAPL migration into the Merrimack River.

System Design and Modeling: The proposed technical design to replace the pump-and-treat system consisted of a funnel-and-gate barrier to manage and physically contain NAPL in conjunction with GZB vertical groundwater circulation systems to enhance product recovery. The funnel-and-gate design consisted of low conductivity cutoff walls (funnels) with one or more openings (gates). The system was designed such that the cutoff wall barriers physically impede the transport of NAPL and diverts the flow of groundwater to the gate areas. A series of GZB wells were placed at strategic locations on the upgradient side of the barrier and in the gate areas to facilitate NAPL recovery.

During the initial design and modeling phase, various configurations of the barrier installation were examined. The purposes of these examinations were to: i) establish a mathematical modeling system to predict efficacy of various system configurations; ii) utilize site-specific data to calculate hydraulic and NAPL flow regimes as they are influenced by barrier installation and groundwater circulation cells; and iii) identify the most efficacious funnel-and-gate/GZB configuration for the Nashua site.

Funnel-and-Gate Modeling. The effective application of a funnel-and-gate system at the Nashua site required an accurate understanding of the aquifer hydraulics. Accordingly, 3-dimensional numerical modeling of groundwater flow analysis was performed. The model used was Frac3DVS (Therrien and Sudicky, 1993), an efficient simulator of saturated-unsaturated groundwater flow and solute transport in porous and fractured porous media (Shikaze *et al.*, 1995). A 3-dimensional, rectangular computational domain of 1150 feet in the x-direction (approximately parallel to the Merrimack River), 400 feet in the y-direction (approximately parallel to the direction of ground water flow), and 90 feet in the vertical z-direction was established. Site specific factors pertaining to hydraulic control, NAPL distribution, and geology were integrated into the modeling initiatives.

GZB System Modeling. Site specific modeling was conducted to define the radius of influence (R) for GZB systems either in the gate areas or those positioned behind the barrier wall. The model assumed both natural groundwater flow, and accelerated groundwater flow through the gate areas (as induced by hydraulics of the barrier wall). The resulting flow fields of single or multiple GZB installations differs from natural groundwater flow fields or accelerated groundwater flow (due to the barrier wall) only in a limited area around the GZB. The 3-dimensional flow field in a defined, limited aquifer region was obtained by superimposing: i) a horizontal uniform flow field, computed in a vertical cross section and representing either the natural groundwater flow or accelerated flow, and ii) radially symmetric, vertical flow fields for each GZB. Superimposition of the different flow fields with their own discretization was achieved by interpolating and adding the different flow vectors at the various nodes of a simple rectangular grid with variable grid distances independently chosen for each Cartesian coordinate. The rectangular grid allowed for some refinements near the wells and their screen sections. More details of the numerical computations are given in Herrling *et al.* (1991).

Integrated System Modeling. To ensure capture of ground water and NAPL, the physical location of the funnel walls considered alignment of the barrier system relative to the principal direction of groundwater flow, accounting for seasonal/tidal variations. To ensure NAPL containment and recovery, multiple construction scenarios were considered which included simulations of various gate locations, gate widths, sheet pile lengths (*e.g.* depth of penetration), *etcetera*.

RESULTS AND DISCUSSION

Funnel-and-Gate Modeling. A total of 11 application options were modeled. The initial models considered various combinations of total wall length (480, 630, or 850 feet), number of wall segments (1, 3 or 4 wall segments), number of gate areas (1 to 3 gates), and depth of wall penetration into the formation (45 or 50 feet). The width of the gate areas was modeled at 30 feet. Rationale for considering these various barrier configurations focused on the analysis of performance and economics of the installation. For example, in the domain used, wall penetration to a depth of 50 feet simulated complete connection with lower permeability bedrock which, in theory, would minimize underflow. Conversely, barrier penetration to a depth of 45 feet simulated a potential underflow component which is perhaps more representative of actual field conditions since complete connection to bedrock *in situ* is unlikely. Moreover, permeability differences between the aquifer and the bedrock confining layer appears to be of gradual transition occurring over a distance of 5 to 10 feet. Thus, in the event that complete connection into bedrock was determined to be unlikely, or unnecessary, to reduce underflow, then the installation of the barrier to a depth of 45 feet would represent a more cost efficient design.

In the final analysis, the most effective means of physical containment was a steel sheet pile wall approximately 650 feet in length installed along the river's edge to a depth of 50 feet (no gates). This wall length was determined efficient to contain completely all areas of potential NAPL migration. At the north and south ends of the wall "wing walls" were placed each 130 feet total length (Figure 1). Each wing wall was designed with a 30 foot gate area in the middle of the wing section. In this configuration, the majority of the modeled streamlines approaching the barrier wall were diverted towards either the north or south gate areas. Complete anchoring of the sheet pile barrier into a lower permeability material helped prevent modeled underflow.

GZB Modeling. The GZB was modeled with an internal flow of $4 m^3/h$ and a saturated thickness of about 23 feet. To ensure sufficient ground water flow and NAPL capture, with particular attention given to units positioned in the gate areas, the calculated practical radius of influence was based on 30% of the stagnation point. This yielded a radius of 25 feet. Based on these calculations, one pore volume sweep of the aquifer circulation cell areas occurs once every 8 days. This flow was determined sufficient to capture NAPL directed toward the areas via the barrier wall.

Integrated Systems Modeling. Integrated modeling solved issues associated with the number, dimensions, and placement of gates, location and number of GZB wells, and diversion of flow either around or beneath the system. For example, the residence time of impacted ground water and NAPL within the gate areas must be sufficient for the necessary capture of the diverted NAPL (*i.e.*, GZB well recovery). Multiple GZB enhanced product recovery systems were placed strategically upgradient of the barrier wall to manage and remove contained product. These four GZB systems were placed in areas of known or suspected NAPL accumulation (*e.g.*, natural bedrock depressions), and considered transitional flow regimes affected by the installation of the funnel-and-gate system. Additional GZB product recovery systems may be integrated at a later date to further enhance NAPL recovery.

Effective management of the flow in the gate areas was critical to control diverted NAPL. Accordingly, in addition to the 4 source management GZB wells, a single GZB unit was placed at the southern gate to recover migrating NAPL. Modeling considering hydraulics and NAPL movement demonstrated no need for a similar unit in the northern gate area. Induced hydraulics through the gates (as a function of funnel construction) was considered in the integrated design. Given the GZB practical radius of 25 feet (30% of the stagnation point) and a gate width of 30 feet, the positioning of a single GZB system in the gate area will be sufficient to recover diverted NAPL, if any.

Permitting Requirements: This project required permits from the following agencies: i) New Hampshire Wetlands Board "Major Impact" Wetlands Permit, NHDES "Site-Specific" (Erosion Control) Permit, and ii) NHDES Groundwater Monitoring Permit (Remedial Action Plan). It was also subject to review by: i) Federal Army Corps of Engineers, ii) NHDES Air Emissions Permit, iii) City of Nashua Conservation Commission, iv) City of Nashua Zoning Board of Adjustment, v) NHDES Shoreland Protection Program, vi) NHDES Section 401 Certification Program, and vii) New Hampshire Department of Safety - Marine Patrol. In addition, this project required two "Special Exceptions" from the City of Nashua Zoning Board: one for work within a Prime Wetland and Buffer Zone, and the second for work within a Designated Floodway. One or more of the permitting agencies listed above solicited comments from the following organizations: i) NHDES Rivers Coordinator, ii) City of Nashua Building Department, iii) Lower Merrimack River Advisory Board, iv) Nashua Regional Planning Commission, v) U.S. Fish and Wildlife Service, vi) National Oceanic and Atmospheric Administration (NOAA), vii) U.S. EPA Region I New England, and viii) U.S. Coast Guard.

TECHNICAL CONCLUSIONS

- Use of Waterloo Barrier™ technology for the GZB/Funnel-and-Gate system allows for barrier installation immediately along the river's edge thus providing complete and immediate *in situ* containment of NAPL entering the Merrimack River.
- GZB systems will collect NAPL from the entire saturated zone.
- GZB/funnel-and-gate design requires no groundwater removal which:
  - i) significantly reduces long-term economic liability associated with the operation and maintenance of groundwater pre-treatment systems;
  - ii) negates the need for groundwater discharge permits; and
  - iii) is not influenced by water infiltration and groundwater or river flow.
- GZB/funnel-and-gate systems represent the most advanced, innovative *in situ* source management and containment technologies with wide-scale applicability to other sites (especially those without existing groundwater treatment plants).
- Significantly reduced O&M costs.

REFERENCES

Herrling, B., Buermann, W. and Stamm, J. 1991. "Hydraulic Circulation Systems for *In Situ* Bioremediation and/or *In Situ* Remediation of Strippable Contamination". *In*. R.E. Hinchee (ed), *In Situ Bioreclamation, Applications and Investigations for Hydrocarbon and Contaminated Site Remediation*. Butterworth-Heinemann, Boston, MA. Pages 173-195.

Shikaze, S.G., C.D. Austrins, D.J.A. Smyth, J.A. Cherry, J.F. Barker and E.A. Sudicky. 1995. "The Hydraulics of a Funnel-and-Gate System: A Three-Dimensional Numerical Analysis". IAH Congress XXVI: Solutions '95, Edmonton, Alberta, Canada.

Smyth, D., J. Cherry and R. Jowett. 1995. "Treat Groundwater in Place: *In Situ* Funnel-and-Gate System Corrals Water for Treatment". *Soil & Groundwater Cleanup*. December Issue:36-43.

Starr, R.C. and J.A. Cherry. 1994. "*In Situ* Remediation of Contaminated Groundwater: the Funnel-and-Gate System". *Groundwater* 32:465-476.

Therrien, R. and E.A. Sudicky. 1993. *User's Guide for FRAC3DVS - an Efficient Simulator for 3-Dimensional, Saturated-Unsaturated Groundwater Flow and Chain-Decay Transport in Porous or Discretely-Fractured Porous Formations.* Waterloo Centre for Groundwater Research, University of Waterloo, Canada.

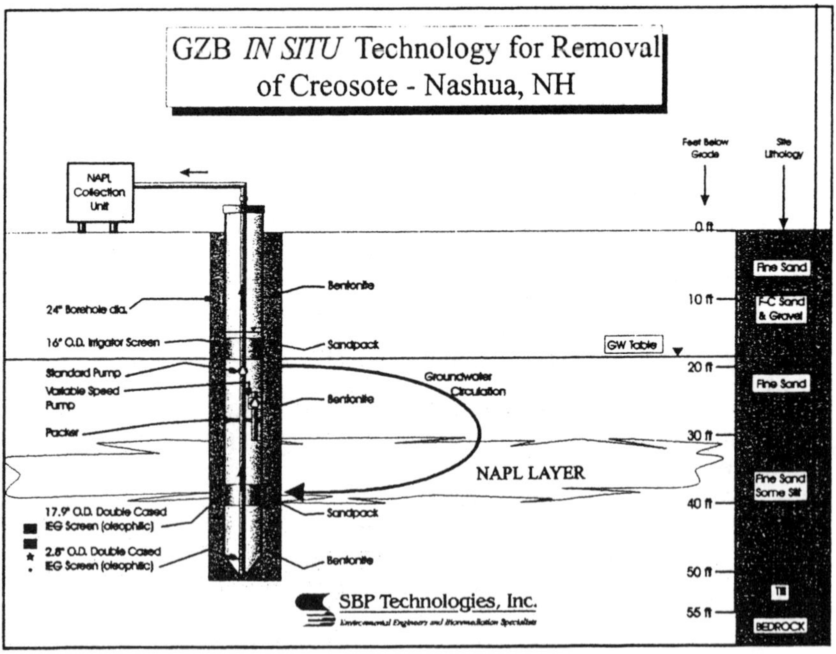

FIGURE 1

BIODEGRADATION AND IMPACT OF PHTHALATE PLASTICISERS IN SOIL

C.D.Cartwright, I.P.Thompson, and R.G.Burns

ABSTACT: Phthalate anhydride esters (PAEs) are used principally as plasticizers and have been implicated widely in the disruption of endocrine and reproductive systems in wildlife and humans. PAEs have been detected in all environments including groundwater (0.052 - 170 μg/l) and soil (0.15 - 1,480 mg kg^{-1}). The degradation of PAEs in soil and their impact on the microbial community was investigated using GC (FID) and selective isolation.

Enrichment cultures (7d, 25°C, 0.1% (v/v) PAEs) with soil (pH 8.1, 46% sand, 50% silt, 4% clay, 3.6% organic content) isolated *Comamonas acidovorans* A, B, and C and a *Xanthomonas* sp. (identified by fatty acid methyl ester analysis).

Degradation of diethyl phthalate (DEP) (100 μg C g^{-1} dw. soil) in soil occurred relatively rapidly at 20°C ($t_{½}$ =24h, t_{90}=7d) and much slower at 10°C ($t_{½}$ =7d, t_{90}=14d). Significant degradation of DEP in soil at <20°C has not been reported previously. Inoculation ($4x10^6$ bacteria g^{-1} dw. soil) with *C. acidovorans* A, B, or C, or *Xanthomonas* sp. did not accelerate degradation of DEP at 20°C.

Only 5% of di(2-ethylhexyl) phthalate (DEHP) (100 μg C g^{-1} dw. soil) was degraded after 40d (20°C). Supplementation with yeast extract (1 mg g^{-1} dw. soil) or N:P:K fertiliser (1:1:3; 1.4 mg g^{-1} dw. soil) failed to accelerate the degradation of DEHP.

DEP (10 mg g^{-1} dw. soil) reduced total culturable bacterial counts by 40% after 24h and 57% after 7d. The pseudomonad count dropped by 60% after 24h and 89% after 7d, under the same conditions. DEHP (10 mg g^{-1}) had no effect on pseudomonad or total bacteria.

BIODEGRADATION OF TRICHLOROETHANE UNDER DENITRIFYING CONDITIONS

Juli L. Sherwood, ***James N. Petersen***, Rodney S. Skeen

ABSTRACT: Previous work has demonstrated that the Hanford Denitrifying Consortium (HDC), cultured from sediments obtained from the DOE's Hanford site in southwestern Washington state, can effectively degrade carbon tetrachloride (CT). Under electron donor (i.e., acetate) limiting conditions, this consortium has been shown to transform CT to CO_2, with minimal production of the chlorinated intermediate chloroform. Others work has suggested that conditions which facilitate CT destruction, including denitrifying environments, are also conducive to 1,1,1-trichloroethane (TCA) transformation. Therefore, we have hypothesized that denitrifying bacteria might be able to degrade TCA under electron donor limiting conditions in a manner similar to CT degradation.

To test this hypothesis, denitrifying enrichments were grown from soil sediments obtained from the Hanford site and from a chlorinated ethylene contaminated aquifer that underlies the DuPont Plant West Landfill near Victoria, Texas. All materials were collected using standard soil sampling techniques. Once removed from the borehole, the material was transferred into sterile jars, shipped to our laboratory, and stored in an anaerobic glove box until use.

A series of fed batch experiments were conducted in which acetate and nitrate were measured and returned to design levels every day. The levels of TCA, as well as the dechlorination intermediates 1,1-dichloroethane, chloroethane, and ethane, in the reactors were tracked to determine the kinetics of the transformation process. In this paper, the results of these experiments are presented.

BIOREMEDIATION OF MINERAL OIL, PAH AND PCB IN DRY SOLID REACTORS

Gemoets, J., Bastiaens, L, Van Houtven, D., Springael, D., Hooyberghs, L. & Diels, L. (Environmental Technology, VITO, Mol, Belgium)

INTRODUCTION

Bioremediation is well recognized to be an economically competitive and ecologically sound remediation technology when the site and/or soil conditions are appropriate. Great research efforts are therefore directed towards defining the scope of its applicability, as well as to enlarging it. Micro-organisms can thrive in a great variety of environmental conditions and can metabolize or cometabolize a wide range of pollutants.

An inconvenience of biological approaches to remediation of contaminated land is that these processes are not always very predictable in terms of achievable residual pollutant levels and required remediation times. This can be due, amongst others, to the sensitivity of biological processes to inhibition by often unknown chemicals, by the pollutant itself which may be present in too high concentrations or by degradation products, or to limited bioavailability of the pollutants.

Performing preliminary treatability studies on a bench-scale is therefore recommended. This can reduce the risk for not achieving the required cleanup levels in an appropriate time and can substantially improve process performance by optimizing its operating parameters. Bench-scale studies are often undertaken after preliminary laboratory screening-studies which are done with simple experimental set-ups (e.g. shaker flask tests) on a very small scale and which therefore can only yield very limited information.

Bioreactors hold great promise for treating soils which are polluted by recalcitrant chemicals which are difficult to treat by conventional bioremediation methods such as landfarming, windrow-turning or in-situ bioremediation. Process parameters can be optimized and controlled easily.

VITO has developped a bench-scale dry solids reactor system (DSR, Carpels et al., 1995) that can be used to simulate bioremediation technologies which operate under non-slurry conditions, such as bioventing, landfarming, dry solid bioreactors, as well as slurry conditions, such as reactor systems for the treatment of contaminated dredging sludges or for polishing fine soil fractions obtained from a soil washing process.

MATERIALS AND METHODS

Three paddle-mixed reactors are used, having a capacity of 50 kg soil each (total volume of 200 liter each). The relatively large scale allows for multiple non-interactive sampling and for realistic extrapolation for design of full-scale bioreactor systems. The system, shown in figure 1, is fully closed and controlled: operating parameters such as temperature and concentrations of O_2

and CO_2 in the off-gas are permanently monitored (by paramagnetic and infrared sensors resp.) and automatically controlled under *Labview*. Volatile hydrocarbons in the off-gas can be monitored as well (on-line GC with FID-detector, PID, FID). Supply of air, nutrients, extra inocula and liquids is automated and soil sampling is performed with minimal perturbation of the system (without process interruption and headspace disturbance).

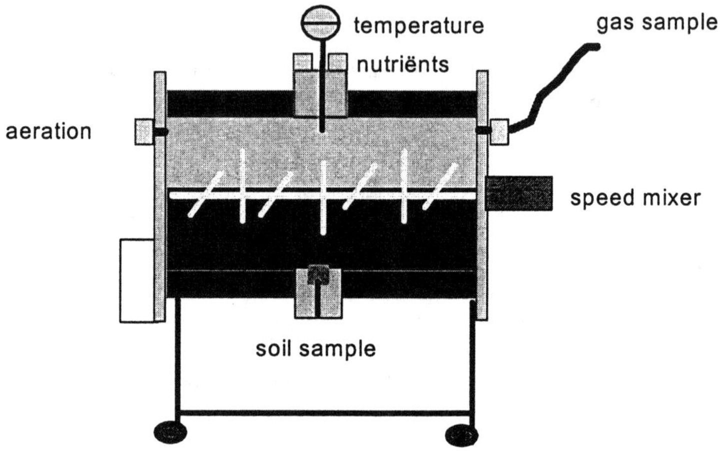

FIGURE 1.: Schematic diagram of a Dry Solid Reactor for bench-scale testing.

Mineral oil in soil samples is analyzed on hexane extracts by GC-analysis with an FID-detector or/and on tetrachloroethane extracts by IR-absorption. PAHs in soil samples are analyzed on hexane extracts by HPLC analysis with a UV detector and by GC/MS analysis.

RESULTS

The DSR reactors are being used to evaluate bioremediation by determining biodegradation kinetics, removal efficiencies and residual concentrations of mineral oil, crude oil, PAHs, PCBs (chloro)aromatics and complexed cyanides. The effects of nutrient and surfactant additions and of inoculation with micro organisms selected at VITO are being investigated.

Degradation experiments with a shaker flask system, using soil slurries with 20% dry solids, are used in conjunction with DSR test runs with 70-80% dry solid contents to get information on the asymptotic residual level of contaminants in the soil after extensive biological treatment. The combined data are used to make (conservative) predictions on biological treatability, required treatment time and residual pollutant levels after completion of soil remediation. The laboratory results are coupled to field data from actual remediation projects to validate and refine the extrapolation from laboratory research to actual full-scale remediation.

Mineral oil biodegradation was studied with contaminated soils from gasoline dispensing stations (historic pollution) and for accidental spills (tests with spiked soils and tests with soils contaminated by recent spills in the field). Inoculation with selected oil degraders (such as *Acinetobacter calcoaceticus* LH168, *Pseudomonas aeruginosa* LH220, unidentified strain LH270 and mixtures thereof) had for most cases of historic pollution little effect on oil degradation rates and efficiencies, but proved helpful for accidental spills. An example is shown in figures 2, illustrating timecourses for degradation of mineral oil in soil from a site with historic pollution by diesel, and in soil which has been recently contaminated by a diesel spill from a tanker truck. Test conditions were a.o. sandy soil, 70% dry solids and 28°C, no nutrient addition.

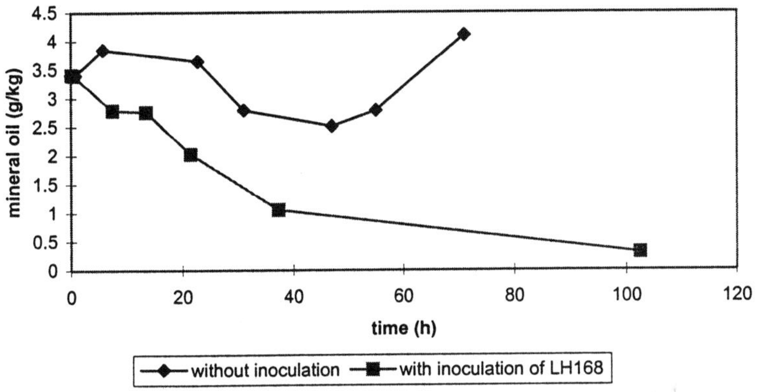

Fig. 2: Effect of inoculation for soil polluted by a recent spill of diesel

Biodegradation of PAH was studied with soils spiked with PAH compounds (fluorene and higher ring compounds) as well as with contaminated soils from a manufactured gas plant site. PAH degrading bacteria have been isolated from a mixture of PAH polluted soils and sediments by enrichment on PAH sorbing surfaces (Teflon®, Zirfon® and polysulfon membranes; Bastiaens L. et al. 1997). The properties of some of these strains are summarized in table 1. The PAH degrading strain *Sphingomonas sp. LB126* is able to utilize fluorene as sole source of carbon and energy and cometabolizes phenantrene, dibenzothiophene, pyrene and anthracene (Bastiaens L. et al. 1997b). In liquid medium with only fluorene, inreased concentrations of the compound led to inreased degradation rates, but also increased lag phases. The presence in the culture medium of non-metabolizable PAHs such as fluoranthene, benza(a)anthracene, benzo(a)pyrene and chrysene, and of PAHs that are only partially cometabolized such as pyrene and anthracene, had no effect on degradation of or growth on fluorene. However, inhibitory effects were observed in the presence of phenantrene and dibenzothiophene. Inhibition seems to be due to accumulation of dibenzothiophene or phenantrene cometabolic products. In non-sterile soil slurries (17% dry matter), rates of fluorene degradation by strain LB126 were found to be higher than in liquid mineral medium cultures.

Inhibitory effects of phenantrene on fluorene degradation were undone when a phenathrene degrading strain and LB126 were added simultaneously, and this as well for liquid and for slurry conditions.

TABLE 1. Properties of some PAH degrading bacteria isolated by VITO

Type	Method of Isolation	Metabilized Products	Cometabolized Products	Identification
LH162	liquid enrichment culture	phenantrene	fluorene, fluoranthene dibenzothiophene (anthracene)	Sphingomonas sp.
LB126	liquid enrichment culture	fluorene	dibenzothiophene phenantrene, (pyrene, anthracene)	Sphingomonas sp.
LB208	liquid enrichment culture	pyrene, phenantrene, fluoranthene	fluorene, dibenzothiophene	Mycobacterium gilvum
LB307T	Teflon membr.	phenantrene	not determined	Mycobacterium sp.
LB501T	Teflon membr.	anthracene	fluorene, dibenzothiophene, phenantrene, fluoranthene, (1.2-benzoanthacene)	Mycobacterium sp.

It was possible to upscale the fluorene degradation process in a DSR system which contained 50 kg of soil (70% dry matter). The results of the latter experiment are summarized in figure 3.

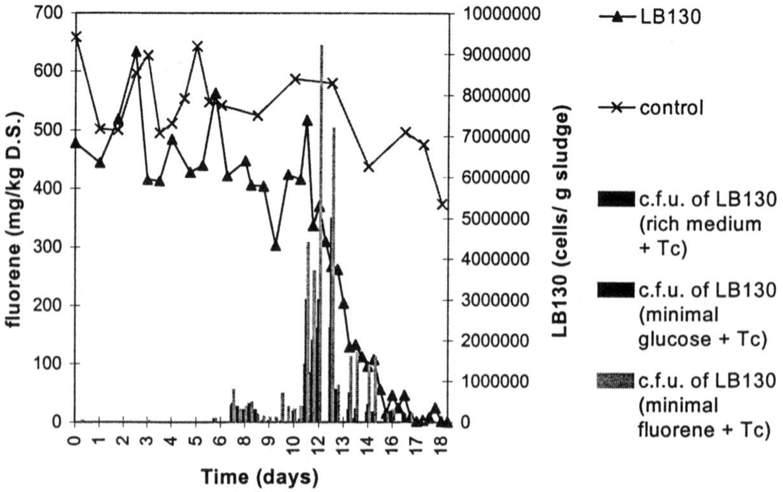

Figure 3.: Effect of inoculation on biodegradation of fluorene.

Bacteria were added to a reactor containing soil spiked with 500 mg/kgdm at the start of the experiment and after 7, 8 and 10 days. It seems that the amount of bacteria after the first three inoculations was insufficient to start fluorene degradation. An inoculum of at least $4.5x10^6$ cells/g soil was required to promote extensive degradation of fluorene. Five days after the onset of degradation, most of the fluorene had been metabolized. During fluorene degradation, the viable count of the fluorene degrader initially increased and declined afterwards to undetectable levels. In the control reactor to which no bacteria were added, no significant fluorene degradation was observed. Strain LB126 was marked with a lux reporter gene to facilitate its detection in the soil samples.

The DSR system is also being used for treatability studies with contaminated soil from manufactured gas plants (MGP). In this study the effects of surfactant addition and of chemical pretreatment on PAH biodegradation are being investigated. Comparison studies are made for PAH biodegradation by selected bacteria and by white rot fungus. Recent results will be presented. They will be used to optimize pilot scale demonstrations of soil washing/bioremediation and windrow-turning to be performed late 1997 at a MGP-site in Lier, Belgium, under the EU-sponsored *Life* program.

Small scale versions of the above reactors have also been built, with a capacity of 3 kg of soil each. They are used for preliminary test runs for optimizing treatment conditions for the larger bench-scale reactors described above. A comparison of biodegradation results for the small and larger size systems will be presented and analyzed with respect to upscaling considerations. The construction of a continuous dry solids bioreactor with a capacity of 1000 kg soil is planned for large-scale pilot demonstrations.

REFERENCES:

Carpels, M., Van Houtven, D., Elslander, H., Diels, L., Hooyberghs, L., Van Roy, S., 1995. "The Use of a New Generation Bench Scale Bioreactor for the Research of Dredged Sludge and Soil Remediation" In Proceedings of the third International Symposium *In Situ and On-Site Bioreclamation, San Diego, California*

Bastiaens, L., Springael, D., Diels, L., Wattiau, P., Verachtert, H. 1997. "Isolation of New Polycyclic Aromatic Hydrocarbon (PAH) Degrading Bacteria by Use of PAH Sorbing Carriers"; to be presented at the *International Symposium on Environmental Biotechnology*, Oostende, Belgium.

Bastiaens, L., Springael, D., Remes, G., Vereecken, J., Diels, Verachtert, H. 1997. "Metabolization and Cometabolization of single PAHs and PAH Mixtures in Liquid Cultures, Soil Slurries and Dry Solid Reactors (DSR)", to be presented at the *International Symposium on Environmental Biotechnology*, Oostende, B.

FEASIBILITY AND COMPATIBILITY OF GROUNDWATER-CIRCULATION-FLOWS

Juergen Stamm (University of Karlsruhe, Karlsruhe, Germany)

ABSTRACT: The groundwater circulation well (Grundwasser-Zirkulations-Brunnen [GZB]) is an in situ remediation technique for the removal of volatiles and microbially degradable contaminants from subsoil and groundwater. The operation of the well is based on the creation of a vertical circulation flow in the well field between two screened sections which are hydraulically separated. The groundwater which is captured by one of the screened sections of the well will be decontaminated within the well column using a suitable remediation procedure and subsequently flow back to the aquifer through the other screen section. Thus GZB-technique can be depicted as a combination of pumping and reinfiltration within the well. The system is used at numerous sites in Germany and the United States.

The contribution will demonstrate the differences between GZB applications in confined and phreatic aquifers. Furthermore it shows the numerically calculated flow field as a combination of vertical and horizontal circulation caused by accompanying pumping wells. The latter can be supported by measured field data at a full-scale installation in Germany.

The practicability of a multiscreened well (e.g., three screened sections for establishing two vertical circulation cells in a vertical line) is demonstrated in a large-scale laboratory investigation in VEGAS (Versuchseinrichtung fuer Grundwasser- und Altlastensanierung Stuttgart, Germany). It reveals the remediation system's performance in layered aquifer conditions.

A NOVEL LIQUID FOAM CARRIER FOR USE IN BIOREMEDIATION

Ashley J. Wilson and W. Bernard Betts (University of York, York, England)

ABSTRACT

Introduction. Biologists at the University of York have developed a novel bioremediation technology in which oil-degrading microorganisms found naturally in the environment are incorporated into a liquid foam carrier similar to dense blanket foams used for fire-fighting purposes. Recent research at York has led to the design of a suitable foam carrier, and the formulation of an optimum foam-microbe system (a 'Bioactive Foam') for enhanced oil biodegradation.

The Foam. The foam employed in a terrestrial application would include the following components:

(1). foam-*forming* protein(s);
(2). foam-*stabilising* proteins [which may be the same as (1)];
(3). a thickening agent to retard drainage of the foam;
(4). nutrients, inorganic (fertilizers);
(5). nutrients, organic (vitamins, etc.);
(6). water;
(7). a microorganism, or a consortium of microorganisms;
(8). enzymes and biosurfactants (produced by microorganisms);
(9). oxygen-enhanced air

Benefits from the Use of a Foam Carrier. The three most important advantages in using a foam carrier are: (1) easier targeting of the site to be remediated including good adhesive properties of the foam to, for example, vertical rock or tank surfaces; (2) a controllable liquid and gaseous environment for containment of microorganisms and continued supply of nutrients, control of pH, etc.; (3) the creation of a large area interface between the microbes and the hydrocarbon they will degrade. The benefits of using foam are manifold but it has been demonstrated in the laboratory that, because of the increased microbial activity occurring when conditions are optimised, rates of hydrocarbon degradation can be significantly increased over equivalent non-foam systems.

How the Foam Works. Shortly after the foam is applied to the hydrophobic surface its structure starts to change - albeit slowly - due to the processes of Ostwald ripening, bubble coalescence, and gravitational separation of liquid from gas. The liquid draining from the foam is rich in oil-degrading microorganisms, enzymes, nutrients, and biosurfactant (protein) and forms a thin layer between the oil and the foam. Since most of the components in the oil are less dense than water they float through this aqueous region and rise through the foam where they are in intimate contact with the bioactive agents (microbes, enzymes, etc.), nutrients, and dissolved gas.

As the oil components rise ('cream') through the aqueous layer below the foam, they emulsify to a complex of oil-in-water and water-in-oil forms which are partially stabilised by the protein surfactant. The interface region between the foam (which acts as a reservoir for gas, microorganisms, and nutrients, etc.) and the oil is, therefore, complex and presents a large surface area at which the biodegradation process can proceed.

Range of Application of the Technology. It is envisaged that the technology described herein has application to hydrocarbon-based pollutants which may be the source of problems in a range of scenarios. Such situations include: (1). cleaning residual oils of a variety of types from storage tanks in ocean- and sea-going oil tankers, after the discharge of their cargo; (2) remediation of oiled beaches and rocky shores following major release of oil from tankers at sea or blow-outs from offshore production; (3) remediation of sites around oil refineries and marine terminals; (4) cleaning polluted soil in proximity to diesel-storage facilities (e.g., farms, building/construction sites, etc.); (5) cleaning polluted ground sites in proximity to oil pipelines; (6) remediation of sites in proximity to terrestrial oil prospecting and drilling platforms; (7) remediation of sites close to natural sources such as seepage from oil deposits; (8) remediation of disused gasometer sites.

A visit to the polluted sites at Milford Haven in South Wales resulting from the *Sea Empress* oil-tanker disaster in February 1996 has demonstrated clearly that while the foam-microbe technology is probably not sufficiently robust to withstand the rigours of the polluted ocean surface under hard weather conditions, it would be ideally suited to the remediation of oiled rocky and sandy beaches.

NATURAL ATTENTUATION AND A MICROBIAL FENCE: RE-ENGINEERING CORRECTIVE ACTION

Jacqueline A. Abou-Rizk (Union Carbide Corporation, S. Charleston, WV)
Maureen E. Leavitt (SAIC, Oak Ridge, TN)

ABSTRACT: Under RCRA, if a site does not meet stringent groundwater quality criteria, a corrective action program must be implemented to bring the site into compliance. This was the case with a closed, capped, remote hazardous waste landfill. Concentrations of various organic contaminants in groundwater exceeded background levels. Considering the low concentrations (averaging approximately 1 ppm) and proximity to receptors, an Alternative Concentration Limit proposal was submitted to the regulatory agency but was not accepted. Left with few options, the site owners considered a pump-and-treat-based system. The multi-million dollar effort to implement such a system at the remote site seemed unreasonable considering the low levels of contaminants and low risk to receptors. It also posed new risks, i.e., potential salt water intrusion and transportation/disposal of hazardous waste. A new option, a "green" solution, satisfied the corrective action goals while avoiding increased exposure to contaminants and considerable operations costs. This paper highlights site characteristics and the path that produced a corrective action program satisfying all stakeholders. The solution was an innovative, wind- and solar-powered, corrective action program, based on natural attenuation combined with a microbial fence. The full-scale system will operate for a fraction of the cost of a pump-and-treat system, yet it will protect the receiving water from exposure to organic contaminants. This project illustrates the benefit that can be realized with public and regulatory participation and "re-engineering" corrective action.

INTRODUCTION

At a coastal Georgia site with a RCRA landfill, groundwater samples from wells along the Point of Compliance statistically exceed background levels, although contaminant concentrations are very low (<5 mg/L). This situation requires a corrective action plan. The landfill is a twenty-acre site with one edge along an estuarine creek and the other edges along a three-thousand acre pine forest, making it very remote. Operations at the site were shut down a decade ago and the landfill was closed and capped.

The site groundwater protection standard was written according to Georgia law; contaminants cannot exceed background except where allowed in 40CFR 264.94. Background samples are obtained at an on-site upgradient well which is adjacent to and downgradient of a pond.

Since the groundwater posed little risk, when the background levels were exceeded, the site owners submitted an Alternate Concentration Limit (ACL) document which was not accepted by the government's environmental agency. The site owners had few alternatives and they decided to design a pump and treat system, while continuing to look for alternatives. The pump and treat system was a multi-million dollar system, an inappropriately costly system based upon the site risks; however, it seemed the only workable technology at the time.

During the regulatory review period in the early 1990s, bioremediation technology was new but looked applicable for the site considering its almost ideal characteristics. The site's aquifer had been characterized: it is sandy, has a known groundwater flow rate and direction and is close to ground surface. Groundwater contaminants were generally known to be biodegradable, although the low levels of contaminants were a concern (i.e., would there be enough carbon for the bacteria to survive?).

The site owners presented a program to investigate bioremediation to the state regulators and were given permission to further investigate this technology as a potential corrective action for the site groundwater. Given the tight deadline to prove bioremediation's ability to clean up the groundwater and the general uneasiness of regulators with natural attenuation in the early 1990s, a thorough aggressive program was developed including field observations, a literature review, a laboratory treatability test, BIOPLUME II modeling, and pilot demonstrations.

Preliminary Biofeasibility Study. To determine the general feasibility of implementing a bioremediation system, a laboratory treatability study was completed (Abou-Rizk, et al., 1995). The study examined replicate water samples that were untreated, enhanced with oxygen and enhanced with both oxygen and nutrients. Over the 21-day test period, oxygen consumption and carbon dioxide production were monitored using a computerized respirometer. The results showed that aeration increased the rate of degradation, but the addition of nutrients did not produce additional rate increases. Major conclusions drawn from the study were:

- A viable bacterial population exists in the subsurface that is capable of degrading the target contaminants at the low levels found at the site.
- Oxygen addition increases the rate of biodegradation.
- Nutrients do not provide a significant increase in the biodegradation rate.
- Anaerobic activity in untreated samples caused some reduction in contaminant concentrations.

The laboratory study results confirmed conclusions from other reports found in the literature. These data provided the foundation to design a prototype bioremediation system. They also provided potential anaerobic biodegradation rates that could be utilized during BIOPLUME II modeling.

Pilot Test. Once the state regulators were convinced of the potential application of bioremediation as a full-scale treatment system, a pilot scale system was set up in the field. Since the site was remote and no utilities were available, a solar-powered system was installed (Figure 1).

The solar-powered pilot system was successful in treating the contaminants, and it was also quite useful in defining operational factors.

Site owners chose oxidation-reduction potential (redox) as the primary performance criterion based on the assumption that a positive redox value represents oxidative conditions. Prior to treatment, redox values from impacted compliance wells were always negative (as low as -300 millivolts). As the study developed, it was determined that positive redox values were directly related to lower contaminant concentrations.

One of the most important observations was that higher dissolved oxygen concentrations could be obtained with intermittent air injection instead of full-time air injection (Figure 2). It was also confirmed that very low quantities of air were needed to treat the contaminants since the air compressors supplied only up to 2 scfm of air. Both of these findings greatly impacted the full-scale design.

During the pilot work the site owners continued to look into alternative energy and oxygen delivery systems. The investigation of alternative energy led the site owners to wind power. A system was found which provides compressed air from a compressor attached to a conventional windmill. The wind-powered system was tested over five months and it met the site needs. It was simple to install, required minimal maintenance, and delivered sufficient air to maintain positive redox values. The intermittent nature of wind in this location was sufficient and compatible with the requirement for intermittent air injection. The wind speed and direction were measured and the data collected were used to develop the full-scale system.

The Role of Intrinsic Bioremediation. During time period of the bioremediation investigation, intrinsic bioremediation gained recognition as a reasonable alternative for low risk sites. Several site characteristics supported the decision to incorporate intrinsic bioremediation as a part of the site corrective action program: it is in a remote location, the landfill is closed, capped and maintained under RCRA, and the contaminant concentrations are very low.

To provide a technical basis for this determination, BIOPLUME II models were constructed using site specific data. Acetone was chosen as the representative target compound since it was the most prevalent throughout the site and it was present at the highest concentrations among all contaminants. The model was run with 1, 5 and 10 year scenarios, using the maximum observed acetone concentration (supplied as a constant source). As a worst case, the models were also run with ten times the maximum observed acetone concentration. Wherever possible within the model input, conservative or worst-case values were used. In each case, the model output showed that acetone would never reach the receiving body of water. These results supported the decision to rely on intrinsic bioremediation to compensate for contamination at two of the three target areas at the site.

Role of the Public. The state regulators welcomed the combined oxygen-enhanced remediation and intrinsic remediation system. However, they were concerned about the reaction of the public. Therefore, they requested a series of public meetings to address questions and concerns. At each meeting, technical

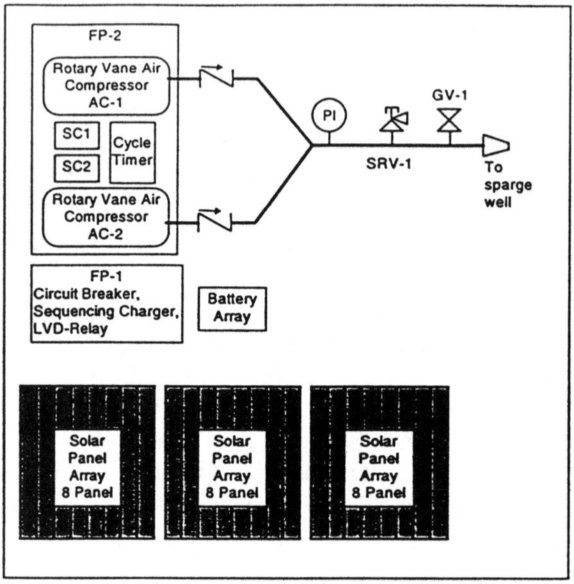

Figure 1: Solar Sparge Unit Schematic

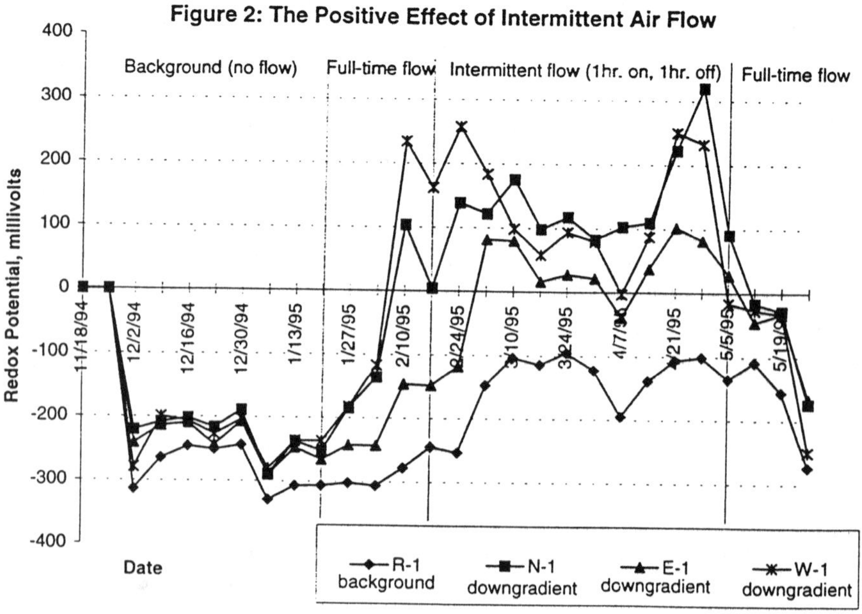

presentations were made to explain the science of bioremediation and its application at the site. Updates as to results and progress were also presented. Further, the specifics of the corrective action system (number of air injection wells, for example) and the RCRA permit modification were discussed. The public was interested in bioremediation and accepted the idea.

Project Status. Currently, the permit modification application is under review by the state regulatory agency. The key components of the corrective action system are two natural attenuation areas and one oxygen-enhanced bioremediation area (Figure 3). The oxygen delivery system is a windmill/compressor system (Figure 4) consisting of up to three windmills and air will be delivered through wells installed with "push" technology. The solar panels used for the solar-powered pilot test will be used to supply instrument power. Once approved, the full-scale system is scheduled to be installed within seven months.

Installation and maintenance costs for this innovative system are low and site safety concerns are close to nil. The windmill/compressor systems cost only approximately $20,000 each (installed) and require once a year maintenance to replace lubricating oil. The savings of this system both in cost and safety concerns are signficant, particularly over time, when compared to: 1)an electric air compressor system (cost of installing overhead power lines, an electric air compressor system, the monthly electric bill and maintenance), 2) a gasoline-powered generator (cost of the generator, fuel and maintenance, potential for spills) or 3) multiple solar panel/air compressor systems (cost of systems, plus battery replacement every 5 to 6 years).

SUMMARY

The re-engineering process brought the site owners a true chance to choose the most reliable, long-term technical solution for the site. Acceptance of the remediation solution by the state regulatory agency and the public was earned through a serious study of the science of bioremediation, proof of its success at the site, and many meetings to address concerns. Success can also be attributed the public's understanding of the groundwater risk at the site. We proudly say that this remedy is "green" through and through.

ACKNOWLEDGEMENTS

The authors would like to thank Arthur M. McClain and Carol L. Dudnick of Union Carbide Corporation and Samuel J. Senn of BASCOR Environmental, Inc. for their help in the preparation of this paper.

REFERENCES

Abou-Rizk, J. A. M., Leavitt, M.E., and Graves, D. A., 1995. "In Situ Aquifer Biosparging of Organics Including Cyanide and Carbon Disulfide." In R. E. Hinchee, et al., *Applied Bioremediation of Petroleum Hydrocarbons*, pp. 175-183. Battelle Press, Columbus, OH.

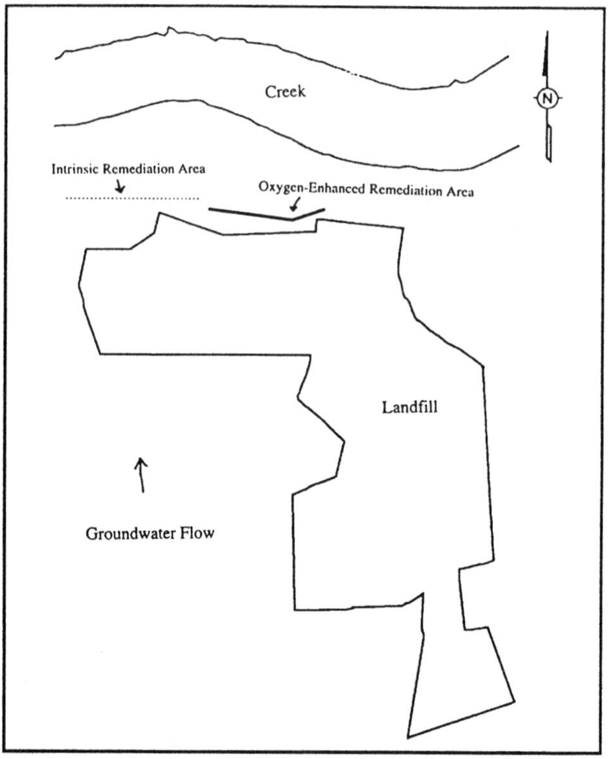

Figure 3: Full-Scale Design Features - Site Plan w/ Intrinsic and Oxygen-Enhanced Remediation Areas

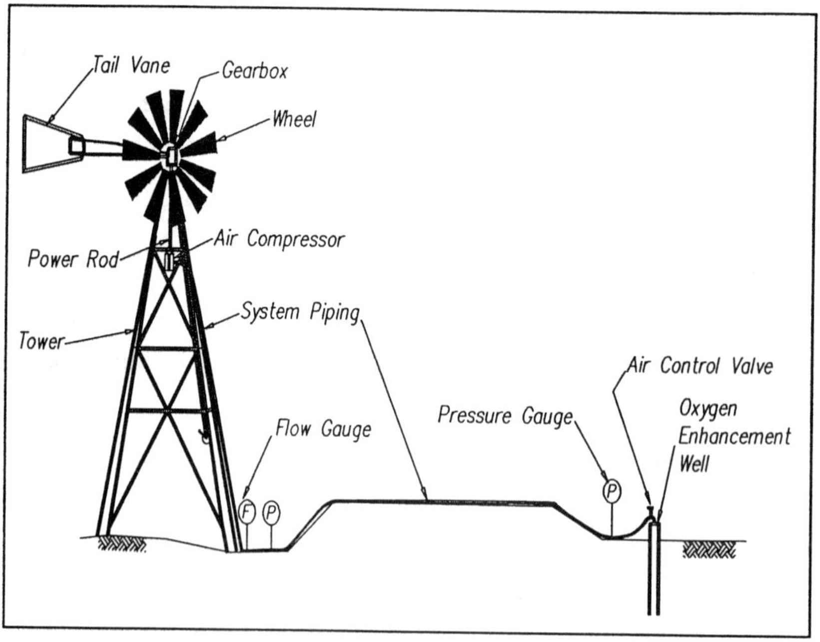

Figure 4: Windmill/Compressor System Schematic

MIGRATION BARRIERS: SPARGING AND BIODEGRADATION IN TRENCHES

Robert D. Norris and David J. Wilson
(ECKENFELDER INC., 227 French Landing Drive, Nashville, TN 37228)

ABSTRACT: Mathematical modeling is used to examine three barrier types bas ed on sparging and biosparging. Efficiency is strongly dependent on design for sparging, less so for biosparging. Other barrier designs are also discussed.

INTRODUCTION

There are several situations in which migration barriers may be the best alternative.. Site owners, regulators, or other stakeholders may not be comfortable with intrinsic bioremediation at a particular site. In some cases intrinsic bioremediation may not adequately control the plume, or may not prevent contaminant migration across a property line or to a receptor. In these cases a migration barrier may provide inexpensive protection of downgradient receptors. Such barriers include air sparging wells, wells containing slow release oxygen compounds, and intercepting trenches with sparged air or some other electron acceptor. In any case, a zone is created within the aquifer which contains elevated levels of electron acceptor through which the contaminated groundwater moves. See Figure 1.

If the barrier is located some distance upgradient of the point of compliance, this provides space and time for intrinsic bioremediation between the barrier and the point of compliance. Also, placing the barrier near the contaminant source permits use of a shorter barrier. However, the barrier must be far enough from the source that the levels of biodegradable compounds are low enough that the barrier, assisted by intrinsic bioremediation, will provide sufficient electron acceptor to reduce contaminant levels to within regulatory limits.

INTERCEPTOR TRENCHES

We address aeration trenches as barriers for plume containment. In these, the water entering the trench is sparged, resulting in stripping of VOCs and provision of oxygen for degradation of biodegradable contaminants. In stripping it may be possible to use the overlying vadose zone as a bioreactor for treatment of the off gas, or, if some of the VOCs are not biodegraded, one may capture the off gas by soil vapor extraction. (Gudemann and Hiller, 1988; Wilson, et al., 1992, 1996)

Three types of aeration barrier trench are shown in Figure 2. All involve injection of air into the saturated zone below the deepest level of the plume, followed by upward movement of the air, during which VOCs are stripped and oxygen transferred to the groundwater for biodegradation. Nutrients may be added, as well. The extent to which bioremediation and/or air stripping takes place is controlled by several factors: These include:

(1) the biodegradability of the contaminant(s)
(2) the volatility (Henry's constant and vapor pressure) of the contaminant(s)
(3) the residence time of the water in the trench
(4) the air flow rate
(5) system design factors (configuration, bubble size, etc.)
(6) the temperature

Both biodegradation and air stripping are enhanced by long residence times, high air flow rates, and high temperatures. Biodegradation competes with stripping, and is reduced if the constituents are highly volatile. Biodegradation rates typically achieve maximum values at modest air flow rates. Generally, therefore, one can maximize biodegradation of VOCs by operating at the minimum air flow at which contaminant removal is satisfactory; much less air is needed to provide oxygen for biodegradation than to air strip most VOCs. Maximizing biodegradation may allow operation without vapor recovery, thus reducing costs.

System design has little effect on oxygen transport; all three systems in Figure 2 provide adequate oxygen transport; see Table 1. For biodegradation alone, therefore, one would avoid the expense of the second and third designs. However, if one must strip recalcitrant VOCs, the third design in Figure 2 is far better than the others. The removal of TCE by these systems was modeled under standard conditions of trench depth, air and water flow, and VOC mass transfer rate coefficient. The first design (cross current) air stripped 88.36 % of the influent TCE; the second (cross current/counter current) removed 96.38 %; the third (cross current/counter current with a counter current section at the bottom) removed 99.97 %. See Table 2. (Mutch, Norris and Wilson, 1995)

In the third type of trench the counter current section need not be long, as seen in Table 3. One meter is ample to give high removals. Small air bubbles increase surface area and bubble residence time, enhancing mass transport of oxygen and VOCs. High air flows result in large bubbles, turbulence, and reduced hydraulic conductivity across the trench, which may lead to bypassing of the plume around the ends of the trench.

The air injection trench must be long and deep enough that it intercepts the plume with some safety margin. It must provide adequate contact time at maximum ground water flow; this depends on both biodegradation and air stripping rates. If recalcitrant VOCs are present, air is required at a rate which insures their adequate stripping; this can be explored by modeling. Provision must be made for off gas recovery and treatment, if needed. If the intent is to minimize air stripping and maximize biodegradation, the air flow must be sufficient to provide a stoichiometric excess of oxygen for oxidation of constituents of concern and non-target biodegradables. The air flow should not be much larger, however, since that may result in excessive stripping of VOCs and a need for off gas collection and treatment.

Monitoring wells should be located short distances upgradient and downgradient of the barrier to check its effectiveness (1) in removing most of the contaminants and (2) in providing oxygen downgradient for destruction of residual biodegradables. This takes place by intrinsic bioremediation between the barrier and the point of compliance. Monitoring wells should be located at the ends of the trench to assure that by-passing of contaminants around the barrier is not occurring. And monitoring wells should be placed well downgradient from the barrier and somewhat upgradient from the point of compliance to assure that the system is operating satisfactorily, or to warn in case that it is not.

RELATED TECHNIQUES

Systems Based on Zero-Valent Metals. Gillham (1994) and coworkers have commercialized zero valence metals to provide permeable barriers for chlorinated solvents. These systems are applicable to metals such as chromium and radio nuclides such as uranium and technetium. Others have developed permeable barriers in which biological processes degrade the constituents of interest or, in some cases, precipitate metals.

Systems Based on Air Sparging Air sparging, in which air is forced into the aquifer through wells screened in a short interval below the zone of contamination, was developed primarily to address source areas. Placing a row of air sparging wells along the barrier provides an aeration zone over several feet along the flow path. Such systems can be effective and require fewer wells than do passive systems, but require power and maintenance, and may result in vapors reaching the ground surface, utility trenches or buildings unless vapor recovery is provided.

Systems Based on Oxygen Release Compounds A more passive system utilizes magnesium peroxide (ORC^R) to slowly release oxygen. The barrier can be constructed by placing socks containing ORC in a row of fairly closely spaced wells. Oxygen, moving by advection, dispersion, and diffusion, creates a zone of increased dissolved oxygen. ORC can also be placed in trenches, horizontal wells, injected using Geoprobe techniques, or augured into the ground.

Method Selection The selection of intrinsic remediation, a migration barrier, or a more highly engineered system can be evaluated through fate and transport modeling.(Norris, Dupont and Gorder, 1996) If a migration barrier is selected, the criteria discussed here in conjunction with site location, the presence of underground utilities, etc., can be used to select the most appropriate barrier.

REFERENCES

Gudemann, H., and D. Hiller. 1988. "In-Situ Remediation of VOC Contaminated Soil and Ground Water by Vapor Extraction and Ground Water Aeration." Proceedings, Third Annual Haztech International Conference, Cleveland, OH.

Mutch, R. D., R. D. Norris, and D. J. Wilson. 1996. "Groundwater Cleanup by In Situ Sparging. XI. An Improved Aeration Curtain Design." Separ. Sci. Technol., in press.

Norris, R. D., R. R. Dupont, and K. Gorder, 1996, "Solection of Remedial Strategies Based on Modeling of Plume Dynamics", Petroleum Hydrocarbons and Organic Chemicals in Groundwater: Prevention, Detection, and Remediation, Houston, TX, Nov. 13-15.

Sandom, M. J., and J. S. Seyfried. 1996. "Enhanced Bioremediation of a Pipeline Release with Time Release Oxygen." I&E Special Symposium, American Chemical Society, Birmingham, AL, September 9-12.

Wilson, D. J., S. Kayano, R. D. Mutch, and A. N. Clarke. 1992. "Groundwater Cleanup by In Situ Sparging. I. Mathematical Modeling." Separ. Sci. Technol., 27,1023.

Wilson, D. J., and R. D. Norris. 1997. "Groundwater Cleanup by In-Situ Sparging. XII. Engineered Bioremediation with Aeration Curtains." Separ. Sci. Technol., in press.

Wilson, D. J., R. D. Norris, and A. N. Clarke, 1996, "Barrier Trench Design and Operation for Air Stripping and Biodegradation", I&EC Special Symposium, American Chemical Society, Birmingham, AL, Sept. 9-12, pp. 787-790.

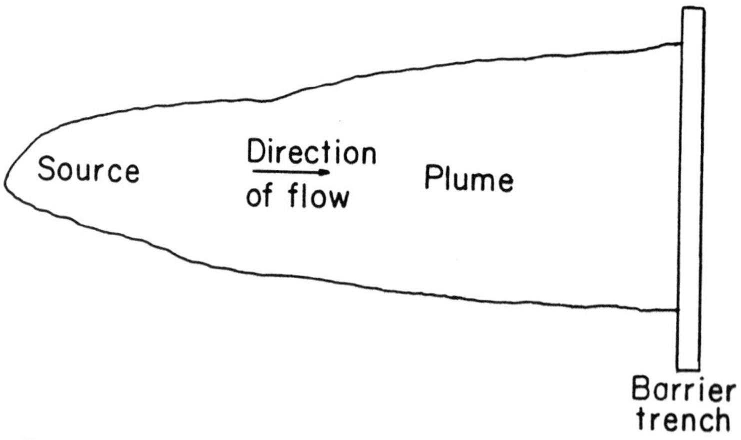

Figure 1. Plan view, plume and aeration barrier trench.

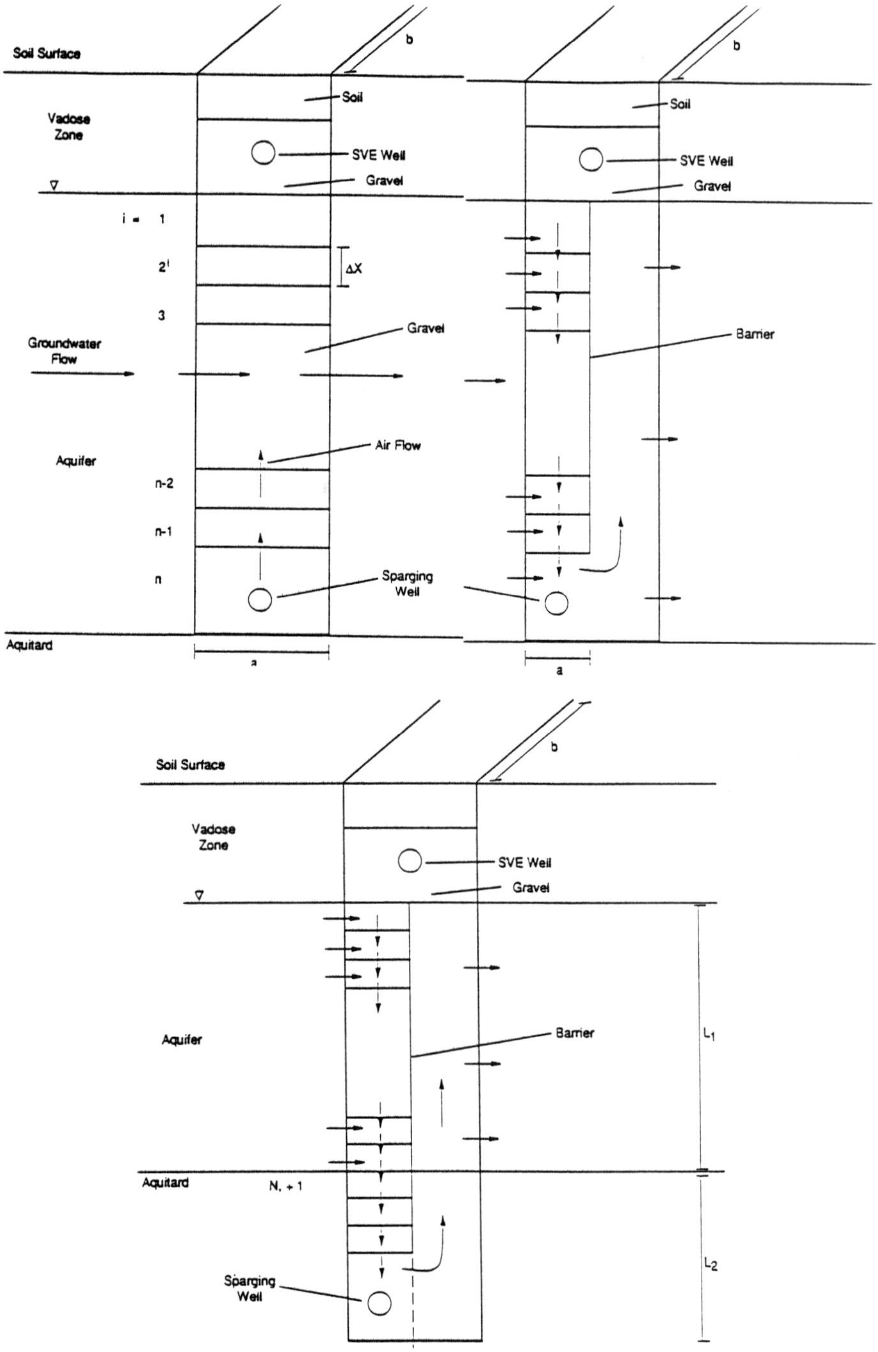

Figure 2. Three configurations of aeration barrier trenches.

TABLE 1. Percent O_2 saturations achieved by the three aeration curtain configurations considered.

Air Flow Rate (m^3/sec)	Cross Current (%)	Cross Current/ Counter Current	Cross Current/ Counter Current with Counter Current Section
0.01	99.8	99.96	100.000
0.001	98.3	99.58	99.997
0.0001	84.7	94.8	99.35

TABLE 2. Percent TCE removal achieved by the three aeration curtain configurations.

Air Flow Rate (m^3/sec)	Cross Current (%)	Cross Current/ Counter Current	Cross Current/ Counter Current with Counter Current Section
0.0050	76.80	91.13	98.03
0.0075	84.84	94.89	99.82
0.0100	88.36	96.38	99.97
0.0125	90.52	97.19	99.993

TABLE 3. Effect of counter current section length on percent TCE removal, third barrier trench configuration.

Length of Counter Current Section (m)	Percent TCE Removal
0.0	96.38
0.5	99.31
1.0	99.87
1.5	99.97
2.0	99.99

PASSIVE AND SEMI-PASSIVE TECHNIQUES FOR GROUNDWATER REMEDIATION

Michael J. Brown
(Komex International Ltd., Calgary, Alberta; formerly University of Waterloo)

Michaye L. McMaster
(Beak International Inc., Guelph, Ontario; formerly University of Waterloo)

J.F. Barker, J.F. Devlin, D.J. Katic, S.M. Froud
(University of Waterloo, Waterloo, Ontario)

ABSTRACT: A pilot-scale demonstration to study passive and semi-passive techniques for in-situ groundwater remediation has been in progress since October 1996. The goal of our research is to evaluate in-situ anaerobic/aerobic sequential remediation schemes, for the treatment of contaminant plumes containing a mixture of gasoline components (BTEX) and chlorinated hydrocarbons. The experiment is operating under controlled conditions in a sand aquifer near Alliston, Ontario. Groundwater flow rates are controlled under a forced hydraulic gradient within 3 parallel sheet piling gates. A contaminant plume of dissolved toluene (TOL), carbon tetrachloride (CTET), and perchloroethylene (PCE) is introduced at the upgradient end of each gate. Each gate is instrumented with multilevel piezometers and stainless steel wells for periodic sampling to monitor the fate of contaminant and tracer solutes. Gate 2 is an experimental control with no remedial technologies installed. Gate 1 is equipped with 2 removable cassette boxes containing zero-valent iron to dechlorinate PCE and CTET abiotically. Another cassette contains an oxygen-releasing compound (ORC®), designed to increase dissolved oxygen levels and promote aerobic biodegradation of TOL. A permeable wall is installed in Gate 3 for periodic nutrient flushes to promote anaerobic biodegradation of PCE and CTET. In-situ aerobic biodegradation of the remaining pollutants is stimulated by an oxygen biosparging gate further downgradient. The treatment capabilities of each technology are currently being evaluated, and preliminary field data is presented.

INTRODUCTION

Objective. Mixed contaminant plumes are a common problem encountered in groundwater remediation projects. Treating mixtures of organic contaminants is a challenging endeavour, since each compound may be treatable with existing remediation technologies, but in many cases the different classes of pollutants are not treatable with any single technology. Recent research in progress at the University of Waterloo (UW) has been aimed at performing in-situ groundwater treatments in a passive or semi-passive fashion. The research project summarized here attempts to combine several semi-passive remediation technologies in sequence to treat a plume of mixed contaminants, in-situ.

Site Description. The experimental site for the test is Canadian Forces Base Borden, about 90 km north of Toronto, Ontario, Canada. The surficial aquifer has been well characterized during previous work by UW researchers, including Devlin & Barker (1994). The surficial unconfined aquifer is approximately 4 m thick at the site, and consists of well-sorted, fine to medium-grained sand. It is underlain by a clayey silt unit on the order of 13 m thick, which serves as a local aquitard. Water table depth varies from 0.5 to 1.5 m below ground surface (bgs). The natural groundwater flow velocity is about 10 cm/day.

MATERIALS AND METHODS

General. The goal of the Borden experiment is to evaluate the performance of previously developed technologies combined in sequence at the pilot scale. This requires the installation and instrumentation of a multiple gate flow control structure composed of Waterloo Barrier® sealable joint sheet piles (Starr & Cherry, 1994). A plan view of the 3 gates and important instrumentation is shown in Figure 1. The sheet piling gates were installed with each pile penetrating between 0.2 to 0.5 m into the clay aquitard. When sheet piling installation was completed, the sealable joints were sealed with a bentonite grout. Each gate structure is open at the upgradient end and closed at downgradient end.

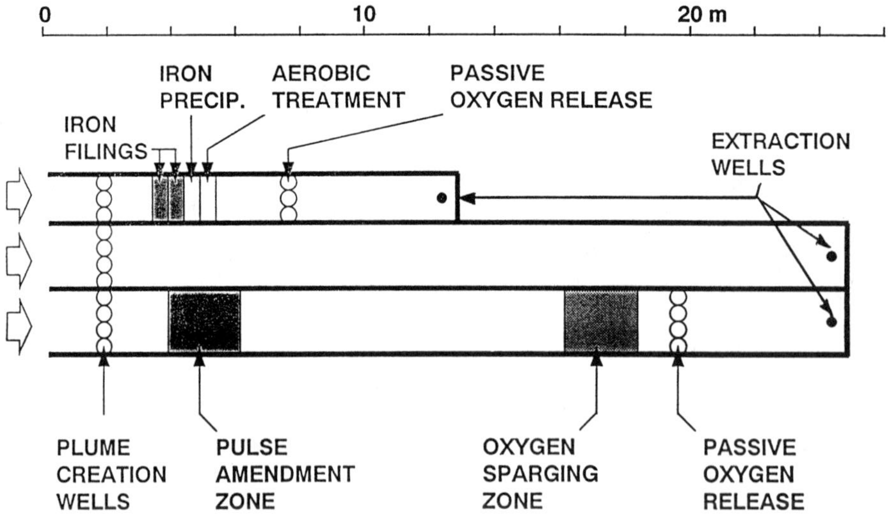

Figure 1 - Plan view of sheet piling gates and selected instrumentation.

A row of pvc source wells (3 or 4 per gate) was installed at the open upgradient end of each sheet piling gate. Each well is 25 cm o.d. casing with screened intervals from about 0.9 to 3.0 m bgs. The source wells have two primary purposes. Initially, a pulse of conservative tracer solution (potassium bromide, KBr) was injected into the wells. Migration of this tracer pulse toward the extraction wells under forced hydraulic gradient conditions was monitored over several months. Tracer arrival times at various sampling fences were analysed in an effort to better characterize groundwater flow at the site, including flow velocities in different parts of each gate, and lateral or vertical heterogeneities in the aquifer.

The second purpose of the large diameter source wells is for long-term controlled addition of an organic contaminant plume, which contains approximately 1 mg/L PCE, 1 mg/L CTET, and 4 mg/L TOL. The three volatile organic compounds (VOCs) of interest are slowly introduced into each source well using diffusive passive release source emitters (Wilson and Mackay, 1995). The diffusive source emitter system is designed to run semi-passively for several months with minimal maintenance. Piezometers upgradient and downgradient of the source wells are sampled regularly to ensure proper system operation.

A fully screened SS well (5 cm o.d.) is installed at the downgradient end of each gate. These wells are pumped continuously, to provide a forced hydraulic gradient controlling groundwater flow. Extraction wells are also intended to capture any VOCs which are not remediated in the treatment gates. These pollutants flow to an above-ground treatment system composed of a zero valent iron canister and a granular activated carbon drum.

Each gate is also instrumented with multilevel piezometers and stainless steel (SS) monitoring wells to monitor plume migration through the gates temporally and spatially. Details of the sampling network are given in Figure 1. The multilevel piezometers each consist of four 0.3 cm o.d. stainless steel (SS) tubes strapped to a central pvc stalk for support. The SS sampling tubes extend to approximately 1.2, 2.0, 2.7, and 3.5 m bgs for discrete point sampling at different depths in the aquifer. In addition, the downgradient side of each cassette in Gate 1 is equipped with nests of ss sampling tubes. Overall, there are 87 multilevel piezometers installed in the 3 gates, representing 348 discrete sampling points. SS monitoring wells (3 cm o.d.) are fully screened from 0.6 to 3.7 m bgs. Nine SS wells are located inside the gates for plume monitoring during the experiment. An additional 8 wells are installed outside the gates, to monitor possible plume leakage through the sheet piling or upgradient of the sources.

Zero valent iron & ORC treatment systems. Gate 1 contains a groundwater treatment module consisting of 4 removable steel cassettes filled with reactive media. The cassettes are supported by a permanent outer casing, which was installed in conjunction with the sheet piling walls. Each cassette includes SS mesh screen on the upgradient and downgradient sides to permit unimpeded groundwater flow. When lowered into place, the cassettes are hydraulically sealed against the outer casing by means of inflatable silicone rubber tubing. Cassette #1 contains a 25% by weight mix of zero valent iron filings and coarse silica sand, while cassette #2 contains only iron filings. Based on previous research reported by Gillham & O'Hannesin (1994), the iron filings are expected to abiotically degrade the chlorinated VOCs (CTET, PCE) in the contaminant plume.

Cassette #3 is backfilled with coarse sand and gravel-sized crushed limestone. The purpose of this material is to minimize the potential for permeability reduction further downgradient, in cassette #4 and the natural aquifer material. Possible precipitates generated by geochemical changes from the abiotic zero valent iron reactions may preferentially accumulate in cassette #3. This cassette may also minimize the impact of biofouling due to proliferation of iron bacteria, which may affect the ORC socks and/or natural aquifer material downgradient.

Cassette #4 contains suspended permeable tubes, or socks, filled with ORC® powder from Regenesis Corporation. A description of previous UW research on BTEX remediation with this product is given by Smyth et al. (1995). It is expected that this material will release sufficient dissolved oxygen (DO) to promote aerobic biodegadation of TOL in the plume. The DO may also promote aerobic biodegradation of PCE breakdown products if they emerge from the iron filings, including dichloroethene (DCE) and vinyl chloride (VC). Three fully-screened large diameter wells installed downgradient of the cassettes may also be used to provide additional DO if cassette #4 is ineffective.

Control gate. Gate 2 is an experimental control, where no effort will be made to remediate VOCs from the source wells. Contaminants will be removed at the extraction well and treated above ground. Plume migration will be monitored to quantify contaminant mass loss due to intrinsic processes such as sorption, dispersion, and biodegradation.

Nutrient flush & oxygen sparging systems. Gate 3 will test anaerobic and aerobic biodegradation in sequence. A nutrient flush system at the upgradient end of gate 3 is designed to promote anaerobic biodegradation of the chlorinated VOCs in the plume. The system is modified from Devlin & Barker (1994). A permeable wall was excavated, instrumented with 6 multilevel piezometers, 3 injection wells and 3 extraction wells, and backfilled with filter sand. Groundwater extracted from the permeable wall is amended with 300 mg/L benzoate and reinjected. For the initial nutrient flush, a 1% Modified Bushnell-Haas (MBH) solution was also included (see Mueller et al., 1991). Approximately 1 pore volume of the wall is amended with the benzoate solution during each flush, completed in approximately 6-8 hours of groundwater recirculation. Periodic short-term flushes are followed by several days to weeks of passive flow, depending on site-specific conditions.

Longitudinal dispersion during the non-pumped period should promote effective mixing of the nutrient "spike" with the contaminant plume as amended groundwater moves beyond the permeable wall. Theoretical calculations and previous field work have shown that if the pulsing cycle is repeated at the appropriate time interval, a stable microbial population can be supported at some distance downgradient from the permeable wall. A soluble carbon substrate (e.g. benzoate) can be consumed by both aerobic and anaerobic bacteria. Thus, injection of benzoate should promote aerobic activity and transform oxidized substances such as DO, nitrate and sulfate. Under the resulting anaerobic conditions, anaerobic bacteria may cause reductive dehalogenation of dissolved CTET and PCE in the contaminant plume, assuming sufficient quantities of other essential nutrients were available in the aquifer (Devlin & Barker, 1994).

The second remedial installation in Gate 3 is an oxygen biosparging gate, located 7.5 m downgradient of the nutrient flush. It is designed to promote aerobic biodegradation of TOL and possible daughter products of the reductive dechlorination of CTET and PCE. The biosparging gate contains 6 L-shaped sparging pipes and 4 piezometers, which were installed in an open excavation later backfilled with a coarse pea gravel. Lab testing showed that the sparger design provided a relatively equal oxygen distribution to all bubbling ports, and thus should provide sparging coverage across the full width of the gate.

The objective of the biosparging system is to add a residual concentration of trapped oxygen bubbles to the pea gravel but minimize volatilization, by cycling short-term oxygen addition periods (minutes) with much longer passive water flow and monitoring periods (days to weeks). Dissolution of trapped oxygen into passing groundwater should promote aerobic biodegradation reactions in situ. A surface cover and gas sampling ports located above the biosparging gate are intended to control off-gases and quantify VOC mass loss due to volatilization. Groundwater samples collected inside and downgradient of the biosparging gate will be used to evaluate system performance, by measuring DO levels, VOC concentrations, and other geochemical data. A row of fully screened pvc wells (25 cm o.d.) is installed just downgradient of the biosparging system. These wells are intended for supplementary DO supply, which may be required if VOCs persist downgradient of the biosparging gate.

DISCUSSION OF PRELIMINARY RESULTS

Some preliminary field results from CFB Borden are available for discussion as of January 1997. Further details are provided elsewhere in this proceedings by Katic et al. (1997) and Froud et al. (1997).

Tracer tests with a conservative tracer solute (KBr) were conducted to characterize groundwater flow velocities under the forced hydraulic gradient conditions imposed by pumping. Extraction pumps were set at 90 to 130 mL/min based on theoretical calculations. Dissolved KBr was injected slowly into each source well, and tracer migration downgradient was monitored temporally by periodic sampling at each piezometer fence. Although data analysis is ongoing, preliminary work indicates that groundwater velocities range from 15-20 cm/day through the gates.

Katic et al. (1997) describes progress in Gate 3. A short-term test of the biosparging gate demonstrated good areal coverage by bubble flow, indicating that most horizontal flow pathways through the gate should encounter oxygenated water. Since periodic oxygen sparging started, dissolved oxygen concentrations just downgradient of the biosparge gate have increased from 0 mg/L background levels to 20 mg/L DO. Periodic nutrient flushes began in October 1996. Decreased redox potential (Eh) downgradient of the flush zone offers preliminary evidence of enhanced bacterial activity due to the nutrient flushes.

Early data from Gate 1 are summarized in Froud et al. (1997). Dissolved oxygen concentrations from the ORC® in cassette #4 have been lower than expected, with a maximum value of about 2 mg/L. The dissolution process is apparently affected by high pH conditions produced by the iron cassettes upgradient. Alternative methods to eliminate the pH problem and/or supplement delivery of dissolved oxygen are currently under consideration.

ACKNOWLEDGEMENTS

Funding for this work was provided wholly or in part by the United States Department of Defense under Grant No. DACA 39-93-1-001, to Rice University for the Advanced Applied Technology Demonstration Facility for Environmental Technology Program (AATDF). This work does not necessarily reflect the position of these organizations and no official endorsement should be inferred.

REFERENCES

Devlin, J.F., and Barker, J.F. (1994). “A semi-passive nutrient injection scheme for enhanced in situ bioremediation”. Ground Water, 32 (3): 374-380.

Froud, S.M., Gillham, R.W., Barker, J.F., Devlin, J.F., Brown, M.J., and McMaster, M.L. (1997). “Sequential treatment using abiotic reductive dechlorination and enhanced bioremediation”. In this proceedings.

Gillham, R.W., and O’Hannesin, S.F. (1994). “Enhanced degradation of halogenated aliphatics by zero-valent iron”. Ground Water, 32(6): 958-967.

Katic, D.J., Barker, J.F., McMaster, M.L., Brown, M.J., and Devlin, J.F. (1997). “Field trial of an in-situ anaerobic/aerobic bioremediation sequence”. In this proceedings.

Mueller, J.G., Lantz, S.E., Blattman, B.O., and Chapman, P.J. (1991). “Bench-scale evaluation of alternative biological treament processes for the remediation of pentachlorophenol- and creosote-contaminated materials: solid phase remediation”. Environ. Sci. Technol., 25(6): 1045-1055.

Smyth, D.J., Byerley, B.T., Chapman, S.W., Wilson, R.D. and Mackay D.M. (1995). “Oxygen-enhanced in situ biodegradation of petroleum hydrocarbons in groundwater using a passive interception system”. Proceed., Symposium on Groundwater & Soil Remediation (GASReP), Toronto, Oct.1995.

Starr, R.C., and Cherry, J.A. (1994). “In situ remediation of contaminated groundwater: the funnel-and-gate system”. Ground Water, 32(3): 465-476.

Wilson, R.D., and Mackay, D.M. (1995). A method for passive release of solutes from an unpumped well. Ground Water, 33(6): 936-945.

ARRAYS OF UNPUMPED WELLS FOR PLUME MIGRATION CONTROL OR ENHANCED INTRINSIC REMEDIATION

Ryan D. Wilson and Douglas M. Mackay (University of Waterloo, Waterloo, Ontario)

ABSTRACT: Arrays of unpumped wells, containing either reactive media or materials or devices that release amendments, can be installed in the path of a contaminant plume to form a discontinuous permeable treatment zone. Treatment within or downgradient of this zone can be designed to either 1) completely control the plume or 2) reduce contaminant flux past some plume cross-section to a level such that intrinsic processes can further reduce concentrations to cleanup goals at the point of compliance. In either case, the degree of treatment is directly related to well spacing in the array. For complete plume remediation, the influence of the treatment array must encompass the entire plume. Using well arrays to enhance intrinsic remediation allows for relaxation of well spacing criteria. The required percentage of treatment in this case will depend on initial contaminant concentration, rate of intrinsic degradation, ground-water velocity, and distance to the compliance boundary.

INTRODUCTION

The new arena of semi-passive in situ plume migration control demands a fundamental shift in the way aquifer restoration is approached. The aggressive pumping-based techniques unsuccessfully employed in the past are confounded by aquifer heterogeneity and rate-limited transport factors of both contaminant and amendment (Mackay and Cherry, 1989; Cherry et al., 1996). Semi-passive techniques shift the focus away from forcing contaminants or amendments into treatment zones, and toward using natural ground water flow to deliver contaminant plumes to treatment zones where they can be immediately destroyed (Gillham and Burris, 1992; Robertson and Cherry, 1995) or allowed to mix with amendments and degrade downgradient (Devlin and Barker, 1994; Byerley et al., 1997).

Creating treatment zones with arrays of unpumped wells installed in the path of a contaminant plume is one concept that may be more broadly applicable than some of the other semi-passive methods recently proposed. Where the water table is deep or the aquifer is very coarse, well installation may be more practical than alternatives such as funnel-and-gate or continuous permeable walls. Wells can serve as in situ reactors or as a means to release amendments that promote biodegradation or other reactions downgradient (Wilson et al., 1997).

Objective. The objective of this study is to investigate the role that treatment by semi-passive well arrays can play in the enhancement of intrinsic remediation in cases where intrinsic processes cannot reduce concentrations to cleanup goals at the point of compliance. The goal is to establish a relationship between intrinsic degradation (average plume concentration, degradation rates, ground water velocity, and distance to the point of compliance) and the percentage of treatment required by semi-passive in situ techniques to achieve overall cleanup criteria.

TREATMENT BY ARRAYS OF WELLS

Complete Treatment. For the case of treatment by amendment release, there are two principal constraints influencing the performance of arrays of wells, both of which are sensitive to well spacing. Firstly, for complete treatment, the flux of amendment released in the array must meet or exceed the demand imposed by the contaminant. Secondly, especially where downgradient biodegradation reactions are intended, natural divergent spreading of discrete amendment plumes from individual wells must be great enough to create a zone of amendment of sufficient concentration to achieve complete lateral treatment. These relationships are examined in detail by Wilson et al. (1997), and will not be presented here. In that work it was shown that if wells are spaced on the order of 3 to 4 times their diameter, plumes of moderate concentration can be completely treated within the zone directly influenced by the array. The tight well spacing required to satisfy the above two constraints may, in some cases, make this approach either technically or financially inappropriate. However, in some cases the site conditions may be such that arrays of wells may be the best approach, regardless of the number of wells needed.

Enhanced Intrinsic Treatment. At many sites, the rate of intrinsic biodegradation may not be sufficient to reduce concentrations to cleanup levels at the compliance point (e.g. property boundary). In these cases some enhancement of intrinsic degradation will be required. Commonly, natural biodegradation is limited by a deficiency of electron acceptors in the core of the plume due either to low natural background levels or prior utilization in biodegradation reactions. Thus intrinsic biodegradation is limited by the mixing of electron acceptors from unimpacted ground water at the outer edges of the plume. Arrays of wells could be employed to passively deliver additional amendments to the plume core with the intention of this "pretreatment" being to partially, rather than completely reduce contaminant concentrations.

There are two end-members of the spectrum of intentionally partial treatment; 1) the plume can be completely captured by the well array and reduced in concentration by a uniform amount across the plume (Figure 1), and 2) the plume can be partially captured but completely treated where it is captured (Figure 2). The first end-member may be useful where there is especially high contaminant loading and/or unacceptably high risk (e.g. a UST spill close to a downgradient receptor). This high loading can be reduced by partial semi-passive treatment to concentrations that can be degraded by intrinsic processes before reaching the compliance boundary. The second end-member is likely the most attractive with respect to practicality and cost. Sufficient amendment is released from wells at

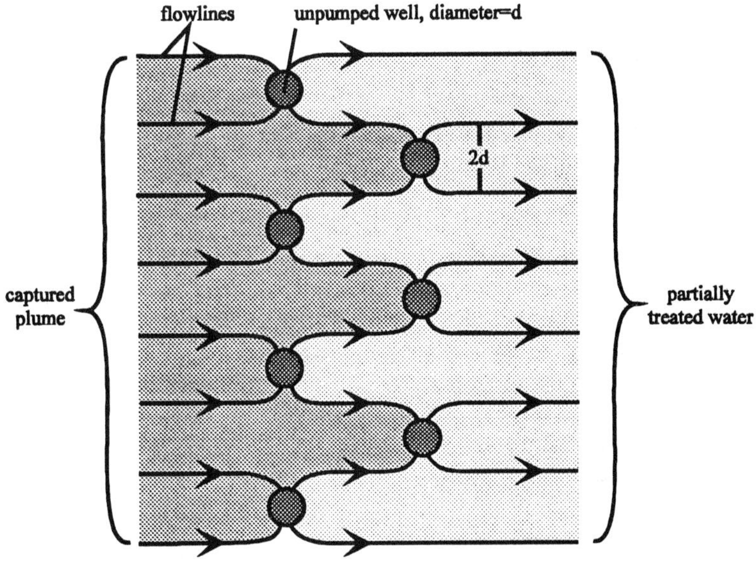

Figure 1. Array of unpumped wells showing complete capture and partial treatment. Schematic flowlines show hydraulic effects caused by wells.

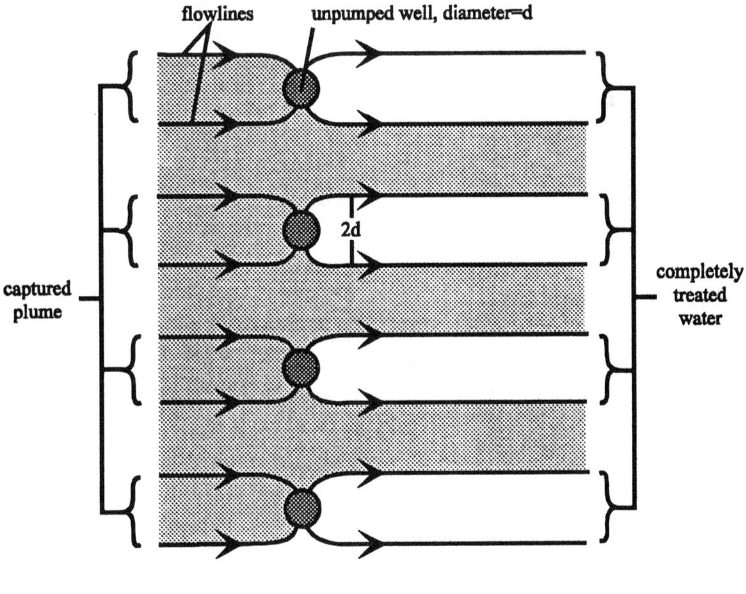

Figure 2. Array of unpumped wells showing partial capture and complete treatment of the captured fraction. Plume surface area exposed to amendment-rich ground water is effectively doubled.

wider spacing to achieve complete treatment immediately downgradient of the wells, but untreated plume passes between the capture zones of the individual wells. This "pinstripe" treatment effectively increases the surface area of the plume and enhances the subsequent intrinsic remediation by 1) maximizing exposure to (and mixing with) naturally occurring electron acceptors, and/or 2) providing excess electron acceptors for eventual mixing with the untreated pinstripes.

Degree Of Intervention. It is, of course, the goal in remediation design to achieve maximum effect with minimum intervention. In the case of arrays of wells, this means releasing the necessary amendments from the fewest number of wells. In order to estimate the required number of wells, it will be necessary to estimate the minimum percentage of plume mass flux that must be treated by the array. Total areal contaminant flux can be estimated from average plume concentration, cross sectional plume area, and ground water velocity. The percentage of this flux removed by intrinsic degradation can be estimated from the total flux and the rate of natural degradation. The difference between total flux and that which can be removed by intrinsic degradation before reaching the point of compliance is the flux that must be treated by the array. Further, dividing this flux by the average plume concentration gives the cross-sectional area that must be treated. The number of wells required can be found by dividing the lateral dimension of required treatment by the impact of a single well (assumed to be a minimum of twice the well diameter).

For example, suppose it is known or reliably estimated that intrinsic degradation processes downgradient of a particular cross-section are capable of removing only 75% of the flux from a 10 mg/l benzene plume (10 m wide by 1 m thick) before it reaches some compliance point. If the ground water velocity is 0.1 m/day, then the total plume flux is 10,000 mg/day and the intrinsically degradable flux is 7500 mg/day. Thus the well array would have to treat a flux of 2500 mg/day. Dividing this by average benzene concentration and ground water velocity gives 2.5 m^2 of cross-sectional area to be treated, which corresponds to 2.5 m of lateral width in this simple case. If 0.25 m diameter wells are used, the capture zone of each will be 0.5 m wide, and the number needed will be 5. It may be possible to use fewer wells if sufficient travel distance from treatment point to compliance is available. In that case the impact of each well would be greater due to enhanced spreading of amendments caused by lateral dispersion.

CONCLUSIONS

Intrinsic remediation is an aquifer restoration option that has gained favor recently. There are, however, some sites where intrinsic processes do not reduce the flux of contaminants at a satisfactory rate and some intervention is required. Installing arrays of wells containing passive amendment delivery devices or materials in the path of a contaminant plume is a method recently proposed to enhance intrinsic degradation. Arrays can be designed to reduce contaminant flux so that natural processes can remove the balance before reaching a compliance point. Effective design of these arrays to minimize the number of wells installed require knowledge of both plume and aquifer characteristics. Subtracting average contaminant flux removed by intrinsic degradation from the total plume flux yields the flux of plume that the well array must treat. The lower the demand on the array, the wider apart the wells can be placed and the fewer wells are needed. Sparse arrays such as this should prove more cost-effective than other methods such as permeable walls and funnel-and-gate systems.

REFERENCES

Byerley, B.T, S.W. Chapman, D.M. Mackay, and D.J.A. Smyth. 1997. "A pilot test of passive oxygen release for enhancement of in situ bioremediation of BTEX-contaminated ground water". in review *Ground Water Monitoring and Remediation.* Submitted May 1996.

Cherry, J.A., S. Feenstra, and D.M. Mackay. 1996. "Concepts for the remediation of sites contaminated with dense non-aqueous phase liquids (DNAPLS)." In J.F. Pankow and J.A. Cherry (Eds.), *Dense Chlorinated Solvents and other DNAPLs in Groundwater.* Waterloo Press, Portland, Oregon, pp. 475-506.

Devlin, J.F. and J.F. Barker. 1994. "A semipassive nutrient injection scheme for enhanced in situ bioremediation." *Ground Water.* v. 32, no. 3, pp. 374-380.

Gillham, R.W. and D.R. Burris. 1992. "In situ treatment walls - Chemical dehalogenation, denitrification, and bioaugmentation." In *Subsurface Restoration Conference, Third International Conference on Ground Water Quality Research,* pp. 66-68, Dallas, Texas, June 21-24, National Center for Ground Water Research, Rice University.

Mackay, D.M. and J.A. Cherry. 1989. "Groundwater contamination: Pump-and treat remediaton." *Environ. Sci. Tech.*, v. 23, no. 6, pp. 630-636.

Robertson, W.D. and J.A. Cherry. 1995. "In situ denitrification of septic-system nitrate using reactive porous media barriers: Field trials." *Ground Water.* v. 33, no. 1, pp. 99-111.

Wilson, R.D., D.M. Mackay, and J.A. Cherry. 1997. "Arrays of unpumped wells for plume migration control by semi-passive in situ remediation." in review *Ground Water Monitoring and Remediation.* submitted October 1996.

BIOLOGICAL REACTIVE WALL/ENHANCEMENT OF INTRINSIC CONDITIONS

Sami A. Fam, Mindi F. Messmer, Anne Lunt, and Keith Marcott
(Innovative Engineering Solutions, Inc. & Safety-Kleen Corporation)

ABSTRACT: This paper will describe a biological reactive wall which is designed to enhance intrinsic remediation at a site impacted by waste oil (dissolved, adsorbed to soil and pure phase). The groundwater at the site has elevated sulfate levels due to historical site use (up to 140,000 ppm). The elevated sulfate levels are fortuitous since sulfate reducing bacteria are capable of anaerobically degrading the hydrocarbon impacts thus oxidizing dissolved hydrocarbons while producing hydrogen sulfide. The reactive wall is designed to provide pure phase oil removal, pH adjustment (due to historical use of sulfuric acid), nutrient addition (nitrogen and phosphorous sources), and possibly enhance biosurfactant producing bacteria. Culturing of sulfate reducing, biosurfactant releasing microorganisms is currently underway.

Due to electron acceptor availability, it was determined to be most beneficial to optimize anaerobic biodegradation (vs. aerobic) within the groundwater. Optimization of anaerobic degradation is most efficient due to the abundance of sulfate (anaerobic electron acceptor) already in the groundwater and the limited water solubility of oxygen. Addition of dissolved oxygen sufficient to achieve remediation goals within reasonable engineering constraints (limited injection points) is unlikely based on the mass of impacted material.

Site data shows that in regions where inorganic nutrients, pH and sulfate levels are present in sufficient amounts, hydrocarbon degradation is underway. In samples collected from areas where conditions are not hospitable to natural degradation, treatability testing showed that naturally occurring microorganisms can degrade the dissolved hydrocarbons after pH and nutrient amendment is performed. Hydraulic data and three dimensional (3D) groundwater modeling were used to design the reactive wall and calculate the required biogeochemistry amendments. It was observed that the funnel and gate system does not lead to full plume capture and that a reactive wall spanning the length of the plume was required. Site data and laboratory treatability testing has confirmed this conceptual site model. Reactive wall construction is scheduled in 1996.

REDUCTION AND IMMOBILIZATION OF MOLYBDENUM BY *DESULFOVIBRIO DESULFURICANS*

Mark D. Tucker, (Sandia National Laboratories)
Larry L. Barton and Bruce M. Thomson (University of New Mexico)

ABSTRACT: Contamination of groundwater from abandoned uranium mill tailings piles is a serious concern in many areas of the western United States. Contaminants of concern include U, As, Cr, Se, V, Mo and other heavy metals. Since traditional groundwater remediation methods, such as pump and treat, have proven to be prohibitively expensive or ineffective, attention has turned to alternative methods such as permeable biobarriers for restoration of these sites. Previous research has shown that *Desulfovibrio desulfuricans* and other sulfate-reducing bacteria may potentially be used in a biobarrier because they have the ability to reduce and precipitate the heavy metals U, Se, and Cr from solution by enzymatic mediated reactions. However, the ability of these bacteria to reduce many other heavy metals is not known. This research reports the enzymatic reduction of Mo(VI) to Mo(IV) by sulfate-reducing bacteria. Mo(VI) was reduced to Mo(IV) when washed cells of *D. desulfuricans* were suspended in bicarbonate buffer solution with either lactate or H_2 as the electron donor and Mo(VI) as the electron acceptor. Mo(VI) reduction by *D. desulfuricans* in the presence of sulfide resulted in the extracellular precipitation of the mineral molybdenite [$MoS_{2(s)}$]. Mo(VI) reduction did not occur in the absence of an electron donor or in the presence of heat-killed cells of *D. desulfuricans*. Attempts to grow *D. desulfuricans* with Mo(VI) as the sole electron donor were unsuccessful. Direct chemical reduction of Mo(VI) by sulfide or by H_2 was also unsuccessful, even when heat-killed cells of *D. desulfuricans* were added to provide a potential catalytic surface for the nonenzymatic reaction. *D. vulgaris* reduced Mo(VI) equally as well. The results indicate that enzymatic reduction of Mo(VI) by sulfate-reducing bacteria may contribute to the accumulation of Mo(IV) in anaerobic environments and that these organisms may be useful for removing soluble Mo from contaminated groundwater in a permeable biobarrier.

GASOLINE BIODEGRADATION IN A PERMEABLE BIOREACTIVE BARRIER

Laleh Yerushalmi, Michelle F. Manuel and Serge R. Guiot
(Biotechnology Research Institute, Montreal, Canada)

ABSTRACT: Aerobic biodegradation of gasoline in a packed-bed biobarrier during batch and continuous modes of operation was investigated in this work. A gasoline-degrading microbial consortium, originally isolated from soil was used to inoculate the biobarrier. Granulated peat moss and protruded stainless steel packings were used as the support materiel for the immobilization of microbial cells. Gasoline removal efficiency of the biobarriers was 100% during the repeated batch operation while ranging from 94% to 100% during the continuous operation. During the repeated batch operation the highest overall gasoline elimination rates were 37.7 mg/l.d and 36.5 mg/l.d obtained at initial gasoline concentrations of 70.3 and 55.5 mg/l for the stainless steel and peat moss-filled biobarriers, respectively. During the continuous operation the elimination rates of the biobarriers ranged from 3.2 to 22.9 mg/l.d and 4.0 to 13.9 mg/l.d for the stainless steel and the peat moss-filled biobarriers, respectively. These results were obtained with the hydraulic retention time (HRT) varying from 17 hours to 6 days, corresponding to liquid velocities of 4 to 35 cm/d, and initial gasoline concentrations of 11.1 to 74.0 mg/l.

INTRODUCTION

This paper discusses the development and performance of a permeable bioreactive barrier (biobarrier) for the remediation of soil and groundwater contaminated with gasoline or its hydrocarbon constituents. Gasoline is among the most common contaminants of soil and groundwater around the world. The toxic and water soluble fraction of gasoline can often mix with drinking water and cause serious health concerns. Natural attenuation processes have been shown to reduce the concentration of gasoline as well as other hazardous organic compounds (Shailubhai, 1986). However, these processes are generally very slow and may take up to several years to completely degrade the contaminants.

Microbial decontamination or bioremediation of soil and groundwater by an in situ biobarrier is a promising and viable alternative. In situ bioremediation techniques are becoming increasingly attractive because they eliminate the need to remove the affected media from the site. This technique has been applied in the remediation of groundwaters contaminated with petroleum hydrocarbons and chlorinated compounds (Hutchins et al., 1995; Thomson et al., 1995).

The biobarrier developed in this work utilizes a fixed support material to immobilize the microbial cells. The packed-bed bioreactor is intended to be used as an in situ device to treat the gasoline-contaminated soil and groundwater. The immobilized microbial cells inside the biobarrier degrade the hazardous organic compounds in the groundwater as the water flows through the system. The kinetics of gasoline biodegradation by the isolated microbial culture has previously been discussed (Yerushalmi et. al., 1996).

MATERIALS AND METHODS

Isolation and Identification of the Microbial Culture. The indigenous microbial culture was isolated from the top layers of a gasoline-contaminated soil sample by enrichment techniques. Commercial gasoline with a concentration of 100 ppm obtained from a Shell gasoline station was used as the source of carbon and energy for the microorganisms.

Medium. A minimal salts medium (MSM) with the following composition was used for culture enrichment as well as the biobarriers' feed, in (g/l): $KH_2\ PO_4$, 0.87; $K_2H\ PO_4$, 2.26; $(NH_4)_2\ SO_4$, 1.1; and $Mg\ SO_4.7H_2O$, 0.097. To this solution was added 1 ml (per liter) of a trace metals solution as discussed before (Yerushalmi et al., 1996). The final pH of the medium was 6.9-7.1. The medium was sterilized by autoclaving at 120° C for 20 minutes.

Cell Immobilization. Microbial cells were immobilized on a support material inside the biobarriers. Two different supports were used in this investigation, protruded stainless steel (Cannon Instrument Company, Pa., USA) and granulated peat moss (Produits Recyclable Bioforet, Quebec, Canada). The Pro-Pak S.S. packings were 0.6 x 0.6 cm in size and had an average particle density of 5.13 g/ml with a free space of 93%. The granulated peat moss had an average particle density of 0.64 g/ml and a free space of 57%. The packings were soaked in MSM medium and sterilized (autoclave at 120° C for 20 minutes) before use.

Biobarrier Operation. The biobarriers consisted of a stainless steel body with a rectangular cross sectional area and a total volume of 5.2 liters (Figure 1). They were initially operated in batch mode in order to promote the attachment of microbial cells to the support material and to develop a cellular biofilm. Gasoline was injected into the system after its depletion in the liquid phase. The feed was aerated before its introduction in the biobarriers in order to provide oxygen for the aerobic biodegradation process. The stainless steel and the peat moss-filled biobarriers were initially 31% and 38% filled, respectively. The liquid volume was gradually increased to 95% of the total reactor volume. Mixing was provided by recirculating the liquid at a rate of 14 to 30 liters per day depending on the liquid content of the biobarriers. Gasoline removal was evaluated by analyzing gas-phase samples taken from a side tower located on the outlet port of the biobarriers. The gas phase in the side tower was in equilibrium with the liquid phase inside the biobarrier.

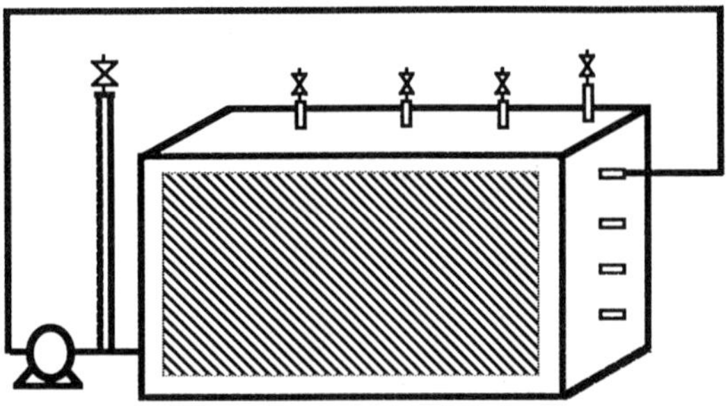

FIGURE 1. Schematic diagram of the fixed-bed biobarrier during repeated batch operations.

The setup for the continuous operation of biobarriers is schematically presented in Figure 2. The feed (MSM) was continuously agitated and sparged with sterile air in order to provide oxygen for the process. A four-channel peristaltic pump model Gilson minipuls 2 (Gilson Medical Electronics, WI, USA) was used to deliver feed to the four inlet ports on the side of the biobarriers. Two peristaltic pumps model Masterflex (Cole Palmer Instrument Company, Illinois, USA) were used to introduce gasoline solution consisting of 370 mg/l gasoline in MSM medium into the biobarriers.

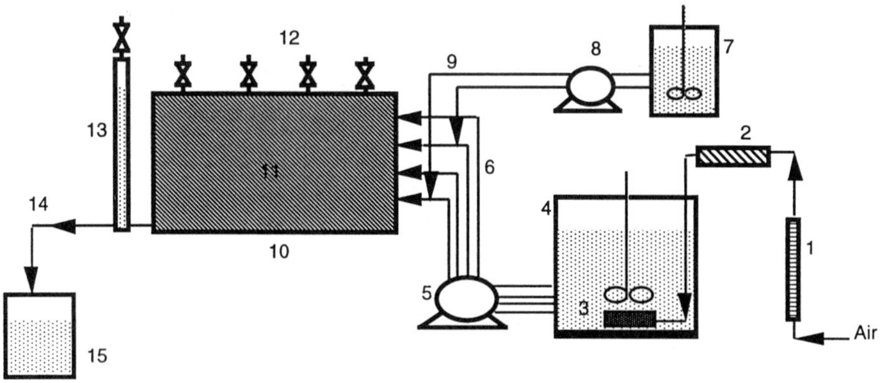

FIGURE 2. Schematic diagram of the biobarrier assembly during continuous operation. 1-Rotameter, 2-Air filter, 3-Air diffuser, 4-Feed tank, 5- Feed pump, 6-Feed lines, 7-Gasoline solution tank, 8-Pump, 9-Gasoline lines, 10-Biobarrier, 11-Packing, 12-Top ports, 13-Side tower, 14-Effluent line, 15-Effluent tank.

The efficiency of biobarriers in the removal of gasoline during the continuous mode of operation was evaluated according to their removal efficiency (RE) and elimination rate (ER) as defined by the following equations:

$$\text{Removal Efficiency (RE)} = (S_i - S_e) * 100 / S_i \quad 1)$$

$$\text{Elimination Rate (ER)} = Q (S_i - S_e)/V_f \quad 2)$$

where S_i is the inlet concentration of compound (mg/l), S_e is the outlet concentration of compound (mg/l), Q is the inlet feed flow rate (l/h) and V_f is the occupied volume of the biobarriers (l).

Analytical Techniques. Gasoline was analyzed by gas chromatography using a Perkin-Elmer, model Sigma 2000 gas chromatograph equipped with a FID detector (250° C). Separation was done on a 1/8", S.S. column (1% SP 1000) packed with 60/80 mesh Carbopak B with temperature programming (50° C, 10° C/min up to 210° C, 210° C for 20 min). Carrier gas was nitrogen at a flow rate of 65 ml/min.

RESULTS AND DISCUSSION

The microbial culture was previously shown to completely degrade gasoline during batch cultivation by free cell suspensions (Yerushalmi et al., 1996). A similar pattern was observed with the immobilized cells during the batch and continuous operations. The dynamics of benzene degradation as a representative of gasoline consumption during the repeated batch operation of the biobarriers is shown in Figure 3.

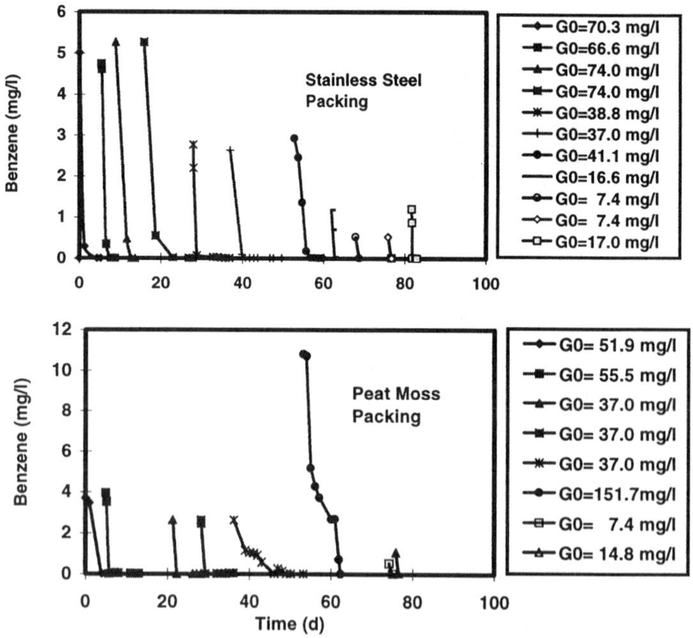

FIGURE 3. Benzene removal in the fixed-bed biobarriers during repeated batch operation.

Initial gasoline concentrations in the liquid phase ranged from 7.4 to 74.0 mg/l and 7.4 to 55.5 mg/l for the stainless steel and the peat moss-filled biobarriers, respectively. Gasoline was completely removed providing a 100% removal efficiency with both of the biobarriers under all the operating conditions examined. The overall gasoline elimination rates based on the occupied volume of the biobarriers ranged from 8.2 to 37.7 mg/l.d and 2.6 to 36.5 mg/l.d for the stainless steel and the peat-moss-filled biobarriers, respectively.

The continuous operation of the biobarriers is currently under investigation. In the peat moss-filled biobarrier the initial gasoline concentration changed from 22.2 to 74.0 mg/l while the hydraulic retention time (HRT) varied from 3 to 6 days, corresponding to liquid velocities of 4 to 8 cm/d. The overall gasoline elimination rate changed from 4.0 to 13.9 mg/l.d resulting in gasoline removal efficiencies of 95 to 100%. The stainless steel-filled biobarrier exhibited 94 to 100% gasoline removal efficiencies with gasoline concentrations of 11.1 to 74.0 mg/l and hydraulic retention times of 17 hours to 6 days, corresponding to liquid velocities of 4 to 35 cm/d. Under these conditions gasoline elimination rate changed from 3.2 to 22.9 mg/l.d. An example of a typical output of the biobarriers in the continuous culture as demonstrated by chromatographic analysis of the reactors' gas phase is depicted in Figures 4 and 5 for the two biobarriers. The operating conditions as well as the resulting removal efficiencies and the elimination rates of the two biobarriers are presented in Table 1.

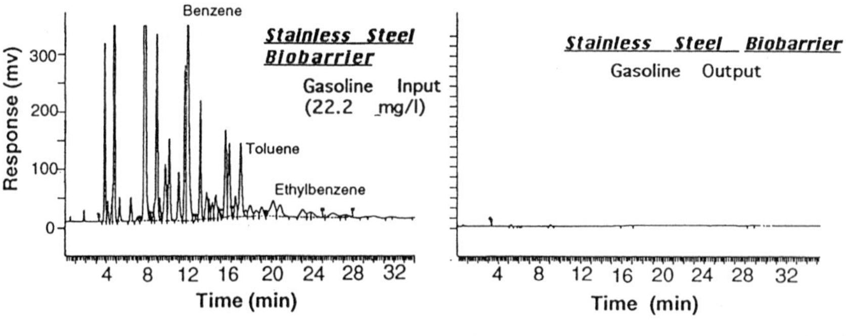

FIGURE 4. Chromatographic analysis of gasoline input and output in the stainless steel-filled biobarrier.

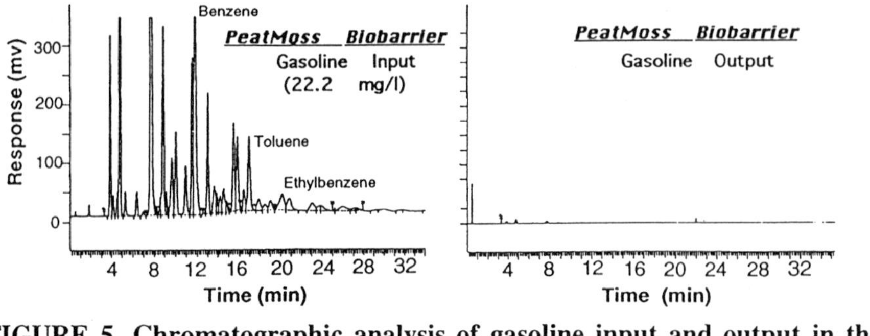

FIGURE 5. Chromatographic analysis of gasoline input and output in the peat moss-filled biobarrier.

TABLE 1. Performance of the biobarriers during continuous operation.

Packing	Hydraulic Retention Time (HRT) (days)	Initial Gasoline Concentration (mg/l)	Gasoline Removal Efficiency (%)	Elimination Rate (mg/l.d)
Peat moss	3	22.2	95	4.0
Peat moss	3	44.4	99	8.3
Peat moss	3	74.0	99	13.9
Peat moss	6	44.4	100	4.2
Stainless steel	0.7	14.8	100	19.4
Stainless steel	1	14.8	100	13.8
Stainless steel	3	11.1	94	3.2
Stainless steel	3	22.2	100	6.9
Stainless steel	3	44.4	100	13.8
Stainless steel	3	74.0	100	22.9
Stainless steel	6	44.4	100	6.9

CONCLUSIONS

Both biobarrier packings have supported almost complete biodegradation of gasoline under the conditions examined. So far, the highest gasoline elimination rates have been 22.9 and 13.9 mg/l.d obtained with the stainless steel and the peat moss-filled biobarriers, respectively. Research is currently in progress to determine the gasoline removal capacities of the biobarriers at higher gasoline concentrations as well as lower hydraulic retention times. Future work will focus on the use of the biobarriers for in situ bioremediation processes.

REFERENCES

Hutchins,S. R., J. T. Wilson, and D.H. Kampbell. 1995. "In Situ Bioremediation of a Pipeline Spill Using Nitrate as the Electron Acceptor." In R. E. Hinchee, J. A. Kittel and H. J. Reisinger (Eds.), *Applied Bioremediation of Petroleum Hydrocarbons*, pp.143-153. Battelle Press, Columbus.

Shailubhai, K. 1986. "Treatment of Petroleum Industry Oil Sludge in Soil.", *Trends in Biotechnol.* 4, 202-206.

Thomson, J. A. M., M. J. Day, R. L. Sloan, and M. L. Collins. 1995. "In Situ aquifer Bioremediation at the French Limited Superfund Site." In R.E. Hinchee, J. A. Kittel and H. J. Reisinger (Eds.), *Applied Bioremediation of Petroleum Hydrocarbons*, pp.453-459. Battelle Press, Columbus.

Yerushalmi, L.; R. Halko; M. F. Manuel and S. R. Guiot. 1996. "Kinetics of Gasoline Biodegradation by Free and immobilized Cells." In C. E. Delisle and M. A. Bouchard (Eds.), *19th International symposium on Wastewater Treatment,* pp. 83-93. Montreal, Canada, November 19-21.

IN-SITU BIOSCREENS

Huub H.M. Rijnaarts, A. Brunia and M. van Aalst (TNO Institute of Environmental Sciences, Energy Research and Process Innovation, Department of Environmental Biotechnology, Apeldoorn, The Netherlands)

ABSTRACT: In-situ bioscreens represent a new and emerging technology for in-situ isolation and remediation of contaminated sites. An in-situ bioscreen is a local zone in a natural porous medium such as soil that has a high contaminant retention capacity (isolation) and an increased biodegradation activity of hazardous organics and/or the immobilisation of dissolved heavy metals. Thus, contaminants are removed from groundwater flowing through such a bioscreen. Laboratory bench scale tests with bioscreens made from activated carbon particles with attached microbes supplied with various electron donors/acceptors are being studied. In one system oxygen (and/or nitrate) is used to degrade oil/BTEX and other mobile aromatic compounds. The data show that i) bioconversion efficiencies can be greatly increased by combining oxygen with nitrate, and ii) hydraulic conductivity performance is at best at biocarrier particle sizes between 1 and 3 mm. Test results with electrochemical and other redox supply systems (electron donors/acceptors) in bioscreens, degrading chlorinated solvents and hexachlorocyclohexanes, show good prospects for application of bioscreens in funnel and gate and other remediation technologies.

INTRODUCTION

In-situ activated bioscreens belong to the group of permeable reactive barriers and represent a new, innovative and emerging technology for long-term in-situ bioremediation of contaminated sites. An in-situ bioscreen is a local zone in a natural soil porous medium that has an increased activity towards the biodegradation of hazardous organics combined with a high capacity for immobilisation and retention of the contaminants. Thus, contaminants are removed from groundwater flowing through such a screen (remediation) , under the influence of natural hydraulic potential gradient. The down-stream soils, aquifers and other possible receptors are so protected against the pollutants. TNO investigates the feasibility, the process optimisation, and the in-situ application of bioscreen technology. Reactive barriers based on complete biological processes or on a combination of physico-chemical and biological processes, especially for chlorinated compounds and oil-related and aromatic compounds, are currently being studied at our laboratory and at various test-sites. Especially a combination of sorption and biodegradation by using sorptive biocarrier particles appears to be beneficial. In this presentation the following topics will be addressed: i) a short overview of possible types of bioscreens, ii) a bioscreen of activated carbon particles with attached microbes, iii) bioscreens achieved by in-situ fouling on soil or aquifer particles, and iv) additional aspects of bioscreen optimisation.

TYPES OF BIOSCREENS

Bioscreens can vary in hydrogeological application, redox-status, and type of biocarrier material.

Hydrogeological application. A bioscreen may form a complete shield around a contaminated site (Figure 1A) or it may be a small part of the isolation through which the contaminated groundwater plume is channelled (Figure 1B).

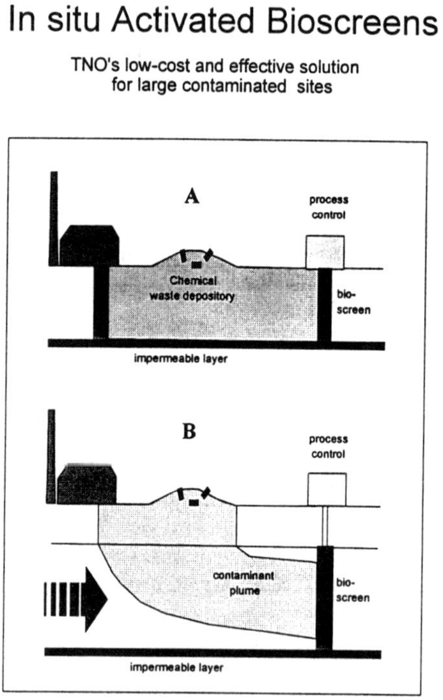

FIGURE 1. Different hydrogeological applications of in-situ bioscreens.

Redox-status. Bioscreens are further characterised by their redox-status and oxygenation level required for the biodegradation of various hazardous organic compounds (Bouwer and Zehnder, 1993). In general, higher chlorinated compounds (PCB's, chlorinated benzenes, PER, TRI, chlorinated methanes and ethanes, β- and δ- hexachlorocyclohexane) can only be dechlorinated and degraded under anaerobic conditions. Often, this anaerobic phase must be followed by an aerobic step to make the bioconversion complete. On the other hand, completely aerobic bioscreens are effective for non-chlorinated contaminants (such as oil, BTEX, and PAH) and for some chlorinated contaminants (some lower chlorinated aliphatics and lower chlorinated aromatics, α- and γ-hexachlorocyclohexane). Bioscreens may also be used for immobilisation of heavy metals. This requires sulphate reducing conditions and pH 7.5. Hence, the most widely applicable bioscreens would combine anaerobic and aerobic conditions.

A recent development is the use of electron-acceptors with a redox-potential in the upper range of the redox-spectrum as an alternative for oxygen. The advantage is that many problems associated with in-situ aeration (e.g. compound volatilisation, pore-clogging due to iron-oxide formation) can be avoided. Electron-acceptors like Fe(III), manganese-oxide, and nitrate, appear to be well suited for application in bioscreens.

Type of biocarrier. Excavation of a part of the soil and replacement with particles coated with microorganisms is likely to be a cost-effective method to construct a bioscreen in field cases where the contamination has not penetrated the subsurface to great depth, i.e., less than 10-15 m. Coarse particle porous media with a strong capacity for pollutant adsorption and microbial attachment, such as activated carbon or sand mixed with compost are suitable to construct the in-situ bioscreen. These materials may be either pre-coated or coated under in-situ conditions with appropriate microorganisms to obtain the required biodegrading activity in the bioscreen. In many cases, excavation will not be feasible (i.e., the example in Figure 1B). Then, an activated bioscreen may be created by stimulating the indigenous microbial populations to adapt to a new and more suitable redox situation and to develop the appropriate contaminant degrading activity. However, the required adaptation times may be long. For example, the development of a complete PER and TRI dechlorinating activity under anaerobic conditions in laboratory soil columns requires about one to two years (De Bruin et al., 1992). Under in-situ conditions this may be even longer. Start-up periods due to adaptation may be strongly reduced by injection of non-indigenous microorganisms with the appropriate attachment and compound-degrading properties. Moreover, a more stabile and predictable bioscreen process-performance may be achieved in this way.

A BIOSCREEN OF ACTIVATED CARBON COATED WITH MICROBES

The feasibility of an aerobic BTEX degrading bioscreen is currently tested and the results will be presented. Laboratory experiments are performed with BTEX-degrading pure cultures (*Rhodococcus* sp. C125, *Rhodococcus erythropolis* A177, *Pseudomonas putida* mt2 (Rijnaarts et al., 1993b), activated sludge/BTEX enrichment cultures, mm-sized activated carbon particles, mixtures of compost and coarse sand, and toluene as the model BTEX compound. The following processes are studied:

Microbial attachment. Attachment of the microorganisms on the activated carbon and sand/compost carriers is studied using a previously developed column test system (Rijnaarts et al., 1993a; Rijnaarts et al., 1993b). Bacterial suspensions were injected in pulses into the columns in order to obtain uniform biocoating profiles. Bacterial attachment reached a level of up to 50 % coverage of the external surface area of the carrier particles.

BTEX-adsorption. The adsorption characteristic of toluene onto the activated carbon was provided by the company NORIT. The adsorption of toluene onto sand compost mixtures was estimated using the partitioning theory, the log K_{OC} value of toluene from literature, and the organic carbon content of the mixture. From these results the BTEX (toluene) adsorptive capacity of the bioscreen can be calculated. The calculations show that a sorptive screen with a width of one m. can intercept a BTEX plume for a period of months (compost) to many years (activated carbon).

BTEX biodegrading activity. Toluene bioconversion by the microorganisms attached onto the activated carbon carriers was tested by measuring oxygen-consumption and toluene removal using column tests. From these results the potential BTEX biodegrading capacity was calculated from the amount of oxygen supplied. For BTEX concentrations >3 mg/l, multi-aeration-points and/or supplementary electron acceptors will be required.

Permeability performance. The hydraulic permeability as a function of bioscreen operation time was studied at laboratory scale in columns. The hydraulic conductivity of the biofilter is important for the appropriate in-situ functioning of the bioscreen. The results indicate that a pre-coating of biocarrier particles with non-clogging BTEX degrading microorganisms prevents extensive biofilm growth and pore clogging.

Overall evaluation. The practical feasibility of BTEX sorption-degrading bioscreen depends on the technical possibilities, needs, and limitations associated with the in-situ implementation at contaminated sites. It may be necessary to develop special devices for the in-situ application of a bioscreen and for the supply of oxygen and supplementary electron-acceptors.

BIOSCREENS ON NATURAL GRAINS BY IN-SITU BIOFOULING

Bioscreens that do not involve excavation are created either by stimulation of a part of the indigenous microbial population or by injection of non-indigenous microbes that are specialised in the biodegradation of certain contaminants. The following factors influence microbial attachment and transport in groundwater: the type of polymers on the exterior of bacterial cells (cell-coating), the ionic strength of the groundwater, and the surface chemistry and size of the porous medium particles (Lindqvist and Enfield, 1992; Rijnaarts et al., 1993; Rijnaarts, 1994; Rijnaarts et al., 1995a; Rijnaarts et al., 1995b; Rijnaarts et al., 1996a; Rijnaarts et al., 1996b). These factors control cell-solid interactions which determine the initial transport behaviour of a microbial cell suspension injected into a porous medium. In Figure 2 this is illustrated in terms of initial penetration depths as before (Rijnaarts et al., 1996a; Rijnaarts et al., 1996b). The parameters mentioned above also influence cell-cell interactions, and thereby the tendency of bacterial cells to attach to each other, to clog the pores and to create a stabile biofilm (Cunningham et al., 1990; Rijnaarts, 1994; Rijnaarts et al., 1995a, Rijnaarts et al., 1995b). As a consequence, these factors strongly affect the porous medium properties after prolonged injection and during bioscreen process operation. Hence, determination of the adhesion and biofilm-forming characteristics of microbial populations prior to their employment in an in-situ bioscreen will be essential for a successful application of such a technique.

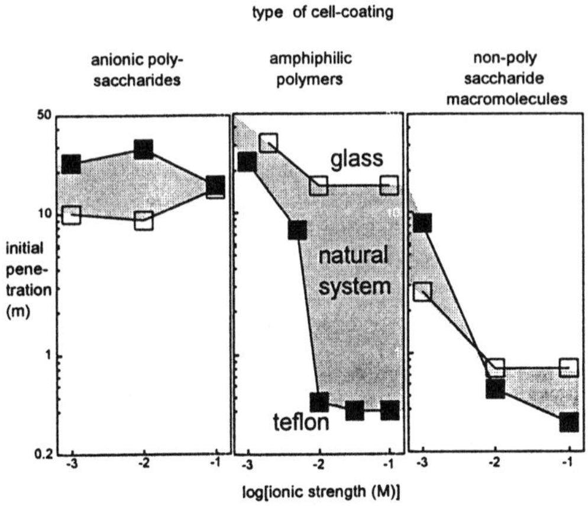

FIGURE 2. Initial penetration depths of injection of bacteria into porous media (depth at which the influent cell concentration is reduced by a factor 100) as a function of ionic strength for various types of cell coating (Rijnaarts et al., 1996a; Rijnaarts et al., 1996b).

ADDITIONAL ASPECTS OF BIOSCREEN OPTIMIZATION.

One of the problems any in-situ treatment has to deal with is a contaminant concentration that is variable: generally, this concentration is initially high and after prolonged operation low. A bioscreen must be capable to function at all concentration regimes. During the initial phase, the concentrations may exceed the toxicity limits for microorganisms. Adsorbents like activated carbon may help to prevent such toxic effects (Rijnaarts et al., 1993). In the lower concentration regime, the biomass in the screen may decay which may result in a less efficient contaminant removal. Another problem may be a clogging of the porous medium, by the biofilm, which may create preferential flow paths and a subsequent reduction in the contaminant removal efficiency of the bioscreen. Hence, further insight into biofilm growth and the prevention of bioclogging during bioscreen operation is essential for a further optimisation of these in-situ isolation and remediation systems.

ACKNOWLEDGEMENTS

This research was partially funded by a grant of NOVEM (Nederlandse Onderneming Voor Energie en Milieu BV) within the framework of the Environmental Technology Programme 1994, The Netherlands.

REFERENCES

Bouwer, E. J. and A. J. B. Zehnder. 1993. "Bioremediation of organic compounds-putting microbial metabolism to work." *Trends in Biotechnology 11*: 360-367.

Cunningham, A. B., E. J. Bouwer and W. G. Characklis. 1990. "Biofilms." In W. G. Characklis and K. C. Marshall (Eds.), *Biofilms in porous media*, pp. 697-732. John Wiley & Sons Inc., New York.

De Bruin, W. P., M. J. J. Kotterman, M. A. Posthumus, G. Schraa and A. J. B. Zehnder. 1992. "Complete Biological Reductive Transformation of Tetrachloroethene to Ethane." *Appl. Environ. Microbiol. 58*: 1996-2000.

Lindqvist, R. and C. G. Enfield. 1992. "Cell Density and Non-equilibrium sorption effects on bacterial dispersal in groundwater microcosms." *Microb. Ecol. 24*: 25-41.

Rijnaarts, H.H.M., W. Norde, J. Lyklema, and A. J. B. Zehnder. 1993a. "Effect of substrate adsorption and bacterial adhesion on Bacterial growth and Activity." In *Soil decontamination using biological processes*, pp. 155-160. DECHEMA, Frankfurt am Main, Germany.

Rijnaarts, H. H. M., W. Norde, E. J. Bouwer, J. Lyklema and A. J. B. Zehnder. 1993b. "Bacterial adhesion under static and dynamic conditions." *Appl. Environ. Microbiol. 59*: 3255-3265.

Rijnaarts, H. M. M. 1994. "Interactions between Bacteria and Solid Surfaces in Relation to Bacterial Transport in Porous Media." Ph.D. Thesis, Wageningen Agricultural University, The Netherlands.

Rijnaarts, H. H. M., W. Norde, E. J. Bouwer, J. Lyklema and A. J. B. Zehnder. 1995a. "Reversibility and mechanism of bacterial adhesion." *Colloids Surf. B: Biointerf. 4*:5-22.

Rijnaarts, H. H. M., W. Norde, J. Lyklema and A. J. B. Zehnder. 1995b. "The isoelectric point of bacteria as an indicator for the presence of cell surface polymers that inhibit adhesion." *Colloids Surf. B: Biointerf. 4*:191-197.

Rijnaarts, H. H. M., E. J. Bouwer, W. Norde, J. Lyklema and A. J. B. Zehnder. 1996a. "Bacterial deposition in porous media related to the clean-bed collision efficiency and substratum-blocking by attached cells." *Environ. Sci. Technol. 30*(10): 2869-2876.

Rijnaarts, H. H. M., E. J. Bouwer, W. Norde, J. Lyklema and A. J. B. Zehnder. 1996b. "Bacterial deposition in porous media: effects of type of cell-coating, substratum hydrophobicity and ionic strength." *Environ. Sci. Technol. 30*(10): 2877-2883.

SEMI-PASSIVE OXYGEN RELEASE BARRIER FOR ENHANCEMENT OF INTRINSIC BIOREMEDIATION

Steven W. Chapman, Brian T. Byerley, David J. Smyth, Ryan D. Wilson and Douglas M. Mackay (University of Waterloo, Waterloo, Ontario, Canada)

ABSTRACT: A pilot scale field demonstration of the use of Oxygen-Releasing Compound (ORC™) was conducted at the site of a former gasoline service station facility. ORC was installed into a barrier consisting of a tight pattern of "treatment wells" placed in two staggered rows across a trial segment of the petroleum hydrocarbon plume relatively near the apparent source of contamination. Monitoring using fencelines of multilevel wells up and down gradient of the barrier indicated significant reductions in BTEX mass flux through the zone impacted by the barrier. Influent BTEX concentrations were highly variable with maximum concentrations exceeding 60 mg/L. Observed reductions in average BTEX concentrations were approximately 70% at 51 days after ORC installation and decreased thereafter, apparently due to decreasing oxygen release from the ORC product. Along flowpaths with average upgradient BTEX concentrations below about 5 mg/L, near complete BTEX treatment was achieved and "excess" dissolved oxygen was observed which would be expected to be available for further BTEX degradation downgradient. Mass flux estimates of BTEX and dissolved oxygen indicated that less than 10% of the oxygen released from the treatment wells contributed to BTEX degradation or was observed as "excess" oxygen. Characterization of other oxygen demands indicated that non-BTEX components of the organic contamination were significant and account for a significant part of this discrepancy.

INTRODUCTION

Generally in plumes of petroleum hydrocarbon contamination, the oxygen demand imposed by biodegradation of BTEX and other compounds exceeds the dissolved oxygen available creating anaerobic conditions within the plume core. Since biodegradation of these compounds is generally faster under aerobic conditions, rates of intrinsic biodegradation will decrease since the rate will be dominated by the slower anaerobic transformations within the core. The faster aerobic transformations will occur only at the plume periphery. Depending on the mass of contaminants released and the site conditions, a plume of hydrocarbon contamination may still migrate unacceptably long distances from the source.

Introduction of oxygen into the core of a hydrocarbon plume will allow aerobic biodegradation of the contaminants to occur to the extent allowed by the additional input of oxygen provided additional sinks for oxygen are also overcome. This will reduce the flux of contaminant mass past the zone where the stimulated biodegradation occurs and may result in a reduction in the distance the plume migrates at concentrations that pose a risk.

A pilot scale field trial was conducted to investigate enhancement of intrinsic biodegradation of BTEX contaminated groundwater by the release of oxygen from unpumped "treatment" wells containing a commercially available solid oxygen releasing product. The following is an overview of the trial and results. More detail are provided by Byerley et al. (1996).

Site Description. The site is a former gas service station located in south-western Ontario. Underground gasoline storage tanks were removed from the site in 1992 and a remediation program undertaken which included excavation of contaminated soils and the operation of a vapor extraction system. However residual gasoline apparently remains in the subsurface and a shallow plume of contaminated groundwater continues to emanate from the site. The pilot test was located in a grass covered area at the site approximately 20 to 30 m downgradient from the apparent source of contamination. The site is located over an unconfined fine to medium sand aquifer with the water table approximately 5 m below ground surface.

Site Characterization. Site characterization efforts included monitoring water table elevations and gradient fluctuations, and hydraulic conductivity estimates using laboratory permeameter tests on cores and short duration pumping tests in several 5 cm monitoring wells in the vicinity of the trial. The water table was noted to vary by about 0.4 m with gradient fluctuations of about 20°, apparently caused by seasonally variable aquifer recharge. Groundwater flow velocities are estimated to range from 10 to 20 cm/day based on the hydraulic conductivity, gradient and porosity data which compare well with results from a tracer test (Byerley et al., 1996).

EXPERIMENTAL DESIGN

The trial was designed to achieve significant degradation of BTEX across a segment of the plume and to provide detailed monitoring data to quantify the results as clearly as possible.

Treatment Barrier. The barrier consisted of seven 20 cm diameter "treatment wells" located in two staggered rows on 0.6 m centers with the well screen extending approximately 1.5 m above and below the water table (Figure 1). The close spacing was used to ensure that groundwater flowing through the trial portion of the plume would be completely intercepted by the treatment wells.

The oxygen source used was ORC™ (Oxygen Releasing Compound, Regenesis Bioremediation Products), which releases oxygen upon contact with water according to the following overall reaction:

$$MgO_2 + H_2O \rightarrow 1/2\ O_2 + Mg(OH)_2$$

The ORC was supplied in filter socks which were about 16.5 cm in diameter and 38 cm in length, each containing approximately 13.5 kg of a 50% mixture of ORC and silica sand. Sets of four ORC socks were enclosed in cylindrical cages of plastic fence meshing and lowered by rope into each of the treatment wells. Assays of the ORC-sand mixture indicated an initial total available (releasable) oxygen content of about 3.4% of the total mass.

Monitoring Installations. Thirty-seven multilevel monitoring wells were installed on 0.3 m centers in three fencelines perpendicular to the initially estimated groundwater flow direction (Figure 1). The multilevels were constructed of six 3.2 mm stainless steel sampling tubes attached to the outside of a 2 cm diameter PVC centerstock. A vertical spacing of 0.15 m was used between sampling points and the multilevels were installed to cover the 0.75 m vertical interval immediately below the water table. Fence 1 was located approximately 0.6 m upgradient from the centerline of the most upgradient row of treatment wells, Fences 2 and 3 were located at 0.6 m and 4.4 m downgradient from the centerline of the most downgradient row of treatment wells, respectively. Additional monitoring included nine conventional 5 cm diameter PVC monitoring wells along with monitoring points placed within several of the treatment wells to evaluate the performance of the ORC.

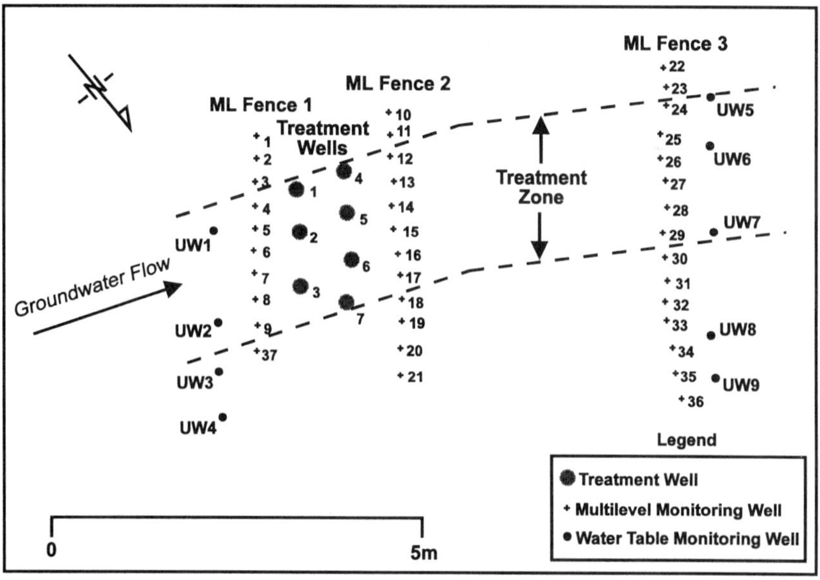

FIGURE 1. Plan view of field trial setup.

Groundwater Sampling and Analysis. Monitoring was carried out in four major sampling "snapshots" over about a 4.5 month period following installation of the ORC on April 19, 1995 (Day 0). During these snapshots the fences were sampled for BTEX and dissolved oxygen (DO). Fence 3 was only sampled during the final 2 snapshots. DO levels in the treatment wells were regularly monitored during the trial and to 6 months after ORC installation Both during and after the trial, samples were collected for analysis of other parameters to allow a more thorough interpretation of results. These included inorganic parameters (such as Fe, Mn), total organic carbon (TOC), total petroleum hydrocarbons (TPH), chemical oxygen demand (COD) and biological oxygen demand (BOD). BTEX analyses were performed by GC/FID methods using a GC equipped with an automatic headspace sampler. BTEX samples were collected using a manifold capable of sampling 12 multilevel points at once. DO was measured in the field using electrodes in flow-through cells.

RESULTS AND DISCUSSION

Summary of Monitoring Results. Following installation of the ORC socks, DO concentrations rose from <0.5 mg/L to greater than 16 mg/L (upper limit of DO probe) in the treatment wells and generally remained above 16 mg/L up to Day 80 and declined thereafter. After 182 days DO levels ranged from 1.4 to 4.3 mg/L in the treatment wells. Elevated DO levels were observed at Fence 2 on Day 36 and coincided with decreased BTEX concentrations. The first major sampling snapshot occurred 51 days after ORC installation when it was presumed that the full effect of the treatment wells propagated beyond Fence 2. Fence plots showing total BTEX and DO contours for the first snapshot are shown in Figure 2. The plume was highly variable both spatially and temporally, with BTEX concentrations ranging from below detection to greater than 60 mg/L influent to the treatment wells. A "hole" was observed in the plume with high background DO levels which was a persistent but mobile feature throughout the trial.

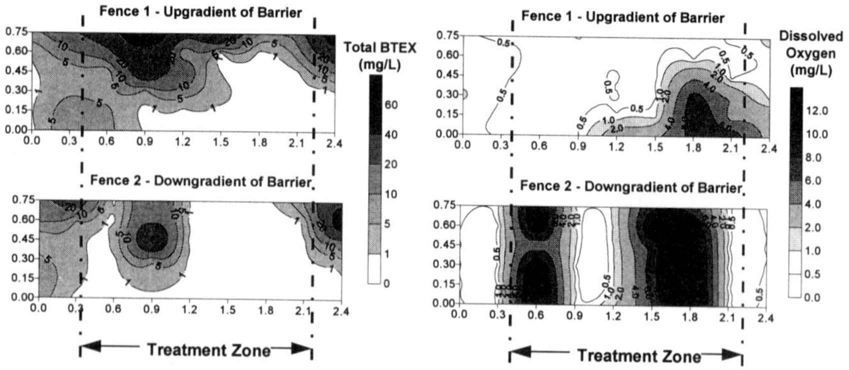

FIGURE 2. Contour maps of (a) total BTEX and (b) dissolved oxygen for snapshot 1 (Day 51).

The "treatment zone" shown on Figures 1 and 2 encompasses the zone captured by the treatment wells based on the groundwater flow direction and similarities in BTEX distribution between the fences. The treatment zone covers an area 1.8 m in width and 0.9 m in depth, assuming samples from each monitoring point are representative of a cross-sectional area of 0.3 m by 0.15 m. Average values of total BTEX were calculated for each multilevel fence in the treatment zone for each snapshot (Figure 3). BTEX levels generally increased influent to the treatment wells during the trial at Fence 1. Immediately downgradient of the treatment wells at Fence 2, BTEX concentrations were generally lower than noted at Fence 1 but also increased with time. Insufficient Fence 3 data was available to confirm a temporal trend.

Estimates of Treatment Efficiency. Treatment efficiency was estimated by comparing the average values of total BTEX within the treatment zone at Fences 1 and 2 (Figure 3). Corrections were not made for the groundwater travel time between the fences since the travel time is relatively short and to avoid uncertainties in data interpolation. BTEX reductions between Fences 1 and 2 ranged from 71% for snapshot 1 (Day 51) to -2% for snapshot 4 (Days 126-132) with absolute reductions in average BTEX concentrations from 7.2 mg/L to -0.3 mg/L. For the zone containing low BTEX concentrations (the vicinity of the "hole" where average input concentrations were below 5 mg/L), reduction in BTEX concentrations ranged from 99.8% for snapshot 1 to 97.7% for snapshot 4. The decline in treatment efficiency over time was attributed to decreasing rates of oxygen release from the treatment wells along with increasing influent BTEX concentrations. "Excess" DO levels were observed at Fences 2 and 3 in the zone where BTEX was completely degraded which presumably would be available for further BTEX biodegradation downgradient from Fence 2.

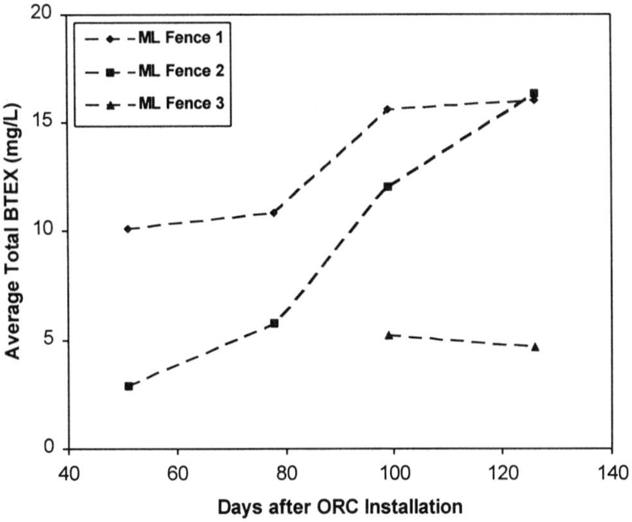

FIGURE 3. Plot of average total BTEX concentrations within the treatment zone at the three fences during the four sampling snapshots.

Oxygen Balance Estimates. To provide an indication of the performance of the ORC barrier, estimates of the amount of oxygen released by the treatment wells were compared to estimates of the mass used for BTEX degradation or observed as "excess" DO. Assay tests were performed on ORC socks removed from treatment well ORC7 following the trial 153 days after installation. Results indicated that approximately 48% of the initially available oxygen was released during that time period. Assuming the results from this well were applicable to the other treatment wells, the total amount of oxygen released to the treatment zone during the trial was estimated.

Mass fluxes of BTEX and DO through the cross-sections within the treatment zone at Fences 1 and 2 were estimated. Assuming each mg decrease of BTEX between Fences 1 and 2 used 3 mg of oxygen, it was possible to estimate the mass of oxygen either used for BTEX degradation or observed as "excess" DO. The amount of oxygen accounted for by observed BTEX degradation or "excess" DO was less than 10% of the total estimated mass of oxygen released to the treatment zone during the same period. Although these estimates are subject to a number of assumptions and uncertainties, they still clearly indicate that there were losses of oxygen other than insitu biodegradation of BTEX. Initial rapid release of oxygen which was not characterized in this study, oxygen loss from treatment wells to the atmosphere or vadose zone, degradation of compounds other than BTEX and other demands such as reduced inorganic species are expected to account for a significant part of this discrepancy.

CONCLUSIONS

Results of the trial demonstrated that intrinsic biodegradation of BTEX in groundwater can be enhanced using oxygen releasing solids in unpumped wells placed across the migration pathway of a plume. Significant reductions in BTEX mass flux were observed in a zone impacted by treatment wells containing a commercially available oxygen releasing product. However the degree of plume remediation was neither expected or found to meet typical regulatory requirements. even with the tight pattern of treatment wells used in the study. Thus complete plume cut-off close to the source of contamination using this approach may not be a realistic goal. However any reduction in mass flux represents a reduction in potential impacts and risk associated with the plume downgradient of the site. With reduction in contaminant mass flux, intrinsic biodegradation would be expected to control downgradient plume migration over shorter travel distances.

REFERENCES

Byerley, B. T., S. W. Chapman, D. J. Smyth, and D. M. Mackay. 1996. "A Pilot Test of Passive Oxygen Release for Enhancement of In-Situ Bioremediation of BTEX Contaminated Ground Water". In submission.

MANAGEMENT OF A HYDROCARBON PLUME USING A PERMEABLE ORC® BARRIER

Jeffrey G. Johnson (GRAM, Inc., Albuquerque, New Mexico)
Joseph E. Odencrantz (TRI-S Environmental, Newport Beach, California)

ABSTRACT: A "permeable oxygen barrier" was formed by depositing 342 Oxygen Release Compound (ORC®) filter socks in a series of 20 six-inch polyvinyl chloride (PVC) source wells. The objective was to increase the dissolved oxygen (DO) levels of the aquifer and to enhance the intrinsic bioremediation of dissolved phase benzene, toluene, ethyl-benzene, and total xylenes (BTEX) contamination. The DO and BTEX levels were monitored by sampling monitor points attached at three depths to the exterior of the source well screens. Additional monitoring points downgradient of the barrier were also sampled. The change in DO concentrations was determined by contouring the distribution of DO at different times in a cross-section view across the front of the barrier and in a plan view to determine the area affect. The same approach was used for estimating changes in the total BTEX in the aquifer. The results of the analysis showed an increase in DO above background levels, both at and downgradient of the barrier, and an overall decrease in BTEX. The initial degradation half-life of benzene after ORC® addition in the vicinity of the barrier was, approximately, 50 days.

INTRODUCTION

At the request of the New Mexico Environment Department (NMED), GRAM, Inc., in conjunction with Regenesis Bioremediation Products, conducted a pilot study from August through December 1994. The purpose of the pilot study was to determine, under field conditions, the ability of ORC® to increase the DO levels in the aquifer. The pilot study consisted of the installation and sampling of one 6-inch PVC ORC® source well, 26 downgradient monitoring points, and three existing monitor wells. Ground water samples were collected at various times to determine changes in DO and BTEX concentrations. The results of the pilot study showed that oxygen was released into, and dispersed through, the subsurface as indicated by an increase in DO at the source well, and downgradient of the source well, and that remediation occurred at various points in the system as indicated by a decrease in BTEX. Based on the results of the pilot study NMED granted permission to proceed with the installation of a full-scale remediation system.

Site Description. The site selected for conducting the investigation was the location of a former filling station in Belen, New Mexico. The site is referred to as the Shell North Main site by NMED and is approximately 50 km south of Albuquerque, New Mexico. There was a release of an undetermined amount of regular gasoline over an unknown period of time at the site. A series of investigations had previously determined that dissolved phase contamination in the ground water extended off site, and some residual sorbed phase soil contamination remained on site. There was no evidence of free-phase product on the site greater than a slight sheen in the existing monitor wells before the start of the investigation. The depth to ground water at the site is approximately 1.5 m below ground surface and the soils at the site are primarily composed of alluvial sands, silts, and clays (GRAM, Inc. 1995).

Product Description. The product selected for use in providing oxygen to the site was ORC®. ORC® is a patented solid form of magnesium peroxide that releases oxygen when hydrated. The product is delivered from the manufacturer in nylon "filter socks." Each filter sock is composed of a tight weave polyester and contains a 1:1 mixture of ORC® and #90 silica sand and weighs 5.7 kg (Regenesis, 1995).

FULL-SCALE SYSTEM

As a result of the pilot study, GRAM, Inc., in conjunction with Regenesis Bioremediation Products, designed and installed a full-scale oxygen distribution system to use ORC®. The purpose of the full-scale system was to evaluate the effectiveness of a full-scale application of ORC® and, if possible, to prevent future off-site migration of dissolved phase contamination. The full-scale system consists of 20 6-inch PVC ORC® source wells installed to form two permeable barriers. The location of overhead power lines prevented the installation of one continuous barrier at the south-eastern most area of the plume. A long barrier, consisting of 16 source wells, and a slightly off-set shorter barrier, consisting of 4 source wells, were installed. The source well used for the pilot study is part of the long barrier. Additional source wells were installed during March 1995. The layout of the full-scale system is shown in Figure 1.

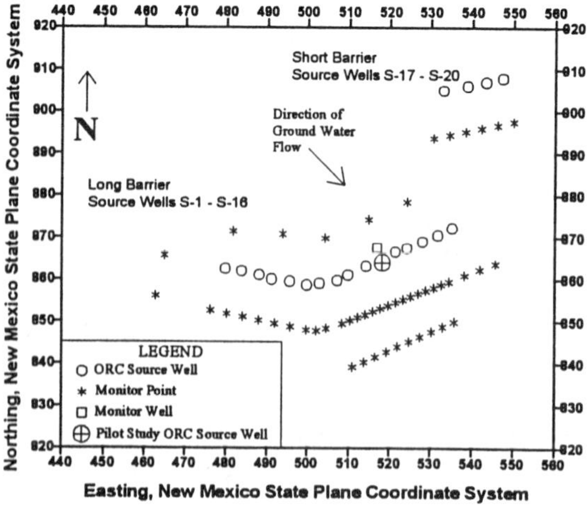

FIGURE 1. Shell North-Main Detailed Site Map

A total of 342 ORC® socks were emplaced in the source wells on April 3, 1995. The system was monitored for 3, 5, 9, 12, 20, 30, 47, 60, 75, 93, 200, 288, and 365 days after ORC® emplacement to determine changes in DO and BTEX concentrations in the ground water. Monitor points were attached at three depths (2.1 m, 4.3 m, and 6.4 m) to the exterior of the source well screens during installation so that DO and BTEX levels could be monitored at the barrier with minimal disturbance of the ORC® in the source wells. Additional up-gradient and down-gradient monitor points and monitor wells were sampled to determine both the horizontal and lateral distribution of DO throughout the aquifer, and the biodegradation of BTEX in the aquifer as a result of the increase of DO.

ANALYTICAL METHODS

Temperature, pH, conductivity, and DO were monitored on site. A Hydac water meter was used for measuring the pH, conductivity, and temperature of all ground water samples collected in the field. To provide accurate information at a reasonable cost, field screening procedures were used for monitoring DO and BTEX concentrations throughout. DO levels were measured in the field using the Hach-Modified Winkler digital titration method. Ground water samples were collected in 40-ml VOA vials, preserved with mercuric chloride, placed on ice, and transported back to the office where BTEX concentrations were measured using Ohmicron Total BTEX RaPID Assays® Immunoassay Test Kits. Duplicate samples, collected for laboratory analysis of BTEX and TPH using EPA Methods 8020 and 8015 Modified, were collected in 40-ml VOA vials, preserved with mercuric chloride, placed on ice, and transported directly to the laboratory.

RESULTS OF ANALYSIS

The collected data were analyzed to determine interactions between the added ORC® and the BTEX plume. The change in DO concentrations was determined by contouring the distribution of initial oxygen and the increase in total oxygen. These calculations were performed to determine background conditions and conditions at monitoring events for operating the full-scale system. The same approach was used for estimating change in the total BTEX in the aquifer.

Areal Analysis. The volume of DO was calculated using Surfer for Windows, a contouring and three-dimensional surface mapping software (Golden Software 1994). The values for the three methods, Trapezoidal Rule, Simpson's Rule, and Simpson's 3/8 Rule, were averaged and then used for determining trends. The same procedures used for the analysis of the DO concentrations were used for BTEX. A portion of each plot upgradient of the source wells was blanked out during the calculations to determine a more realistic area of consideration. The blanking set the values for the upgradient concentrations to zero. This prevented values to be extrapolated into areas without monitoring points (no data to support). In a previous publication (Odencrantz et al., 1996) the blanking was not performed which showed a considerable different "macroscope" behavior of the system. The source well monitor points at 4.3 m below ground level were used.

The results of the analysis showed that there was an increase in DO downgradient of the barrier and a change in BTEX concentrations. Upon ORC® addition, there was evidence of dispersion of oxygen downgradient by Day 9. There was also a marked decrease in the size of the BTEX plume. It was apparent from the contour plots that the plume was retreating and was in the process of being "cut-off" from any further downgradient migration. The BTEX levels showed an initial drop until Day 20. This was followed by an increase in BTEX levels until Day 75. The initial decrease was during a period of a rising ground water table and the increase of BTEX concentrations was during a period of a falling ground water. The water table elevations for monitor well SH-5, located 1 m upgradient of the barrier are shown in Figure 2.

The results of the analysis, presented as normalized mass curves for DO and BTEX, for the areal analysis over a one-year period are summarized in Figure 3.

Cross-Sectional Analysis. The same methods used for determining the volumes of DO and BTEX for the areal analysis were used for the cross-sectional analysis. Contour plots were made of the DO and BTEX concentrations in a cross-section across the face of each of the barriers.

The average DO background concentration was 1.4 mg/L and 1.8 mg/L for the long barrier and the short barrier, respectively. The results of the data analysis indicated there was an increase of DO above background levels during the first nine days. The average concentration for DO reached a maximum of 13.1 mg/L and 14.8 mg/L on Day 9 for each of the barriers. Individual maximum levels greater than 20 mg/L DO were achieved at the long barrier and there was a general decrease in DO after the initial increase. Levels of DO were maintained greater than background levels (1 to 2 mg/L) for more than 200 days after the emplacement of the ORC®. The ORC® was replaced in half of the source wells 275 days after initial emplacement. DO levels in the vicinity of the wells where the ORC® was replaced again increased to above 10 mg/L. The levels in the vicinity of wells that did not have the ORC® replaced were still above background levels at 365 days after initial emplacement.

The analysis of the BTEX data showed an overall decrease in BTEX levels across the barrier. Average background BTEX concentrations were 2.3 mg/L and 1.4 mg/L for the long and short barriers, respectively. There was a slight increase in the average BTEX concentrations to 2.6 mg/L and 2.1 mg/L on Day 3. There was an overall decrease in BTEX concentrations through Day 93. There was a slight increase in BTEX concentrations measured on Day 47 and just before the partial replacement of the ORC® on Day 190. There was an overall decrease following the partial replacement.

The results of the analysis, presented as normalized mass curves for DO and BTEX, for the two barriers over a one-year period are summarized in Figure 4 and Figure 5.

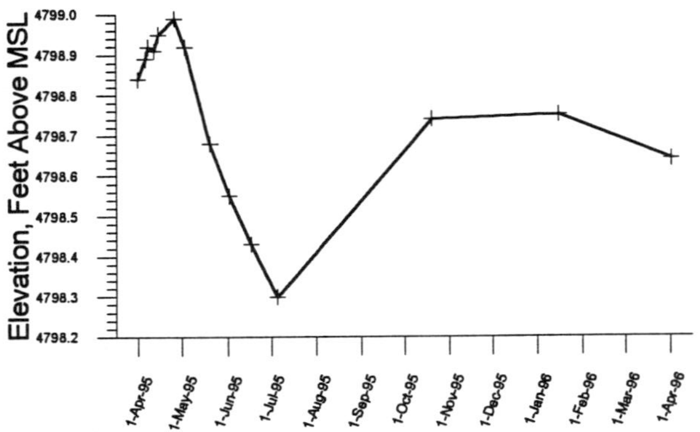

FIGURE 2. Monitor Well SH-5 Water Table Elevations

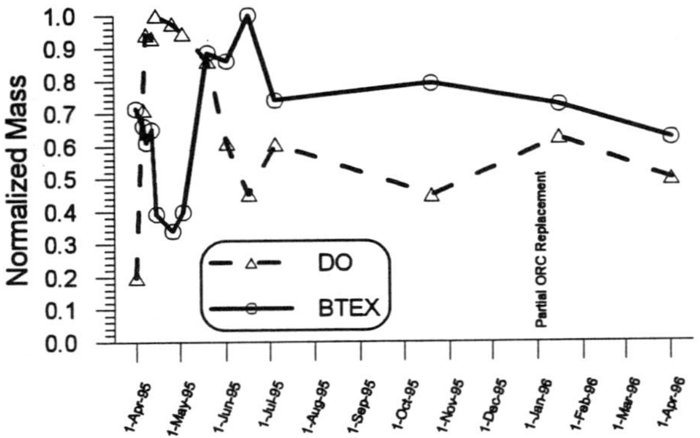

FIGURE 3. Normalized Mass DO & BTEX Areal Analysis

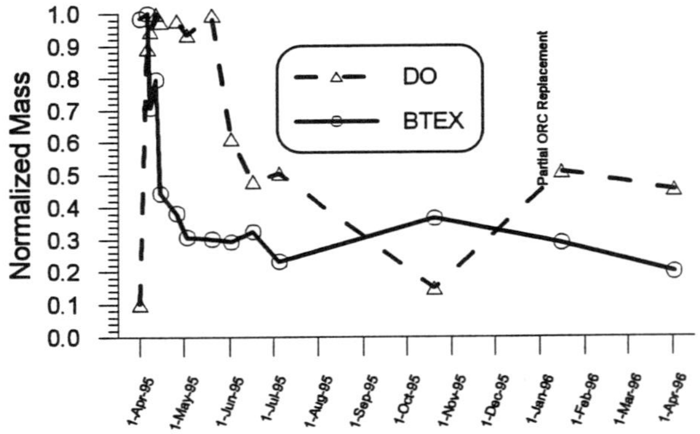

FIGURE 4. Normalized Mass DO & BTEX Long Barrier

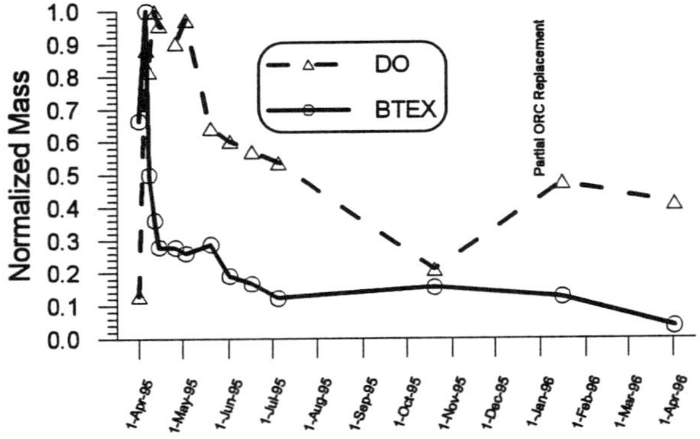

FIGURE 5. Normalized Mass DO & BTEX Short Barrier

VARIABLES

There were two events that occurred during the course of this study that had unquantified impacts on the site. The first was the partial excavation of contaminated soil upgradient of the barriers. During the backfilling of the excavation 500 pounds of pure ORC®, which equates to twenty pounds of available oxygen, was mixed in with the clean fill to act as an additional source of oxygen. The second event was a dewatering activity downgradient of the barriers. The dewatering consisted of pumping approximately 12 million gallons of water over a four week period from multiple off-site locations. The excavation was completed in March 1995 and the dewatering occurred during November 1995.

CONCLUSIONS

There are three indicators of bioremediation (National Research Council, 1993):

(1) Decrease in the volume of contamination.
(2) The presence of microbes capable of degrading the contaminant of concern.
(3) Actual indication that biodegradation occurred.

The results of monitoring have shown an increase in DO levels and a decrease in BTEX levels along the barriers and varying degrees downgradient of the barriers. There is clearly far field (Figure 3) and near field (Figure 4 and Figure 5) behavior demonstrated by the areal (away from barrier) and cross-sectional (near barrier), respectively. There was limited biological analysis done that showed that there were indigenous hydrocarbon degrading microbes present in sufficient quantities to degrade the contaminants of concern. The objective to determine if ORC® was able to increase DO levels was successful. The secondary objective to show evidence of increased biodegradation occurring at the site was also successful.

REFERENCES

Golden Software. 1994. *Surfer for Windows User's Guide*. Version 5. Golden Software, Inc., Golden, CO.

GRAM, Inc. 1994. *ORC Pilot Study Report, Belen, New Mexico*. GRAM, Inc., Albuquerque, NM.

National Research Council, Committee on In Situ Bioremediation. 1993. *In Situ Bioremediation, When Does it Work*. National Academy Press, Washington, D.C.

Odencrantz, J. E., J. G. Johnson, and S. S. Koenigsberg. 1996. "Enhanced Intrinsic Bioremediation of Hydrocarbons Using an Oxygen-Releasing Compound." *Remediation*. 6(4): 99-114.

Regenesis Bioremediation Products. 1995. *Oxygen Release Compound (ORC®) Product Description and Price List*. Regenesis Bioremediation Products, San Juan Capistrano, CA.

CHARACTERIZATION OF A NEW SUPPORT MEDIA FOR *IN SITU* BIOFILTRATION OF BTEX-CONTAMINATED GROUNDWATER

Dominique Forget, Louise Deschênes, Dimitar Karamanev, ***Réjean Samson***

NSERC Industrial Chair in Site Bioremediation, BIOPRO Research Centre, Chemical Engineering Department, École Polytechnique de Montréal, P.O. Box 6079, Downtown, Montreal, Quebec, H3C 3A7, Canada.

ABSTRACT: A new filtering media, granular peat moss (GPM), was studied for its possible utilization in an *in situ* biofilter for the remediation of groundwater contaminated with gazoline. This study showed that GPM has excellent microbiological and mechanical properties for its utilization as a support media in an *in situ* biofilter. Batch microcosms using ^{14}C-toluene demonstrated that the indigenous microflora was able to mineralize this compound under both aerobic and denitrifying conditions. Tests were conducted on a small scale biofilter of 0,5 L. A porosity of 80,8% and a hydraulic conductivity of $1{,}23*10^{-2}$ cm/s were measured. Sorption of toluene on GPM was evaluated and a retardation factor of 22,6 was observed. The hydrodynamics of the column were studied and the results indicated that the GPM filtering media allowed a plug flow regime with dispersion without dead zones or preferential flow paths. When operated in continuous mode without substrate addition (absence of toluene), oxygen consumption within the filter was low with a measured value of 0,50 mg O_2/L.hour. This suggests that carbon sources within the GPM are not readily mineralized, therefore reducing the risk of development of anaerobic zones in the reactor. Kinetics of aerobic toluene biodegradation in the filter were studied. A first order biodegradation rate of 0,41 min^{-1} was observed. Moreover, the filtering media did not show any signs of clogging after more than 6 months of operation. The biofilter was shown to operate in stable conditions over a 5 month period with an average toluene concentration of 450 µg/l at the entrance and a concentration below the detection limit at the exit. These results show that granular peat moss has great advantages as a filtering media for the *in situ* bioremediation of BTEX-contaminated groundwater.

INTRODUCTION

In situ biotreatment of BTEX-contaminated groundwaters is usually achieved through the injection of nutrients into the aquifer to stimulate biodegradation by indigenous microflora. Unfortunately, this approach often proves to be problematic due to the heterogeneity of the subsurface geological materials which causes poor distribution of the nutrients throughout the contaminated zones. Moreover, clogging of the injection wells is often observed due to the preferential microbiological growth which occurs near the nutrient source (Taylor, 1993).

A new approach to *in situ* bioremediation of contaminated groundwaters is the funnel-and-gate system (Starr and Cherry, 1994). This technique consists of cutoff walls with openings that contain *in situ* biofilters in which the biodegradation of BTEX compounds is optimized. In order to be efficient, the funnel-and-gate technology requires a filtering media that offers great advantages both at the microbiological and mechanical levels. In view of that fact, this study evaluated a new support media for *in situ* biofiltration of BTEX-contaminated groundwater

MATERIAL AND METHODS

Chemicals. Toluene was chosen as a model compound (Allen, 1991). Radiolabeled toluene (specific activity 9,7 mCi/mmol, purity > 98%) was obtained from Sigma Chemicals (St-Louis, MO) and used in all microcosm studies. All other experiments were carried out with toluene (purity > 99,9%) obtained from Anachemia Chemicals (Montreal, Canada).

Synthetic groundwater was used throughout this study. Its composition, inspired by the groundwater found in the Montreal region (Canada), was the following (mg/l): $FeCl_2$: 1,4; NaCl:617,3; Na_2CO_3: 678,4; KNO_3: 6,1; K_2HPO_4: 10,7; $CaCO_3$:15,0; $MgSO_4$: 18,8. Concentrated H_2SO_4 was used to adjust the pH at a value between 6,8 and 7,2.

Filtering media characteristics. Granular peat moss BB2-95 was provided by Premier Tech (Riviere-du-Loup, Canada). Granules had diameters ranging from 0,64 and 0,95 cm. Porosity, density and hydraulic conductivity of the filtering media were measured with a falling head permeameter (ASTM D-2434 rev. 74). Sorption of toluene on granular peat moss was evaluated in batch experiments as described by Stuart and al. (1991).

^{14}C-toluene mineralization in granular peat moss. The mineralization of ^{14}C-toluene in granular peat moss was studied in 110 ml serologic bottles under both aerobic and denitrifying conditions. Twenty milliliters of filtering media was introduced in each bottle as well as 80 ml and 90 ml of synthetic groundwater for microcosms under aerobic and denitrifying conditions respectively. Air was injected during one hour in aerobic microcosms while nitrogen was injected for 30 minutes in denitrifying microcosms. A concentration of 100 000 dpm of radiolabeled toluene was then injected in all bottles. Unlabeled toluene was added to reach a final concentration of 4,4 mg/l of synthetic groundwater. To trap the ^{14}C-CO_2, a 4 ml glass tube containing 1 ml of 1N KOH was placed in each bottle. The bottles were sealed with Teflon-lined valves (type Mininert, Supelco inc., Bellefonte, PA) and placed on a rotary shaker at 100 RPM (10°C). Each experiment was carried out in triplicates with abiotic controls containing 2% (w/w) NaN_3. At intervals of approximately 48 hours, KOH was removed for analysis and replaced by fresh alcali. KOH samples were counted for trapped $^{14}CO_2$ with a liquid scintillation counter (Wallac 1409, Turku, Finland).

Four series of microcosms under aerobic conditions were studied. In the first, the filtering media consisted of granular peat moss alone. In the second, the filtering media was inoculated with 20% (v/v) of contaminated soil provided by Hydro-Quebec (Pointe-aux-Trembles, Quebec). In the third series, 10% (v/v) of composted chicken manure was added. Finally, synthetic groundwater was replaced with a mineral salt medium (Greer and al., 1990) in the fourth series. Three series of microcosms under denitrifying conditions were prepared. A concentration of 800 mg/l of $NaNO_3$ was added in the first series. A slow release nitrogen fertilizer (HIGHN 22-4-6, Scotts, Marysville, OH) was introduced in the second series. Finally, no amendment was made in the third series. Concentration of NO_3 and NO_2 was followed in all denitrifying microcosms. Filtered samples (0,45 μm Millex-HV filter) were analyzed by HPLC (Dionex Co.). The ions were separated on a 250x4 mm IONPAC AS4A-SC chromatographic column with a guard column (IONPAC AG4A-SC) and an ion self suppressor. Aqueous bicarbonate buffer was used as eluent.

Biofiltration of toluene in an 0,5 L column. Experiments were conducted in a column of 39 mm of internal diameter and 44 cm of height with sampling ports placed at the entrance and at the exit. Synthetic groundwater was contained in a tank where air was continuously injected to reach oxygen saturation (11,7 mg/l). Toluene was injected in the inflow stream with a syringe pump (Orion M365) and mixed with the water by means of a static mixer. The entire experimental setup was made out of glass, stainless steal and Teflon to avoid adsorption problems and was placed in a 10°C chamber. A polarographic probe (Cole Parmer, Vernon Hills, IL) was used to measure dissolved oxygen concentration at the entrance and the exit of the biofilter.

Residence time distribution within the biofilter was studied with a NaCl tracer. For residence times ranging from 4 minutes to 8 hours, spikes were induced at the entrance of the column and chloride concentrations at the exit were followed in time with a probe (Accumet, Pittsburgh, PA). Kinetics of toluene degradation within the filter was evaluated. For residence times in the reactor ranging from 2 minutes to 2,4 hours, concentrations in toluene were measured both at the entrance and at the exit of the biofilter at steady-state. Samples were analyzed by GC/MS (Hewlett-Packard 5890 Gas Chromatograph connected to a HP 5971 Mass Spectrometer) with a capillary column HP-624 of 25 m x 0,20 mm and helium as the carrier gas. The system was equiped with a purge and trap device (Hewlett-Packard 7675).

RESULTS AND DISCUSSION

Physical properties of the media. In order to optimize the efficiency of the biofilter, different organic filtering media were studied. Among them, granular peat moss was chosen for its excellent microbiological and mechanical properties. The density and the porosity were evaluated at 0,144 g/cm^3 and 80,8% respectively. This latter value is slightly inferior to the characteristic value of peat moss porosity of 92 % usually reported in literature (Todd, 1980). This can be explained by the presence of the polymeric material found in granular peat moss which obstructs a certain portion of the pores. The hydraulic conductivity was very high with a measured value of $1,23*10^{-2}$ cm/s which is comparable to a sand or a gravel.

Sorption of toluene on granular peat moss was studied and a retardation factor of 22,6 was calculated. This relatively important adsorption is partly attributable to the polymeric agent in the granular peat moss.

^{14}C-toluene mineralization in granular peat moss. In order to assess the microbiological properties of the granular peat moss filtering media, batch microcosms were prepared using ^{14}C-toluene as a model compound. After a lag period of 8 days, 70 % of the ^{14}C-toluene was mineralized within 17 days in microcosms containing granular peat moss alone (figure 1a). In contrast, a lag phase of 5 days was observed within microcosms containing granular peat moss inoculated with contaminated soil and 70 % of mineralization occurred after 14 days (figure 1a). After 21 days of experimentation, both series of microcosms were spiked again with ^{14}C-toluene. For both experiments, no lag phase was observed and 70 % of ^{14}C-toluene mineralization occurred within 6 days indicating that the indigenous microflora of the granular peat moss had well adapted to toluene. Furthermore, addition of a mineral salts medium (MSM) and of composted chicken manure did not increase the rate of mineralization within granular peat moss (figure 1b) which shows that the indigenous microorganisms were not limited in nutrients during the tested period of time.

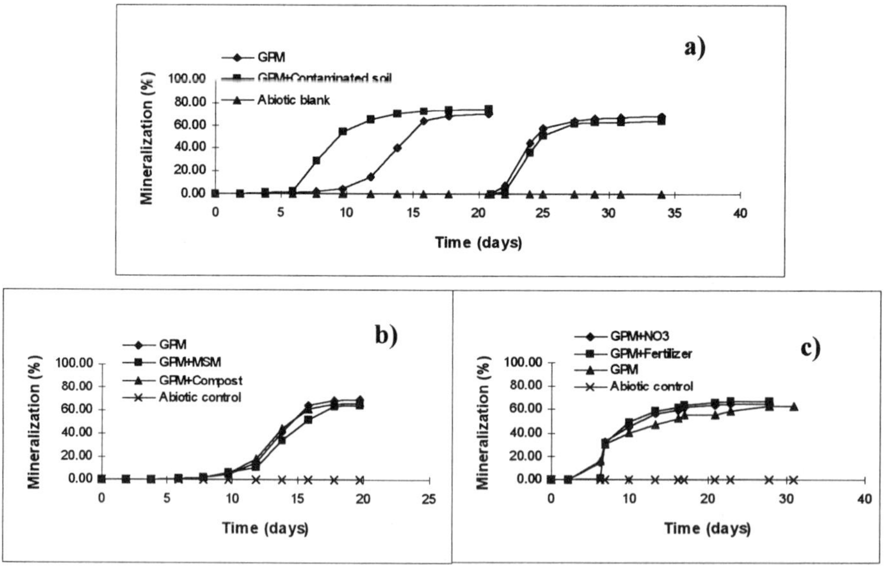

FIGURE 1. Mineralization of ^{14}C-toluene at 10°C a) under aerobic conditions b) under aerobic conditions with nutrient amendments c) under denitrifying conditions

Batch microcosms studies were also performed under anaerobic conditions in order to find out if the indigenous microflora had the genetic capacity to mineralize toluene under denitrifying conditions. Biodegradation was observed after a lag period of 6 days and 65 % of mineralization occurred after 25 days (figure 1c) in the three experiments carried out with different nitrate sources. The similarity of the three profiles shows that the nitrate concentration present in the synthetic groundwater provides a sufficient level of electron acceptors. Furthermore, concentration of nitrates was found to decrease in the microcosms confirming that denitrifying conditions did prevail during mineralization (results not shown).

Biofiltration of toluene in an 0,5 L column. Residence time distribution experiments within the column provided Peclet numbers ranging from 9,5 to 17,8 for residence times of 4 minutes to 8 hours respectively (results not shown). Therefore, the biofilter operated under a plug flow regime with dispersion. Only one maximum was observed on each distribution curve which permits to conclude that no preferential flow paths are present within the biofilter. Comparison of the theoric and observed dynamic volumes showed that no dead zones prevailed in the bioreactor. Furthermore, the filtering media did not show any signs of clogging after more than 6 months of operation.

In absence of toluene, oxygen consumption within the filter was quite low , in the order of 0,5 mg/L.hour indicating that carbon sources associated to the granular peat moss are not easily mineralized. This reduces considerably the risk of development of anaerobic zones in the bioreactor.

Kinetics of toluene biodegradation within the biofilter were studied at steady state for an average concentration of 1,2 mg/l of toluene at the entrance. Exit toluene concentrations ranged from 3,3 μg/l to 380 μg/l from residence time in the biofilter ranging from 2,4 hours to 2,1 min respectively. A first order kinetic constant of 0,41 min^{-1} was calculated. A wide divergence exists between the kinetics of toluene degradation rates reported in the literature (Alvarez, 1991) due to the fact that these constants are system specific. Furthermore, no data concerning kinetics of toluene degradation within peat moss has been found in literature.

The long term behavior of the biofilter was studied over a 5 month period for an average toluene concentration of 450 μg/l at the entrance (results not shown). After a lag phase of 7 days, the efficiency of toluene biodegradation in the reactor stabilized at a value near 100 % and remained stable through the experiment. Oxygen consumption in the filter fluctuated between a value of 2,5 and 3,0 mg/l. At startup, biomass concentration in the granular peat moss was of the order of $5{,}0*10^6$ MPN/g GPM (dry wt). Initially, a growth phase was observed over a 30 day period and concentrations peaked at $2*10^7$ MPN/g GPM (dry wt) at the entrance and $1{,}7*10^7$ MPN/g GPM (dry wt) at the exit. No compaction of the filtering media was observed. Nutrient concentrations were followed at the entrance and at the exit of the filter and the system was found to be self-sufficient.

This study has shown that granular peat moss presents excellent mechanical and microbiological properties and therefore has great advantages as a filtering media for the *in situ* bioremediation of BTEX-contaminated groundwater. It is thus possible to foresee an on site application of the system within a short term period.

ACKNOWLEDGMENTS

This work was supported by Alcan, Analex, Browning-Ferris Industries, Cambior, Hydro-Quebec, Petro-Canada, Premier Tech, SNC-Lavalin Environnement inc., the Centre québécois de valorisation des biomasses et des biotechnologies and the Natural Sciences and Engineering Research Council of Canada (NSERC). The authors would also like to ackwnoledge the technical assistance of Manon Leduc.

REFERENCES

Allen, R. M. 1991. "Fate and Transport of Dissolved Monoaromatic Hydrocarbons During Steady Infiltration through Unsaturated Soil." Ph.D. thesis. University of Waterloo, Waterloo, Ontario.

Alvarez P. J. J., P. J. Anid and T. M. Vogel. 1991. "Kinetics of Aerobic Biodegradation of Benzene and Toluene in Sandy aquifer Material." *Biodegradation.* 2: 43-51.

Greer, C. W., J. Hawari and R. Samson. 1990. "Influence of Environmental Factors on 2,4-D Dichlorophenoxyacetic Acid Degradation by *Pseudomonas Cepacia.*" *Archives of Microbiology.* 154: 317-322

Starr, R. C. and J. A. Cherry. 1994. "In Situ Remediation of contaminated Groundwater: The Funnel and Gate System." *Ground Water.* 32: 465-476.

Stuart, B. J., G. F. Bowlen and D. S. Kosson. 1991. "Competitive Sorption of Benzene, Toluene and the Xylenes onto Soil". *Environmental Progress.* 10: 104-109.

Taylor, R. T., M. L. Hanna, N. N. Shah, D. R. Shonnard, A. G. Duba, W. B. Durham, K. J. Jackson, R. B. Knapp, A. M. Wijesinghe, J. P. Knezovich and M. C. Jovanovich. 1993. "In Situ Bioremediation of Trichloroethylene-contaminated Water by a Resting-Cell Methanotrophic Microbial Filter." *Hydrological Sciences.* 38: 323-342.

Todd, D. K. 1980. *Groundwater Hydrology.* Second edition. Wiley. New York 535 p.

PHYSICO-CHEMICAL OPTIMIZATION OF BIOFILM DEVELOPMENT IN FRACTURED ROCK AQUIFER CONDITIONS

Nathalie Ross[1] ([1]NSERC Industrial Chair on Site Bioremediation, BIOPRO Research Centre, École Polytechnique de Montréal, Québec, Canada), Louise Deschênes[1], Bernard Clément[2] ([2]Industrial Engineering and Mathematical Department, École Polytechnique de Montréal, Québec, Canada) and Réjean Samson[1]

ABSTRACT: Microbial barriers formed with indigenous biomass and exopolymeric matter, produced in a fractured rock aquifer, offer an excellent potential as a mean to prevent spreading of a contamination plume. The objective of this study was to optimize physico-chemical conditions for biofilm formation on a porous medium in groundwater. Mixed microbial populations, cultivated from natural groundwater, were used to inoculate semi-continuous reactors which contained a ceramic plate immersed in synthetic groundwater (10°C, in darkness). Statistical design was used to study the effects of aeration, carbon source, carbon source feeding rate, C:N:P ratio, calcium ions, and pentachlorophenol (PCP) as a contaminant, on suspended biomass, suspended exopolysaccharide (EPS), and biofilm production. Results indicated that high carbon source feeding rate (20mg/m^2.min) and addition of calcium ions (100mg/L) provided optimum conditions for suspended biomass production (1.8g/L after 8 days) while presence of PCP (10 mg/L) slightly reduced it. High concentration of EPS (1.6g/L) was measured in water when the carbon source feeding rate was 20mg/m^2.min. Biofilm thickness was significantly increased (250 µm after 6 days) with aeration and molasses high feeding rate (20mg/m^2.min).

INTRODUCTION

The complexity of fracture patterns and contaminant transport in fractured rock aquifer are well recognized obstacles that limit the application of remediation technologies (Fetter 1992). A novel method for controling more adequately the aquifer treatment area, consists of stimulating indigenous microorganisms to produce EPS and form an underground biobarrier. Physico-chemical conditions of the selected area will determine the efficiency of the biofilm development, its thickness, and stability (Characklis and Marshall 1990).

Work has been carried out on lab-scale biobarrier formation, which consisted of injecting starved cells and a carbon source in different porous media (glass bed, sand or gravel) for the containment and the biodegradation of contaminants (Taylor et al. 1993) (Characklis and Marshall 1990) (Bellamy et al. 1993). The objective of the present study was to determine the effects of important physico-chemical parameters and presence of PCP as a model contaminant, on biofilm development on porous media, in aquifer conditions. The hypothesis ar: that groundwater indigenous microbial population contains EPS producers organisms, and that injection of a carbon source and oxygen would be sufficient to promote biofilm formation.

MATERIALS AND METHODS

Groundwater and microbial enrichment. Natural groundwater was obtained from an observation well operated at *Le Centre de tri et d'élimination des déchets* (CTED, Montreal, Canada). The groundwater sample provided the microorganisms used to colonize the ceramic. Chemical analysis of the gropundwater sample was used to compose a synthetic water with the following composition (mg/L): Na_2CO_3 (679), NaCl (617), $MgSO_4$ (19), $CaCO_3$ (15), K_2HPO_4 (11), KNO_3 (6) and $FeCl_2$ (1) (pH≅8,3). Synthetic water was autoclaved at 121°C and 103.4kPa (20 min.). The culture of indigenous microorganisms was accomplished in an Enlenmeyer (1L) enriched with molasses (5g/L) until microbial concentration reached 10^7 heterotrophic microorganisms/ml (200rpm, placed at 10°C in darkness). Heterotrophic microorganisms concentration was determined by the most probable number (MPN) method (APHA et al. 1992). The method of Dubois et al. (1956) was used to determine carbohydrate concentration. The culture was maintained at -20°C with glycerol (15g/L) until use (Cargill et al. 1992).

Biofilm development. Experiments were conducted in semi-continuous reactors (1L) with a ceramic coupon attached to the top (10°C, in darkness). Reactors were filled with synthetic groundwater (500ml) and inoculated with a microbial culture (10^5 heterotrophic microorganisms/ml). The physico-chemical conditions tested are listed in TABLE 1. The carbon source feeding was carried out once a day with a syringe at 20ml/d (concentrations 4,09 and 40,9 g/L). Calcium chloride solution (100mg/L) was used for calcium ions enrichment. Sodium pentachlorophenol (NaPCP, 10mg/L) was added to evaluate the effect of contamination on biofilm development. Statistical fractionated experimental design (Plackett-Burman design) was used to determine the effects of physico-chemical conditions on biofilm development (Clément 1990). Each experiment was performed in duplicate.

TABLE 1 - Physico-chemical conditions for the optimization of biofilm development on porous media in synthetic groundwater

Designation	Condition	Modality (-)	Modality (+)
A	Feeding Rate	$2mg/m^2.min$	$20mg/m^2.min$
B	Aeration	Agitation 200rpm	Sparging $8,3cm^3/s$
C	Carbon Source	Saccharose	Molasses
D	PCP	0mg/L	10mg/L
E	Calcium Ions	0mg/L	100mg/L
F	C: N: P ratio	50: 10: 1	200: 10:1

Microbial production was analyzed by mesuring the following parameters: suspended biomass concentration, suspended EPS concentration, and biofilm thickness on ceramic (Hacking et al. 1983) (Trulear and Characklis 1982). Water pH, oxydoreduction potential (ORP), and carbohydrate concentration were also measured daily during 192 hours (ASTM 1990) (ASTM 1993). The phenol-sulfuric method was used to obtain carbohydrate concentration (Dubois et al. 1956). Residual PCP was measured with high performance liquid chromatography (HPLC) (Barbeau 1996).

RESULTS AND DISCUSSION

Microbial growth and biofilm development. The growth of groundwater indigenous microorganisms was represented by a typical batch growth curve (Characklis and Marshall 1990) (FIGURE 1A). Microorganisms developed up to a concentration of 1369 mg/L after 192 hours, at high carbon source feeding rate (20 mg/m^2.min) and without PCP as a contaminant. On the other hand, addition of PCP (10mg/L) and a carbon source feeding rate of 2 mg/m^2.min lead to a biomass concentration of 90 mg/L after 192 hours. The EPS concentration curve was similar to a typical logistic curve like those obtained in batch culture (Characklis and Marshall 1990) (FIGURE 1B). The EPS production under high feeding rate (20mg/m^2.min) was 37% higher than under low feeding rate (2mg/m^2.min). The EPS concentration reached 780mg/L after 192 hours.

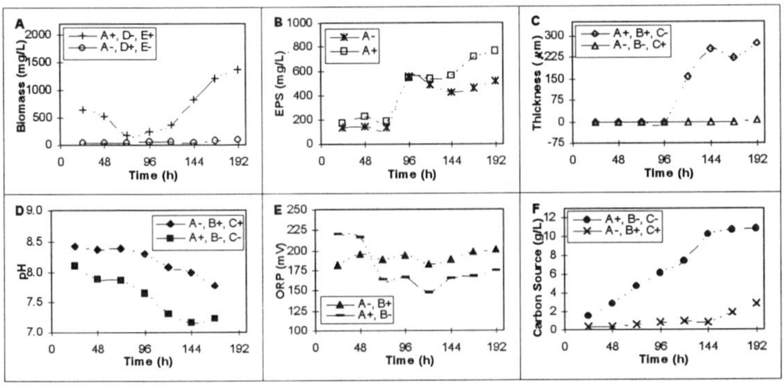

FIGURE 1. Microbial growth and biofilm development on ceramic coupon immersed in synthetic groundwater. A: Suspended biomass (mg/L), B: Suspended EPS (mg/L), C: Biofilm thickness (µm), D: Water pH, E: ORP (mV), and F: Carbon source concentration (g/L).

Biofilm thickness was the most sensitive variable. Aeration and feeding with molasses at 20mg/m^2.min contributed to develop the biofilm within 96 hours (FIGURE 1C). The biofilm thickness reached 250µm after 144 hours. A biofilm thickness of only 6,5 µm was obtained on ceramic when agitation (200rpm) and saccharose feeding (2mg/m^2.min) were applied. Conditions that led to a large production of biomass contributed to a decrease in pH (FIGURE 1D). ORP was higher (195mV in average) at low feeding rate and when air was sparged (FIGURE 1E). Carbohydrate accumulated in the water up to 10,9 g/L after 192 hours when the feeding rate was high and agitation was provided. This accumulation was correlated with the biomass production and decay (FIGURE 1F). When saccharose was used at 2mg/m^2.min, accumulation was 3g/L after 192 hours.

Effects of physico-chemical conditions on microbial growth and biofilm formation. Biofilms, which developed well when molasses was used as the carbon source, were 5,6 times thicker as compared to biofilms developed in saccharose fed reactors: 50 versus 9µm respectively (p=0,0001) (FIGURE 2C). The carbon source played several roles in the system. First, the carbon source promoted the growth of a microbial population, and conditioned the ceramic before cell attachment (Honig 1953). The molasses composition probably contributed to an effective conditioning of the ceramic. In fact, inorganic salts (≅ 8%) and amino acids (≅ 2%) in molasses could adsorbed on ceramic with ionic interactions (Characklis and Marshall 1990). Second, cell adhesion on ceramic could be promoted by the presence of divalent ions in molasses (≅3%) (Characklis and Cooksey 1983). Third, the hydrogen bonds, hydrophobic and ionic interactions between molasses, cells and EPS could explain a thicker biofilm (Bryan et al. 1986). Saccharose, which contained less than 0,02% of ashes and no trace of amino acids, was probably not effective in conditioning the ceramic surface and promoting cell attachment.

On average, a high carbohydrate feeding rate allowed production of 9 times more suspended biomass than low carbohydrate feeding rate: 360 mg/L compared to 40 mg/L respectively (p=0,0001) (FIGURE 2C). The same conditions lead to an increase of 40% in EPS production: 391 to 543 mg/L on average for 192 hours (p=0,0094) (FIGURE 2B). The biofilm was also 20 times thicker: 3 versus 57μm when the reactor was fed at a high rate (p=0,0001) (FIGURE 2A). Microbial diversity is related to the feeding rate. Indeed, filamentous microorganisms could have developed when feeding rate was as low as $2mg/m^2$.min and thus lead to a low density biofilm (Bellamy et al. 1993). On the other hand, dense biofilm could have develop when carbon source feeding was $20mg/m^2$.min because colonies were more dense and numerous.

Oxygen concentration and ORP were directly proportional to biofilm thickness (Characklis and Cooksey 1983). Air sparging ($8,3cm^3/s$) favored a biofilm 5,6 thicker than with agitation at 200rpm (50 versus 9μm respectively, p=0,0001). The aeration type caused different degree of water turbulence, which could explain the variation in biofilm thickness. Transport of organic compounds and cells onto the ceramic were more effective when air was sparged. Transport of cells and molecules from the liquid to the surface is the major factor involved for biofilm development, particularly when microbial population is low (Characklis and Marshall 1990). Groundwater mixed microbial population usually contains a large fraction of facultative aerobic bacteria (Fetter 1992). This could explain the fact that no significant difference in biomass concentration was observed between reactors aerated with air sparging and those aerated with mechanical agitation.

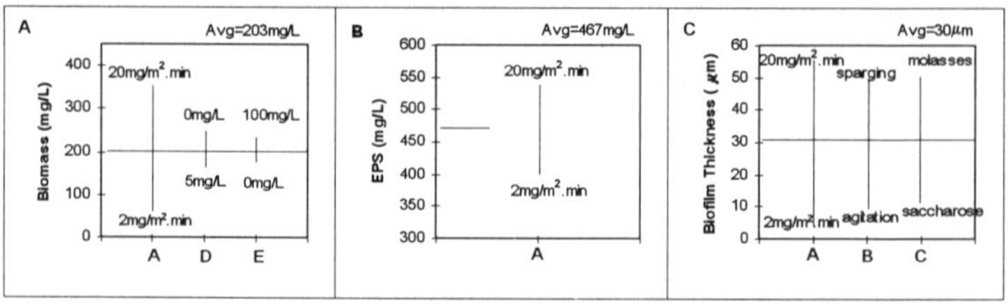

FIGURE 2- Significant effects of physico-chemical conditions on microbial production and biofilm development: A: Suspended Biomass (mg/L), B: Suspended EPS (mg/L) and, C: Biofilm Thickness (μm)

Enrichment with calcium chloride (100mg/L) contributed to an increase in microbial population (p=0,0131) which suggested that microorganisms in biofilm increased as well (Huang and Pinder 1994). An increase of microbial population in biofilm is usually associated with a densification of the biofilm instead of an increase in thickness (Ramsay et al. 1989). The addition of PCP, to simulate groundwater contamination, affected microbial growth. Over a period of 192 hours, the average microbial population was 159mg/L in PCP-contaminated water compared to 247 mg/L in synthetic groundwater (p=0,0001) (FIGURE 2A). The repressive effect on microbial growth could be associated to the accumulation of chloride ions and reduction of pH. Results also showed that a decrease in microbial population was obtained after 120 hours (data not shown).

CONCLUSIONS

Results confirmed that indigenous microbial population of an aquifer can produce EPS rapidly and promote the formation of a biofilm on a porous surface in groundwater conditions. The development of a biobarrier seems to be achievable without microorganisms injection. Stimulation of the production of EPS by indigenous population was obtained with the injection of molasses and air sparging. Results from reactors containing PCP suggested that *In situ* biobarrier could be tolerant to groundwater contaminated with PCP. The inhibition effect of PCP on microbial growth was probably due to the accumulation of chlorine ions. Such accumulation would be avoided in aquifer because of dispersion.

ACKNOWLEDGMENTS

The authors would like to acknowledge the support from the Chair partners: Alcan, Ltée, Analex, Inc., Browning-Ferris Industries, Cambior, Centre québécois de valorisation de la biomasse, Hydro-Québec, Natural Science and Engineering Research Council, Petro-Canada, SNC-Lavalin.). This research was also supported by the "Fondation québécoise de recherche en environnement".

REFERENCES

APHA, AWWA, and WEF. 1992. *Standard methods for the examination of water and wastewater*, .

ASTM. 1990. "Standard Test Method for pH of Water." , 350-358.

ASTM. 1993. "Standard Practice for Oxidation-Reduction Potential of Water." , 345-349.

Barbeau, C. 1996. "Bioaugmentation de sols contaminés au pentachlorophénol par la méthode des sols activés," Mémoire, École Polytechnique de Montréal, Montréal.

Bellamy, K. L., de Lint, N., Cullimore, D. R., and Abiola, A. 1993. "In-situ Intercedent Biological Barriers for the Containment and Remediation of Contaminated Grounwater." *P.90-1T12-6001/CI Revised*, Droycon Bioconcepts Inc., Regina.

Bryan, B. A., Linhardt, R. J., and Daniels, L. 1986. "Variation in Composition and Yield of Exopolysaccharides Produced by *Klebsiella* sp. Strain K32 and *Acinetobacter calcoaceticus.*" *Applied and Environmental Microbiology*, 51(6), 1304-1308.

Cargill, K. L., Pyle, B. H., Sauer, R. L., and McFetters, G. A. 1992. "Effects of culture conditions and biofilm formation on the iodine susceptibility of *Legionella pneumophila.*" *Can. J. Microbiol.*, 38, 423-429.

Characklis, W. G., and Cooksey, K. E. 1983. "Biofilms and Microbial Fouling." *Advances in Applied Microbiology*, 29, 93-138.

Characklis, W. G., and Marshall, K. C. 1990. *Biofilms*, John Wiley & Sons, Inc., New York.

Clément, B. "La qualité par la planification d'expérience (méthode Taguchi)." *Congrès Association québécoise de la qualité*, régionale de Québec, 17.

Dubois, M., Gilles, K. A., Hamilton, J. K., Rebers, P. A., and Smith, F. 1956. "Colorimetric Method for Determination of Sugars and Related Substances." *Analytical Chemistry*, 28(3), 350-353.

Fetter, C. W. 1992. *Contaminant Hydrogeology*, Maxwell Macmillan Canada, Toronto.

Hacking, A. J., Taylor, I. W. F., Jarman, T. R., and Govan, J. R. W. 1983. "Alginate Biosynthesis by *Pseudomonas mendocina.*" *Journal of General Microbiology*, 129, 3473-3480.

Honig, P. 1953. "Principles of Sugar Technology." , Elsevier Press, Inc., Housron, 767.

Huang, J., and Pinder, K. L. 1994. "Effects of Calcium on Development of Anaerobic Acidogenic Biofilms." *Biotechnology and Bioengineering*, 45, 212-218.

Ramsay, J. A., Cooper, D. G., and Neufeld, R. J. 1989. "Effects of Oil Reservoir Conditions on the Production of Water-Insoluble Levan by *Bacillus lichenformis.*" *Geomicrobiology Journal*, 7, 155-165.

Taylor, R. T., Hanna, M. L., Shan, N. N., and Shonnard, D. R. 1993. "In situ bioremediation of trichloroethylene-contaminated water by a resting-cell methanotrophic microbial filter." *Hydrological Sciences - Journal des Sciences Hydriques*, 38(4), 323 - 342.

Trulear, M. G., and Characklis, W. G. 1982. "Dynamics of biofilm process." *Journal WPCF*, 54(9), 1288-1301.

ACTIVE BIOFILM BARRIERS FOR WASTE CONTAINMENT AND BIOREMEDIATION:LABORATORY ASSESSMENT

Matthew J. Brough, Dr Abir Al-Tabbaa and Dr Robert J. Martin
(School of Civil Engineering, University of Birmingham, B15 2TT, U.K.)

ABSTRACT: Laboratory assessment of the introduction and maintenance of a biofilm barrier utilising activated sludge micro-organisms (A.S.M.O.) within sand columns was performed, examining bioimpedance (permeability), viability (heterotrophic plate counts) and biodegradative potential(COD removal). Various methods of activated sludge introduction into the sand column were looked at, with in situ sand/activated sludge mixing identified as the most promising in terms of column clogging. Increasing Mixed Liquor Suspended Solids (MLSS) of the A.S.M.O. caused a greater clogging of voids, with smaller average diameter sands exhibiting a greater extent of pore clogging. Batch and continuous feeding resulted in a maximum 28.1% and 78.7% decrease in sand column permeability respectively in test runs ranging from 1 to 5 days. Associated with this permeability decrease were increases in COD removal and colony forming units/g of dry weight of sonicated sand (CFU/g). Continuous feeding of biologically activated sand columns causes permeability decreases with associated increases in biofilm viability and improvements in biodegradative potential. However, sand columns do not become completely clogged, allowing continuing flow of the synthetic sewage feed and maintenance of the biofilm barrier. The permeability of biologically activated sand columns could be restored with a 1% solution of sodium hypochlorite.

INTRODUCTION

Engineering of physical barriers to isolate contaminated soil and groundwater from the local environment is well developed. However, these impermeable barriers, once breached cannot react to the rupture : they are passive. Recent isolation technologies undergoing research, accelerate natural microbial degradation processes in soil. In-situ stimulation of bacteria, will not only cause permeability decreases in porous media (Kalish et al., 1964, Shaw et al., 1985 and many others) but provide an area of increased contaminant biodegradation. Sub-surface biofilms can bring about the transformation of many trace organic groundwater contaminants with biofilm growth kinetics, and hence biotransformation, strongly influenced by transport characteristics which govern mass transport (Cunningham et al., 1990). The active biofilm barrier will biodegrade any inflow of contaminant with a corresponding increase in biomass (further reducing permeability).

The feasibility of implementation of a biofilm barrier is being studied. A.S.M.O.s due to their natural affinity for surface adhesion, and diverse species population, are suited to biofilm plugging of porous media and contaminant acclimatisation. Providing the correct environmental conditions are supplied to the A.S.M.O.s, their biodegradative potential will be maintained in the sub-surface, causing plugging due to bacterial and exo-polysaccharide accumulation.

Objective. The bench study involves optimisation of a biofilm barrier within sand columns in terms of reductions in permeability and contaminant treatment efficiency (COD removal). Varying permeability sand columns are saturated with the A.S.M.O.s and permeability testing carried out, investigating the effects of method of sludge introduction, nutrient addition and toxic shock (dosing with sodium hypochlorite). Varying contaminant concentrations are passed through the biologically activated sand columns to examine their biodegradative capability in terms of COD removal.

This work will form the basis for implementation of a biofilm barrier, utilising various support media inoculated with A.S.M.O.s. The most successful method of bioimpedance and biodegradation identified, will be applied into a 3-d soil tank using the in situ method of deep soil mixing with special augers.

MATERIALS AND METHODS

Test Cultures Return activated sludge samples were taken exhibiting MLSS from 372 mg/l to 4694 mg/l with a mean MLSS of 920 mg/l. Samples were stored in covered glass reactors to inhibit algal growth, and sparged with air to ensure sufficient dissolved oxygen. Cultures were batch fed on a synthetic sewage feed (1.5625g Glucose, 0.3125g KH_2PO_4, 3.1250g bacteriological peptone/litre of distilled water = 5g/l COD) of 1.1 litres/day and monitored for MLSS variation and viability.

Porous Media Two Leighton Buzzard sands, a coarse and a medium with particle sizes of 0.6-1.18 and 0.3-0.425mm, were used in this study. The sand was washed, autoclaved for 25 minutes and dried at 105° prior to the inoculation of columns with the A.S.M.O.'s. Sand was introduced in layers (10cm thick) into a water filled column, stirred for 1 minute to remove entrapped air bubbles and tamped down 40 times across the sample surface. The average void ratio of both sands was found to be the same and equal to 0.56. The initial permeability of the sands (K_{free}) was 2.611, 0.863 x 10^{-3}m/s for the coarse and medium sands respectively.

Experimental Rig The rig consisted of three columns, 7cm in diameter and 36cm high. The permeability was measured using the constant head test (BS1377, 1991). Tap water used for the permeability experiments was continually pumped around the constant head system ensuring a maximum static head of 1.5m oxygen saturated water to the column outflow. Nutrients/sodium hypochlorite were supplied by dosing pumps, into the tap water percolating upwards through the columns. The permeability was measured along the full length of the columns using four equally spaced, 7cm, standpipes. Diametrically opposite ports were used for liquid phase and sand sampling.

Experimental Procedures All permeability testing was carried out at room temperature with appropriate water viscosity corrections. Activated sludge samples were taken for MLSS determination and either locked in or injected into the column. Locked-in procedure simply involved, sludge addition during sand placement and tamping, with typically 1 hour of run-time for displacement of excess 'free' bacteria. Injection of bacteria simply involved the pumping of activated sludge flocs through the column.

Nutrient feeding was performed on a batch or continuous basis. Batch feeding involved displacement of pore water with the synthetic sewage. After 24 hour saturation , permeability was measured with time. Continuous feeding involved pumping of nutrients into the tap water entry line. When no more change in permeability was noted, a 1% solution of sodium hypochlorite was passed through the column, to assess the barriers resistance to a biocide.

During column runs, permeability testing, biofilm viability and COD sampling was performed along the length of the column. Permeability was measured at regular intervals, with sand and liquid samples taken daily for biofilm viability and COD determination. Sand samples were taken from the centre of a single column during a 3 column parallel run, to make comparisons of permeability and CFU/g. Sand samples were washed three times in 1/4 strength Ringers solution, and sonicated for 30s to remove any biofilm, with 0.5 ml of the resulting suspension plated out, in duplicate, onto Nutrient Agar at 37°c for 48 hours. The sand sample was washed with distilled water and the dry weight measured. 50ml Liquid samples were taken for COD determination.

RESULTS AND DISCUSSION

Results presented here are typical examples of over 35 test runs made, the time requirements for each run ranging from 1-5 days. Test runs were performed either in single, duplicate or triplicate columns with similar initial permeabilities.

Permeability was measured at time i (K_i), relative to the baseline permeability of a clean sand column (K_{free}) and initial baseline permeability of the inoculated sand column (K_{base}). This gives the contribution to permeability change due to an initial clogging by Activated Sludge flocs (K_i/K_{free}) and reductions due to the method of operation/maintenance of the biofilm barrier (K_i/K_{base}).

For locked-in systems, a range of MLSS Activated Sludges were used to produce an initial physical clogging of voids between sand particles. On the basis of the data presented in Figure 1, an increase in MLSS caused increased initial pore plugging prior to a system run. This initial pore plugging is more pronounced in medium sand, with consistently lower K_i/K_{free} values over the range of MLSS.

Following locking-in of activated sludge the system was run for 1 hour to allow flow through of 'excess' free bacteria. Initial column flow rate was kept constant, although this was difficult due to flow instability caused by sloughing of

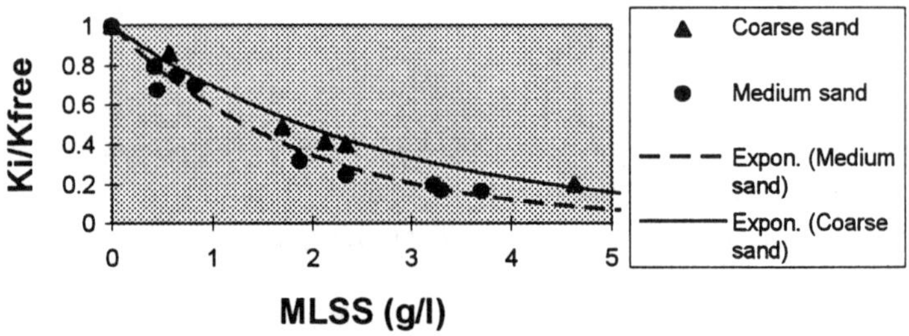

FIGURE 1. Initial clogging of sand due to activated sludge inoculation.

bacteria. A maximum 17% increase in permeability (K_i/K_{base}) was recorded due to the sloughing of Activated Sludge. For low initial MLSS (<1g/l) there was a relatively high sloughing of activated sludge with permeability decay rates decreasing for higher initial MLSS. Highly loaded systems contain a greater proportion of flocs and are therefore less susceptible to permeability decay.

Of five batch fed locked-in test runs, one resulted in a decreased permeability (K_i/K_{base}=0.72). The unsuccessful tests showed a gradual increase in permeability, as activated sludge was washed out of the column. As batch systems are static during nutrient saturation, adsorption processes of biofilm accumulation are greatly reduced. This highlights the importance of flow-related shear stress upon permeability reduction.

Continuous nutrient addition to locked-in systems caused an average 39.4% decrease in permeability, with a maximum 78.7% decrease obtained. Five test triplicate-column runs were performed to analyse COD removal and biofilm accumulation during formation of a biofilm barrier, with continuous nutrient addition. In all cases, decreases in permeability caused increases in the magnitude of COD removal/m of assumed biologically activated sand column. This is shown in Figure 3 with minimum permeability occurring at the same point as maximum COD removal. Results would suggest that with increased influent COD there is an improved reduction in permeability (see figure 2). Furthermore with increased COD influent in different test runs, there is no appreciable change in the magnitude of COD removal per m. Although the COD is not completely removed by the action of the biologically activated sand column, the effluent could be recycled through the biofilm barrier or the biobarrier thickness increased.

Plate counts of sonicated biofilm from sand particles show consistently increases in CFU/g with decreasing permeability (see Figure 2). This infers that permeability reductions are as a result of increases in viable biofilm, and not just due to an accumulation of exo-polysaccharides. Furthermore the increase in COD removal/m with decreasing permeability suggests an improvement in the biofilm barrier's biodegradative potential. However, as all test runs show, sand columns cannot become completely plugged. This ensures viability of the biofilm barrier as the synthetic sewage is allowed to flow through the column, supplying required nutrients (i.e. C,N and P).

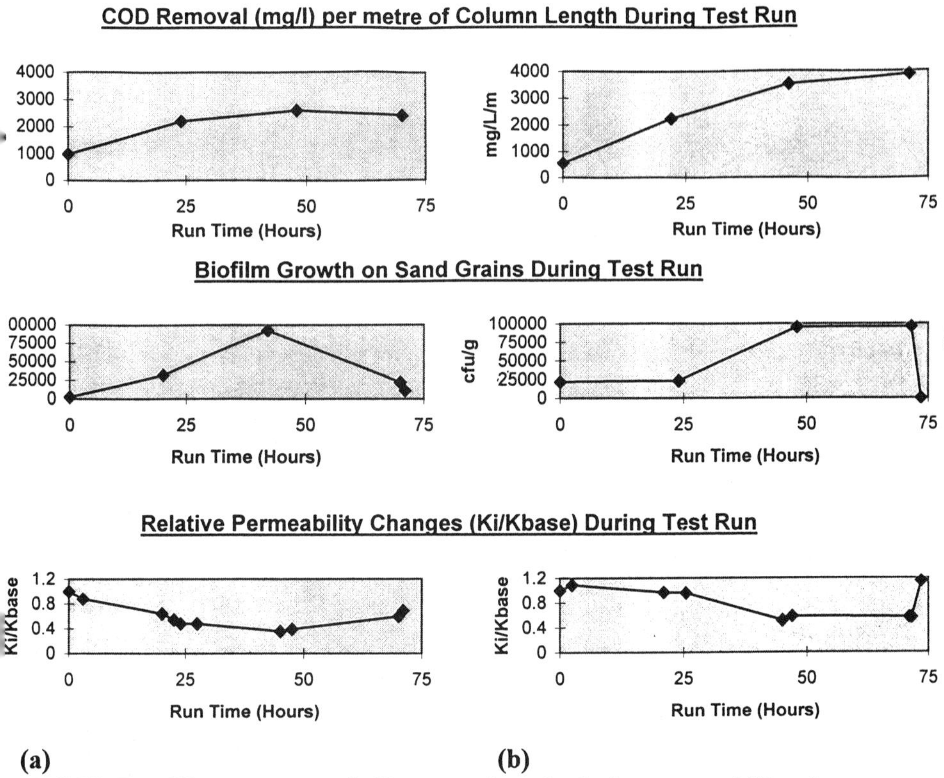

FIGURE 2. COD removal, biofilm growth and relative permeability changes for test run with (a) Influent COD 2160 mg/l (b) Influent COD 528 mg/l.

Several runs were performed injecting sludge into the columns. Permeability reductions were noted during sludge injection, however, head changes were constant throughout the column length, suggesting a clogging at the inlet. As continuous feeding proceeded, after sludge injection, permeability decay rates decreased in the direction of flow (see Figure 3). This inferred Activated Sludge at the inlet was sloughing through the column and clogging pore spaces in the direction of flow. As we move along in the direction of flow, more and more sloughed bacteria are entrained.

Once a biofilm barrier was established, a 1 % solution of sodium hypochlorite was passed through the column. In many cases permeability was restored to the column within 1 hour of chlorine dosing, often above the value of K_{base}, causing a black turbid effluent to leach from the columns outlet.

Comparisons with plate count of sonicated biofilm from sand particles, shows a decrease in CFU/g of dry weight of sand with increases in permeability. At the end of test runs (see Figure 2b), a dramatic decrease in CFU/g can be seen with increases in permeability due to sodium hypochlorite contact.

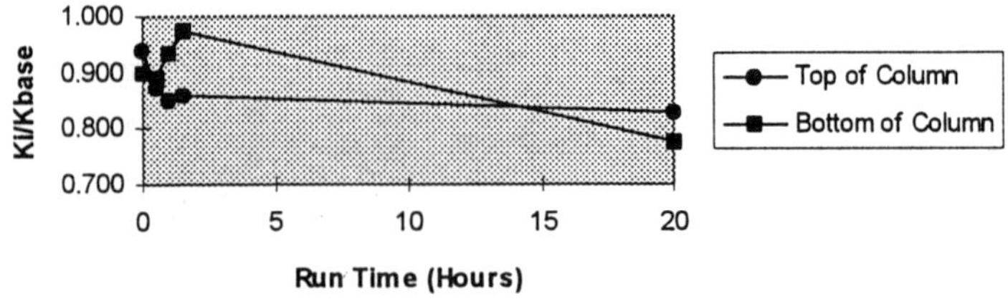

FIGURE 3. Permeability decay after sludge injection and continuous feeding.

CONCLUSIONS

- Locking in of activated sludge caused an initial sand column plugging, with increased MLSS causing higher plugging and a lower initial permeability decay.
- Injection of activated sludge was unsuccessful due to inlet clogging, however, permeability reductions were noted during subsequent continuous feeding.
- Batch feeding of activated sludge cultures is not suitable for column clogging due to the importance of hydrodynamic factors upon biofilm kinetics.
- Continuous feeding of locked in systems proved the most effective technique for the plugging of sand columns. Associated with this permeability reduction were improvements in the viability (heterotrophic plate counts) and biodegradative potential (COD removal per m). Complete plugging never occurred, therefore allowing continued feeding and maintenance of the biofilm barrier.
- 1% sodium hypochlorite percolation (1 hr contact) through plugged columns restored initial permeability. High contaminant concentrations may have a similar effect..

ACKNOWLEDGEMENTS
The authors gratefully acknowledge the financial support of Rockwell (U.K.) Ltd.

REFERENCES

BS1377, 1991. *Methods of Test for Soils for Civil Engineering Purposes*, British Standards Institution, London.

Cunningham A.B., E.J. Bouwer, W.G. Characklis, 1990. "Biofilms in Porous Media." In.W.G.Characklis and K.C.Marshall (Eds.), *Biofilms*, pp. 697-732. John Wiley and Sons Inc., New York.

Kalish P.J., J.E. Steward, W.F. Rogers, E.O. Bennett. 1964. "The Effect of Bacteria on Sandstone Permeability." *J. of Pet. Technol.* 16: 805-814.

Shaw J.C.,B. Bramhill, N.C. Wardlaw, J.W. Costerton. 1985. "Bacterial Fouling in a Model Core System." *Appl.Environ.Microbiol.* 49(3): 693-701.

BIOREMEDIATION OF PENTACHLOROPHENOL-CONTAMINATED GROUND WATER: A PERMEABLE BARRIER TECHNOLOGY

Jason D. Cole, ***Sandra L. Woods***, Peter J. Kaslik,
Kenneth J. Williamson, David B. Roberts

ABSTRACT: Pentachlorophenol (PCP) is a common environmental contaminant due to its use in wood preservation, agriculture, and manufacturing. This poster describes the development of a down-borehole permeable barrier technology for the sequential anaerobic/aerobic bioremediation of pentachlorophenol-contaminated ground water at a wood preserving facility. Under anaerobic conditions, reductive dechlorination of PCP produces lesser chlorinated congeners. Dechlorination pathways and rates depend on environmental conditions and the history of the microbial consortium, and substrate selection.

Substrate evaluation has centered around compounds that are Generally Recognized as Safe (GRAS) by the Food and Drug Administration. In batch tests, several potential electron donors including imitation vanilla flavoring were evaluated for the ability to support PCP degradation in anaerobic and aerobic environments. Under anaerobic conditions, imitation vanilla flavoring resulted in the immediate biotransformation of PCP by the *para* dechlorination pathway to form 2,3,5,6-tetrachlorophenol. In similar aerobic bioassays, complete removal of PCP's anaerobic metabolites, 3,4-dichlorophenol and 3,5-dichlorophenol, was observed with imitation vanilla flavoring. Laboratory findings have provided the basis for nutrient requirements in a field-scale demonstration of a sequential anaerobic/aerobic biological treatment technique for PCP-contaminated ground water.

A 24" diameter borehole was constructed in a PCP-contaminated aquifer at a wood preserving facility. A permeable barrier reactor containing anaerobic and aerobic zones has been constructed for the field demonstration. The site has been characterized, regulatory permission has been granted for the study, and process optimization studies are in progress.

STUDIES OF BIOCLOGGING FOR CONTAINMENT AND REMEDIATION OF ORGANIC CONTAMINANTS

Colin D. Johnston and John L. Rayner (Centre for Groundwater Studies, CSIRO Division of Water Resources, Perth, Western Australia)
D. Stanley De Zoysa and Santo R. Ragusa (Centre for Groundwater Studies, CSIRO Division of Water Resources, Adelaide, South Australia)
Michael G. Trefry and Greg B. Davis (Centre for Groundwater Studies, CSIRO Division of Water Resources, Perth, Western Australia)

ABSTRACT: Laboratory studies of the feasibility of bioclogging to produce barriers to the flow of contaminated groundwater are presented. The ability of indigenous bacteria at a contaminated site to produce polysaccharides which would clog the pore space was investigated in aquifer slurries using glucose as a carbon source. Aerobic conditions using atmospheric oxygen and H_2O_2 as well as anaerobic conditions with nitrate as an electron acceptor were tested. The effect on saturated hydraulic conductivity (K_{sat}) was determined in laboratory columns repacked with aquifer material. Results from the aquifer slurries showed that most polysaccharides were produced by microbiota grown under atmospheric O_2 conditions while H_2O_2 showed limited potential in producing polysaccharides. On the other hand, anaerobic conditions did not appear to favour polysaccharide production. However in one case, slurries of material from immediately below the water table produced much greater polysaccharide concentrations than other treatments with aquifer material from different locations and depths. Amendment of a laboratory column of this material under anaerobic conditions reduced K_{sat} by a factor of 0.07. Factors affecting delivery of the carbon source and other amendments at the field site were investigated with groundwater models.

INTRODUCTION

Bacteria may be stimulated to produce slimes which clog the pore spaces in aquifers and reduce their hydraulic conductivity. Such bioclogging may potentially be harnessed to engineer subsurface barriers that can be used to control the flow of contaminated groundwater and effect degradation of the contaminants. The objective of the work described here is to produce a biologically active, low permeability zone or barrier *in situ* that provides temporary containment of both dissolved organics and organic non-aqueous phase liquids. Laboratory work is aimed at determining the extent to which indigenous bacteria from a site contaminated by BTEX (benzene, toluene, ethylbenzene, xylene) compounds can produce slimes (extra-cellular polysaccharides) as well as the most favourable energy and nutrient conditions for this to occur. Also, a pilot-scale demonstration of bioclogging will be carried out at the contaminated site by stimulating the indigenous bacteria to produce polysaccharides *in situ*. This differs from the work of Cunningham et al. (1991) who injected ultramicrobacteria into the porous media, which were subsequently resuscitated. Although the focus is on gasoline and dissolved BTEX, techniques developed as part of this work could be applied to groundwater contamination by a wide range of pollutants.

The Field Site. A site 15 km north-west of the city of Adelaide in South Australia was chosen for a pilot-scale trial of bioclogging. Leaks from oil pipelines at the site have contaminated the aquifer with diesel and gasoline. Dissolved BTEX from the gasoline has formed a plume in groundwater that is moving towards an estuary 200 m away. Stratigraphy at the site consists of 1.5 m of dredged fill material above a clayey layer approximately 0.3 m thick. This is underlain by approximately 7 m of loose, grey fine to medium grained sand. Dry bulk density is typically around 1.5 Mg/m^3. Pump tests indicate K_{sat} is in the range 3 - 10 m/day (Trefry 1996). Significant amounts of peat and seagrass occur in the upper part of the sand. Organic carbon contents ranged from 0.06 to 7.00% with a median value of 0.18%. Nineteen percent of the samples analysed had organic carbon contents greater than 1%. A stiff brown clay underlays the sand sequence. The water table is around 1.8 m below ground surface. The dissolved BTEX contamination extended from the water table to a maximum 2.5 m below the water table. Concentrations are high (benzene ranged up to 30 mg/L) in the vicinity of the residual gasoline. Benzene is persistent in the aquifer and concentrations of toluene, ethylbenzene and xylene are greatly reduced relative to benzene down gradient of the source. The aquifer is anoxic with dissolved oxygen less than 0.6 mg/L except at the water table and E_h is generally less than -250 mV. Further details of the site are presented by Johnston et al. (1996a,b).

MATERIALS AND METHODS

Aquifer Slurries. Aquifer slurry experiments were used to measure the production of polysaccharides in the presence of different electron acceptors, for different forms and concentrations of carbon source and for aquifer material taken from different depths and locations at the field site. The combinations tested are summarised in Table 1. Oxygen was supplied as atmospheric O_2 and as aqueous H_2O_2. Nitrate was used as an alternative electron acceptor at 730 mg/L. Experiments were done in 500 mL Erlenmeyer flasks with 200 g of aquifer material and 150 mL of growth medium containing organic and inorganic nitrogen, phosphate and trace elements (see Johnston et al., 1996a). A reducing agent was also added to the anaerobic flasks to establish E_h at -130 mV. After adding the carbon source, the flasks were incubated in duplicate at 21°C to match site conditions. Periodic monitoring of the flasks determined polysaccharide concentration, E_h and pH. Polysaccharide concentration was determined by difference between total carbohydrates and residual glucose. Controls with no growth medium or glucose were also incubated. Initial preparations and periodic sampling were carried out in an anaerobic chamber.

TABLE 1. Summary of aquifer slurry experiments

Aquifer material		Electron acceptor	Carbon source	
Location	Depth (m below water table)		Form	Concentration (mg/L)
site a	2 - 3	O_2	glucose	1000 - 50000†
site a	2 - 3	H_2O_2‡	glucose	20000
site a	2 - 3	NO_3^-	glucose	1000 - 50000†
site b	0 - 1	NO_3^-	glucose	20000
site c	0.0 - 0.5	O_2	glucose	10000, 20000
site c	2.0 - 2.5	O_2	glucose	10000, 20000
site c	3.5 - 4.0	O_2	glucose	10000, 20000
site c	0.0 - 0.5	NO_3^-	glucose	10000, 20000
site c	2.0 - 2.5	NO_3^-	glucose	10000, 20000
site c	3.5 - 4.0	NO_3^-	glucose	10000, 20000
site d	3.5 - 4.0	NO_3^-§	glucose	10000, 20000
site d	3.5 - 4.0	NO_3^-	molasses	10000 - 30000¶

‡ concentrations 170, 340, 680, 1360, 2040 and 2720 mg/L
§ concentrations 73, 182, 365, 730 mg/L
† concentrations 1000, 5000, 10000, 15000, 20000, 50000 mg/L
¶ concentrations 10000, 20000, 30000 mg/L equivalent carbohydrates

Column Studies. Column studies were used to determine the effect of polysaccharide production on hydraulic conductivity. An initial study used a 0.5 m-long, 80 mm-internal diameter perspex tube with ports for monitoring hydraulic pressures at 0.02, 0.08, 0.20, 0.32 and 0.48 m along the column. The column was repacked to a dry bulk density of 1.6 Mg/m^3 with material from site b, 0 - 1 m below the water table. A solution of the growth medium, 20000 mg/L glucose and 730 mg/L nitrate (as $NaNO_3$) was passed through the column for 3 days at a constant volumetric flux density of 0.11 m day^{-1}. Pumping ceased and K_{sat} was then determined every 3 days by briefly resuming pumping.

A second series of column tests using 0.3 m-long, 50 mm-internal diameter acrylic tube was conducted with aquifer material from site d, 3.5 - 4.0 m below the water table. The columns had ports at 0.03, 0.09,0.15, 0.21 and 0.27 m. In this series of tests, peaty organic material was segregated and only the fine-medium sand was used in the columns. This series was run under anaerobic conditions with and without nitrate and with different glucose concentrations. The growth medium was not used but NH_4Cl was added at 1600 mg/L as a nitrogen source. Groundwater from the field site was the solvent.

RESULTS AND DISCUSSION

Aquifer Slurries. In the aerobic slurries in contact with atmospheric O_2, polysaccharides concentrations peaked after 24 - 96 hours then declined over the rest of the 7- to 12-day incubation. There was a non-linear response to glucose concentrations - 10000 mg/L was required to get substantive (> 2500 mg/L) peak polysaccharide concentrations, while 15000 - 20000 mg/L was required for sustained polysaccharide concentrations greater than 1000 to 2000 mg/L (Table 2). For example, after 7 - 12 days, in the slurries with 20000 mg/L glucose, polysaccharides ranged from 2080 - 3940 mg/L. The polysaccharides produced depended on the depth below the water table (Table 3), with highest concentrations from the material 0.0 - 0.5 m below the water table and lowest concentrations from the intermediate depths, 2.0 - 2.5 m below the water table.

TABLE 2. Polysaccharide concentrations in aquifer slurries using material from site a, 2 - 3 m below the water table.

Glucose	Polysaccharide concentration (mg/L)			
concentration	Aerobic slurries		Anaerobic slurries	
(mg/L)	maximum	after 7 days	maximum	after 9 days
1000	70	50	230	20
5000	150	90	-	90
10000	2560	360	780	150
15000	3830	1170	-	230
20000	4000	2290	-	450
50000	7400	4320	2760	530

TABLE 3. Polysaccharide concentrations in aquifer slurries after 12-days incubation using material from different depths at site c

Depth interval	Polysaccharide concentration (mg/L)			
	Aerobic slurries		Anaerobic slurries	
(m below water table)	10000 mg/L glucose	20000 mg/L glucose	10000 mg/L glucose	20000 mg/L glucose
0.0 - 0.5	2350	3940	2880	3040
2.0 - 2.5	1140	2080	710	870
3.5 - 4.0	1580	3860	1120	1360

Difficulties were envisaged in creating aerobic conditions with atmospheric O_2 in the aquifer so H_2O_2 was trialed at a range of concentrations. Unfortunately the H_2O_2 was not efficacious - peak polysaccharide concentrations only increased to 2420 mg/L compared to 1510 mg/L in the control and after 6 days of incubation, concentrations were less than 430 mg/L and were little different to the control of 270 mg/L. The limited utility stemmed from the consumption of the H_2O_2 during reduction of organic matter in the aquifer material. The H_2O_2 may have also broken down the polysaccharides. At the higher concentrations, it may have been toxic to the bacteria.

Overall, anaerobic conditions are not as favourable as aerobic conditions for polysaccharide production (Tables 2 and 3). However, interpreting results from anaerobic slurries is somewhat confused due to differences between material taken from different locations and different depths at the field site. A comparison can be made for the slurries using 20000 mg/L glucose with 730 mg/L NO_3^-

(1000 mg/L $NaNO_3$). At site a, 2 - 3 m below the water table polysaccharide concentrations were 450 mg/L after 9 days and at site d, 3.5 - 4.0 m, concentrations were 290 mg/L after 14 days after attaining a maximum of 1170 mg/L. In fact for all nitrate concentrations tested with this material, peak polysaccharide concentrations and concentrations after 14 days were similar to controls with no nitrate. However, at site c, polysaccharide concentrations after 12 days incubations varied from 870 - 3040 mg/L for the three depths (Table 3). The site b, 0 - 1 m slurry gave a spectacularly different result with a maximum polysaccharide concentration of 14100 mg/L and the concentration was still 4440 mg/L after 30 days incubation.

Molasses was investigated as an alternative carbon source because of cost benefits and to avoid the addition of supplements. With the supplementary growth medium, polysaccharide concentrations were similar to those for glucose in anaerobic slurries of the same aquifer material. Greater polysaccharide concentrations were observed without the growth medium. However polysaccharide concentrations after 9 days of incubation were very modest - 240 mg/L for the 10000 mg/L carbohydrate, 570 mg/L for the 20000 mg/L carbohydrate and 1270 mg/L for the 30000 mg/L carbohydrate treatments.

Column Studies. For the initial column with material from site b, 0 - 1 m below the water table, K_{sat} of the interval of the column between 0.02 and 0.32 m (the 0.48 m port was not measured after 3 days) decreased from an initial value of 0.33 m/day to 0.023 m/day after 390 hours. The most rapid change in K_{sat} was observed in the first 18 hours and further reductions continued slowly after 50 hours. Pressures showed that K_{sat} had been reduced throughout the column. At 48 hours, K_{sat} within the column varied by a factor of 2.4 compared to the factor 14 reduction for the column as a whole. The relatively low initial K_{sat} compared to pump test estimates is attributed to the inclusion and distribution of peaty material during homogenising and repacking of the column.

In the second series of tests with material from site d, 3.5 - 4.0 m, the trial with 10000 mg/L glucose, 1600 mg/L NH_4Cl and no nitrate did not produce any significant reduction in K_{sat}. There was production of copious amounts of CO_2 (80% by volume) and N_2 (20% by volume) in the column indicating fermentation reactions. All the glucose was being utilised within the column and organic acid production is presumed to be the cause of lowered pH.

The other trial in the second series which used 20000 mg/L glucose, 1600 mg/L NH_4Cl and 730 mg/L nitrate, showed large reductions (factor 0.02) in the K_{sat} of the column as a whole. This was due to clogging in the inlet endplate of the column and within the first 0.03 m of the column. Changes in K_{sat} over the rest of the column (0.03 - 0.27 m) were hard to identify because of the very large head losses across the inlet but data indicate a reduction from 12 m/day to 4 m/day over 18 days. Evolution of N_2 (80-90% by volume) and CO_2 (10-20% by volume) indicated denitrification and some fermentation in the column.

Implementation at the Field Site. The objective in bioclogging at the field site is to produce a pilot-scale barrier of reduced hydraulic conductivity 10 m long by at least 2 m wide extending to 4 m below the water table. To engineer such a barrier, the appropriate amendment solutions need to be determined and a delivery system for the amendments need to be designed. An objective has been to use a once-off addition of amendment solution. Using nitrate as an electron acceptor seems a logical choice in these circumstances. Dissolved atmospheric oxygen would not persist in the anoxic, high organic carbon aquifer at the field site and the laboratory tests have shown H_2O_2 to be ineffective. However, the response of polysaccharide production in the laboratory tests with nitrate call for further investigation to elucidate the variable and sometimes contradictory results. Column studies are continuing to provide data for the design of the field trial.

Challenges faced in the design of a cost-effective delivery system include efficient delivery to the specified volume of the aquifer, distribution of the amendments before significant clogging takes place and minimising disturbance to existing contamination. Extensive 3-D groundwater flow and transport modelling has suggested that a linear, five-well, balanced injection/extraction system could effectively deliver the amendment solution to the required volume of the aquifer.

ACKNOWLEDGEMENTS

This work was made possible with part funding by South Australian Department of Environment and Natural Resources Office of the Environment Protection Authority, BP Australia Ltd, Minenco and CRA.

REFERENCES

Cunningham, A.B., W.G. Characklis, F. Abedeen, and D. Crawford. 1991. "Influence of biofilm accumulation on porous media hydrodynamics." *Environ. Sci. Technol.*, *25*: 1305-1311.

Johnston, C.D., J.L. Rayner, D.S. De Zoysa, S.R. Ragusa, M.G. Trefry, G.B. Davis, and C. Barber. 1996a. "Bioclogging of aquifers for containment and remediation of organic contaminants: Preliminary laboratory and field studies." In Proc NATO/CCMS 1996 Pilot Study, 11-16 February 1996, Adelaide, South Australia.

Johnston, C.D., J.L. Rayner, M.G. Trefry, D.S. De Zoysa and S.R. Ragusa,. 1996b *Bioclogging of Aquifers for Containment and Remediation of Organic Contaminants: A Report on Progress March - October 1996*. CSIRO Division of Water Resources, Consultancy Report 96-42

Trefry, M.G., 1996. *Pump Test Analysis for the Largs North Aquifer, South Australia*. CSIRO Division of Water Resources, Technical Memorandum 96.21

THE USE OF OXYGEN RELEASE COMPOUND (ORC®) IN BIOREMEDIATION

Stephen Koenigsberg, Craig Sandefur and William Cox
(Regenesis Bioremediation Products, San Juan Capistrano, CA)

ABSTRACT: ORC® is a unique formulation of magnesium peroxide that releases oxygen slowly when hydrated. The compound is insoluble and releases oxygen while being converted to ordinary magnesium hydroxide which is also insoluble. ORC® is packaged in exchangeable filter socks and is contacted with contaminated groundwater via an array of wells or trenches. ORC® can also be made into a slurry for permanent applications in the saturated zone, or dispersed as a free powder for the in situ or ex situ treatment of soil. These methods help optimize the natural bioremediation of aerobically degradable compounds and are being used on over 700 sites nationwide as a low-cost, passive bioremediation protocol.

ORC®-mediated oxygenated zones generally last four months to a year as a function of contaminant flux. The objective of the "oxygen barrier" is plume cut-off; however, any significant reduction of contaminant mass will bring the control point back to the source and reduce risk. A broad array of treatment points, in which ORC® is backfilled or injected, has been shown to be an effective source treatment. The points can be implemented with low-cost, small-bore technologies to achieve full remediation or risk reduction objectives.

An overview of the history of the environmental applications of ORC®, and the evolution of the strategies for site closure involving direct treatment of contaminant sources and control of plume migration, will be presented. This will include 1) a compendium of oxygen barrier field results featuring several large-scale, fully monitored demonstrations, 2) results from a several source placement applications, and 3) results from a variety of in situ soil applications. All of these projects were directed at BTEX remediation and showed that ORC® is a highly effective bioremediation protocol. Preliminary evidence for the use of ORC® in the remediation of vinyl chloride, PCP, PAHs, and MTBE will also be presented.

SEQUENTIAL TREATMENT USING ABIOTIC REDUCTIVE DECHLORINATION AND ENHANCED BIOREMEDIATION

Susan M. Froud (University of Waterloo, CANADA),
R.W. Gillham, J.F. Barker, J.F. Devlin (University of Waterloo, CANADA),
M.J. Brown (Komex International Ltd., CANADA)
and M.L. McMaster (Beak International Inc., CANADA)

ABSTRACT: Remediation of contaminated groundwater can be complicated by the presence of several, often very different, contaminants, requiring very different treatment technologies. For this reason, the concept of sequential treatment is being considered and a controlled field evaluation of a reduction-oxidation sequence is currently underway at CFB Borden, Ontario. A mixed chlorinated solvents (PCE and CT) and aromatics (toluene) dissolved-phase plume is being introduced at concentrations of 1 to 5 mg/L for all components. The semi-passive treatment system consists of removable cassettes with zero valent iron, used for abiotic reductive dechlorination of chlorinated solvents; followed by oxygen release compound (ORC™), used to enhance natural aerobic biodegradation of petroleum aromatics. Preliminary results indicate that degradation of the chlorinated compound in the presence of iron is proceeding as expected, with transport of the toluene being uninhibited through this zone. Evaluation of the secondary phase of the experiment involving enhanced bioremediation (particularly of toluene) will be possible as the plume continues to develop. The earliest data indicate that the aerobic biodegradation stage of treatment may be limited by the slow release of oxygen from the ORC™ socks due to the high pH water emanating from the iron. Further investigations into this problem are currently being undertaken.

INTRODUCTION

Groundwater plumes frequently comprise more than one type of contaminant, and often, the contaminants can not be treated by a single method of remediation. For this reason, the possibility of combining several technologies, of proven effectiveness for individual components of the plume, is being considered. The question remains, however, as to what inherent difficulties might exist in this type of sequential treatment system.

In addition, the ineffectiveness and inefficiencies associated with "Pump and Treat" technologies have been realized (Mackay and Cherry, 1989) and a move toward more passive alternatives is being made (Gillham, 1995; Barker et al, 1995). With this change in strategy it is expected that remedial efforts will become more cost effective and less labour intensive.

The Advanced Applied Technology Demonstration Facility (AATDF) has provided funding for the critical evaluation of several semi-passive or passive technologies for the remediation of such multi-contaminant plumes. A controlled field experiment is currently underway at CFB Borden, Ontario.

EXPERIMENTAL DESIGN

Multiple Gate System The Borden experimental site is shown in Figure 1 to consist of three channels (referred to as gates) to control groundwater flow in the natural sand aquifer. Sealable joint sheet-piling (Waterloo Barrier^{TM}), keyed into the underlying clay aquitard at depths of 3.5 to 4 m below ground, form the walls of the gates. The up-gradient ends are open to allow capture of the natural groundwater flow, while the down-gradient ends of the gates are closed. The hydraulic gradient within the system is controlled by pumping from extraction wells, at a rate sufficient to achieve the desired groundwater flow velocity of between 10 and 15 cm/day. The site is heavily instrumented with multi-level piezometers and standard US EPA performance wells. Also, a row of eleven 25.4 cm diameter PVC source wells screened from depths of 1 to 4 m are located at the open end of the gates. The locations of all wells are shown in Figure 1.

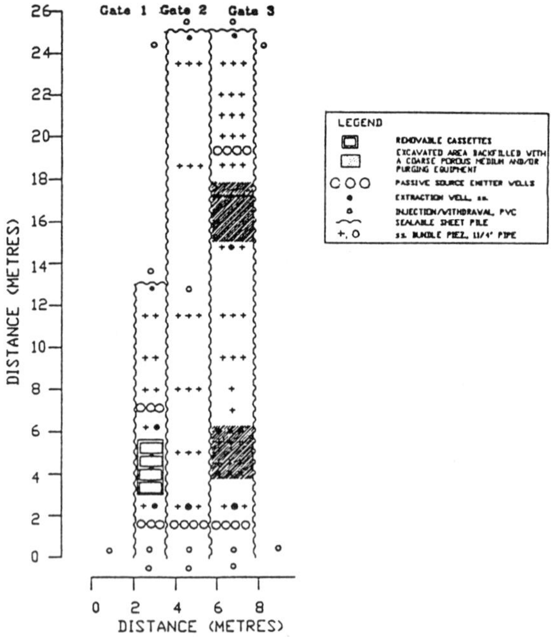

FIGURE 1: Site layout for experimental gates at CFB Borden, Ontario

In June and July, 1996, a pulse of conservative tracer solution (KBr) was released from the source wells and was carried under the imposed hydraulic gradient conditions of the system. Detailed monitoring was conducted to characterize the groundwater flow regime, including flow velocities and heterogeneous zones.

The creation of a mixed organic plume comprised of perchloroethylene (PCE), carbon tetrachloride (CT), and toluene, at concentrations of 1 to 5 mg/L for each component, commenced on October 17, 1996. The diffusive release source emitters were designed based on work by Wilson and Mackay (1995) and will operate for a several months in a semi-passive fashion.

The sequential treatment technologies, zero valent iron and enhanced bioremediation using oxygen release compound (ORC™), being used in Gate 1 for the remediation of the mixed organic plume are the focus of this paper. Gate 3 is being used to investigate the effectiveness of sequential enhanced anaerobic and aerobic bioremediation zones, while Gate 2 functions as a control gate in which only natural attenuation processes occur.

Gate 1 Design In Gate 1, a cassette module containing four removable cassettes was installed and sealed within the sheet-piling walls. Multi-level piezometers were attached to the down-gradient side of the first three cassettes to permit groundwater monitoring. Conceptually, these cassettes can be used for remediation materials for this experiment (zero valent iron and ORC™) and re-used again in subsequent experiments to evaluate other combinations of materials. Details of the cassette system are shown in Figure 2.

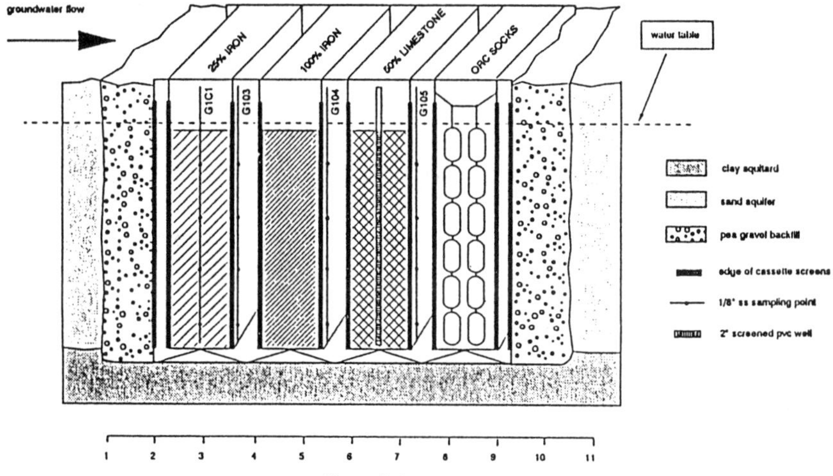

FIGURE 2: Schematic of the removable cassette system

On April 23, 1996, cassette C1 was backfilled with a 25 wt% granular iron filings and 75 wt% silica sand mixture, while cassette C2 was backfilled with 100 % granular iron. Previous work by Gillham and O'Hannesin (1994) has demonstrated zero valent iron to be effective in the abiotic reduction of chlorinated contaminants including PCE and CT; however, the iron material is ineffective in the degradation of non-halogenated aromatics. The use of two iron cassettes will permit rate monitoring of the dechlorination process and will facilitate the identification of breakdown products resulting from incomplete degradation.

On June 10, 1996, cassette C3 was backfilled with a 50 % by volume mixture of washed limestone screenings and silica sand to promote the precipitation of any dissolved constituents emanating from the iron cassettes.

Cassette C4 contained standing water into which fifty-six ORC™ socks (approximately 10 cm diameter, 30 cm length) were suspended on November 15, 1996. Dissolved oxygen (DO) concentrations and other parameters (pH, Eh, conductivity and temperature) were monitored on a regular basis in the vicinity of the cassette to determine the performance of the system. It was anticipated that increased DO concentrations in the groundwater through the use of ORC™ would stimulate the aerobic biodegradation of toluene, the only contaminant that should remain in the plume after treatment with zero valent iron.

RESULTS AND DISCUSSION

The results of the conservative tracer test indicate that the groundwater flow velocity in the natural sand aquifer is approximately 13 cm/day. Through the cassette system, however, the average flow velocity is 20 cm/day, likely due to an increased porosity. Tracer data also provided insight into the presence of some heterogeneities within the natural sand, but through the engineered materials in the cassettes good lateral and vertical mixing of the tracer solution was observed, a positive feature for effective remediation.

Preliminary data indicate that degradation of the chlorinated solvents (PCE and CT) is being achieved in the iron cassettes, while concentrations of toluene have not decreased significantly. Analyses for possible degradation products from the chlorinated solvents are underway, but results are not yet available. The plume has not yet reached the monitoring points down-gradient of the ORC™ cassette and thus data concerning toluene removal are not yet available.

Figure 3 provides plots of pH values and DO concentrations along the length of Gate 1 at all sampling depths. pH values were determined in the field using an Orion Model 260 pH meter. DO measurements were determined in the field using both an Orion Model 835 DO meter and oxygen Chemetrics kit for comparison. Results shown are for the DO probe only, however, they agree reasonably well with those obtained by the Chemetrics method. All probe measurements were conducted using a "flow-through" cell.

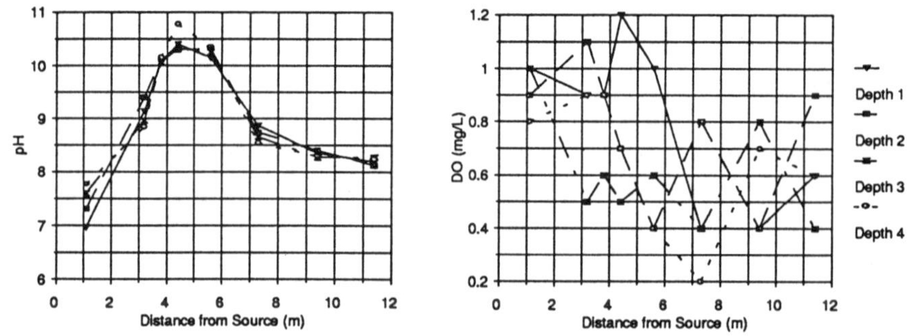

FIGURE 3: pH and dissolved oxygen (DO) profiles on December 16, 1996

Sharp increases in pH in the vicinity of the iron cassettes (2.8 to 3.8 m from source), to values greater than 10, are evident and are comparable to previous results (Gillham and O'Hannesin, 1994). A gradual decrease in pH is observed further down-gradient after contact with the natural aquifer material, indicating the presence of some buffering capacity in Borden sand.

The DO profiles appear to be quite variable, however, a marked increase beyond the ORC™ (5 m from source) is not evident. Previous studies using ORC™ have shown initial increases in DO concentrations to values exceeding 20 mg/L in the immediate vicinity of the socks, over the short-term (several weeks). A DO plume with concentrations up to 10 mg/L was sustained down-gradient of ORC™ source wells (Smyth et al., 1995). Discussion with the product manufacturer (Regenesis Bioremediation Products, Inc., San Juan Capistrano, California) and research into the components of ORC™ (magnesium hydroxide, oxide and peroxide) has lead to the hypothesis that the product is relatively stable at high pH and consequently releases less oxygen under these conditions.

FUTURE WORK

The high pH environment (greater than 10) down-gradient of the iron cassettes may be limiting the release of oxygen from the ORC™ socks. By decreasing the pH of the water contacting the ORC™, the release of oxygen may be enhanced. Various alternative designs to achieve this are under consideration.

Prior to the implementation of any design variations, the following questions remain to be answered:

- Is pH the variable controlling the performance of the ORC™?
- What is the buffering capacity of the in-situ aquifer material?
- What is the desirable pH value to activate the ORC™?

These questions will be addressed through simple laboratory tests.

If the ORC™ limitations in the cassette system can not be resolved, large diameter contingency wells (SW's 12, 13 & 14; Figure 1), approximately 2.5 m down-gradient of the iron zone, may provide a more suitable environment for the slow-release of oxygen from the socks. Inherent to the performance of the ORC™ is the availability of sufficient natural buffering capacity of the aquifer to ensure that the desired pH exists at this new location (currently the pH is approximately 8.5).

If no suitable means of promoting the release of oxygen from the ORC™ socks can be devised, alternative methods to increase the DO concentration for use in the cassette system could be evaluated. These techniques might include oxygen gas diffusion through low-density polyethylene tubing (LDPE), or bio-sparging.

The introduction of oxygen into the system using ORC™ or an alternative means does not guarantee the effectiveness of the system. Other problems that might hinder remediation efforts could include, the development of an unfavourable environment for microbial populations in the aerobic zone (ie. high pH), or the production of precipitates in the iron or aquifer that may have adverse effects on the hydraulic properties of the system. Further testing and analyses will be performed to evaluate the importance of these potential problems

CONCLUSIONS

Results to date indicate that zero valent iron is performing effectively in removing the chlorinated hydrocarbons; however, the low DO values suggest that toluene will not be degraded in the present system. If it is confirmed that the performance of the ORC™ is a consequence of the elevated pH, then modifications, possibly the addition of a buffer material, may be required in order for the iron-ORC™ treatment materials to be compatible.

ACKNOWLEDGEMENTS

Funding for this work was provided wholly or in part by the United States Department of Defense under Grant No. DACA 39-93-1-001, to Rice University for the Advanced Applied Technology Demonstration Facility for Environmental Technology Program (AATDF). This work does not necessarily reflect the position of these organizations and no official endorsement should be inferred.

REFERENCES

Barker, J., C. Austrins, J. Gorman, J. Devlin, D. Smyth, and J. Cherry. 1995. "Controlled In-situ Bioremediation of Groundwater." Published in the proceedings of the 5th Annual Symposium on Groundwater and Soil Remediation (GASReP), Toronto, Ontario, Oct. 2-6, 1995.

Gillham, R.W. 1995. "In-situ Treatment of Groundwater: Metal-Enhanced Degradation of Chlorinated Organic Contaminants." Published in the proceedings of the Recent Advances in Ground-Water Pollution Control and Remediation conference, A NATO Advances Study Institute, Kemer, Antalya, Turkey, May 20-June 1, 1995.

Gillham, R.W. and S.F. O'Hannesin. 1994. "Enhanced Degradation of Halogenated Aliphatics by Zero-Valent Iron." *Ground Water 32*(6): 958-967.

Mackay, D.M. and J.A. Cherry. 1989. "Groundwater Contamination: Pump-and-Treat Remediation." *Environmental Science and Technology. 23*(6): 630-636.

Smyth, D.J.A, B.T. Byerley, S.W. Chapman, R.D. Wilson and D.M. Mackay. 1995. "Oxygen-Enhanced In-situ Biodegradation of Petroleum Hydrocarbons in Groundwater Using a Passive Interception System." Published in the proceedings from the 5th Annual Symposium on Groundwater and Soil Remediation (GASRep), Toronto, Ontario, Oct. 2-6, 1995.

Wilson, R.D. and D.M. Mackay. 1995. "A Method for Passive Release of Solutes from an Unpumped Well." *Ground Water. 33*(6): 936-945.

FIELD TRIAL OF AN IN SITU ANAEROBIC/AEROBIC BIOREMEDIATION SEQUENCE

Dennis J. Katic (University of Waterloo, Waterloo, Canada)
John F. Devlin (University of Waterloo, Waterloo, Canada)
James F. Barker (University of Waterloo, Waterloo, Canada)
Michaye L. McMaster (Beak International Inc., Guelph, Canada)
Michael J. Brown (Komex International Ltd., Calgary, Canada)

ABSTRACT: A pilot scale experiment at Canadian Forces Base Borden is evaluating the performance of a combined anaerobic/aerobic bioremediation sequence within a funnel-and-gate flow control structure to treat a mixed plume of dissolved organic contaminants consisting of chlorinated solvents and aromatic hydrocarbons. Initial induction of anaerobic conditions are accomplished by a permeable wall injection system intended to introduce benzoate, a labile carbon substrate, into the flow system. Further downgradient, a biosparging system aerates the contaminated groundwater. Both systems are operated in a pulsed manner. Preliminary geochemical data indicate dissolved oxygen and redox potential in the anaerobic zone as low as 0.15mg/L and -242.6mV respectively (relative to background values of 0.25mg/L and -103.1mV), increasing to as high as 20.9mg/L and 169.2mV immediately downgradient of the biosparge system.

INTRODUCTION

Difficulties in the remediation of some contaminated groundwater can arise due to the presence of mixtures of different classes of organic compounds, such as chlorinated solvents and aromatic hydrocarbons. Combining multiple in situ technologies in tandem can increase the effectiveness of remediation of these groundwaters. Cometabolic biotransformation of dissolved chlorinated solvents under anaerobic conditions (McCarty and Semprini, 1994) and aerobic degradation of aromatic hydrocarbons (Bianchi-Mosquera et. al., 1994) have been demonstrated in field studies. A combined anaerobic/aerobic bioremediation sequence thus has the potential to treat a mixture of organic contaminants in situ.

A pilot scale experiment at Canadian Forces Base Borden attempts to induce geochemical conditions in the aquifer to effect in situ bioremediation of an artificially created plume containing 1 to 5 mg/L of mixed organic contaminants. Anaerobic transformation of tetrachloroethene and tetrachloromethane will be followed by aerobic transformation of toluene and residual by-products of the anaerobic process. Geochemical indicators used to establish the redox environment include redox potential, dissolved oxygen, nitrate, nitrite, sulfate, ferrous iron, methane, ethane and ethene.

MATERIALS AND METHODS

The treatment system is confined within a sheet piling funnel (Figure 1), located within a shallow sandy aquifer. Groundwater enters the upgradient end and is drawn through by extraction at the downgradient end.

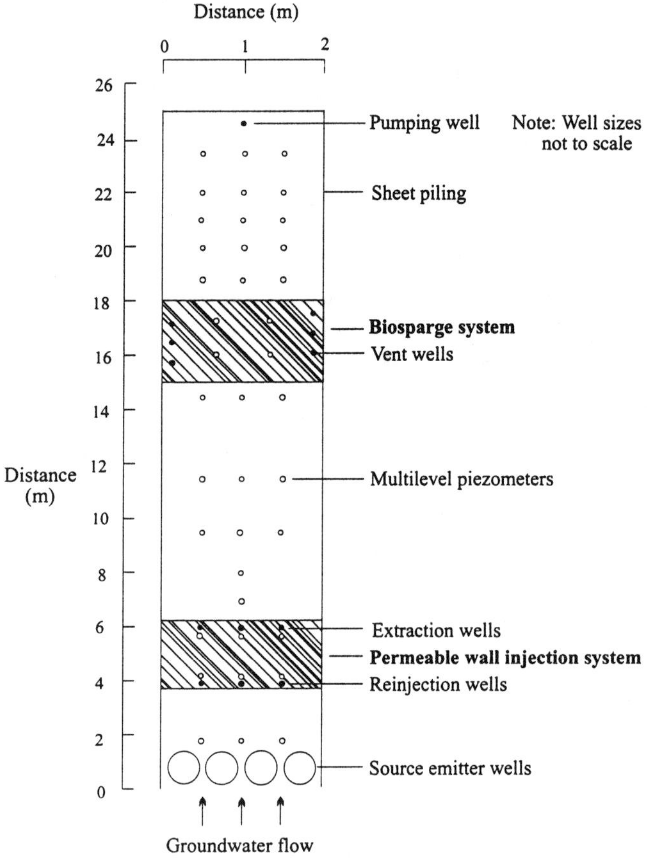

FIGURE 1. Plan view schematic layout of funnel-and-gate treatment system.

Anaerobic conditions are achieved in the first stage by a permeable wall injection system (Figure 2), as described by Devlin and Barker (1994, 1996). The system is designed to intermittently circulate and amend native groundwater within the wall with a labile carbon substrate (sodium benzoate). Advection and dispersion facilitate mixing between the added nutrients and contaminants downgradient of the injection zone. Pulse intervals are set to create a continuous, strongly reducing zone approximately 5m downgradient of the injection system within which chlorinated contaminants are remediated. The in situ design consists

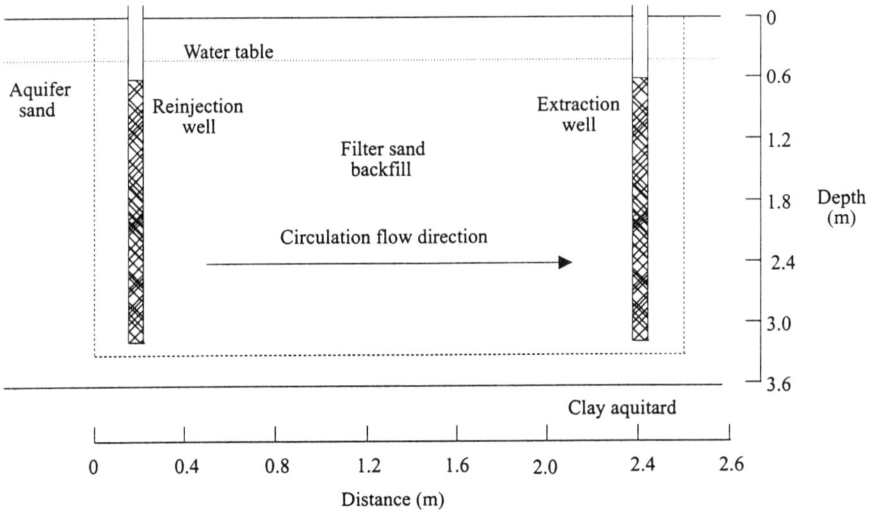

FIGURE 2. Longitudinal cross-section of permeable wall injection system.

of filter sand filling the wall, three 5.1cm diameter PVC extraction wells at the downgradient end, three identical reinjection wells at the upgradient end, a circulation pump, and an in-line nutrient amendment system consisting of a water-driven proportional feeder and reservoir.

Further downgradient, aerobic conditions are induced via a biosparge system (Figure 3) which, at pulsed intervals, delivers gaseous oxygen up to the residual capacity of a gravel filled formation. Oxygen delivery is achieved by six L-shaped 5.1cm diameter PVC aeration wells, with vent holes located laterally across the bottom of the biosparge zone. Four manifolded 0.63cm o.d. polyethylene tubes extend to equally spaced lengths along the horizontal section of each well, with all wells fed by a single pressurized oxygen cylinder.

Non-sealed sheet piling extending from above ground surface to the water table at the up and down gradient ends of the formation help control lateral migration of contaminants that may have been volatilized during the sparging process. It is expected that additional biodegradation of volatilized organics will occur in this headspace zone and this process will be monitored. Narrow diameter (0.64cm o.d., 0.3cm i.d.) stainless steel soil gas probes fitted with Mininert caps are used to monitor off-gas concentrations immediately above the saturated zone.

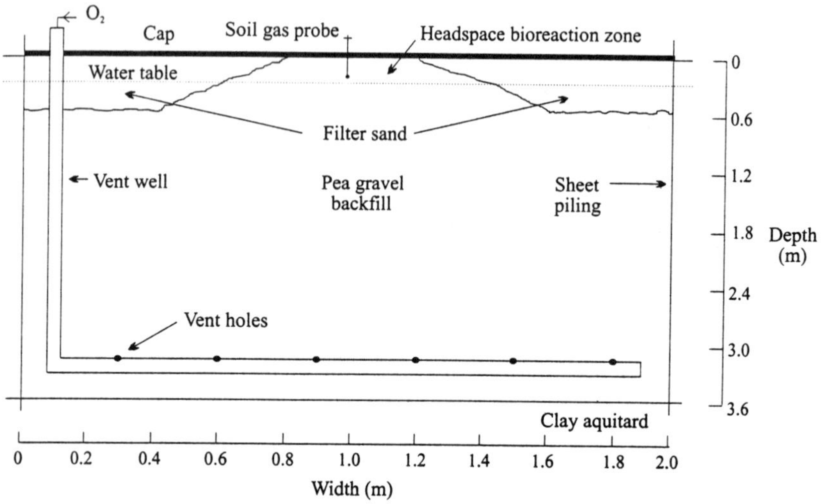

FIGURE 3. Transverse cross-section of biosparge system.

PRELIMINARY RESULTS AND DISCUSSION

Trends in dissolved oxygen (DO) and redox potential (E_h) with distance along the length of the treatment systems are shown in Figure 4. These were obtained after both systems were operated, although before contaminants were introduced.

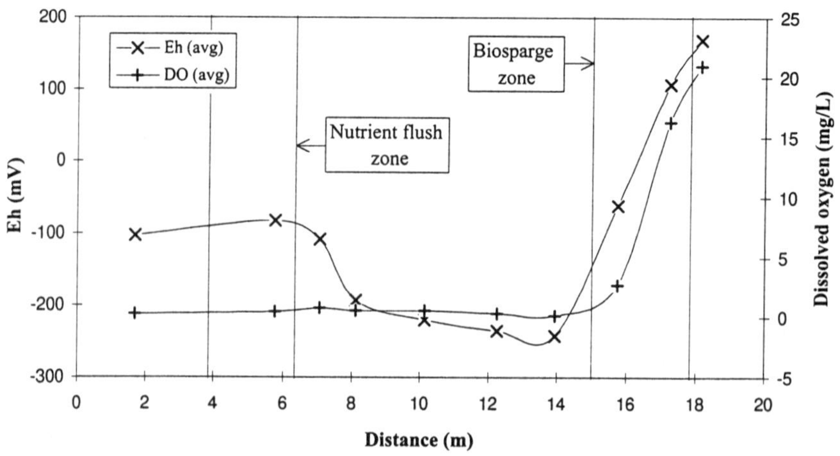

FIGURE 4. Depth averaged DO and E_h trends longitudinally along the length of the treatment system. Measurements are over a one month period.

Preliminary results indicated that highly reducing conditions can be maintained in the anaerobic zone via pulsed nutrient delivery, with DO and E_h values as low as 0.15mg/L and -242.6mV respectively, representing depth-averaged values between 1.26 to 3.36m below ground surface. A slight upward trend in Eh at the tail end of the permeable wall and into the aquifer as well as slightly elevated DO (within 0.5mg/L above background) suggested that aeration of the groundwater had occurred during nutrient circulation. However, all DO values remained below 1mg/L for the entire length of the anaerobic treatment zone. Further downgradient E_h values declined, reaching a minimum prior to the biosparge system. This suggests that establishing and maintaining geochemical conditions conducive to reductive dechlorination of solvents is possible using such a system. A plume containing dissolved tetrachloroethene and tetrachloromethane is currently entering the system, with concentration profiles expected to decline with distance downgradient.

Relatively high dissolved oxygen contents within and immediately downgradient of the biosparging system (see Figure 4) were obtained. This indicates reasonably effective mass transfer of gaseous oxygen to the groundwater using the delivery system. Whether conditions remain aerated in the presence of up to a 200mg/L biochemical oxygen demand (primarily as benzoate) has yet to be determined. It is anticipated the oxygen content will be sufficient to permit aerobic degradation of the combined flux of nutrient, toluene and lower order breakdown products of the anaerobic system prior to reaching the end of the treatment system.

The extent of contaminant degradation is currently being assessed, with the amount and rate of mass loss (including the potential for volatilization from the biosparging system) to be quantified over the next four months.

ACKNOWLEDGMENTS

Funding for this work was provided wholly or in part by the United States Department of Defense under Grant No. DACA 39-93-1-001, to Rice University for the Advanced Applied Technology Demonstration Facility for Environmental Technology Program (AATDF). This work does not necessarily reflect the position of these organizations and no official endorsement should be inferred.

REFERENCES

Bianchi-Mosquera, G.C., R.M. Allen-King, and D.M. Mackay. 1994. Enhanced Degradation of Dissolved Benzene and Toluene using a Soild Oxygen Releasing Compound. *Ground Water Monitoring and Remediation*. Winter: 120-128.

Devlin, J.F., and J.F. Barker. 1994. "A Semipassive Nutrient Injection Scheme for Enhanced In Situ Bioremediation." *Ground Water*. 32(3): 374-380.

Devlin, J.F., and J.F. Barker. 1996. "Field Investigation of Nutrient Pulse Mixing in an In Situ Biostimulation Experiment." *Water Resources Research*. 32(9): 2869-2877.

McCarty, P.L., and L. Semprini. 1994. "Ground-water Treatment for Chlorinated Solvents." In R.D. Norris, R. Brown, P.L. McCarty, L. Semprini, J.T. Wilson, D.H. Kampbell, M. Reinhard, E.J. Bouwer, R.C. Borden, T.M. Vogel, J.M. Thomas and C.H. Ward, *Handbook of Bioremediation*, pp. 87-116. Lewis Publishers, Boca Raton, FL.

SITE CHARACTERIZATION METHODS FOR THE DESIGN OF IN-SITU ELECTRON DONOR DELIVERY SYSTEMS

Steven D. Acree (US-EPA, NRMRL-Ada, OK), Mike Hightower (Sandia National Laboratory, Albuquerque, NM), Randall R. Ross (US-EPA, NRMRL-Ada, OK), **Guy W. Sewell** (US-EPA, NRMRL-Ada, OK), Brent Weesner (Lockheed Martin Specialty Components, Largo, FL)

ABSTRACT: The Department of Energy and the U.S. Environmental Protection Agency have been involved in designing and evaluating a pilot field demonstration of reductive anaerobic biological in-situ treatment technologies (RABITT) for use as a standard remedial technology for chloroethene contamination. Innovative site characterization techniques have been utilized to identify the hydraulics of the site and in particular the vertical distribution of relative hydraulic conductivities. Direct extraction of intact frozen cores has been utilized to determine the vertical distribution of contaminants in the pore spaces and on the solid matrix of site material. The combination of these techniques along with standard site characterization methods has been used the develop a three-dimensional picture of the site with vertical resolutions down to 0.5 ft (15 cm). This information has then been used to evaluate different scenarios for nutrient/electron donor delivery at the site, and when used with appropriate transport and flow codes was used to exclude designs which did not allow for significant mixing of donor and contaminants, or which did not efficiently deliver nutrients/donors to all contaminated zones. It is felt that the use of site characterization data in this manner is critical to the effective and appropriate design and implementation of RABITT and other in-situ treatment technologies.

INTRODUCTION

Reductive anaerobic biological in-situ treatment technologies for the remediation of ground water contaminated with chloroethenes is a promising approach to an all too common environmental problem. The design of these treatment systems is a complex environmental engineering challenge requiring a clear understanding of the contaminant distribution, the hydrogeologic setting and the geochemistry. To an even greater extent than with pump-and-treat systems or aerobic-catabolic-bioremediation, the application of RABITT requires that this site conceptual model be a detailed three-dimensional representation, incorporating flow/time dynamics to ensure interaction of electron donor, contaminants and active microorganisms under appropriate conditions.

The Department of Energy, the U.S. Environmental Protection Agency, The State of Florida, and Industry Partners have been involved in the design and implementation of a pilot-scale field demonstration of RABITT to support the DOE's Innovative Treatment Remediation Demonstration (ITRD) Program. The study involves creation of an in-situ circulation cell and the injection of electron donor/nutrients to stimulate biological transformation processes. Constraints on system design included a potentially short half life of the injected solutions and the need for controlled mixing of donor and contaminants. Travel time through the contaminated media was required to be no longer than approximately 100 days. Design of an effective and efficient system under such constraints requires three-dimensional characterization of contaminant distribution, hydrology, and geochemistry.

Detailed, three-dimensional site characterization is seldom performed due, in part, to a lack of appreciation of the potential effects of heterogeneity on remedial design and effectiveness. Estimates of bulk or "average" parameters obtained from traditional monitoring wells and aquifer tests generally have been used in design. Techniques for obtaining detailed hydraulic information, such as extensive laboratory permeameter testing and multi-level slug tests (Molz and others, 1990), have been available but are often costly, difficult to apply, and may not be representative at the field scale. Sensitive borehole flowmeters suitable for detailed characterization have only recently become commercially available. Despite availability, there still exists a general lack of recognition regarding uses for these tools and the value of detailed data in defining contaminant transport and fate processes/rates and in remedial design. The following case study illustrates the potential value of detailed hydraulic parameter and contaminant distribution data in cost-effective designs.

BACKGROUND

The site of the pilot study is located in central Florida and was used in the 1960's for disposal of drums of waste and construction debris. Subsurface contamination, as indicated by contaminant concentrations in ground-water samples from monitoring wells, is heterogeneously distributed in the shallow aquifer. Geology of the upper 30 ft (9.1 m) of the saturated zone is predominantly fine sands with varying fractions of silt and clay. Fill material, construction debris, and lagoon sediments are present in the upper few feet of the subsurface. The top of the Hawthorn Formation is encountered at a depth of about 30 ft (9.1 m) in this area and consists of clay and limestone.

The water table at the site is located at depths of less than 10 ft (3 m) below land surface and varies seasonally. Ground-water flow at the site is strongly influenced by a ground-water extraction system that is currently in operation. Bulk hydraulic conductivity of aquifer materials was estimated using data from a 72-hour multi-well pumping test. Estimates of horizontal hydraulic conductivity clustered in a relatively narrow range from approximately 1 ft/d to 3 ft/d (0.3-0.9 m/d). Estimates of vertical hydraulic conductivity were less certain. The most reliable data set indicated that a horizontal to vertical anisotropy ratio of about 10:1 may be representative of bulk conditions in the shallow saturated zone of interest. Based on the potential for significant heterogeneity in hydraulic conductivity and contaminant distribution, detailed characterization of site conditions was undertaken.

MATERIALS AND METHODS

Borehole Flowmeter Description and Methodology. A sensitive electromagnetic borehole flowmeter was used to define the relative hydraulic conductivity distribution of aquifer materials screened by a test well. The study consisted of measuring the vertical component of ground-water flow at several depths in the well under undisturbed (ambient) conditions and during constant-rate ground-water extraction. Measurements made during constant-rate ground-water extraction or injection indicate the distribution of flow to the well and allow interpretation of the relative hydraulic conductivity distribution of materials within the screened interval (Molz and others, 1994).

Test well NEBIOTW-1 was installed in the vicinity of the proposed bioremediation pilot study site. The borehole was drilled using a wash rotary technique whereby a casing is driven in advance of the rotary bit and materials are washed from the hole. The objective of this technique was to minimize formation damage during drilling so that more representative data regarding hydraulic properties could be obtained. The well was installed using an approximately 1-inch (0.4 cm) thick artificial sand pack and screened from approximately 5 ft (1.5 m) below the water table (i.e., 9 ft or 0.7 m below land surface) to the top of the

Hawthorn Formation at approximately 30 ft (9.1 m) below land surface. Casing and screen were standard 2-inch Schedule 40 PVC materials. The investigation was performed using the following general protocol:

- Ambient vertical flowrates (undisturbed conditions) were measured from total depth to the top of the well screen at 1 ft intervals using the 0.5-inch (0.2 cm) ID probe.

- A peristaltic pump was used to stress the aquifer and establish a stable flow field under pumping conditions. Total flowrates were measured using graduated cylinders and a stop watch at routine intervals. Tests were performed at two different extraction rates (i.e., approximately 2.7 l/min and 4.8 l/min).

- After conditions in the well stabilized, the flowmeter was used with the 1.0-inch (0.4 cm) ID probe to measure vertical flowrates at each of the elevations occupied during the ambient flow profile. Measurements were also repeated at different times following the start of extraction to ensure a stable flow distribution was maintained.

Measurements of flowrates under ambient and constant-rate pumping conditions were analyzed using methods described by Molz and others (1994). Flow to the well from each interval is assumed to be horizontal and proportional to the transmissivity of the formation after an initial stabilization period. The relative hydraulic conductivity profile was estimated using Equation 1 developed by Molz and others (1990), which relates the dimensionless ratio K_i/K to the net induced flow from each interval, interval thickness, total flow from the well, and aquifer thickness influenced during the test.

$$\frac{K_i}{K} = \frac{(\Delta Q_i - \Delta q_i)/\Delta z_i}{Q_P/b} \quad ; \quad i = 1, 2, 3....n \qquad (1)$$

where:

K_i = horizontal hydraulic conductivity of interval i,
K = average hydraulic conductivity of screened materials,
Q_i = induced flow from interval i,
q_i = ambient flow from interval i,
z_i = interval i thickness,
Q_P = total extraction rate, and
b = aquifer thickness influenced by the test.

Determination of Contaminant Distribution. Three-dimensional characterization of contaminant distribution was also performed. Chemical analysis of monitoring well samples was used to identify the areal extent of contamination at the site and to identify an area for the pilot demonstration. However, greater vertical resolution of contaminant distribution was needed to refine the design of the delivery system to ensure appropriate mixing. Hollow-stem augers in combination with an impact driven core barrel were used to collect continuous sleeved cores of aquifer materials. The sleeved cores were sectioned, sealed, and frozen in the field for transport to the laboratory. Subsections of the frozen sleeved cores were obtained and used for various characterization studies. Subsections representing the combined pore water and soil matrix were analyzed by GC/Mass Spectrometry for types and relative concentrations of contaminants. Vertical resolutions of as little as 15 cm (0.5 ft) were achieved.

RESULTS AND DISCUSSION

Flowmeter Results. The majority of the induced flow entered the well in two zones located near the middle and bottom of the screened interval. Approximately 25% of the flow entered the well within the bottom 0.5 foot (15 cm) of the well screen. This indicates that the top of the Hawthorn Formation is much more conductive in this area than originally conceptualized, resulting in a significant flow contribution from depths below this interval. Therefore, the value of hydraulic conductivity estimated for this zone is considered relatively uncertain.

The hydraulic conductivity profile (Figure 1), estimated from this study and the results of the previous multiwell pumping test, indicates that a stratified hydraulic structure exists. A zone of relatively high conductivity exists between approximately 15 ft and 22 ft (4.6-6.7 m) below TOC. This interval is bounded by zones of lower hydraulic conductivity. Aquifer materials in this zone are as much as approximately one-half order of magnitude more conductive than the bulk hydraulic conductivity in the screened interval. Hydraulic conductivity estimated for aquifer materials near the well screen ranged from less than 0.1 ft/d to approximately 27 ft/d (0.03-8.2 m/d), spanning over two orders of magnitude.

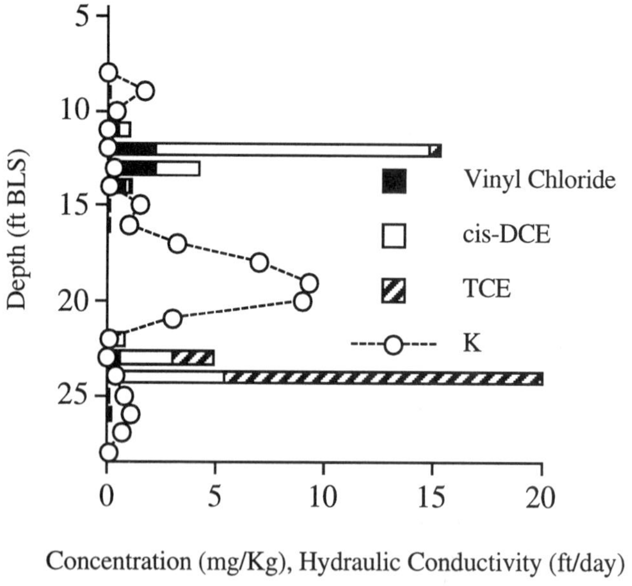

FIGURE 1. Absolute hydraulic conductivity of each measured flow interval estimated using a bulk hydraulic conductivity of 2 ft/d obtained near well NEBIOTW-1, and organic contaminant concentrations in soil/pore water samples from composite site cores (PS1 and PS2), vs depth. Depth is relative to top of casing which is approximately land surface.

Contaminant Distribution. Results of analyses from initial cores (Figure 1) indicate contaminants at these locations were heterogeneously distributed and predominantly associated with materials of lower hydraulic conductivity identified in the flowmeter survey. The profile of the contaminants detected in the core material suggests that reductive biotransformations are occurring in-situ and that the potential for augmenting those transformations with the addition exogenous electron donor is high. It is also felt that recoveries of vinyl chloride in the mg/kg range argue that the sample collection and handling procedures are robust and appropriate for the characterization activities.

System Design. Detailed, three-dimensional characterization data combined with more conventional information were used to define the conceptual model for contaminant distribution and transport at this site. A three-dimensional ground-water flow model was developed from these data and used to screen various injection and extraction scenarios. Scenarios were evaluated with respect to well configurations for delivery of these solutions and potential travel times of injected solutions through contaminated zones. Potential designs that did not provide sufficient transport of injected solutions through target zones, which included materials with relatively low hydraulic conductivity, in approximately a 100 day time frame were excluded from consideration.

Detailed design considerations such as these would not have been possible without the three-dimensional characterization data to identify contaminated zones and the hydraulic conductivity distribution. The design chosen for the pilot study (Figure 2) incorporates infiltration galleries and horizontal wells for fluid circulation which may be scaled up in a cost-effective manner.

ACKNOWLEDGMENTS

We would like to thank the other ITRD members who have contributed to this effort, in particular we would like to acknowledge efforts of Todd McAlary (BEAK Consultants Limited), Hal Koechlein (Lockheed Martin Specialty Components), Frank Beck and Mike Cook (US-EPA, NRMRL-Ada).

Although the research described in this paper is supported in part by the US-Environmental Protection Agency through an in-house research program, it has not been subjected to Agency review and therefore does not necessarily reflect the views of the Agency, and no official endorsement should be inferred.

REFERENCES

Molz, F.J., G.K. Boman, S.C. Young, and W.R. Waldrop, 1994. Borehole flowmeters: field application and data analysis, J. Hydrology, 163:347-371.

Molz, F.J., O. Güven, J.G. Mellville, I. Javandel, A.E. Hess, and F.L. Paillet, 1990. A new approach and methodologies for characterizing the hydrogeologic properties of aquifers, U.S. Environmental Protection Agency, EPA/600/2-90/002, 205 pp.

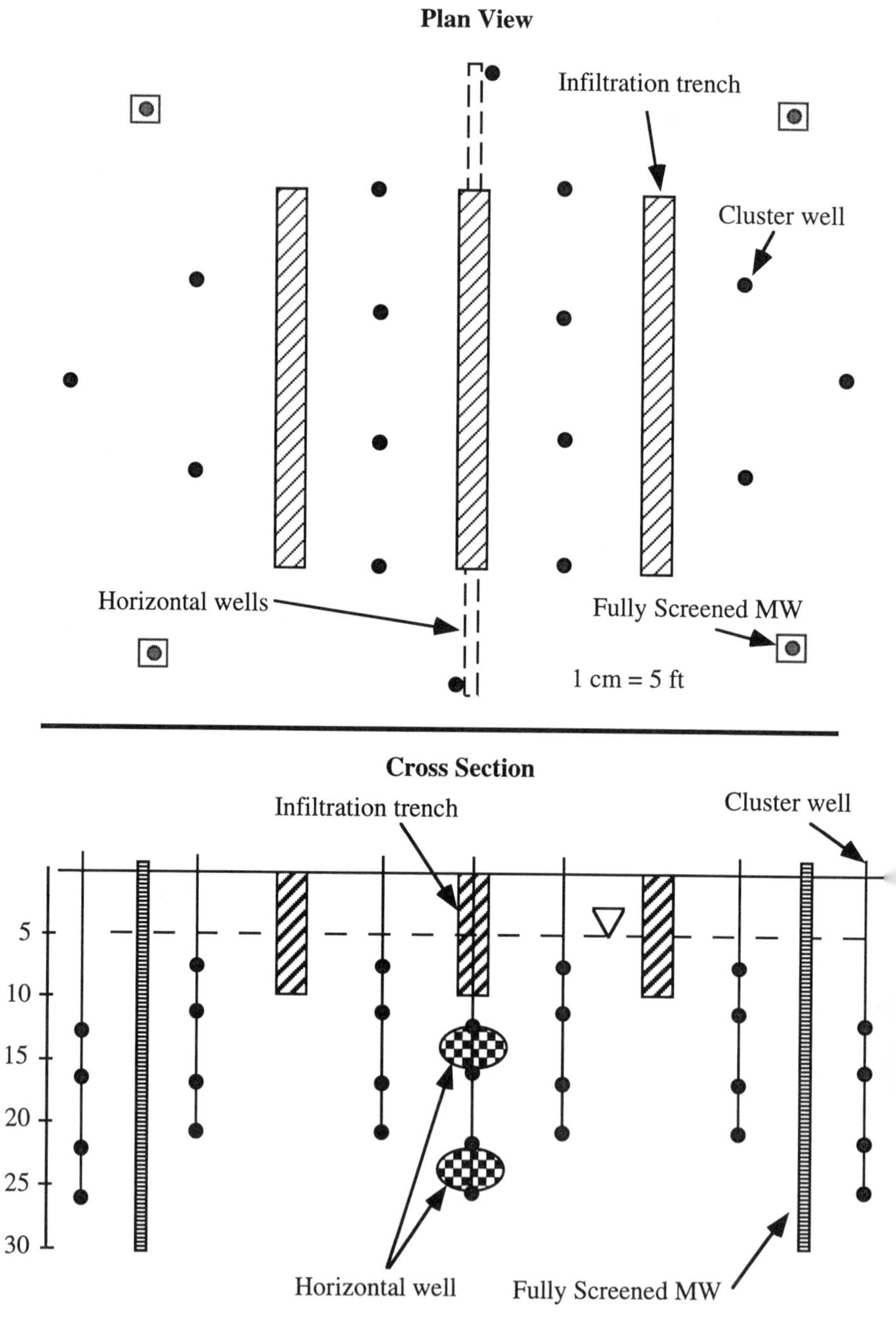

FIGURE 2. Schematic of pilot biotreatment system design using infiltration galleries and horizontal wells for extraction/injection, and monitoring system.

NUTRIENT TRANSPORT DURING BIOREMEDIATION OF CRUDE OIL CONTAMINATED BEACHES

Brian A. Wrenn (Environmental Technologies & Solutions, Rochester, NY)
Michel C. Boufadel and Makram T. Suidan (Univ. of Cincinnati, Cincinnati, OH)
Albert D. Venosa (U.S. EPA, Cincinnati, OH)

ABSTRACT: The effect of wave energy on transport of dissolved nutrients in the intertidal zone of sandy beaches was studied by comparing the washout rates of a conservative tracer (lithium) on two beaches in Maine. The physical characteristics of the two beaches were similar, and they were subjected to the same tidal influences, but the wave energies were very different. Scarborough Beach is a high energy beach that faces southeast toward the Atlantic Ocean, whereas Ferry Beach is in a protected harbor. This difference in wave energy caused lithium to be washed out of Scarborough Beach much more rapidly than from Ferry Beach. The higher wave energy at Scarborough Beach also appears to have increased the amount of lithium that was diluted directly into the water column. These differences in transport rate and mechanism have important implications for the feasibility of bioremediation for cleanup of oil-contaminated shorelines.

INTRODUCTION

The growth rate of oil-degrading bacteria on contaminated shorelines is often limited by the availability of nutrients, such as nitrogen and phosphorus (Pritchard and Costa, 1991; Bragg et al., 1993; Lee et al., 1993; Venosa et al., 1996). Effective bioremediation requires nutrients to remain in contact with the oiled beach material, and the concentrations should be sufficient to support the maximal growth rate of the oil-degrading bacteria throughout the cleanup operation. Contamination of coastal areas by oil from offshore spills usually occurs in the intertidal zone, where the washout of dissolved nutrients can be extremely rapid. Lipophilic and slow-release formulations have been developed to maintain nutrients in contact with the oil (Atlas and Bartha, 1992), but most of these rely on dissolution of the nutrients into the aqueous phase before they can be used by hydrocarbon degraders (Safferman, 1991). Therefore, design of effective oil bioremediation strategies and nutrient delivery systems requires an understanding of the transport of dissolved nutrients in the intertidal zone.

Transport through the porous matrix of a beach is driven by a combination of three main factors: tide, waves, and the flow of freshwater from coastal aquifers. The focus of this research was on the effects of tide and wave activity. Tidal influences cause the groundwater elevation in the beach, as well as the resulting hydraulic gradients, to fluctuate rapidly (Nielsen, 1990; Wrenn et al., in press). Wave activity affects groundwater flow through two main mechanisms. First, when waves run up the beach face ahead of the tide, some of the water percolates vertically through the sand above the water line and flows horizontally when it reaches the water table (Riedl and Machan, 1972). Waves can also affect groundwater movement in the submerged areas of beaches by a pumping mechanism that is driven by differences in head between wave crests and troughs (Riedl et al. 1972).

The relative effects of tide and waves on nutrient transport in the intertidal zone of sandy beaches was investigated by comparing the washout of a conservative tracer, lithium, on two beaches in southern Maine. Scarborough Beach is a high energy beach that faces the Atlantic Ocean, whereas Ferry Beach is in a sheltered harbor at the mouth of the Scarborough Marsh. Lithium transport at

Ferry Beach was driven almost exclusively by tidal effects, whereas tide and waves both affected transport at Scarborough Beach.

EXPERIMENTAL DESIGN

Site Description. The two beaches used in this study are subjected to very different wave energies, but in other respects they are quite similar. Both are composed primarily of medium to fine sand with relatively narrow particle size distributions. Differences in the composition of the two beaches suggest that the hydraulic conductivity of Scarborough Beach might be slightly larger than Ferry Beach, but the small permeability differences were expected to have much less influence on solute transport than the differences in wave energy. The tide was identical at both sites.

Plot Setup and Sample Collection. The tracer was applied to the beach in discrete areas called "plots." Each plot was 5 m wide (i.e., parallel to the shoreline), and they were either 10 m (Ferry Beach) or 12 m (Scarborough Beach) long (i.e., perpendicular to the shoreline). Although the plots on Ferry Beach were shorter than those on Scarborough Beach, the difference in elevation between the tops (i.e., the landward edges) and the bottoms (i.e., the seaward edges) of the plots was approximately the same on both beaches. The plots were set up such that the landward edges were at the elevation that was expected for the highest tide that would occur during the study.

A transect consisting of six multi-port sample wells was installed perpendicular to the shoreline through the center of each plot. The layout of these transects and the elevations of the tops and bottoms of the plots on both beaches are shown in Figure 1. Three of the six sample wells were installed inside the plots, one well was installed landward of the plots, and two were installed seaward of the plots. Figure 1 also shows the locations of the sample ports for each well.

Sprinklers were used to apply the tracer to the beach surface inside the plot boundaries at low tide. Lithium nitrate (>99.7%; Cyprus Foote Mineral Co., Kings Mountain, NC) was dissolved in 100 gallons of fresh water to a final concentration of 33 g/L, which gave it a density approximately equal to the local seawater. Water samples were collected from the multi-port wells periodically for about two weeks.

Water Level Measurement. The water levels in the beaches were measured with transects of six piezometer wells that were installed perpendicular to the shoreline. Piezometer wells were installed at the top, bottom, and middle of the plots. One well was landward of the top, and two were seaward of the bottom of the plots. The most seaward well, which was screened over a four-foot interval above the beach surface, was used primarily to measure the level of the tide whenever it was high enough to submerge any part of the sample well transects. Vibrating wire piezometers (RocTest, Inc., Plattsburgh, NY) were used to measure the water level at each well position. Three readings were usually taken for each piezometer every 15 minutes. These three readings were averaged to smooth out the effect of waves on the water level measurements.

RESULTS AND DISCUSSION

Hydraulic Gradients. The two main forces that drive solute transport in sandy beaches are waves and tidally induced hydraulic gradients. Although no quantitative measurements of the wave activity at the two beaches used in these studies are available at this time, a qualitative comparison can be made by inspection of Figure 1. Whereas the water level changed fairly smoothly at Ferry Beach in

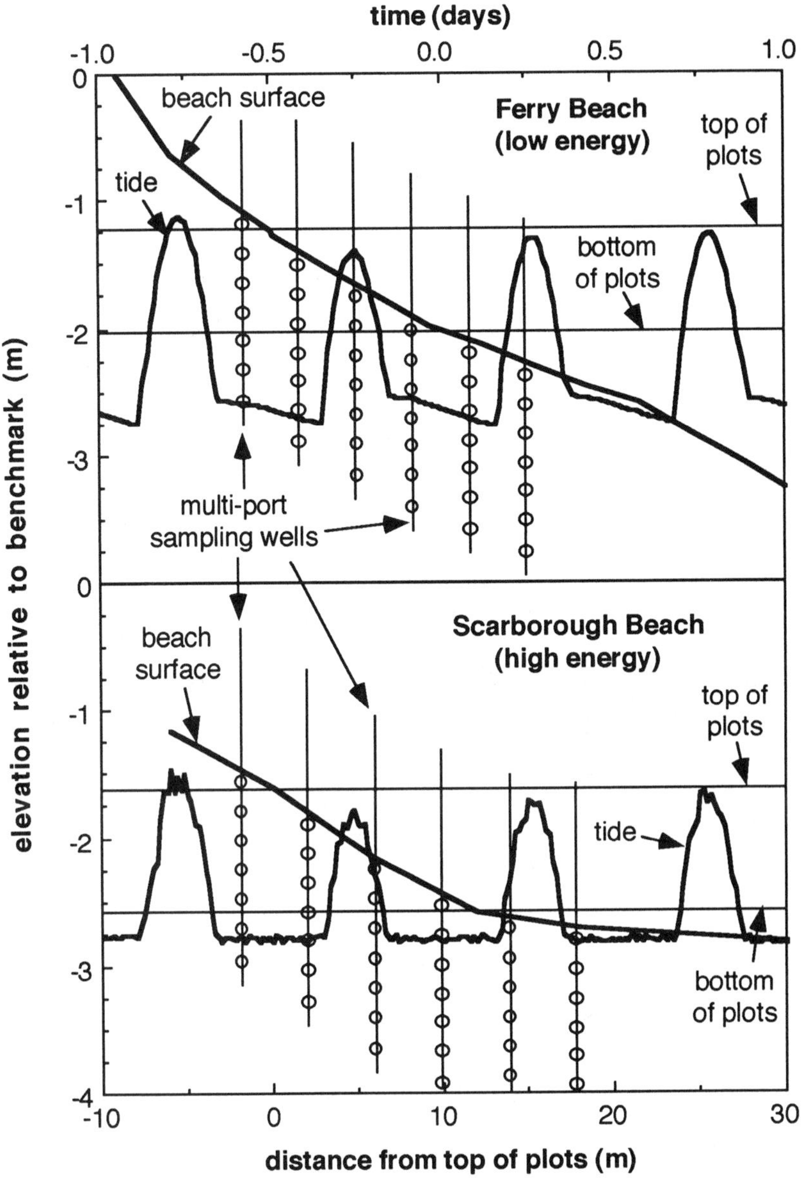

FIGURE 1: Beach profiles showing well positions and the elevations of the tops and bottoms of the experimental plots (i.e., the areas to which the tracer was applied). The circles on each well mark the depths of the sample ports. All elevations were measured relative to a benchmark, but the absolute elevations of the benchmarks on the two beaches were not the same. The tide measurements show that the absolute elevations of the plots were similar on both beaches. Time is measured relative to the beginning of the experiment (i.e., when the tracer was applied).

response to the tide, the response was quite jagged at Scarborough Beach. Although multiple readings were taken whenever water level measurements were made, it was not possible to completely eliminate variations due to waves from the Scarborough Beach data.

The effects of waves can also be seen in Figure 2, which shows the hydraulic gradients in the bottom (seaward) half of the plots for both beaches. The response at Ferry Beach was relatively smooth, whereas the gradient fluctuated rapidly at Scarborough Beach. Wave run up and subtidal pumping probably both contributed to these abrupt changes in the hydraulic gradient. In general, the responses of the hydraulic gradients to the tide were similar in both beaches. For example, landward-directed (i.e., positive) hydraulic gradients developed only briefly in this region of both beaches. (Landward-directed gradients persisted much longer in the top half of the plots, however.) Most of the time, the hydraulic gradients were directed seaward (i.e., negative), which is consistent with previous observations (Nielsen, 1990; Wrenn et al., in press).

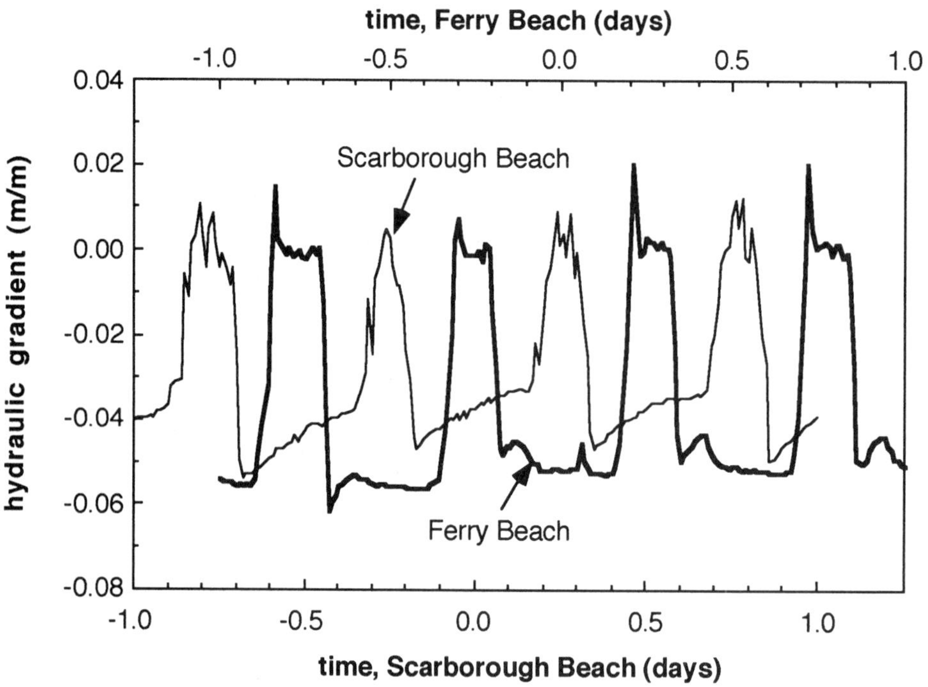

FIGURE 2: Hydraulic gradients in the bottom half of the plots at Ferry and Scarborough Beaches. Positive values indicate landward-directed gradients and negative values indicate gradients that are directed seaward. The time is measured relative to the beginning of the experiment, and the time scales for the two beaches are offset by 6 hours to improve readability.

Tracer Washout. Lithium was removed from Scarborough Beach much more rapidly than from Ferry Beach. At Scarborough Beach, less than 15% of the added lithium was recovered at the first high tide following tracer application, whereas about 60% was recovered at Ferry Beach. Tracer was essentially completely removed from the experimental domain at Scarborough Beach within four days, but at Ferry Beach tracer was still present after two weeks. It seems likely that most of the tracer that was lost during the first high tide was diluted into the water column. Following the initial large decrease in tracer concentration, washout proceeded more slowly as the lithium moved through the beach subsurface. Therefore, one of the main effects of wave activity appears to be to change the amount of tracer (or nutrient) that is immediately diluted into the water column relative to that which penetrates into the beach to be transported by subsurface flow.

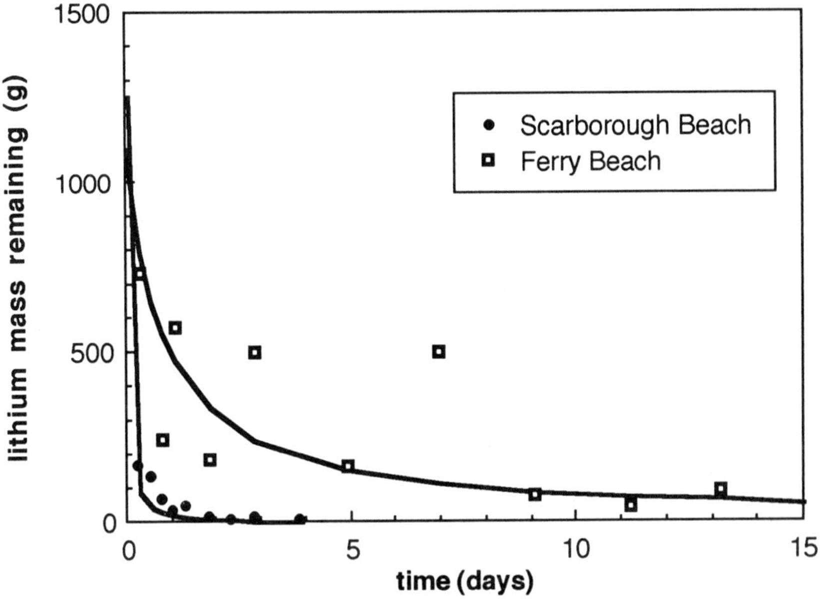

FIGURE 3: Washout of lithium from high energy (Scarborough) and low energy (Ferry) beaches following application of the tracer to the beach surface at low tide during the full-moon spring tide. The mass of lithium remaining was determined using water samples collected from all six multi-port sample wells in each plot.

These results suggest that surface application of nutrients will be ineffective on high energy beaches, because most of the nutrients will be lost to dilution at high tide. On low energy beaches, however, this might be an effective and economical bioremediation strategy. Nutrients that are released from slow-release or lipophilic formulations will probably behave similarly to the dissolved tracer that was used in this study. Therefore, they will not be effective on high energy beaches unless the release rate is high enough to achieve adequate nutrient concentrations while the tide is out. Subsurface application of nutrients might be more effective on high energy beaches. Since crude oil does not penetrate very deeply into most beach matrices (Gundlach, 1987), however, nutrients must be present near the beach surface to effectively stimulate bioremediation. Nutrients move downward and seaward during transport through the intertidal zone of sandy beaches (Wrenn et al., in press). Therefore, nutrient application strategies that rely on subsurface introduction must provide some mechanism for insuring that the nutrients reach the oil-contaminated area near the surface.

ACKNOWLEDGMENTS

We are grateful for the cooperation of the citizens and town council of Scarborough, ME for allowing us to use Ferry Beach for this research and to the Sprague Corporation for granting us permission to use Scarborough Beach. The cooperation and assistance of Greg Wilfert and the employees of Scarboro Beach Park is also gratefully acknowledged. Finally, we are indebted to Dave Corbeau, Marine Resources Officer for the Town of Scarborough, and the Scarborough Police and Fire Departments for their valuable assistance during this study.

REFERENCES

Atlas, R. M. and R. Bartha. 1992. "Hydrocarbon biodegradation and oil spill bioremediation." In K. C. Marshall (Ed.) *Advances in Microbial Ecology, Vol. 12*, pp. 287-339. Plenum Press, New York, NY.

Bragg, J. R., R. C. Prince, E. J. Harner, and R. M. Atlas. 1993. "Bioremediation effectiveness following the *Exxon Valdez* spill." *Proc. 1993 International Oil Spill Conference*, pp. 435-447. American Petroleum Institute, Washington, DC.

Gundlach, E. R. 1987. "Oil-holding capacities and removal coefficients for different shoreline types to computer simulate spills in coastal waters." *Proc. 1987 International Oil Spill Conference*, pp. 451-457. American Petroleum Institute, Washington, DC.

Lee, K., G. H. Tremblay, and E. M. Levy. 1993. "Bioremediation: Application of slow-release fertilizers on low-energy shorelines." *Proc. 1993 International Oil Spill Conference*, pp. 449-454. American Petroleum Institute, Washington, DC.

Nielsen, P. 1990. "Tidal dynamics of the water table in beaches." *Water Resources Research 26*: 2127-2134.

Pritchard, P. H. and C. F. Costa. 1991. "EPA's Alaska oil spill bioremediation project." *Environmental Science and Technology 25*: 372-379.

Riedl, R. J. and R. Machan. 1972. "Hydrodynamic patterns in lotic intertidal sands and their bioclimatological implications." *Marine Biology 13*: 179-209.

Riedl, R. J., N. Huang, and R. Machan. 1972. "The subtidal pump: A mechanism of interstitial water exchange by wave action." *Marine Biology 13*: 210-221.

Safferman, S. I. 1991. Selection of nutrients to enhance biodegradation for the remediation of oil spilled on beaches. *Proc. 1991 International Oil Spill Conference*, pp. 571-576. American Petroleum Institute, Washington, DC.

Venosa, A. D., M. T. Suidan, B. A. Wrenn, K. L. Strohmeier, J. R. Haines, B. L. Eberhart, D. King, and E. L. Holder. 1996. "Bioremediation of an experimental oil spill on the shoreline of Delaware Bay." *Environmental Science and Technology 30*: 1764-1775.

Wrenn, B. A., M. T. Suidan, K. L. Strohmeier, B. L. Eberhart, G. J. Wilson, and A. D. Venosa. (in press) "Nutrient transport during bioremediation of contaminated beaches: Evaluation with lithium as a conservative tracer." *Water Research*

PULSED NUTRIENT INJECTION FOR IMPROVED BIOMASS DISTRIBUTION

Brent M. Peyton, Brian S. Hooker, Michael J. Truex, and Mark G. Butcher

ABSTRACT: Plugging of nutrient injection wells is problematic for co-metabolic in situ bioremediation because substrate for microbial growth must be distributed from the well. In addition, improving the spatial or volumetric biomass distribution has the potential to reduce the number of wells required for a given site and to improve the overall effectiveness of the remediation system. Pulsed injection of electron donors and acceptors has been suggested as a method for evenly distributing biomass for more effective in situ bioremediation.

Laboratory tests were used to help define field-scale nutrient addition strategies for co-metabolic in situ biodegradation of tetrachloromethane under denitrification conditions using acetate as the substrate. Separate laboratory soil column experiments were conducted with either continuous or pulsed acetate and nitrate addition; however, in all experiments the overall injected mass was equal. The longitudinal profiles of biofilm that developed under each nutrient addition strategy were determined by protein and cell enumeration of post-test sediment cores. Compared to continuous nutrient addition, a pulsing strategy resulted in more evenly distributed biomass profiles in soils. The pulsing strategy also resulted in an order of magnitude less biomass formed near the point of nutrient addition, thus demonstrating a reduction in near-well biofouling.

The field demonstration was completed in the spring of 1996 using a two-well injection/extraction groundwater recirculation system. Pulsed nutrient injection strategies were successfully used to distribute biomass away from the injection well to maximize the zone of active contaminant destruction. Co-metabolic tetrachloromethane destruction was sustained for 4 months without significant increases in the pressure required to inject groundwater.

This research was supported by the Department of Energy, Office of Science and Technology. Pacific Northwest National Laboratory is operated for the DOE by Battelle Memorial Institute under Contract DE-AC06-76-RLO 1830.

LABORATORY AND FIELD-SCALE TCE BIODEGRADATION IN GROUNDWATER UNDER AEROBIC CONDITIONS

L. T. LaPat-Polasko and N. R. Chrisman Lazarr
(Woodward-Clyde Consultants, Phoenix, AZ)

ABSTRACT: Several bench-scale studies were conducted prior to the design of a field-scale remediation system for trichloroethylene (TCE)-contaminated groundwater in Tucson, Arizona. These semi-continuous soil column studies evaluated potential cometabolic conversion of TCE under bulk aerobic conditions in site groundwater. Tyrosine, methanol, and vanillin were supplied as potential cometabolic inducers. The results of the bench-scale study indicated that the highest removals of TCE from groundwater occurred in a soil column amended with methanol, hydrogen peroxide, and nutrients. Comparable (although slightly lower) removals of TCE occurred in the absence of methanol under bulk aerobic conditions.

Based on the results of bench-scale testing, an in situ bioremediation amendment system was designed for implementation in the field. Ongoing operation of the bioremediation system includes the amendment of methanol (at 25 mg/L), hydrogen peroxide (at 25 mg/L), and diammonium phosphate (at 12 mg/L as nitrogen) to treated groundwater. An injection/extraction wellfield is used to distribute the amended groundwater to the subsurface. The groundwater zone of interest consists of unconsolidated, dense reddish clay to sandy clay with some thin lenses of fine sand. The groundwater zone lies above a regional aquifer and is impacted by surface water infiltration and unrelated regional aquifer injection activities. The estimated hydraulic conductivity of the groundwater zone ranges from 0.015 to 8.5 feet per day with an average bulk hydraulic conductivity of approximately 0.5 foot per day. The groundwater zone starts at approximately 60 to 85 feet below ground surface (bgs) and extends to approximately 110 to 120 feet bgs. The screened interval of treatment wells is approximately 50 feet long and is located within the groundwater zone of interest. Monitoring wells are used to assess the efficacy of amendment distribution and extent of TCE degradation.

FEASIBILITY SCREENING STUDY FOR *IN SITU* BIOREMEDIATION OF HEAVY GAS OIL IN REFINERY SOILS

Dominic Meo, III (CH2M HILL)
W.T. Frankenberger, Jr. (Center for Environmental Microbiology)
M. Hillyer and B.L. McFarland (Chevron Research & Technology Company)

ABSTRACT: A twelve-week bench-scale treatability study screened *in situ* bioremediation options for weathered hydrocarbons in a refinery soil contaminated with heavy gas oil. The study was conducted with five columns packed with soil in three layers collected from the field to simulate soil profiles at the site. The rate of intrinsic, static bioremediation was compared to rates achieved with the addition of: (i)nutrient amendments; (ii)nutrient amendments and a sufactant; (iii)nutrient amendments, a surfactant and hydrogen peroxide; and (iv)nutrient amendments, a surfactant and air venting. Bioremediation progress was monitored within three depth horizons (0-3", 3-24", and 24-48") utilizing: EPA 418.1 (TRPH); EPA Modified 8015 (TPH: $<C_{10}$, C_{10}-C_{20}, C_{20}-C_{30}, C_{30}-C_{40}, and $>C_{40}$); ammonium; nitrate and orthophosphate; and heterotrophic and hydrocarbonclastic microbial population measurements. A graphical, analytical technique utilizing "trellis plots" facilitated understanding the relationship between treatment efficiency and soil depth. This technique is analogous to developing data relationships in an analysis of variance (ANOVA).

INTRODUCTION

A sandy soil at a refinery was contaminated near the surface with weathered, heavy gas-oil hydrocarbons. In the first 3 inches of soil, the concentration of heavy gas oil, as measured by EPA 418.1, averaged 130,000 mg/kg of soil. This concentration fell down to 19,000 mg/kg of soil at a depth of approximately 4 feet. The depth of contamination was limited to the first few feet. As the soil was directly beneath a low-lying pipe rack, access was severely restricted, and excavation of the contaminated soil would be both difficult and expensive. The refinery elected to assess the feasibility of *in situ* bioremediation as a cost-effective alternative. Given both the high molecular weight and high concentration of the heavy gas oil, it was decided to evaluate the effectiveness of different biostimulant treatments, but it was not clear which type or combination of types would be most effective.

Objective. The objective of this study was to assess whether *in situ* bioremediation was feasible and whether it could be accomplished more cost effectively than excavation of the contaminated soil. Although the concentration of heavy gas oil was very high on the surface, the lead agency monitoring soil conditions at the refinery had agreed to a clean-up level of 7,000 mg/kg of soil. The most effective treatment or combination of biostimulants would be selected by using "trellis" plots to analyze the data.

MATERIALS AND METHODS

Soil-Column Setup. Soil samples were collected at the site using a methodology which would facilitate duplicating field conditions in the lab. Using a separate container for each contamination "horizon," soil was collected at depths of 0-3", 3-24" and 24-48". In the lab, soil from each depth horizon was sieved and mixed. The soil was sandy in texture, had a moisture content of 4-7% and a pH of 7.2-7.6. Then, the soil was carefully packed into five clear, plastic columns with an inside diameter of 3 inches (7.6 cm) and an overall length of 72 inches (182.9 cm). To approximate the heavy gas-oil concentration profile at the site, the deepest soil was placed in the columns first and the surface soil last. A total of 9 kg of soil was used in each column.

Soil Treatments. One column served as the static, control column, receiving no treatment, not even water. The other four columns were treated as follows: (i) nutrients consisting of 100 mg N as $(NH_4)_2SO_4$ plus 50 mg P as K_2HPO_4 per kg of soil; (ii) same nutrients as (i) plus a non-ionic surfactant, applied as a 2% solution of Witconol SN-70, an ethoxylated alcohol; (iii) same nutrients and surfactant as (ii) plus 250 mg of H_2O_2 per liter of water added; and (iv) same nutrient and surfactant as (ii) plus venting with oil-free air flowing at the rate of 5 ml per minute. All columns were incubated in a vertical position at ambient temperature (23°C ± 2°C) for twelve weeks. Each treatment was applied a total of three times: once on September 20, 1994, when the study was initiated, then again on November 11 and December 7, 1994.

Soil Sampling and Analysis. Subsamples collected from each contamination horizon within the soil columns were analyzed for TRPH (total, recoverable petroleum hydrocarbon), using EPA 418.1 and TPH (total petroleum hydrocarbon), using EPA Modified 8015 at three-week intervals. The hydrocarbon content as measured by the EPA modified 8015 method was fractionated into $<C_{10}$, C_{10}-C_{20}, C_{20}-C_{30}, C_{30}-C_{40}, and $>C_{40}$ units to monitor the microbial preference in degrading weathered heavy gas oil. The hydrocarbon-oxidizing population was also monitored at 0, 4, 8, and 12 weeks. Inorganic N (NH_4-N and NO_3-N) and orthophosphate-P were determined at 0, 6 and 12 weeks to follow the fate of added nutrients with time of incubation.

RESULTS AND DISCUSSION

Soil TRPH Analysis. This study shows that there was a rapid decline in TRPH over 12 weeks of incubation. The most effective treatment was the application of nutrients alone, producing a 72% decline in TRPH, while the other treatments promoted a 56 to 65% decline in TRPH. Aeration was not a limiting factor in the upper soil layer (0-3") which was initially sieved and well mixed, then exposed to the atmosphere throughout the study.

Soil TPH Analysis. In addition to monitoring the heavy hydrocarbons (>C_{40}), the EPA modified 8015 method was used to detect hydrocarbons from <C_{10} to C_{40} (TPH). It is obvious that this fraction of hydrocarbons is much more susceptible to biodegradation. In comparing the overall drop in TRPH (EPA 418.1) vs. TPH (EPA modified 8015) produced by all 5 treatments over 12 weeks of incubation, the former (TRPH) averaged a 36.4% drop in hydrocarbons while the latter (TPH) averaged a 64.2% drop. Table 1 summarizes the decline in TRPH and TPH produced by nutrients (N and P) alone, the most effective amendment.

TABLE 1. Summary of decline in TRPH and TPH produced by addition of nutrients (N and P) only.

STATIC CONTROL

Soil Depth	Starting TRPH (ppm)	Starting TPH (ppm)	% TRPH Decline After 12 Weeks	% TPH Decline After 12 Weeks
0-3"	130,000	84,300	40	72
3-24"	29,000	35,200	14	61
24-48"	19,000	11,500	0	40

ADDITION OF NUTRIENTS (N and P) ONLY

Soil Depth	Starting TRPH (ppm)	Starting TPH (ppm)	% TRPH Decline After 12 Weeks	% TPH Decline After 12 Weeks
0-3"	130,000	84,300	72	78
3-24"	29,000	35,200	62	78
24-48"	19,000	11,500	42	52

Microbial Population and Nutrient Status. The hydrocarbon-oxidizing microbial population was monitored in the soil columns throughout this study. In almost all cases, the microbial-population count increased 1 to 4 log units over the 12-week incubation period, starting from 23-49 x 10^2 most probable number (mpn)/g of soil. This increase in the hydrocarbon-oxidizing microbial population upon the addition of nutrients and/or surfactant confirmed the most effective treatments predicted by monitoring the TRPH and TPH content in the soil, as described above.

The nutrient status (NH_4-N, NO_3-N and orthophosphate-P) was assessed to determine the fate and transport of nutrients in the soil columns. Most of the N was retained in the upper layers of the soil columns. Over the term of the study, the NH_4-N became immobilized, or concentrated by the biomass. After 12 weeks of incubation, the added NH_4-N content was similar in concentration to the background levels observed in the static, control column, indicating that the added N was readily used by the hydrocarbon-oxidizing population. The NO_3-N content was also concentrated by the hydrocarbon-degrading microorganisms. The added phosphorous was mainly found in the upper layers where it was utilized in the most contaminated soil fraction. The orthophosphate-P content generally declined upon degradation of the heavy gas oil.

Trellis Plot Analytical Approach. Resource limitations, especially cost, frequently limit bioremediation-feasibility, or screening-study experiments which ideally would be done with replication and subjected to an ANOVA analysis. For example, the experiment described above would best be run and analyzed as a repeated-measurements ANOVA, with at least two (preferable three) replicates of each condition in order to calculate the appropriate error term needed for a quantitative assessment of the different treatments at each soil depth. Failing a quantitative analysis, a graphical analytical technique along the lines suggested by Cleveland utilizing "trellis plots" enables graphical comparisons to be made between treatments and treatments at different depths (analogous to testing the significance of the main effect of a treatment and the treatment/depth interaction in an ANOVA).

By ratioing the amount of hydrocarbon present in the different soil samples at a particular time to their initial concentration, comparisons can be made between samples having widely different initial hydrocarbon contamination levels. A robust "loess" (local regression) smoothing method was used to minimize the influence of outliers (anomalies) by employing a local second-degree regression line to develop a smoothed, fitted curve. Residuals (deviations) from the loess fit are used to ensure that the data are normally distributed prior to analysis. To assess the amount of variation in the data, a residuals-fitted (r-f) data-spread plot is used to determine the distribution functions for adjacent r-f values, corrected for the mean. This process is similar to the role of an R^2 statistic in data regression.

Using trellis plots, we plotted the results, e.g. TRPH concentration as a function of time for all five treatments and three depths on the same plot, using the same axis scalings. These trellis plots enable quick, visual comparisons between the overall efficiency of the treatments and also between treatments at each depth (Figure 1). Since the starting TRPH concentrations were different at different depths, the ratios of TRPH concentration at a particular time to the zero-time concentrations are plotted. The five columns of plots are for the five treatments, coded as: C = static control, N = nutrients added; NS = nutrients + surfactant; NSA = nutrients + surfactant + air venting; NSP = nutrients + surfactant + peroxide treatment. The three rows of plots are for the three soil-column depths; the shallowest is at the top, with a loess, smoothed curve added to each plot.

CONCLUSIONS AND RECOMMENDATIONS

As shown in Table 1, the most effective treatment in promoting the overall biodegradation of the heavy gas oil, as indicated by the decline in TRPH (EPA 418.1), was the application of nutrients (N and P) alone. Based on Figure 1 and other trellis plots which are not included here, the following general conclusions were reached:

FIGURE 1. Decline in Total Recoverable Hydrocarbons Using Various Treatment

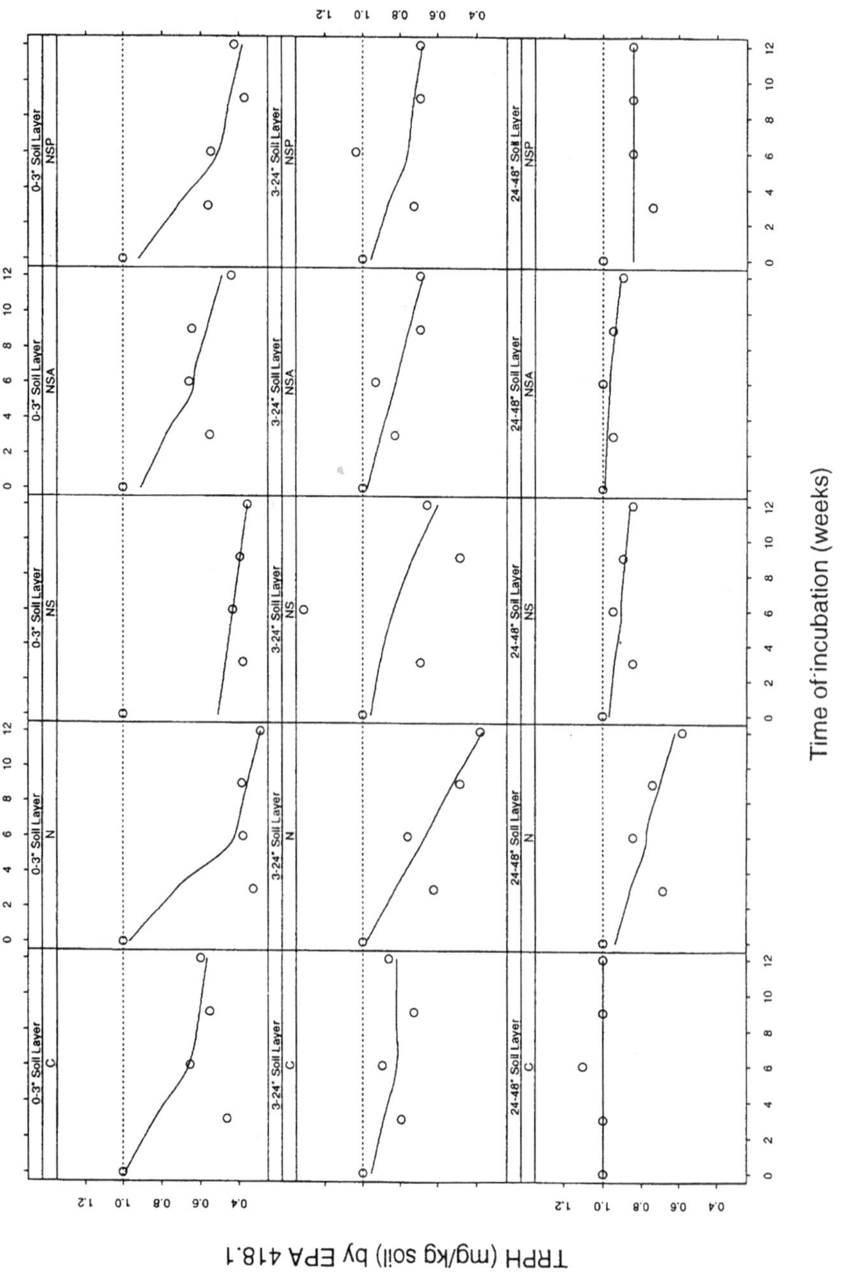

- <C_{10} fraction: declined more rapidly at all depths than all other fractions.
- C_{10}-C_{20} fraction: no difference in decline, regardless of treatment; the decline may have become slower with depth.
- C_{20}-C_{30} fraction; no difference in decline, regardless of treatment, and the decline decreased with depth; declined at same rate as C_{10}-C_{20} fraction.
- C_{30}-C_{40} fraction: no difference in decline within the top soil layer, regardless of treatment.
- >C_{40} fraction: no difference in decline within the top soil layer, regardless of depth.
- With the exception of a minor improvement in the decline in TRPH at lower depths, venting offered no significant benefit.

The success of this study is attributed in part to the porous nature of the sandy soil which supports a high microbial activity and permits the rapid and uniform introduction of treatments. Sieving and mixing of the soil, which was necessary to obtain uniform TRPH and TPH readings for each soil layer, before the soil was packed into the columns, also contributed to the success of this study. Despite these factors, the study clearly predicts that the heavy gas oil found at the site can be remediated using a combination of the treatments described above and careful field-monitoring and management practices.

After only 12 weeks, TPH had met the clean-up level of 7,000 mg/kg in the upper (0-3") and middle (3-24") layers and approached this level in the lower (24-48") layer. Given the increasing trend by lead agencies to base clean-up levels in soils on leaching potential, the concentration of hydrocarbons at the site with chain lengths greater than C_{40} (as measured by EPA 418.1) should not be of concern. This favors using EPA modified 8015 as a test for verifying compliance with the clean-up level. The only point to be confirmed before recommending that the refinery proceed with *in situ* bioremediation is that the cost of bioremediation is less than that of excavation. Based on past experience, this should be the case.

In the event *in situ* bioremediation does proceed, the recommended treatment is nutrients (N and P), which proved most effective in decreasing TRPH, plus a non-ionic surfactant to enhance the decline of TPH. Moisture content should be maintained between 12 and 15% by weight, and a series of wells connected to a mechanical blower should be used to aerate the soil. Additionally, the surface of the soil which is effectively sealed by the very high concentration of weathered, heavy gas oil must be mechanically broken up prior to the start of the remediation program.

REFERENCE

Cleveland, W.J. 1993. *Visualizing Data.* Hobart Press.

THE ROLE OF SOIL NITROGEN CONCENTRATION IN BIOREMEDIATION

J.L. Walworth (Univ. of Alaska Fairbanks, Palmer Research Center, Palmer, AK)
C.R. Woolard (Univ. of Alaska Anchorage, Anchorage, AK)
J.F. Braddock (Univ. of Alaska Fairbanks, Fairbanks, AK)
C.M. Reynolds (USA-CRREL, Hanover, NH)

ABSTRACT: The relationship between soil nitrogen (N) levels, soil water content, and petroleum biodegradation were studied by adding various levels of N to a sand, a sandy loam, and a silt loam. Biodegradation was related to soil N expressed as a fraction of soil water (N_{H2O}, mg N/kg soil H_2O) better than N expressed as a fraction of soil dry matter (N_S, mg N/kg soil). In a second study, petroleum-contaminated loamy sand was wetted to various water contents and treated with a range of N levels. Total soil water potential and O_2 consumption were both closely related to N_{H2O}. A third experiment testing effects of N fertilizer and NaCl on total soil water potential suggests that observed responses to over-fertilization were caused by depressed soil water potential. The expression of N_{H2O} accounts for both nutrient status and N fertilizer contribution to soil water potential. An optimum N_{H2O} level of approximately 2,000 mg N/kg H_2O is tentatively identified.

INTRODUCTION

Microorganisms responsible for petroleum biodegradation require N, although methods determining optimum N application levels have not been perfected. Many studies report the beneficial effects of N addition, however several studies indicate no benefit, or even deleterious effects from added N fertilizer (Table 1). There is no consensus on the optimum level of N for microbial petroleum biodegradation.

Nitrogen requirements for bioremediation systems are often expressed as C:N ratios. Waksman (1924) related the N required to degrade organic substrates to substrate carbon (C) content, composition of microbial biomass, and efficiency of substrate degradation. This approach has been adopted for design of bioremediation systems, but a broad range of C:N ratios have been recommended. Generally accepted optimal ratios are between 20:1 and 10:1, although values ranging from 200:1 to 9:1 have been reported (Huesemann, 1994).

The optimum C:N ratio may vary with substrate load. Applying enough N to satisfy the stoichiometric requirement in highly contaminated soils (establishing a C:N ratio of 10:1) can reduce the rate of biodegradation. Brown et al. (1983) reported an optimum C:N ratio of 9:1 in soil containing 3,500 mg/kg of refinery sludge, but the optimum ratio was 124:1 in soil with 21,000 mg/kg of petrochemical sludge, and biodegradation was reduced when the C:N ratio was adjusted to 23:1. Genouw et al. (1994) reported that a C:N ratio of 170:1 stimulated microbial CO_2 production in a soil with 44,000 mg hydrocarbons/kg soil, whereas respiration was reduced with a C:N ratio of 11:1.

Fertilizer N recommendations for bioremediation may be based on soil mass (g N/kg soil), rather than on substrate C levels. Huesemann (1994) recommended adding no N if inorganic soil N exceeded 50 mg N/kg soil, and adding no more than 250 mg N/kg soil at one time to avoid leaching and toxicity. Reports of N-induced petroleum biodegradation inhibition are not uncommon (Table 1). Reported inhibitory N fertilization levels range from just over 100 to 4000 mg N/kg soil, although the reported levels do not necessarily represent the minimum inhibitory level as most experiments include only a few levels of N application. A more accurate measure of critical N levels would be helpful for providing adequate N for bioremediation, while avoiding N inhibition.

TABLE 1. Published reports of N-induced inhibition of bioremediation.

Reference	Soil Texture	Inhibitory N Level (mg N/kg soil)
Brown et al., 1983	sandy clay	913
Dibble and Bartha, 1979	sandy loam	1667
Huntjens et al., 1986	sand	400
Genouw et al., 1994	loamy sand	4005
Morgan and Watkinson, 1992	"sandy soil"	1750
Zhou and Crawford, 1995	"sandy subsoil"	119

An original concept is presented in this paper for measuring N for bioremediation systems: inorganic N can be related to soil water (N_{H2O}), rather than to substrate level (C:N), or to soil dry mass (N_S). Nitrogen salts used in fertilizer formulations are highly soluble and partition into the soil solution, decreasing soil osmotic potential. Over-fertilization, which can depress activity of petroleum-degrading microbes that are sensitive to soil water potential, is tempered by soil water content. Wet soil provides a larger volume of water for dilution of fertilizer N than dry soil, so hazards of over-fertilization are less in moister soil. Using C:N ratio or N_S ignores this relationship, whereas use of N_{H2O} accounts for both nutrient level and fertilizer contribution to soil water potential, regardless of soil texture or water content.

MATERIALS AND METHODS

Experiment 1. Soil water-holding capacity (WHC, water held at approximately -33 kilopascals (kPa)) is related to soil texture, so comparing N response in different types of soils provides a means of evaluating effects of soil water content (while maintaining a constant matric water potential). Three soils were compared: 1) silt loam with a WHC of 30% and 8,000 mg #2 diesel fuel/kg soil, 2) loamy sand with a WHC of 9%

and 7,400 mg crude oil/kg soil, and 3) sand with WHC of 2.5% and approximately 3,000 mg JP-5/kg soil. The silt loam and loamy sand were each fertilized with 0, 400, 800, or 1200 mg N/kg soil (as NH_4NO_3) (Walworth and Reynolds, 1995). The sand was treated with 0, 100, 200, or 300 mg N/kg soil, (20-20-20 fertilizer composed of NH_4-polyphosphates, KNO_3, and urea) (Braddock et al., in review). Soils were wetted to their WHC's and incubated at 10°C. Extent of biodegradation, expressed either as petroleum loss (silt loam and loamy sand) or CO_2 production (sand) was normalized by expressing each treatment as a percentage of the top treatment with the respective soil.

Experiment 2. In a single soil, varying N levels and soil water contents results in a range of soil solution N concentrations. Knik subsoil (sand) was dried to 2% moisture and mixed with field-moist (20% H_2O) Knik silt loam in a 10:1 ratio, resulting in a loamy sand with pH 6.3, 1% organic C, and 11.5% WHC. It was treated with 800 mg $Ca(H_2PO_4)_2 \cdot H_2O$/kg soil and 5,000 mg of #2 diesel fuel/kg soil, and amended with 0, 250, 500, or 750 mg N/kg soil added as NH_4NO_3. Water was added to reach 5.0, 7.5, or 10.0% water content (by weight). Total (matric plus osmotic) soil water potential was measured with a thermocouple psychrometer. Head space O_2 and CO_2 were measured once per week.

Experiment 3. To evaluate soil water potential and N level independently, salinity was controlled by adding NH_4NO_3 or NaCl in various combinations. NH_4NO_3 (0, 150, or 300 mg N/kg soil) and NaCl (0, 300, or 600 mg/kg soil) were added to the sand:silt loam mix described above. Soil water content was adjusted to 7.5%. Water potential was measured via thermocouple psychrometry; head space O_2 and CO_2 were measured twice per week.

RESULTS AND DISCUSSION

Data collected from Experiment 1 are given with N expressed as both N_S and N_{H2O} (Table 2). The latter expression (N_{H2O}) was determined by dividing soil mass-based N concentration by the soil water content:

$$N_{H2O} = \frac{mg\ N}{kg\ H_2O} = \frac{mg\ N}{kg\ soil} \times \frac{kg\ soil}{kg\ H_2O}$$

When N was expressed as a function of soil dry weight (N_S), optimum and inhibitory levels of N were between 10 and 20-fold different for the three soils. Expressing N as a fraction of soil water (N_{H2O}) reduced the variability to a factor of 2 to 3. Note that the sand, the driest soil, was most sensitive to over- fertilization, and the moist silt loam was least sensitive.

TABLE 2. Effect of soil texture and N level on petroleum biodegradation.

	-------- Optimum Level --------		------- Inhibitory Level -------	
Soil texture	**N_S** (mg N/kg soil)	**N_{H2O}** (mg N/kg H_2O)	**N_S** (mg N/kg soil)	**N_{H2O}** (mg N/kg H_2O)
sand	22	914	61	2514
loamy sand	259	2910	596	6697
silt loam	399	1330	756	2520

TABLE 3. Relationship between soil water, nitrogen, and water potential.

Water Content	Linear Regression Equation	r^2
5.0%	$\Psi_T^* = 123.3 + 2.35 \times N_S$	0.97
7.5%	$\Psi_T = 116.7 + 1.33 \times N_S$	0.98
10.0%	$\Psi_T = 121.0 + 0.95 \times N_S$	0.95
All water contents	$\Psi_T = 85.7 + 0.1113 \times N_{H2O}$	0.96

$^*\Psi_T$ = total (osmotic + matric) water potential measured in -kPa.

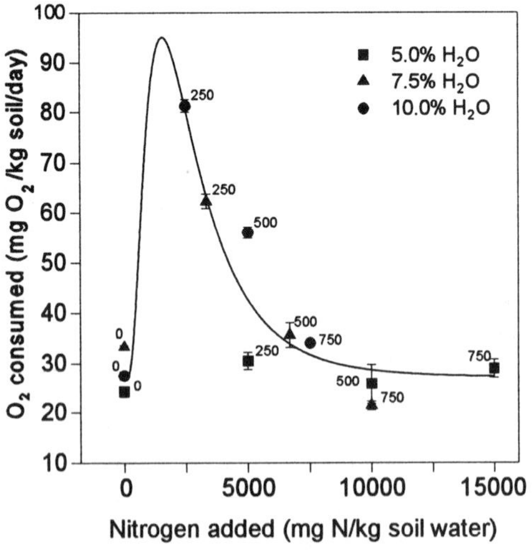

FIGURE 1. N_{H2O} versus O_2 consumption by petroleum-contaminated soil. Vertical bars are standard errors; line is a log-normal regression curve. Subscripts are levels of added N.

In Experiment 2, soil water potential (Ψ_T) was much more closely related to N_{H2O} than to N_S, reflecting partitioning of fertilizer salt into soil solution. With N_S distinct relationships are described for each water content (Table 3). Note that changes in soil water potential are much greater if N is added to dry soil than to moist soil. Use of N_{H2O} results in a single equation describing the relationship between soil N and soil water potential.

The relationship between biological O_2 consumption and N_{H2O} is shown in Figure 1. There was little activity where no N was added (N_{H2O} was 0). Oxygen

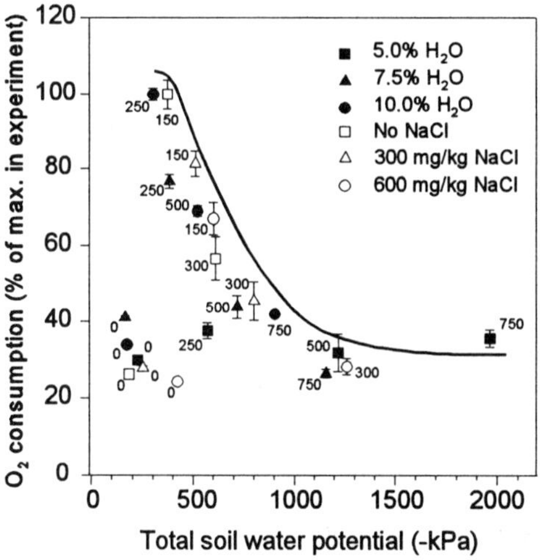

FIGURE 2. Effect of water potential on O_2 consumption by petroleum-contaminated soil. Vertical bars are standard errors. Subscripts indicate N fertilizer level.

consumption was maximized with an N_{H2O} application of 2500 mg N/kg H_2O, and decreased where N_{H2O} exceeded that level. From these data, it is impossible to more precisely identify an optimum N_{H2O} because 2500 mg N/kg H_2O was the lowest N_{H2O} level in the experiment.

The effects of soil water potential on biodegradation are shown in Figure 2, which contains data from Experiments 2 (solid symbols) and 3 (hollow symbols). Soils that received no N addition had favorable soil water potential (close to 0), but there was very little microbial activity because the system was N-limited. In soils that were not N-limited, microbial respiration was inhibited by soil water potential below approximately -400 kPa (further to the right in Figure 2). Inhibition of microbial respiration caused by decreased soil water potential was independent of soil water content, level of fertilization, or source of the potential (NH_4NO_3 or NaCl).

CONCLUSIONS

Biodegradation was more closely associated with N_{H2O} than N_S. In a given soil, total soil water potential is closely related to N_{H2O} because soluble inorganic N salts are partitioned into the soil solution. Correspondingly, dry soil is easily over-fertilized, whereas wet soil is much less sensitive to excess N. In dry soil soluble N salts are partitioned into a small volume of soil water; in moist soil N is partitioned into a larger volume of water and effectively diluted. Nitrogen response is closely related to soil moisture content through soil water potential, and data should not be directly transferred across soil conditions. For example, costly errors could result if one tried to transfer data derived from a slurry reactor (very high water content) to a land-farm or a non-slurry bioreactor (low water content).

N_{H2O} (mg N/kg soil H_2O) accounts for both nutritional and osmotic aspects of N fertilization. The optimum N_{H2O} level in the soils studied was approximately 2,000 mg N/kg H_2O. N_{H2O} should be a useful expression in the formulation of fertilizer management plans for bioremediation systems.

ACKNOWLEDGEMENTS
Funding for this research was provided by the Army Environmental Quality Technology (EQT) Program, Work Unit AF25-RT-005, Investigation of the Feasibility of Low Temperature Biotreatment of Hazardous Wastes and Strategic Environmental Research and Development Program, (SERDP Project 712-94 ARMY) Enhancing Bioremediation Processes in Cold Regions.

REFERENCES

Braddock, J. F., M. L. Ruth, P. H. Catterall, J. L. Walworth, and K. A. McCarthy. In review. "Enhancement and Inhibition of Microbial Activity in Hydrocarbon-Contaminated Arctic Soils: Implications for Nutrient-Amended Bioremediation." Submitted to: *Environ. Sci. Technol.*

Brown, K. W., K. C. Donnelly, and L. E. Deuel, Jr. 1983. "Effects of Mineral Nutrients, Sludge Application Rate, and Application Frequency on Biodegradation of Two Oily Sludges." *Microb. Ecol.* 9: 363-373.

Dibble, J. T. and R. Bartha. 1979. "Effect of Environmental Parameters on the Biodegradation of Oil Sludge." *Appl. Env. Microbiol.* 37(4): 729-739.

Genouw, G., F. De Naeyer, P. Van Meenen, H. Van de Werf, W. De Nijs, and W. Verstraete. 1994. "Degradation of Oil Sludge by Landfarming - A Case-Study at the Ghent Harbour." *Biodeg.* 5: 37-46.

Huesemann, M. H. 1994. "Guidelines for Land-Treating Petroleum Hydrocarbon-Contaminated Soils." *J. Soil Contam.* 3(3): 229-318.

Huntjens, J. L. M., H. De Potter, and J. Barendrecht. 1986. "The Degradation of Oil in Soil." In: J.W. Assink and W.J. vander Brink (Eds.), *Contaminated Soil*, pp. 121-124. Martinus Nijhoff Publishers, Dordrecht, The Netherlands.

Morgan, P. and R. J. Watkinson. 1992. "Factors Limiting the Supply and Efficiency of Nutrient and Oxygen Supplements for the *In Situ* Biotreatment of Contaminated Soil and Groundwater." *Wat. Res.* 26(1): 73-78.

Waksman, S. A. 1924. "Influence of Microorganisms upon the Carbon:Nitrogen Ratio in the Soil." *J. Agric. Sci.* 14: 555-562.

Walworth, J. L. and C. M. Reynolds. 1995. "Bioremediation of a Petroleum-Contaminated Cryic Soil: Effects of Phosphorus, Nitrogen, and Temperature." *J. Soil Contam.* 4(3): 299-310.

Zhou, E. and R. L. Crawford. 1995. "Effects of Oxygen, Nitrogen, and Temperature on Gasoline Biodegradation in Soil." *Biodeg.* 6: 127-140.

REMEDIAL STRATEGY FOR PETROLEUM HYDROCARBONS: ENHANCED INTRINSIC BIOREMEDIATION

Christopher H. Nelson, Curtis S. Wright, James E. Goetz, and Karen Van Rijn

ABSTRACT: Normal refueling operations at the City of Tucson's Thomas O. Price Service Center resulted in petroleum hydrocarbon losses over an extended period. Primarily gasoline was released into the vadose zone beneath the pump islands where it migrated vertically 100 feet until it encountered the uppermost aquifer, resulting in the lateral migration of dissolved-phase hydrocarbons over a distance of 1,800 feet.

Soil vapor extraction with thermal oxidation off-gas treatment was selected as the remedial approach for the highly impacted soils near the source zone. The depth of water, lateral extent of the plume, low hydraulic conductivity of the upper aquifer, and long-term operating costs eliminated pump and treat approaches for groundwater. Air sparging and biosparging approaches were considered but eliminated due to excessive drilling and installation costs. Intrinsic bioremediation became the optimal alternative for groundwater remediation.

A review of historical data indicated that significant quantities of naturally occurring dissolved oxygen were being transported into the contaminant plume through existing groundwater flow, resulting in the natural bioattenuation of the contaminated groundwater at the fringes of the plume. Additionally, low levels of nitrate were being transported on-site and consumed within the plume. Therefore, it was reasonable to assume both dissolved oxygen and nitrate were serving as terminal electron acceptors in the ongoing bioremediation of the hydrocarbon plume.

Laboratory tests were conducted confirm the role of oxygen, nitrate, and other inorganic nutrients in the bioremediation of the BTEX and TPH contaminants in groundwater from the site. Over 90% reduction in BTEX and TPH was observed in the oxygen and nitrate-supplemented systems during bench-scale testing. Field pilot tests were also conducted to confirm the effectiveness of nitrate-amended nutrients which included identifying specific bacterial strains collected from the site.

Based on these results, a full-scale nutrient injection system was designed for installation in late 1996. The system is designed to effectively disperse nitrate in the source area without allowing nitrate to migrate offsite. This is an enhanced intrinsic bioremediation approach using nitrate-amended nutrients. The concept promotes naturally occurring denitrifying bioremediation within the center portions of the hydrocarbon plume and allows residual dissolved oxygen to collapse the plume over time. This paper will present data from the laboratory, pilot-, and full-scale stages of the project.

CASE STUDY OF BIOVENTING INCLUDING NUTRIENT ADDITION AT KINCHELOE AFB

Mary Katherine O'Mara (U.S. Army Corps of Engineers, Buffalo District)

INTRODUCTION

Biological treatment will be used at the former Kincheloe Air Force Base (KAFB) in Michigan at an area used for fire training exercises. The area is contaminated with petroleum hydrocarbons and it will be remediated using bioventing. The site is amenable to bioremediation but the biodegradation rates are slow. A long term pilot study was initiated to decide if nutrient addition will accelerate the biodegradation rate.

Biological treatment offers a cost-effective alternative to traditional site remediation technologies. Many experts regard biotreatment as the technology of the future (Zappi et. al., 1992). Biological treatments use microorganisms to transform contaminants into non toxic products. These treatments are attractive remediation options because they are cost effective and often destroy the contaminants. Recent usage of the terms in situ bioremediation in both scientific and popular literature implies that knowledge of biodegradation has produced one or more reliable technologies that are fully operable. However, the fundamental progression from rigorous pure science to full scale engineering has rarely been completed for in-situ biodegradation (Madsen 1991).

Site description. KAFB is located in Kinross Township, Chippewa County, on the eastern side of the Upper Peninsula of Michigan. The installation was a Strategic Air Command base closed in 1977. This site contains a fire training area (FTA) which was used for training in fire-fighting techniques using JP-4 fuel, diesel, gasoline, waste oil, or solvents. The FTA encompasses approximately three acres of flat, unpaved land with light brush and scattered small trees.

The site contaminants are generally hydrocarbon compounds, which is typical of an area used for fire-training activities. Contamination was primarily found in the vadose zone soil, although contaminants were detected in the groundwater. Total recoverable Petroleum Hydrocarbons (TRPH) concentrations in the soil samples varied from below detection levels to 12, 000 mg/kg. TRPH concentrations in the groundwater samples varied from below detection levels to 3 mg/l. The FTA is underlain by a light brown, fine- to medium-grained sand. The total depth to bedrock is approximately 163 feet below ground surface (BGS). Groundwater at KAFB occurs in the unconsolidated glacial outwash sediments. The depth to groundwater was gauged between 62.81 and 63.24 BGS. The groundwater flow direction is toward the southeast.

BV is a potential remediation technology based on several factors. The identified contaminants were primarily hydrocarbons that are amenable to bioremediation. The subsurface conditions were nearly ideal for this technology, including a sandy homogeneous soil, and a deep vadose zone. However, based on preliminary feasibility studies the initial bio-consumption rates were low. A long-term bioventing test was designed to further investigate the possible applications of BV and to learn if nutrient addition can accelerate the bio-consumption rates in the nutrient impoverished soils.

Prior field investigations. The entire KAFB was assessed during three studies before the fire training area was singled out for further study and remediation. Parsons Engineering Science (PES) conducted the third field investigation at the former KAFB in 1991. Based on the findings of the 1991 investigation, it was concluded that the FTA was a source of groundwater contamination and should be further characterized and remediated.

PES conducted a fourth field investigation at the FTA in 1994. Twenty five soil gas samples and ten confirmatory laboratory samples were collected. The field samples were analyzed for BTEX, TCE, and PCE. This sampling program showed that soil contamination was very localized. The highest contamination was found near the center of the FTA and decreased as the radial distance from this point increased. The vertical contamination profile indicated high levels of hydrocarbons near the surface to 25 feet BGS, then decreasing to just above the water table where the level increased in the source area. Plume boundaries were more fully defined during the long term bioventing study conducted in 1995-1996.

Site preparation for the in-situ bioventing preliminary pilot test included the construction of one central vent well (VW), one background well (BW), five multi-depth soil vapor monitoring points (VMPs), and an air injection system. The air injection VW was installed in the center of the contamination plume. The BW was installed approximately 200 feet northwest of the VW. The purpose of the BW was to monitor background conditions at the site. Five VMPs, were installed at varying distances away from the VW. Each VMP was constructed with three individual monitoring probes at different depths. In-situ respiration and air permeability preliminary pilot tests were performed at the FTA. The respiration and permeability tests followed the AFCEE protocol. (Hinchee et. al., 1992) Before any air injection activities the soil gas in the VW, BW and each VMP was monitored for initial oxygen, carbon dioxide, and total volatile hydrocarbon (TVH) levels. All soil gas samples collected from the VW and all VMP intervals exhibited depleted oxygen levels, elevated carbon dioxide levels, and TVH concentrations varying between 840 ppm and 11,200 ppm. These sample results showed that the indigenous microorganisms have depleted much of the naturally available oxygen supply.

The in-situ respiration test determined the initial biodegradation rate of the soil contaminants. Data from this test was used to estimate the required cleanup time for the site based on initial contaminant levels. Generally the biodegradation rates measured during the in-situ respiration test represent those for a full scale bioventing system.

Three points were chosen for the respiration test, the soil gas at these points was oxygen deficient, and had elevated TVH levels. A BW was monitored to assess background levels of soil gas O_2. The O_2 was at atmospheric level in the BW. The respiration test consisted of aerating the site for 24 hours until the O_2 in the soil gas was at atmospheric concentrations. Each point chosen for the respiration test was connected to a blower and flow meters were attached to each point to measure the flow. Helium was delivered to each point to act as an inert tracer to measure air diffusion. The aeration was then stopped and the rate of O_2 uptake was measured. After five days of monitoring the test was terminated. The respiration test showed enhanced biodegradation rates of 156 to 305 mgTPH/kg/yr. Using the calculated biodegradation rates, it was approximated that it will take 15 years to clean up the site without the addition of nutrients.

The air permeability test was conducted to determine the effective radius of oxygen influence during air injection at a single vent well (VW). This was a primary design parameter necessary to determine required well spacing for a full-scale BV remediation system. The soil permeability test was conducted by injecting air at a constant rate from a single venting well while measuring the pressure changes over time at the vapor monitoring points. Based on the pressure response observed the estimated empirical radius of oxygen influence was greater than 200 feet.

Based on the results of the bioventing feasibility test it was determined that no additional vent wells or blowers would be necessary for full scale operation.

METHODS AND MATERIALS

Description of Long Term Bioventing Project: The initial biodegradation rates calculated from the Bioventing Feasibility Test (1994) were low. A long-term bioventing pilot test was designed to assess whether biodegradation is feasible for remediation of this site and to determine the potential for nutrient addition to accelerate bio-consumption rates. Soil samples were collected before, during, and after the test to quantify the extent of remediation and the effects of nutrient addition.

The sampling protocol monitored the entire contaminated area to detect the rate at which the contamination was disappearing, and to monitor the growth of microorganisms. Several sampling events over a one year period were conducted to get a temporal distribution of contaminant disappearance at the site. The site sampling protocol was designed to quantify the amount of contamination that was being removed by biodegradation and not by physical forces. It is very difficult to prove that biodegradation is occurring in the field. Most studies that have shown that biodegradation is occurring were highly controlled bench scale studies that allowed a mass balance to be completed. It is difficult if not impossible to perform a mass balance in a field scale study so other methods must be employed to show that the disappearance of contamination can be attributed to biodegradation. The methods employed to prove that biodegradation was occurring included the measurement of the disappearance of contamination, quantifying the amount of contamination lost by physical forces such as volatilization, and monitoring the growth of hydrocarbon degrading microorganisms. The sampling protocol included observing the differences in the respiration rates of microorganisms, and contaminant disappearance in areas treated with nutrients as compared to areas not treated with nutrients. The differences observed in the area treated with nutrients could be attributed solely to biodegradation, however this method could only prove that biodegradation was occurring it could not absolutely quantify the amount of contamination being removed by this mechanism. Several persistent compounds were chosen to act as internal tracers, these compounds can be expected to undergo the same abiotic processes as the BTEX compounds. If the ratios of persistent compounds to BTEX decreases, the net loss of BTEX can be attributed to a reductive mechanism such as bioremediation. Soil samples taken before beginning bioventing, and after the system had been operating for approximately one year quantified the extent of remediation.

Nine inner matrix samples were taken in pre-designated areas near the center of the plume. The inner matrix soil borings were drilled to a depth of 50 feet. The inner matrix samples were analyzed for VOCs, metals, ammonia-nitrogen, TKN, nitrate, ortho-phosphate, sulfate, COD, pH, alkalinity, total cyanide, percent moisture, grain size distribution, and TPH degrader count. Ten outer matrix soil samples were taken in areas determined from field screening. The outer matrix soil borings were drilled to a depth of 30 feet. The outer matrix samples were analyzed for VOCs, and TPH degrader count. One background sample was taken in an uncontaminated area. The background soil boring was drilled to fifty feet. The background sample was analyzed for VOCs, metals, ammonia-nitrogen, TKN, nitrate, ortho-phosphate, sulfate, COD, pH, alkalinity, total cyanide, TPH degrader count, percent moisture, and grain size distribution.

In order to determine the possible effects of nutrient addition on the rate of biodegradation in the FTA, two cells of the testing area were further treated prior to starting the bioventing system. One of these cells (SB15) was treated with an aqueous solution containing 500 ppm reagent grade ammonium nitrate and 500 ppm reagent grade ammonium phosphate. Five hundred gallons of this solution were applied topically to the soil. The solution completely saturated a conical section to a depth of approximately 8 feet. The second cell was a control cell, treated with water and no additional nutrients. Five hundred gallons of water were applied topically to this spot. This cell was expected to be of similar size and shape as the first cell. VMPs were installed six feet deep in the center of each cell to allow for later respiration testing. Several extra samples were taken from the top ten feet of SB15 during the first sampling event. In addition three samples were collected by a hand boring from SB15 approximately one month after the start-up of the bioventing system. The samples were collected from the same depth intervals as the samples collected from SB15 during the first sampling event. This sample quantified the initial increase in nutrient levels from nutrient addition and assessed the infiltration and migration of nutrients. The samples were analyzed for ammonia-nitrogen, TKN, nitrate, cyanide, ortho-phosphate, and percent moisture.

During the first sampling event the soil cores were screened to determine the area of highest contamination, two samples were taken from each soil boring location. During the second and third sampling events the soil borings were placed as close as possible to the first event locations and taken from the same depths. The samples were analyzed for the same parameters as selected during the first sampling event. Taking the samples in the same location during every sampling event allowed a point to point comparison of the effects of bioventing and nutrient addition on site.

Air samples were collected from a flux chamber during the first sampling event and approximately one month after the first sampling event. The air sample was collected in the area of highest contamination. This sample was collected to quantify the hydrocarbon content of air emissions before and during the operation of the bioventing system.

After nine and twelve months of operation confirmatory soil samples were collected from the fire training area. Three sampling events were chosen because at least three points were needed to give enough data to make meaningful conclusions.

During the second and third sampling events respiration tests were performed. The purpose of the respiration tests were to measure the increase, if any, in the instantaneous biodegradation rate in the nutrient and water enriched cells as compared to the untreated areas. The respiration tests were performed by shutting down the bioventing system and taking periodic readings of O_2, CO_2, and TVH from each VMP.

RESULTS

The first sampling event took place in September 1995, the second event was held in June 1996, and the final sampling event was completed in September 1996. The nutrient addition was successful in raising soil nitrogen concentrations to a depth of eight feet. The addition of phosphorus was only successful in raising soil concentrations of this element for two feet BGS. This is because the phosphorus reacted with the calcium in the soil rendering it immobile. The nutrient concentrations over the entire site were depleted by the second sampling round. The respirations tests were conducted after the nutrients had been depleted so no conclusions can be drawn from the results of these tests. The air sample taken before the bioventing system was started showed no ambient volatilization. The sample taken after the system was started showed no hydrocarbons venting from the site. The ratios of BTEX compounds to persistent compounds decreased throughout the study. The number of microorganisms on site increased with the addition of oxygen but decreased when the available supply of nutrients was depleted. The number of microorganisms increased more in the area treated with nutrients than in the untreated areas. The area impacted by VOCs expanded after the bioventing system was started however by the third sampling event the area had returned to less than initial conditions due to contaminant removal.

CONCLUSIONS AND RECOMMENDATIONS

- The addition of air did significantly increase the number of aerobic microorganisms on site which were capable of utilizing the contamination during respiration.
- The decrease in the number of microorganisms on site during the third sampling round may be related to the depletion of soil nutrients throughout the site.
- The concentrations of total hydrocarbons and BTEX compounds decreased in the area treated with nutrients at a greater rate than in areas not treated with nutrients.
- Air samples collected on site showed negligible VOC emissions, this indicates that the contaminants are not being removed by volatilization, which supports the hypothesis that the disappearance of contamination is due

to biodegradation.

- The reduction of the ratio of BTEX to persistent compounds indicates that the BTEX compounds are being removed by a reductive mechanism such as biodegradation.
- Bioventing is an applicable technology to remove BTEX compounds. The total amount of BTEX on site was decreased by approximately 20% in the areas not treated with nutrients over a one year period. This indicates that the site could be remediated in five years, however the rate of biodegradation at this site will slow considerably because all of the available soil nutrients have been depleted. Also, the rate of biodegradation will begin to slow as the amount of contamination decreases. These factors will raise the five year estimate, a conservative estimate of the amount of time needed to remediate the site without the addition of nutrients is 10 to 15 years. Approximately 50% of the contamination was removed in the area treated with nutrients. This indicates that the site could be cleaned up in two years with the addition of nutrients, however the rate of removal would begin to slow as the contamination decreases. A conservative estimate of the amount of time needed to remediate the site with the addition of nutrients is approximately 5 years.
- For future studies all measurements of the effects of nutrient addition should be made one or two days after application. This includes conducting respiration tests, sampling to determine if the nutrient application was successful, and enumerating bacteria.
- Nutrients should be applied to the test cell more than once during the study.

REFERENCES

Hinchee, R., Ong, S., *Test Plan and Technical Protocol for a Field Treatability Test for Bioventing* Prepared for U.S. Air Force Center for Environmental Excellence, 1992.

Madsen E.L. (1991). "Determining in situ biodegradation, facts and challenges." *Environmental Science and Technology*, Vol. 25, No. 10, 1663-1672.

Zappi, Mark E., Gunnison, Douglas, Pennington, Judith, Cullinane, M. John, Teeter, Cynthia L., Brannon, James M., Meyers, Tommy E., Banerji, Shankha, and Sproull, Robert. 1992. *Technical Approach for In Situ biological Treatment Research Bench-Scale Studies*, Miscellaneous Paper IRRP-92-XX, US Army Waterways Experiment Station, Vicksburg, MS.

ASSESSMENT OF NUTRIENT-CONTAMINANT CARBON RATIOS FOR ENHANCING IN SITU BIOREMEDIATION

Richard B. Coffin (Environmental Quality Sciences, Naval Research Laboratory, Washington, DC); Michael T. Montgomery (Geo-Centers, Inc., Ft. Wash., MD); Cheryl A. Kelley (Environmental Quality Sciences, Naval Research Laboratory, Washington, DC); and Luis A. Cifuentes (Texas A&M University, College Station, TX)

ABSTRACT: Through the SERDP CU-030 bioremediation development program, groundwater circulation systems have been installed at sites contaminated with BTEX and PAH hydrocarbons to inhibit transport of contaminants to adjacent ecosystems and stimulate bioremediation of the soil. This approach may enhance contaminant bioavailability and increase degradation rates by the natural bacterial assemblage. With recirculation of the groundwater through the contaminated zone, there is potential to enhance degradation with the addition and mixing of nutrients. Balancing microbial C/N/P ratios can increase the net carbon flow through the microbial assemblage. This study presents a comparison of concentrations of nutrients and contaminants in the groundwaters over four sampling events during nine months, at the two contaminated sites. Distinctly different patterns in nitrogen cycling were observed between the sites. In one system vitrification appeared to be a major component of the total nitrogen cycle. In the other system, because of hypoxia, nitrate appeared to support microbial requirements for electron acceptors. These results are important for determining nitrogen speciation and concentrations required for enhancing contaminant degradation.

Another interesting comparison of nitrogen pools in the two ecosystems was the large difference in concentration of dissolved organic nitrogen (DON) in the systems. In the PAH-contaminated system, DON was a major component of the total dissolved nitrogen (TDN) pool . If the components of the DON pool are labile, this nitrogen source may be preferentially assimilated and inorganic nitrogen additions may not enhance the carbon mineralization. In both of the contaminated environments, phosphorus appeared to be a limiting factor in balancing C/N/P ratios required for efficient bacterial growth. The data from this study provide a survey for requirements of nitrogen and phosphorus amendments that will be used to enhance microbial degradation of contaminants.

ENHANCED BIOREMEDIATION OF SOIL AND GROUNDWATER USING NUTRIENT INJECTION

Regina S. Porter (Southeastern Technology Center, Augusta, Ga), Marc Ghetti (Freeman & Vaughn Engineering, Inc., Augusta, Ga), William S. Anderson III (S&ME, Savannah, Ga) and Sherri Johnson (DOE-SR, Aiken, SC)

ABSTRACT: The efficacy of using a nutrient injection system to remediate soil and groundwater contaminated with petroleum products has been demonstrated in two underground storage tank sites. The first site presented only soil contamination and the second site presented soil and groundwater contamination. Bioremediation at the first site lasted for 131 days while bioremediation lasted 51 days at the second site. Benzene was degraded to below detectable levels in all analyzed samples and groundwater analyses showed high biodegradation rates for all BTEX constituents. A third demonstration has been in progress for soil and groundwater contamination to confirm the encountered enhanced biodegradation rates for indigenous microorganisms.

INTRODUCTION

The in-situ bioremediation of soil and groundwater of underground storage tanks (UST) sites contaminated with petroleum products is significantly improved using a mixture of nutrients in the gas phase. One of the characteristics of UST contaminated sites is the presence of an unbalanced nutrient ratio. Microbial growth requires oxygen : nitrogen : phosphorus in the molar ratio of 20 : 12 : 1 in addition to a carbon source (Looney, B. B., 1995). If these nutrients are readily available, biodegradation rates can be significantly improved. This concept was investigated at two sites using pilot demonstrations. The results of these investigations are discussed here. A third site has been selected for an additional demonstration and is in progress.

The nutrient injection technology utilized in the pilot demonstration is patented as PHOSter™ and was developed by a research team from Westinghouse Savannah River Technology Center, Oak Ridge National Laboratory, and Ecova Corporation (Hazen, T.C., 1992). Freeman & Vaughn Engineering, Inc. was responsible for the engineering design and set-up of the pilot demonstrations. The project was funded by the Southeastern Technology Center (STC) located in Augusta, Georgia. STC has a Cooperative Agreement with the Department of Energy (DOE) that establishes guidelines to deploy and demonstrate innovative technologies developed in DOE laboratories.

Objective. The pilot demonstrations discussed in this abstract had three main objectives:

- To measure the biodegradation rate of indigenous microorganisms stimulated by the addition of nutrients.
- To determine the major parameters that affect the in-situ bioremediation process using PHOSter™.
- To determine the efficiency of PHOSter™ when compared with biosparging alone.

PHOSter^{TM} can be used to improve existing or new biosparging/bioventing systems. The nutrients delivered via injection wells consist of air, nitrous oxide, and vapor phase triethylphosphate (TEP). A reliable system for adding the gaseous mixture in a controlled and uniform manner is the key element in reducing operating time and remediation costs.

Site Description. PHOSter^{TM} was demonstrated at two UST sites. The first site is located in the City of Aiken, South Carolina, and the second site is located in Augusta-Richmond County, Georgia. The Aiken site, currently owned by the City of Aiken, was abandoned by the previous owner in 1992 when a leakage in one of the gasoline underground tanks was detected. Three USTs installed to depths of approximately 12 feet were in place when the project started. The native soil at the site consist predominantly of hard clay. Backfilled sandy soil is also present at the site. No contamination of groundwater had been detected at the Aiken site. Soil contamination ranged from 5 μg BTEX/Kg to 12,366 μg BTEX/Kg.

The Augusta-Richmond County site is an active site and has been used as the county public works facility for over forty years. Refueling of city and county owned vehicles is conducted on-site. There are currently five USTs in place at the site. Two 12,000 gallons USTs that are currently in service contain diesel and gasoline products. The site is at an elevation of approximately 140 feet within the City of Augusta and is adjacent to the Augusta Canal. The site is not, for the most part, a natural site since it has been backfilled over the years with coarse construction materials, such as bricks and concrete blocks, and sands or clayey sands. The characteristics of both sites are described in Table 1.

PROCESS DESCRIPTION

The flow diagram of the nutrient delivery system is presented in Figure 1. Air, nitrous oxide and TEP are mixed prior to the injection wells. The system delivered the nutrient mixture at a molar ratio of 20:12:1 of air : nitrous oxide : TEP. A pilot test to determine the radius of influence for an injection system was performed prior to installation of the wells. The radius of influence is dependent on the hydraulic conductivity of the soils and compressor capacity. The selected number of wells was based on the determined radius of influence and on the characteristics of the contaminated plume. The processes for the two sites are summarized in Table 2. Soil samples were obtained through the use of Direct-Push Technology (DPT), a very effective technology for obtaining soil samples. The use of DPT creates no soil cuttings for disposal, and it requires closure of only a small diameter hole (1").

Table 1. Properties of the two sites where PHOSter™ was demonstrated.

	Soil Type	K[3]	Groundwater Level (ft)	Microbial Count (cfu/g[2])	Total Kjeldahl Nitrogen (mg/Kg)	Total Phosphorus (mg/Kg)	Moisture content (%)
Aiken, SC	sandy clay and hard plastic clay	15×10^{-3} ft/day	> 50[1]	1.5×10^{3} to 1.6×10^{6} at 11 to 15 ft	47	20	7 - 24
Augusta, Ga	sandy clay and backfilled soil	not determined	~15[1]	9,000 to 1×10^{4} (soil)	62 to 140	170 to 1700	20.7 to 24.6

[1]No water supply wells are located within one-half mile of the site.
[2]Colony forming units per gram.
[3] Estimated hydraulic conductivity

Table 2. Process description for the two sites where PHOSter™ was demonstrated.

	Radius of Influence (ft)	Injection Cycles (per day)	Number of Wells	Depth of Wells (ft)	Delivery System	Number of Samples Analized	Benzene (maximum concentration)
Aiken, SC	5	8 (3 hours per cycle)	11	20 (2″ diameter PVC)[1]	4 scfm @ 30 psi	4	1,200 µg/Kg
Augusta, Ga	15	8 (3 hours per cycle)	4 (2 soil and 2 groundwater)	10 to 20 (2″ diameter PVC)[1]	1 scfm @ 10 psi	4	2,400 µg/L

[1]The screened section of the wells is 10 ft long

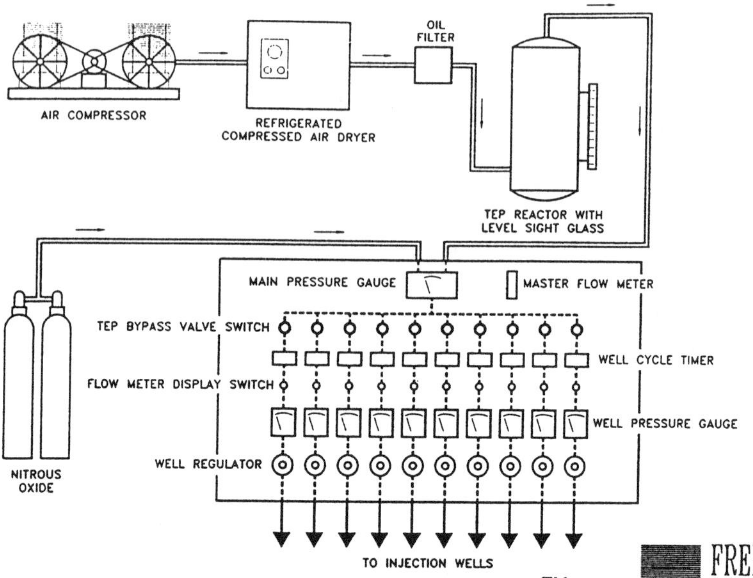

FIGURE 1. Flow diagram for PHOSter^{TM}.

RESULTS

Aiken, South Carolina. The system was in operation for 131 days. BTEX and Naphthalene analyses (EPA Method 8260) showed significant decrease in the levels of contamination. Significant results are presented in Table 3. Benzene was degraded to below detectable levels in all analyzed samples.

Augusta-Richmond County, Georgia. The system was in operation for 55 days. Soil and groundwater samples were analyzed for BTEX by EPA Method 8260. Groundwater analyses showed enhanced biodegradation for all BTEX constituents (Table 4). Soil analyses showed a 50% reduction in the benzene levels after 55 days.

CONCLUSIONS AND RECOMMENDATIONS

The need for more efficient and cost effective technologies to remediate soil and groundwater is well documented (Brockman et al., 1995). The currently used approach for vadose zone remediation is excavation of the contaminated soil followed by incineration and/or disposal in a landfill. This process stimulates site disruption and only transfers the contamination from one location to another. Additionally, landfilling and/or incineration of contaminated soil may not be the most cost effective alternative. Soil vapor extraction is also used for vadose zone remediation; however, the use of vapor extraction requires a volatile organic carbon (VOC) capturing device that substantially increases the remediation cost. Soil vapor extraction systems are usually used in conjunction with air sparging systems. PHOSter^{TM} presents several advantages when compared to these technologies. The results obtained from the demonstrations summarized here indicate that PHOSter^{TM} is a more efficient and cost effective

Table 3. Significant results for the site in Aiken, South Carolina. Results in μg/Kg; location GP6, depth 11′ to 15′.

Constituent	Baseline Concentration	24 days of treatment	43 days of treatment	131 days of treatment
Benzene	1,200	1,100	BDL[1]	BDL[1]
Toluene	11,000	9,200	140	380
Ethylbenzene	6,500	6,400	740	690
Total Xylenes	21,000	12,000	840	3,300
Naphthalene	8,400	17,000	1,100	730

[1]BDL = Below Detection Limits

Table 4. Significant groundwater results for the site in Augusta-Richmond County. Results in μg/L; location IW1 and IW2.

Constituent	Baseline Concentration IW1/IW2	6 days of treatment IW1/IW2	38 days of treatment IW1/IW2	55 days of treatment IW1/IW2
Benzene	130/2400	12/BDL	BDL/BDL	BDL/BDL
Toluene	240/8200	BDL/BDL	BDL/BDL	BDL/BDL
Ethylbenzene	390/2100	BDL/BDL	BDL/BDL	BDL/BDL
m and p-Xylenes	650/6600	15/BDL	BDL/BDL	BDL/BDL
o-Xylenes	140/2400	BDL/BDL	BDL/BDL	BDL/BDL

BDL = Below Detection Limits.

technology. In addition, since the injection system operates at low pressures and flow rates, VOC discharges to the atmosphere are negligible. Furthermore, pulsed operations like the PHOSter™ set-up minimize volatilization of VOCs.

The most common approach for saturated zone remediation is still pump-and-treat. Conventional groundwater physical treatment involves pump and treat systems where groundwater is pumped out of the ground, treated and either discharged or re-injected. However, these systems rarely achieve typical cleanup level goals. In some cases, pump and treat systems have been in operation for many years with little reduction in dissolved phase hydrocarbon concentrations and very little overall removal. Clean-up levels of groundwater using PHOSter™ in the Augusta-Richmond County site were extremely encouraging but more data is necessary to confirm those results. A third demonstration has been set up to check the results previously obtained. A significant increase in the microbial population just above the water table in the vadose zone was encountered at the end of the demonstration and in-situ respirometry indicates that the groundwater contamination was impacted by biological oxidation and not by a physical process. However, the release of carbon dioxide was not monitored during the process and biodegradation rates could not be established. Thus, more information is necessary to draw a reasonable mass balance. The significant observation of the pilot test at this point is that remediation has occurred in a very expedient manner.

Early reports from industrial sites indicate approximate biodegradation rates in the vadose zone of 0.5 mg BTEX/Kg of soil per day using biosparging alone and 5 mg BTEX/Kg of soil per day using PHOSter™ (Looney, B.B., 1995). Thus, the addition of the PHOSter™ process resulted in a substantial increase in contaminant mineralization. Based on the soil characteristics of the two sites, hydraulic conductivity and soil moisture content seem to be the major parameters that will determine the effectiveness of PHOSter™.

REFERENCES

Brockman, F.J.; Payne, W.; Workman, D.J.; Soong, A.; Manley, S. and Hazen, T.C. 1995. "Effect of Gaseous Nitrogen and Phosphorus Injection on In Situ Bioremediation of a Trichloroethylene-Contaminated Site." *Journal of Hazardous Materials 41*: 287 - 298.

Hazen. T. C. 1992. *Test Plan for In Situ Bioremediation Demonstration of the Savannah River Integrated Demonstration Project DOE/OTD TTP No.: SR 0566-01 (U)*. Prepared for the Department of U.S. Energy under Contact No. DE-AC09-89R180035.

Looney, B.B. 1995. Technical Evaluation of PHOSter™ to Optimize Early Biosparge Operation and Minimize Fugitive Emissions. Internal Report, Savannah River Technology Center.

MARKET OVERVIEW OF THE BIOREMEDIATION MARKET

Olin R. Jennings (The Jennings Group, Inc.)

ABSTRACT: The bioremediation industry continues to grow as this technology gains acceptance by regulators, customers, and remediation companies. However, good market information and demographics generally are not available.

This session will present the results of The Jennings Group's annual survey of bioremediation consultants, remediation contractors, and equipment suppliers. The survey will include information on the number of projects started and completed, size of the market, number of suppliers, industry profitability, and other market and business related trends and information. The survey will cover both the historic market, the market for 1994, and the future outlook for bioremediation. The Jennings Group conducted the first comprehensive demographic survey of the bioremediation industry in 1993, published the first bioremediation industry resource guide in 1994, and is a co-author of a comprehensive market and business analysis of the U.S. bioremediation industry which was completed in 1993. The Jennings Group made a similar presentation at the last symposium in San Diego.

INTERNATIONAL BIOREMEDIATION: RECENT DEVELOPMENTS IN ESTABLISHED AND EMERGING MARKETS

David J. Glass (D. Glass Associates, Inc., Needham, Massachusetts)
Thomas Raphael (Umweltberatung Dr. Raphael, Schwerte, Germany)
Jacques Benoit (AGRA Earth & Environmental, Calgary, Alberta, Canada)

ABSTRACT: The world market for bioremediation products and services in 1996 was perhaps as high as a half billion U.S. dollars, at least US $250 million of which was outside the U.S. Significant markets for bioremediation exist in several European countries, notably Germany and the Netherlands, with smaller existing markets in the Nordic countries and, outside Europe, in Canada. Bioremediation is also becoming more widely utilized in several nations around the world, including other European countries, such as France, Spain, and the U.K. We expect biological treatment methods to be used in the rapidly developing environmental markets in Eastern Europe, the Pacific Rim and Latin America.

OVERVIEW: WORLD REMEDIATION MARKETS

Historically, the U.S. has made up almost half the world market for environmental products and services, however, in the early years of the next century the U.S. will probably represent only 35-40% of the world market. Europe will claim about 30% of the market, but substantial growth will be seen in Asian and South American markets (OECD 1996b). For the most part, bioremediation activities have been seen in countries where there is an existing remediation industry (Glass et. al. 1995). The U.S. has been the world leader, with about half the world bioremediation market, but the overall European market could match or surpass the U.S. market by the turn of the century, with each in the approximate range of US $350-600 million (Glass et. al. 1995).

BIOREMEDIATION IN THE UNITED STATES

Although overall U.S. remediation markets have matured and have seen little or no growth in recent years (OECD 1996a), bioremediation and other innovative technologies have continued to grow within the overall market, albeit not necessarily as rapidly as in previous years. Bioremediation is being chosen more frequently for *in situ* projects, often in combination with technologies like soil vapor extraction. We estimate the 1996 U.S. bioremediation market to be $200-250 million, the vast majority of which is attributed to services rather than product sales.

We anticipate that this sector will see only single-digit growth in coming years, primarily due to changing pressures in the marketplace. Government funding cuts leading to decreased regulatory enforcement and the growing importance of programs like Brownfields are causing economic driving forces to become more important than regulatory forces. Increased adoption of natural attenuation will lead to reduced markets for "interventional" bioremediation, and while this may be environmentally beneficial, it will lead to lower revenues for the bioremediation industry. We therefore predict substantially slower market growth following the turn of the century, perhaps as low as 2.5-5% per year, and feel the market in 2000 will be $300-500 million, and $400-700 million in 2005.

BIOREMEDIATION IN EUROPE

Europe represents the second largest bioremediation market in the world, with the greatest opportunity for short-term growth. Although the better established markets have grown only slowly in recent years, viable environmental industries are arising in several European countries, where bioremediation is well positioned to play a major role.

Germany. The situation for bioremediation in Germany has been presented in detail (Raphael and Glass 1995), but there have been many changes in the bioremediation market and the technologies practiced since that time. The most significant trend is the importance of central remediation stations, whose large capacities have made it economically unattractive to treat soil on-site. Germany's soil remediation stations had estimated 1996 total capacity of about 3 million metric tons per year, with total bioremediation capacity of about 2 million metric tons per year (Schmitz 1995). We estimate that these facilities were used to only 50-60% of capacity, which together with soil treated in on-site projects, created a total of about 1.6 - 2.0 million metric tons of soil treated by bioremediation. At an average cost of DM 110 (US $75), this translates to a 1996 market of US $120-150 million.

The German bioremediation market has slowed somewhat in recent years. The availability of large capacities in some of the Länder for landfilling and declining waste production due to recycling have depressed prices of all treatment technologies. Furthermore, the overall economic weakness has reduced the amount of remediation activities, and bioremediation has faced competition from other methods such as fixation or encapsulation, and from the lowered prices of incineration and thermal desorption. In the wake of these changes, many companies that have specialized in on-site treatment, including several small bioremediation companies, are no longer able to compete against the large central remediation stations and have given up their business.

Another reason for this downturn is the lack of a fixed federal legislative framework for soil remediation. This situation is expected to change: a new federal waste and recycling law, the "Recycling and Waste Management Act", was enacted in October 1996 (Haznews 1996b). Under this law, the only alternatives allowed for contaminated soil would be removal or recycling. The Soil Protection Act was still in preparation in late 1996 and should cause additional changes in the handling and selection of remedial technologies.

Umweltschutz Nord, with 16 bioremediation stations as of December 1996, is Europe's leading bioremediation company, and is involved in several bioremediation projects at military sites, including a field trial for bioremediation of TNT at a former ammunition factory site.

The Netherlands. As previously discussed (Glass et al. 1995), growth of the bioremediation industry in the Netherlands will largely be dependent upon relaxation of the target values for petroleum hydrocarbons, which is not expected for several years. However, the Netherlands Research Program on In Situ Soil Bioremediation (NOBIS) awarded research grants amounting to about US $4.4 million to eight projects in 1995 (Haznews 1996a). The municipal waste treatment and disposal firm VAM has established a joint venture with Heijmans Milieutechniek to run a bioremediation facility with a capacity of 20,000 metric tons of soil in Wijster, while Heijmans is joining forces with three other companies, Ecotechniek Bodem, Mourik-Groot-Ammers, and Tauw Milieu, to form a joint venture, SoilNeth, to apply a variety of techniques, including bioremediation, to soil cleanup.

Sweden. Legislation for site remediation remains unclear, and there may be no responsible party for acts of pollution which occurred before 1989. The Swedish Environmental Protection Agency has launched a program for remediation of contaminated sites, which received funding of only 10 million kronor (approx. US $1.5 million), far less than requested. Among companies active in Sweden are VBB Viak, ANOX and its affiliate Marksanering i Sverige, EkoTec, Gotthard Nilsson, SAKAB, and the Swedish Geotechnical Institute (P. Englöv, L. Larsson, A. Persson, personal communications).

Finland. Bioremediation has been well accepted by government regulators and site owners, and was used for about 25% of the contaminated soil cleaned in 1995. Several local soil treatment plants with bioremediation capability are expected to come on line in 1997. An estimated 1,177 contaminated sites will need remediation within 20 years, at an estimated cost of US $850 million. New regulatory guidelines for contaminant endpoints are in preparation, and although risk assessment will be used, the new guidelines are expected to be stricter than at present. However, opportunities for alternative technologies will be created as new permit requirements and stricter disposal standards begin to limit the use of landfilling. Government funding is now available for remediation of old or abandoned gasoline stations, as well as other orphan sites. (R. Valo, personal communication).

Table 1. Military Sites in Europe.

Germany
West Group of Soviet Army: 14,000 contaminated sites, 4,048 to be remediated.
Federal Republic of Germany Armed Forces sites: 3,285 suspected to be contaminated.
Poland
Former Soviet military sites: Approximately 35 sites to be cleaned, estimated US $2 billion cost.
Hungary
Former Soviet military sites: At least 150 sites still to be cleaned, estimated US $1 billion cost.
Sources: TerraTech 1996 and USDOC 1996.

United Kingdom. The U.K. market for land remediation is US $350 million (USDOC 1996), and there may be 5-20,000 contaminated sites in England, Scotland and Wales, of which as many as 1,000 may be priority sites (Haznews 1996c). The Department of the Environment issued a draft guidance document to implement the contaminated land regulations under the 1995 Environment Act, that explains how contaminated land will be identified and who should be responsible for any required remedial action. A recently enacted landfill tax will make techniques like bioremediation more economically competitive. In 1996, Viridian Bioprocessing, which develops microbial processes for effluent treatment, was acquired by International Bioremediation Services, an international producer of microbial products and processes, while BioTal Limited merged its Land Remediation Division with the Dutch firm HMVT to form Telluric Limited (J. McCluskey, personal communication).

Austria. The 1989 "Contaminated Sites Act" has been revised, with the major impact being a change in the financing of projects for site remediation through fees for wastes. From an initial 1989 inventory of 2,500 suspected sites, 122 sites have been as mandated for remediation. At 23 sites, remediation or preventive actions have been started, and four sites have already been remediated.

Switzerland. The Swiss firm Ebiox has almost completed soil cleanup using its vacuum heap bioremediation system at a former gas work site in Winterthur. NUVAG, a subsidiary of Umweltschutz Nord, will be using a mobile soil washing plant for the remediation of a mineral oil contaminated site in Winterthur.

France. Although France has no specific law to deal with contaminated sites, the federal Ministry of the Environment issued a report in 1994 that listed 669 polluted sites. This is an ongoing activity which, coupled with regional efforts, is expected to identify tens of thousands of sites of which thousands will be found to be polluted. Responsible owners will be required to investigate the sites and to complete risk assessments, classifying each site into one of three categories of contamination. Under a directive of January 1989, sites where the responsible party fails to pay for a cleanup can become orphan sites, where remediation will be managed by the Agence de l'Environnement et de la Maîtrise de l'Energie (Ademe), financed by a Superfund-like tax established in 1995 (P. Jacquemin, personal communication). A few bioremediation projects in France are known, most of which have involved hydrocarbon contamination (Glass et al. 1995).

Spain. The State Secretariat for the Environment has identified 4,532 contaminated industrial sites about 5% of which are considered dangerously polluted. Roughly half these sites are in the highly industrialized Basque and Catalonia regions (USDOC 1996). Bioremediation technologies have not yet been used in Spain to any appreciable degree.

Italy. Italy's environmental regulations are just beginning to be strengthened to meet EU standards, but the country has an estimated soil remediation market of US $75 million (OECD 1996a). Castalia has treated hydrocarbon contamination *in situ* at a landfill, and has developed an air-venting process to treat perchloroethylene from nonsaturated soils. The U.S. company EBASCO used bioremediation at the AGIP-Trecate site, where an oil well exploded in February 1994.

BIOREMEDIATION IN CANADA

Acceptance of bioremediation as a remedial option continues to be slow in Canada, primarily because of the extremely low costs of landfill disposal, long timeframes for bioremediation due to colder climate, and weak government enforcement. In Eastern Canada (especially in Quebec) there are about a dozen large regional and commercial bio pads that have had some success in treating UST soils in the last 5 years. Landfarming is not an accepted practice in Quebec, but biopiles have recently gained popularity. In Ontario, landfarming has been an accepted practice for many years and continues to be popular. The provincial government has been reviewing and consolidating environmental regulations and has moved to a more risk-based process for contaminated sites which, although well received by industry, has resulted in a slowdown of the overall remediation market. In Western Canada, the oil and gas industry continues to prefer landfarming for the treatment of hydrocarbon contaminated soils. Biopiles are being used more frequently although there are very few *in situ* bioremediation projects. Leading companies include Biogenie and Sanexen in Quebec, Grace Bioremediation Technologies, Beak Consultants and Biorem in Ontario, and AGRA Earth & Environmental, O'Connor Associates Environmental and Golder Associates in Western Canada.

EMERGING REMEDIATION MARKETS

Central and Eastern Europe. The former Eastern bloc countries have severe environmental problems, but have not seen significant remediation activity. However, several nations have begun to adopt more realistic approaches to environmental protection. Much of the money being spent on environmental problems will be devoted to the more immediate concerns of air pollution, sewage and wastewater treatment and drinking water pollution. However, most of these countries have significant soil and groundwater contamination problems arising from the mining industry, former Soviet military bases, misuse of agricultural chemicals, and the continuing problem of improper disposal of hazardous wastes, so site remediation is at least a mid-term priority. The countries with the most promising situation for site remediation are shown in Table 2.

Table 2. Emerging Eastern European Markets for Bioremediation.

Czech Republic
Will spend US $5.3 billion for solid wastes and "damages from the past"; identified remediation of high-risk sites as a short-term priority.
Poland
US $200 billion next 25-30 years for environmental problems; restoration of environmentally damaged lands named a long-term priority.
Hungary
US $50-100 billion needed next 20 years for all environmental problems; new environmental law in 1995.
Bulgaria
Conducted 1993 inventory of polluted agricultural lands, heavy metals most common contaminant.

Source: USDOC 1996.

Japan. Japan offers a promising market for contaminated soil and water remediation and bioremediation because of its long history of industrialization and high land values. Several driving forces will affect the development of bioremediation in Japan.

Economics. Because of the limited land for expansion, construction activities mostly use existing industrial sites, which are often found to be contaminated. Since there is limited availability of suitable landfills, site owners must develop cost effective remediation techniques to allow soils to be reused. As a result, soil remediation is primarily driven by cost, time and the availability of suitable remediation options.

Culture. The Japanese culture is an integral part of all business transactions, with the highest quality standards demanded in all activities. Site owners thus require extensive research and demonstration of efficacy before initiating remediation activities, possibly hindering increased use of bioremediation. Also, the fear of public disclosure has caused many owners to delay remediation projects and to make bioremediation projects highly confidential.

Technology Acceptance. The Japanese are a technology-based society and require solutions using the most up to date technologies. Bioremediation, as a natural phenomenon not relying on sophisticated technologies, may be slow to gain acceptance. Also, tight construction schedules often cause contaminated soils to be excavated and stored off-site, which may lead to a greater focus on *ex situ* bioremediation techniques.

Regulatory Framework. In 1970, the Basic Law for Environmental Pollution Control was amended to include soil pollution, but it was only in 1991 that the law was further amended to include environmental quality standards for 25 substances. In 1996, the Japanese Environment Agency proposed new contaminated site legislation that would provide enforcement powers in cases where liable parties cannot be determined. However, environmental enforcement in Japan is often done through consensus and compromise, reducing the importance of strong regulations. The Japanese Environment Agency recognizes the importance of bioremediation and plans to issue guidelines for owners wishing to use this option.

Publicly available information on bioremediation projects is difficult to obtain. To date there have been very few large scale bioremediation projects in Japan, but there has been a considerable amount of research activity. In 1991, 67 companies, mostly including construction and water treatment companies, established the Japan Forum for Soil Environmental Remediation (now known as the Geo-Environmental Protection Center) to study soil reclamation technologies, with some emphasis on bioremediation. Among the many universities and institutions involved in bioremediation research, the most active has been the Japan Research Institute (JRI), which has united industry, government and academia to develop bioremediation technologies for application to soil and groundwater. It has conducted a number of symposia and undertook a pilot test on *in situ* bioremediation in 1994. In 1996 JRI formalized a partnership with Walsh Environmental (formerly ECOVA Corporation) for the application of bioremediation to chlorinated organics.

Some of the larger construction companies have set up their own bioremediation research programs, including Obayashi, Kumagai, Takenaka, Kajima and Konoike Construction. Obayashi Corporation has also conducted a full scale bioremediation project in Kuwait as a result of the Gulf War. Among other active companies, Kurita Water Industries is developing a vapor-phase bioreactor for trichloroethylene.

The potential market for site remediation in Japan is good, although prospects for bioremediation are uncertain. The lack of available land will make large scale *ex situ* landfarming unattractive for many owners, and the length of time and uncertainty of the final clean-up levels may hinder the wider use of *in situ* bioremediation. We therefore do not expect large scale use of bioremediation for at least another 3 to 5 years.

Other Asian Markets. Environmental industries are beginning to emerge in the rapidly-developing countries in Asia, as environmental spending becomes a national priority. Although most of this money will go towards municipal solid waste and wastewater treatment and air pollution control, it will begin to create an infrastructure for environmental protection.

Table 3. Emerging Asian Environmental Markets.

Country	Estimated or Planned Environmental Spending (U.S. dollars)
China	$5.0 billion per year
Hong Kong	$2.6 billion (1995-2000)
Philippines	$400 million per year
South Korea	$11.0 billion (1991-95)
Taiwan	$12 billion (1991-96)
Thailand	$600 million per year

Sources: SGS 1996, USDOC 1996.

Table 3 shows promising Asian environmental markets. In India, the government has commissioned an inventory of hazwaste disposal sites to be complete near the end of the century, and Malaysia has a need for oil pollution and spill control technologies (SGS 1996). Among Singapore's tough environmental laws are requirements for petrochemical and chemical companies to conduct baseline site assessments and to be responsible for increases in pollution (SGS 1996). South Korea has enacted a series of laws including the Soil Protection Act of 1995, under which a roster of sources of contamination and interventional concentration levels will be established (Bae 1995)

Latin America. Significant opportunities may exist in Mexico and Latin America. Mexico's soil remediation market may be as high as US $1 billion (USDOC 1996) although other estimates are much lower, and several U.S. engineering companies and manufacturers of microorganisms are known to have established operations there. Argentina, although lacking a comprehensive national environmental law, had an estimated 1992 market for pollution control products and services of US $170 million (USDOC 1996). Much of the ongoing remediation has focused on petroleum contamination, with some interest in using biological techniques. There are the beginnings of a remediation industry in other countries such as Chile, where the legal authority is lacking but where some remediations have taken place.

ACKNOWLEDGMENTS

We would like to thank the following individuals for providing information on bioremediation activities in their countries: Peter Englöv, VBB Viak (Malmö, Sweden); Brian Herner, Biorem (Guelph, Ontario, Canada); Patrick Jacquemin,

Ademe (Angiers, France); Lennart Larsson, Swedish Geotechnical Institute (Gothenburg, Sweden); John McCluskey, Telluric Limited (Cardiff, U.K.); Guiliano Mortola, Ramoco (Genova, Italy); Anders Persson, ANOX AB (Lund, Sweden); and Risto Valo, Soil and Water Ltd. (Helsinki, Finland).

REFERENCES

Bae, W. K. 1995. "Korea's Environmental Protection and Remediation Programs", presentation at Superfund XVI, Washington, DC, November 7, 1995.

Glass, D. J., T. Raphael, R. Valo, and J. Van Eyk. 1995. "International Activities in Bioremediation: Growing Markets and Opportunities." In R. E. Hinchee, J. A. Kittel, H. J. Reisinger, (Eds.), *Applied Bioremediation of Petroleum Hydrocarbons*, pp. 11-33. Battelle Press, Columbus, OH.

Haznews. 1996a. "Dutch Bioremediation Research Funding in '95." *Haznews* No. 101 (August), pp. 8-9.

Haznews. 1996b. "New German Waste Law in Force." *Haznews* No. 104 (November), pp. 12-13.

Haznews. 1996c. "UK Contaminated Land: Local Authorities Burdened." *Haznews* No. 105 (December), pp. 10-11.

Organization for Economic Cooperation and Development. 1996a. *The Environmental Industry: The Washington Meeting*. OECD, Paris.

Organization for Economic Cooperation and Development. 1996b. *The Global Environmental Goods and Services Industry*. OECD, Paris.

Raphael, T. and D. J. Glass. 1995. "Bioremediation in Germany: Markets, Technologies and Leading Companies." In R. E. Hinchee, J. A. Kittel, H. J. Reisinger, (Eds.), *Applied Bioremediation of Petroleum Hydrocarbons*, pp. 35-45. Battelle Press, Columbus, OH.

Schmitz, H. J. 1995. "Die Jagd nach dem Boden geht weiter (The run on the soils goes on)." *TerraTech* No. 3, pp. 34-47.

SGS (Thailand) Co. Ltd. 1996. *Environmental Markets Asia: 1996-1997*. SGS Co. Ltd., Bangkok.

TerraTech. 1996. "Russische Militärs aus ökologischem Dornröschenschalf waschgeküsst." *TerraTech* No. 1, pp. 28-31.

U.S. Department of Commerce. 1996. *Global Export Market Information System*, searchable index of envirobusiness documents, including the Central and Eastern European Business Information Center, available at http://www.itaiep.doc.gov/.

COST EVALUATION OF ANAEROBIC BIOREMEDIATION VS. OTHER *IN SITU* TECHNOLOGIES

Gary E. Quinton (DuPont, Wilmington, Delaware)
Ronald J. Buchanan (DuPont Environmental Remediation Services, Wilmington, Delaware)
David E. Ellis (DuPont, Wilmington, Delaware)
Stephen H. Shoemaker (DuPont, Houston, Texas)

ABSTRACT: DuPont has developed a method to compare *in situ*, substrate-enhanced anaerobic reductive dehalogenation with other *in situ* technologies. The methodology employs a template site with a subsurface aquifer 300 m (1000 ft) long by 133 m (400 ft) wide. The template can incorporate variable depth, depending on the number of requisite subsurface treatment zones; for example, 10 m (30 ft), 20 m (60 ft), and 30 m (90 ft) can be used to approximate a particular site. Variables considered in the estimate for bioremediation include estimated engineering and flow/transport modeling costs; equipment costs; and operation, maintenance, and monitoring costs. The methodology allows the user to evaluate, on a consistent economic basis, substrate-enhanced anaerobic bioremediation, intrinsic bioremediation, *in situ* permeable reactive barriers, and pump-and-treat systems. Cost metrics presented include present cost in US dollars (USD), USD/lb of contaminant removed, and USD/1000 gals treated, using a discounted cash-flow analysis. Evaluation of the results indicates that intrinsic bioremediation is the cost-effective alternative at equivalent levels of protection.

COST ANALYSIS OF RISK-BASED CORRECTIVE ACTION

Kimberly L. Davis, P.E. (University of Tennessee, Knoxville)
Christian Kiernan (University of Tennessee, Knoxville)
Gregory D. Reed, Ph.D., P.E. (University of Tennessee, Knoxville)

ABSTRACT: The University of Tennessee's (UT) Waste Management Research and Education Institute (WMREI) has been tracking remediation costs for Tennessee underground storage tank (UST) site cleanup for the past four years, in order to study the cost-effectiveness of cleanup methodologies for different site characteristics. A model is being developed from collected site information that predicts costs of assessment, cleanup and monitoring. One way that Tennessee and other states are attempting to encourage cost-effective cleanups, while also safeguarding human health and the environment, is through the implementation of risk-based corrective action (RBCA) plans. Tennessee requires site owners to input geological parameters, exposure pathway parameters and contaminant concentrations into a formula that assigns a numerical score to each site. For sites that score below 500, owners are allowed to bypass cleanup and monitor the plume only. The cost model being developed by UT will be used to determine the efficacy of Tennessee's RBCA system in promoting cost-effectiveness in the management of these contaminated sites. Also, the current and potential future impact of RBCA on the composition (e.g. site characteristics, location) of sites undergoing cleanup was examined, as well as the cleanup methods being used.

INTRODUCTION

The Waste Management Research and Education Institute (WMREI) has been working with the Tennessee Department of Environment and Conservation (TDEC), Underground Storage Tank (UST) Division over the past four years in the study of the cost-effectiveness of cleanup at UST sites in Tennessee (Davis et al., 1995). This research has culminated in the identification and analysis of 76 UST sites throughout Tennessee, for which numerous site parameters (e.g. site characteristics, costs of various phases of work, and cleanup technology(s) implemented) have been documented. The original goal of this work was to compare the cost-effectiveness of cleanup for technologies using biotechnology (e.g., landfarming of soil) with "traditional" cleanup technologies (e.g., excavation of soil and incineration).

In order to satisfy federal financial responsibility requirements, Tennessee, along with a majority of states, established a Petroleum UST Fund ("Trust Fund") in 1988, from which UST site owner/operators could request reimbursement for certain costs of cleaning up a site. Data tracked for reimbursements from this Trust Fund served as a major source of information for WMREI researchers, as it breaks out cost information for assessment, water treatment, soil treatment and monitoring. The amount of financial assistance projected to be provided nationwide by states with

UST Trust Funds is significant - over $16 billion (Gurr, T.M. and R.L. Hommann, 1996). The amount of money being reimbursed was not anticipated by most states when they first established these funds, including Tennessee, and as a result, Trust Fund balances have dropped dangerously low. These problems are being addressed by Tennessee by removing certain "inefficiencies" in how money is reimbursed. For example, limits are placed on the number of manhours for specific personnel working on reports, corrective action design, and assessment activities.

One way that many states are attempting to encourage cost-effective cleanups, while also safeguarding human health and the environment, is through the implementation of risk-based corrective action (RBCA) plans. This decision-making process has been developed for the past 2 ½ years by an American Society for Testing and Materials (ASTM) task group, and has been adopted by ASTM as Emergency Standard ES38-94 (ASTM, 1994). The use of RBCA is not a new concept; the U.S. Environmental Protection Agency has been developing it as part of the Superfund program for years. RBCA establishes a process that incorporates the qualitative analysis of a site for contaminant pathways and receptors before an extensive site assessment is undertaken.

Tennessee currently uses a site "ranking system" which allows UST site owners/operators to score their sites using the parameters listed in Table 1. If a site scores under 500 points, the site owner/operator previously had the option to choose long-term monitoring over corrective action, based on a perceived lack of risk to surrounding receptors. Beginning August 1, 1996, sites ranking under 500 could no longer obtain money from the Trust fund for cleanup, but could be reimbursed for monitoring activities.

This ranking system loosely follows what is referred to as "Tier 1" of RBCA, where sites are classified by their need for immediate corrective action (triggered by the presence of free product, for example) and "Tier 2" of RBCA, which involves a decision-making process for establishing a more realistic set of corrective action goals. However, the ranking system is bypassed if there is an explosion hazard, drinking water is affected, or a receptor, such as a sewer line, is affected.

Another element of RBCA which is incorporated in TDEC's guidelines is the option for the site owner/operator to apply for a site-specific cleanup standard based on the completion of a risk-based exposure assessment.

OBJECTIVES

The previous data that WMREI has already collected on Tennessee UST sites contains many of the parameters in Table 1 that are being used to aid regulators in ranking sites. An opportunity was present to take this baseline data, add more current data, then assess the impact of individual, or groups of "ranking" parameters on the cost-effectiveness of cleanup. Since one of the goals of implementing RBCA is to save money being sought from the Tennessee Trust Fund, this data was analyzed to determine the efficacy of the ranking system in promoting the desired cost-effectiveness in the management of the contamination at these sites.

TABLE 1. Site ranking parameters used by the TDEC UST Division.

Type of Parameter	Parameter Description	Range
Geologic and Hydrogeologic Factors	Minimum depth to the water table	<5 to 100 feet
	Minimum distance between water table and contaminated soil	<5 to 100 feet
	Soil Permeability	$<10^{-6}$ to $>10^{-4}$ cm/sec
	Calculated Ground Water Flow Rate	<10 to >260 feet/day
Receptor Factors (distance from source)	Basements	<50 to 300.1 feet
	Sanitary Sewers	<50 to 300.1 feet
	Storm Water Sewers	<50 to 300.1 feet
	Other subsurface utilities	<50 to 300.1 feet
	Public water supply source	<0.1 to >0.51 miles
	Private water supply source	<0.1 to >0.51 miles
	Distance to surface water	<0.1 to >0.51 miles
Contaminant Factors	Benzene in groundwater	<1.0 to >500.1 ppb[1]
	TPH[2] in groundwater	<1.0 to >500.1 ppb
	BTX[3] in soil	<1.0 to >50.1 ppm[4]
	TPH in soil	<1.0 to >50.1 ppm

[1] parts per billion
[2] total petroleum hydrocarbons
[3] benzene, toluene and xylenes
[4] parts per million

METHODOLOGY

A total of 400 sites were targeted for addition to the 76 site database compiled in 1994. These sites were selected because they underwent the ranking process (which began in early 1994) and because they were represented on the Trust Fund database, which documents all costs for assessment, corrective action and monitoring. Data from the "UST Site Ranking Form" for each site were entered into our database, along with the Trust Fund tracking data. Additional data regarding site characteristics were also collected from TDEC files and personnel. These sites represent a variety of geographical locations, management options (e.g., corrective action or monitoring only), types of contamination, and corrective action technologies. Figure 1 shows a possible relationship among independent parameters

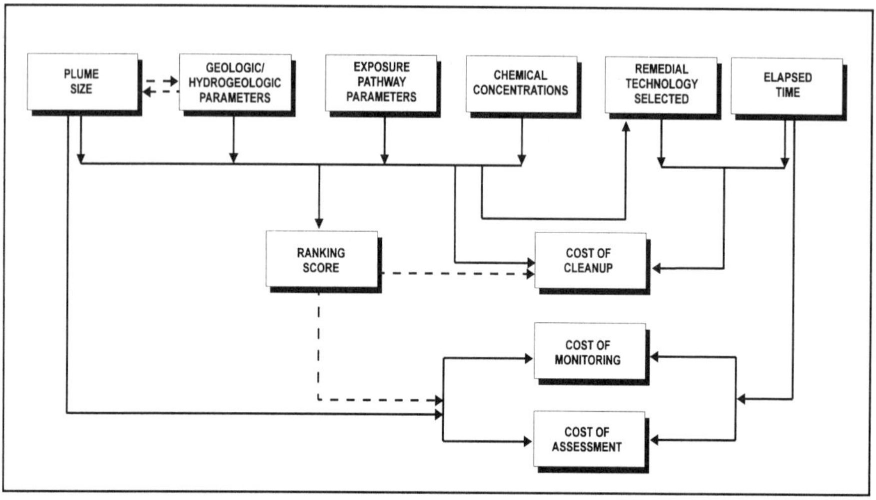

FIGURE 1. Proposed relationship among parameters describing contaminated UST site management.

(plume size, geologic and hydrogeologic parameters, chemical concentrations, remedial technology selected, duration of site management), the ranking score and the cost of site management.

The first phase of work will examine the key factors that influence the ranking score and characteristics of sites that "rank out" (score under 500). The current and potential future impact of the ranking system on the composition (e.g., site characteristics and location) of sites undergoing cleanup versus monitoring only will also be studied, along with the cost and the cleanup methods being used. It is probable that intrinsic bioremediation will be implemented more frequently now in lieu of groundwater pump and treat schemes.

Next, the cost drivers will be identified. It is possible, for example, that site characteristics will have less influence on corrective action cost than the cleanup technology selected. A cost function will be developed that will predict monitoring and cleanup costs based on these cost drivers. The difference in cost between (1) placing a site the monitoring-only status for two years and (2) performing corrective action will be determined for sites with similar characteristics.

PRELIMINARY RESULTS

WMREI is completing the process of collecting data on Tennessee UST sites that will be adequate to test the hypothesized relationships among the parameters, as shown in Figure 1. Initial tests of the data indicate that groundwater contaminant concentrations are a good indicator of a site's ranking score, and to some degree, cost of cleanup. Further analysis of the proposed model, using statistical techniques such as multivariable least square regression, will hopefully illuminate the key drivers of cost, as well as "risk drivers" as defined by TDEC's ranking system. The results of this work will not only be of use to cost estimators of UST site cleanup, but also to

TDEC in estimation of the cost savings promoted by the site ranking system. Also, analysis of the site ranking system will help clarify how TDEC's formula is working in capturing the "less risky" sites (those that score under 500) for monitoring only. Thus far, TDEC personnel have indicated that the 500 point threshold is successful in targeting these sites, which are amenable to natural attenuation.

ACKNOWLEDGMENTS

The authors would like to express their appreciation to the following TDEC UST Division personnel for their continued support on this project: Mr. Chuck Head, (Director), Mike Langreck (Environmental Protection Specialist) and Blake Evans (Accounting Manager).

REFERENCES

American Society for Testing and Materials. 1994. *Emergency Standard Guide for Risk-Based Corrective Action Applied at Petroleum Release Sites*. ES 38-94, Philadelphia, PA.

Davis, K.L., G.D. Reed, and L. Walter. 1995. "A Comparative Cost Analysis of Petroleum Remediation Technologies." In R.E. Hinchee, J.A. Kittel and H.J. Reisinger (Eds.), *Applied Bioremediation of Petroleum Hydrocarbons*. Battelle Press, Columbus, Ohio.

Gurr, T.M. and R.L. Hommann, R.L. 1996. "Managing Underground Storage Tanks," *Pollution Engineering*, May, pp. 40-44.

NEW APPROACHES TOWARDS PROMOTING THE APPLICATION OF INNOVATIVE BIOREMEDIATION TECHNOLOGIES

Gary G. Broetzman (Colorado Center for Environmental Management (CCEM)), Michael J. Chacón (New Mexico Environment Department), Paul W. Hadley (California Department of Toxic Substances Control), Dawn S. Kaback Ph.D. (CCEM), Roger W. Kennett (Arizona Department of Environmental Quality), Nettie J. Rosenthal, J.D. (CCEM)

ABSTRACT: During 1995-96, the Interstate Technology and Regulatory Cooperation Working Group (ITRC), in conjunction with the Colorado Center for Environmental Management (CCEM), performed a series of case studies in six states regarding regulatory acceptance of innovative technologies. The ITRC, consists of 22 states and was formed as a working group of a federal advisory committee. ITRC selected *in situ* bioremediation (ISB), involving injection of additives to ground water (i.e., engineered ISB), as the example innovative technology. Interviews conducted jointly by ITRC member state and CCEM representatives revealed legislative, regulatory and administrative approaches used by the states to overcome regulatory barriers to demonstration and deployment of engineered ISB. Successful state approaches included: innovative technology support groups within state cleanup agencies; state interagency committees to encourage specific innovative technologies; concurrent evaluation of conventional and innovative remedies; selection of innovative remedy with a conventional backup technology; state financial guarantee in event of failure of innovative technology; waiver of permit requirements for injection; flexible permits; and centralization of cleanup authority and ground water protection into one state agency. A detailed report of the case studies is available through CCEM (ITRC/CCEM 1996).

INTRODUCTION

Interstate Technology and Regulatory Cooperation. Under the auspices of the Federal Advisory Committee to Develop On-site Innovative Technologies, a group of representatives from 22 states gathered in Denver, Colorado during February 1995 to initiate a cooperative effort that became known as the ITRC. The ITRC mission is to facilitate cooperation among the states in the implementation of innovative technologies that will clean up contaminated sites safely, economically, effectively, and quickly. This is accomplished by conducting projects, staffed by members from various states, regarding subjects related to innovative technologies.

Objectives of the Case Studies. The primary objective of the ITRC, in performance of the case studies, was to document and report how state regulatory agencies encourage use of innovative technologies for environmental restoration. The intent was not to focus on barriers to implementation of new environmental technologies, but to emphasize approaches to resolving barriers.

The secondary objective of the ITRC was to determine whether the case study approach would have merit. The adopted approach involved state-to-state interviews and a biased selection of case studies in favor of states that successfully deployed a selected innovative technology. Collection of anecdotal data from a limited sample, rather than a more comprehensive or statistical approach, was evaluated by the ITRC to assess its usefulness to other states.

The ITRC was assisted by CCEM in performance of the original case studies, drafting of the original report (ITRC/CCEM 1996), and this article.

Selection of ISB. Engineered ISB was selected by the ITRC as the vehicle for the case studies, because many members of the ITRC believe that ISB can provide cost-effective, safe, and successful cleanup, yet the technology is not widely utilized. The ITRC perceived that the major impediments to implementation of ISB were institutional and regulatory -- not technical.

CASE STUDY ISSUES AND METHODOLOGY

Institutional/Regulatory Barriers to Innovative Technologies. The major incentives for use of innovative technologies for environmental restoration include their promise of faster, better, safer, and more cost-effective cleanups. Yet, institutional/regulatory barriers to the use of innovative technologies often arise by virtue of (1) the lack of cost and performance data, and (2) an inflexible institutional/regulatory framework.

In a 1993 study prepared for CCEM, Stone & Webster Environmental Services identified risks associated with uncertain performance of innovative technologies, including the following (CCEM 1993):

- Risk aversion. Regulators may be unwilling to assume the risk of an innovative remedy that may not prove to be either safe or effective. Regulators must be accountable to both their management and the public when assuming risks.
- Desire to expedite cleanup. Remedial site managers (i.e., regulators) must adhere to scheduled milestones. Responsible parties benefit from getting the site out of the media, the public eye, and regulatory scrutiny. Public opinion generally favors immediate action.
- Desire to maintain a projected budget. Remedial site managers are under pressure to maintain projected budgets and responsible parties have a significant incentive to minimize cleanup costs. Yet, the study of an innovative technology may drive up costs in the short-term. If the technology should not perform as expected, budgets may be overrun.

Regulatory drivers also may impede the implementation of innovative technologies. Such regulatory concerns include the following:

- Regulatory standards. Frequently, regulatory standards can actually impede, rather than facilitate a cleanup. For example, land disposal restrictions may prohibit removal of contaminated media containing listed hazardous wastes.
- Permitting procedures. Permitting of innovative technologies is often a lengthy process that is unfamiliar to many regulators. Additionally, there is potential for interagency friction when one agency is ready to approve use of an innovative technology, but another agency feels compelled to adhere to permitting processes used for conventional technologies.

Institutional/Regulatory Barriers to ISB. Institutional barriers to implementation of ISB are typical of most innovative technologies. Because of performance uncertainties associated with ISB, the ITRC anticipated that some regulators would not be willing to assume the risk of approving ISB. The biochemical mechanisms causing the degradation of hydrocarbons and solvents are complex. On a site-specific basis, subsurface conditions (such as the presence and activity of the organisms), generally, are also unknown.

Regulatory barriers arise because ISB technologies crosscut multiple state and federal environmental programs and agencies. The ITRC focused on ISB technologies that involved injection of additives into the ground water, because of the tension commonly found between state agencies responsible for remediation versus agencies responsible for ground water quality. Ground water quality standards can inhibit or prohibit injection of additives that accelerate biodegradation of chemicals in ground water. In some states, the lengthy time required for obtaining a permit for discharge to ground water discourages the use of ISB at non-Superfund sites (i.e., permits are not required for Superfund cleanups occurring entirely on-site).

Case Study Selection Criteria. The ITRC initially screened candidate states by whether a state had successfully implemented ISB at one or more remediation sites. The criteria for ultimate selection of case studies included: (1) injection into the ground water of additives; (2) selection of at least one case study each that involved petroleum, chlorinated solvents and wood preservative contaminants; (3) diversity in site conditions, geography, and regulatory framework; and (4) local and state cooperation with the interviewers.

Sites in six states were selected for case studies. These states were Illinois, (petroleum), Massachusetts (petroleum), Montana (wood preservative), New York (chlorinated solvents), Oregon (wood preservative), and South Carolina (chlorinated solvents).

Case Study Approach. The ITRC reviewed key documents and interviewed key people (e.g., regulators, site owners, and consultants) associated with each case study. Interviews were conducted by teams of two, consisting of a representative from an ITRC member state and a CCEM representative. The ITRC selected Massachusetts as the pilot case study. Data were presented at an ITRC meeting where member states provided constructive feedback and encouragement of the case study approach. Based upon the feedback, data were collected from the remaining five states and the results presented in a report (ITRC/CCEM 1996).

RESULTS AND DISCUSSION

Approaches that States Have Used to Encourage Use of Innovative Technologies. The ITRC confirmed that concern regarding potentially adverse impacts to ground water quality was a significant barrier to state acceptance of ISB. A related barrier was the requirement for a federal underground injection control (UIC) or state-required ground water discharge permit that involved a lengthy permitting procedure and/or rigid permit conditions that restricted or inhibited the use of ISB. In some of the states, these barriers created interagency conflicts, pitting a ground water protection agency against hazardous waste or Superfund agencies. Table 1 identifies and explains approaches intentionally adopted by the states to address the regulatory barriers.

Possible Consequences of Broad-based State Actions on Innovative Technologies. During the interviews, state representatives shared their views about the impacts of legislation, rulemaking, and administrative polices that were not specifically directed toward the use of ISB or other innovative technologies, but impacted those technologies, nonetheless. Table 2 summarizes those observations.

TABLE 1. State Approaches to Encourage Use of Innovative Technologies

Approaches	Explanation
Discrete innovative technology or technical support group within state environmental agency to support remedial programs.	A dedicated group of technical experts, who are not responsible for site management, can focus on keeping abreast of new technical approaches. This group is typically centralized, whose mission is to provide technical support to remediation agencies.
Interagency committee set up to address specific innovative technology.	In Massachusetts, a committee identified issues inhibiting implementation of ISB and specified data or procedural reforms that were required for resolution. In South Carolina, a ground water quality committee is set up to address the implications of ground water requirements on the use of innovative technologies used in a variety of programs.
Evaluation of dual remedies (conventional and innovative) concurrently during feasibility study phase of Superfund cleanup.	In New York, dual evaluation assures that innovative technologies are given fair consideration without extending the duration of the feasibility study.
Selection of dual remedies (innovative with backup conventional remedy or enhancement).	In New York and Oregon, selection of dual remedies in a Superfund Record of Decision reduces the risks associated with an innovative technology failure, including the need to reevaluate remedial alternatives years later.
State financial guarantee of innovative technology.	Illinois legislation covers the full cost of a replacement technology if the innovative technology did not meet cleanup goals.
Waiver of permit requirement for injection of additives into ground water.	In Massachusetts, the permit waiver is specific to ISB. In New York, permits are waived at many cleanup sites (not just Superfund sites), potentially encouraging ISB use.
Flexible permits that anticipate occasional amendments to cover innovative technology demonstration results.	In South Carolina, permits allow amendments to closure plans to accommodate innovative technology results and provide the framework for obtaining other types of permits that may be required for implementing or demonstrating the innovative technology.
Flexible regulations that allow site-specific waiver or variance from ground water quality standards and points of compliance.	Several states have procedures that allow for the short-term exceedances of ground water quality standards, caused by injection of additives, without any actual endangerment of human health. Downgradient points of compliance and mixing zones also are used to encourage injection to enhance ISB.
Centralization of cleanup authority into one agency.	In Massachusetts, one agency is responsible for assuring a safe, effective, and timely cleanup, and implementation of the ground water protection standards.

TABLE 2. Broad State Actions That May Impact The Use Of Innovative Technologies

State Actions	Potential Impact
Deletion of preference in state Superfund statute for treatment in favor of co-equal consideration of containment, institutional controls and treatment.	This action is a major disincentive for a responsible party to attempt treatment, much less any innovative remedial technologies, even if an innovative treatment technology may prove cheaper, safer, better, and faster in the long-term.
Amendments in state cleanup statutes that make cleanup levels in soil and/or ground water health risk-based.	Frequently, these type of statutory amendments cause cleanup levels to be less stringent. Although this shift will likely lead to less costly cleanup (or no cleanup), it may reduce the initiative for pursuing innovative technologies, if a conventional technology can achieve a less stringent cleanup level.
Indirect state oversight through licensed professionals, who are not state employees, but are state-certified.	State certification of environmental professionals responsible for cleanup can discourage or encourage use of innovative technologies, depending on the level of risk avoidance exercised by the certified professional. If not risk adverse, the additional resource burdens presented by innovative technologies (e.g., education, monitoring, review) are absorbed by the certified professional, rather than state regulators.

Approaches to Change. A state that is considering how to enhance its support for innovative technologies, or ISB in particular, can obtain guidance from the case studies. The approaches to change include legislative, regulatory, organizational, and policy changes. Legislation is a powerful means of creating incentives or disincentives for the use of innovative technologies. But, change may be well within the administrative powers of the regulatory agencies, in the form of regulatory, organizational or policy reform, without the necessity of legislative amendment.

Usefulness of the Case Study Approach. The case study approach could be used to guide states regarding resolution of institutional/regulatory barriers with respect to other troublesome issues (e.g., incineration and cost containment of remediation). The ITRC discovered that the approach of state colleagues interviewing one another created an opportunity for trust; the interviewed states shared with candor the process of resolving institutional/regulatory barriers. Also, the ITRC lent credibility to the process as being the sponsoring organization and provided the opportunity to showcase each state's success story.

CONCLUSIONS

- The degree to which a state maintained a flexible approach to implementation of ground water protection requirements (e.g., using risk-based cleanup levels, downgradient points of compliance, or temporary waivers or variances), determined the degree of implementation and support of ISB in that state. A trend toward flexibility in meeting ground water requirements was observed among the states.
- Most states that have approved use of ISB technologies have resolved fundamental differences between the water quality and site cleanup programs regarding injection of additives into ground water. Because EPA does not regulate ground water (except under the UIC program), each state has independently developed its approach to resolving internal conflicts without national guidance.
- State agency managers determine the degree to which ISB and other innovative technologies, in general, are demonstrated and deployed in their state. Managers initiated innovative technology programs or were responsive to regulated community or staff requests to do what was necessary to test ISB.
- Experience gained through use of ISB has allayed many of the concerns of water quality protection staff. Also, proactive education of regulators by experts has helped gain more ready acceptance by regulators of innovative technologies.

ACKNOWLEDGMENT

The Case Studies were supported by the Department of Energy Office of Science and Technology.

DISCLAIMER

The views expressed in this article are those of the authors only and do not represent the views or policies of the states of Arizona, California, or New Mexico.

REFERENCES

Colorado Center for Environmental Management (CCEM). 1993. *Methods for Assuring the Use of Innovative Technologies.* CCEM, Denver, CO.

Case Studies Task Group of the Interstate Technology and Regulatory Working Group (ITRC) and Colorado Center for Environmental Management (CCEM). 1996. *Case Studies of Regulatory Acceptance; In Situ Bioremediation Technologies.* Prepared for the Federal Advisory Committee to Develop On-Site Innovative Technologies. ITRC/CCEM, Denver, CO.

PUBLIC PERCEPTION OF ENVIRONMENTAL BIOTECHNOLOGY: THE CANADIAN PERSPECTIVE

Kate Devine (DEVO Enterprises, Inc./ Biotreatment News, Washington, DC).

Terry McIntyre (Environment Canada, Ottowa, Ontario)

ABSTRACT: A 1996 Canadian public perception survey of environmental biotechnology indicates that public priorities are focused on the benefits of biotechnology and tended to show a more favorable view towards environmental and health applications, viewing food biotechnology as more of a profit-garnering endeavor. However, those surveyed did not favor use of genetic engineering techniques for advancement of any applications. Providing information on the risks and benefits of research and commercial activities as well as inclusion of the government in such educational operations as well as funding, research and standards enforcement were suggested as means of advancing these technologies to maximum commercial fruition.

INTRODUCTION

This study was precipitated in recognition of a lack of empirical data in the literature regarding examination of how the public feels, is likely to feel, understands or even cares about environmental applications of biotechnology products and processes. Such data is critical if the Canadian federal government is to continue to promote the advancement of innovative biotechnology products and processes. It is also considered invaluable to environmental biotechnology practitioners particularly those who contemplate field releases of microbial products as a prelude to development of communication strategies with affected communities. Therefore, in the winter of 1996, the Clean Technology Advancement Division (CTAD) of the Environmental Technologies Advancement Division (ETAD) of Environment Canada conducted a public survey to assess perceptions and attitudes regarding environmental biotechnology applications.

Objectives. The project was designed to explore:

- the public's understanding of the concept of biotechnology and awareness of specific applications:
- awareness of potential benefits and risks of biotechnology;
- perceptions of current use of these applications;
- acceptability of specific applications within a given community;
- perceptions of trade-offs and willingness to make them;
- credibility of alternative messages and information sources to calm fears; and
- the role for government agencies in funding, encouraging, regulating and undertaking biotechnology applications.

MATERIALS AND METHODS

Eight "focus" groups, intended to divulge opinions and not to establish definitive conclusions, were surveyed. The survey was carried out by consultants with no federal government representatives in attendance during the focus group meetings so as not to influence the outcome of the respondents' comments. These groups consisted of nine or ten people each, who were assembled and asked several questions in a two-hour session to which they responded in writing. The survey was conducted in the cites of Montreal, Toronto, Saskatoon and Vancouver, large cities in four Canadian provinces that currently account for about 90% of the country's biotechnology activity. Coincidental to the selection of these cities were imminent small scale experimental field tests of a genetically engineered *Pseudomonas aureofaciens*, designed to examine the ability of a microcosm to predict microorganism survival in the field. Focus groups were formed on the bases of gender and education (high school or less vs. university). People were a mix of socio-economic groups and age was restricted to those between 25 and 55. Those excluded were environmental activists, market researchers, the media, those in advertising and federal government employees.

RESULTS AND DISCUSSION

Public's Attitudes on Technology/Biotechnology and Applications. Survey results showed that people have experienced angst associated with the rate of development and complexity of science and technology. The term "biotechnology" brought health- and food-related impressions to most respondents with oil-spill consuming bacteria and non-chemical pesticides being the only environmentally-related applications mentioned by very few of those surveyed. When presented a list of biotechnologies (specialty chemicals - e.g., bioplastics; biosensors; bioremediation; bioleaching; biofiltering; biologically-produced fuels; biological pesticides; phytoremediation; and composting), the survey participants identified biologically-produced fuels, biological pesticides and composting most often. Few people were aware of any other environmental biotechnologies except composting and biologically-produced fuels although respondents assumed other applications were being used. After brief introductions to the various environmental biotechnologies, participants were generally supportive, with those that were university-educated generally more comfortable with such applications.

Views on Risks and Benefits. After a brief description of select biotechnology applications, survey respondents were asked to rate the applications on a scale of 1 to 10, with 1 being very opposed to its use and 10 being very supportive. Biological pesticides scored much lower than the other applications (Table 1).

TABLE 1. Canadian public perception survey: acceptance rating of select environmental biotechnologies.

Application	Rating [(a)]
specialty chemicals (e.g., bioplastics)	7.9
biosensors	7.5
bioremediation	8.8
bioleaching	7.7
biologically-produced fuels	8.6
biological pesticides	6.5
phytoremediation	8.2

(a) On a scale of 1 to 10, 1 being very opposed and 10 being very supportive.

Additionally, environment-related and health-related biotechnology applications were considered more important than food-related applications, thereby diffusing the notion that people fear biotechnology products and processes. Most of the perceived benefits focused on technologies' abilities to clean up difficult problems. Many respondents indicated resignation to the inevitability of the introduction of biotechnologies although they had reservations about future impacts. Respondents also felt that advances in cleanup technology would promote a disincentive to address the issue of waste generation. Almost all participants expressed discomfort with technologies involving genetic alteration of organisms. Reasons for this included expected effects, such as food chain alteration and biodiversity loss.

Communication. Respondents indicated they would be unlikely to protest environmental biotechnology applications in their neighborhood, it they were kept informed, although providing very little information concerning the application would raise suspicions. Most wanted to learn of the disadvantages of an application, including long term consequences, as well as the advantages. Sources that were considered to be relatively reliable were the government, some of the media, worldwide organizations (such as the World Health Organization), Greenpeace, advocacy groups, the Internet and watch groups (such as the Better Business Bureau). Participants gave verbal ratings of 1 to 10 on their confidence in various sources of biotechnology information. Opinions varied greatly, but the broadest ratings were seen for environmental groups while religious leaders, corporations and industry associations had the lowest credibility rating.

Government Involvement. No one was aware of specific government activity as pertains to research and regulatory activities, although they assumed such was taking place. Participants felt that the government should be involved in all aspects of technological development, such as funding, research, setting and enforcing standing and public education. Respondents suggested the establishment of independent watch groups indicating more confidence in an independent body overseeing biotechnology than the government.

Implications. Several implications can be drawn from the survey results.

- Knowledge of environmental biotechnology applications is minimal, even among college educated people, and lack of knowledge generally leads to suspicion about such applications. Public support for environmental biotechnologies will require a deliberate educational endeavor.

- People feel that biotechnological applications in the health and environmental markets are a societal priority over food applications that, in general, tend to be viewed more as profit-making endeavors. This attitude should have a positive effect on future environmental biotechnological commercialization efforts.

- Suspicion concerning environmental biotechnology applications is probably higher than that associated with food biotechnology given that people can avoid engineered foods (if labeled) but people may be subject to environmental applications and the effects of such without prior knowledge.

- Terminology has an effect on public attitudes in that the term "genetic" has negative connotations while terms with identifiable words within them, such as biofiltration or biorestoration, are viewed as more acceptable.

- Comfort levels seem to increase when known technologies, such as composting, are associated with the new, unknown technologies. That is, people tend to accept unknown applications when discussed in the same context as those with which they are familiar.

- While knowledge of environmental biotechnology was generally very low, comfort levels concerning future applications of new environmental biotechnologies was higher among those with more education. The more educated may feel that they are better able to judge information's validity.

- People register apprehension about being asked to trust researchers they do not know. They also feel that such researchers would not necessarily be aware of the consequences of their own research actions.

- Education is needed to introduce the public both to those performing the applications and to the specific objectives of the applications. The benefits as well as the costs need to be discussed as acknowledging potential risks will hold more credibility in the long run for gaining public acceptance. While most feel that there is a benefit to be derived from environmental biotechnology, many also feel that there should be limitations of some sort placed upon its utilization, possibly in the form of ethics code or guidelines. Such endeavors should include the public.

CONCLUSIONS

The importance of prior knowledge of applications is a necessary precursor to successful development of a communication strategy in dealing with the public prior to proposed release. For example, the information regarding the public's perception of environmental biotechnology applications was used with respect to three province experimental field trials (British Columbia, Saskatchewan and Ontario) that occurred soon after the focus group information collection. Open houses at select universities were held to introduce the public to the concept and objectives of those trials.

The implicit support for environmental biotechnology shown by this survey adds heightened awareness of the need to continue to acknowledge, emphasize and build upon informed public participation and consent in developing Canadian biotechnology programs. As well, for the conduct of similar future surveys, the angst exhibited over science and technology must be distinguished from specific societal concerns unique to environmental applications of biotechnology.

In closing, the mechanism that supports the funding of such projects in Canada, the National Biotechnology Strategy, is 15 years old and is currently being subject to detailed analysis to reflect current realities and future priorities. Elements identified so far from this survey analysis considered integral to the continued success of the revitalized National Strategy include: emphasis on consumer awareness, the role of government in addressing ethical issues associated with biotechnology, importance of public input to the decision-making process; and the need for development of a National Biotechnology Advisory Council to be more responsive to national concerns.

ACKNOWLEDGEMENTS

In addition to the co-authors, the project's advisory team consisted of Dr. Genevieve Bechard (Natural Resources Canada - BIOMINET) and Dr. Terry Leung (Industry Canada's Office of Consumer Affairs).

DISCLAIMER

This paper has not undergone peer review within Environment Canada and, as such, the views presented reflect those of the authors only who acknowledge any errors of omission or fact.

REFERENCES

Environment Canada and Industry Canada. 1996. *Environmental Applications of Biotechnology: Focus Groups.* Final Report. March.

PRACTICAL EXPERIENCE IN LANDFARMING OF GASOLINE-CONTAMINATED SOILS

David P. Dunn, Larry L. Schneider, Duane D. Dubrock

ABSTRACT: Landfarming is a process in which soil is treated ex situ by application onto an existing noncontaminated soil base and subsequently integrated into that soil base. The primary remedial action is volatilization of petroleum during and after application. Soil integration enhances the bio-remediation process.

This land disposal facility in southern Illinois was permitted to accept three thousand five hundred forty (3,540) cubic yards of gasoline-contaminated soil. The generator site is located approximately fifteen miles from the permitted treatment facility. Site selection was determined by a series of criteria established by the Illinois Environmental Protection Agency (IEPA). Some of these requirements include: 200-foot setback from residences, bodies of water, intermittent streams, 20-foot setback from property lines and areas with slope less than five percent, to name a few.

Initial permitting through the IEPA, Bureau of Land was streamlined by requesting a copy of a previously accepted permit through the Freedom of Information Act (FOIA). Modeling the permit after one that is approved will reduce greatly the number of additional information requests. With appropriate permitting in place, transfer of contaminated soil to the treatment site was initiated. Cost reductions were realized by dropping the initial lift in pre-positioned locations throughout the application area. Soil spreading was accomplished for the first lift utilizing a D-450 bulldozer. Subsequent lifts were spread utilizing lime spreaders. Permitted lift thickness was one half inch per lift, one inch total.

Random sampling of the application area indicated soil levels dropped from an average 35,569 parts per billion (ppb) BTEX to less than 9.4 ppb BTEX upon closure. Monitoring of lysimeters during the treatment stages ensures that the migration of petroleum to the water table does not occur. Clean closure of this facility was received after submission of laboratory analyses confirming the absence of contamination.

GROWING MARKETS OF ENVIRONMENTAL PROTECTION IN CHINA

Zheng Yuan—Yang and Lian—Kai Wen
(University of Petroleum, Beijing, China)

ABSTRACT: In the ten odd years past, the GNP of China increased with years by 9. 5%, it is also bringing pressure on environmental protection. For implementing the sustainable development strategy, China will pay more attention to enevironmental control at the course of economic development. The expenses of environmental protection, which accounted for 0. 8% of GNP at 1995, will be increased to 1. 5% at 2000. The investments of environmental protection in next five years will amount to 320 billion Yuan. There is a large capacity of markets in environmental protection industry including technologies, products, and services in China.

BACKGROUND

Since 1979 China's economics has been seeing a repid development, the GNP increase rate reaching 9. 5% per year. In the following years, China's economic development will continue to keep up a rather high speed. Table 1 shows the situation and prediction about China's economic development from 1995 to 2000.

TABLE 1. The situation and prediction about China's economic development

Item	1990	1995	2000
Grain Crops (million t)	440	465	490
Coal (million t)	1070	1298	1400
Steel (million t)	66	94	105
Grude oil (million t)	138	149	155
Natural gas (billion m^3)	15. 2	17. 4	
Electric power (billion kWh)	620	1000	1400

Chemical fertilizer(million t)	19	25	28
Automobile(million)	0.5	1.5	2.7
Ethylene(million t)	1.6	2.4	4.2

The rapid developing industries in China will certainly result in an increase in the effluent of pollutants. As an example, table 2 gives the anticipated information about the atmospheric pollution. It is bringing about great pressure on China's environmental protection work.

TABLE 2. The estimated effluent of atmospheric pollutants

Item	1992	2000
SO_2(million t)	17	21—23
Smoke and powdery dusts (million t)	16	21

ACT

In practising the sustainable development strategy, China is laying emphasis on environmental protection and thinking of it as a basic state policy.

In the near future years, China's environmental protection work will focus on that follows.

Industrial pollution prevention. In this field will popularize clear production, develop clear coal technology, improve the structure of energy resources, widely promote the reuse of industrial wastes, and turn the end treatment of industrial effluents to the pollution control at overall production processing.

Municipal pollution control. In a comprehensive way, will pay equal attention to control pollution out of daily life, industry and transportation. Will spare no efforts to build up central treatment facilities for urban wastewater and solid waste.

Global environmental protection. China will make great efforts to protect atmosphere, both for China's sustainable development and for making contribution to the international cooperation in global environmental protection. The stress is on saving energy resources, carrying out afforestation, and gradually eliminating the ozone-depleting substances.

MEASURE

For not only to solve the environmental problems left over by history, but also to control the new ones in the development, in the coming five years, China will take two important mearsements in environmental protection.

The plan for controlling the total effluents. The purpose of this plan is, while maintaining the rapid growth of economic development, to preserve the polluted effluents at the level of 1995's. For strictly pollution controlling, the policy of combing the standard of effluent concentration with the quota system of effluent will be pursued, and the advanced technologies for the clear production, the resources conservation, and the waste treatment will be developed and imported.

Green engineering program for transcentury. This program aims at improving the environmental quality of some key regions as a focal point work, and will be fulfilled in 15 years. By the end of the century, the emphasis is put on dealing with the water pollution prevention of Liao River (Northeast China), Hai River (North China), Huai River (Middle East China), Tai Lake (Jiangsu Province), Chao Lake (Anhui Province), and Dian Lake (Yunnan Province), and atmospheric pollution prevention in acid rain control areas and sulphur dioxide control areas.

MARKET

In order to attain the target, China will increase the expenses of environmental protection from 0.8% to 1.5% of GNP. It is estimated that, from 1996 to 2000, China will invest in environmental protection about 320 billion Yuan (see table 3) with a 20% increase rate per year. Besides, in these five years, 180 billion Yuan will be invested in the "Green engineering program for transcentury"

TABLE 3. China's investment for environmental protection (1996—2000)

Item	**Estimated investment (billion Yuan)**
Atmospheric pollution control	150
Water pollution Control	130
Solid waste, noise and others	40

In recently years, China has seen a rapid development in the market of environmental protection. In 1995, China has about 7000 enterprises serving environmental protection on various scales with an output value of 13 billion Yuan per year. These can still hardly meet the needs for China to invest heavily in environmental protection in the future. Now China is making great efforts to develop advanced technologies and products and wishes to carry out cooperation in technology and economics with all countries for her environmental protection plan. Nowadays, there are great market potentialities of environmental industries including technologies, processes, products, services, and so on in China.

COST EFFECTIVENESS OF SELECTED REMEDIATION TECHNOLOGIES AND DESIGN PROTOCOLS

Jack C. Parker and Mesbah Islam
(Environmental Systems & Technologies, Inc., Blacksburg, Virginia USA)

ABSTRACT: A case study is presented to compare the total present-value cost of various technologies and design protocols to meet specified cleanup criteria for a gasoline spill in sandy soil. Five remedial options are investigated: 1) free product recovery (FPR) followed by natural attenuation with quarterly monitoring, 2) FPR followed by soil vapor extraction with dewatering, 3) FPR followed by bioventing, 4) FPR followed by air sparging, and 5) bioslurping. Two design protocols are considered for each option: a conventional engineering design method and a computer-aided design optimization method. Estimated costs were lowest for bioslurping and highest for FPR with natural attenuation. Total costs for the computer-optimized systems were 6 to 16% lower than those of conventionally designed systems.

INTRODUCTION

A variety of methods may be used to clean up soil and groundwater contaminated by nonaqueous phase liquids (NAPLs). If floating free product occurs, conventional practice has been to remove any recoverable free product before addressing mitigation of residual contamination. Commonly used technologies for remediation of residual NAPL contamination include soil vapor extraction (SVE), soil bioventing (SBV) and in-situ air sparging (IAS). More recently interest has developed in concurrent vacuum enhanced free product recovery and bioventing (VER/BV), referred to as bioslurping. The biodegradability of many hydrocarbons has also led to increasing interest in natural attenuation (NA) as a remediation strategy.

Selection of the most appropriate technology for a site is commonly based on subjective estimates of the effectiveness and costs of various options. Computer-aided remedial design (CRD), in contrast, involves quantitative analysis of system effectiveness to select the most cost-effective option. The objective of this paper is to describe a CRD methodology for hydrocarbon contaminated sites and to present an illustration of the method.

The hypothetical case study considered involves a gasoline spill in a sandy soil. A water table occurs at a depth of 12 meters and free product is observed over an area approximately 80 m long and 50 m wide. A slug test yielded an aquifer conductivity of 25 m d^{-1}. An in-situ respiration test and short duration pilot tests were also performed.

The CRD approach and conventional design methods (designated hereafter as the "SOP" approach) will be described and compared.

COMPUTER-AIDED REMEDIAL DESIGN PROTOCOLS

Free Product Recovery. An estimate of the total hydrocarbon volume in the subsurface (56,000 L) was obtained by interpolating the maximum well product thickness from each monitoring well following the method of Parker et al. (1994), using grain size data to estimate capillary parameters. The current free product volume (27,000 L) was similarly estimated from the most recent well gauging data. Residual hydrocarbon in the unsaturated zone at the source was regarded to be negligible relative to the total mass. The methodology of Parker et al. (1994), implemented in the program SPILLCAD (ES&T, 1996), was used to evaluate free product recovery options. The results indicated that water extraction at 0.5 L s^{-1} at an existing monitoring well would control the free phase plume and recover approximately 10,000 L of free product in a period of 430 days. The area and volume of contaminated soil were estimated to be 4,000 m^2 and 4,200 m^3, respectively.

Natural Attenuation. The objective of the analysis of natural attenuation was to estimate the time required for natural processes to attenuate dissolved benzene concentrations throughout the aquifer to less than 1 μg L^{-1}. This level was used as the remedial goal for all analyses in the case study. The time to attenuate dissolved benzene was estimated using the program VADSAT (API, 1995), which models dissolved transport from a nonaqueous phase liquid (NAPL) in the unsaturated zone (1-D) and/or saturated zone (3-D). The contaminant source was assumed to be entirely below the water table, which will tend to *under*estimate the time for natural attenuation. Following free product recovery, the volume of hydrocarbon was estimated to be 46,000 L (initial less removal by FPR). The mass fraction of benzene in the NAPL was assumed to be 1%, which is typical for weathered gasoline. The hydraulic gradient was estimated from well gauging data to be 0.2%, aquifer organic carbon content was assumed to be 0.1%, and dispersivities were estimated from literature correlations with travel distance. Biodecay was modeled by oxygen superposition assuming dissolved oxygen in the aquifer of 5 mg L^{-1}, a stoichiometry of 3g O_2 per g benzene, and an average of 10% of total oxygen consumption for benzene biodecay. The results indicated natural attenuation of benzene would require at least 5 years.

Vapor Extraction, Bioventing, Air Sparging and Bioslurping. The program BIOVENTING^{plus} (ES&T, 1997) was used to evaluate and optimize the design of SVE, SBV, IAS and VER/BV systems. BIOVENTING^{plus} includes modules for airflow, mass recovery and cost estimation. An analytical airflow model determines air pressure, flow rate, pore volume turnover rate and extraction efficiency as functions of distance from a well. The air sparging flow model considers the effects of soil capillary properties and injection pressure on air permeability. Mass removal rates and cumulative mass removal are computed with a multicell, multiphase, multicomponent model, which solves mass balance equations for a system of recovery wells considering mass removal due to concurrent vapor extraction, biodecay and free product recovery.

The initial contaminant volume prior to initiating SVE, SBV and IAS remediation was assumed to be the amount remaining following FPR (46,000 L). The maximum biodecay rate was determined from an in-situ respiration test (5 mg kg^{-1} d^{-1}). The horizontal air permeability (32 darcies) and horizontal-to-vertical anisotropy ratio (10.0) were estimated from airflow pilot tests. Gasoline composition was modeled as a 22 species mixture. Extraction efficiency, characterized by a pore volume turnover time for 50% efficiency, was estimated from pilot test off-gas concentrations. BIOVENTING^{plus} was used to compute the time to reach the cleanup criteria of 1 µg L^{-1} dissolved benzene in equilibrium with residual NAPL in soil or groundwater.

SVE wells were assumed to be screened for 3 m above the static water table, and well vacuum was varied from 0.5 to 1.5 m H_2O. Calculations with SPILLCAD indicated that a water pumping rate of 5 L s^{-1} was needed per meter of vacuum to prevent upwelling.

The air flow rate for SVE was calculated to yield twice the oxygen demand, based on an in-situ respiration test. Preliminary analyses indicated that air emission rates for total hydrocarbons and benzene would be too high to allow air injection. Therefore, vacuum systems only were considered and water pumping was assumed to prevent upwelling as described above. BIOVENTING^{plus} was used to investigate various combinations of vacuum and pulsing frequency.

For VER/BV, free product recovery parameters were estimated from SPILLCAD, and vacuum was varied between 0.5 and 1.5 m H_2O head to assess optimum performance. Water pumping was assumed to vary with vacuum as described above to maintain zero net change in the liquid levels.

For the IAS system design, the air pump was assumed to be capable of a maximum air flow rate of 300 L s^{-1} and a maximum gauge pressure of 10 m H_2O. Depth to the sparge point and flow rate per well were treated as design variables.

CONVENTIONAL DESIGN PROTOCOLS

Free Product Recovery. Using the analytical capture zone method of USEPA (1996) a single well with a water pumping rate of 0.5 L s^{-1} was determined to be adequate for FPR.

Vapor Extraction. Based on a pilot air extraction test, it was determined that an air flow rate of 60 L s^{-1} would yield a vacuum at the water table corresponding to 1 m H_2O at the well and 0.01 m H_2O at a distance 38 m from the well. The later was defined as the pressure radius of influence (PROI) and was taken as an operational measure for well spacing. As the PROI under these conditions is approximately half the maximum plume dimension, a single well with this flow rate was selected for the SOP-design SVE system.

Bioventing. The SBV air flow rate was computed as the flow needed to deliver twice the oxygen needed to meet the maximum aerobic decay rate as follows

$$Q_{air} = \frac{2\rho VBS}{C_{ox}}$$

where ρ is the soil density (1,700 kg m^{-3}), V is the contaminated soil volume (4,200 m^3), B is the maximum biodecay rate from an in-situ respiration test (5 mg kg^{-1} d^{-1}), S is the mass stochiometric ratio for oxygen consumption (3 g g^{-1}), and C_{ox} is the concentration of oxygen in the atmosphere (273 mg L^{-1}). The computed flow rate of 9 L s^{-1} was used as the average design rate. However, at this flow rate, the PROI is only 16 m. From the SVE pilot test, a PROI of 38 m was determined with a flow rate of 60 L s^{-1} and a vacuum of 1 m H_2O. Pulsed flow at this pressure, 15 percent of the time, will achieve the desired average flow rate.

Air Sparging. A pilot test with a sparge well 4 m below the water table was performed with an air injection rate of 58 L s^{-1} yielding a gauge well pressure of 8.7 m H_2O and a PROI of 16 m. Based on the spill area and PROI, five sparging wells were estimated to be needed. This design is within the equipment limits, i.e., total air flow rate less than 300 L s^{-1} and a maximum gauge pressure of 10 m H_2O.

Bioslurping. A pilot test with an induction tube at the level of the water table and with a vacuum at the well screen of 1 m H_2O yielded a PROI of 38 m, indicating a single well operated in this manner is sufficient to remediate the site.

COST ESTIMATION

For a given system design, BIOVENTING[plus] was used to compute the total cost, discounted to present value (PV), as

$$PV = C_o + C_1N + FP_1Q_wNT + FP_2Q_aNT + FP_3T$$

where C_o is the fixed cost independent of the number of wells, C_1 is the fixed cost per well, N is the number of wells, F is a time discount factor, P_1 is the cost per volume of water flow, Q_w is the water flow rate per well, T is the remediation time in years, P_2 is the cost per volume of air flow, Q_a is the air flow rate per well, and P_3 is the cost per unit time for operating costs not related to flow rates. The discount factor is calculated as

$$F = \frac{1-(1+D/100)^{-T}}{T\,D/100}$$

where D is the discount rate expressed as an annualized percentage. Cost parameters used in the analyses are summarized in Table 1. No capital costs per well were allocated to FPR on the assumption that an existing monitoring well would be used for FPR. Costs per volume of water pumped were assumed to be 2 times higher for FPR and VER/BV systems due to the presence of free product. Based on estimates of air emissions from BIOVENTING[plus], off-gas treatment was assumed for all air-based technologies. Therefore, costs per unit air flow were assumed to be the same for all technologies. For all scenarios, costs for quarterly monitoring and reporting were assumed to be $20,000 per year and a discount rate of 5 percent per year was employed. For scenarios preceded by FPR, costs for the secondary technology were discounted to the beginning of the FPR operation.

TABLE 1. Unit Costs Used in Cost Calculations.

Cost Parameter	FPR	SVE	SBV	IAS	VER/BV
Fixed cost per well, $/well	-	5,000	5,000	5,000	5,000
Other fixed cost, $	10,000	10,000	10,000	10,000	10,000
Cost per vol. water, $/m^3	0.5	0.25	0.25	0.25	0.5
Cost per vol. air, $/m^3	-	0.00025	0.00025	0.00025	0.00025
Other oper. cost, $/year	20,000	20,000	20,000	20,000	20,000
Discount rate, %/year	5	5	5	5	5

RESULTS

A summary of estimated remediation times and costs for each option, using the conventional design approach (SOP) and the computer-aided remedial design methodology (CRD), is presented in Table 2. Costs of serial technologies (i.e., FPR followed by NA, SVE, etc.) represent the total remediation cost, discounted to the value at the beginning of the remediation effort. The estimated cost for FPR alone for the illustration was $31,500.

Both SVE systems utilize a single well. The CRD design operates at a vacuum of 0.5 m, whereas the SOP system operates at 1.0 m. The higher pressure for the SOP design results in an air flow rate 100% higher than the CRD rate. Due to reduced extraction efficiency at the higher flow rate, recovery time only decreases 20%, whereas the cost (for the SVE system alone) increases 15%.

The CRD-based SBV system utilizes 2 wells at 0.5 m vacuum with 30% pulsing, while the SOP system uses 1 well at 1.0 m vacuum and 15% on-time. The 2-well system reaches the clean up goal one year sooner, which reduces operating costs sufficiently to compensate for higher capital costs. SBV and SVE systems have almost identical performance and cost. Model results indicate that 80 percent of the mass loss for the SVE system is actually due to biodecay. Clearly, SVE and SBV are not really distinct technologies.

The CRD-based IAS system utilizes 4 wells at a depth of 5 m with an injection pressure of 10 m and a flow rate of 58 L s^{-1} per well. The SOP system has 5 wells at a depth of 4 m with an injection pressure of 8.7 m and a flow rate of 58 L s^{-1} per well. Placing sparge points deeper yields essentially identical remediation performance with 20% lower total air flow. Inspection of the impacts of sparge point depth for a constant air pressure indicates that air flow rate decreases and remediation time increases with increasing sparge point depth. However, radius of influence exhibits a maximum value at a sparge point depth below the water table equal to approximately half of the injection pressure (in equivalent water head). Costs are minimized at about the same depth.

TABLE 2. Comparison of Remediation Time and Cost for Various Options and Design Protocols.

	Remediation Time (yrs)		Total Cost ($x1000)	
	SOP	CRD	SOP	CRD
FPR-NA	6.2	-	118	-
FPR-SVE	2.0	2.3	92	84
FPR-SBV	3.3	2.4	93	86
FPR-IAS	1.9	1.9	78	73
VER/BV	0.6	0.8	70	59

The optimum CRD-based VER/BV system utilized a single well operating at a vacuum of 0.5 m and a liquid pumping rate of 2.5 L s^{-1}, while the SOP system utilized 1 well operated at 1 m vacuum and 5 L s^{-1} liquid flow.

For the assumed unit cost parameters, VER/BV was the least costly option, followed by FPR-IAS, FPR-SVE, FPR-SBV and FPR-NA. The high cost of NA reflects long term monitoring expenses. The lowest cost alternative is 50 percent of the highest cost option, emphasizing the importance of remedy selection. Remediation time generally follows the same trend as cost and varies by a factor of 10 between the fastest to slowest options.

For the technologies considered, CRD produced total cost savings of 6 to 16 percent over conventional design methods. Cost savings for individual technologies (e.g., SVE independent of FPR) ranged from 10 to 16%. The level of effort required to achieve these savings is minor. All of the analyses discussed in this paper were performed in less than 10 hours, which represents a cost of about 1% of the remediation effort.

REFERENCES

American Petroleum Institute. 1995. *VADSAT A Vadose and Saturated Zone Model for Assessing the Effects of Groundwater Quality from Subsurface Petroleum. User's Guide to Version 3.0.* Washington DC.

ES&T, Inc. 1996. SPILLCAD *Data Management and Decision Support for Hydrocarbon Spills. User and Technical Guide*, Blacksburg, Virginia.

ES&T, Inc. 1997. *BIOVENTING^{plus} A Program to Evaluate Soil Vacuum Extraction, Bioventing, Air Sparging and Bioslurping Technologies, User and Technical Guide,* Blacksburg, Virginia.

Parker, J. C., D. W. Waddill, and J. A. Johnson. 1994. *UST Corrective Action Technologies: Engineering Design of Free Product Recovery Systems*, USEPA Risk Reduction Laboratory, Edison, New Jersey.

USEPA. 1996. *How To Effectively Recover Free Product At Leaking Underground Storage Tank Sites*. EPA 510-R-96-001. OUST/OSWER.

INTERSTATE ACCEPTANCE OF IN SITU BIOREMEDIATION TECHNOLOGIES

Paul W. Hadley, (California Dept. of Toxic Substances Control, Sacramento, CA)
Louis C. Rogers, P.E., (Texas Natural Resource Conservation Commission, Austin, TX)
Steve R. Hill, (Coleman Research Corporation, Boise, ID)

ABSTRACT: The In Situ Bioremediation Technology Specific Task Group (ISB Group), a subgroup of the state-led Interstate Technology and Regulatory Cooperation (ITRC) Working Group, determined that given appropriate conditions in situ technologies can remediate contaminants more cost-effectively than conventional technologies. The Department of Energy, Office of Science and Technology (EM-50) provided the major portion of the funding to assist states in identifying mechanisms to encourage the demonstration and use of innovative approaches to site remediation. The ISB Group developed a "General Protocol" for the class of in situ bioremediation technologies to guide technology developers through the complicated matrix of multiple state regulatory approvals and requirements.

The members of the ISB Group (representing states, federal agencies, industry and stakeholders) acknowledge, based on experience, that broad acceptance of a technology requires an organized collective process. The group's "General Protocol" defines "roles and responsibilities" for participants in a multi-state demonstration. Each have clearly defined responsibilities during three phases of the technology demonstration process. The ISB Group identified four groups of issues particularly pertinent to deployment of in situ bioremediation technologies.

INTRODUCTION

In situ bioremediation technologies rely on the capabilities of indigenous or introduced microorganisms to degrade, destroy or otherwise alter objectionable chemicals in soils or ground water. These technologies can be applied to soils or deep sediments and in arid or wet regions. In situ bioremediation is a class of technologies as variable as the subsurface itself. The In Situ Bioremediation Technology-Specific Task Group (ISB Group), a subgroup of the Interstate Technology and Regulatory Cooperation (ITRC) Working Group, recognizes that given appropriate conditions, in situ technologies can remediate contaminants more cost effectively than conventional technologies.

In February of 1995, the ISB Group was established to focus on in situ bioremediation (ISB) technologies. The purpose was to review the existing use of ISB technologies, identify barriers to their effective use and recommend actions which, if implemented, will promote multi-state acceptance[1] of data obtained during the demonstration[2]. To promote state agency, federal agency, industry, user, tribal and stakeholder cooperation, the ISB group set out to develop products, processes and recommendations which would accelerate the safe and effective development and deployment of site cleanup technologies using ISB techniques which are as good or better and cost less than conventional technologies.

PURPOSE AND SCOPE OF THE GENERAL PROTOCOL DOCUMENT

The General Protocol (Interstate Technology and Regulatory Cooperation Working Group, 1996) provides guiding principles and a standard approach for conducting safe and appropriate demonstrations of in-situ bioremediation techniques to foster interstate acceptance of the test results from a variety of in-situ bio demonstrations. This document emphasizes the establishment of objectives, criteria and measures so that work plans can be designed consistent with those measures, and results can be verified.

The General Protocol presents an outline containing the essential elements the proponent of an in situ bioremediation test must address when initiating a demonstration. The outline represents a compilation of concerns gathered by the ITRC states. A process is also presented which defines the parties responsible for verifying demonstration results and transferring those results to other states for acceptance.

In addition, as a guide to the proponent, The General Protocol contains examples of recommended technology-specific protocols which have been developed by industry and tested in field applications of intrinsic bioremediation and bioventing of hydrocarbons. These technology-specific protocols have been evaluated by members of the ISB Group. Use of these protocols will increase the likelihood that the essential information required by the states has been included in the design of the demonstration and test plan.

GENERAL OUTLINE

The essential elements of an ISB demonstration proposal are contained in the General Outline section of the document. This provides guidance to the

---

1

Acceptance for this application is defined as "States will accept the results of the demonstration as if they had overseen the demonstration themselves."

2

In the context of this document, the term "demonstration" refers to field scale deployment of an in situ bioremediation technique or technology to show (demonstrate and validate) the performance of that technique or technology. Full-scale application of in situ technique or technology may in fact be used to measure, record and document performance. The central element of a demonstration is the determination of performance.

technology proponent during development of the initial proposal for a demonstration. The proposal should contain enough detail so that the other parties can identify the applicable regulatory requirements for the project, the innovative nature and scope of the project, the advantages this technology might have over conventional technologies and the concerns the participants might have with this technology.

RESPONSIBILITIES OF THE PARTIES

Many "parties" need to voluntarily accept responsibility during the implementation of a General Protocol based demonstration. These parties include Host States[3], Participating States[4], Proponents[5], Tribes and Community/Other Public Stakeholders. In some cases, Tribes with regulatory authority may act as Host or Participating States in this process. The responsibilities of each party are discussed below in phases of a demonstration program.

Phase 1: Identification of the Parties. The first phase in the process describes the opportunity for the Proponent to identify a Host State in which to conduct a demonstration, and select those states where the technology could be transferred following successful demonstration. The Host State will coordinate the activities of the Participating States. Each Participating State must identify and be responsible for any intra-state coordination steps necessary to fully evaluate a demonstration proposal.

The Proponent, with the assistance of the Host State, should also identify other Stakeholders who may be impacted or have a vested interest in the outcome of the demonstration. Participating States should also identify interested community/public Stakeholders to include in the review of technology demonstration proposals.

Phase 2: Demonstration Design. The second phase of the process is the submission of the demonstration proposal, prepared by the Proponent consistent with the General Outline. Upon receipt each State or Tribal environmental agency will identify the applicable regulatory requirements for the technology demonstration. These will be compiled into a single document and provided to

---

[3] The State which receives the demonstration proposal and oversees the implementation.

[4] States which are projected to receive an application to use the successfully demonstrated technology in their state in the future.

[5] The company or organization (public or private) submitting a proposal to the host state to conduct a demonstration.

the Proponent for use in developing the demonstration plan. Community/public Stakeholders in turn should identify any concerns or sensitivities they have with the demonstration proposal or technology and deliver these to the Proponent.

This phase validates the common concerns and regulatory requirements among Participating States as well as identifying unique concerns of one or more Participating States or other Stakeholders that may be incorporated in the demonstration in the Host State. This will guide the development of the performance objectives for the demonstration

Once the performance objectives are established, the Proponent and the States must establish verification criteria and the verification measures for this demonstration. These are the elements which the Host State will use to verify the performance of the technology and report that success to the other Participating States.

Phase 3: Implementation and Reporting Results. Phase three of this process is the actual implementation of the demonstration plan. During this phase the Host State is responsible for issuing any applicable permits or approvals from their agency(ies) and overseeing the demonstration. The Proponent must collect and validate the results of the demonstration and submit those results to the Host State for verification of the demonstration according to the predefined verification criteria and measures, and report those results to the other Participating States.

The Host State, in collaboration with Participating States, will develop a summary evaluation describing the level of acceptance of the demonstrated technology based on the predefined verification criteria and measures.

ISSUES

Through its' collective experience and depth of representation the ISB Group found several issues pertinent to in situ bioremediation. These issues, as summarized by the ISB Group, are as follows:

- Cleanup levels, and the approaches used by various jurisdictions to derive those numerical criteria, vary among state and federal agencies. Although a single set of concentration based cleanup levels cannot be developed to apply to all jurisdictions, it is recommended that a work group be established to formulate recommendations for changes that encourage consistency in approach, if not numerical criteria.
- Factors beyond the jurisdiction of the state regulatory agencies often dictate the type of remedial technology that is deployed. These factors include addressing the concerns of participants in real estate transactions and the financial institutions lending on such transactions and the public's opposition and fear of a technology. These pressures often discourage the deployment of cost-effective techniques and technologies, particularly natural attenuation and bioventing, and thus reduce the potential market for affordable remedial measures. It is necessary to address the concerns of these non-regulatory entities in order to broaden market acceptance of many affordable remedial options, as well as encourage the free market to

continue to develop remediation techniques and technologies.

- Natural attenuation for petroleum hydrocarbons, particularly benzene, toluene, ethyl benzene and xylene, is well demonstrated as a remedial option for groundwater. It is recommended that for all sites where remediation is deemed necessary, particularly fuel tank sites, the appropriate agencies should evaluate natural attenuation as a remedy, referencing their agencies to consider the ITRC work-product concerning this topic and the various technical guidance documents and references now available in the literature.
- Bioventing is a cost-effective in situ technology which reduces petroleum hydrocarbon contamination by accelerating natural biological conversion processes. Where remediation of soils is deemed necessary, particularly for leaking underground fuel sites overseen by state agencies, the use of bioventing should be encouraged as a remedial measure.

CONCLUSIONS

The members of the ISB Group know, through experience, that acceptance of a technology by multiple states and state agencies, tribes (as regulators or stakeholders), and local communities requires an organized collective process. In the "General Protocol" the ISB Group defined categories of responsibilities for participants in a multi-state demonstration.

The "General Outline" to the General Protocol contains the essential elements of an in situ bioremediation demonstration. It provides guidance to the proponent during the development of the initial demonstration proposal.

The ISB Group of 15 states, the Federal government, industry, and representatives from an environmental and other non-profit groups along with members of the larger ITRC working group, which include 26 states, industry community and tribal representatives, offered comments and guidance throughout the development of the General Protocol. Thus, the "Protocol Binder and Resource Document" reflects the input of many sectors of our society interested in site cleanup.

DISCLAIMER

The views expressed in this article are those of the authors only and do not represent the views or policies of the state of California or Texas.

REFERENCE

Interstate Technology & Regulatory Cooperation Working Group. 1996. *Protocol Binder & Resource Document of ITRC's In Situ Bioremediation Task Group.* Prepared for the Western Governors' Association by the Interstate Technology & Regulatory Cooperation Working Group.

RISK-BASED IN SITU BIOREMEDIATION DESIGN

J. Bryan Smalley (University of Illinois, Urbana-Champaign, Illinois)
Barbara Spang Minsker (University of Illinois, Urbana-Champaign, Illinois)

ABSTRACT: Risk-based corrective action (RBCA) is rapidly becoming an accepted approach to remediating contaminated sites. Under a RBCA approach, the risks to human health and the environment associated with a contaminated site are evaluated and appropriate corrective measures are taken as needed to reduce risk to acceptable levels. Evaluation of natural attenuation and in situ bioremediation strategies will likely be an important part of RBCA, particularly for petroleum releases. Such evaluations will usually involve the use of predictive models to assess risks. The development and implementation of a management model which will simultaneously predict risk and propose cost-effective options for reducing risk to acceptable levels is described. This model will use coupled computer simulation and optimization models to assist in making decisions. In addition, the uncertainty associated with particular model parameters is incorporated into the model in order to produce a more reliable remediation design. Such models have been shown to be effective for reducing costs associated with more traditional remediation approaches, but have never been applied within the RBCA framework. This note presents the model formulation and describes the solution methodology.

INTRODUCTION

Due to the adverse impacts of petroleum hydrocarbons on human health and the environment, considerable resources have been expended to restore sites contaminated with petroleum. In the past, cleanup goals were often established without regard to risk, mandating remediation of groundwater to background or non-detection levels, to maximum contaminant levels, or to some level of total petroleum hydrocarbons. Such practices have produced goals that are often difficult or impossible to achieve and have made site restoration prohibitively expensive.

Numerous states, such as Michigan, have implemented or are in the process of implementing RBCA standards for petroleum-release sites based on procedures developed in 1994 by the American Society for Testing and Materials (ASTM, 1994). ASTM describes RBCA as a "tiered" approach to risk assessment with movement from a lower to a higher tier only when necessary. Within each tier, risk assessment and remedial action are appropriately tailored to the extent of available site assessment data. A Tier 1 evaluation involves comparing conservative, generic screening-level concentrations, or Risk-Based Screening Levels (RBSL's), to site conditions. Higher tiers (Tiers 2 or 3) involve a greater degree of sophistication and expense for data collection and modeling but may allow overall cost savings because site-specific target levels (SSTL's) are established as remediation goals. Progression to a higher tier is warranted when the assumptions made to develop the RBSL's are inappropriate relative to site conditions or when the expenditures associated with higher tier evaluation and subsequent corrective action are lower than the preceding tier remediation costs.

Tier 2 and Tier 3 evaluations will likely involve the use of predictive models to assess risks. In this work, a management model is being developed which can simultaneously predict human health risk and propose a cost-effective strategy for reducing this risk to an acceptable level. This model will use coupled computer simulation and optimization models to assist in making recommendations. Such models have been shown to be effective for reducing costs associated with more traditional remediation approaches, but have never been applied within the RBCA framework. Optimization models are more useful for design than simulation models alone. For example, it is unlikely given a virtually unlimited number of possible design options (well locations and pumping rates) that trial-and-error simulation will achieve a least-cost design. In addition, optimization allows accurate and informed comparisons to be made between remediation alternatives (e.g. enhanced in situ bioremediation vs. pump and treat) because both designs are optimal for the given site conditions.

In situ bioremediation will likely be an important part of RBCA, particularly for petroleum releases; therefore, this work will focus on in situ bioremediation as the first remediation technology to be modeled. The model can evaluate the potential for natural attenuation to reduce risks without the expense of engineered remediation measures, or if natural attenuation is not sufficient, identify cost-effective measures to increase the rate of in situ bioremediation.

A final and key aspect of this coupled model is the capability to account for uncertainty or variability in model parameters. Accounting for parameter uncertainty and variability will support the development of a risk reduction strategy that is both cost-effective and sufficiently but not overly conservative. This note presents the formulations for the model and describes the solution methodology. Results for various representative scenarios will be presented at the Bioremediation Symposium.

FORMULATION OF THE SIMULATION MODEL

The function of the simulation component of the coupled model is to predict the extent of natural attenuation or to enable evaluation of the effectiveness of a bioremediation design. The simulation model employs mathematical representations of contaminant fate and transport due to advection, dispersion, adsorption, and degradation by microbial populations. Design of an in situ bioremediation strategy usually involves determining the locations and pumping rates for injection and extraction wells. For petroleum hydrocarbons, injection wells are used primarily to stimulate microbial growth and to accelerate degradation of the contaminant, known as the electron donor or primary substrate, by injecting an electron acceptor, such as oxygen. Degradation of key petroleum constituents, benzene, toluene, ethylbenzene and xylene (BTEX), has been most successful under aerobic conditions with oxygen as the electron acceptor.

The simulation model is a two-dimensional, depth-averaged finite element biodegradation model called BIO2D (Taylor, 1993). The model predicts hydraulic heads and contaminant, oxygen, and microbial biomass concentrations in a confined aquifer that would result from pumping rates selected by the optimization component of the model. There are four governing equations that characterize this model. The first describes steady-state groundwater flow and is solved once for a given set of pumping locations and rates. After the first equation is solved, groundwater velocities are determined by employing Darcy's law. The three remaining equations describe time-dependent changes in contaminant, oxygen, and biomass concentrations. The rate of substrate degradation is modeled using the Haldane variant of the Monod equation. This equation includes a term which allows inhibition of microbial growth to occur at high substrate (contaminant) concentrations. BIO2D uses the Galerkin finite element method in space and a variably weighted finite difference approximation in time to find solutions to the four governing equations.

FORMULATION OF THE OPTIMIZATION MODEL

General Formulation. In order to find the least-cost strategy in terms of well locations and pumping rates, the simulation model described above is coupled with an optimization model. The optimization model consists of three main components: the objective function, the simulation model, and the constraints. The objective function describes the goal of the optimization model which is to minimize the total cost of the bioremediation strategy and is presented below as equation (1).

$$\text{Min } C_{TOT} = \sum_{i=1}^{I} (c_i^{fix} * x_i + c_i^{var} * |u_i| * t_{cu}) + c_{lab}^{var} * t_{cu} + \text{monitoring costs} \quad (1)$$

where i is a potential injection/extraction well site, c_i^{fix} is the capital/fixed cost associated with well installation ($), x_i is a well installation indicator variable (0 or 1), c_i^{var} is the variable cost associated with operating a well ($/m^3), u_i is a decision variable for the pumping rate for a well (u_i is positive for extraction and negative for injection) (m^3/day), t_{cu} is the cleanup time period [a calculated variable where $t_{cu} = f(u_i)$], and c_{lab}^{var} is the variable cost associated with labor to manage/operate the site ($/day). Monitoring is required at performance wells to ensure that standards are being met. The monitoring costs are a function of the duration, frequency, and extent of monitoring and will accrue during the cleanup period and for a period afterwards when closure is being verified. The exact form of these costs is under development, but they will include monitoring costs for both intrinsic and active remediation.

As described previously, the simulation model predicts hydraulic heads and contaminant, oxygen, and microbial biomass concentrations in the aquifer over time for a set of pumping rates, u_i, at well locations, i. Given these physical relationships, the optimization model will select well locations and pumping rates to minimize costs. A notable feature of the optimization model is that in its implementation, each u_i has four possibilities: extraction, injection, injection of electron acceptor (oxygen), or a combination of injection and electron acceptor injection. The final possibility allows the model not only to select the rate of injection but also to determine implicitly the concentration of electron acceptor in the injection water. In making these decisions, the model must meet the following constraints:

$$-u_{max} \leq u_i \leq 0 \quad \text{all injection wells} \tag{2}$$

$$u_{max} \geq u_i \geq 0 \quad \text{all extraction wells} \tag{3}$$

$$x_i = \begin{cases} 0 & \text{if } u_i = 0 \\ 1 & \text{otherwise} \end{cases} \quad \text{all i} \tag{4}$$

$$Risk_{j,r} = Risk_{j,r}^{ing}(c_{j,r}^{avg}) + Risk_{j,r}^{derm}(c_{j,r}^{avg}) + Risk_{j,r}^{inhal}(c_{j,r}^{avg}) \leq TR \quad \text{all j, all r} \tag{5}$$

$$h_{i,r} - h_{max} \leq 0 \quad \text{all injection wells, all r} \tag{6}$$

$$-h_{i,r} + h_{min} \leq 0 \quad \text{all extraction wells, all r} \tag{7}$$

where u_{max} is the maximum pumping rate at a given well site (m^3/day), j is a well site with potential for human use, r is a realization set (defined later), $Risk_{j,r}$ is the total individual lifetime cancer risk, $Risk_{j,r}^{ing}$, $Risk_{j,r}^{derm}$, and $Risk_{j,r}^{inhal}$ are the risk due to ingestion of well water, dermal absorption of well water, and inhalation of volatiles from well water, respectively, $c_{j,r}^{avg}$ is the time-averaged contaminant concentration over the exposure duration (calculated from the simulation model) (mg/L), TR is the target individual lifetime cancer risk, and $h_{i,r}$, h_{max}, and h_{min} are the hydraulic head (m), the maximum hydraulic head (m), and the minimum hydraulic head (m) at a given well site, respectively.

Many of the constraints are simply physical limits on certain variables. Equations (2) and (3) are equipment-based limits on pumping rates and equations (6) and (7) prevent the model from allowing unrealistically high hydraulic heads at well sites. The one for x_i, equation (4), is more functional. This constraint ensures that the fixed cost of a well at a given site is incurred (i.e. installation occurs) if pumping is selected by the optimization model. The constraint essential to a risk-based design is one that requires the total individual lifetime risk calculated by the model due to contaminant exposure to be less than a target risk level. Risk is a unitless number representing increased individual cancer risk (e.g., 10^{-6} or 1 in 1,000,000) and is a function of the modeled contaminant concentration in the groundwater at a well site with potential for human use. The three exposure pathways considered are ingestion, dermal absorption, and inhalation of volatiles from contaminated well water. Depending on site conditions, numerous exposure routes are possible, but this model is currently restricted to human receptor exposure to contaminated groundwater from a well. Future work will examine other exposure routes and risk to the environment (ecological risk).

Accounting for Modeling Uncertainty. One of the most difficult problems associated with the simulation-optimization approach to groundwater remediation is incorporating the effects of modeling uncertainty into the optimization process. The parameters that contribute to this uncertainty belong to one of two categories: physical, chemical, or biodegradative parameters associated with the simulation model equations or risk assessment parameters that are part of the equation for calculating lifetime human health risk.

Much uncertainty in groundwater modeling results from a lack of knowledge regarding the natural spatial variability of porous media parameters, such as hydraulic conductivity. The heterogeneity of hydraulic conductivity can be a key factor in determining the transport and ultimate fate of groundwater contaminants. Because sampling for such parameters is expensive and often difficult, it is impossible to fully characterize the spatial variability. Based on limited sample data, numerous valid or possible maps (realizations) of the spatial distribution of groundwater parameters can be estimated and incorporated into the optimization process (Wagner and Gorelick, 1989).

Similarly, uncertainty associated with various components of the risk equation leads to uncertainty in the ultimate estimation of risk. Current regulatory guidance under the Comprehensive Environmental Response, Compensation, and Liability Act (CERCLA) and the Resource Conservation and Recovery Act (RCRA) requires only single-point estimates of risk using rigid and usually conservative values in the calculation. It is possible, though, to define distributions of values (probability density functions) for various exposure input variables (Kangas, 1996). As above, numerous realizations of a given exposure input variable can be generated by repeatedly sampling from the appropriate distribution.

Uncertainty is incorporated into the coupled model by first generating numerous (e.g. 30) realization sets in which each set includes one realization for each user-specified groundwater or risk assessment parameter. The optimization procedure, then, is required to produce a solution (pumping locations and rates) that is valid for all realization sets. The reliability of the design produced by the coupled model is evaluated by applying the design to a much greater number of newly-generated realization sets. The reliability of a given design is the percentage of realization sets for which successful remediation (no constraint violations) is achieved.

Optimization Procedure. The optimization of groundwater management models can be a highly nonlinear and nonconvex mathematical programming problem, particularly in the case of remediation design with contaminant transport constraints. Alternative combinatorial optimization methods, such as genetic algorithms and simulated annealing, have been successfully applied to groundwater management problems and confer certain advantages over more traditional nonlinear programming methods (McKinney and Lin, 1994). A genetic algorithm method is implemented for this coupled model. The general operation of such a method applied to this risk-based bioremediation problems is as follows.

First, initial sets of pumping rates at specified well locations (solution sets) are generated randomly. For all realization sets and for all solution sets, the simulation model, BIO2D, is run and total lifetime risk is predicted for a human receptor exposed to contaminated groundwater. The performance of each solution set for all realizations is then evaluated based on the value of the objective function and any penalties that are assessed due to constraint violations. The worst solution sets are discarded, and the process is repeated by generating new solution sets from those that are retained until the procedure converges.

CONCLUSION

The development and implementation of a risk-based groundwater management model has been described. The strength of the coupled model presented in this note is that it provides a method for developing cost-effective, reliable in situ bioremediation designs within a risk-based framework. Risk-based corrective action is becoming a widespread approach for sites contaminated by petroleum spills; ASTM is currently developing risk-based standards that can be applied at any chemical release site. As such guidelines take hold, tools such as the one described will be a valuable component of the RBCA process.

REFERENCES

ASTM. 1995. *Standard Guide for Risk-Based Corrective Action Applied at Petroleum Release Sites*. E 1739-95. American Society for Testing and Materials, West Conshohocken, PA.

Kangas, M. 1996. "Probabilistic Risk Assessment." *ASTM Standardization News*. 24(6): 28-33.

McKinney, D.C., and M.-D. Lin. 1994. "Genetic Algorithm Solution of Groundwater Management Models." *Water Resources Research*. 30(6): 1897-1906.

Taylor, S.W. 1993. "Modeling enhanced in-situ biodegradation in groundwater: Model response to biological parameter uncertainty." *Proceedings: 1993 Groundwater Modeling Conference*. International Ground Water Modeling Center, Golden, CO.

Wagner, B.J., and S.M. Gorelick. 1989. "Reliable Aquifer Remediation in the Presence of Spatially Variable Hydraulic Conductivity: From Data to Design." *Water Resources Research*. 25(10): 2211-2225.

RATES OF HYDROCARBON BIODEGRADATION IN THE FIELD COMPARED TO THE LABORATORY

Albert D. Venosa and John R. Haines (U.S. EPA, Cincinnati, OH)
Edith L. Holder (University of Cincinnati, Cincinnati,OH)

ABSTRACT: A field study conducted on the shoreline of Delaware in 1994 showed that the first-order rate constants calculated from the field study were nearly identical to rate constants calculated in sealed laboratory flasks, using the same microbial populations from the study site, when the rate data were normalized to the highest alky-substituted homologue in a given polycyclic aromatic hydrocarbon (PAH) series. A laboratory experiment was conducted to determine how widespread this relationship was among a diverse series of microbial consortia. Eight undefined mixed cultures isolated from various U.S. marine shorelines were incubated for a period of one month in quadruplicate shake flasks in the presence of artificial seawater containing weathered crude oil. Flasks were sacrificed at periodic intervals and the contents analyzed for depletion of crude oil constituents by gas chromatography/mass spectrometry (GC/MS). Rates of biodegradation were calculated by nonlinear regression analysis of the analyte depletion data and compared to rates measured in the Delaware field study. Results affirmed the findings from Delaware. Six of the eight mixed cultures isolated from different parts of the U.S. behaved similarly in closed flasks (with respect to the relative rates of biodegradation of crude oil hydrocarbons) to the microbial consortium that degraded the light crude oil in the open field on the beach of Delaware Bay. Results suggested that one can use the relationship between degradation rate and substrate structure or molecular weight as a reliable indicator of biological activity.

INTRODUCTION

In the summer of 1994, researchers from the U.S. Environmental Protection Agency's National Risk Management Research Laboratory and the University of Cincinnati conducted a field study on the shoreline of Delaware Bay to investigate the bioremediation of light crude oil experimentally released onto a sandy beach (Venosa et al., 1996). One of the important findings from that study was that the first-order rate constants calculated from the field study were nearly identical to rate constants calculated in sealed laboratory flasks when the rate data were normalized to the highest alky-substituted homologue in a given polycyclic aromatic hydrocarbon (PAH) series. In other words, even though the rate constants in the field were substantially lower than those measured in the laboratory in an absolute sense, when ratios of lower alkyl-substituted compounds within a homologous series to the highest alkyl-substituted homologue were compared, field and laboratory ratios were identical, except for naphthalene (nap) and methylnaphthalene (C_1-nap). This finding suggested that the rates of disappearance of the PAHs were indeed due to biodegradation as opposed to dissolution. Since nap and C_1-nap are substantially more volatile than the other PAHs in crude oil, it would not be expected that they follow the same loss rate pattern as the other less volatile compounds. Other researchers have reported that the degradation of PAHs is lower as more alkyl groups are added to the parent compound (Cerniglia, 1992; Bayona et al., 1986); Prince, 1993; MacGillivray and Shiaris, 1994).

Results from the above observations prompted an interest to determine how widespread these relationships are among different microbial consortia. An experiment was conducted in which eight undefined mixed cultures isolated from various U.S. marine shorelines were incubated in replicate shake flasks in the presence of artificial seawater containing weathered crude oil for a period of one month. Flasks were sacrificed at periodic intervals and the contents analyzed for depletion of crude oil constituents by gas chromatography/mass spectrometry (GC/MS). Rates of biodegradation were calculated by nonlinear regression analysis of the analyte depletion data and compared to rates measured in the Delaware field study.

MATERIALS AND METHODS

Cultures. Undefined mixed cultures were obtained from several geographically distinct marine shorelines of the United States (3 from Alaska, 1 from Texas, 3 from Maine, and 1 from Delaware). The Alaska cultures were labeled 4ORG (Disk Island, Prince William Sound), SNG (Knight Island, Prince William Sound), and KN213 (Knight Island, Prince William Sound); the Texas culture was labeled BS48 (Big Shell Beach, south of Corpus Christi); the Maine cultures were designated CV, HLS, and SPT (all isolated from the central coast of the state); and the Delaware culture was FB (Fowler Beach, Delaware). Each was originally grown on Alaska North Slope crude oil (ANS521) weathered by heating at 272°C (521°F) (NETAC, 1993). After several passages, the cultures were collected by centrifugation, washed with sterile saline, and then stored frozen in glycerol for later use. On the day of the experiment, the cultures were thawed, centrifuged, and resuspended in sterile artificial seawater. A 1.0 mL volume of each culture was used for inoculating the test flasks.

Experimental Design. Shake flasks (250 mL capacity) were filled with 100 mL of artificial seawater (Spotte et al., 1984) containing 0.5 g of ANS521 crude oil and a nutrient mixture consisting of 2.88 g/L KNO_3, 300 mg/L $Na_5P_3O_{10}$ (sodium tripolyphosphate), and 50 mg/L $FeCl_3{\cdot}6H2O$. A total of 228 flasks were set up (4 replicates of the 8 cultures for sacrificing at days 1, 2, 4, 8, 12, 20, and 28 plus 4 uninoculated flasks for T_0 oil determinations). The flasks were incubated at 20° C and shaken on orbital shaker tables at 200 rpm.

Analysis of Crude Oil Components. Appropriate quadruplicate flasks were sacrificed at the designated times, and the entire contents were extracted with dichloromethane (DCM) followed by solvent exchange into hexane for GC/MS analysis of the saturate and PAH components of the oil. A 1.0-µL aliquot of the hexane extract was injected into a Hewlett-Packard 5890 Series II gas chromatograph equipped with an HP 5971A Mass Selective Detector (MSD). The MSD was operated in the selected ion monitoring (SIM) mode for quantifying specific saturated hydrocarbons, PAHs, and sulfur heterocyclic constituents. Details and operating conditions of the GC/MS have been described elsewhere (Venosa et al., 1996).

Regression Analysis. All data were analyzed by nonlinear regression analysis, assuming the first-order ralationship $C_i = C_0\exp(-kt)$, where C_0 = analyte concentration at time t_0 and C_i = analyte concentration at time t_i.

RESULTS AND DISCUSSION

Overall Biodegradation Rates. Table 1 summarizes the slopes (initial rates after a 1-day lag period) and r^2 values of the nonlinear regression analyses performed on all 8 cultures in the laboratory shake flask experiment. Data for both saturates (normal alkanes) and 2- and 3-ring PAHs are shown in the table. Summary statistics for the same analytes from the Delaware field study (Venosa et al., 1996) are also included for comparison. All 8 cultures contained highly competent alkane degraders (biodegradation rate coefficients and coefficients of determination were high). Greater differences among the cultures showed up in the PAH degradation data.

TABLE 1. Summary of Rate Coefficients and Coefficients of Determination for the Total Normal Alkane and Total 2&3-Ring PAH Data.

	n-Alkanes		2- & 3 Ring PAHs	
Culture	**k, day^{-1}**	***r^2**	**k, day^{-1}**	***r^2**
KN213	-0.480	0.939	-0.186	0.908
4ORG	-0.425	0.958	-0.274	0.938
BS48	-0.377	0.963	-0.026	0.581
CV	-0.500	0.904	-0.076	0.823
HL	-0.519	0.922	-0.075	0.832
FB	-0.550	0.941	-0.047	0.650
SNG	-0.170	0.876	-0.081	0.883
SPT	-0.379	0.878	-0.105	0.867
Field	-0.089	0.871	-0.033	0.829

*n = 32, except for the field data, where n = 40

The biodegradation rate coefficients for all but one (SNG) of the cultures were about 5-fold higher (range was 4.2 to 6.2, 1.9 for SNG) for the saturates than that measured in the field, whereas for the PAHs differences were wider, ranging from slower for BS48 to about 8.3-fold higher for 4ORG.

Figure 1 compares the biodegradation rates of the Alaska (4ORG) culture to the field data for the individual saturated (Figure 1a) and aromatic (Figure 1b) oil components. The data are generally representative of the other 7 cultures. With a few exceptions, the patterns of loss were indicative of biodegradation: for the alkanes, rates declined as carbon number increased, and for the PAHs, rates declined as alkyl groups on the parent structure increased in number. The rates of loss of all compounds were close to an order of magnitude lower in the field compared to the lab. This difference was especially marked for the branched alkanes (pristane and phytane) and the higher molecular weight alkanes (n-$C_{33\text{-}35}$) compared to the lower and intermediate molecular weight normal alkanes. When Figure 1c and d are examined, where the biodegradation rates of the field data and the lab data are depicted as a ratio, the decline in the rate of loss of all compounds in the field *relative* to the decline in the laboratory was generally quite similar in magnitude. The error bars represent the upper and lower 95% confidence limits for each ratio. The confidence limits overlap for virtually all analytes except the 3 highest *n*-alkanes and the branched alkanes. The branched alkanes, like the PAHs, belong in a class by themselves and should not be included in the normal alkane group for purposes of comparing rates of biodegradation, since the metabolic pathways are significantly different.

These observations are especially noteworthy considering that the highly weathered medium crude oil used in the lab study was substantially different from the slightly weathered light crude oil used in the Delaware field study. Six of the eight cultures isolated from different parts of the U.S. behaved similarly in closed flasks (with respect to the relative rates of biodegradation of crude oil hydrocarbons) to the microbial consortium that degraded the light crude oil in the open field.

Although it is well known that the rates of biodegradation of hydrocarbons decrease with increasing molecular weight (higher carbon numbers in the case of alkanes, more substituted alkyl groups in the case of PAHs) (Cerniglia, 192; Bayona et al., 1986; Prince, 1994), it is of special interest that such close agreement between laboratory and field results has been observed even with different crude oil types and different microbial consortia. One could make the argument that, in the field, the decreasing rates of disappearance of compounds within a homologous PAH series as the number of alkyl groups increase could be at least partially explained by differences in solubility (i.e., the parent and lower alkyl-substituted compounds washing away faster than the higher substituted ones). However, the data from this paper suggest that solubility differences were not high enough to explain the rate differences observed in the Delaware field study. In a closed laboratory flask, no washout of compounds is possible because the biological reactions occur in a closed vessel. Thus, rates of loss can only be explained by volatilization or biodegradation. In an earlier laboratory study, the results of which were reported as part of the Delaware field study, sealed respirometer flasks were used to calculate loss rates, and comparability between

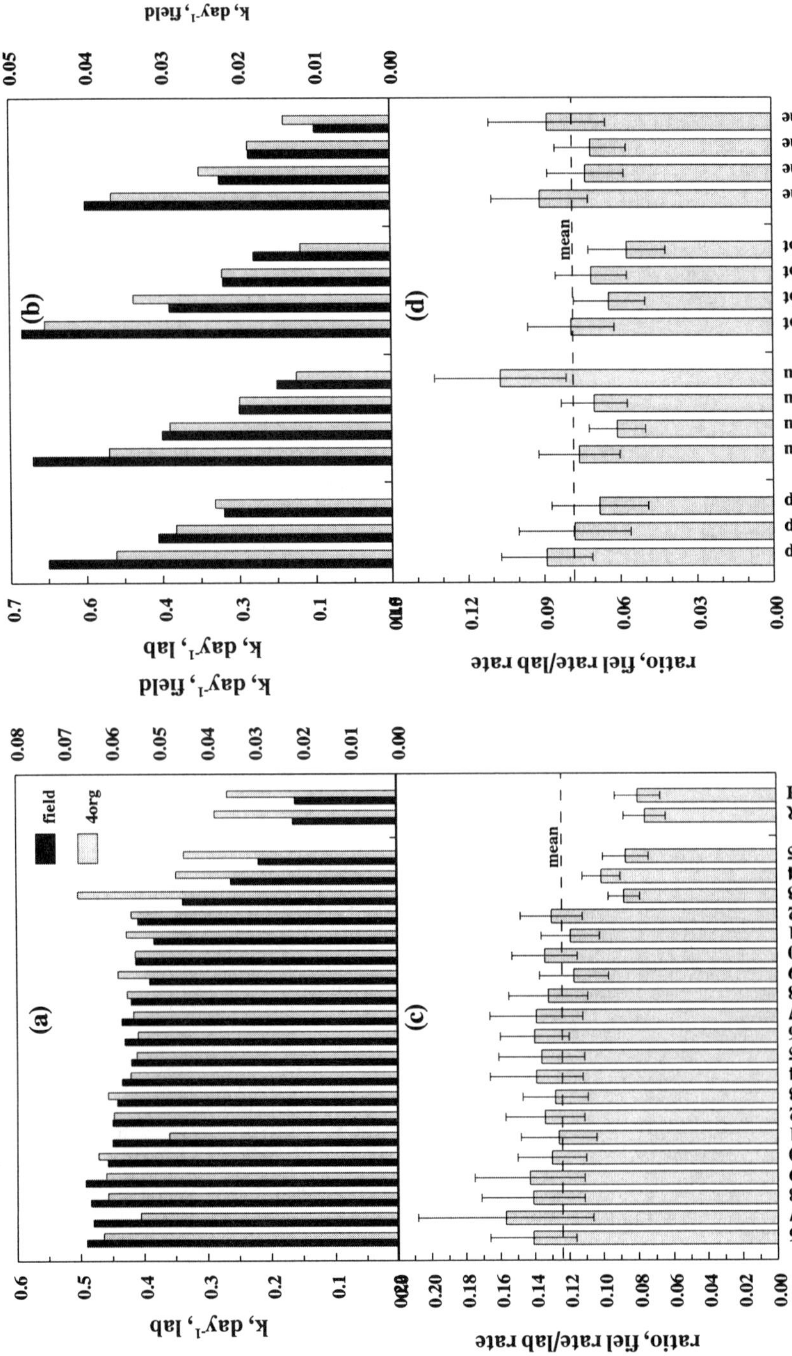

FIGURE 1. Biodegradation rates of alkanes (a) and PAHs (b) in the lab and field and the ratios of field/lab rates for alkanes (c) and PAHs (d).

lab and field rate ratios was also affirmed. This suggested that volatile losses, which are nonexistent in a sealed flask, were negligible compared to biodegradative losses (with the exception of the highly volatile nap and C_1-nap, where the ratios of field rates to lab rates were significantly higher than the mean ratios). Thus, first-order decay rates of hydrocarbon analytes in the Delaware field investigation were due primarily to biodegradation.

Results from this paper and others suggest that one can use the relationship between degradation rate and substrate structure or molecular weight as a reliable indicator of biological activity. This relationship is further strengthened when it has been derived from data that have been normalized to a nonbiodegradable biomarker, which eliminates much of the effect due to washout and solubilization.

REFERENCES

Bayona, J.M., J.Albaiges, A.M.Solonas, R.Pares, P.Garrigues, M.Ewald. 1986. "Selective Aerobic Degradation of Methyl Substituted Polycyclic Aromatic Hydrocarbons in Petroleum by Pure Microbial Cultures." Int. J. Environ. Anal. Chem. 23: 289.

Cerniglia, C.E. 1992. "Biodegradation of Polycyclic Aromatic Hydrocarbons." Biodegradation 3: 351.

MacGillivray, A.R. and M.P.Shiaris. 1994. "Microbial Ecology of Polycyclic Aromatic Hydrocarbon (PAH) Degradation in Coastal Sediments." In: G.R.Chaudry (Ed.), "Biol. Degradation and Bioremediation of Toxic Chemicals." Chapman and Hill, London, pp. 125-147.

National Environmental Technology Applications Center. 1993. "Evaluation methods manual, Oil spill response bioremediation agents." University of Pittsburgh Applied Research Center, Pittsburgh, PA.

Prince, R.C. 1993. "Petroleum Spill Bioremediation in Marine Environments." Crit. Rev. Microbiol. 19(4):217-242.

Spotte, S., G. Adams, and P.M.Bubucis. 1984. "GP2 Medium Is an Artificial Seawater for Culture or Maintenance of Marine Organisms." Zoo. Biol. 3:229-240

Venosa, A.D., M.T.Suidan, B.A.Wrenn, K.L.Strohmeier, J.R.Haines, B.L.Eberhart, D.King, and E.Holder. 1996. "Bioremediation of an Experimental Oil Spill on the Shoreline of Delaware Bay." Environ. Sci. Technol. 30(5): 1764-1775.

MARINE OIL SPILLS: ENHANCED BIODEGRADATION WITH MINERAL FINE INTERACTION

Kenneth Lee (Fisheries and Oceans, Mont-Joli, Quebec, Canada)
Andrea M. Weise (Fisheries and Oceans, Mont-Joli, Quebec, Canada)
Tim Lunel (AEA Technology, Culham, Oxfordshire, United Kingdom)

ABSTRACT: Interactions between oil and mineral fines in seawater (clay-oil flocculation) may stimulate the physical dispersion of stranded oil from the intertidal environment into the sea. Oil droplets stabilized by mineral fines do not adhere strongly to sediments. Surf-washing procedures to accelerate this natural process were successfully used to clean-up an oil spill contaminated site on the coast of Wales (U.K.). Chemical analysis of field samples and laboratory shaker flask studies uphold the assumption that oil-mineral fine interactions in seawater will accelerate the natural rate and extent of hydrocarbon degradation.

INTRODUCTION

Clay-oil flocculation has recently been used to explain the natural recovery of oiled shorelines, especially sheltered low-energy environments (Bragg and Owens, 1995; Bragg and Yang, 1995). A significant quantity of stranded oil may be dispersed by tides and currents when the adhesion of oil to sediments is reduced by its surface interaction with micron-sized mineral particles (Bragg and Yang, 1995).

On February 15, 1996, the *Sea Empress* ran aground on the coast of Wales (U.K.), spilling 70,000 tonnes of Forties Blend and 370 tonnes of Heavy Fuel Oil. A large proportion of oil stranded on the shoreline was successfully recovered within a few weeks by the use of traditional methods. Of the remaining sites, including cobble and shingle beaches where oil became stranded as the high tide receded, there was an indication that the residual oil did not adhere strongly to some sediments. High turbidity in some of the nearshore waters suggested the occurrence of clay-oil flocculation. This was confirmed by direct microscopic observations at Amroth Beach when samples of oiled sediment were agitated with seawater from the surf zone (Lee et al., 1997). On the basis of this evidence, recommended oil spill cleanup methods for this site included surf washing, to accelerate clay-oil flocculation processes, as proposed by Owens et al. (1994). Over several tidal cycles, tracked excavators were used during the period of low tide to move material from the oiled zone at the high water mark (the berm) towards the middle of the intertidal zone. The physical/chemical interaction between oil and mineral fines was promoted with each rising tide, as the surf redistributed the cobbles on the beach. Comparison of the treated to untreated sectors of the beach conclusively demonstrated the efficacy of the surf washing operations to disperse oil stranded on the shore into the sea.

As an oil spill countermeasure, some environmentalist are concerned that surf washing will only move residual oil from one compartment to another within the ecosystem. However, we postulate that oil within the environment is biodegraded at a higher rate when dispersed at sea by the mechanism proposed.

MATERIALS AND METHODS

Samples of sediment and seawater from Amroth Beach, collected during the rising tide after berm relocation, were shipped to the laboratory for detailed chemical analysis by gas chromatography/mass spectroscopy (GC/MS). The samples were dried, extracted with dichloromethane (DCM), concentrated, and solvent-exchanged into hexane prior to injection into a gas chromatograph equipped with a mass selective detector (MSD). The instrument was operated in the selected ion monitoring (SIM) mode to quantify specific saturated hydrocarbons, polynuclear aromatic hydrocarbons (PAHs), and sulphur heterocyclic constituents.

To investigate the mechanism and significance of mineral fine interactions with oil on biodegradation rates, replicate 500 mL shaker flasks (3 for each experimental treatment) with and without mineral fines (38% quartz, 22% feldspars, 20% illite, and 10% chlorite; 25 ppm at a mean particle diameter of 1.3 μm), containing nutrient amended seawater (300 mL) and oil (75 ppm Terra Nova Crude Oil weathered to a density of 0.85 mg/L) were monitored over a 56 day period. Flasks were shaken for 24 hours on an horizontal shaker to stimulate the physical/chemical interaction between oil and the mineral fines, prior to incubation on an orbital shaker (100 rpm, 10°C) for the remainder of the experiment. Flasks were sampled on Days 1, 7, 14, 28, and 56 for microbiological and chemical parameters. For GC/MS analysis, oiled controls were liquid-liquid extracted in DCM while clay-oil flocs were collected on glass fibre filters and extracted by sonicating with DCM. Microbiological parameters monitored during the study included the number of heterotrophic and oil-degrading bacteria, *n*-[1-^{14}C]hexadecane mineralization rates, and methyl-[^{3}H]thymidine incorporation rates into DNA.

RESULTS AND DISCUSSION Removal of emulsified oil from the cobble on the beach was quantified by GC/MS analysis for resolved n-alkanes (C_{12}-C_{35}) and 2-4 ring aromatic compounds (Figure 1, Figure 2). Evidence for enhanced oil biodegradation is apparent in the comparison of data on the distribution of n-alkane and aromatic components within clay-oil flocs recovered from the water (Figure 3a, 3b) with that of the oil emulsion on untreated sectors of the beach (Figure 1a; Figure 2a). Based on the known solubility of individual oil components, the difference observed, cannot be attributed to dissolution alone. Furthermore, there is an observed decrease in the C_{17}/pristane and C_{18}/phytane ratios between the oil emulsion on the cobble and that of the suspended oil recovered from the water column.

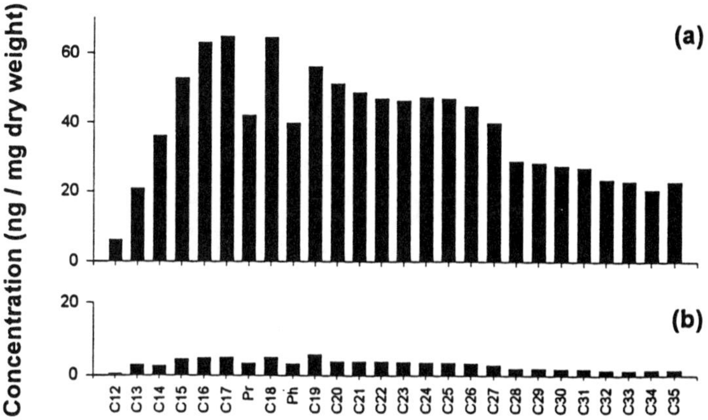

Figure 1. Distribution and concentration of *n*-alkanes and the isoprenoids pristane (Pr) and phytane (Ph) in untreated (a) and surf washed (b) sectors of the beach.

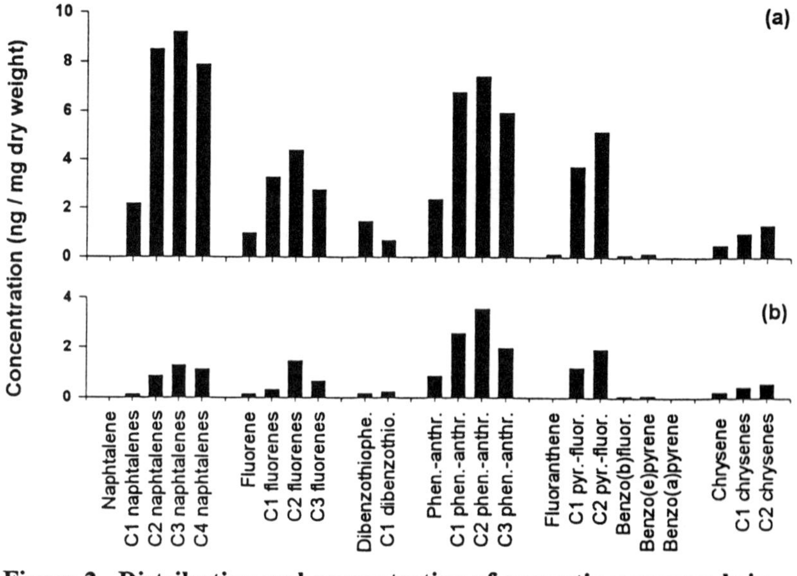

Figure 2. Distribution and concentration of aromatic compounds in untreated (a) and surf washed (b) sectors of the beach.

Time-series shaker flask experiments conclusively demonstrated that the addition of mineral fines accelerated the rate and extent of oil degradation (Figure 4; Figure 5). The mineral fines prevented the oil from adhering strongly to the surface of the flasks. Much of the oil interacting with the mineral fines was maintained in the water column as discrete droplets. An increase in the oil's surface/volume ratio rendered it more accessible to oxygen, nutrients, and bacterial attack. Conclusions drawn from the chemical analysis were in agreement with microbiological data on the numbers of oil-degrading bacteria.

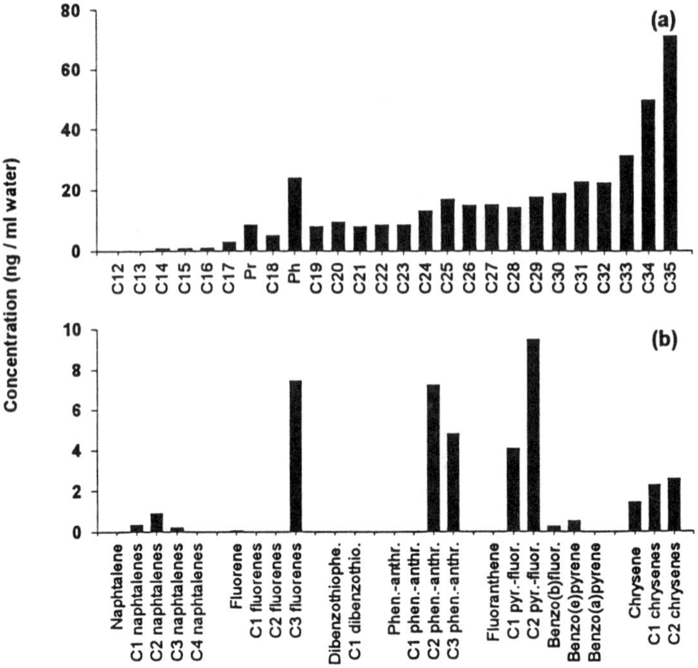

Figure 3. Distribution and concentration of *n*-alkanes (a) and aromatic compounds (b) in nearshore waters.

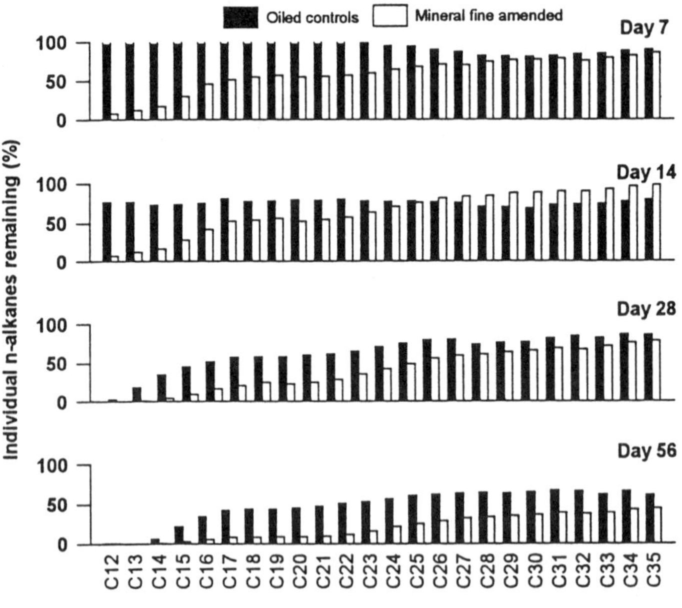

Figure 4. Percent individual *n*-alkanes remaining, relative to Day 1, in oiled controls and mineral fine amended samples on Days 7, 14, 28, and 56.

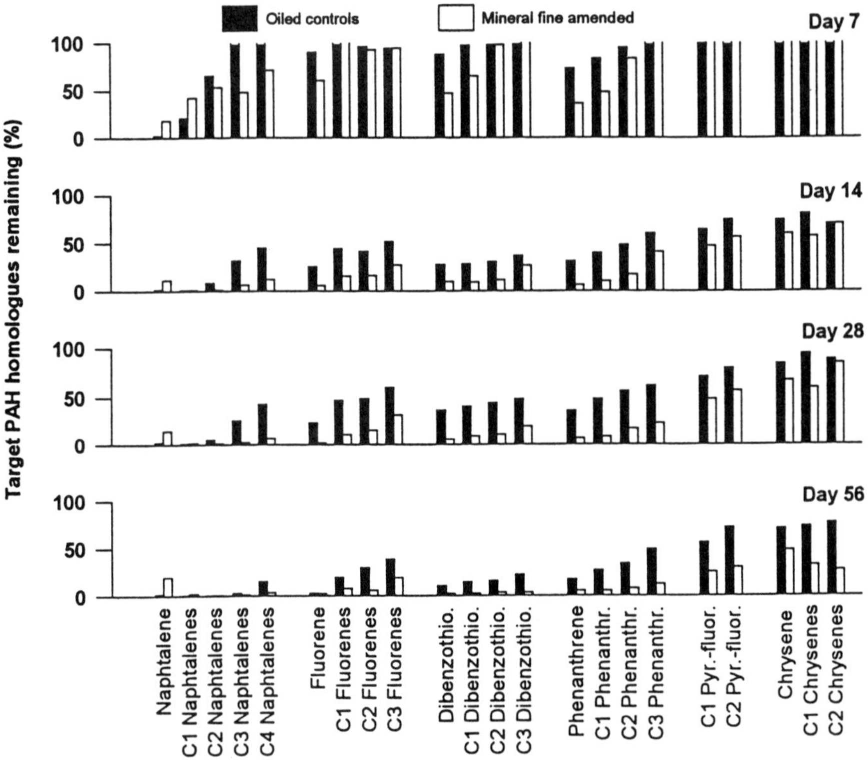

Figure 5. Percent PAHs remaining, relative to Day 1, in oiled controls and mineral fine amended samples on Days 7, 14, 28, and 56.

CONCLUSIONS

Detailed studies on the persistence of the oil released into the sea were conducted to fully evaluate the ecological impacts of surf washing as an oil spill countermeasure. In the field trial described, chemical evidence indicated that oil found within the particulate fraction of nearshore waters was biodegraded to a greater extent than the oil stranded on the shore. However, it cannot be proven that the bulk of this residual oil fraction was released from the shore, as a direct result of the surf washing operations. Conclusive evidence that a stimulation of biodegradative activity occurs with the interaction between oil droplets and mineral fines in seawater was obtained with controlled laboratory shaker flask experiments. Surf washing should now be considered one of the oil spill countermeasures that effectively enhance the rate of oil removal from the ecosystem.

ACKNOWLEDGMENTS

This work was supported by the Panel of Energy Research and Development (PERD), Canada and the Marine Pollution Control Unit (MPCU) of the United Kingdom Department of Transport.

REFERENCES

Bragg, J.R., E.H. Owens. 1995. "Shoreline Cleansing by Interactions between Oil and Fine Mineral Particles." *Proceedings of the 1995 Oil Spill Conference*, pp. 219-227. American Petroleum Institute, Washington, D.C.

Bragg, J.R. and S.H. Yang. 1995. "Clay-Oil Flocculation and its Role in Natural Cleansing in Prince William Sound following the Exxon Valdez Oil Spill." In P.G. Wells, J.N. Butler, and J.S. Hugues (Eds.), *Exxon Valdez Oil Spill: Fate and Effects in Alaskan Waters*, pp.178-214. American Society for Testing and Materials, Philadelphia.

Lee, K., T. Lunel, P.Stoffyn-Egli, P. Wood and R. Swannell. 1997. "Shoreline clean-up by acceleration of Clay-Oil Flocculation Processes." *Proceedings of the 1997 Oil Spill Conference*, (In Press). American Petroleum Institute, Washington, D.C.

Owens, E.H., J.R. Bragg and B. Humphrey. 1994. "Clay-Oil Flocculation as a Natural Cleansing Process following Oil Spills - Part 2: Implications of Study Results in Understanding Past Spills and for Future Response Decisions." *Proceedings of the 17th Artic and Marine Oil Spill Program (AMOP) Technical Seminars*, pp1149-1167. Environment Canada, Ottawa.

APPLICATION OF WASTEWATER SLUDGE TO MICROBIAL DEGRADATION OF CRUDE OIL

Hideaki Maki, Masami Ishihara, Shigeaki Harayama
(Kamaishi Institute, Marine Biotechnology Institute, Kamaishi, Iwate, Japan)

ABSTRACT: We investigated the effects of the application of wastewater sludge on microbial degradation of crude oil. Seawater media supplemented with either 10% (v/v) sludge (corresponding to ca. 200 mg/l of total nitrogen) or inorganic nutrients (1 g/l of NH_4NO_3 and 0.2 g/l of K_2HPO_4) were prepared under the following conditions: (a) sterilized seawater + non-sterilized sludge, (b) non-sterilized seawater + sterilized sludge, (c) non-sterilized seawater + non-sterilized sludge, and (d) non-sterilized seawater + inorganic nutrients. The rates of biodegradation of linear alkanes and alkylnaphthalenes in the sludge-supplemented cultures, i.e. cultures (a), (b) and (c), were higher than that in culture (d) in which inorganic nitrogen and phosphate were supplemented. However the degradation of other polycyclic aromatic hydrocarbons (PAHs) did not proceed rapidly in cultures (a) and (d). This observation indicated that indigenous microorganisms in seawater play an important role in the degradation of these PAHs. Even though the bacterial density in cultures (b) and (d), the degrees of oil degradation were found to be greater in culture (c) than in cultures (b) and (d). This result implies that bacterial density does not necessarily correlate with the degree of oil degradation.

INTRODUCTION

Many experimental results have shown that the supply of nitrogen and phosphorous is the most critical factor to determine the rate of microbial degradation of crude oil (Atlas and Bartha, 1972; Murakami et al., 1987). In fact, the application of fertilizers for clean-up of spilled oil has been shown to be effective (Pritchard and Costa, 1991). However, elaborate fertilizers cost, and less expensive substitutes are highly anticipated.

Approximately 1.5 million tons of excess sludge are produced in 1992 in Japanese wastewater treatment plants, most of them being dumped in land and/or ocean as shown in Table 1. Wastewater sludge contains abundant nitrogen and phosphorous, and thus has a great potential to be used as a fertilizer in bioremediation. We thus investigated the effect of the application of excess sludge on biodegradation of crude oil in batch cultures.

TABLE 1. Annual utilization of wastewater sludge in Japan. (1992)

Landfill	1,149 t
Land farming	192 t
Construction Materials	108 t
Ocean Dumping	15 t
Others	32 t
Sum	1,496 t

MATERIALS AND METHODS

Materials. Crude oil was heated at 230°C before being used for biodegradation tests. The composition of modified natural seawater liquid medium (NSM) was as follows: (per liter) *N*-2-hydroxyethylpiperzine-*N'*-ethanesulfonic acid 23.8 g; and 80% (v/v) fresh seawater. pH of the medium was adjusted to 7.8 with KOH. Wastewater sludge was sampled from a sludge thickener at a municipal wastewater treatment plant.

Oil degradation test. Five hundred-ml Erlenmeyer-flasks containing 100 ml of NSM media supplemented with 0.2% (w/v) of heated oil and either 10% (v/v) of sludge (corresponding to ca. 200 mg/l of total nitrogen) or inorganic nutrients (1 g/l of ammonium nitrate, 0.2 g/l of dipotassium hydrogenphosphate, and 0.2 g/l of iron (III) citrate *n*·hydrate,) were incubated on a rotary shaker at 20°C.

Analysis. Value of suspended solid (SS), the concentrations of total nitrogen and phosphate of sludge were determined according to Standard Methods (1995). Oils in batch cultures were extracted with dichloromethane, and resulting dichloromethane extracts were heated up to 50°C to evaporate dichloromethane in an incubator overnight. The dried samples were subsequently subjected to the thin layer chromatography coupled with flame ionization detection (TLC-FID, Iatron Co., a model of IATROSCAN MK-5) or to the gas chromatography coupled with the mass spectrometery (GC-MS, Shimadzu Co., a model of GCMS-QP5000) without any clean-up treatment of analytes. TLC-FID and GC-MS analysis were done according to Goto et al., (1994) and Wang et al., (1994), respectively. All values obtained by the instrumental analysis were normalized by that of hopane (Prince et al., 1994). Viable counts of bacteria were determined on Marine Agar plates. (Difco Co.)

RESULTS AND DISCUSSION

Properties of sludge. SS, total nitrogen concentration and total phosphate concentration of the sludge used for oil biodegradation tests were shown in Table 2.

TABLE 2. Properties of sludge used.

MLSS	20 g/l
Total nitrogen	2,340 mg/l
Total phosphate	590 mg/l

Bacterial growth. Colony forming units of the four cultures are shown in Figure 1. The bacterial density between days 4 and 10 was the highest in condition of non-sterilized seawater + sterilized sludge (= culture (b)), followed by that of non-sterilized seawater +

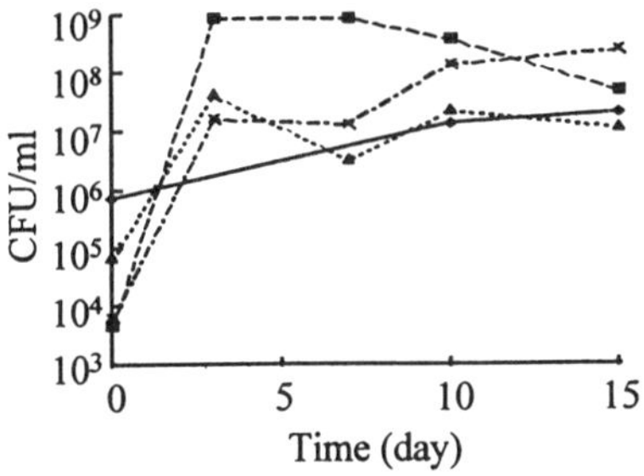

FIGURE 1. Bacterial growth on crude oil. Sterilized seawater + non-sterilized sludge (a, ◆), non-sterilized seawater + sterilized sludge (b, ■), non-sterilized seawater + non-sterilized sludge (c, ▲), non-sterilized seawater + inorganic nutrients (d, ×)

inorganic nutrients (= culture (d)). The bacterial density in condition of sterilized seawater + non-sterilized sludge (= culture (a)) and that of non-sterilized seawater + non-sterilized sludge (= culture (c)) were lower than those of the above two conditions. However, the degrees of oil degradation were found to be greater in culture (c) than in cultures (b) and (d) (see below). These observations imply that bacterial density does not necessarily correlate with the degree of oil biodegradation.

Effect of sludge supplementation on oil degradation. Rapid biodegradation of both saturates and aromatics was observed in culture (c) (non-sterilized seawater + non-sterilized sludge), whilst the biodegradation in culture (d) (non-sterilized seawater + inorganic

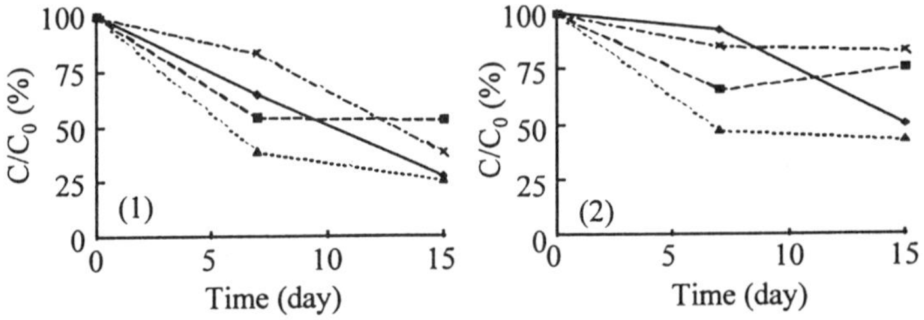

FIGURE 2. Degradation of saturates (1) and aromatics (2). The concentration of these fractions were determined by TLC-FID, and normalized by the concentration of hopane. Sterilized seawater + non-sterilized sludge (a, ◆), non-sterilized seawater + sterilized sludge (b, ■), non-sterilized seawater + non-sterilized sludge (c, ▲), non-sterilized seawater + inorganic nutrients (d, ×)

nutrients) was the lowest irrespective of the highest bacterial growth at day 15. The biodegradation of saturates and aromatics in culture (a) (sterilized seawater + non-sterilized sludge) proceeded significantly, indicating that indigenous microbes in the sludge are also capable of degrading saturates and aromatics of crude oil (Figure 2).

The rates of biodegradation of linear alkanes and alkylnaphthalenes in the sludge-supplemented cultures (cultures (a), (b) and (c)) were much faster than that in culture (d) in which inorganic nutrients were supplemented (Figure. 3 (1), (2)). The results showed that

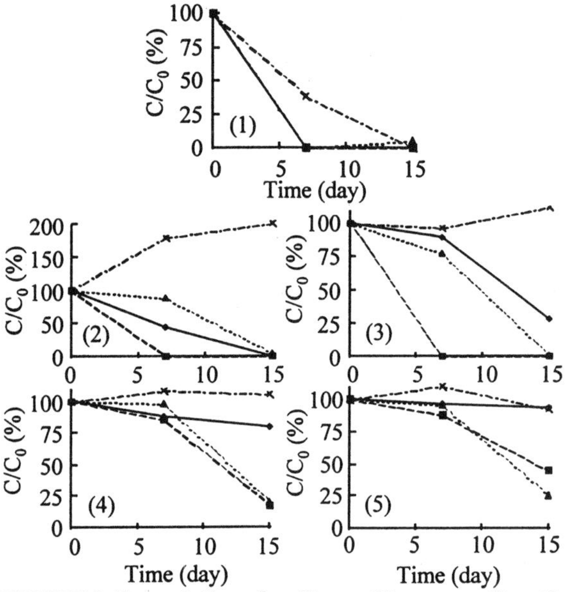

FIGURE 2. Degradation of *n*-alkanes (1), non- and/or alkyl ($C_{0\text{-}4}$) - naphthalenes (2), ($C_{0\text{-}2}$) - fluorenes (3), ($C_{0\text{-}4}$) -dibenzothiophenes (DBTs) (4), and ($C_{0\text{-}7}$) -phenanthrene/anthracene (5). The concentrations of these compounds were determined by GC-MS , and normalized by the concentration of hopane. Sterilized seawater + non-sterilized sludge (a, ◆), non-sterilized seawater + sterilized sludge (b, ■), non-sterilized seawater + non-sterilized sludge (c, ▲), non-sterilized seawater + inorganic nutrients (d, ×)

indigenous microbes in the sludge degraded linear alkanes and alkylnaphthalenes. However, the biodegradation of dibenzothiophene (DBT), phenanthrenes/anthracene and their alkylated derivatives occurred in cultures (a) and (d) but not in culture (a). (Figure. 3 (4), (5)). This observation demonstrated that indigenous microorganisms in seawater might play an important role in the degradation of these PAHs.

As a summary, the degradations of crude oil in the sludge-supplemented cultures were found to be rather efficent in comparison with that in the culture with inorganic nutrients.

Acknowledgment This work was performed as a part of the Industrial Science and Technology Frontier Program supported by New Energy and Industrial Technology Development Organization (NEDO).

REFERENCES

Atlas, R. M., and R. Bartha. 1972. "Degradation and Mineralization of Petroleum in Seawater, Limitation by Nitrogen and Phosphorus." *Biotechnol. Bioeng. 14*: 309-317.

Eaton, A. D., L. S. Clesceri, and A. E. Greenberg. (Eds.). 1995. *Standard Methods*. 19th Edition. APHA, AWWA, and WEF.

Goto, M., M. Kato, M. Asaumi, K. Shirai, and K. Venkateswaran. 1994. "TLC/FID Method for Evaluation of the Crude-Oil-Degrading Capability of Marine Microorganisms." *J. Mar. Biotechnol. 2*: 45-50.

Murakami, A., H. Okumura, S. Omi, A. Yamane, and T. Sugihara. 1987. "Microbial treatment of crude oil on the sea surface using nutrient microcapsules." *Hakkokogaku* (in Japanese) *65*: 113-120.

Prince, R. C., D. L. Eimendorf, J. R. Lute, C. S. Hsu, C. E. Haith, J. D. Senius, G. J. Dechert, G. S. Douglas, and E. L. Butler. 1994. "17α(*H*), 21β(*H*)-Hopane as a Conserved Internal Marker for Estimating the Biodegradation of Crude Oil." *Environ. Sci. Technol. 28*: 142-145.

Pritchard, P. H., and C. F. Costa. 1991. "EPA's Alaska Oil Spill Bioremediation Project." *Environ. Sci. Technol. 25*: 372-379.

Wang, Z., M. Fingas, and L. Ken. 1994. "Fractionation of a Light Crude Oil and Identifcation and Quantitation of Aliphatic, Aromatic, and Biomarker Compounds by GC-FID and GC-MS, Part II." *J. Chromatogr. Sci. 32*: 367-382.

TRANSFORMATIONS OF POLYAROMATIC HYDROCARBONS IN SULFATE- AND NITRATE-REDUCING ENRICHMENTS

Karl J. Rockne (Civil Engineering, University of Washington, Seattle, WA)
H. David Stensel (Civil Engineering, University of Washington, Seattle, WA)
Stuart E. Strand (Forest Resources, University of Washington, Seattle, WA)

ABSTRACT: Bacteria from creosote-contaminated marine sediments were enriched for two years in fluidized bed reactors (FBR) with nitrate or sulfate as the sole potential terminal electron acceptor and with PAH as the sole source of carbon and energy. Influent and effluent analysis showed removal of the PAH naphthalene, biphenyl, and phenanthrene in the FBRs. To confirm these results, cells were harvested from the FBR and used in serum vial tests. Tests using strict anaerobic techniques confirmed the transformation of phenanthrene (specific removal rate, $k = 1.12 \pm 0.20$ μg PAH/mg VSS/day) with stoichiometric removal of nitrate. Biphenyl was removed in sulfate-reducing enrichments ($k = 0.517 \pm 0.033$ μg PAH/mg VSS/day) with stoichiometric amounts of sulfide produced. These results contradict the view that PAH are recalcitrant to biodegradation without oxygen and suggest that there may be the potential to bioremediate PAH-contaminated sediments *in situ* with anaerobic electron acceptor amendment.

INTRODUCTION

Polycyclic aromatic hydrocarbons (PAH) are of great concern as environmental pollutants because of their toxicity to wildlife and humans and their recalcitrance to biodegradation (Demuth et al., 1993). In marine environments PAH are predominantly found in near-shore sediments around harbors and shore-side facilities which produce and store PAH-containing materials. Due to their high hydrophobicity, PAH tend to sorb to suspended particulates and remain sorbed to organic matter in sediment where they are buried by ongoing sedimentation.

Most PAH less than 5 rings can be degraded by bacteria aerobically. However, the oxic zone of polluted sediments typically makes up only the top 1-10 mm and compounds spend 90-99% of the time under anoxic conditions (Aller, 1992). Under anaerobic conditions, PAH are difficult to biodegrade because of the stability of the ring structures. Recent research has suggested, however, that unsubstituted polyaromatic compounds can be degraded under anaerobic conditions (Mihelcic and Luthy, 1988; Coates et al., 1996).

Our goals in this research were to 1) enrich for anaerobic PAH-degraders under nitrate- and sulfate-reducing conditions from PAH-contaminated marine sediment and 2) determine whether PAH-removal was strictly anaerobic and concurrent with stoichiometric electron acceptor reduction.

MATERIALS AND METHODS

Fluidized bed reactors (FBRs) were developed with either sulfate or nitrate as the sole potential electron acceptor and PAH as the sole carbon source. FBRs were inoculated with coal tar creosote-contaminated sediment obtained from Eagle Harbor in Puget Sound, Washington. Sediment was sampled with a box-coring device in April 1993 as described previously (Rockne et al., 1997).

FBR Enrichments. Approximately 25 g of sub-surfacial sediment was placed in an erlenmeyer flask (1L) and diluted with 1 L de-oxygenated artificial seawater (ASW) medium (Rockne et al., 1997) amended with either 3.5 mM $NaNO_3$ (nitrate-reducing enrichment) or 28 mM $NaSO_4$ (sulfate-reducing enrichment). The flasks were placed on a shaker table (100 rpm) for two weeks prior to inoculation of the FBRs.

Following 20 minutes settling, supernatant from the initial enrichment was used to fill the FBRs. The FBRs consisted of a bed of diatomaceous earth biocarrier (Celite Corp., Lompoc, CA) fluidized by recirculating ASW media as described previously (Melin et al., 1996). The FBRs were maintained at room temperature (20°-27° C).

Fresh anaerobic ASW with the appropriate electron acceptor was pumped into the FBR, resulting in a one day hydraulic residence time. The feed was pumped through a glass column filled with glass beads (Fisher Scientific, Bellafonte, PA) coated with phenanthrene, biphenyl, dibenzofuran, and naphthalene. These PAH dissolved into the aqueous phase and were pumped into the reactor. PAH and electron acceptor were monitored in the reactor, influent, and effluent. PAH were quantified using gas chromatography (GC) or high pressure liquid chromatography (HPLC) (Rockne et al., 1997). Sulfide was measured by iodometric titration (APHA, 1995). Nitrate was measured using a cadmium reduction colorimetric kit (Hach Co., Loveland, CO) or by HPLC-UV spectroscopy (Schroeder, 1987).

PAH Degradation Experiments. To confirm the anaerobic PAH removal seen in the FBR, experiments with subsamples of the FBR were performed in serum bottles under strictly anaerobic conditions. Cellmass and biocarrier were anoxically transferred to a serum bottle (30 mL) from the FBR through a side port. Cells were removed from the carrier by vigorous shaking on a vortexer for approximately one minute. Following settling of the biocarrier (1 minute), the supernatant was resuspended in fresh media and used in the degradation experiments described below.

Media were prepared by heating over de-oxygenated N_2 with a strong reductant (0.2 mM Ti-citrate) to ensure no trace oxygen was present. Bottles were prepared using Hungate's technique (Holdeman et al., 1977). Briefly, bottles were gassed with de-oxygenated N_2 using a canula and filled with anaerobic media. N_2 was de-oxygenated by passage through a column of reduced copper fillings at 300° C. Following the experimental setup, the bottles were transferred to an anaerobic glove box (Coy Industries) containing a gas composition of N_2 80%, CO_2 18%, H_2 2%.

PAH degradation experiments were conducted using 30 mL glass bottles with 20 mm, teflon-lined, thick butyl rubber stoppers (West Biodirect, Lionville, PA) in duplicate or triplicate. PAH was dissolved in methylene chloride (MC) and distributed evenly around the inside of the serum bottles. The MC was allowed to evaporate leaving sorbed PAH on the vial wall. Cell suspensions were added after PAH were sorbed to the bottles. Bottles with filter-sterilized, de-gassed, de-ionized water (DI) were spiked with identical amounts of MC/PAH to determine the initial PAH concentration. PAH was measured following a one week incubation to allow for complete solubilization of the PAH in the DI. Killed controls were spiked (5% v/v) with formaldehyde solution (37%, J. T. Baker).

The experimental bottles were sampled inside the glove box using a syringe flushed with degassed N_2 in a serum bottle fitted with a thick butyl rubber stopper. The bottle contained $TiCl_3$ solution (20 mL, 2 M) to further scrub trace oxygen and was transferred to the glove box immediately prior to sampling.

Samples (1 mL) were placed in vials (4 mL) fitted with teflon-lined septa and diluted 1:1 with methanol. The vials were centrifuged and the supernatant used for determination of PAH by HPLC. Nitrate and nitrite were measured from these samples by HPLC-UV (Schroeder, 1987). Samples for sulfide analysis (1 mL) were drawn into a N_2-purged syringe and pushed into a butyl rubber stopper for transfer out of the glove box and immediately measured by iodometric titration allowing no exposure to oxygen. Cell mass was measured as volatile suspended solids (VSS) according to Standard Methods (APHA, 1995).

RESULTS AND DISCUSSION

All PAH were initially removed by both FBRs (Figure 1). Naphthalene, biphenyl, and phenanthrene continued to be nearly totally removed in both FBRs after 500 days of operation. Continued removal of these PAH beyond 100-200 days was attributed to biodegradation. In contrast, effluent dibenzofuran concentrations increased following a lag of 100-200 days and eventually reached the feed concentration. Initial dibenzofuran losses were attributed to sorption to biosolids, tubing, and bed material in the FBR. Eventually sorption sites were saturated allowing breakthrough. Because the early results with biphenyl, naphthalene, and phenanthrene could also have been due to adsorption, the acclimation period (if any) could not be determined. To confirm that the PAH removal was strictly anaerobic and to determine electron acceptor stoichiometry, biomass from the reactors was removed and challenged with individual PAH in serum bottle tests.

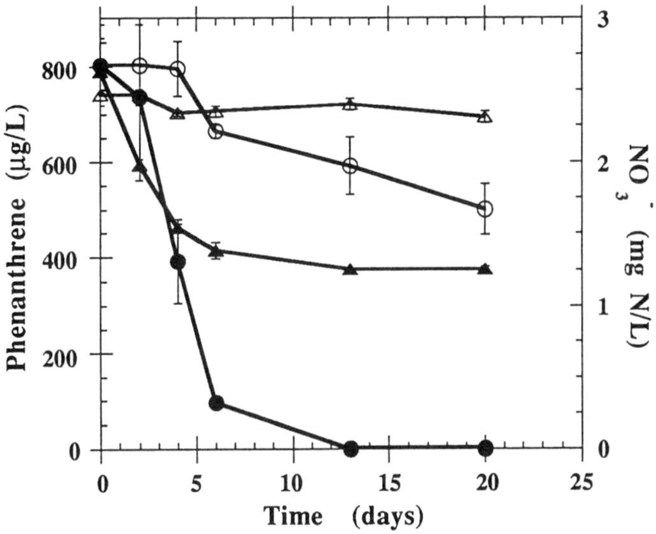

Figure 2. Anaerobic biodegradation of phenanthrene (●) with concomitant stoichiometric reduction of nitrate (▲) by a nitrate-reducing enrichment culture. Corresponding killed controls are shown by open markers. Average of triplicates ± std. error of the mean.

Nitrate Reducers. Phenanthrene was completely removed by the nitrate-reducing culture with an initial specific removal rate, $k = 1.12 \pm 0.20$ µg PAH/mg VSS/day (Figure 2). Phenanthrene losses (corrected for loss to the stopper seen in the killed control) were concomitant and stoichiometric with nitrate removal (0.95 ± 0.059 of stoichiometric) assuming denitrification to N_2O. Nitrate was not

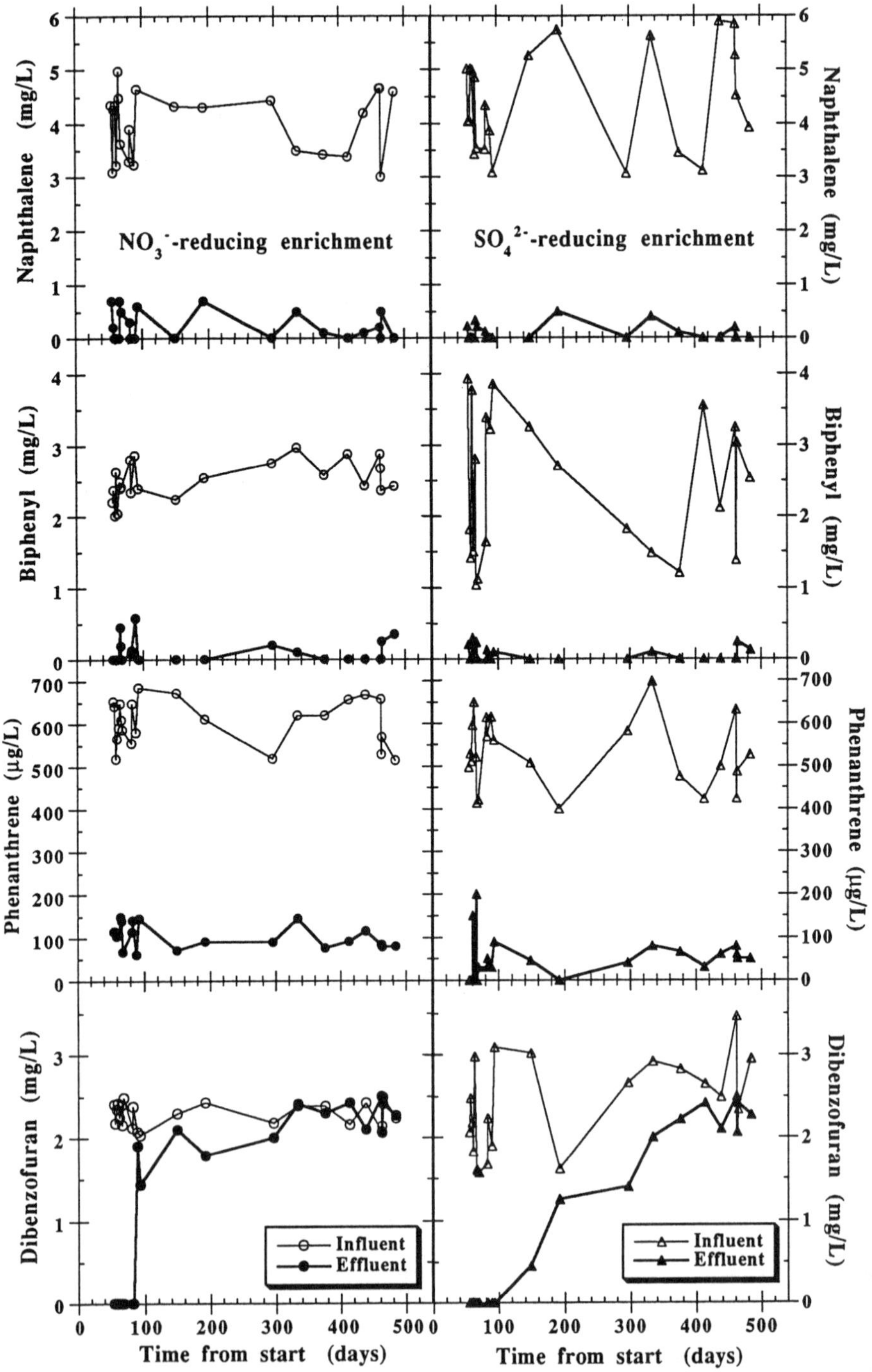

Figure 1. Feed (open markers) and effluent (closed markers) PAH concentration in nitrate-reducing (●, left) and sulfate-reducing (▲, right) FBRs showing biodegradation of biphenyl, naphthalene, and phenanthrene.

transformed significantly, either in parallel cultures with no PAH (data not shown) or in killed controls. Similar results were seen with naphthalene and biphenyl. Dibenzofuran was not degraded, as expected from the FBR results.

Sulfate Reducers. Sulfate reduction was dependent on the presence of PAH. In experiments with biphenyl present in excess of solubility, sulfide was continually produced at a specific rate of 1.12 ± 0.08 μg S/mg VSS/day. In contrast, very little sulfide was produced without biphenyl present. The biphenyl-exposed culture was then resuspended in fresh media and fed biphenyl below solubility (Figure 3). Biphenyl was removed at a specific rate of 0.517 ± 0.033 μg PAH/mg VSS/day. Biphenyl removal was near stoichiometric with sulfide production (1.03 ± 0.066 of stoichiometric), corrected for killed and no PAH controls. In contrast to the rapid removal of PAH by the denitrifiers, PAH removal rates were much slower in the sulfate reducing enrichment. Specific removal rates were 50-80% less than the denitrifiers, which would be expected because of the lower energy yield from sulfate-reduction.

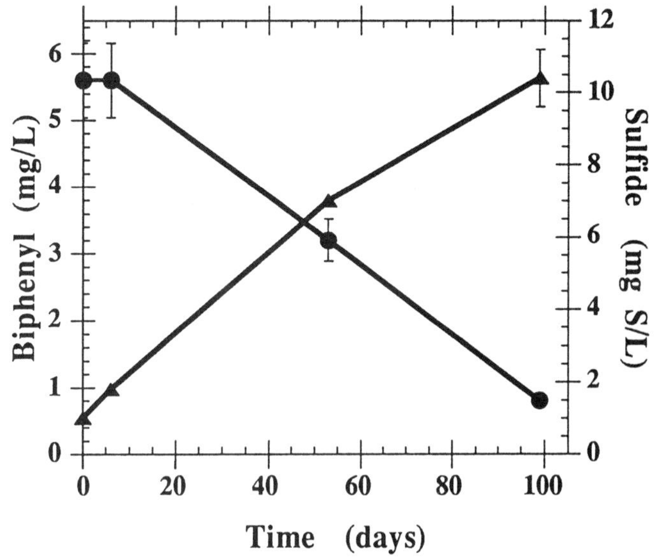

Figure 3. Anaerobic biodegradation of biphenyl (●) with concomitant stoichiometric production of sulfide (▲) by a sulfate-reducing enrichment. Average of duplicates ± std. error of the mean.

CONCLUSIONS

Current research has shown that aromatic compounds once thought to be recalcitrant to biodegradation under anaerobic conditions are degraded anaerobically. Naphthalene, acenaphthene, and phenanthrene are the only PAH reported to be biotransformed under anaerobic conditions. Our results are the first to show phenanthrene biodegradation under denitrifying conditions and the first report showing clear evidence of anaerobic biphenyl biodegradation. Further, this is the first enrichment culture able to biodegrade PAH under sulfate-reducing conditions (previous results were from bulk sediment incubations). These results indicate that the long held view that PAH are recalcitrant to biodegradation without oxygen is not accurate.

Given the cost and difficulty in introducing oxygen into sediment, alternative strategies are needed to make *in situ* biotreatment a viable strategy. These results suggest that there may be the potential to remediate PAH-contaminated sediment using an anaerobic treatment strategy.

ACKNOWLEDGMENTS

We thank Dr. Jim Staley, Dr. John Leigh, Dr. Robert Sanford, and Dr. Joanne Chee-Sanford for helpful input on this research. This work was supported by grant ONR N00014-92-J-1578 from the Office of Naval Research.

REFERENCES

Aller, R. C. 1992. "Bioturbation and remineralization of sedimentary organic matter: effects of redox oscillation". Presented at *Organic Geochemistry Division Symposium*, Geochemical Society of America, 1992.

APHA. 1995. *Standard Methods for the Examination of Water and Wastewater.* 19th ed. American Public Health Association. Washington, DC.

Coates, J. D., R. T. Anderson, and D. R. Lovley. 1996. "Oxidation of Polycyclic Aromatic Hydrocarbons under Sulfate-Reducing Conditions." *Appl. Environ. Microbiol.* 62(3):1099-1101.

Demuth, S., E. Casillas, D. A. Wolfe, B. B. Mccain. 1993. "Toxicity of saline and organic solvent extracts of sediments using the Microtox(r) bioassay." *Archives of Environmental Contamination and Toxicology* 25(3): 377-386.

Holdeman, L. V., E. P. Cato, and W. E. C. Moore. 1977. *Anaerobe Laboratory Manual.* Anaerobe Laboratory, Virginia Polytech. Inst. State Univ., Blacksburg, Virginia.

Holland, P. T., C. W. Hickey, D. S. Roper, T. M. Trower. 1993. "Variability of organic contaminants in inter-tidal sandflat sediments from Manukau harbour, New-Zealand." *Archives of Environmental Contamination and Toxicology* 25(4): 456-463.

Mihelcic, J. R., and R. G. Luthy. 1988. "Microbial Degradation of Acenaphthene and Naphthalene under Denitrification Conditions in Soil-Water Systems." *Appl. Environ. Microbiol.* 54(5):1188-1198.

Melin, S. E., J. A. Puhaaka, S. E. Strand, K. J. Rockne, and J. F. Ferguson. 1996. "Fluidized-Bed Enrichment of Marine Ammonia-to-Nitrite Oxidizers and their Ability to Cometabolically Oxidize Chloroaliphatics." *Internat. Biodeterioration and Biodegradation.* 38(1): 9-18.

Rockne, K. J., H. D. Stensel, and S. E. Strand. 1997. "Cometabolic Enhancement of PAH Degradation by Marine Methanotrophic Enrichment Cultures." Submitted to *Bioremediation Journal.*

Schroeder, D. C. 1987. "The Analysis of Nitrate in Environmental Samples by Reversed-Phase HPLC." *J. Chromatogr.* 25(9): 405-408.

EFFECTS OF CRUDE OIL CONTAMINATION AND BIOREMEDIATION IN A SOIL ECOSYSTEM

Kevin Lawlor, ***Kerry Sublette***, Kathleen Duncan, Estelle Levetin, Paul Buck, Harrington Wells, Eleanor Jennings, Susan Hettenbach, Scott Bailey (University of Tulsa, Tulsa, OK); J. Berton Fisher (Amoco Technology Center, Tulsa, OK); Timothy Todd (Kansas State University, Manhattan, KS)

ABSTRACT: Analyses of samples taken from three experimental soil lysimeters demonstrate marked effects on the soil chemistry and on bacterial, fungal, nematode, and plant communities three years after the application of crude oil. The lysimeters are located at the Amoco Production Research Environmental Test Facility in Rogers County, OK, and were originally used to evaluate the effectiveness of managed (application of fertilizer and water, one lysimeter) vs. unmanaged bioremediation (one lysimeter) of Michigan Silurian crude oil compared to one uncontaminated control lysimeter. Five 2-foot-long soil cores were extracted from each lysimeter, each divided into three sections, and the like sections mixed together to form composited soil samples. All subsequent chemical and microbiological analyses were performed on these nine composited samples.

Substantial variation was found among the lysimeters for certain soil chemical characteristics [% moisture, pH, total Kjeldahl nitrogen (TKN), ammonia nitrogen (NH_4-N), phosphate phosphorous (PO_4-P), and sulfate (SO_4^{-2})]. The managed lysimeter had 10% the level of total petroleum hydrocarbons (TPH-IR), as did the unmanaged lysimeter. Assessment of the microbial community was performed for heterotrophic bacteria, fungi, and aromatic hydrocarbon-degrading bacteria (toluene, naphthalene, and phenanthrene) by dilution onto solid media. There was little difference in the number of heterotrophic bacteria, in contrast to counts of fungi, which were markedly higher in the contaminated lysimeters. Hydrocarbon-degrading bacteria were elevated in both oil-contaminated lysimeters. In terms of particular hydrocarbons as substrates, phenanthrene degraders were greater in number than naphthalene degraders, which outnumbered toluene degraders. Levels of sulfate-reducing bacteria seem to have been stimulated by hydrocarbon degradation. Nematodes were extracted from soil samples, identified to genus, and classified according to their mode of nutrition. All vegetation and roots were removed from each lysimeter after the soil samples were taken, representative plants were pressed for identification, and the dry weight of all plants (total biomass) for each lysimeter was determined. The plant species were predominantly those found in disturbed habitats. The greatest number of species was found in the control lysimeter, while the total biomass was highest in the managed lysimeter.

CRUDE OIL BIODEGRADATION IN SATURATED AND VADOSE SOIL: LABORATORY SIMULATION

Paolo Carrera (Eniricerche, S.Donato Milanese, Italy)
Pasquale Sacceddu (Eniricerche, Monterotondo, Italy)
Andrea Robertiello (Eniricerche, Monterotondo, Italy)

ABSTRACT: An accidental spill of 200 m^3 of Marsa-el-Brega crude oil, caused a contamination of 4,000 m^2, in a vineyard cultivated area. The oil infiltrated to a depth of 4.5 m, with concentration raising to 60,000 mg/L. In order to investigate the hydrocarbons fate and to measure the biodegradation rate, laboratory trials were undertaken using both indigenous bacteria and commercially available hydrocarbon degraders bacterial cultures. Evidence from flask experiments recommended the profitable stimulation of the native microflora and confirmed the uselessness of allochtonous microbial starters for the degradation of crude oil. To evaluate a bioremediation option for both saturated and vadose soil, different laboratory mesocosms simulating respectively biosparging and bioventing treatment were setted up. Under laboratory conditions, mesocosms simulating *in situ* bioventing showed approximately 70% TPH degradation in 80 days. Interesting results refer to hydrocarbon degradation trials in the saturated zone operating at 13 C, simulating the actual groundwater kinetic and without other external addition of final electron acceptors. Under these conditions, a degradation rate of approximately 50% was monitored using both soluble and slow-release-fertilizers. Taking into account the indication coming from lab activities, full-scale bioremediation of the site is currently under way.

INTRODUCTION

A spill of approximately 200 m^3 of light paraffinic crude oil from a pipeline failure in Northern Italy, produced a contamination of 4,000 m^2, in an area cultivated with prized vineyard very closed to a river. The site, characterized by gravelly and sandy soil, showed a high transmissivity value. The oil polluted vadose and saturated soil to a depth of 4.5 m, with concentration up to 80,000 mg/L. After two weeks from the accident, samples of contaminated soil were collected to perform laboratory experiments with the aim to assess the feasibility of bioremediating the site. The specific objectives of the experimental activities were: (i) to verify the ability of autochtonous microorganisms to degrade the crude oil contamination by flask experiments; (ii) to determine the feasibility of using bioventing and biosparging to remediate respectively the vadose and saturated soils by mesocosms.

MATERIALS AND METHODS

Soil sampling. A set of 20 samples was collected at 15-20 cm depth from the contaminated area for microbiological purposes. The soil samples were excavated using a spoon and pulled into polystirene sterile tubes for microbiological characterization.
For lab bioremediation simulation trials, soil samples were collected at 2.5 m depth by using a mechanical excavator. Samples were than removed from coarse material and pulled into 20 L plastic containers.

Sample analysis. Total Petroleum Hydrocarbons (TPH) were determined in soil samples according to EPA 3540. To analyzed the composition of Marsa El Brega crude oil, the technique suggested by the Imperial Oil Research Laboratories (Jobson A., et al., 1972) has been used.

Nickel measurement. Nickel was adopted as biodegradation marker of crude oil in saturated soil (Robertiello, A. , et al. 1983).
Soil aliquots of 20 g have been extracted with 10 ml methylene chloride, dried at room temperature with nitrogen and solubilized in 1 ml of methyl isobuthyl ketone.
Nickel has been determined using an Atomic Absorption Spectrophotometer, at 232 nm.

Microbiological characterization. For eterotrophic bacteria counts, aliquots of 0.1 ml of soil extracts (10 g of soil into 100 ml Ringer solution), after serial dilution, were plated onto Tryptone Soya Agar for direct counting of colony forming units per gram of soil (CFU/g).
To determine the most probable number (MPN/g) of hydrocarbon oxidizing bacteria, diluted samples of soil extracts were used to inoculate Bushnell-Haas Broth added with 1 mg/l resazurine as indicator. After the addition of crude oil and 30 days incubation at 20 C, the results were calculated according to the MPN tables of Standard Methods.
To characterize the bacteria isolated from the site, procedures according to Bergey's Manual of Systematic Bacteriology have been followed.

Flask biodegradation tests. A set of 500 ml Erlenmeier flask filled with 100 ml of saline medium was added by 500 mg of Marsa el Brega crude oil and 2 ml of cell suspension (10^8 CFU/ml). Total hydrocarbon residual after biodegradation, were measured by FT-IR Spectroscopy after CCl_4 extraction (Olivieri et al. , 1976).

Simulation apparatus One of the apparatus realized for simulating the remediation of vadose soil is schematized in Figure 1. The analytical programme included: total eterotrophyic bacteria, hydrocarbon oxidizing bacteria and hydrocarbon degradation in the soil (by Nickel as internal marker).

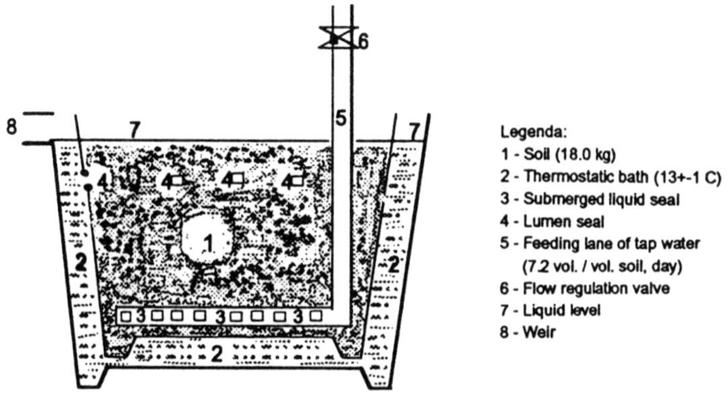

FIGURE 1. Scheme of an experimental simulation unit for the biotreatment of hydrocarbon contaminated soils in saturated environment.

With the aim to simulate the efficacy of bioventing in the vadose soil, apparatus as schematized in Figure 2, were set up and managed at room temperature.

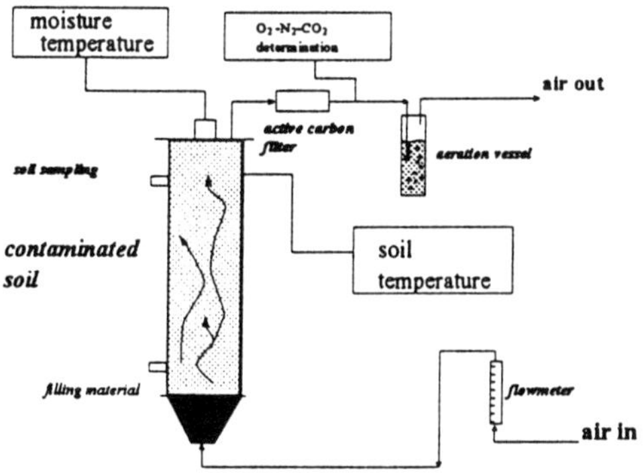

FIGURE 2. Scheme of a laboratory apparatus for the simulation of a bioventing treatment in vadose soil.

Monitoring parameters included moisture, temperature, redox potential, on line respirometry, hydrocarbon degradation and microflora analysis.
All the simulation equipments were managed and monitored over a 3-months period.

RESULTS

Microbial findings.

The microbiological preliminary findings revealed a conspicuous and well adapted microbial population apparently unaffected by toxic injury, consisting prevalently of hydrocarbon-oxidizing bacteria. Microbial counting showed average values of 3×10^6 CFU/g of etherotrophic and > 1×10^6 MPN/g hydrocarbon oxidizing bacteria.
Gram positive strains, mainly belonged to *Arthrobacter, Brevibacterium, Corynebacterium* and *Mycobacterium* genera, gram negative bacteria isolated from the soil, represented more than 40% of the total microflora mainly represented by *Pseudomonas putida, Acinetobacter calcoaceticus* and *Xanthomonas maltophilia.*

Flask biodegradation tests.

In order to investigate the hydrocarbons fate and to measure the biodegradation rate, laboratory trials were undertaken using both indigenous bacteria and commercially available hydrocarbon degraders bacterial cultures.
In particular (see Figure 3), the biodegradation efficiency of different commercial starters (A-E), a degradative inoculum isolated in Eniricerche from refinery oily sludges (F) and natural bacteria coming from the spill site (G), were compared.

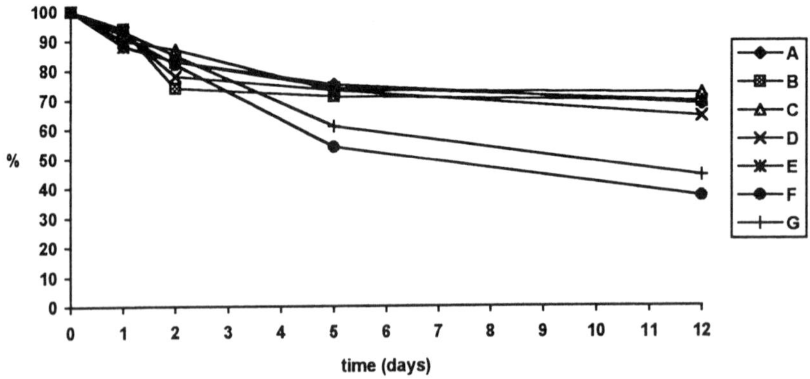

Figure 3. Comparison of crude oil degradative efficiency of different microbial inocula

Bioremediation of saturated soil.

Mesocosms as described in Materials and Methods, were amended with: (A) C:N:P=150:1:0.25 water soluble fertilizers (ammonium nitrate, potassium phosphate); (B) slow-release-fertilizers (ammonium magnesium phosphate); (C) no nutrients. In the following Table 1 the hydrocarbons abatement, using Nickel as internal biodegradation marker, is reported.

Table 1. Contaminants abatement

Apparatus	Nickel (day 0) ppm	Nickel(day80th) ppm	Biodegradation %
A	5.7	25.9	45
B	5.7	30.8	54
C	5.7	12.5	21

Bioremediation of vadose soils.
The activity was addressed to verify the efficiency of a bioventing approach to be realized after soil vapour extraction in order to remove volatile fraction of contaminants. The analysis, referred to the TPH abatement are reported in the following Table 2 as a mean of three determinations.

Table 2. Variation in the composition (%) of residual contaminants during bioventing experiments.

Components	starting point	36 days	80 days
TPH	6.15	3.43	2.11
Asphaltenes	11.2	18.6	21.9
Saturated	63.4	35.2	21.2
Aromatics	19.1	22.4	24.5
NSO polar compounds	6.3	23.8	32.4

Redox potential values, always over 130 mV, as well as pH data around 6.5, measured during the 80 days of experimental trials, were always adequate to microbial development and activity.
In agreement with data above mentioned, on line respirometry, showed an oxygen consumption reaching 70% during the first week of experiments.
Table 3 resumes the biodegradation kinetics data obtained during bioventing simulation trials as a mean of three determinations.

Table 3. Biodegradation kinetic data under bioventing conditions (mg/kg/day)

days 0-36	80.4
days 37-60	60.0
days 61-80	52.0

CONCLUSIONS

Evidences from flask experiments recommend the profitable stimulation of the native microflora and confirmed the uselessness of allochtonous microbial starters for the degradation of crude oil in soil characterized by a conspicuous and well adapted microbial population. Microbial activity measured in the unsaturated contaminated soil treated under bioventing conditions is able to assure an initial biodegradative kinetic of contaminants in the order of 80 mg/kg of soil per day.

Both nutrient formulations used give encouraging results in order to remediate the saturated soil. In this respect at least 50% of total contaminants have been biodegraded after 80 days of treatment at 13 C. Nevertheless, the easiest applicability of diluted solutions containing potassium phosphate and ammonium nitrate suggests the use of such formulation instead of ammonium magnesium phosphate.

The experimental results achieved at laboratory level allowed the implementation of an *in situ* integrated bioremediation system running succesfully from 1.5 year. The field activity were conducted by Ambiente and Aquater, Companies active in Environmental Services within the ENI Group.

REFERENCES

Jobson A., Cook, F.D. and Westlake, W.S. 1972. "Microbial utilization of crude oil". *Appl. Microbiol.* 23: 1082-1089.

Robertiello, A. , Petrucci, F., Angelini, L. and Olivieri, R. 1983. "Nichel and Vanadium as biodegradation monitors of oil pollutants in aquatic environment". *Water. Res.* 17: 497-500.

Olivieri, R., Bacchin, P., Robertiello, A., Oddo, N., and Tonolo, A. 1976 " Microbial degradation of oil spills enhanced by a slow-release fertilizer". *Appl. Environ, Microbiol.* 31: 629-634.

BIOLOGICAL TREATMENT OF HIGHLY WEATHERED CRUDE OIL AFFECTED SOILS

Bruce M. Haikola, Mark Q. Henley and ***Randolph M. Kabrick***

ABSTRACT: During the early years of crude oil discovery and development in Texas, produced crude oil was stored in large earthen pits. This practice was discontinued by legislation which was enacted in the 1930's, thereby requiring all produced crude oil to be stored in above-ground tanks. However, many of these old earthen storage pits remained open but out of service well into the 1970's and were subsequently closed by pushing in the earthen berms to fill the pits. As part of a redevelopment effort, several of these old pits (total area of 19 acres) located in east Texas were remediated using biological land treatment.

Soils collected from the site were subject to chemical analyses and biological treatability testing, including respirometry and pan studies. The soils contained total petroleum hydrocarbon (TPH) levels, as measured by USEPA Method 418.1, that ranged from four to 14 percent by weight. Analysis by thin layer chromatography showed that the hydrocarbon was 50 percent asphaltic, 10 percent polar, 20 percent aromatics and 20 percent aliphatics. Respirometry and pan treatability testing showed that only the aliphatic and polar fractions were biodegraded. Furthermore, significant biodegradation occurred only after the affected soils were amended (diluted) with a clean soil. Maximum biodegration occurred at 50 percent dilution with clean soils resulting in over 30 percent degradation of TPH in eight weeks of treatment.

A human health and environmental risk evaluation conducted subsequent to the treatability testing indicated that a final treatment level of three percent TPH was protective of human health and the environment. A large pilot-scale study was conducted to further demonstrate that biological treatment of these crude oil affected soils could effectively meet the treatment goal of three percent TPH.

The pilot-scale system consisted of an engineered land treatment system (clay lined, drainage system, run-off and run-on control, storm water detention pond, and irrigation system) with a treatment area of 5.2 acres. Crude oil affected soils were loaded to the land treatment unit (LTU) with conventional excavation equipment and blended with clean sand (up to 50 percent by volume). Agricultural nutrients were applied in excess of stoichiometric amounts (based on expected TPH degradation of 30 percent) and the treatment zone soil moisture was maintained at 70 percent of field capacity. The LTU was tilled on a daily basis using a combination of conventional agricultural tillage equipment and a road stabilizer. Two lifts of 5800 cubic yards each were treated in this manner. Initial TPH concentrations were reduced from five percent to below three percent within 12 weeks of treatment for each lift.

DEMONSTRATED COST EFFECTIVENESS OF BIOVENTING AT A LARGE CRUDE-OIL IMPACTED SITE

H. J. Reisinger, S. A. Mountain, V. Owens, J. Godfrey, D. Arlotti, G. Andreotti, and G. DiLuise

ABSTRACT: A large-scale bioventing system covering approximately 15 hectares (37 acres) was constructed to address vadose-zone hydrocarbon resulting from an oil well blowout. Results of soil sample analyses showed that the hydrocarbon in deeper soils occurs in a patchy distribution and that, in some locations, it extends from ground surface to the water table up to 11-m below grade. A two-phased bioventing pilot test that was conducted in two locations showed that a zone of remediation of approximately 76-m could be developed around a single air-injection point; it also showed that the subsurface was, in general, very permeable and amenable for distributing large volumes of air at low injection pressures. The second phase of testing (in situ respiration measurement) showed that an active consortium of heterotrophic organisms exists in the subsurface and that they are capable of degrading the hydrocarbon present through aerobic metabolic pathways.

The bioventing system consists of five independent high-flow blower stations housed in steel cages in excavated zone fields that deliver air to 26 air-injection points. Vacuum extraction is not incorporated into the bioventing system due to the relatively low overall vapor pressure of the remaining crude oil fraction in the subsurface. Oxygen concentrations are maintained between 5 and 20 percent in the bioventing area to support aerobic biodegradation. Performance monitoring forms the basis of the biovent optimization process; it includes subsurface pressure measurements, vapor monitoring (oxygen, carbon dioxide, total volatile hydrocarbon), and in situ respiration testing. During the first six months of operation, respiration rates have averaged 5 mg/kg/d in the monitoring points. It is estimated that over one million kilograms of crude oil (over 500,000 L) have been mineralized as a result of the bioventing operations. The cost of this remediation is approximately \$2.00 per m^3, which is very economical relative to other remediation technologies.

LABORATORY SIMULATION OF THE BIOREMEDIATION OF OIL POLLUTED SAND BEACHES

Christine Dalmazzone (IFP, Rueil-Malmaison, France)
Daniel Ballerini (IFP, Rueil-Malmaison, France)

ABSTRACT: The aim of this study, which is a part of the EUREKA BIOREN programme, was to simulate in the laboratory the bioremediation of sand beaches polluted by an oil spill. An experiment was set up at Institut Français du Pétrole (IFP) in order to compare the efficacy of two types of additives (a slow release fertilizer and a BIOREN formulation) in semi-closed basins filled up with fine sand and polluted with a weathered crude oil. The other objective was to achieve a mass balance by assessing the different ways of oil degradation or disappearance. It was shown that the residual amount of oil is lower with the BIOREN formulation after two months. Furthermore, it was possible to achieve a satisfying mass balance by quantifying the hydrocarbons content in the sediment and in the aqueous phase, as well as the CO_2 content in the gaseous phase.

INTRODUCTION

Bioremediation consists in adding materials to contaminated environments to cause an acceleration of the natural biodegradation process. This procedure has received most attention, notably after the Exxon Valdez incident (Bragg et al., 1992; Atlas and Cerniglia, 1995).

The main objective of the european EUREKA BIOREN programme was to develop commercial formulations of nutriments able to enhance hydrocarbons biodegradation on contaminated shorelines (partners: ELF AQUITAINE, IFP, ELF Norway and NAT). The specific aim of our laboratory study in IFP was to simulate the bioremediation of sand beaches polluted with a weathered crude oil and to assess the efficacy of a slow release fertilizer and a formulation of nutriments developed during the previous phases of the BIOREN programme. Over the comparison of the nutriments action, the other objective of the study was to achieve a mass balance by assessing the different ways of degradation of the crude oil: partial biodegradation, complete mineralization and washing of hydrocarbons by the action of surfactants or biosurfactants. Two closed basins were therefore designed and filled up with fine sand and a tidal movement of synthetic seawater was set up with a system of pumps. The pollutant was an Arabian Light crude oil topped at 250°C. Sediment was sampled for chemical and microbiological analyses. The aqueous phase was also sampled for determination of the hydrocarbons content. The air flow was controlled in each basin and analysed for CO_2 concentration. Biodegradation was essentially assessed by chemical analyses, as suggested by a standardized experimental protocol which was developed by an international working group (Merlin et al., 1994).

solid-liquid chromatography on silica-gel columns and quantified by gravimetry. The saturated and aromatic fractions were then analysed by computerised capillary gas-chromatography (GC-FID). Some sediment samples were also collected in order to count total heterotrophic and oil degrading bacteria. Finally, total organic carbon (TOC) was determined in the aqueous phase every day, as well as the amount of ammoniacal nitrogen (NH_4).

At the end of the experiment, the residual amount of hydrocarbons in the sediment and in the water was extracted and quantified.

Nutriments. A minimum medium was used in the first experiment: synthetic seawater (distilled water + 28 g/L sea salt for aquarium) + 0.5 g/L $(NH_4)_2SO_4$ + 0.1 g/L yeast extract. In the second experiment, two different solid nutriments were tested. The first one was a slow release fertilizer, Max Bac®, derived from the "Customblend" used during the Exxon Valdez incident, and sold by Grace Sierra (USA). It consists of fertilizer granules coated with an organic resin material, supplemented with vitamins. It contains nitrate and ammoniacal nitrogen (220 g N/kg) and phosphoric acid. This additive was compared to a formulation which was developed during the first phases of the BIOREN programme and which consists of extrudates.

RESULTS AND DISCUSSION

First experiment. The aim of the first experiment was to test the experimental setting and the overall procedure. The inoculum used was a consortium of microorganisms from soils in the first basin (B1) and a consortium of microorganisms from seawater in the second one (B2). The synthetic seawater supplemented with ammonium sulfate was changed every two weeks. The duration of the experiment was 42 days. Figure 2 shows the evolution of the composition of the oil extracted from the sediment. O-terphenyl was used as biomarker, because of its resistance to biodegradation. The ratio nC18/o-terphenyl is a good indicator of the extent of biodegradation (Table 1).

The oil extracted from the sediment samples (5 cm depth maximum) was submitted to biodegradation as shown by the decrease of the nC18/o-terphenyl ratio. However, the relative evolution of the different chemical families is characteristic of a weak level of biodegradation. It was confirmed by the experimental values of the percentage of Total Extractible Organic Matter (Table 1). Furthermore, a mass balance was performed at the end of the experiment. The difference between the initial mass of BAL 250 and the mass of organic matter extracted after 42 days was about 9 g in both cases (5 % of the initial amount of oil). The quantity of carbon calculated from the experimental values of the CO_2 concentration in the gaseous phase was between 8 and 13 g.

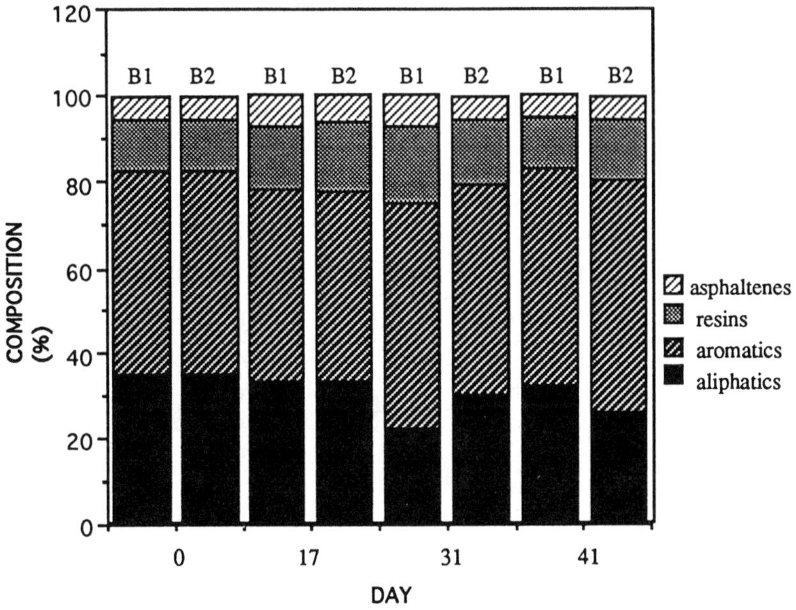

FIGURE 2. Evolution of the composition of the oil extracted from the sediment (%)

TABLE 1. Evolution of the Total Extractible Organic Matter (TEOM) and the nC18/o-terphenyl ratio during the experiment

Day	0	17		31		41	
		B1	B2	B1	B2	B1	B2
TEOM (g/kg sediment)	nd	27.4	33.8	26.7	51.1	31.2	28.6
nC18/o-terphenyl	0.889	0.593	0.573	0.381	0.358	0.337	0.175

Second experiment. The objective of the second experiment was to compare our formulation (Basin 2) with Max Bac® (Basin 1). The inoculum was similar for both basins and contained a consortium of microorganisms from sea water and soils. Synthetic sea water was distilled water with 33 g/L of aquarium salt. This experiment was conducted during two months. The pollutant was applied as previously described and nutriments were applied one week after the oil application (day 7). The amount of applied nitrogen was calculated in order to theoretically ensure a complete mineralization of oil. The evolution of the composition of the oil extract (%) is represented in Figure 3.

Results from chemical and microbiological analyses as well as mass balances are presented.

MATERIALS AND METHODS

Materials. Two basins made of plexiglass (400 x 400 x 300 mm) were designed and filled up with several layers of sand washed with clear water: gravel , medium sand (about 1 mm diameter) and fine sand (about 0.3 mm diameter). Synthetic sea water (distilled water and sea salt for aquarium) was pumped from a glass tank in order to simulate tide cycles (4 tide cycles per 24 hours). The inoculum was a consortium of microorganisms from different origins. The pollution was conducted at high tide, by using a confinement ring (33 cm diameter) in order to avoid the contamination of the container walls. The oil was an Arabian Light crude oil topped at 250°C (BAL 250) to simulate a weathered crude oil. O-terphenyl was initially added to the crude oil (0.5 % w/w) to be used as an extraction yield indicator as well as a biomarker. The amount of oil that was applied was calculated on the basis of about 2 L/m^2 on a depth of 5 cm (about 180 g of oil). A general view of the experimental apparatus is given in Figure 1.

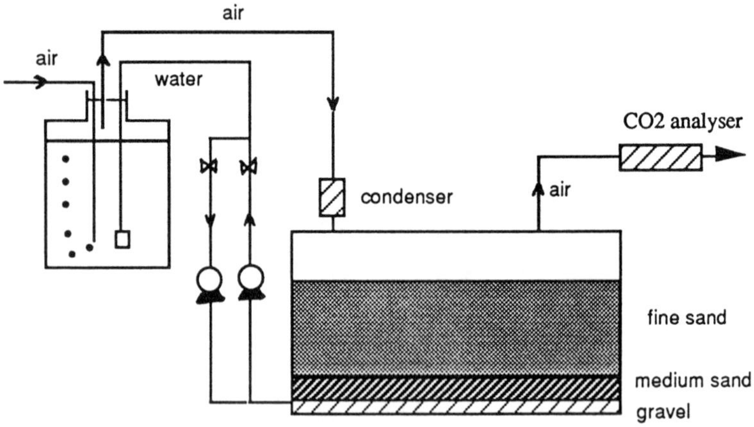

FIGURE 1. Schema of the experimental apparatus

Sampling and analyses. The CO_2 concentration in the gaseous phase was continuously analysed with a CO_2 analyser COSMA Rubis 3000. Sediment samples were regularly collected in order to perform hydrocarbons analyses. Total extractible organic matter (TEOM) was extracted with dichloromethane in a soxhlet (6 hours) and quantified by gravimetry after evaporation of the solvent. The oil extract was then dissolved again in a solvent and separated into asphaltenes, aliphatics, aromatics and resins. Asphaltenes were determined as the hexane insoluble fraction and aliphatics, aromatics and resins were separated by

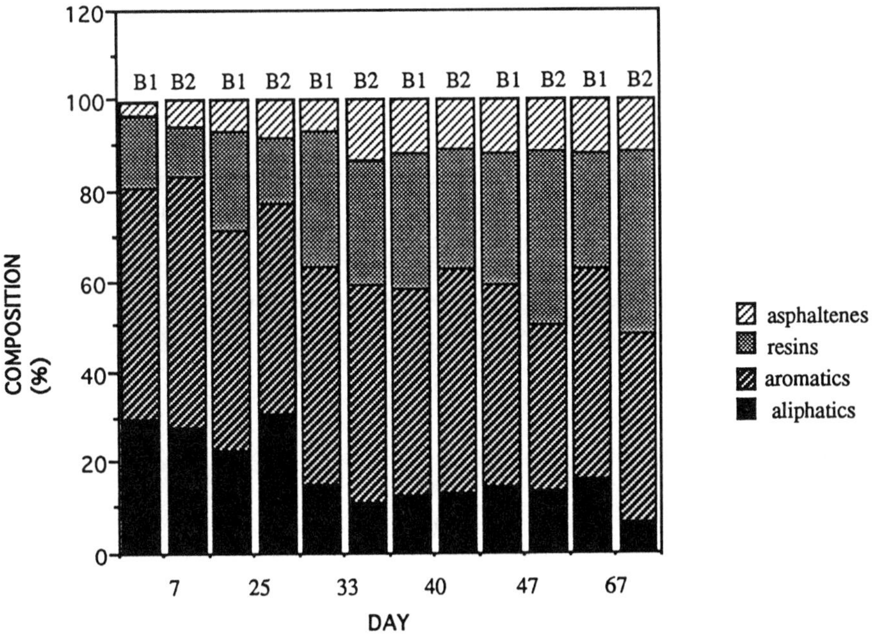

FIGURE 3. Evolution of the composition of the oil extracted from the sediment (%); B1: Mac Bac, B2: Bioren formulation

The aliphatic fraction was strongly degraded after two months, particularly with the Bioren formulation. The aromatic fraction tends also to decrease, but much more slowly. The polar fraction regularly increases during the experiment, especially in basin 2. It is noteworthy that the Bioren formulation tends to work better than Max Bac at the end of the experiment. It seems to be confirmed by the microbiological analyses on oil degrading bacteria: the number of oil degrading bacteria per gram of sediment is much greater in basin 2 at day 67 (8.0×10^5 against 8.0×10^4), when it was the contrary in the middle of the experiment (1.4×10^3 against 8.0×10^4 in basin 1). However, the number of total heterotrophic bacteria was always higher with the Bioren formulation: 2.5×10^6 against 8×10^5 in the middle of the experiment and 1.7×10^7 against 5×10^5 at the end.

The results of the evolution of the Total Extractible Organic Matter (g/kg dry sediment) and of the nC18/o-terphenyl ratio are gathered in Table 2. A rapid decrease of the TEOM is observed during the first month and a stabilization is then observed around 20 g/kg. The evolution of the nC18/o-terphenyl ratio follows exactly the same behaviour. Biodegradation occurs very quickly after application of nutriments.

Table 3 shows the main results of the mass balance that was performed at the end of the experiment. The residual organic matter was extracted from the sand and quantified. All the aqueous effluents contained an oil-in-water emulsion for both basins. It is noteworthy that the emulsion formed in the case of basin 2 (BIOREN formulation) was finer and more stable. Water effluents were systematically recovered during the experiment and the emulsified oil was extracted after breaking of the emulsions. The amount of oil in the aqueous phase was similar for both basins. The amount of carbon from the CO_2 analyses was also quantified: the extent of mineralization seems to be higher in the case of the BIOREN formulation. These results clearly show that the residual amount of oil in the sediment is lower with the BIOREN formulation.

TABLE 2. Evolution of the Total Extractible Organic Matter (TEOM) and the nC18/o-terphenyl ratio during the second experiment

Day	7		25		40		67	
	B1	B2	B1	B2	B1	B2	B1	B2
TEOM (g/kg sediment)	39.3	43.6	21.2	19.8	16.6	20.6	18.0	17.1
nC18/o-terphenyl	0.670	0.870	< 0.1		< 0.1		< 0.1	

TABLE 3. Results of the mass balance from the Total Organic Matter extracted from sediment and water and from the gas phase analyses (CO_2)

	Initial spilt crude oil BAL 250 (g)	TEOM from sediment (g)	TEOM from water (g)	Difference (g)	Carbon from CO_2 analyses (g)
Max Bac	162	101	13	48	51
Bioren	173	69	9	95	55<C<135*

* The mineralization of the carbon source from the formulation has to be taken into account (total C from CO_2: 135 g; total C from the nutriment: 80 g)

REFERENCES

Atlas, R. M., and C. E. Cerniglia. 1995. "Bioremediation of Petroleum Pollutants." *BioScience. 45*(5): 332-338.

Bragg, J. R., R. C. Prince, J. B. Wilkinson, and R. M. Atlas. 1992. *Bioremediation for Shoreline Cleanup Following the 1989 Alaskan Oil Spill.* Exxon Company, Houston, TX, USA.

Merlin, F. X., K. Lee, R. Swannel, J. Oudot, A. Bassères, T. Reilly, C. Chaumery, C. Dalmazzone, and P. Sveum. 1994. "Protocol for Experimental Assessment of Bioremediation Agents on a Petroleum Polluted Shoreline.", in *Proceedings of the 17th AMOP*, pp. 465-478. Environment Canada, Ottawa, Ontario, Canada.

FIELD EVALUATION OF BIOREMEDIATION TO TREAT CRUDE OIL ON A MUDFLAT

Richard P.J. Swannell & David J. Mitchell (AEA Technology plc, National Environmental Technology Centre, Oxfordshire, UK)
D. Martin Jones (University of Newcastle, Tyne and Wear, UK)
Alyson Willis (University of Sunderland, Tyne and Wear, UK)
Kenneth Lee (Fisheries and Oceans, IML, Québec, Canada)
Joe E. Lepo (University of West Florida, CEDB, Pensacola, Florida, USA)

ABSTRACT: A field evaluation of the use of bioremediation to treat oiled fine sand in the intertidal zone of Stert Flats (Somerset, UK) was conducted and the use of *in situ* respirometry to monitor bioremediation success was evaluated. Previous experimental studies had shown that superficial oil is rapidly removed from Stert Flats with tidal action removing or depositing 0.05 - 0.10 m of fine sand in single tidal cycle. Thus, only the oil found at depth, as a result of penetration or burial by sediment deposition is persistent. To evaluate the feasibility of bioremediation to treat this stranded sub-surface oil, a field trial was conducted using inorganic sources of nitrogen and phosphate. Arabian light crude oil (weathered and emulsified with 25% seawater) was added to selected plots a coverage of 4 $l.m^{-2}$. Regular addition of nutrients (sodium nitrate and potassium dihydrogen orthophosphate) were made throughout the three month experiment, beginning 1 week after oil application. The application rate was determined by a separate laboratory study. The success of the bioremediation strategy was determined by chemical analysis of the residual hydrocarbons and monitoring CO_2 evolution *in situ*. The results suggest that inorganic fertiliser does stimulate the biodegradation and mineralisation of oil buried in the aerobic zone of fine sediments.

INTRODUCTION

Bioremediation has now been shown to be effective on a range of shoreline types (Prince 1993, Swannell *et al* 1996). Field studies have shown that it can be used successfully to clean rocky, cobble (Bragg *et al* 1994) and coarse sand (Venosa *et* al 1996) shorelines. These studies have lead to the formulation of some operational guidelines on bioremediation for responders (MPCU 1995, Swannell *et al* 1996). Much less attention, however, has been given to fine sediments such as those found in the upper parts of mudflats around the UK coast. These areas often have poor access and are difficult to clean using conventional methods. Moreover the field experiments carried out to date have also concentrated on the ability of bioremediation to treat surface contamination of shoreline sediments (for example Rosenburg *et al* 1992), and less consideration has been given to the potential of bioremediation to treat buried oil. In order to fill these important gaps in our current understanding of the potential of bioremediation, our field experiment was designed to ascertain the feasibility of using bioremediation on the upper part of a mudflat to treat oil contamination buried at a depth of 15 cm. The experiment utilised mesh enclosures to retain sediment in this aerobic zone, and evaluated *in situ* respirometry (Swannell *et al* 1994) as a technique for studying bioremediation success.

METHODS AND MATERIALS

Specification of Field Site and Placement of Plots. A field site in the South-West of England (Stert Flats: 51° 12.3' North, 03° 03.9' West) was chosen for the experimental work. This site has been previously used for bioremediation field experiments and tidal and sediment movements on this site have been well characterised. A total of 12 plot areas were marked on an 80 m stretch of sand (mud content 3.2%, 80% of particles in range 125-180 μm) using stainless steel poles, which also served to anchor the mesh enclosures in the beach. Twelve mesh enclosures were manufactured using Nitex (pore size = 200 μm) material (Lee and Levy, 1992), measuring 0.4 (L) x 0.4 m (W) x 0.05 m (D). These were filled with beach material from the field site and buried at each of the delimited plot locations at a depth of 0.15 m, 6 m apart.

Preparation of Oil. Arabian Light Crude Oil was chosen as the test oil for the experiment as it is know to contain a high proportion of biodegradable components (Swannell *et al* 1995). The oil was weathered by agitation with air at room temperature until a constant weight was achieved. This process removed 20% of the oil by volume. The oil was then emulsified with artificial seawater (Instant Ocean), using a mechanical mixer (Silversan, Bucks) to form a 25% water-in-oil emulsion. The weathering and emulsification were used to simulate oil spilled at sea and washed ashore (Lee *et al* 1995).

Experimental Design. The plots were divided into three blocks of four randomly assigned treatments (Table 1).

TABLE 1. Allocation of Treatments in the Field Experiment

Plot Number	Block Number	Treatment
1	Block 1	Oil and Fertiliser
2		Oil Only
3		Control
4		Fertilised Control
5	Block 2	Control
6		Oil Only
7		Fertilised Control
8		Oil and Fertiliser
9	Block 3	Oil and Fertiliser
10		Fertilised Control
11		Oil Only
12		Control

For the oiled plots oil was applied at a rate of 4 $l.m^{-2}$ bag area. Inorganic fertiliser (sodium nitrate and potassium dihydrogen orthophosphate) was applied at fortnightly intervals. This application rate was determined in laboratory studies using Stert sediment (Swannell *et al* 1995) at 100:2:0.2 (oil:nitrate:phosphate). Fertiliser was applied in seawater to the mesh containers after partial excavation.

Monitoring. The success of the bioremediation treatment was monitored using a variety of methods. Samples were extracted a random points within the enclosures at Day 0, 42 and 101 and analysed for residual hydrocarbons using a method described elsewhere (Swannell *et al* 1995). In addition to a range of aliphatic and aromatic hydrocarbons, the analysis process identifies a range of geochemical biomarkers including 17α(H), 21β(H)-hopane which were used to accurately determine the extent of biodegradation independent of any physical removal of oil that may have occurred (Bragg *et al* 1994). Microbial activity as a result of oil biodegradation was also monitored using a method of *in situ* respirometry. Carbon dioxide evolution rates from the bags were determined before the addition of oil to the microcosms and at 14 day intervals thereafter. The method used is a modified version of that described in Swannell *et al* (1994). A flux box (as illustrated in Figure 1) was placed over each of the microcosms after they had been excavated and placed on the beach surface nearby. The air within the box was circulated through the cell of an Infra-Red Gas analyser (Servomex, UK). Readings were taken at minute intervals for 5 minutes. The analyser was calibrated at the beginning of the sampling and after every 3 subsequent measurements, using CO_2-free air and a standard gas (236 ppm CO_2 , Linde Gas Ltd, Manchester, UK).

FIGURE 1. Diagram of Equipment Used to Determine *in situ* CO_2 Evolution Rates

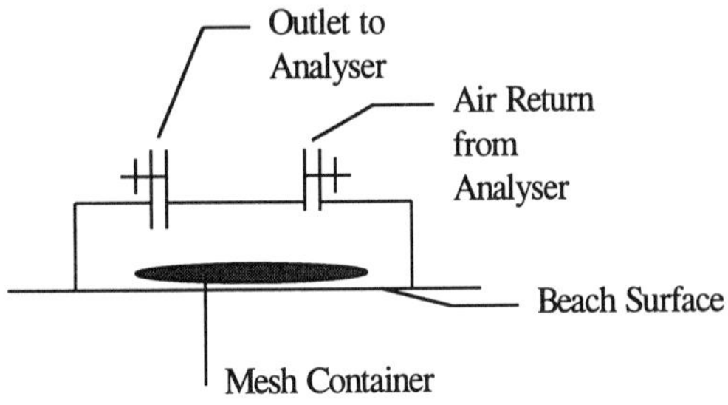

RESULTS AND DISCUSSION

Residual Hydrocarbon Analysis. At the end of the trial the analysis suggested that the oil was more degraded in the fertilised plots than in the oiled, unfertilised controls (Table 2).

TABLE 2. Crude Oil Component/Hopane Ratios at Day 101

Ratio	Plot 1 (O/F)	Plot 2 (O)	Plot 8 (O/F)	Plot 6 (O)	Plot 9 (O/F)	Plot 11 (O)
n-C18/Phy	1.18	3.38	3.46	3.58	1.74	3.52
TGCD (AL)	3422	3876	4252	4516	3060	4815
TGCR (AL)	600	1171	1134	1385	700	1296
TGCR (AR)	20.5	27.1	34.9	28.1	25.5	30.1

O = Plots Treated With Oil Only
O/F = Plots Treated With Oil and Fertiliser
n-C18/Phy = n-c18 / Phytane
TGCD (AL) = Total GC Detectable Aliphatics / Hopane
TGCR (AL) = Total GC Resolvable Aliphatics / Hopane
TGCR (AR) = Total GC Resolvable Aromatics / Hopane

A statistical analysis by a three-way, factorial analysis of variance of this data (for effect of treatment, block and date) was completed. The results for the effect of block and treatment are reported here. This analysis showed that observed differences between the oiled only plots and the plots treated with oil and fertiliser were highly significant for the ratios of resolvable and detectable aliphatic hydrocarbons against hopane. There appeared to be no significant effect of treatment on the degradation of aromatic hydrocarbons. The p-values calculated by this analysis are given in Table 3. A further effect was seen in that there was a significant difference in the rates of degradation in Block 1 & 3 and Block 2, with a much slower rate seen in the four plots positioned in the middle of the experimental area.

TABLE 3. p-Values for Observed Differences Treatments

Oil Component	p - Value for Comparison	
	Treatment	Block
n-C18/Phytane	0.0001	0.0001
TGCD (AL)	0.0001	0.062
TGCR (AL)	0.0001	0.0001
TGCR (AR)	0.715	0.050

The result of the *in situ* respirometry measurements (Figure 2) support the conclusions from the oil chemistry. Whilst this elevation in gas generation seen in the oiled, fertilised plots is substantial, we are currently analysing the results statistically to determine whether this difference is significant. Note that the addition of fertiliser to unoiled beach material also stimulated CO_2 evolution. Chemical and microbiological analysis showed that this was not a result of oil mineralisation but was probably caused by the degradation of other organic compounds.

FIGURE 2. Levels of Carbon Dioxide Evolution from Each Plot

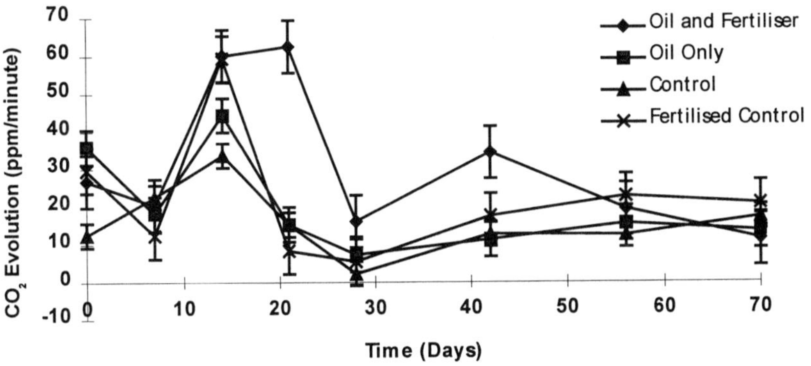

CONCLUSIONS

The residual hydrocarbon analysis clearly showed that the addition of inorganic fertilisers to the oiled aerobic fine sediment significantly enhanced the level of biodegradation in comparison to untreated oiled sediment. This indicates that bioremediation may be considered for the treatment of buried oil in aerobic fine sediments. Oil can become buried on sandy shorelines following tidal and sediment movements, or even as a result of conventional physical methods used to clean oiled fine sediment substrates. Both these phenomena were noted during the response to the recent *Sea Empress* incident (Colcomb *et al* 1997) and hence our findings expand the range of conditions under bioremediation may be considered as a response to an oil spill incident. Moreover bioremediation of buried oil is less intrusive and may be less damaging to the environment than traditional physical techniques. The work also clearly showed that *in situ* measurements of CO_2 evolution can be used as an accurate guide to the success of bioremediation treatments in the field. The results obtained through this work have shown that this data correlates with the other chemical data, and has the additional advantages of being both cost-effective and rapid.

REFERENCES

Bragg, J. R., R.C. Prince, E.J. Harner and R.M. Atlas. 1994. "Effectiveness of Bioremediation for Exxon Valdez Oil Spill." *Nature:* 413-418.

Colcomb, K., D. Bedborough, S. Shimwell, T. Lunel, K. Lee, R. Swannell, P. Wood, N. Bailey, C. Halliwell, L. Davies, M. Sommerville, A. Dobie, D. Mitchell and M .McDonagh. 1997. "Shoreline Clean-Up and Waste Disposal Issues During the Sea Empress Incident." To be presented at the *1997 Oil Spill Conference*, Fort Lauderdale, FL., USA, April 7-10,1997.

Lee, K.,& E.M. Levy. 1992. "Microbial Degradation of Petroleum in an Intertidal Beach Environment - in situ Sediment Enclosure Studies 1987." In *Marine Ecosystem Enclosed Experiments,* Proceedings of a Symposium held in Beijing, People's Republic of China. Published by International Development Centre, Ottawa, Canada, ISBN 0-88936-543-1 pp 140-155.

Lee, K., R.P.J. Swannell, P. Sveum, M. Guillerme, F. Merlin, T. Reilly, J. Oudot, J. Ducreux. 1995. "Protocol for the Assessment of Bioremediation Strategies on Shorelines." *Proc. 1995 Oil Spill Conference* API, Washington, D.C.

MPCU. 1995. "Scientific, Technical and Operational Advice (STOP 2/95)." In *Operational Guidelines for the Application of Bioremediation Agents* Marine Pollution Control Unit, Coastguard Agency, Southampton, UK.

Prince, R.C. 1993. "Petroleum Spill Bioremediation in Marine Environments." *Crit. Rev. Microbiol. 19:*217-242.

Rosenberg, E.R., R. Legmann, A. Kushmaro, R. Taube, E. Adler and E.Z. Ron. 1992. "Petroleum bioremediation - a multiphase problem." *Biodegradtion 3*:337-350.

Swannell, R.P.J., K. Lee, and M. McDonagh. 1996. "Field Evaluations of Oil Spill Bioremediation." *Microbiol. Rev. 60* (2):342-365.

Swannell, R.P.J., B.C. Croft, A.L. Grant and K. Lee. 1995. "Evaluation of Bioremediation Agents in Beach Microcosms." *Spill Science and Technology Bulletin 2*(2/3):151-159.

Swannell, R.P.J., K. Lee, A. Bassères and F.X. Merlin. 1994. "A Direct Respirometric Method for in situ Determination of Bioremediation Efficacy." In *Proceedings of the 17th Arctic and Marine Oil Spill Program Technical Seminar*, Environment Canada, Ottawa, Canada, pp 1273 - 1286

Venosa, A.D., M.T. Suidan, B.A. Wrenn, K.L. Strohmeier, J.R. Haines, B.L. Eberhardt, D. King and E. Holder. 1996. "Bioremediation of an Experimental Spill on the Shoreline of Delaware Bay." *Eviron. Sci. Technol 30:*1764-1775.

ON-SITE BIOREMEDIATION OF OIL SLUDGE/ CRUDE OIL-CONTAMINATED SOIL

Banwari Lal and Sunil Khanna
(Tata Energy Research Institute, New Delhi, India)

ABSTRACT: Petroleum refining unavoidably generates a huge volume of oil sludges constituting a disposal problem. Currently oil sludge is being disposed of in open fields in refinery areas and, with continuous disposal of oil sludge in soil, there is serious threat of groundwater contamination. On-site bioremediation of oil sludge/crude oil-contaminated soil with a developed bacterial consortium at Mathura refinery (North India) has been conducted in 24 plots (plot size 1 X 1 m) with 6 treatments (4 replicates/treatment), covering a wide range of contamination levels (3.7% to 11.3%).

In control plots the average concentration of sludge/crude oil was 0.572 g/10 g soil on day zero (when the experiment was initiated) and it decreased to 0.498 g/10 g soil in 90 days after the experiment was initiated, of a 12.9% loss of oil sludge/crude oil in control plots which may be due to various abiotic factors. However on day zero, the average concentration of oil sludge/crude oil was 0.918 g/10 g soil in a plot inoculated with the bacterial consortium along with nutrients and the concentration decreased to 0.565 g/10 g soil in 90 days which showed a 38.4% loss of sludge/crude oil. Similarly in plots inoculated by pure bacterial isolates of *Acinetobacter calcoaceticus* S30 plus *Alcaligenes odorans* P20 (in combination), the average concentration of oil sludge/crude oil was 0.899 g/10 g soil and it reduced to 0.579 g/10 g soil in 90 days which accounted for a 35.5% loss of oil sludge/crude oil. The oil sludge/crude oil concentration in soil (up to 1 foot) also decreased quite significantly (31.4% loss) in plots inoculated by bacterial consortium alone (without addition of nutrients). However the addition of nutrients in plots inoculated with pure bacterial isolates (S30 and P20) could not enhance the degradation of oil sludge/crude oil. The qualitative analysis of oil sludge/crude oil degradation indicated that among the n-alkanes the major degradation was of docosane, tetracosane, hexacosane, octacosane, triacontene, dotriacontane, and tritriacontane. GC finger printing of the aromatic fraction revealed that more than 78% of the aromatic fraction was degraded in 90 days and the major degradation was of fluorene, dibenzothophene, cyclopentaphenanthrene, pyrene, benzanthracene, and perylene.

IN SITU BIOREMEDIATION OF OIL SLUDGES

C. Infante, M. VialeRigo, M. Salcedo, J. Rodríguez, A. Melchor (CORPOVEN, Caracas, Venezuela); E. Bilbao and R. Arias (INTEVEP, Caracas, Venezuela)

ABSTRACT: Bioremediation is a promising technology for treatment of oil sludges generated during refining process and production by the Venezuelan Oil Industry. In order to evaluate this technology, studies were conducted at laboratory scale (160-L bioreactors) and in situ (a lagoon of 1,000 m^3 of oil sludge). The sludge constituents were sediments resulting from the decanting process of oil sludges generated during tank and water/oil separator cleaning procedures and those from accidental spills of crude oil. Disposal of these sludges in open lagoons has been a common practice for a long time, allowing the action of weathering to take place.

Indigenous microorganisms were stimulated by amendment with water, nutrients (C/N=60, C/P=800), and in situ mixing of the sludge. The extent of biodegradation was estimated by the reduction in hydrocarbons (saturate and aromatic fractions, w/w) over 40 days of treatment. Microbial activity was evaluated measuring the CO_2 production by respirometric methods.

A mean biodegradation value of 52, 40 and 41%, expressed as mass of saturates and aromatic compounds per mass of sludge, was registered after 40 days in the laboratory and in situ, respectively. The microbial activity remained between 0.5-1.8 mg CO_2/gr sludge, which is considered an adequate value for the resulting biodegrading processes. These results suggest that in situ bioremediation can be used to remediate oil sludges from waste lagoons.

BIOREN: RECENT EXPERIMENT ON OIL POLLUTED SHORELINE IN TEMPERATE CLIMATE

Stéphane LE FLOCH, François Xavier MERLIN (CEDRE, France)
Michel GUILLERME (ELF AQUITAINE, France)
Pierre TOZZOLINO (ELF AKVAMILJO, Norway)
Daniel BALLERINI, Christine DALMAZZONE (IFP, France)
Tore LUNDH (NAT, Norway)

ABSTRACT: In order to assess the efficiency of two bioremediation formulations designed in the scope of BIOREN project, it has been decided to conduct a field experiment in a temperate climate, in Brittany (France) at the end of 1996.

The first part of the study was aimed at finding an appropriate site for the experiment: a sheltered sandy beach large enough and without external disturbance factors such as nutrient extra supply from agricultural activities.

On the chosen site the experiment has been conducted on four large plots, oiled on 25 m², reproducing the following conditions: oil and formulation A, oil and formulation B, oil and Inipol EAP 22 taken as a well known reference product, and oil only for a control plot; in addition an uncontaminated plot has been considered to assess the background levels (oil, nutrient, microflora).

Bioremediation has been monitored for four months on a periodic basis by documenting the changes in the quantity and chemical composition of the residual hydrocarbons.

Interstitial water has been periodically taken to monitor the nutrient levels in the plots.

Sediment sampling strategy was done following particular statistic considerations.

This paper describes the methodology of this experiment.

INTRODUCTION

In the scope of Eureka technical projects, BIOREN is an international program dealing with the development of bioremediation products to be used on marine oil pollution.

This program is splitted into four successive steps to define and validate formulations able to enhance the in situ biodegradation of hydrocarbons: after a bibliographical and theoretical approach to define products potentially efficient (I), laboratory studies were carried on to check the efficiency of these products (II), then the convenient formulations were assessed in mesocosms in controlled conditions (III) and, at last, the more promising formulations were validated by field experiments carried out in different climate conditions (IV).

In this respect, a field experiment was conducted on the Southern coast of Britanny, France, between October 1996 and March 1997, to compare the efficiency of two bioremediation formulations.

PREPARATION OF THE EXPERIMENT: SELECTION OF THE EXPERIMENTAL SITE

Previous field trials have shown the need to carefully choose the experiment site in order to avoid any adverse condition which could hide or prevent the oil bioremediation (Merlin, 1995a); this requirement is even more important because field trials are small scale pollution models, very sensitive to the environmental conditions.

- the site should be very sheltered in order to avoid that wave actions naturally clean the sediment; in this respect bays and estuaries are the most amenable locations for bioremediation (Lee and Levy, 1991; Merlin et al., 1992),
- the background level of nutrients must be sufficiently low to see the effect of a nutrient enrichment strategy; locations exposed to agricultural or urban effluents have to be avoided; it is noteworthy that biostimulation can no longer be demonstrated above an ambient level of nitrogen of 1 to 2 mg/L (Merlin et al., 1995b; Venosa et al., 1995),
- the sediment should be fine and homogeneous to simplify the sampling and analytical protocols and should be permeable enough to allow aerobic conditions of biodegradation; fine to medium sands are therefore the most convenient sediments (Lee and Levy, 1987).
- the site location should be large enough to avoid any cross contamination between the different experimental plots; few hundred meters of shoreline offering similar environmental conditions (exposure, sediment, profile...) are requested.
- at last, the human socioeconomic environment should be compatible with the experiment itself; the recreational areas, protected sensitive areas, maricultural areas are not suitable to run an experiment involving in situ oil release.

A preliminary study was conducted on the whole coast of Brittany (around 500 km long), to find sites which could be acceptable according to the requirements listed above.

Proper conditions were only found on two sites; one of them was chosen: the LETTY, on the Southern coast of Brittany. The experimental site is located on the internal side of a sandy spit closing partially a small estuary (see fig 1); it is composed of medium to fine sand (600 ɡm - 1450 ɡm), subjected to tides and is protected from ocean waves; a preliminary survey showed that the interstitial water contains low concentration of nitrogen (NO_3 < 0.5 - 0.8 mg/L).

EXPERIMENTAL DESIGN

According to the experimental protocol defined few years ago by the working group on bioremediation leaded by Cedre (Merlin et al., 1994), experimental plots of sediment were set up on the sandy beach: four different conditions were compared: oil and formulation A (O + A), oil and formulation B

(O + B), oil and Inipol EAP22 (O + Inipol EAP 22, as a well known reference product) and oil alone (O, no treatment). An additional plot was left unoiled in order to monitor the background level of the site; bioremediation has been assessed for four months on a periodic basis by documenting the changes in the quantity and chemical composition of residual hydrocarbons (fractionating into saturates, aromatics, resins and asphaltenes and GC-FID on the saturate and aromatic fraction).

The treatment conditions were randomly assigned to the different plots. All the plots (5 m x 5 m) were set at high water neap level. Each one was covered and enclosed in a fine mesh stretched on a 6 m x 6 m frame in order to retain the oily sediment, and also the tested bioremediation products and to restrict any external disturbance. In addition, drain pipes were buried above and on the side of the plots to prevent the interstitial water running in the plots during the ebb time.

From a scientific viewpoint, the lack of replica (triplication of the treatment conditions) was one of the main criticism of this experimental design: due to logistic reasons and to the available space on the beach, it was not possible to triple the number of plots to take into account the site self-variability. Four additional small control plots (2 m x 3 m) which received only oil were therefore set up along the experimental zone between the larger plots (see fig 2).

The oil was an Arabian Light crude oil topped at 170°C to simulate a weathered oil.

The oil was applied homogeneously on the plots at 2 L/m² using small pierced booms (0.5 m wide); this oil application rate was previously optimized in the laboratory on samples of sand in order to warrant the permeability of the sediment after oiling.

MONITORING AND SAMPLING STRATEGY

Surface sediment (0 cm - 5 cm) were periodically taken over in the central part of the plots (the sampled zone, 3 m x 3 m itself divided into nine 1 m² squares, see. fig 3).

To optimize the analyse works, four series of samples were taken according to statistical considerations; the co-ordinates of each sample were determined by using random numbers.

- series A: one sample taken in the center of the plot to get a basic information on the progress of the experiment (sampling at 0, 3, 6, 8, 10, 15 weeks after the beginning of the experiment),
- series B: four samples taken at each corner of the sampled zone, at four dates, (0, 3, 8, 15 weeks after the beginning of the experiment) on the oiled plots (O, O+A, O+B, O+Inipol EAP 22); sixteen of these samples were selected according to the square latin method and analysed to identify potential spatial variations of the oil content in the plot,
- series C: the largest series, six samples in each plot (2 in the upper part, 2 in the middle and 2 in the lower), at ten dates (0, 1, 2, 3, 4, 5, 6, 8, 10, 15 weeks after the beginnig of the experiment). All these samples were archived; then according to the information gained from series A and B, the most interesting

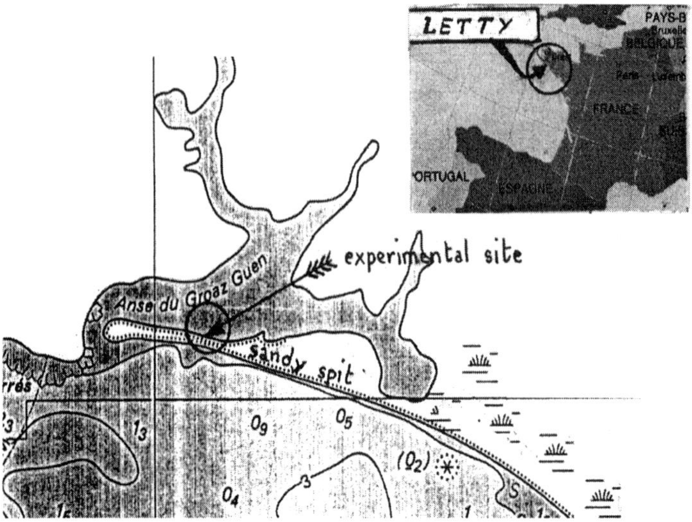

FIGURE 1: Location of the experimental site in Brittany.

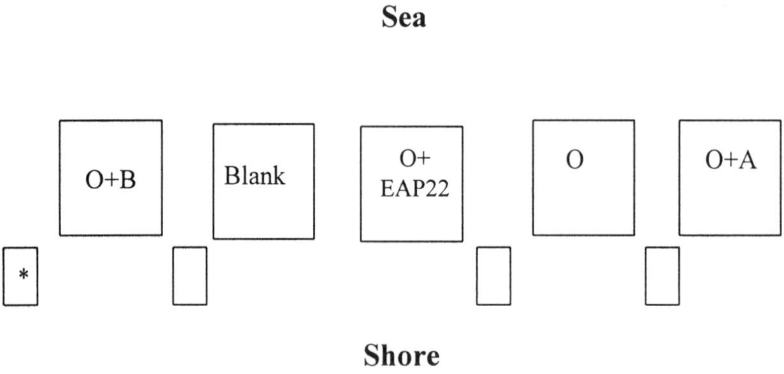

* : 4 small oiled control plots.

FIGURE 2: Arrangement of the plots on the shoreline.

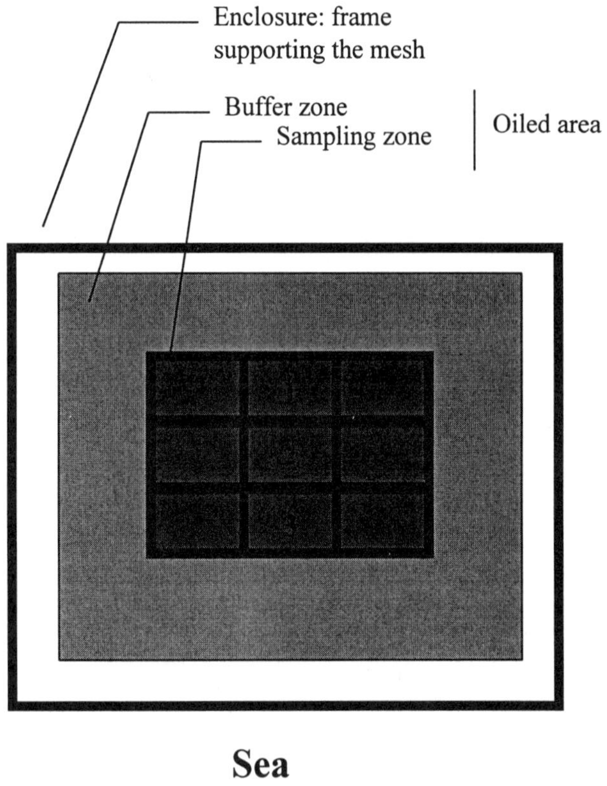

1 : upper sampled zone
2 : medium sampled zone
3 : lower sampled zone

FIGURE 3 : Description of an experimental plot

will be selected and possibly pooled to be analysed in order to get the required information on the bioremediation products efficiency,

- series D: additional sediment sampling was carried out in the four small oiled control plots in order to assess the variability of the site itself.

Additional measurements, sampling and analyses were also peridically performed:

- recording of the environmental conditions,
- analysis of the nutrient content in the interstitial water (nitrogen: NO_3, NO_2, NH_4, Kjeldahl, and phosphorus: PO_4),
- permeability of the surface sediment to ensure that aerobic conditions remain even after oiling of the sediment (this assessment is done by measuring the time needed by a known volume of water to get through the sediment in an open cylinder half buried in the sand),
- microbiological assessments on sediment samples.

PROGRESS OF THE EXPERIMENT (Jan, 97)

This experiment started in November 1996 is still running by the time this paper is written: as the analyses are not yet performed, very few results are available; this does not allow to draw actual conclusions on the efficiency of the products.

The observations of the sediment show that the oil did not penetrate deeper than 4 cm to 7 cm in the sediment.

The first extractions performed on the sediment samples (series A at the beginning of the experiment) show that all the plots sediment were polluted up to 14 000 ppm to 21 000 ppm (total organic matter) while the ambient level (blanck plot, unoiled) was below 100 ppm of organic matter.

After the 3 first weeks the amount of organic matter remained in the order of 10 000 ppm to 20 000 ppm, which shows that the oil leaching by the waves and currents was relatively low.

At last, the permeability measurements showed that permeability decreased gradually in the oiled plots (oil slowly plugs the sand interstices); the surface of all the plots was therefore raked in order to restore proper aerobic conditions.

By the time this poster will be presented at the Bioremediation Symposium, additional data will be available.

REFERENCES

Lee, K., and E. M. Levy. 1991. « Bioremediation : Waxy Crude Oils Stranded on Low-Energy Shorelines. » *Proceedings, 1991 International Oil Spill Conference*, San Diego, CA, 4-7 March 1991, pp. 541-547.

Lee, K., and E. M. Levy. 1987. « Enhanced Biodegradation of a Light Crude Oil in Sandy Beaches » *Proceedings of the 1987 Oil Spill Conference,* Baltimore, MD, 6-9 April 1987, pp. 411-416.

Merlin, F. X. 1995a. « Devising an Experimental Protocol to Evaluate the Effectiveness of Bioremediation Procedures. » *Paper presented at the Second International Oil Spill Research and Development Forum,* London, United Kingdom, 23-26 May 1995, pp. 37-44.

Merlin, F. X., P. Pinvidic, C. Chaumery, J. Oudot, R.P.J. Swannell, A. Basseres, C. Dalmazzone, J. Ducreux, K. Lee, and T. Reilly. 1995b. « Bioremediation : Results of the Field Trials of Landevennec (FRANCE). » *Proceedings of the 1995 International Oil Spill Conference,* American Petroleum Institute, Long Beach, California, February 27- March 2, pp. 917-918.

Merlin, F. X., K. Lee, R. P. J. Swannell, J. Oudot, A. Basseres, T. Reilly, C. Dalmazzone, C. Chaumery, and P. Sveum. 1994. « Protocol for Experimental Assessment of Bioremediation Agents on a Petroleum Polluted Shoreline. » *Proceedings of the 17th Marine Oil Spill Program (AMOP) Technical Seminar,* Vancouver, British Columbia, June 8-10, pp. 465-478.

Merlin, F. X., C. Chaumery, and J. Oudot. 1992. « Elaboration of an Experimental Method to Assess Biodegradation Agents : Bioremediation Trials on Oil Polluted Beach. » *Proceedings of the 15th Marine Oil Spill Program (AMOP) Technical Seminar,* Edmonton, Alberta, 10-12 June, pp. 723-730.

Venosa, A. D., J. R. Haines, B. A. Wrenn, K. L. Strohmeier, B. L. Eberhart, M. T. Suidan, and B. Anderson. 1995. « Bioremediation of Crude Oil Released on a Sandy Beach in Delaware. » *Paper presented at the Second International Oil Spill Research and Development Forum,* London, United Kingdom, 23-26 May 1995, pp. 68-77.

ASSESSMENT OF MIXED CULTURES FOR BIOREMEDIATION PRODUCT TESTING

J.R. Haines, United States Environmental Protection Agency, Cincinnati, Ohio; E.L. Holder, University of Cincinnati, Ohio; and A.D. Venosa, United States Environmental Protection Agency, Cincinnati, Ohio

ABSTRACT: Samples of beach sediments from many locations in the United States were used to enrich for hydrocarbon degrading microorganisms. Sediments were incubated with mineral nutrients and an aromatic enriched fraction of crude oil to enrich for aromatic hydrocarbon degrading bacteria. A shake flask experiment was designed to assess the ability of the various cultures to degrade crude oil as described in the NETAC protocol (NETAC, 1991). Each flask contained 100 mL of artificial seawater, nutrients, and weathered Alaska North Slope crude oil. Eight enrichment cultures were used to inoculate four replicate flasks for each sample time. The sample times were 0, 1, 2, 4, 8, 12, 20, and 28 days. After incubation at 20°C, alkane and aromatic hydrocarbon degrading bacteria were measured by MPN and residual oil was extracted with methylene chloride, solvent exchanged with hexane, and analyzed by gas chromatography-mass spectrometry. A total of 28 alkanes and 40 aromatic hydrocarbons were measured. Alkanes were extensively degraded by all cultures with approximately 80% reduction occurring within 8 days of incubation. The cultures were variable in their ability to degrade aromatic hydrocarbons. The greatest reduction in aromatic content (>90%) was achieved by two cultures by day 20. Substituted aromatics were poorly degraded by some cultures. Based on the results of these experiments a positive control culture will be chosen for bioremediation product testing.

INTRODUCTION

The production and marketing of bioremediation products has become an active business due to the need for environmentally benign materials for oil spill cleanup. Responsible officials desire a cost effective, useful product. A manufacturer needs the ability to ensure that their product is effective both from developmental aspects and from marketing aspects. The public wants to know that products used to clean up spills are effective and innocuous in terms of further ecological damage. In all cases, a bioremediation product is inherently attractive because it takes advantage of natural processes to clean up a petroleum spill. The ability of microorganisms to degrade petroleum is well known and used in many venues to degrade unwanted petroleum (Atlas, 1984; Leahy and Colwell, 1990). Use of bioremediation products in a spill situation is designed to enhance the natural rate of oil degradation either by adding limiting nutrients or adding microorganisms with broader metabolic capability. In the event of a spill, rapid deployment of remedial action is desirable. In the case of bioremediation products, The National Environmental Technology Application Center (NETAC) has developed a series of protocols for evaluation of the effectiveness of products. The protocols are designed in tiers to provide decision points for further testing. All tiers include an effectiveness component and a toxicity component. Tier 2 is a flask test at laboratory scale. Tier 3 is a bench-scale continuous flow microcosm test. In Tier 2 testing, a series of controls are used to evaluate the effectiveness of bioremediation products. A negative control is used to show changes mediated by the applied product compared to no applied product. Blanks are used to show if any effects are attributable to the materials used in setting up the flask test. In biological testing, a positive control is useful to establish that the test method is working as it should. For the purposes of the Tier 2 protocol, a positive control will take the form of a culture or mixed culture that shows significant degradation of oil in the test procedure. Manufacturers of bioremediation products will be able to use the positive control culture to ensure that the test is working properly and that their products are performing as they intended. This paper reports the results of our development of a positive control culture for Tier 2 protocol testing.

MATERIALS AND METHODS

Development of mixed cultures for use in the Tier 2 protocol began by enriching the oil degrading populations of microorganisms from a variety of sediments. Sediment samples were collected from marine shorelines around the United States. Samples were collected from Alaska, Delaware, Maine, and Texas. Each enrichment consisted of 10 g of sediment added to 100 mL of GP2, an artificial seawater solution (Spotte et al, 1984), together with 0.5 g of aromatic hydrocarbon enriched crude oil. The crude oil was enriched using silica gel chromatography resulting in an aromatic fraction that was significantly depleted of alkanes. Cultures were transferred after 28 d of incubation at 20° C on a shaker at 200 rpm. After several transfers and growth cycles, the cultures were transferred into GP2 with Alaska North Slope 521 (weathered under vacuum at 521°F) oil as the carbon source. After incubation for 28 d as above, the entire contents of the duplicate flasks were sacrificed and extracted with methylene chloride for hydrocarbon analysis. The extracts were solvent exchanged into hexane and analyzed by gas chromatography-mass spectrometry (GC/MS) (Venosa et al, 1996). A total of 28 alkanes and 40 aromatic hydrocarbons were determined. Data were normalized to the nonbiodegradable biomarker hopane (Peters and Moldowan, 1993). Cultures demonstrating the most extensive degradation of aromatic hydrocarbons were then used for further development.

Eight of the best cultures from the screening trial were used in an experiment to determine the pattern of degradation over the incubation period. Four replicate flasks of each culture were set up for each sampling time (0, 1, 2, 4, 8, 12, 20, and 28 days). The incubation conditions and oil chemistry analysis was performed as before. Alkane and aromatic hydrocarbon degrader MPNs were determined as described in (Haines et al, 1996; Wrenn and Venosa, 1996).

The three best cultures were chosen based on the ability of the cultures to degrade a wide range of parent PAHs and their substituted homologs. A followup experiment was performed to determine if the dramatic population changes observed in the prior experiment were repeatable. The cultures were inoculated into triplicate flasks using the same conditions as before. Daily sampling was done for 20 d, then on days 22, 24, 26, and 28 using different flasks at each sampling point to avoid population changes induced by changing the surface to volume ratio.

RESULTS AND DISCUSSION

Of the eight cultures chosen from the initial screening test, three were from Maine sediments, three were from Alaska, one from Texas, and one was from Delaware. All eight of the cultures reduced aromatic hydrocarbons in ANS521 oil by more than 60%. Figure 1 shows a bar chart comparing the sum of aromatic hydrocarbons (normalized to hopane) remaining after 28 days incubation.

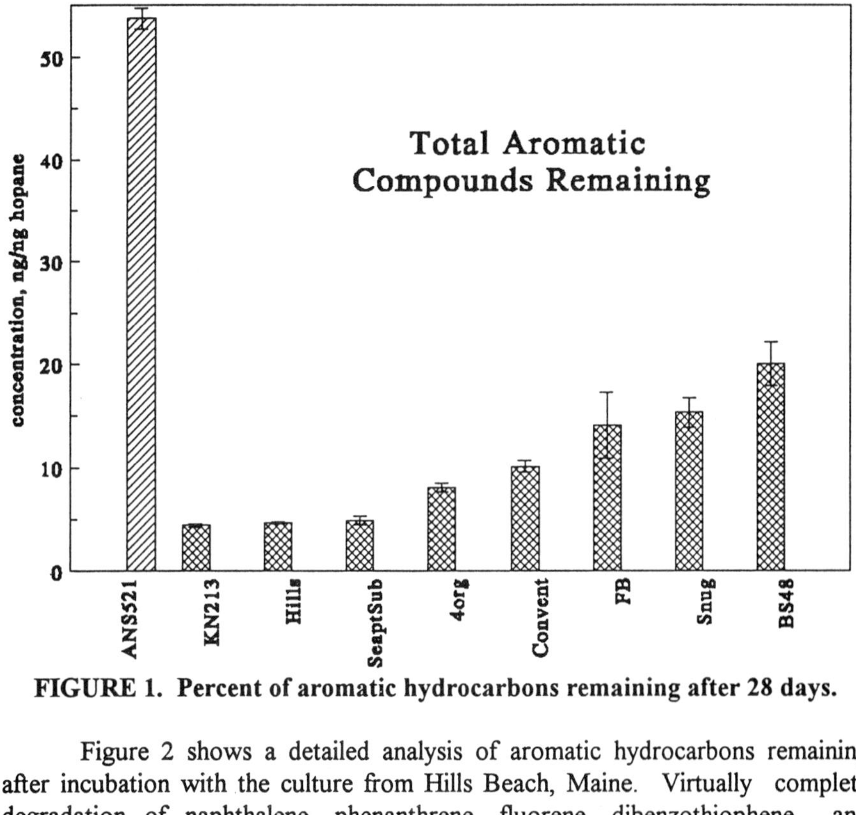

FIGURE 1. Percent of aromatic hydrocarbons remaining after 28 days.

Figure 2 shows a detailed analysis of aromatic hydrocarbons remaining after incubation with the culture from Hills Beach, Maine. Virtually complete degradation of naphthalene, phenanthrene, fluorene, dibenzothiophene and substituted homologs is evident. Higher molecular weight aromatics were also being degraded. Naphthobenzothiophenes, pyrenes, and chrysenes show significant degradation compared to the undegraded oil. This pattern of essentially complete removal of lower molecular weight and less substituted compounds was typical for all cultures.

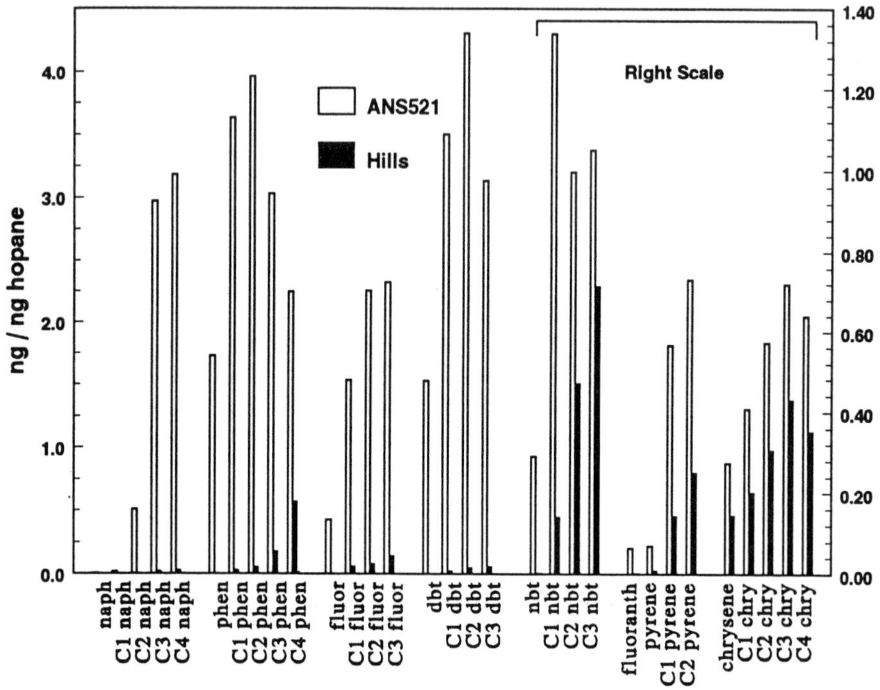

FIGURE 2. Degradation of aromatic hydrocarbons by Hills Beach culture compared to source oil.

The microbiological results showed the populations changing dramatically over the course of the experiment. Both alkane and aromatic degraders exhibited large changes in numbers over relatively short times. The results of the MPN determinations for the follow-up experiment are shown in Figure 3. All points represent the mean of three determinations, and the error bars represent ± 1.0 standard deviation. In the first ten days of incubation significant and frequent population shifts occur among the C16 degraders. The population is relatively stable from day 10 to day 19 and then shows wide swings in numbers again, repeating the pattern observed in the previous experiment. The most likely explanation for this observation is that different groups of microorganisms grow at the expense of different substrates at different times. The source oil contains a wide variety of alkanes other than the straight chain compounds. The changes observed in alkane degraders probably reflects the shifting of the population to those organisms most able to use the different compounds as they become the available substrate.

In contrast, the MPNs of aromatic degraders show relatively smooth increases until day twenty. The most stable population growth is shown by culture 4ORG and the most variable one is KN213 both of which are originally derived from Alaskan sediments. Both KN213 and 4ORG are effective with populations around 1.0 x 10^5 organisms per mL. The Hills Beach culture developed a population between 1.0 x 10^6 and 1.0 x 10^7 per mL. After 20 days of incubation all of the cultures showed large changes in population on different sampling days. The most likely explanation for this observation is that the organisms have exhausted readily available substrates and were beginning to utilize the more resistant materials. The changes we observed reflected the shifting of the population to those organisms most able to use the available substrates.

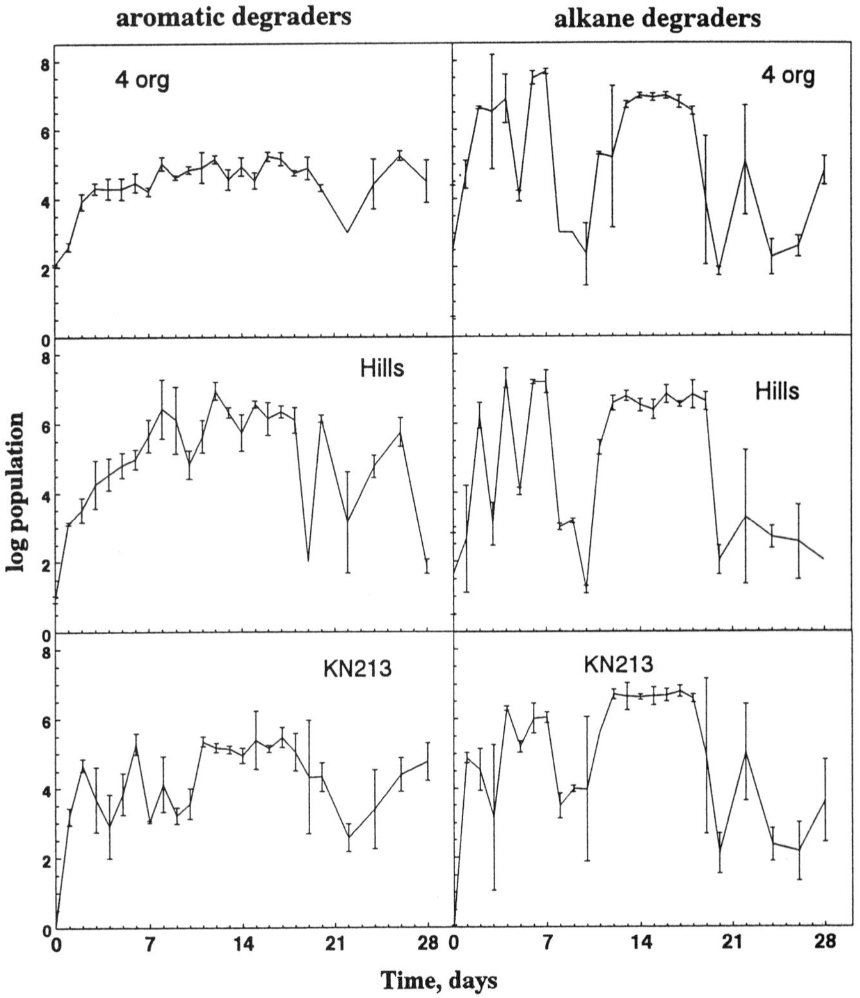

FIGURE 3. Hydrocarbon degrading bacteria (MPN).

A conclusion that can be drawn from the results of these experiments is that the relatively simple environment of a shake flask is more complex than commonly believed when examining mixed cultures with complex substrates such as crude oil. Ongoing work in our laboratory seeks to clarify population changes in mixed cultures over an incubation cycle to understand this complexity and to determine optimum sample time and harvest time to preserve the largest metabolic diversity. The mixed cultures chosen for these experiments have proven to be effective in degradation of aromatic and alkane hydrocarbons removing about 90% of the target aromatics in 28 days of incubation including the substituted homologs. The organisms have not lost this ability on multiple transfers as commonly happens. Although large shifts in populations occurred at different times, the agreement among the four replicates was high (small error bars) signifying good reproducibility. Based on the results of this work, any one of the three cultures, 4ORG, Hills, or KN213 would be a suitable positive control culture for inclusion in future revisions of the NETAC protocol.

REFERENCES

Atlas, R.M. 1984. Microbial Degradation of Petroleum Hydrocarbons: An Environmental Perspective. Microbiol. Rev. 45:180-209.

Haines, J.R., B.A. Wrenn, E.L. Holder, K.L. Strohmeier, R.T. Herrington, and A.D. Venosa. 1996. Measurement of Hydrocarbon-Degrading Microbial Populations by a 96-Well Plate Most-Probable-Number Procedure. J. Ind. Microbiol. 16:36-41.

Leahy, J.G. and R.R. Colwell. 1990. Microbial Degradation of Hydrocarbons in the Environment. Microbiol. Rev. 54:305-315.

National Environmental Technology Corporation. 1991. Oil Spill Bioremediation Products Testing Protocol Methods Manual. Univ. Pittsburgh Appl. Res. Center. Pittsburgh, PA.

Peters, K.E. and J.M. Moldowan. 1993. The Biomarker Guide. Prentice Hall Englewood Cliffs NJ 363 pp.

Spotte, S., G. Adams, and P.M. Bubucis. 1984. GP2 Medium is an Artificial Seawater for Culture or Maintenance of Marine Organisms. Zoo Biol. 3:229-240.

Venosa, A.D., M.T. Suidan, B.A. Wrenn, K.L. Strohmeier, J.R. Haines, B.L. Eberhart, D. King, and E. Holder. 1996. Bioremediation of an Experimental Oil Spill on the Shoreline of Delaware Bay. Environ. Sci. Technol. 30:1764-1775.

Wrenn, B.A. and A.D. Venosa. 1996. Selective Enumeration of aromatic and Aliphatic Hydrocarbon degrading bacteria by a most-Probably-Number Procedure. Can. J. Microbiol. 42:252-258.

SORPTION AND BIODEGRADATION INTERACTIONS DURING PHENANTHRENE REMOVAL IN MARINE SEDIMENT

H. David Stensel, Tom Poeton, and Stuart Strand
(University of Washington Seattle, WA)

ABSTRACT: Polyaromatic hydrocarbon (PAH) contamination of marine sediments is common due to the generation and transport of pollutants related to industrial activity and use of petroleum products. This paper reports on research investigating desorption and biodegradation of phenanthrene by aerobic bacteria in a marine sediment. A model was developed that described desorption of phenanthrene, solids-liquid equilibrium partition coefficients for phenanthrene and a marine sediment, and biodegradation kinetics and growth of a marine PAH-degrading enrichment. Individual components of the model were quantified using a synthetic sediment with a fixed carbon content and Blakely Harbor sediment from Puget Sound. Desorption kinetics, partitioning coefficients, and biokinetic coefficients for phenanthrene by the PAH-degrading enrichment were determined. Biodegradation kinetics were determined for liquid phenanthrene concentrations below solubility concentrations using radiolabelled phenanthrene. The coefficients for the fundamental reactions were then used in the mechanistic model to predict the phenanthrene removal in experiments with phenanthrene-contaminated sediment and the PAH-degrading bacteria.

The experimental results showed that phenanthrene removal and degradation was much faster than predicted by the model. Furthermore phenanthrene degradation in the presence of sediment was much faster than phenanthrene degradation without sediment at the same aqueous phenanthrene concentrations. Liquid surface tension was evaluated from samples taken during phenanthrene degradation and epifluorescence microscopy was used to assess bacteria attachment to the phenanthrene-contaminated media. The results suggest that significant bacterial attachment occurs when phenanthrene is available on the sediment surface. The bacterial attachment may explain the improved phenanthrene degradation kinetics. The results of this research help to improve our understanding of mechanisms that effect bioremediation of marine sediments.

AN ASSESSMENT OF ORGANIC CONTAMINANT BIODEGRADATION RATES IN MARINE ENVIRONMENTS

Barbara Krieger-Brockett (U.Washington Chem. Eng., Seattle, WA)
Jody W. Deming (U.Washington Oceanography, Seattle, WA)
Russell P. Herwig (U.Washington Fisheries, Seattle, WA)

ABSTRACT: Determining *in situ* rates of biodegradation of organic contaminants in submerged marine sediments remains a significant challenge to the assessment of site remediation strategies. This paper summarizes the results of experimental measurements of contaminant mineralization rates by natural microbial populations in the upper 10 cm of sediments at several sites in Puget Sound, WA. These waterways are currently in being treated or in the design phase as Superfund Sites. Recalcitrant (slowly degraded) organic contaminants under investigation are bis(2-ethylhexyl) phthalate (BEP) and benzo(a)pyrene (BaP) and several other polyaromatic hydrocarbons (PAHs). To the authors' knowledge, no site-specific rate data are available for the biodegradation of these compounds in Puget Sound sediments. What little information is available from other marine environments suggests that the rates are extremely slow. We have therefore used a sensitive radioactive isotope method to be able to monitor the evolution of CO_2 from the mineralization of the radiolabeled compounds in a reasonable period of time, with killed controls to distinguish abiotic losses from the microbial action of interest. Labeled compounds were added in tracer amounts to replicate samples of agitated sediment-seawater slurries and also injected into undisturbed sediment subcores. Owing to dilution and agitation, oxygen and other nutrients were more available to the bacteria in slurries than under the more nearly *in situ* conditions simulated in the injected subcores. Thus, the slurry method provided maximum estimates of biodegradation rates, while the injection core method provided minimum rates. An ability to set bounds on the biodegradation rates should improve the design process for pollution cleanup or containment strategies. The relationship between these site-specific *in situ* biodegradation rates and capping of contaminated marine sediments is discussed.

INTRODUCTION AND BACKGROUND

The remediation strategies for contaminated submerged sediments (Superfund sites) in Puget Sound and other coastal areas have become important research subjects and engineering activities over the past five or more years. The danger presented to large fisheries and the food chain, the impact on relatively pristine benthic ecosystems, and the potential effects on recreational activities in Puget Sound have stimulated a number of innovative remediation strategies some of which have already been implemented. One of these, sediment capping, provides immediate containment of the pollutant. However, the continued protection of the water column depends on estimated long breakthrough times for particular organic contaminants to move (partition and diffuse) through the cap. Indefinite containment requires that the cap remain intact despite tidal action, ship movement, and mesoscale sediment disturbances by benthic invertebrates.

The long term consequences and fundamental transformation processes that occur in contaminated marine sediment have been examined by an interdisciplinary research effort at the U. of Washington - the Marine Bioremediation Program (MBP). Our goal is to make fundamental advances in understanding how microscopic organisms native to marine sediments act to degrade organic compounds, and how the physical and chemical transport and rate processes compete with and affect the contaminant transformation. We have taken a "site-directed" research approach where the physical, chemical, and biological data and samples from local contaminated site investigations have focused our research enquiry and provided the opportunity for quantitation and specificity.

In-situ biodegradation and contaminant transport rates in two submerged marine Superfund sites in Puget Sound are the main subjects of this paper. One site, Eagle Harbor, has been contaminated by wood preservation chemical treatment and ship building activities for over 100 years. The other site, Thea Foss, has been contaminated by effluent from large sewer outfalls that seasonally have very large runoff flows from nearby chemical manufacturing industrial areas and port activities. In these two sites the contaminants of interest are the general class of PAHs including creosote (the former site) and plasticizers such as low molecular weight phthalate concentrations (the latter site) that represent a greater danger to marine invertebrates and fish than to humans. Typical sediment concentrations of PAHs in Eagle Harbor greatly exceed sediment quality standards and solubility (depending on composition). Although metals are present, we have not addressed their transformation. The Eagle Harbor site was capped in September 1993 with clean dredged material from a nearby waterway. Currently the Thea Foss waterway is in the remediation design phase (Thornburg, et al, 1996, currently under review).

MATERIALS AND METHODS

The practice of capping contaminated sites in Puget Sound has been advanced by the need to dispose of dredge materials that periodically must be removed from the passages within our ports and transportation waterways. The Puget Sound Dredge Disposal Authority has been involved in developing standards, disposal methods and characterization data for these sediments. The materials and methods used to characterize sediments are described elsewhere. If the dredged materials are relatively "clean", costs are minimized if the dredge spoils can be used as capping material. One such application of a cap was in Eagle Harbor as described by the US EPA (1995).

Meter-thick caps of "clean" dredged sediments have been proposed and demonstrated to be successful barriers to contaminant transport from polluted aquatic sediments into overlying waters (e.g., Truitt, 1986). Thick caps over subaqueous sediments are believed to greatly retard or eliminate natural microbial processes at work aerobically on the degradation of the contaminants; thus, undesirable compounds are not only contained by this treatment approach but effectively preserved until such time as break-through into the water column occurs due to eventual cap disturbance and erosion. Monitoring activities have examined the cap integrity and thickness over time and a typical distribution of cap thickness in Eagle Harbor is given in Fig. 1. It can be seen that much of the site at the time of the measurement contains less than 1 meter of cap material.

A simple modeling exercise leads us to hypothesize that a thin cap over contaminated sediments provides an interfacial region that functions as an *in situ* chemical reactor for PAH and other species degradation over relatively short time scales. Although the methods and details related to this exercise cannot be presented here, a brief overview is given. Sediment characterization data exist for both the contaminated sites and cap material (US EPA, 1995; Thoma, et al, 1993; Thibodeaux and Bosworth, 1990; Krieger-Brockett and Wirth, 1995). Pertinent sediment characterization data consist of organic content, particle composition and size distribution, and porosity or porewater fraction. Together with estimated or measured molecular diffusivities of the contaminants in water, one can approximate the apparent diffusivities or retardation coefficients for specific contaminants as they are transported through the uncapped contaminated sediment as well as through the cap (Krieger-Brockett and Wirth, 1995). The apparent pollutant species' diffusivity or retardation can be approximated by Eq. 1

$$D_{\tau,PAH} = \frac{D_{H_2O,PAH}\ \varepsilon^{4/3}}{(\varepsilon + K_p\rho_B)} \qquad \text{Eq. 1.}$$

In Eq. 1, $D_{H2O,PAH}$ describes the PAH's diffusivity in water, $D_{\tau,PAH}$ describes the PAH's partitioning behavior through the organic-laden sediment matrix, ε describes the porewater fraction for the particular sediment size distribution, K_p describes the water-organic layer partitioning of the contaminant, and ρ_B describes the bulk density of the sediment (for example, as in Wu and Gschwend, 1986; Wang, et al, 1991). The apparent time to breakthough can then be approximated for a particular sediment or cap material by the time estimate in Eq. 2

$$t_{breakthru} \approx \frac{L^2}{4D_\tau} \qquad \text{Eq. 2.}$$

In Eq. 2, L is the diffusion or retardation path length, for example cap thicknesses in Fig. 1. Bioturbation, the mixing done by organisms as they colonize or move through the sediment, has been studied by Aller (1980) and coworkers. They estimate that bioturbation in the top 10 cm of marine sediment can increase the apparent diffusivity by 2 to 10 fold.

Very simplified and approximate breakthrough times were calculated for several sediments examined in previous work (Formica, et al, 1988) and sediments characteristic of the toxic sediments and capping materials used in Puget Sound (USEPA, 1995; Krieger-Brockett and Wirth, 1995). The breakthrough times for cap thicknesses of 1 meter are displayed in Fig. 2. Also given in Fig. 2 is an approximate breakthrough time for sediments in which bioturbation is an important process. This bioturbation process both accelerates the apparent diffusion of species through the sediment and enhances the mixing between the contaminated sediment and overlying cap material and thus dilutes the contaminant at the interface.

RESULTS AND DISCUSSION

Examination of Fig. 2 suggests for less than 1 meter thick caps, the time to breakthrough of toxic organic compounds such as PAH and phthalates can be considerably less than 10 years especially if bioturbation occurs to any great extent. We can see also that the breakthrough time is dramatically changed by the sediment character, i.e., composition (clay or sand), size distribution which affects porosity in a global way, organic content, and the particular pollutant and its partition coefficient. Bioturbation decreases the breakthrough time by up to an order of magnitude. If the sediment and cap are mixed by bioturbation, then an additional question is whether in this mixed zone, biodegradation can occur at a rate fast enough to transform the pollutant before it leaves the cap.

We have used a sensitive radioactive isotope method to monitor the evolution of CO_2 from the microbial biodegradation and mineralization of the radiolabeled recalcitrant pollutants (such as BaP) in a reasonable period of time (Deming, et al, 1997). Killed controls allowed us to distinguish abiotic losses from the microbial action of interest. Labeled compounds were added in tracer amounts to replicate samples of agitated sediment-seawater slurries and also injected into undisturbed sediment subcores. Owing to dilution and agitation, oxygen and other nutrients were more available to the bacteria in slurries than under the more nearly *in situ* conditions simulated in the injected subcores. Thus, the slurry method provided maximum estimates of biodegradation rates, while the injection core method provided minimum rates (Thornburg, et al, 1996, currently under review). The apparent biodegradation rates for the compounds of interest in the two Puget Sound sites depended on sediment physical and chemical character, pollutant concentration, bacteria concentration, specific location in the contaminated site (horizontal heterogeneity), depth in the undisturbed injection cores (vertical heterogeneity), and incubation temperature. However, measured simulated*in situ* rates of marine microbial biodegradation of the pollutants studied exhibit half lives of a few months for BEP and phenanthrene to tens of years for BaP. Although under current verification, substantial biodegradation rates prevail even under anaerobic conditions at depth in the undisturbed sediment (Deming et al, 1997).

Thus it appears probable that *in situ* biodegradation of a pollutant can occur at a rate that is competitive with pollutant diffusion and breakthrough an emplaced sediment cap. Of course the prospect is dependent on the particular microbial consortia, sediment character, and pollutant concentrations that prevail at a particular site. This gives credence to our hypothesis that for some compounds, the cap can act as a chemical reactor in which biodegradation continues to occur even though the cap's principal function is to isolate the contaminated sediment from the water column. Since we were prevented from sampling the emplaced cap in Eagle Harbor, we have no direct evidence of continued biodegradation in the cap.

Because of the roles sediment organic content and particle size distribution play in determining breakthrough (Eq. 1-2), our simple calculations suggest that it may be possible to design a sediment cap in such a way that it does not completely prevent colonization by invertebrates, nor extinguish bacterial activity. One barrier to the rapid biodegradation remains the aging of the pollutant sediment (Hatzinger and Alexander, 1995), and corresponding reduction in bioavailability (Landrum and Robbins, 1990). This topic is currently under investigation.

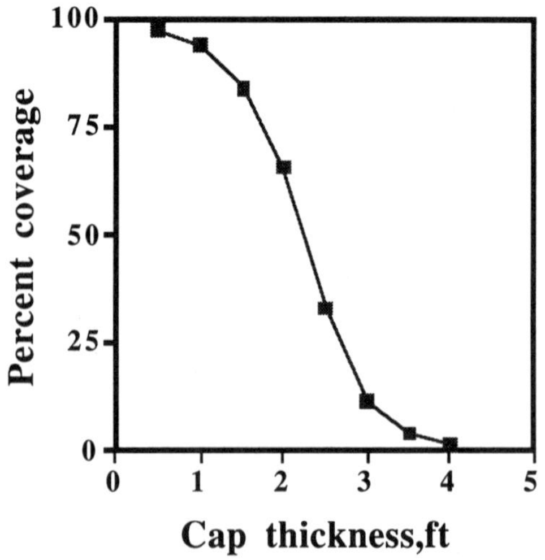

Figure 1. Measured approximate cap thickness in Eagle Harbor (US EPA 1995)

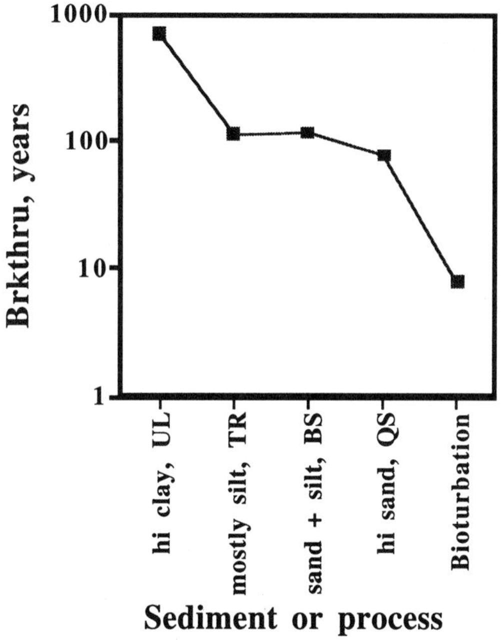

Figure 2. Approximate breakthrough times for typical sediments and PAH type pollutants compared to the time for bioturbation.

ACKNOWLEDGMENTS

The authors would like to acknowledge the financial support of the Office of Naval Research - University Research Initiatives, and Hart Crowser, Inc.

REFERENCES

Aller, R.C., "Quantifying Solute Distributions in the Bioturbated Zone of Marine Sediments by Defining an Average Microenvironment", *Geochimica et Cosmochimica Acta, 44*, 1980, 1955-1965.

US Environmental Protection Agency and US Army Core of Engineers, *Operations, Maintenance and Monitoring Plan for the Wyckoff-Eagle Harbor Superfund Site*, Draft proposal, April, 1995.

Deming, J.W., B. Krieger-Brockett, S. Carpenter, manuscript in preparation, 1997.

Formica, S.J., Baron, J.A., L.J. Thibodeaux, and K.T. Valsara, PCB Transport into Lake Sediments, Conceptual Model and Laboratory Simulation, *Env. Sci. Technol.* 1988, *22* 1435-1440.

Goltz, M.N., and Oxley, M.E., Analytical Modeling of Aquifer Decontamination by Pumping When Transport is Affected by Rate-Limited Sorption, *Water Resources Research, 27*, 4, 1991, 547-556.

Goltz, M.N., and Roberts, P.V., Interpreting Organic Solute Transport Data From A Field Experiment Using Physical Nonequilibrium Models, *Journal of Contaminant Hydrology, 1*, 1986, 77-93.

Hatzinger, P.B., and M. Alexander, Effect of Aging of Chemicals in Soil on their Biodegradability and Extractability, *Env. Sci. Technol* 1995 29 537-545.

Krieger-Brockett, B and T. Wirth, unpublished sediment characterization studies, 1995.

Landrum, P.F., and Robbins, J.A., Chapter 8 - Bioavailability of Sediment-Associated Contaminants to Benthic Invertebrates, in: *Sediments: Chemistry and Toxicity of In-Place Pollutants*s: (Baudo, R., Giesy, J.P., and Muntau, H., Eds.), Leswis Publishers, Inc., 1990, 237-263.

Murphy, E.M., et al, Interaction of Hydrophobic Organic Compounds with Mineral-Bound Humic Substances, *Env. Sci. Technol* 1994 28 1291-1299.

O'Connor, J.M., O'Connor, S.G., Evaluation of the 1980 Capping Operations at the Experimental Mud Dump Site, New York Bight Apex: U.S. Army Waterways Experiment Station,, Vicksburg, MS, 1983

Thibodeaux, L.J. and W.S.Bosworth, 1990, a Theoretical Evaluation of the Effectiveness of Capping PCB Contaminated New Bedford Harbor Bed Sediment, Final Report; Hazardous Waste Research Center, Louisiana State University, Baton Rough, LA.

Thoma, G.J., D.D. Reible, K.T. Vaisaraj, and L.J. Thibodeaux, Efficiency of Capping Contaminated Sediments in Situ. 2. Mathematics of Diffusion-Adsorption in the Capping Layer, *Env. Sci. Technol.* 1993, *27* 2412-2419.

Thornburg, T., B. Krieger-Brockett, J. Deming, R. Herwig, "Rapid Determination of Site Specific Biodegradation Rates in Marine Sediments", under review, 1996.

Truitt, C.L., The Duwamish Waterway Capping Demonstrtion Project: Engineering Analysis and Results of Physical Monitoring; Technical Report D-86-2; U.S. Army Engineer Waterways Experiment Station: Vicksburg, MS, 1986.

W.P. Johnson and G.L. Amy, Facilitated Transport and Enhanced Desorption of PAH by Natural Organic Matter in Aquifer Sediments , *Env. Sci. Tech* 1995 29 807-817. Considers effect of colloids

Wang, X.Q., L.J. Thibodeaux, K.T. Valsaraj, and D.D. Reible, Efficiency of Capping Contaminated Sediments in Situ. 1. Laboratory Scale Experiments on Diffusion-Adsorption in the Capping Layer, *Env. Sci. Technol.* 1991, *25* 1578-1584.

Wu, S. and P.M. Gschwend, Sorption Kinetics of Hydrophobic Organic Compounds to Natural Sediments and Soils, *Env. Sci. Technol*, 1986, 20, 717-725.

IN SITU BIOREMEDIATION UNDER SALINE CONDITIONS

Jeff Raumin (Kleinfelder, Inc., San Diego, California)
Bruce Bosshard (EnviroPacifica, Inc., San Diego, California)
Mike Radecki (NAVFACENGCOM SWDIV, San Diego, California)

ABSTRACT: A Treatability Study was conducted to assess the effectiveness of enhanced *in situ* bioremediation using indigenous heterotrophic bacteria to remediate petroleum hydrocarbons under the saline conditions at two former underground storage tank (UST) sites at the Salton Sea Test Base (SSTB) in Imperial County, California. The low levels of macro nutrients in the subsurface required the addition of metabolic nitrogen and ortho-phosphate in addition to oxygen through air sparging. Unique monitoring of system operation was conducted to track remediation progress, including:

- Dissolved oxygen uptake rates
- Carbon dioxide production rates
- Heterotrophic plate counts

After six months of system operations the hydrocarbon plume was estimated to have been reduced to approximately 20 percent of its original size. Most of the reduction occurred during the last three months of operations following the addition of macro nutrients.

INTRODUCTION

Activities associated with the treatability study were performed under the Comprehensive Long-Term Environmental Action Navy (CLEAN) II Program, by Bechtel National, Inc. (BNI), on behalf of the Southwest Division Naval Facilities Engineering Command (SWDIV).

Two former UST sites, SS2C and SS2E were addressed with *in situ* bioremediation. For purposes of brevity, discussions of Site SS2C only will be presented in this paper. Over the past 40 years the groundwater table has risen approximately 15 feet (ft) (4.6 meters(m)) in the study area in response to a rise of the Salton Sea. During the treatability study, depth to groundwater was approximately 1.5 to 4 ft (0.5 to 1.2 m) below ground surface (bgs) at Site SS2C.

Electrical conductivity measurements of groundwater samples ranged from 4,500 to over 20,000 Mho (2,700 to over 12,000 mg/L salinity) during the operating period. Conductivity generally increases toward the Salton Sea. Presently the Salton Sea averages approximately 45,000 mg/L TDS. Although the measured salinity concentrations during the study were less than expected, high concentrations of Mg, Mn, Ca caused some precipitation of bio-available phosphates.

The former UST at Site SS2C had a capacity of 12,000 gallons (gal) (45,500 liters (L)) and stored diesel fuel. During the removal of the UST in 1993, floating product was observed within the excavation which warranted further investigation.

The treatability study was conducted in three phases:

1. Limited Site Investigation (LSI);
2. Treatability Study;
3. Technical evaluation.

The purpose of the treatability study was to assess the viability and limitations of *in situ* bioremediation of hydrocarbon-contaminated soil and groundwater under saline conditions. The field investigation was designed to evaluate the ability of the natural environment to support *in situ* bioremediation utilizing indigenous aerobic hydrocarbon-consuming microorganisms. Field measurements were collected throughout the project to assess the effectiveness of the bioremediation system design.

The scope of the Treatability Study included the following tasks:

- construction of an *in situ* bioremediation system that included 34 air sparging wells, 28 horizontal vapor extraction wells, air sparging solenoid control valves, and a nutrient addition system;
- groundwater monitoring well installation, development, and sampling;
- assessment of the physical, chemical, and biological characteristics of the site soils and groundwater;
- analysis of the percentage CO_2 in the extracted vapor stream and the dissolved oxygen (DO) uptake rates in the onsite groundwater monitoring wells; and
- collection of soil and groundwater samples at the conclusion of six months of operation to evaluate progress.

Limited Site Investigation: The LSI (BNI 1995a) concluded that the approximate areal extent of the TPH plume was 220 by 90 ft (67 by 27.5 m). Contamination appeared to be concentrated between 4 and 7 ft (1.2 to 2.1 m) bgs, with contamination extending up to 14 ft (4.3 m) bgs. The highest total recoverable petroleum hydrocarbon (TRPH) concentration reported in soil samples collected during the LSI field activities was 17,000 milligrams per kilogram (mg/kg) and 2.5 milligrams per liter (mg/L) in the groundwater. The estimated volume of TPH-contaminated soil was 2,500 cubic yards (yd^3) (1900 m^3) with approximately 8,000 pounds (3600 kilograms) of hydrocarbons present at this site.

The site stratigraphy consists of predominantly silty sand, with interbedded layers of clay and silt at 20 to 30 ft (6 to 9 m) bgs. A 1- to 2-ft (.3- to .6-m) thick continuous silt layer was generally present between 5 and 14 ft (1.5 to 4.2 m) bgs dipping to the east.

Treatability Study: The second phase of the treatability study was the construction, operation, and maintenance of an *in situ* bioremediation system. Construction of the system occurred between 24 April and 5 June 1995. A central equipment compound provided vacuum, pressurized air, and electricity to operate the bioremediation system for both sites.

The compound consisted of a 60-kilowatt (kW) electrical generator, a 15-horsepower (hp) vacuum blower, an air compressor with twin 10-hp blowers, and 120-volt-alternating current (VAC) electrical distribution panel (rated for outdoor use).

The 34 sparge points were operated as 6 sparge groups. Each sparge group header was controlled with a solenoid control valve. Vapor samples for CO_2 analysis were collected from sample ports installed at each of the 28 horizontal vapor extraction wells. A nutrient addition tank was installed at the site, including a venturi which entrained the nutrients into the air stream being injected into the subsurface. The conceptual model of the air sparge system is shown in Figure 1.

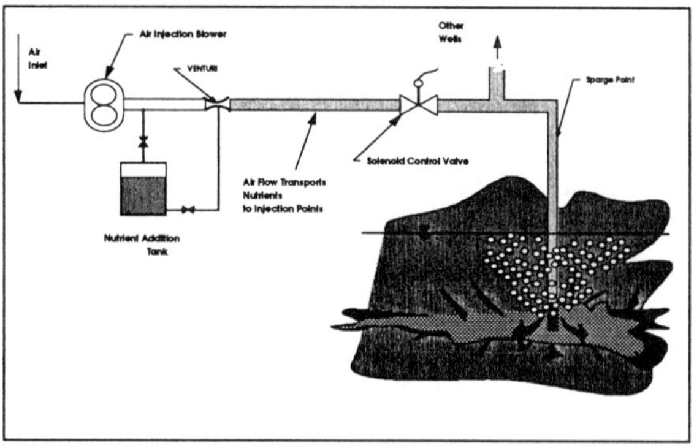

FIGURE 1 - Air Sparge System

TECHNICAL APPROACH AND STUDY FINDINGS

Bioremediation of hydrocarbons requires microorganisms (heterotrophs) that utilize hydrocarbons as a food and energy source. After a release of hydrocarbons into the soil, environmental conditions favor indigenous organisms that naturally evolve under the new conditions. Successful *in situ* bioremediation relies on the creation of a subsurface environment, through engineering means, that accelerates the production of these indigenous organisms. Two types of hydrocarbon biodegradation can occur, aerobic (oxygen is the primary electron acceptor) or anaerobic (an alternate electron acceptor such as nitrate is utilized). Aerobic degradation is the preferred reaction as it is faster and more efficient than the anaerobic degradation reaction. To establish aerobic conditions at the site, 34 sparge wells were installed to inject air (oxygen) and macro nutrients (nitrogen and phosphate).

Typical ratios of macro-nutrients required to support microbial activity range from 100:5:1 for carbon (from petroleum hydrocarbons), nitrogen, and phosphate, respectively, (Jackson and Zenobia 1994), and as low as 300:10:0.05 (Tropy et al. 1989).

The salinity averaged 3.9 mg/L from the groundwater samples collected at the site. The treatability study was designed to assess the effect of the elevated salinity concentrations on the ability to increase the heterotrophic biopopulation to degrade the hydrocarbon contamination at the site. It was uncertain whether the salinity at the site would inhibit the growth of the aerobic microorganisms and limit the rate of hydrocarbon degradation to the extent of increasing the necessary operational time, resulting in higher remedial costs.

BNI operated the treatability system from 9 June to 18 December 1995. The system design enhanced *in situ* bio-activity through the delivery of oxygen and nutrients to the subsurface via the sparge points. The remedial progress of the site was monitored indirectly using dissolved oxygen (DO) uptake rates, TPH concentration, heterotrophic plate counts and CO_2 concentration in the vapor extraction streams.

Dissolved Oxygen Uptake Rates: The DO uptake rate is the change in the DO concentration in the groundwater over time (between air sparging cycles in the vicinity of the well). Microorganisms in the subsurface consume oxygen available in the oxygen-rich (elevated DO) groundwater during the respiration process that degrades petroleum hydrocarbons. The decrease in DO concentration during the biodegradation of the hydrocarbon compounds is the DO uptake rate. The change in the DO uptake rate over time is directly proportional to the change in heterotrophic activity in the subsurface.

Air was intermittently sparged into the groundwater, causing the DO concentration to increase. Chemical oxygen demand and DO off-gassing contribution to the DO uptake rates were assumed to be minimal after the initial system start-up period and to be nearly constant over the course of the study. Therefore, the changes in DO uptake rates over time were attributed solely to biochemical demands resulting from bio-activity. In this manner the DO uptake rates were used as an indirect measurement of bio-activity in the subsurface.

The DO uptake rate was calculated at individual groundwater monitoring wells by collecting DO concentrations and plotting concentrations versus time. These DO curves were generated both by manually collecting DO levels and recording the concentration and the associated time in the field log book and by real-time collection using a strip chart recorder to record the continuous drop of DO over time.

Nutrients were added to stimulate bio-activity, thereby increasing the DO uptake rates. Delays in DO uptake rate increases were observed which was associated with mixing of the nutrients in the subsurface. DO uptake rates in some wells did not increase which may be a result of inefficient mixing or may also be attributed to a general reduction in hydrocarbon contaminants in that specific area.

Baseline DO uptakes rates recorded in June ranged from 0.013 to 0.021 mg/L/min. The DO uptake rates increased each month until August when the rates peaked at 0.093 to 0.290 mg/L/min. The recorded DO uptake rates decreased during September and October to lows ranging from 0.035 to 0.140 mg/L/min. The decreased bio-activity was associated with nutrient limitations at the site. Nutrients were added in September and DO uptakes rates increased from October to December. The Nutrients were added again in November to maintain and increase the DO uptake rates.

Percentage CO_2 Results: Carbon dioxide samples were collected to qualitatively assess heterotrophic microbial activity in the subsurface. CO_2 is a byproduct of aerobic biodegradation and the percentage of CO_2 recorded in the extracted subsurface vapor was associated with microbial activity. Thus, an increase in the reported CO_2 percentage was attributed with an increase in microbial activity.

The field data collected allowed for general assessment of the bioremediation process and appeared to correlate fairly well with the observed hydrocarbon plume reduction. Rather than using the absolute values of CO_2 percentages to predict increases or decreases in microbial activity from month to month, the CO_2 values were used in a more relative manner to assess specific areas of higher microbial activity. These areas of presumed elevated microbial activity (defined as greater than 0.5% CO_2) were compared from sampling event to sampling event to assess an overall increase or decrease in CO_2 production across each site.

Plate Count Levels: Biopopulations in the subsurface were evaluated through the analysis of groundwater samples using the Spread Plate and Membrane Filter Methods (standard methods 9215C (APHA 1992)). The method was further modified by: incubation temperature, incubation period, and toxicity testing. The method was used to estimate the number of live heterotrophic bacteria in the groundwater. Biopopulations in the groundwater decreased during the first three months of operations. Nutrients were added in the third month of operations and the plate counts increased in two of the wells. Plate counts remained elevated in the samples collected in the fifth month but fell off unexpectedly in the samples collected after six months. This may be attributed to reduction in hydrocarbon concentrations. Figure 2 presents an example of the plate count data collected during the six months of system operation.

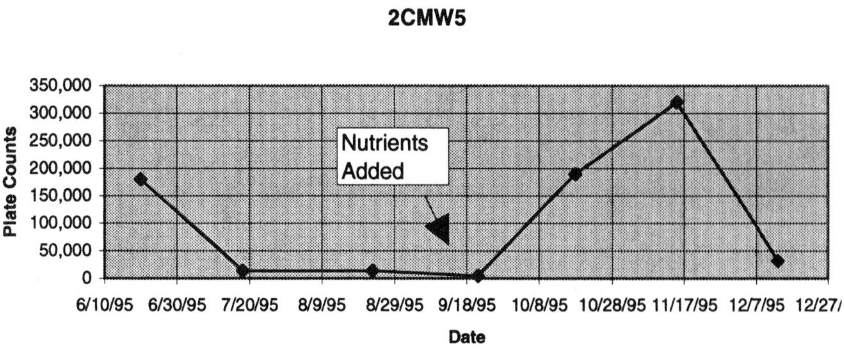

FIGURE 2 - Plate Count Results From a Typical Onsite Well

Reduction of Hydrocarbon Contamination: Hydrocarbon contamination concentrations in the soil were tracked through the collection of soil samples over time from the same vicinity. Minimal hydrocarbon reductions were noted after three months of operations and significant hydrocarbon concentrations (approximately 10,000 mg/kg TPH-diesel fuel) were still present in the subsurface soil. Nutrients, in the form of a nitrogen/phosphate fertilizer, were added to the site in September. Soil samples collected in October indicated a reduction in the TPH levels from 10,000 mg/kg to approximately 1,000 mg/kg. Nutrients were added a second time during November to continue the accelerated biodegradation. The overall concentrations of hydrocarbons in the subsurface were reduced further in samples collected at the completion of the study (December).

CONCLUSIONS

Several parameters were monitored in the field during the treatability study to assess the progress and/or the effectiveness of the system. Of the real-time field monitoring methods, the collection of dissolved oxygen uptake rates appeared to be yield the best information to track the bio-activity at the sites. CO_2 concentrations in the vapor stream were useful, producing general information which appeared to correlate the bio-activity at the site with the overall plume reduction. Plate count analyses on groundwater samples produced quantitative results on the number of heterotrophs present near the wells.

The saline conditions at the site did not inhibit *in situ* bioremediation. The essential component in the subsurface appeared to be the presence of nitrogen and phosphate (macro nutrients) in the subsurface. During the treatability study most of the reduction in contaminants occurred after macro nutrients in the form of an aqueous nitrogen- and phosphate-based fertilizer were injected into the subsurface via the sparge points. A concern of nutrient addition was the possibility of precipitation of the phosphate with calcium in the subsurface. Evidence of precipitation was monitored through the observation of the sparge point injection pressure. Increased pressures were not noted and therefore precipitation was calculated to be minimal.

The hydrocarbon contamination plume was reduced to approximately 20 percent of the original size in six months of operations demonstrating that *in situ* bioremediation was not adversely affected by the elevated saline concentrations in the subsurface.

REFERENCES

American Public Health Association (APHA). 1992. "Standard methods for the examination of water and wastewater." 18th ed.

Bechtel National, Inc. 1995a. "Limited Site Investigation Report." September.

Jackson, J.D. and Zenobia K. 1994. "Using Microbial Kinetics in the Bioremediation of Contaminated Soils." Wise, D.L., editor, *Remediation of Hazardous Waste Contaminated Soils*. pp.681-689.

Tropy, M.F., H.F. Stroo, and G. Bruebaker. 1989. "Biological treatment of hazardous wastes." *Pollution. Engineering*. 21:80-86.

BIODEGRADATION AND VOLATILIZATION OF SAUDI ARABIAN CRUDE OIL

Steven E. Whiteside (Shell Offshore Inc., New Orleans, Louisiana)
Sanjoy K. Bhattacharya (Tulane University, New Orleans, Louisiana)

ABSTRACT: Aerobic degradation of Saudi Arabian crude oil was conducted under various conditions representative of a coastal marine environment. Both bench-scale and laboratory-scale aerobic reactors were utilized to study the changes of intertidal soil contaminated with crude oil. Weathering experiments of the crude oil were conducted to gain insight into the degree of volatilization losses, and to quantify abiotic losses from biodegradation losses of the crude oil. The bench-scale aerobic column studies indicated that Gulf of Mexico conditions were nutrient-limited during short term studies, and the total petroleum hydrocarbon (TPH) content of the soil was reduced sixty to seventy percent in five weeks with supplemental nutrients. Aerobic laboratory-scale batch shaker flask studies were utilized for long-duration (seventy-eight week) experiments. Gravimetric loss of hydrocarbons ranged from eighty to eighty-nine percent under various conditions. Based on weathering tests, approximately twenty percent of this loss was attributed to volatilization. Supplemental nutrient additions did not increase the ultimate gravimetric loss of crude oil in the long-duration experiments. Significant degradation of the asphaltene fraction of the crude oil was observed in the long-duration studies.

INTRODUCTION

Both natural and engineered in situ biodegradation processes play a significant role in alleviating environmental problems associated with organic pollution. In spite of this assurance, the in situ biodegradation of certain organic contaminants is difficult to verify due to the relative degree of both microbiological and abiotic contributions to contaminant loss. This problem is of particular concern when considering appropriate remediation and impacts of accidental crude oil spills in a marine environment.

Bioremediation is time-consuming and not considered as a rapid primary response countermeasure for removal of stranded crude oil on shorelines. Bioremediation can serve as an important process to be utilized in conjunction with other recovery or remediation methods.

A research project was conducted to evaluate the abiotic and biological losses of a specific imported crude oil (Whiteside, 1996).

OBJECTIVES. The specific objectives of the research were to:

1. evaluate the overall "losses" of Saudi Arabian light crude oil, in a marine environment, under controlled laboratory conditions,
2. separate and quantify the various sources of crude oil loss under specific environmental conditions,

3. determine the rates and mechanisms affecting crude oil losses or degradation, and
4. evaluate methods to enhance crude oil degradation.

MATERIALS AND METHODS

The experimental plan was divided into several phases:

1. Tests were performed to characterize the crude, soil, and seawater.
2. Weathering experiments were conducted to characterize and quantify natural abiotic losses of the crude oil under specific environmental conditions.
3. Preliminary bench-scale column experiments were conducted to evaluate TPH degradation of oil- contaminated soil under specific aerobic environmental conditions. These experiments were designed to compare a control experiment (without additions), with a nutrients test, and a seed microorganism (plus nutrients) test.
4. Additional laboratory-scale aerobic batch reactor studies were performed to evaluate the long-term degradation under near-optimum D.O. conditions. The GC-MS studies were performed on the samples subjected to various environmental conditions to identify the decomposition products and residuals.

Saudi Arabian Extra Light crude oil (36 degree API, 1.42% sulphur) was utilized for the experiments. The soil and seawater utilized in the biodegradation testing and other experiments were obtained from the Louisiana coastline. The sandy soil was extremely well sorted with a D_{10} grain size of 0.2 mm and a permeability of approximately 0.1 cm/sec. The seawater was characterized by a pH of 8.2, 18,300 mg/l chlorides, and a total Kjeldahl nitrogen of 1.4 mg/l.

CRUDE OIL WEATHERING EXPERIMENT

A weathering experiment was designed to provide a quantitative evaluation of natural (presumably abiotic) losses of both weight and volume of Saudi Arabian crude oil under specific environmental conditions.

The apparatus utilized in the weathering experiment consisted of a 1000 ml graduated cylinder with a vent. A glass aeration tube extended to the base of the cylinder to provide a full aerated column of oil with constant agitation and mixing. The apparatus was designed to provide extremely accurate and convenient monitoring of both weight and volume changes of the crude oil over an extended time period.

BENCH-SCALE BIOLOGICAL REACTOR DESIGN

Three identical bench scale biological reactor columns were constructed from 6-inch (15.2 cm) schedule 40 PVC pipe illustrated in Figure 1. Aeration was equally supplied to each reactor via ports in the base of each reactor. Seawater was supplied through an overhead tube to provide necessary moisture to each reactor. A cassette pump drive was used to operate individual peristaltic cassette pumps for each of the three reactor columns.

Reactor No. 1 was a control column. Reactor No. 2 evaluated the effect of supplemental nutrients on the growth rate of indigenous bacteria, in an effort to increase the rate of conversion of the hydrocarbon substrate.

Seed microorganisms were added only to Reactor No. 3 to evaluate the effect of non-native bacteria on the rateof TPH degradation. The seeding of Reactor No. 3 was accomplished with naturally-occuring culture of *Pseudomonas* and *Bacillus* genera bacteria.

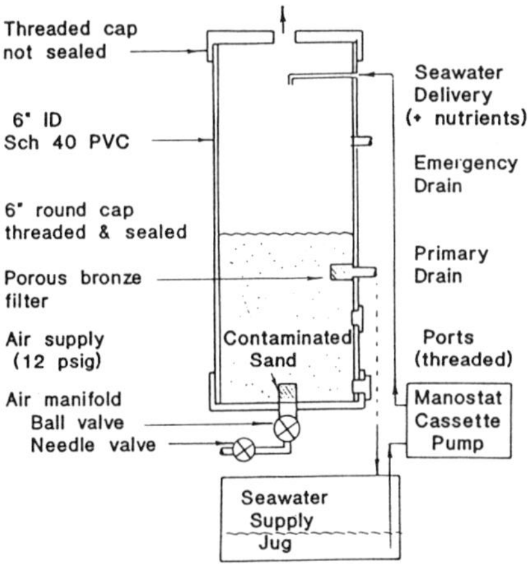

FIGURE 1. Bench-Scale Reactor Design

SOIL ANALYSES

The crude oil contaminated soil samples were tested for total recoverable petroleum hydrocarbon (TPH) concentration using EPA Method 418.1 (Spectrophotometric, Infrared). Reagent grade Fluorocarbon-113 (1,1,2-trichloro-1,2,2-trifluroethane), b.p. 48°C, was used for the hydrocarbon extraction. A Perkin Elmer 1600 Series Fourier Transform Infrared (FT-IR) Spectrophotometer was used for the TPH measurements.

SLURRY REACTOR DESIGN

The laboratory slurry reactor experiments were designed to simulate the long-term results of the Saudi Arabian crude oil in a marine seawater environment under aerobic conditions. The experiments represent, in the author's opinion, near-optimum dissolved oxygen conditions for the biodegradation process to proceed.

The slurry reactor experiments were conducted using 250 ml Erlenmeyer flasks which had been sterilized at 250°F to remove any contaminating microorganisms. Each flask was filled with 100 ml of natural seawater and contained 10 gm of natural beach soil. Each flask was then spiked with 0.500 gm of Saudi Arabian AXL crude oil which had not been subject to prior weathering. Three environmental cases were examined:

1. Control reactor (no supplemental nutrients).
2. Nutrient enriched reactor (nutrients containing 35% total nitrogen, 9% phosphorus pentoxide, P_2O_5, and 3% soluble potash, K_20, determined on a weight basis).
3. Nutrient enriched reactor, but shielded from light.

All slurry reactor flasks were installed on a rotary shaker table for constant agitation and mixing. Each flask was fitted with a glass aeration tube to maintain saturated aerobic conditions throughout the duration of the experiment. The contents of each shaker flask was extracted by liquid- liquid extraction, then analyzed by GC-MS using a procedure based upon SW-846 Method 3510. The laboratory-scale shaker flask experiments included a gravimetric analysis to evaluate the weight loss of the hydrocarbon material recovered from the methylene chloride extraction.

The preparation of samples prior to GC-MS analyses is very important to the analytical results. Asphaltenes were removed by *n* hexane precipitation, and polar compounds were separated using an activated alumina column. GC-MS was utilized for identification of hydrocarbon compounds in samples, utilizing the procedures of U.S. EPA Method 8270. QA/QC internal standards of 1,4 - Dichlorobenzene, d8 - Napthalene, d10 - Acenapthene were utilized. The asphaltene fraction was further analyzed with a GC-TCD elemental analyzer.

RESULTS

WEATHERING EXPERIMENT

Measurements of the crude oil weight and volume were recorded over a 1,400 hour duration. The weight change as a function of time is illustrated in Figure 2. The loss of weight exhibited zero-order kenetics up to approximately 50 hours and thereafter approximated a first-order decay model. The computed API gravity declined from 36.5° API to 29.1° API during the experiment. During the duration of the experiment, an 18% decrease in weight, and a 22% decrease in volume was recorded.

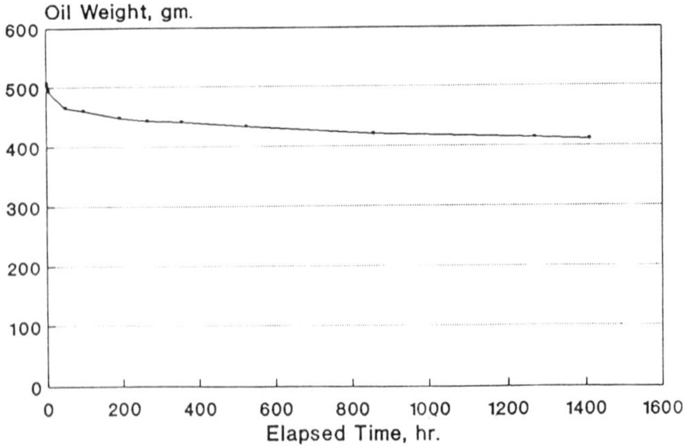

FIGURE 2. Crude Oil Weathering Experiment, Saudi Arabian Crude Oil.

BENCH-SCALE COLUMN REACTOR OBSERVATIONS

The changes in TPH over time are illustrated in Figure 3. Samples taken after seven days varied from 1,600 mg/kg to 2,000 mg/kg, indicating no statistically significant changes from the initial TPH values. Samples taken after 21 days indicated clear evidence of TPH degradation.

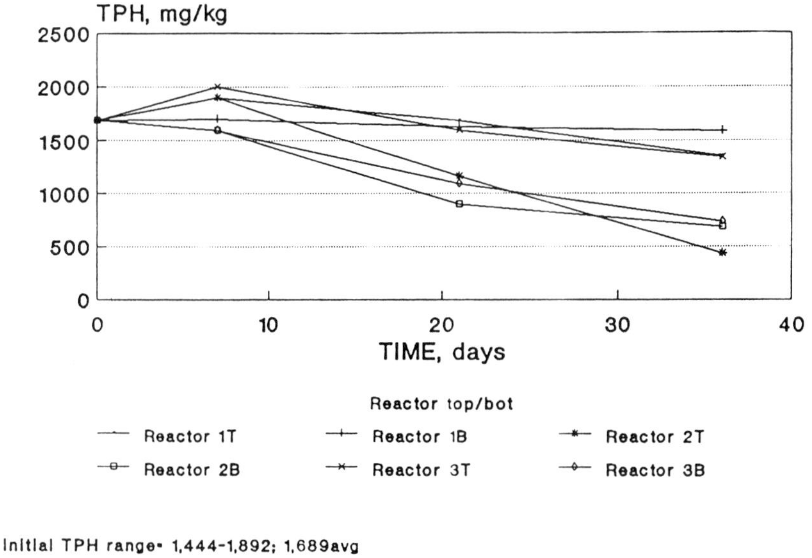

FIGURE 3. Total Petroleum Hydrocarbons vs Time, Bench-Scale Reactors

LABORATORY-SCALE REACTOR RESULTS

The laboratory-scale flasks were subjected to long- term run time on the shaker table and were constantly aerated to maintain oxygen saturated conditions in each reactor flask. Samples were run for a total of 547 days, then hydrocarbons were extracted for gravimetric and GC-MS analyses.

CONCLUSIONS

1. Saudi Arabian AXL crude oil experienced significant weight and volume losses and compositional change as a result of abiotic weathering. Weathering losses measured in the laboratory are believed to be conservative due to limited surface area and lack of photolysis, relative to field conditions. Weathering losses in the laboratory are approximated by a zero-order kinetic model up to fifty hours, then a first-order kinetic model beyond fifty hours. Abiotic weathering of the crude oil was responsible for removal of C_1 through C_{10}, although these

compounds may also undergo simultaneous transformation via biodegradation under the biological experiments.

2. The bench-scale column experiments indicated overall TPH reductions in the range of sixty to seventy percent are possible within a five-week period under aerobic conditions, under the constraints of the environmental conditions studied. Approximately seventeen percent of the TPH reduction is attributed to weathering losses.

3. The addition of supplemental nutrients caused improvement in the rate of TPH degradation during the five-week bench-scale evaluation period. This response is dependent upon the availability of nutrients in the seawater and soil, however, the Gulf of Mexico environment at this location was considered nutrient-limited for the shorter duration of bench-scale studies.

4. The addition (seeding) of naturally-occurring microorganisms did not affect the TPH degradation under the combination of environmental conditions and specific seed bacteria evaluated in the bench-scale studies.

5. Under the relatively optimum aerobic conditions experienced in the laboratory-scale reactors, a total removal of eighty to eighty-nine percent, by weight, of the original live crude oil was recorded during the extended 547 day experiments. Biodegradation effectively mineralized or transformed approximately forty-five to fifty-four percent of the original live crude oil, by weight, over a period of 547 days. Organic compounds identified in the recalcitrant residue following the experiment correspond to hydrocarbon degradation pathways documented by other researchers.

6. Nutrient additions under the long-duration laboratory-scale studies do not conclusively improve the ultimate hydrocarbon losses. Although the transformation of certain hydrocarbons was likely accelerated in the short term by nutrient additions (consistent with bench-scale studies), the residual recalcitrant compounds at the end of 547 days were gravimetrically similar. This data suggests that under long-term biodegradation, the recalcitrance of the remaining organic compounds became the controlling factor, rather than the availability of nutrients.

7. The relatively low asphaltene content of the residual hydrocarbons at the end of 547 days indicated that even the asphaltene fraction had been significantly degraded in all samples. The low level of nitrogen contained in the asphaltene fraction did not indicate significant change during the experiment.

8. The laboratory-scale aerobic experiments are representative of relatively optimum field conditions of dissolved oxygen content in an intertidal surf. Field measurements indicated a rapid decrease in dissolved oxygen below the intertidal sediment surface. Use of the hydrocarbon loss data from this research should be restricted to oil contamination on or near the surface of the intertidal sediment, where comparable dissolved oxygen conditions are present.

REFERENCE

Whiteside, S. E. 1996. "Biodegradation and Volatilization of Saudi Arabian Crude Oil." Ph.D. Dissertation, Tulane University, New Orleans, LA.

MESOCOSM ASSAYS OF OIL SPILL BIOREMEDIATION

Regas Santas (OikoTechnics Institute, Athens, Greece)
Athanassia Korda and Angela Tenente (OikoTechnics Institute, Athens, Greece)
Evangelos Gidarakos (Battelle, Germany)
Kurt Buchholz (Battelle, Arlington, VA)
Photeinos Santas (OikoTechnics Institute, Athens, Greece)

ABSTRACT: Three mesocosms were constructed simulating a sloping intertidal zone in fiberglass tanks (3×1×1 m). Substrate, water, and living organisms from wild unpolluted ecosystems were transferred into the tanks as a natural inoculum for mesocosm development. One-kilogram portions of sediment samples collected from uncontaminated beaches were immersed into Iranian light and placed at two depths (40 and 80 cm below water level) for a total of 12 sediment samples per depth per tank. In each tank, Iranian light was added at a 1:1000 v/v ratio relative to water and installed pumps provided water recirculation (5 m^3/h). The effects of fertilizer type on biodegradation were examined by pouring 330 mL Inipol EAP-22 into the first tank, spreading 330 g of F1 (modified fishmeal) at the water surface of the second one, and no fertilizer to the third one (control).

The effect of various fertilizer additions on hydrocarbon biodegradation was assessed by GC monitoring of n-C_{17}/pristane and n-C_{18}/phytane ratios on days 0, 1, 3, 7, 15 and 30 after mesocosm commissioning, on two replicate samples per mesocosm per depth and at the water surface. Physical and chemical parameters (temperature, salinity, etc.) were also monitored throughout the experiment.

Bioremediation depended on fertilizer added, depth, and sampling date. Inipol produced the most visually dramatic reduction of crude oil: on day 30, the Inipol-assigned mesocosm had the clearest water column of all installed mesocosms. However, based on measurments of the n-C_{17}/pristane and n-C_{18}/phytane ratios, F1 had a better performance. By day 30, F1 increased biodegradation to about 3x the initial value. Similar but less pronounced differences were observed at 40 cm, while no differences were observed at 80 cm for any fertilizer and any sampling day.

INTRODUCTION

Bioremediation, the use of microorganisms to break down hazardous chemicals into nontoxic compounds, is a developing cleanup technology. This technology is most effective in reducing concentration of petroleum products at oil polluted areas (Orzech et al., 1991).

The effectiveness of bioremediation depends on a variety of environmental conditions, including oxygen concentration, indigenous bacteria populations, temperature and nutrient concentration

Many bioremediation techniques have been developed since this technology first appeared in 1967 (Santas et al., 1994). Among them, bioaugmentation, the direct application of microorganisms isolated from the contaminated site or from an off-site vendor, adapted to the specific contaminants and site conditions, cultured, and enhanced. Bioremediation usually involves biostimulation: the addition of oxygen, water and mineral nutrients (combinations of nitrogen, phosphate, and maybe surfactants and trace metals) to accelerate the reproduction of organisms as well as their metabolic activity. Oleophilic additives are preferred in seawater applications because they are dissolved into the oil, facilitating bacterial growth at the oil-water interface (Lacotte et al., 1995).

This paper assesses the bioremediation effectiveness of representative oleophilic fertilizers (Inipol EAP-22, F1) in mesocosms operated outdoors.

MATERIALS AND METHODS

Iranian light was used for all experiments. Table 1 lists the additives used, with their relevant components.

TABLE 1. Additives used and their relevant components in beach simulation experiments

Name	Description	Contents	Source
Inipol EAP-22	Commercially available oleophilic fertilizer	Oleic acid (26.2%), lauryl phosphate (23.7%), 2-butoxy-1-ethanol (10.8%), urea (15.7%), water (23.6%) (Sveum et al., 1994)	Elf Aquitaine, France
F1	Commercially available fertilizer	Modified fish meal	Elf Aquitaine, France (BIOREN project, EUREKA Program)

Mesoscale experiments (Table 2) were performed in continuous-flow 3 m^3 seawater tanks (3×1×1).

The seawater was recirculated at 5 m^3/h, and the average seawater temperature in the tanks was approximately 15°C. Additives were applied immediately after crude oil application. Only bacteria indigenous to the seawater and the sediment were used in these tank experiments.

TABLE 2: Description of experiments performed in continuous-flow tanks

Tank	Crude oil	Additive
1	3 L	Inipol EAP-22 (330 mL)
2	3 L	F1 (330 g)
3	3 L	none

Samples were collected from three different points in each tank: water surface, shore sediments(0.40m), and open water sediments (0.80m).

Samples used for gas chromatographic analyses were processed as follows: water samples (~20 mL) were swirled and oil was extracted with 10 mL n-hexane (Merck, *proanalysi*; >99%). Hydrocarbons in sediment samples were extracted with 10 mL n-hexane. Water was removed from the oil/n-hexane solution by sodium sulfate. The supernatant was separated into saturated and aromatic fractions through a silica-gel packed column (2-25 μm particle size). The saturated hydrocarbons fraction was eluted with 1 mL n-hexane and analyzed on a Hewlett Packard 5890

Series II GC, equipped with flame ionization detector (FID) and a splitless injector. Oil biodegradation was evaluated from the n-C_{17}/pristane and n-C_{18}/phytane ratios.

RESULTS

Data from GC analysis for the n-C_{17}/pristane and n-C_{18}/phytane ratios were subjected to a three-way ANOVA. The sources of variation were fertilizer type (F1; Inipol; no fertilizer), depth and sampling day.

n-C_{17}/pristane. The three factor interaction was significant ($F=10.68$; $df=20, 54$; $P<0.05$). Means analysis was performed, 95% Just Significant Confidence Intervals were placed around the means, and the results for each combination of fertilizer type, depth and sampling day are presented in Figures 1-3. On day 15, the surface sample of the Inipol-treated mesocosm was significantly lower than any other mean (Figure 1).

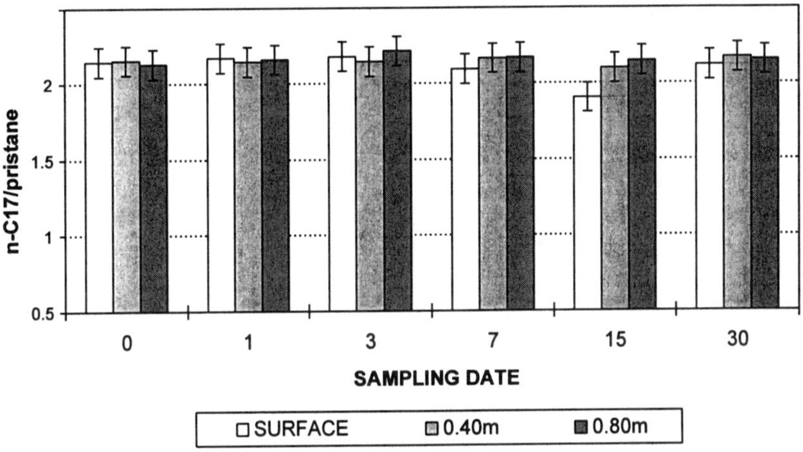

FIGURE 1. Effects of depth and time on biodegradation using Inipol

No significant differences were observed in the F1-treated mesocosm within the first week after commissioning (Figure 2). The n-C_{17}/pristane ratio at the water surface on day 15 was significantly lower than day 7, and the n-C_{17}/pristane ratio on day 30 was significantly lower than any other value. At mid-depth, the n-C_{17}/pristane ratio on day 15 was not significantly different from day 7, while on day 30 this ratio was significantly lower than on day 15. At the deep layer, no significant changes were observed throughout the project.

No significant differences were observed in the control mesocosm (Figure 3).

n-C_{18}/phytane. The three factor interaction including fertilizer type, depth and sampling day was significant ($F=24.84$; $df=20, 54$; $P<0.05$), meaning that the effects of fertilizer type on the n-C_{18}/phytane ratio depend on the combination of depth and sampling day.

No significantly different n-C_{18}/phytane values were observed on days 0-7 across depths for the F1 treated mesocosms (Figure 5). On day 15, the surface and

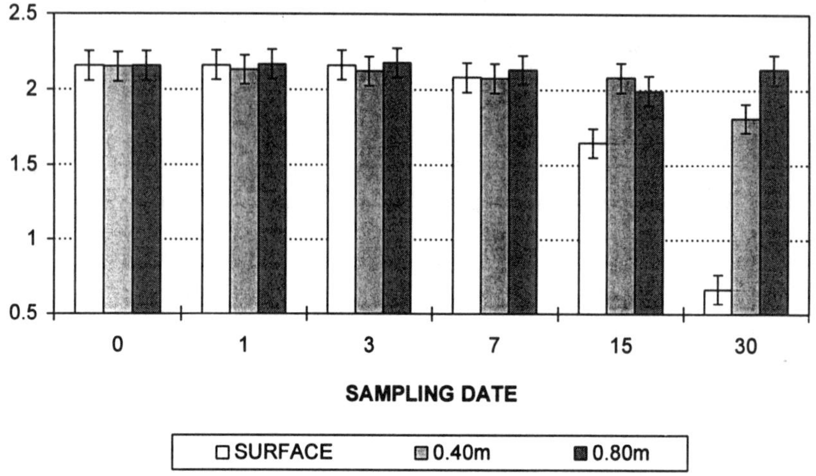

FIGURE 2. Effects of depth and time on biodegradation in F1-treated mesocosms.

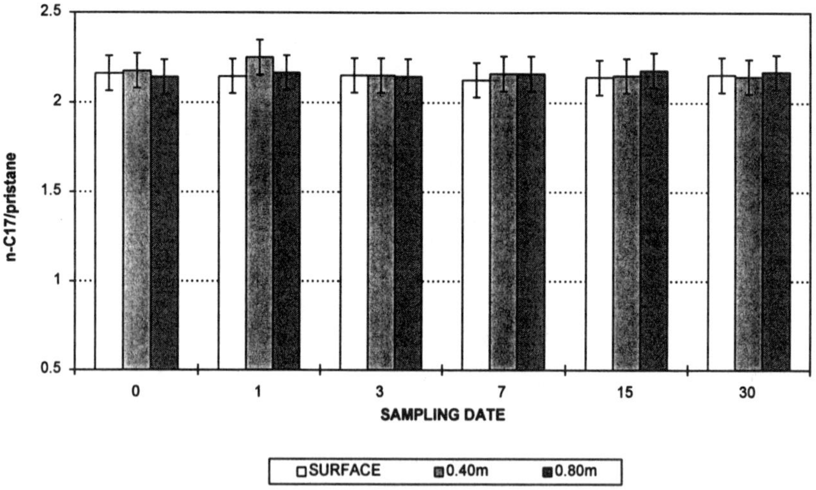

FIGURE 3. Effects of depth and time on biodegradation in the control mesocosm.

deepest waters had significantly lower n-C_{18}/phytane values than the respective ones of day 7, while there was no significant difference in the n-C_{18}/phytane values at mid-depths. On day 30, the surface waters had the lowest values recorded (0.65), while the mid-depth value was significantly lower than the respective one on day 15. In contrast, the n-C_{18}/phytane value at the deepest water was significantly higher than the respective one on day 15, and not significantly different from the values of days 0-7.

In the control tank the n-C_{18}/phytane value for the sediment-water interface was significantly higher on day 1 than the same value for the water surface on day 15 (Figure 6).

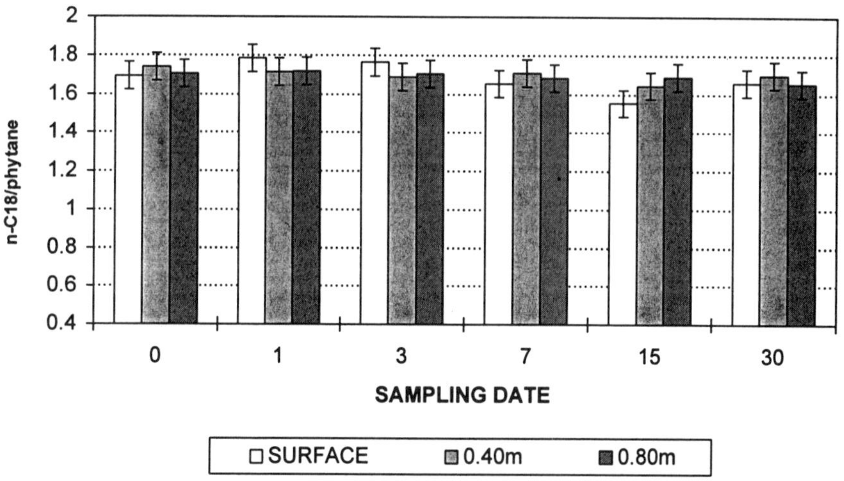

FIGURE 4. Effects of depth and time on biodegradation using Inipol.

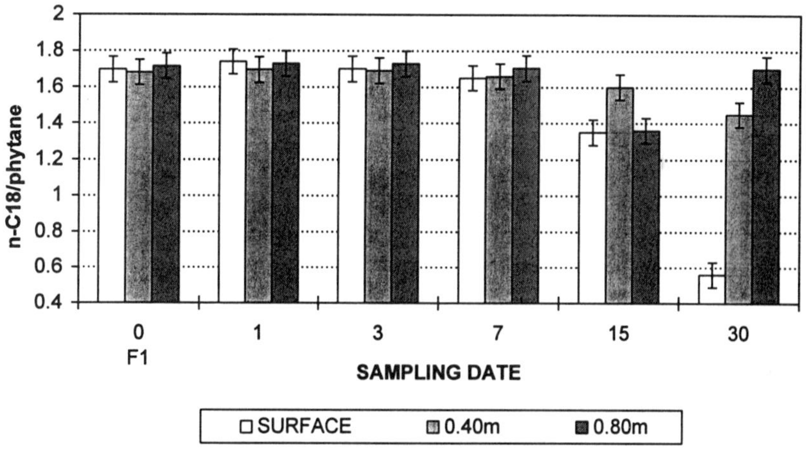

FIGURE 5. Effects of depth and time on biodegradation using F1.

DISCUSSION

F1 was the most effective and fastest-acting treatment (Figures 2 and 5). The n-C_{17}/pristane and the n-C_{18}/phytane ratios at the water surface on day 15 were dramatically lower than the previous days. Within 30 days, the original concentrations of n-C_{17} and n-C_{18} had been reduced by 75% and 65%, respectively, compared to the initial values.

On day 30, visual inspection of the Inipol-assigned mesocosm revealed the clearest water column of all installed mesocosms. After mesocosm decommissioning, it was observed that some of the oil had moved from the water surface towards the beach and had been permanently attached on the cinder blocks used as support for the beach gravel. With regard to the n-C_{17}/pristane and n-C_{18}/phytane ratios,

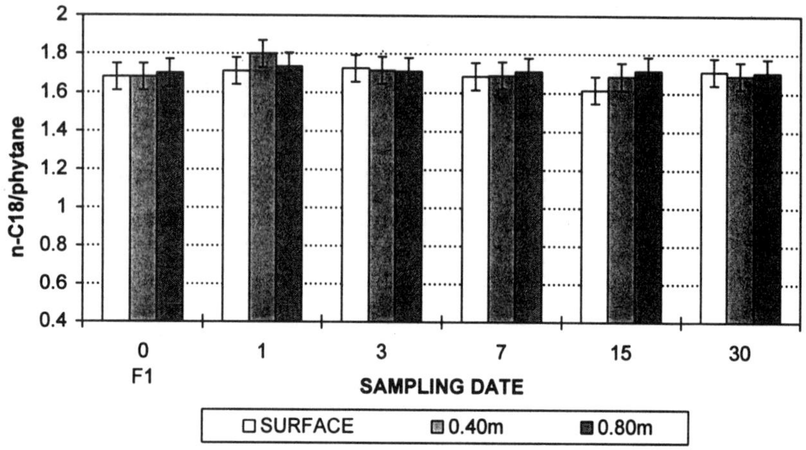

FIGURE 6. Effects of depth and time on biodegradation in the control mesocosm.

Inipol application had less dramatic effects. A possible explanation for this may be that Inipol was sprinkled on the water surface, as opposed to the usual application on the beach.

Unlike the Inipol mesocosm, the degraded oil on the water surface of the F1-treated mesocosm had a non-uniform distribution: there were patches of a thin, clear crust of undegraded material, while some other areas were covered by dark, thick oil residues. A thick crust of petroleum fractions resistant to biodegradation floated on the water surface.

ACKNOWLEDGMENTS

Work supported by EU grant PM/XI.C.4/9517 (DG-XI) to RS, OikoTechnics.

REFERENCES

Lacotte, D. J., G. Mille, M. Acquaviva, and J-C. Bertrand. 1995. "*In vitro* biodegradation of Iranian Light 250 by a marine mixed culture using fertilizers as nitrogen and phosphorous sources." *Chemosphere. 31*(11/12): 4351-4358.

Orzech, M., J. Solyst, and P. Thompson. 1991. *State's Use of Bioremediation: Advantages, Constraints, and Strategies.* Natural Resources Policy Studies. Center for Policy Research. National Governors' Association.

Santas, Ph., and R Santas. 1994. "Status of sea-born bioremediation technologies." In D. L. Wise, D. J. Trantolo (Eds.), *Remediation of Hazardous Waste Contaminated Soils,* pp. 459-479. Marcel Dekker, Inc., New York. NY.

Sveum, P., L. G. Faksness, and S. Ramstad. 1994. "Bioremediation of oil-contaminated shorelines: the role of carbon in fertilizers." In R. E. Hinchee, B. C. Alleman, R. E. Hoeppel and R. N. Miller (Eds.), *Hydrocarbon Bioremediation,* pp. 163-174. Lewis Publishers, Boca Raton, FL.

BIOLOGICAL REMEDIATION OF OIL SPILLS OF THE SARONIKOS GULF, GREECE

Athanassia Korda (OikoTechnics Institute, Athens, Greece)
Angela Tenente and Photeinos Santas (OikoTechnics Institute, Athens, Greece)
Evangelos Gidarakos (Battelle, Germany)
Michel Guillerme (ELF Aquitaine, France)
Regas Santas (OikoTechnics Institute, Athens, Greece)

ABSTRACT: The biodegradation of Iranian Light crude oil is examined in beach sediments of the Saronikos Gulf, Greece. Two oleophilic fertilizers (Inipol and F1) were applied at two concentrations (high, low) on beach sediments. The fertilizer treatments were combined with indigenous and introduced bacteria populations (DB 19). Control treatments received no fertilizer or introduced bacteria. Biodegradation was assessed by GC monitoring of the n-C_{17}/pristane and n-C_{18}/phytane ratios on days 0, 1, 3, 7, 15 and 30 after the addition of crude oil to the sediments.

Regardless of bacterial treatment, additions of high concentrations of F1 (modified fishmeal) resulted in significant hydrocarbon degradation: On day 30, the n-C_{18}/phytane ratio of the high F1 application was significantly lower than all other treatment combinations on any day of the experiment. Although Inipol application had less dramatic effects on the n-C_{18}/phytane ratio, on day 30, the Inipol-assigned beach sediments were the least contaminated at visual inspection.

Inoculation with bacteria did not affect bioremediation significantly throughout the trials. On days 15 and 30, treatments receiving fertilizers and inoculated with bacteria showed a gradual reduction in the n-C_{17}/pristane ratio (1.90 vs. 2.20 on day 1).

INTRODUCTION

In situ bioremediation is a low cost and effective method for the remediation of shorelines contaminated with petroleum hydrocarbons. Previous experiments (Sveum et al., 1994) indicate that the application of nitrogen and phosphorous nutrients enhances the biodegradation of crude oil. Fertilizer application provides the necessary elements for the metabolism of oil-degrading microorganisms. In addition, bioremediation in the field may be stimulated by inoculation with allochthonous microorganisms cultured and adapted for specific contaminants and site conditions (Santas et al., 1994).

This paper studies the capabilities of bioremediation as a primary tool for beach decontamination in an attempts to optimize the field implemention of bioremediation.

MATERIALS AND METHODS

Iranian light was used for all experiments. Table 1 lists the additives used, with their relevant components.

TABLE 1. Additives used in Saronikos Gulf beach simulation experiments

Name	Description	Contents	Source
Inipol EAP-22	Commercially available oleophilic fertilizer	Oleic acid (26.2%), lauryl phosphate (23.7%), 2-butoxy-1-ethanol (10.8%), urea (15.7%), water (23.6%)	Elf Aquitaine, France
F1	Commercially available fertilizer	Modified fish meal	Elf Aquitaine, France (BIOREN project, EUREKA Programme)
DB 19	Commercially available bacterial cleaner	Blend of bacteria, enzymes and nutrients	Recherche Exploitation Produits, France

Beach pebbles (diameter 5-10 mm) from the coast of Saronikos Gulf were used to fill 18 crates (0.15 m^2). A basket (0.035 m^2, 8-9 cm deep) was filled with pebbles and placed in each of the 18 crates after the following treatment combinations:

TABLE 2. Experimental design and treatment combinations.

	Low fertilizer dose		High fertilizer dose	
	native bacteria	introduced bacteria*	native bacteria	introduced bacteria
Inipol EAP-22	33 mL-native	33 mL-introduced	330 mL-native	330 mL-introduced
F1	33 gr-native	33 gr-introduced	330 gr-native	330 gr-introduced
No fertilizer	control (native)	introduced	control (native)	introduced

*100 mL of bacterial solution were added in each basket.

Hydrocarbon degradation was monitored for 30 days following treatment application, throughout the duration of the experiment (November-December 1996).

Approximately 10 gr samples were collected from each basket on days 0, 1, 3, 7, 15 and 30, hydrocarbons extracted with 10mL n-hexane (Merck, *proanalysi*; >99%), and water removed from the extract by sodium sulfate. The supernatant was separated into saturated and aromatic fractions through a silica-gel packed column (2-25 μm particle size). The former fraction was eluted with 1 mL n-hexane and analyzed on a Hewlett Packard 5890 Series II GC equipped with flame ionization detector (FID) and a split/splitless injector. Oil biodegradation was evaluated from the n-C_{17}/pristane and n-C_{18}/phytane ratios.

RESULTS

GC data were analyzed by a four way ANOVA, using fertilizer type (F1, Inipol), fertilizer quantity (high, low), bacteria type (native, native+introduced) and sampling day as the sources of variation.

n-C_{17}/pristane. The four way interaction was not significant ($F=0.64$; df=5.48; $P>0.05$). The only significant three way interaction was the fertilizer by quantity by sampling day ($F=3.82$; df=5.48; $P<0.05$).

Means analysis was performed on the n-C_{17}/pristane data, 95% Just Significant Confidence Intervals were placed around the means and the results are presented in Figures 1 through 3. The high Inipol concentration did not result in significantly higher hydrocarbon degradation than the lower concentration (Figure 1). The high concentration of F1 reduced significantly the n-C_{17}/pristane ratio on day 30 (Figure 2) compared to days 0 through 3. For the low concentration of F1, no significant differences were observed between days 15 and 30; however, the n-C_{17}/pristane ratio on day 30 was significantly lower than on days 0 through 3. No differences were observed between the controls inoculated with natural and introduced bacteria populations (Figure 3).

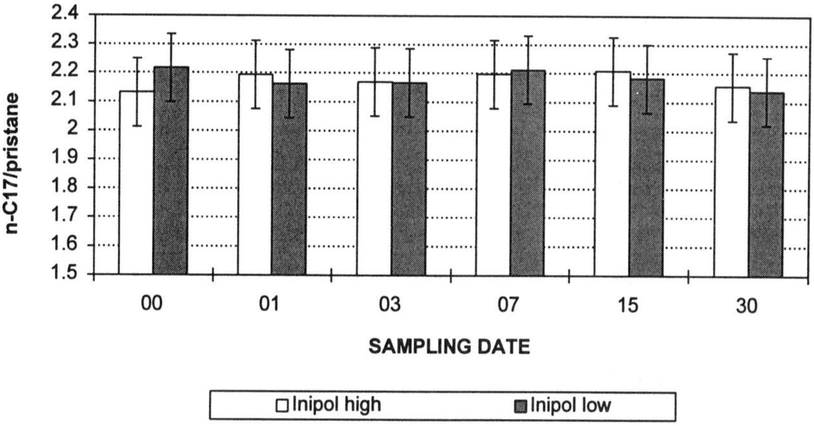

FIGURE 1. Effects of Inipol concentration on biodegradation in beach sediments.

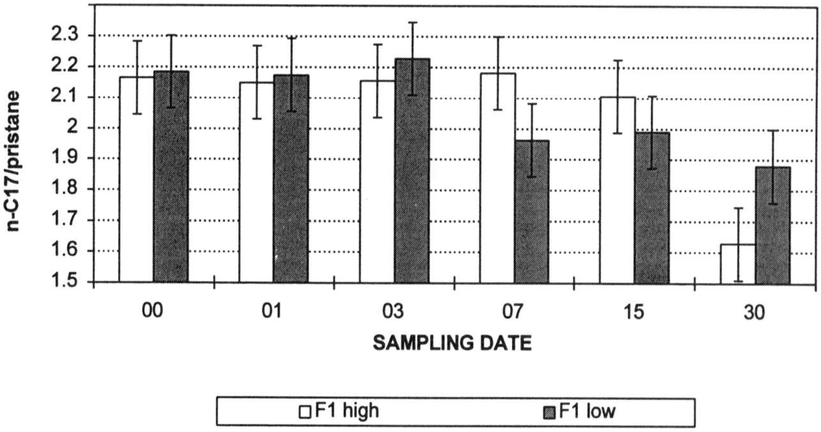

FIGURE 2. Effects of F1 concentration on biodegradation in beach sediments.

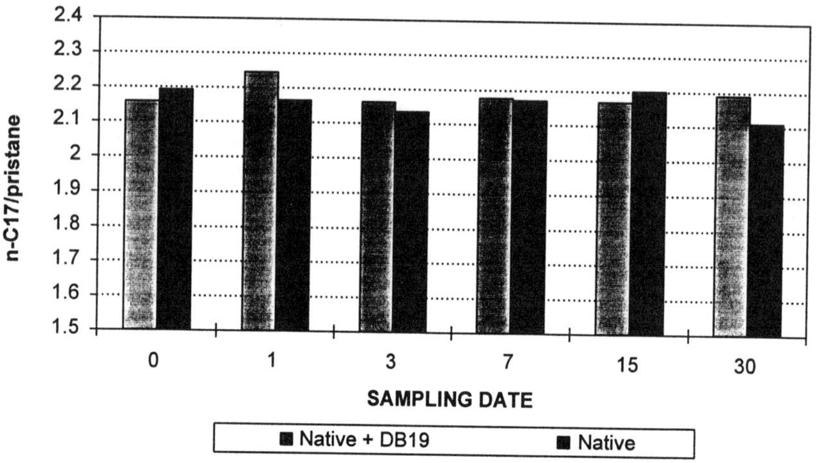

FIGURE 3. Effects of bacteria type on biodegradation without any added fertilizers.

n-C_{18}/phytane. The four way interaction was not significant (F=0.88; df=5.48; P>0.05). The three way interaction including fertilizer type, fertilizer quantity and sampling day was significant (F=2.48; df=5.48; P<0.05).

Means analysis for each fertilizer treatment is presented in Figures 4 through 6. The high Inipol concentration did not result in significantly higher hydrocarbon degradation than the lower concentration (Figure 4). High concentrations of F1 reduced the ratio significantly on day 30 compared to all other sampling days (Figure 5). Low concentrations of F1 reduced the ratio significantly on day 30 compared to days 0, 1, 3, but this value was not significantly different from the respective values on days 7 and 15. Limited variation was detected in the control baskets (Figure 6).

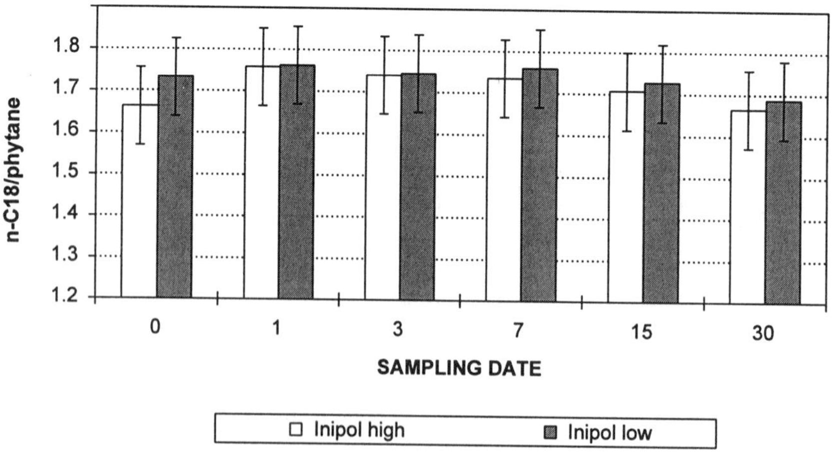

FIGURE 4. Effects of Inipol concentration on biodegradation in beach sediments.

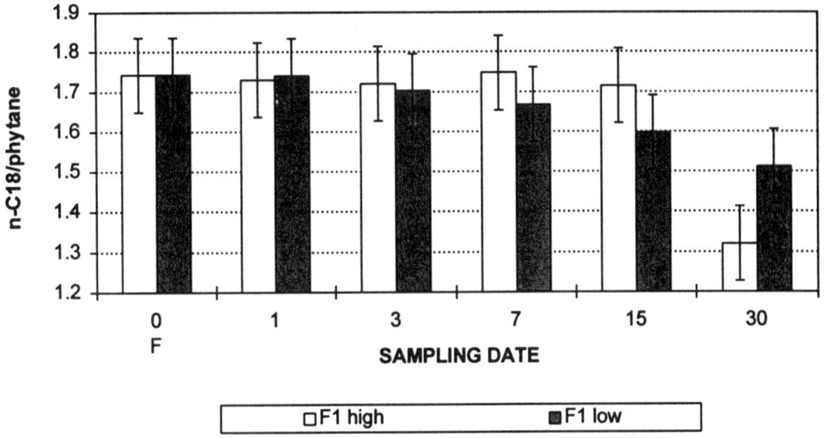

FIGURE 5. Effects of F1 concentration on biodegradation in beach sediments.

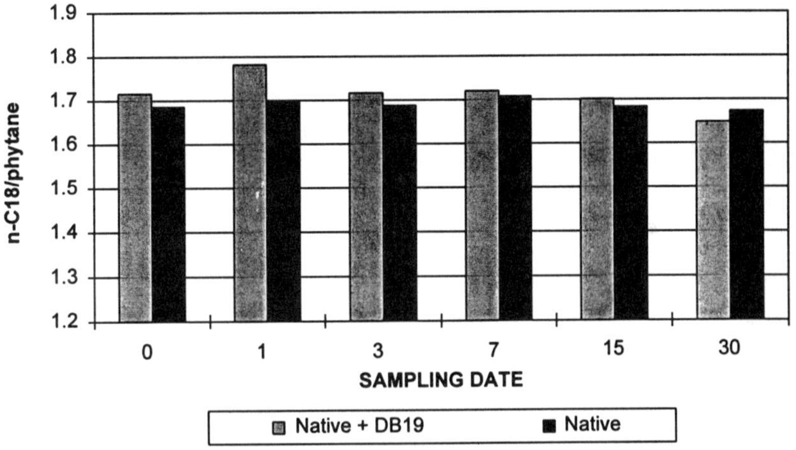

FIGURE 6. Effects of bacteria type on biodegradation without any added fertilizers.

DISCUSSION

The most effective treatment was the high quantity of F1 fertilizer (day 30). It may be interesting to assess the biodegradation of Iranian light in beach sediments for intervals exceeding 30 days.

Introduced bacteria do not seem to enhance biodegradation under the conditions of the present experiment and this agrees with results obtained by Venosa et al. (1991). This could be attributed to either of two factors: the oil-degrading bacteria remained dormant, or they are not as effective in degrading Iranian light as native populations (Venosa et al. 1991). However, these results should be interpreted with caution because control baskets were not replicated and therefore can not be rigorously compared to the rest of the results obtained.

There was no wave action and energy applied on the simulated beach experiment; the only water input was natural precipitation. Temperature conditions were favorable for bacterial activity (average daily temperature 15 °C).

ACKNOWLEDGMENTS

Work supported by EU grant PM/XI.C.4/9517 (DG-XI) to RS, OikoTechnics Institute.

REFERENCES

Santas, Ph., and R. Santas. 1994. "Status of Sea-Born Bioremediation Technologies." In D. L. Wise and D. J. Trantolo (Eds.), *Remediation of Hazardous Waste Contaminated Soils,* pp. 459-479. Marcel Dekker, Inc., NY.

Sveum, P., L. G. Faksness, and S. Ramstad. 1994. "Bioremediation of oil-contaminated shorelines: the role of carbon in fertilizers." In R. E. Hinchee, B. C. Alleman, R. E. Hoeppel, and R. N. Miller (Eds.), *Hydrocarbon Bioremediation,* pp. 163-174. Lewis Publishers, Boca Raton, FL.

Venosa, A. D., J. R. Haines, W. Nisamaneepong, R. Govind, S. Pradhan, and B. Sidique. 1991. "Screening of commercial bioproducts for enhancement of oil biodegradation in closed microcosms." *17th Annual Hazardous Waste Rsearch Symposium*, April 9-11, Cincinnati, OH.

COMBINING OXIDATION AND BIOREMEDIATION FOR THE TREATMENT OF RECALCITRANT ORGANICS

Richard A. Brown, Ph.D., (Fluor Daniel GTI, Trenton, NJ)
Chris Nelson, (Fluor Daniel GTI, Golden, CO)
Maureen Leahy, Ph.D., (Fluor Daniel GTI, Windsor, CT)

ABSTRACT: High molecular weight PAHs are generally resistant to biodegradation processes. Two technologies are being developed as means of treating these types of contaminants. These are bioremediation and oxidation. Both have drawbacks. Bioremediation is a relatively inexpensive but fairly slow process, controlled by the low solubility and/or reactivity of these compounds. Oxidation is a faster process than bioremediation but is fairly expensive. A new approach to the remediation of recalcitrant materials is to combine oxidation and bioremediation. Oxidation can be used to pre-condition substrates for biodegradation, or be used as a "polishing" system. The sequential treatment by *in situ* bioremediation followed by ozonation shows promise as an extremely effective approach.

INTRODUCTION

Semivolatile organic compounds, such as pentachlorophenol (PCP), polyaromatic hydrocarbons (PAHs) and phthalates, represent a challenge to many *in situ* treatments. These compounds have low volatility and solubility and are slow to biodegrade. Soil vapor extraction and air sparging are ineffective without thermal enhancement, and *in situ* bioremediation works slowly, if at all, on many of these compounds. Chemical oxidation, is effective in destroying many of these compounds but is expensive to employ. For these reasons, stand-alone applications of these technologies are not effective for site remediation.

A new approach is to combine chemical oxidation and bioremediation. This combination can result in both effective treatment and substantial cost savings.

There are two questions that need to be answered with respect to combining these technologies. The first is whether or not they are compatible. Will biological activity survive an oxidizing environment? The second question is how these technologies should be combined - with oxidation as the first step or as polishing step, or perhaps intermittently throughout the process.

CHEMICAL OXIDATION

A number of oxidation systems can be employed with bioremediation including ozonation and Fenton's Reagent (hydrogen peroxide plus iron). The types of applications are driven by chemical reactivity and physical properties of these oxidants.

Ozone treatment is a proven and commonly applied method of waste water treatment. The powerful oxidizing effects of ozone (O_3) have been successfully applied for many years in municipal and industrial waste water treatment. A primary reaction mechanism of ozone oxidation involves the attack at carbon-carbon double bonds contained in PAHs and many other organic compounds. Complete oxidation yields carbon dioxide as the final product. Because ozone is a gas and can be easily generated on site, it is amenable to use for *in situ* remediation using air sparging technology (Nelson and Brown, 1994).

Fenton's reagent is a mixture of hydrogen peroxide with ferrous iron (generally ferrous sulfate). This combination produces hydroxyl radicals that attack and degrade many organic compounds. Like ozone, Fenton's reagent has been successfully used for many years for the treatment of waste waters. Application of Fenton's Reagent's to environmental problems has focused on its uses in soil/water slurry reactors and land farming. (Lewis, 1993; Srivastava et al., 1994; Lewis et al., 1995).

BIOREMEDIATION OF PAHs

High molecular weight polyaromatic hydrocarbons (PAH) such as benzo(a)pyrene drive remediation at sites since many are extremely carcinogenic. However, bioremediation is not always effective in degrading these compounds. The rate and extent of biodegradation of PAHs is a function of the number of aromatic rings in the structure. As shown in Figure 1, PAHs with 2 or 3 rings biodegrade rapidly, while high molecular weight PAHs with 5 or 6 rings degrade slowly. This resistance of 5- and 6-ring PAH is due, in part, to the fact that PAH solubility decreases with increasing number of rings. Benzene (a one-ring aromatic compound) has a solubility of 1760 milligrams per liter (mg/L); naphthalene (2 rings), 31 mg/L; chrysene (3 rings), 5 mg/L; and benzo(a)pyrene (5 rings), 5 μg/L. This lower solubility limits the availability of these molecules to the bacteria. Additionally, high molecular weight multi-ring PAHs such as benzo(a)pyrene may be too large to be easily transported through the bacterial cell wall. As a result of these limitations, the conventional biodegradation of PAH mixtures is limited in effectiveness.

ENHANCED BIOREMEDIATION OF PAHs

Because high molecular weight PAHs are generally resistant to biodegradation processes, their presence in soil and groundwater may limit the successful closure of coal tar or creosote sites by conventional *in situ* bioremediation. Also, in states which regulate soil cleanup goals for petroleum products by specific compounds, carcinogenic PAHs such as benzo(a)pyrene present residual problems even at diesel or fuel oil sites. Therefore, the integration of bioremediation and chemical oxidation is of interest as an effective treatment approach for these sites.

The effect of chemical oxidation on the biodegradation of PAHs is illustrated by comparing Figures 1 and 2. As shown in Figure 1, PAH with 2, 3 or 4 rings are readily degraded during aerobic bioremediation, while 5- and 6-ring PAH are much more resistant (Fluor Daniel GTI, 1995). Chemically-assisted bioremediation, such as using Fenton's reagent, Figure 2, is much more effective in treating the full range of PAHs than is conventional bioremediation (Srivastava, et al., 1994).

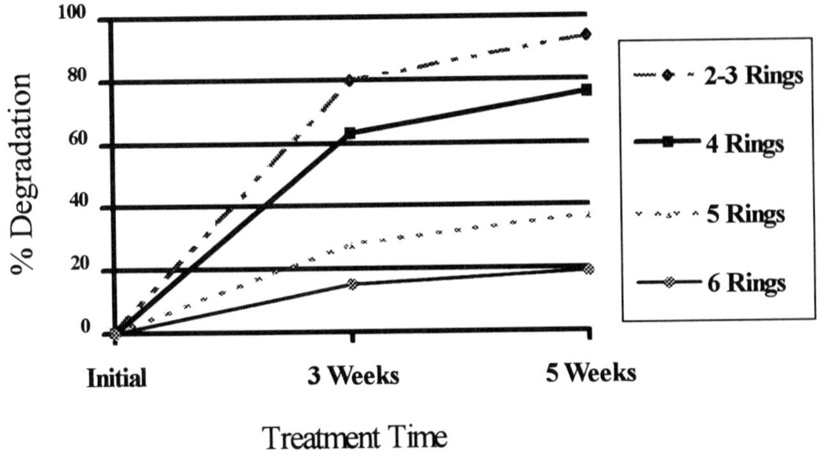

Figure 1: Aerobic Biodegradation of PAHs

With chemically-assisted bioremediation, the biological response is much less perturbed by the number of rings than it is with conventional bioremediation. While the higher molecular weight PAHs are initially slower to biodegrade even with the addition of chemical oxidation, they are removed to a greater extent with chemical pretreatment than without. This suggests that the chemical pretreatment addresses

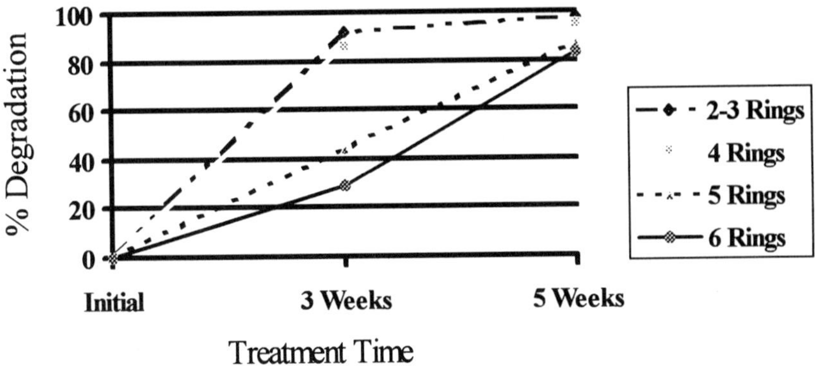

Figure 2: Chemically Enhanced PAH Biodgradation

one or both of the limitations postulated for conventional bioremediation of PAHs - lack of solubility or too large of a molecular structure. Comparing the data in Figures 1 and 2 would suggest that the effect of chemical pretreatment is to both increase the solubility of the PAH and to break the ring structure.

There are several reasons for this supposition. First, the enhancing effect of chemical pretreatment is observed regardless of the number of rings in the PAH. In the chemical-biological treatment (CBT) process developed by the Institute of Gas Technology (IGT), presented in Figure 2, oxidant is added to only partially oxidize the PAHs. Therefore, the complete degradation regardless of number of rings suggests that chemical pretreatment does not destroy the molecular structure, but may create soluble products more available for biodegradation. There is, however, some degradation of the PAH structure with partial opening of the ring(s). There is rate differentiation for PAHs: the 5- and 6-ringed compounds take longer to biodegrade than smaller 2-, and 3-ring PAHs. Smaller PAHs are more susceptible to oxidative attack and show more rapid degradation.

Biological Activity after Chemical Oxidation. An important factor in integrating chemical oxidation and biodegradation is the impact of an oxidizing environment on the microbial population. This is a genuine concern: oxidants such as ozone and hydrogen peroxide have been used as disinfectants. Moreover, in completely mixed systems with relatively long contact times such as slurry reactors, oxidants have been shown to dramatically reduce biological activity. Thus, reintroduction of a bacterial inoculum is often recommended for optimum success of subsequent biological treatments after a chemical oxidation step in the CBT process. However, for *in situ* applications, reinoculation is not practical.

Biological activity following ozone sparging was evaluated at a (pentachlorophenol) PCP site. PCP, while biodegradable, presents problems for *in situ* bioremediation. High concentrations of PCP in soil or groundwater are toxic to bacteria (Ruckdeschel et al., 1987; USEPA, 1991). Therefore, the extent of remediation is unpredictable and often above the regulatory cleanup goals. Since high concentrations of PCP were detected in both soil and groundwater at this site, chemical treatment using ozone sparging was proposed as a conditioning to bioremediation.

Air mixed with ozone was injected below the water table over a six-week test period. During that time, ozonation reduced PCP concentrations in groundwater from 50% to greater than 99%. Ozonation reduced the potentially toxic concentrations of PCP, but the question remained of whether *in situ* biological activity would survive ozonation.

In situ respirometry testing was conducted after ozone sparging to measure biological activity. Immediately after ozonation, biological activity was suppressed as indicated by the production of only small amounts of carbon dioxide within the first 24 hours. However, five to seven days after cessation of ozone sparging, the rate of carbon dioxide production greatly increased indicating a rebound in biological activity. *In situ* oxidation will reduce biological activity, but this effect is only temporary. Sufficient bacteria survive *in situ* ozonation to resume biodegradation after oxidation.

Integrating Oxidation and Biodegradation. The relative efficacy chemical attack and chemically enhanced bioremediation is seen by comparing the maximum removal observed over a five-week period as illustrated in Figure 3. As can be seen, the chemically enhanced bioremediation and direct ozonation give very similar results. Since the direct ozonation takes the PAHs to carbon dioxide, it may be concluded that the continued addition of an oxidant leads to the complete degradation of the PAH structure.

Since both methods are effective in treating PAHs, the question is how may they be effectively combined. Oxidation can be used to pre-condition substrates for biodegradation. Pre-conditioning involves partial oxidation which can sufficiently change the chemical properties of these molecules to increase their availability to biological degradation. Also as in the case of the PCP example discussed above, *in situ* ozonation can be used to reduce toxic concentrations in 'hot spots' to levels at which biological activity can act. While this approach has always been viable for *ex situ* application, the rebound of biological activity after *in situ* ozonation demonstrates that this approach is also applicable to *in situ* remediation.

Oxidation may also be used as a "polishing" system, to remove residual contaminants and achieve regulatory standards. Since chemical oxidants are generally non-specific in their reactions, they will attack both easily biodegraded and recalcitrant molecules. Removal of easily biodegradable molecules by bioremediation will reduce the mass of contaminant to be chemically treated. Lower doses of oxidant can then be cost effectively used to degrade the residual recalcitrant compounds.

A third approach, alternating biodegradative and oxidative conditions over the

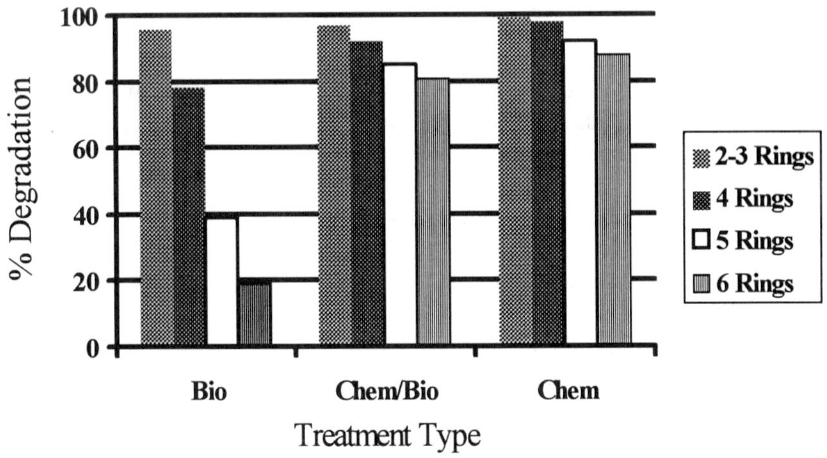

Figure 3: Comparison of Treatment after 5 Weeks

life-time of the remediation, is also feasible. Initial biological degradation will reduce contaminant mass and can increase availability of residual contaminants. Application of chemical oxidants will then pre-condition these residuals for a final treatment by bioremediation. This approach will also be advantageous in cases where toxic intermediates are produced biologically. PCP degradation, for example, produces transient intermediates which can suppress biological degradation. Alternating chemical and biological treatment can overcome this problem.

CONCLUSIONS

The integration of chemical oxidation with bioremediation is proving to be an effective technology for the treatment of many contaminants both *in situ* and *ex situ*. The wide range of chemicals susceptible to chemical oxidation by either ozone or Fenton's reagent, combined with the cost-effective mass reduction of bioremediation makes this a practical approach for site remediation.

REFERENCES

Fluor Daniel GTI, Inc. (1995) Unpublished data.

Lewis, R. F. (1993) "SITE demonstration of slurry-phase biodegradation of PAH contaminated soil." *Air & Waste* 43:503-510.

Lewis, F., R. Kelley and J.P. Moreau (1995) "Combined Chemical-biological Treatment of Soils and Sediments Contaminated with Polyaromatic Hydrocarbons (PAHs)." International Symposium and Trade Fair on the Clean-up of Manufactured Gas Plants. Prague, Czech Republic. Sept. 19-21, 1995.

Nelson, C.H. and R.A. Brown (1994) "Adapting Ozonation for Soil and Groundwater Cleanup." *Chemical Engineering*, November 1994.

Ruckdeschel, G., G. Renner and K. Schwarz. (1987) "Effects of Pentachlorophenol and Some of its Known and Possible Metabolites on Difference Species of Bacteria." *Appl. Environ. Microbiol.* 53(11):2689-2692.

Srivastava, V.J., R.L. Kelly, J.R. Paterek, T.D. Hayes, G.L. Nelson and J. Golchin. (1994) "A field-scale demonstration of a novel bioremediation process for MGP sites." *Applied Biochemistry and Biotechnology* 45/46: 741-756.

United States Environmental Protection Agency. (1991) *On-Site Treatment of Creosote and Pentachlorophenol Sludges and Contaminated Soils.* Office of Research and Development. Washington, DC. EPA/600/2-91/019. Pp. 1-222.

DECHLORINATION BY METHANOGENS OR CO-FACTORS IN THE PRESENCE OF IRON

Paige J. Novak and Gene F. Parkin

ABSTRACT: Preliminary results show that the rate of carbon tetrachloride (CT) and chloroform (CF) dechlorination is enhanced with methanogens and elemental iron (Fe^0)are incubated together. The enhanced dechlorination is thought to be due to H_2-utilizing organisms using the increased pool of electron donor (H_2) to dechlorinate CT and CF more rapidly and completely. Pure culture work with both H_2-utilizing and non-H_2-utilizing methanogens was carried out to test this hypothesis. *Methanosarcina barkeri* and *Methanosarcina thermophila* were used for these studies. *Ms. barkeri* was able to degrade CT and CF much more rapidly when in the presence of Fe^0. Under similar conditions, *Ms. thermophila* also carried out enhanced dechlorination. This could possibly be due to H_2 serving as an electron donor for dechlorination even when the cells are not able to grow on H_2; the internal cell machinery still functions for dechlorination. Work is also being carried out using purified co-factor F_{430} and vitamin B_{12} to investigate and compare the degradation pathway of CT when iron is present.

ENHANCED BIOTRANSFORMATION OF CARBON TETRACHLORIDE BY AN ANAEROBIC ENRICHMENT CULTURE

Syed A. Hashsham (Michigan State Univ., Center Micro. Ecol., East Lansing, MI)
David L. Freedman (Clemson Univ., Environ. Systems Engin., Clemson, SC)

ABSTRACT: In-situ bioremediation of carbon tetrachloride (CT) faces a number of significant challenges, including a relatively slow rate of transformation at CT concentrations above several mg/L and substantial accumulation of lesser chlorinated methanes. The objective of this study was to evaluate the effectiveness of using cyanocobalamin (Cbl) to overcome these limitations. A methanol-grown methanogenic enrichment culture with no prior exposure to halogenated organics was used at CT concentrations ranging from 4-17 mg/L. In the presence of 10 μM Cbl, the overall rate of CT transformation was 5 times higher than in cultures that received none of the coenzyme. CT transformation was also accelerated by Cbl doses as low as 0.1 μM. At the same time, the presence of Cbl minimized accumulation of chloroform, to as little as 0.3% of the total CT consumed. CS_2 was initially a significant abiotic transformation product (30-60%); however, formation ceased as sulfide levels declined. Transformation of CT was sustainable in Cbl-supplemented cultures for over 247 days with H_2 as an electron donor. The adsorption of Cbl to aquifer materials was also found to be marginal, suggesting that minor losses would be encountered during in situ application.

INTRODUCTION

Transformation of carbon tetrachloride (CT) has been studied extensively in the past. Recently this subject has gained interest because of the need to economically remediate hazardous waste sites contaminated with CT. Anaerobic processes have often been characterized as relatively slow and accompanied by accumulation of chloroform (CF) and dichloromethane (DCM). Evidence suggests that the microbial transformation of CT is cometabolic and related to the activity of porphinoids and corrinoids.

One possible way to alleviate the problem of slow rates and accumulation of reductive halogenated products is to augment the microbial community with cyanocobalamin (Cbl), a commercially available form of corrinoid. This approach has been proven successful with a DCM-degrading enrichment culture (Hashsham, et al., 1995) and a pure culture of the acetogen *Acetobacterium woodii* (Hashsham and Freedman, 1996). A 5-10 fold increase in rate and almost complete elimination of CF and DCM, with a concomitant increase in CO_2 were observed when catalytic amounts of a cobalamin were added to live cultures in the presence of a suitable electron donor.

However, before field trials of this approach are warranted, additional questions need to be addressed, including the effect of inoculum source, the potential for adsorption of cobalamin to aquifer materials, and the effect of sulfide levels on CT transformation rates and products. The objective of this study was to address each of these issues, using a methanol-grown enrichment culture having no prior exposure to CT.

MATERIALS AND METHODS

A mixed methanogenic culture was developed using inoculum from a municipal anaerobic digester (Urbana, Illinois), diluted in a basal medium previously described (Freedman and Gossett, 1989) and enriched for methanol degrading organisms. At the start of the CT transformation studies, the culture was actively degrading 4.65 mM of methanol every 2 days, producing nearly stoichiometric amounts of methane.

The sand and sandy soil used for the adsorption studies had the following characteristics; sand: nominal diameter 0.25-0.35 mm, porosity 0.36, bulk density 1.63 g/cm^3, organic content 0.14%; sandy soil: bulk density 1.37 g/cm^3, porosity 0.39, sand 52.5%, silt 28.1%, clay 19.4%, organic content 2.49%, cation exchange capacity 21 meq/100 g, and pH 7.31. Tritiated water, 0.58 x 10^6 disintegrations per minute (dpm) per mL, was obtained from the Radioisotope Laboratory of the University of Illinois at Urbana. The source of all other chemicals used in this study has been described previously (Hashsham, et al., 1995).

Analytical Techniques. CT, CF, DCM, CH_4, and CS_2 were measured by gas chromatography using a flame ionization detector, and CO was analyzed using a thermal conductivity detector, as previously described (Hashsham, et al., 1995). ^{14}C volatile compounds were determined using a GC-combustion tube technique, while $^{14}CO_2$ was analyzed by acidifying and purging 20 mL of the liquid sample with N_2 and trapping it in NaOH (Freedman and Gossett, 1989). A similar procedure involving acidification, purging, and filtration (0.22 μm Whatman) gave the soluble and cell-associated fractions (Freedman and Gossett, 1989). Cbl was analyzed on a 250-mm x 4.6-mm RP-318 column (Bio-Rad Labs) and a liquid chromatograph (Hewlett-Packard 1090) equipped with a diode array detector (Hashsham, 1996). All the products reported in this study are in μmol per bottle unless otherwise specified.

Transformation Studies. CT transformation studies were conducted in 160-mL serum bottles (in duplicate) sealed with slotted gray butyl rubber septa containing 100 mL culture under a 30% CO_2/70% N_2 atmosphere. The effect of Cbl was tested in live culture, autoclaved culture, and autoclaved basal medium, at several Cbl doses. An additional set of live cultures with the above Cbl doses but no CT served as the controls. Methane production from these controls was monitored to determine the extent of possible cyanide toxicity (released from the added Cbl), and to determine if Cbl served as an electron donor. CT doses between 6-19 μmol (4-17 mg/L) were used over the study period (247 days). The bottles were covered with aluminum foil (to minimize photodegradation of the Cbl) and were incubated on a shaker table at 35 °C, with the liquid in contact with the septum.

On day 235, some of the bottles received approximately 1.5x10^6 dpm of [^{14}C]CT along with some unlabeled CT, resulting in approximately 2 μmol/bottle. After complete CT disappearance in most bottles on day 247, they were analyzed for ^{14}C-products. Percent recovery during ^{14}C analysis (defined as the total dpm recovered in all components {CT + CF + DCM + CS_2 + CO + CO_2 + soluble products + cell associated fraction} divided by the total dpm present in a serum bottle at the time of analysis) was 93% (one standard deviation = 3.5%).

Adsorption Studies. Continuous flow adsorption experiments were conducted in a 450-mm x 51-mm liquid chromatography column (Ace Glass, Inc.) packed with approximately 850 g of sand or a sandy soil. After passing through at least 30 pore volumes of 0.01 M $CaCl_2$, the experiment was started by pumping 10 μM Cbl solution spiked with ^3H_2O (resulting activity, 675 dpm/mL) at a flow rate of 0.2 mL/min (pore water velocity of 0.14 m/day). Effluent fractions (20-50 min intervals) were analyzed for ^3H activity and Cbl, as previously described (Hashsham, 1996).

RESULTS

Addition of 10 μM Cbl enhanced the rate of CT transformation by live cultures (Figure 1). The enhancement in rate was sustainable over a period of 247 days when hydrogen was also added. In general, the higher the dose of Cbl added to live cultures, the faster the rate of CT transformation (Table 1). Based on two distinct ranges of rates observed, the total study period was divided into two phases; 0-33 days, and 34-247 days. Autoclaved culture and basal medium also showed an increase in the rate of CT transformation with Cbl addition (Table 1). Initially, this increase was approximately one-third of that observed in live culture with Cbl. The results with basal medium were very similar to the autoclaved culture, hence they are not included. The transformation in autoclaved culture and basal medium, both with 10 μM Cbl added, was sustained for only five weeks, consuming a total of 80 μmol CT (Table 1). After that, the transformation completely stopped, presumably due to the depletion of sulfides. The addition of hydrogen did not help the autoclaved culture or basal medium to transform more CT, as expected.

TABLE 1. Cumulative CT transformed and rate of transformation by live and autoclaved cultures with various Cbl doses

		Cbl dose added to culture (μM)				
Parameter	**Culture**	**0.0**	**0.1**	**1.0**	**5.0**	**10.0**
Cumulative CT transformed (0-33 days, μmol/bottle)	live:	9.9	17.1	46.9	102.9	117.6
	autoclaved:	9.2	9.0	29.0	65.1	76.1
Cumulative CT transformed (34-247 days, μmol/bottle)	live:	39.8	47.3	53.7	82.1	113.6
	autoclaved:	14.6	42.4	33.6	18.6	9.0
Rate of CT transformation (0-33 days, μmol/bottle-day)	live:	0.30	0.52	1.4	3.1	3.6
	autoclaved:	0.28	0.27	0.88	2.0	2.3
Rate of CT transformation (34-247 days, μmol/bottle-day)	live:	0.20	0.24	0.27	0.38	0.57
	autoclaved:	0.08	0.21	0.17	0.05	0.05

The main products of CT transformation that were routinely measured by gas chromatography included CF, CS_2, and CO, with only traces of DCM. A complete distribution of products was obtained in live cultures with various doses of Cbl by administering [^{14}C]CT on day 235 at a CT dose of approximately 2 μmol/bottle (Table 2). CO_2 production increased from approximately 29% with no Cbl to 63-66% with the two highest doses of Cbl. There was very little or no CF or DCM in these two sets. CS_2 levels were also very low, amounting to less than 2.0%. Some CO (0-9%) was also observed in these bottles. The soluble products ranged between 17-31%, with lower percentages corresponding to higher Cbl doses. Between 6-8% of the recovered ^{14}C activity was associated with biomass (nonstrippable residue).

Similar amounts of methane (μmol/bottle) with various Cbl doses were observed in methane controls on day 247: 185±4.8, 188±5.3, 185±5.9, 180±7.1, and 201±4 μmol methane for 0, 0.1, 1, 5, and 10 μM added Cbl, respectively.

The breakthrough curves for Cbl in a column containing sand or a sandy soil using distilled water are shown in Figure 2. The retardation factor (calculated as the ratio of the areas projected on the y-axis by the breakthrough curves of Cbl and ^3H_2O) was found to be 1.9 for sand and 2.4 for sandy soil. The results

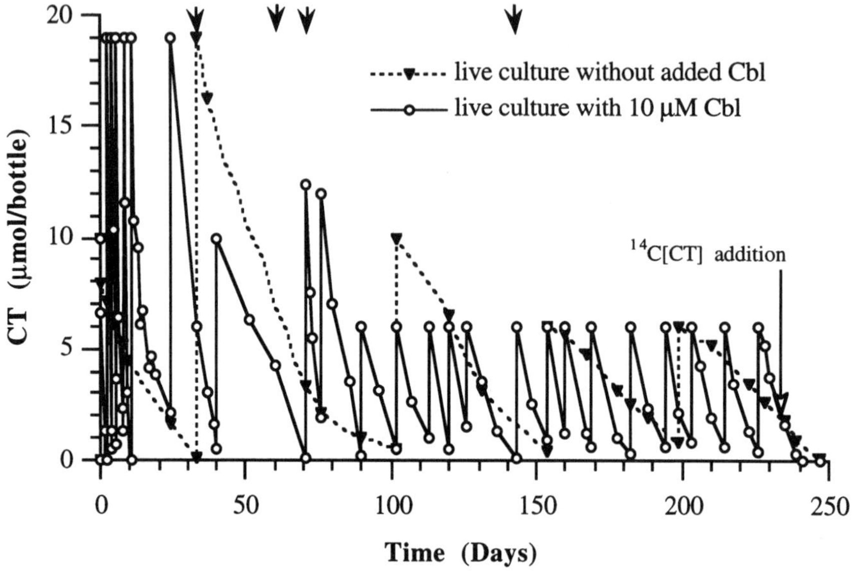

FIGURE 1. Transformation of CT by live cultures with and without Cbl added. Each arrow represents the addition of 80 μmol of H_2. Results for duplicate bottles were very similar.

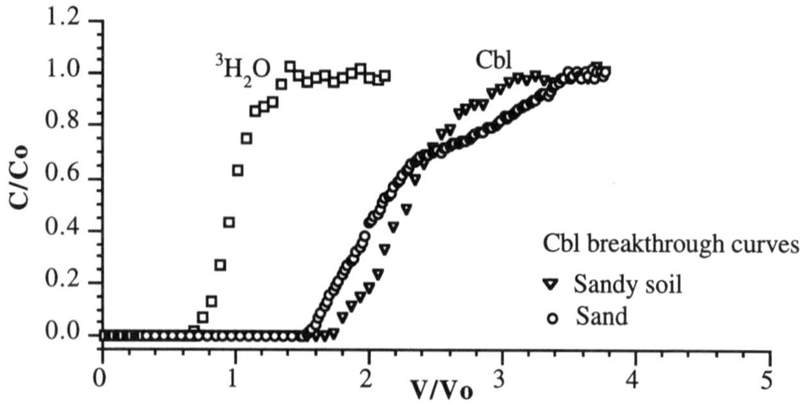

FIGURE 2. Breakthrough curves for ^3H_2O and Cbl in sand and soil.

obtained by batch tests gave a similar retardation factor of 4.0 for sand and 2.3 for kaolinite clay. Adsorption to alumina was insignificant.

TABLE 2. Distribution of ^{14}C (% dpm) from transformation of [^{14}C]CT[a] in live cultures with Cbl

Products	Cbl dose added to live culture (µM) 0.0	0.1	1.0	5.0	10.0
CT	6.8 (±0.57)	5.6 (±1.1)	5.8 (±0.57)	0	0
CF	12 (±0.88)	5.3 (±0.15)	1.6 (±0.84)	0	0
DCM	2.0 (±0.09)	0	0	0	0
CS_2	18 (±1.0)	16 (±0.62)	9.0 (±0.54)	1.1 (±0.30)	2.0 (±0.70)
CO	0	0	3.3 (±0.55)	7.7 (±0.19)	9.0 (±0.40)
CO_2	29 (±1.5)	35 (±1.1)	51 (±2.4)	66 (±0.85)	63 (±2.0)
Soluble NSR[b]	26 (±0.83)	31 (±0.38)	22 (±0.39)	17 (±0.69)	19 (±0.74)
Nonfilterable NSR	6.2 (±0.11)	7.1 (±0.34)	7.3 (±0.45)	8.2 (±1.0)	7.0 (±0.54)

[a] Values in parentheses represent one standard deviation for duplicate bottles.
[b] NSR = Nonstrippable residue.

DISCUSSION

This study demonstrated that enhancement of CT transformation by addition of Cbl to live cultures is not specific to the inoculum source. Cbl increased the formation of CO_2 and decreased CF accumulation. It was also shown that enhanced CT transformation was achieved at very low doses of Cbl. Moreover, the adsorption of Cbl to aquifer material was marginal, indicating that adding cobalamins may be a viable option for in situ bioremediation of CT.

These results confirm our previous observations with a DCM degrading enrichment culture (Hashsham et al., 1995) and a pure acetogenic culture (Hashsham and Freedman, 1996), which achieved a 5-10 fold increase in transformation rate and a significant decrease in CF and DCM. The increase in rate was lower (1.2 to 3-fold) with the methanol enrichment culture, depending on the Cbl dose. However, formation of CF and DCM remained negligible. These results indicate that the extent to which Cbl enhancement occurs may vary with the type of culture, CT and Cbl dose, and redox conditions. The initial high rates observed in this study (during 0-33 days) were most likely due to both nonenzymatic reaction with sulfides, and microbial transformations. Once the sulfides were depleted, CT transformation was dominated by H_2-supported microbial activity.

Only a few previous studies using mixed cultures have reported the distribution of products from CT. From pure culture studies the transformation of CT to CF, CO, CO_2, CS_2, soluble products and cell associated material is well established, under several electron acceptor conditions. The results obtained in this study are very similar to those previously reported by us with a DCM degrading enrichment culture (Hashsham, et al., 1995), except the lower amount of CS_2, which was presumably related to lower sulfide levels. CO is one of the major intermediates of nonenzymatic transformations of CT in the presence of corrinoids. High concentrations of CT may also have contributed to the accumulation of CO, due to its toxicity to carbon monoxide dehydrogenase.

Methanogenesis was completely inhibited in all the live cultures receiving CT. Results from controls that received Cbl but no CT indicated that release of cyanide from Cbl did not contribute to inhibition of methanogenesis. These results also showed that Cbl was not biodegraded to methane and therefore did not enhance CT transformation by serving as an electron donor.

Chemical reactions of CT with sulfides and Cbl resulting in the formation of CS_2 were significant during the initial phase of the experiments (days 0-33). CS_2 formation is a concern because of its neurotoxicity, but its formation can be minimized by avoiding high sulfide levels. If sulfides are present along with CT and Cbl addition increases CS_2 formation, it may be necessary to degrade the CS_2 in an aerobic zone, which has been demonstrated (Smith and Kelly, 1988).

The low retardation factor for Cbl indicates that its movement in aquifers will not be drastically altered due to adsorption. Studies conducted with large molecules including DNA, dyes and proteins, and oligomers indicate that pH, ionic strength, and mineralogy (clay content) rather than organic carbon content of soils play a major role in the adsorption of large molecules. However, the effect of soil organic content on Cbl adsorption needs further investigation, preferably using aquifer soil with varying mineralogical characteristics and organic content.

The results of this study demonstrated that the two major problems encountered during microbial transformations of CT, i.e. a slow rate and accumulation of reductive halogenated products, can be successfully overcome by Cbl additions. Our finding that Cbl-enhanced biotransformation of CT is independent of inoculum source is significant for the application of this approach to bioremediation. Use of relatively low Cbl doses and a marginal adsorption to sand indicate that supplying Cbl to contaminated aquifers may be feasible. Tests under field conditions are needed to more firmly establish the effectiveness and economics of in situ Cbl addition. It is expected that the relatively high cost of Cbl will be counterbalanced by faster in-situ biotransformation, without the accumulation of hazardous products.

ACKNOWLEDGMENTS

This research was supported in part by a grant from the U.S. Department of Energy. The assistance of Mark Lambrü with the adsorption tests is appreciated.

REFERENCES

Freedman, D. L. and J. M. Gossett. 1989. "Biological Reductive Dechlorination of Tetrachloroethylene and Trichloroethylene to Ethylene under Methanogenic Conditions." *Appl. Environ. Microbiol. 55* (9): 2144-2151.

Hashsham, S. A. 1996. "Cobalamin-Enhanced Anaerobic Biotransformation of Carbon Tetrachloride." Ph.D. Thesis, University of Illinois at Urbana-Champaign, Urbana, IL.

Hashsham, S., R. Scholze and D. L. Freedman. 1995. "Cobalamin-Enhanced Anaerobic Biotransformation of Carbon Tetrachloride." *Environ. Sci. Technol. 29* (11): 2856-2863.

Hashsham, S. A. and D. L. Freedman. 1996. *Enhanced biotransformation of carbon tetrachloride by Acetobacterium woodii using hydroxocobalamin.* Proceeding of the 96th General Meeting of the American Society for Microbiology, New Orleans, Louisiana, May 19-23. American Society for Microbiology.

Smith, N. A. and D. P. Kelly. 1988. "Oxidation of Carbon Disulfide as the Sole Source of Energy for the Autotrophic Growth of *Thiobacillus thioparus* strain TK-m." *J. Gen. Microbiol. 134:* 3041.

BIO-COMPATIBILITY OF THE VITAMIN B_{12}-CATALYZED REDUCTIVE DECHLORINATION OF TETRACHLOROETHYLENE

Kelly Millar and Suzanne Lesage
(National Water Research Institute, Burlington, Ontario, Canada)

ABSTRACT: The effectiveness of vitamin B_{12} as a remediation treatment for tetrachloroethylene (PCE) in systems containing microbial activity was examined in soil microcosms and columns. Reductive dechlorination by vitamin B_{12} was dependent upon the availability of the reducing agent, titanium(III) citrate. Bacterial dechlorination of PCE to *cis*-1,2-dichloroethylene (*cis*-DCE) was rapid after an initial acclimation period, however, the dechlorinating consortium consumed the citrate component of the reducing agent. Without citrate to chelate the titanium, it precipitated and the reduced conditions required for the vitamin B_{12}-catalyzed reaction were lost. The addition of glucose as an alternative substrate prevented citrate degradation but resulted in the enrichment of a microbial population that was unable to dechlorinate PCE.

INTRODUCTION

Reductive dechlorination of tetrachloroethylene (PCE) is observed in anaerobic bacterial cultures. The same activity is often evident at contaminated sites, but can be limited by high solvent concentrations. Vitamin B_{12} reduced with titanium(III) citrate is a rapid and effective means of achieving essentially the same reaction at any substrate concentration. A remediation treatment with vitamin B_{12}, coupled with an active bacterial community may, therefore, effectively restore highly contaminated sites. While remediation is often very effective in laboratory microcosms, *in-situ* application of this treatment can be problematic because delivery of the remedial solutions is subject to precipitation, sorption and metabolism. For instance, citrate can serve as a carbon source for bacteria and cause the precipitation of titanium. The addition of alternative carbon sources, for the preservation of citrate, was studied.

In systems containing soil and an active microbial population, the challenge is to maintain a low enough E_h for vitamin B_{12}-catalyzed dechlorination of chlorinated ethylenes to occur. Titanium(III) citrate in concentrations of 4-10 mM can be used in soil microcosms to reduce vitamin B_{12}, without permanent effects on the existing anaerobic microbial species (Millar and Lesage, 1995). Vitamin B_{12} can have initial inhibitory effects on methane production which are reversible after an acclimation period (Lesage et al., 1996). The objectives of the experiments were to study the interactions between a vitamin B_{12} system and a PCE-dechlorinating population and to assess whether a combined system would result in enhanced rates of dechlorination.

MATERIALS AND METHODS

Aquifer Material. Aquifer sand cores were obtained from the Canadian Forces Base Borden, Ontario, in a cell where controlled spill experiments of chlorinated solvents had been previously conducted.

Microcosms. Batch experiments were conducted in 160-mL serum vials, sealed with Teflon-lined butyl rubber stoppers and aluminum crimp seals, containing 50-100 g of Borden sand, various amounts of PCE and vitamin B_{12}, a headspace of $N_2/H_2/CO_2$ (85/10/5%) and 100 mL of titanium(III) citrate (4-10 mM, pH 7-7.5), prepared according to Holliger et al. (1992). Titanium chloride (20%, stabilized) was obtained from Fisher Scientific (Fair Lawn, NJ). As additional substrates, methanol, Bacto-yeast extract (Difco Laboratories, Detroit, MI) and powdered 'glucose solids' (Grain Processing Enterprises, Scarborough, ON) were used. The latter, containing dextrose (15-19%) and maltose (11-15%), is an inexpensive sweetener purchased in bulk for field studies.

Column Study. Two glass columns, 4.8 cm (I.D.) x 30 cm in length, each with 6 sampling ports consisting of ⅛" stainless steel tubing and stainless steel porous frits extending into the centre of the columns, were slurry packed with Borden core material. Titanium(III) citrate (4 mM) was pumped upward through the sand at a rate of 0.5 mL/min for a residence time of 6 hours. Vitamin B_{12} (30 mg/L) was added to the titanium(III) citrate solution of one of the columns. PCE, in methanol, was fed through a 'T' using a syringe pump, yielding a final concentration of 300 μg/L when diluted with the titanium(III) citrate influent.

Analytical Methods. PCE and its dechlorination products were measured in the headspace of the microcosms by using a model 10S Photovac gas chromatograph equipped with a 10.6 eV lamp. A SRI model 8610A dual column GC with an ECD and FID was used for analysis of chlorinated compounds and methane, respectively. For the columns, aqueous samples were removed by syringe and placed into 300-μL vials, from which a 50-μL headspace sample was then removed and analyzed.

Organic acids were analyzed by ion exclusion chromatography using an IC-Pak column (Waters Chromatography, Milford, MA) and a Waters Model 430 conductivity detector. Eluent was 1 mM HCl at 1 mL/min.

Glucose consumption was monitored using a Glucometer Encore blood glucose meter (Bayer Inc., Etobicoke, ON). It should be noted that this method measured only the glucose portion of the 'glucose solids' and not the maltose component.

RESULTS AND DISCUSSION

Microcosms. To study the effect of vitamin B_{12} on bacterial reductive dechlorination, an active population had to be established. Initially dechlorination was slow, but once the cultures became actively methanogenic, dechlorination was rapid with PCE reduced to trichloroethylene (TCE) within one day, followed by further dechlorination to *cis*-DCE within 8 days (Figure 1A). Dechlorination of PCE

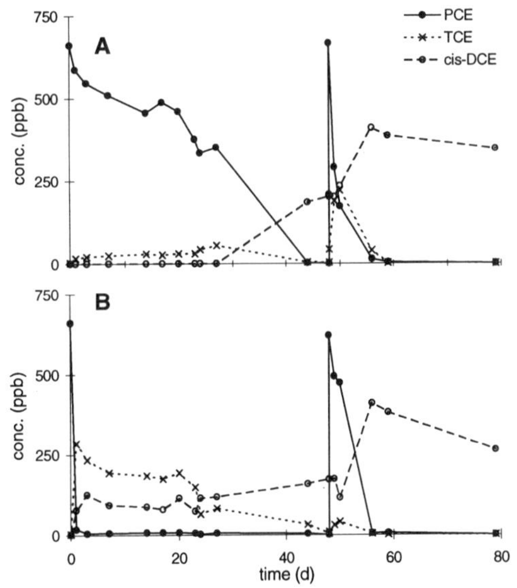

FIGURE 1. Dechlorination of PCE in Borden sand in the absence (A) and presence (B) of vitamin B_{12} (20 mg/L). Microcosms consisted of 100 g sand, 10 mM titanium(III) citrate and 1.5 mM Na_2MoO_4.

in microcosms containing vitamin B_{12} but without aquifer material, proceeded immediately to *cis*-DCE (data not shown). With the introduction of aquifer material, dechlorination to TCE was rapid initially, however, TCE was persistent for over 20 days until a methanogenic population was established, then TCE was dechlorinated to *cis*-DCE (Figure 1B). The persistence of TCE suggested that the vitamin B_{12} was no longer reduced enough to carry the dechlorination reaction further. Dechlorination of a second addition of PCE was therefore likely due to bacterial processes, regardless of the presence of vitamin B_{12} in the system. Analysis of the organic acids in the microcosms indicated that citrate was being degraded to acetate as the predominant product (Figure 2). Without citrate, titanium precipitated and the redox conditions could no longer sustain the vitamin B_{12}-catalyzed reaction.

Column Studies. Dechlorination was also assessed in columns which, through application of a continuous source of remedial solution, would be more representative of an *in-situ* treatment. At the onset of pumping, dechlorination was rapid with approximately 50% of the influent PCE dechlorinated to TCE by the time it reached port 1, a contact time of less than 1 hour (Figure 3). A further 30% was transformed within the 5 hours of reaching port 6. Conversely, dechlorination in the unamended column (no B_{12}) was poor, with less than 10% of the PCE dechlorinated to TCE by port 6 (data not shown). It should be noted that the objective here was not to optimize bacterial degradation, but to look at the difference between dechlorination rates in the presence and absence of vitamin B_{12}.

Dechlorination rates remained consistent for the first

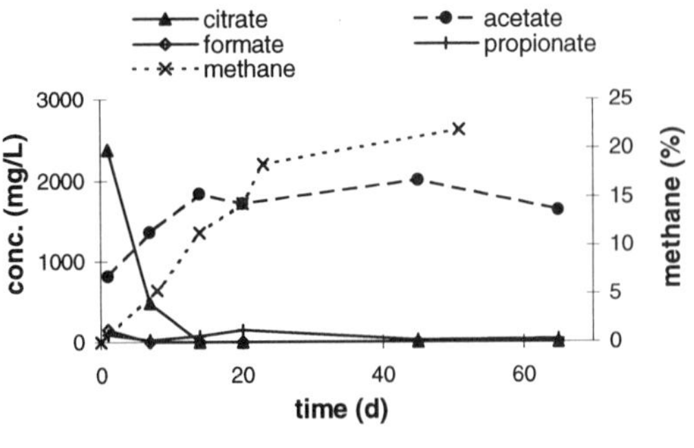

FIGURE 2. Citrate degradation and CH_4 production in Borden sand microcosms. Conditions as in figure 1.

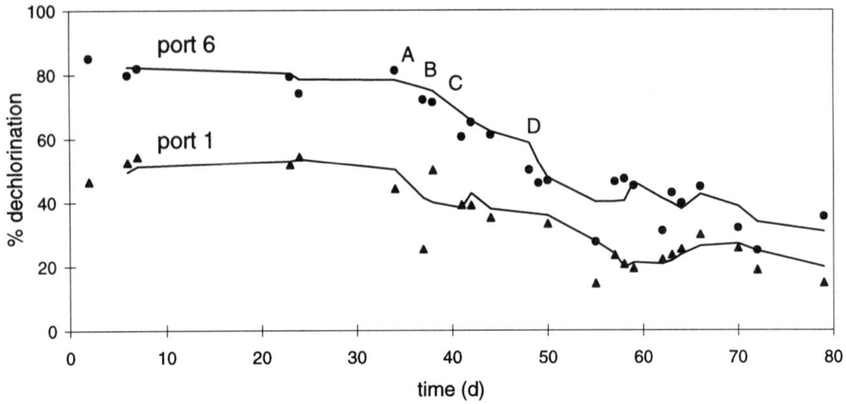

FIGURE 3. Reductive dechlorination of PCE in a Borden sand column containing vitamin B_{12}. Solid lines represent moving averages. A - FeS accumulation, Ti +ve in effluent, B - Na_2MoO_4 added, C - Ti no longer mobile, D - CH_4 production and decreased flow.

30 days of pumping with titanium being mobile throughout the column, as was verified by a simple colorimetric test using HCl and H_2O_2. An accumulation of black ferrous sulfide precipitate in the effluent lines and waste vessel suggested that sulfate-reducing bacteria were flourishing. Sodium molybdate (Figure 3,'B') was added to control this growth. Almost immediately following this addition, the microbial population shifted from sulfate-reducing to methanogenic. Methane production was accompanied by an increase in back pressure, decreased flow, and a drop in dechlorination in the vitamin B_{12} column. In addition, titanium could no longer be measured in either column past port 2. An analysis of the organic acid profile of the columns on day 62 indicated that citrate, necessary for chelating the titanium for transport through the columns, was being degraded to acetate and small amounts of formate and propionate (Figure 4). Pumping rates were increased in efforts to force titanium(III) citrate further along in the column, however, bacterial metabolism increased such that citrate was utilized as fast as it was being introduced.

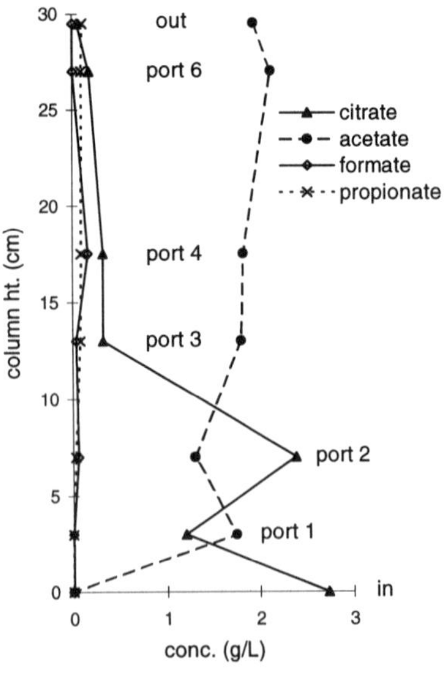

FIGURE 4. Organic acid profile in Borden sand column on day 62.

Alternative Substrates. In order to prevent citrate consumption, microcosms were supplemented with glucose, as potentially a preferred substrate over citrate. Previous work on citrate metabolism indicated that glucose, when added at concentrations

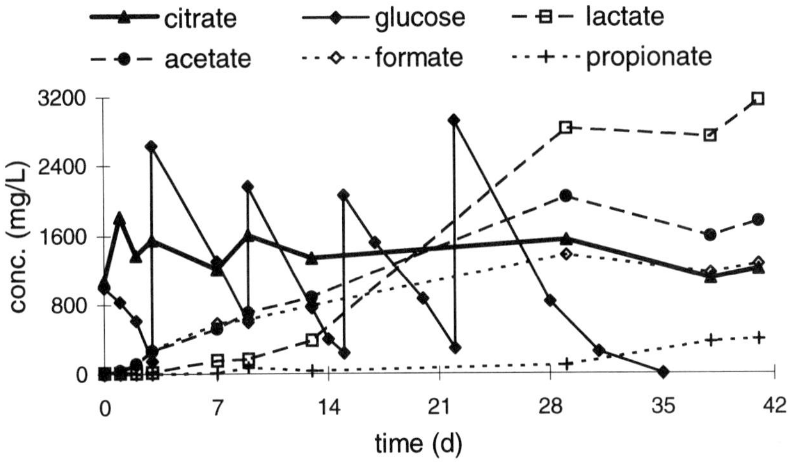

Figure 5. Prevention of citrate utilization by repeated additions of glucose solids. Microcosms contained 50 g Borden sand, 4 mM titanium(III) citrate and glucose solids.

greater than 25 mM, could result in a reduction of citrate consumption (Bellengier et al., 1994). In addition, glucose has been reported to be a sufficient substrate for sustaining reductive dechlorination (Wu et al., 1995; Freedman and Gossett, 1989). Microcosms were established with Borden core material (50 g), titanium(III) citrate (4 mM) and 'glucose solids'. It was observed that by replenishing the glucose before it was exhausted, the citrate remained untouched (Figure 5). Predominant products of glucose utilization were lactate, acetate, formate and propionate. The addition of glucose resulted in the development of a different anaerobic consortium, producing large quantities of CO_2 instead of methane. In addition, the microorganisms present were incapable of dechlorinating PCE (data not shown). In separate experiments with pure glucose, citrate was degraded once all the glucose was consumed, however, the population did not shift to methanogenesis nor to a PCE-dechlorinating consortium (data not shown).

Yeast extract and methanol were also tested as alternative substrates. Methanogenic populations developed faster when yeast extract was added and a shorter initial lag in dechlorination was noted (data not shown). Citrate, in fact, was degraded faster when yeast extract was present with almost complete removal within one day. Methanol also enhanced the onset of methanogenesis, however did not promote rapid dechlorination nor prevent citrate from being utilized. The tendency for cultures to shift from dechlorination to methanogenesis when amended with methanol has been reported (Smatlak et al., 1996; Fennell et al., 1995; Wu et al., 1995).

CONCLUSIONS

The success of an *in-situ* application of reduced vitamin B_{12} for PCE remediation is dependent upon adequate delivery of the remedial solution to the contaminated zone. In situations where citrate may be rapidly consumed, it may be necessary to supplement the system with additional carbon sources, such as glucose, to preserve the integrity of the reducing agent. Whereas it would be desirable to incorporate the benefits of the vitamin B_{12} reaction with an actively dechlorinating bacterial population, amendments with glucose may promote the growth of strains incapable of carrying out reductive dechlorination. Column studies are continuing to further examine the delicate balancing of conditions for optimal dechlorination.

REFERENCES

Bellengier, P., D. Hemme, and C. Foucaud. 1994. "Citrate Metabolism in 16 *Leuconostoc mesenteroides* subsp. *mesenteroides* and subsp. *dextranicum* Strains." *Journal of Applied Bacteriology.* 77: 54-60.

Fennell, D. E., M. A. Stover, S. H. Zinder, and J. M. Gossett. 1995. "Comparison of Alternative Electron Donors to Sustain PCE Anaerobic Reductive Dechlorination." In R. E. Hinchee, A. Leeson, and L. Semprini (Eds.), *Bioremediation of Chlorinated Solvents*, pp. 9-16. Battelle Press, Columbus, OH.

Freedman, D. L. and J. M. Gossett. 1989. "Biological Reductive Dechlorination of Tetrachloroethylene and Trichloroethylene to Ethylene under Methanogenic Conditions." *Applied and Environmental Microbiology.* 55(9): 2144-2151.

Holliger, C., G. Schraa, E. Stupperich, A. J. M. Stams, and A. J. B. Zehnder. 1992. "Evidence for the Involvement of Corrinoids and Factor F_{430} in the Reductive Dechlorination of 1,2-Dichloroethane by *Methanosarcina barkeri.*" *Journal of Bacteriology.* 174(13): 4427-4434.

Lesage, S., S. Brown, and K. Millar. 1996. "Vitamin B_{12}-Catalyzed Dechlorination of Perchloroethylene Present as Residual DNAPL." *Ground Water Monitoring and Remediation.* 16(4): 76-85.

Millar, K. and S. Lesage. 1995. "Dechlorination of Tetrachloroethylene by Anaerobic Bacteria and Vitamin B_{12} for Groundwater Remediation." Presented at the 30th Central Canadian Symposium on Water Pollution Research, Feb. 6-7, Burlington, ON.

Smatlak, C. R., J. M. Gossett, and S. H. Zinder. 1996. "Comparative Kinetics of Hydrogen Utilization for Reductive Dechlorination of Tetrachloroethylene and Methanogenesis in an Anaerobic Enrichment Culture." *Environmental Science and Technology.* 30(9): 2850-2858.

Wu, W. -M., J. Nye, R. F. Hickey, M. K. Jain, and J. G. Zeikus. 1995. "Dechlorination of PCE and TCE to Ethene Using an Anaerobic Microbial Consortium." In R. E. Hinchee, A. Leeson, and L. Semprini (Eds.), *Bioremediation of Chlorinated Solvents*, pp. 45-52. Battelle Press, Columbus, OH.

APPLICATION OF CHEMICAL PROCESSES FOR THE TREATMENT OF LEACHATE FROM SOLID WASTE LANDFILL

Soon Haing Cho, Young Soo Choi
and Je Yong Yoon (Ajou university., Suwon, Korea)
Hee Chan Yoo and Eui Sin Lee (Daewoo Corporation, Suwon, Korea)

ABSTRACT: For the on-site treatment of non-biodegradable or refractory organics contained in the leachate from solid waste landfill, the feasibility of combining chemical treatment processes to the biological treatment was studied. Treatability of coagulation and catalyzed H_2O_2 oxidation process were tested for the treatment of non-biodegradable organics. Analyses of molecular weight distribution of leachate were conducted to investigate the effectiveness of chemical treatment and biological treatment. Major constituents of leachate to be treated were relatively non-biodegradable organics and nitrogen compounds. The overall removal efficiency of coagulation process accompanied by catalyzed H_2O_2 oxidation process was about 48 % (COD basis) which suggests these chemical treatments are necessary processes for the removal of non-biodegradable organics contained in the leachate. Biological process was effective for removing the organics below molecular weight of 500 while chemical process showed opposite result. This fact also suggests chemical treatment process is preferable to treat the leachate containing the organics above molecular weight of 500.

INTRODUCTION

The organics of leachate from newer landfill are relatively biodegradable, while the main organic components from aged landfill are microbially refractory substances (Chian and Dewalle, 1976; Christensen et al, 1992). Biological treatment alone cannot meet the Korean stringent effluent guideline of leachate generated from solid waste landfill, if it is to be treated on-site, because of these non-biodegradable or refractory substances. Accordingly, combining physico-chemical treatment might be necessary. The objective of this study is to access the feasibility of combining chemical treatment processes to the biological treatment for the removal of refractory organics of leachate from solid waste landfill.

MATERIALS AND METHODS

Leachate Characteristics. The leachate used in this study was collected from the Metropolitan Landfill which is the largest solid waste landfill site in Korea. All of the analytical procedures were conducted in accordance with the method prescribed in Standard Methods for the Examination of Water and Wastewater (1995).

Coagulation. Main purpose of coagulation process was to remove the organics contained in the leachate. This process also can serve as an auxiliarly process to minimize the organic loading to chemical oxidation process, which brings the reduction of oxidant and catalyst dosage, and/or to reduce the organic loading to the biological treatment system which is followed after chemical processes. Coagulation with $FeCl_3$ was showed the best treatment efficiencies among the various coagulants tested. Optimum conditions for coagulation process, such as optimum dosage of $FeCl_3$ and optimum pH range, were investigated by bench-scale experiment.

Chemical Oxidation. Oxidation with catalyzed H_2O_2 was selected as the most feasible and effective process among three different oxidation methods such as oxidation by ozone, oxidation by combination of O_3 and UV, and oxidation by catalyzed H_2O_2. Operational parameters, such as catalyst and H_2O_2 dosage, optimum pH and optimum reaction time were investigated. Removal efficiencies of COD were investigated at the determined optimum operational conditions.

Biodegradability Test. Change of biodegradability test was conducted to validate the application of chemical processes before biological processes. Biodegradability change of organic contained in the leachate by each chemical process was measured by using electrolytic respirometer(BI-1000, Bioscience, Inc.). For these purposes, COD and accumulated oxygen uptake(AOU) of leachate before and after introducing to each chemical process were compared.

Molecular Weight Characterization. Organics in the leachate were also analyzed by means of molecular weight basis using ultrafiltration (Slater et al. 1985) methods(YCO5, YM1, YM3). The investigated molecular weight cutoff were 500, 1000, 3000. The ultrafiltration cell was operated in a batch model with nitrogen gas applied to pressurize the system. All samples for characterizing the molecular weight of leachate organics were prefiltered with 0.45 μm membrane filter. Small quantities of permeate were collected and analyzed for TOC. UV-persulfate method was employed for TOC analyses.

RESULTS AND DISCUSSION

Leachate Characteristics. Major constituents contained in the leachate to be treated were nitrogen compounds and organics. More than 99 % of nitrogen compound was existed as NH_3-N form. The BOD_5/COD was in the range of 0.2 to 0.3. This ratio suggests that major portion of organic was considered to be non-biodegradable organics. This ratio also suggest that the application of chemical treatment processes is required to treat the non-biodegradable organics. pH of leachate was in the range of 7 to 8. More than 99 % of total solid existed as dissolved solid. The alkalinity was in the range of 3,500~6,000 mg/L. The concentrations of heavy-metals were negligible.

Coagulation. Optimum pH condition for coagulation was found to be around 5(Figure 1). Optimum $FeCl_3$ dosage was in the range of 1000 to 1400 mg/L(Figure 2). This dosage variation was changed by the characteristics of leachate. Settleability of sludge generated by $FeCl_3$ coagulation was better and the quantity of sludge generated by $FeCl_3$ coagulation was less than that of sludge generated either by alum or by lime.

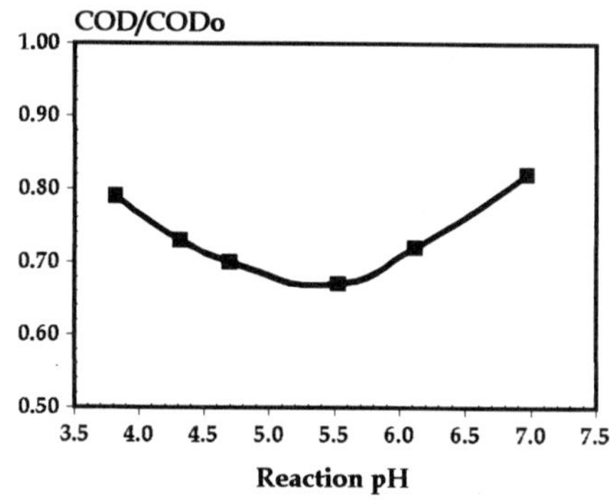

FIGURE 1. Change of removal efficiencies by pH variation; $FeCl_3$.

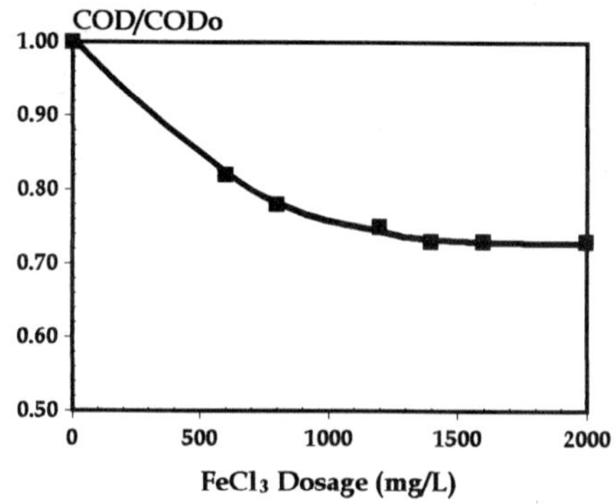

FIGURE 2. Change of removal efficiencies by various dosage of $FeCl_3$.

Chemical Oxidation. Oxidation reaction was rapidly occurred for the first 30 minutes after introducing the H_2O_2 and Fe^{2+} to the leachate. The increment of removal efficiency after 30 minutes was less than 5 %. From these results, it could be concluded that the optimum oxidation time would be from 0.5 to 1 hr. Fig. 3 shows the COD removal efficiencies changes by H_2O_2 dosage and H_2O_2 /Fe^{2+} ratio variations. The optimum H_2O_2 dosage and H_2O_2/Fe^{2+} was found to be 1,500mg/L and 1:1(weight basis), respectively. More than 35 % of COD removal was achieved at these conditions.

Biodegradability Test. The accumulated oxygen uptake(AOU) change with respect to the chemical pretreatment applied to the raw leachate was investigated to access the effectiveness of chemical treatment application. According to this result(Figure 4) the combination with coagulation and catalyzed H_2O_2 oxidation showed the highest AOU which implies that the application of chemical pre-treatments is an effective process to improve the biodegradability of organics contained in the leachate. According to COD removal efficiencies change investigation(Figure 5), coagulation process alone showed good performance of organic removal which suggests that the combination of coagulation with catalyzed oxidation process might not be always required.

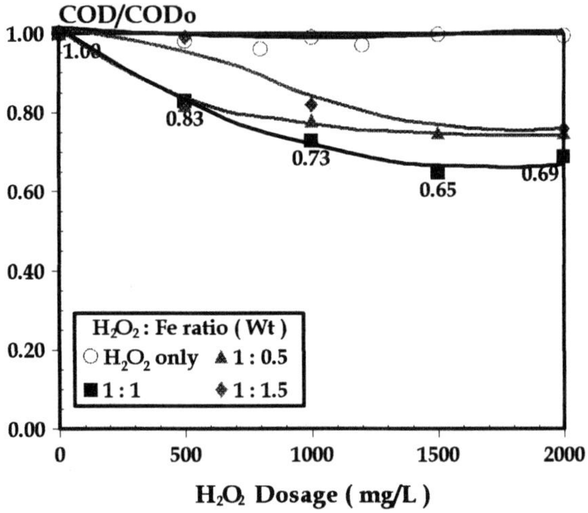

FIGURE 3. Change of COD removal efficiencies change by various H_2O_2 dosage and H_2O_2/Fe^{2+} ratio variation.

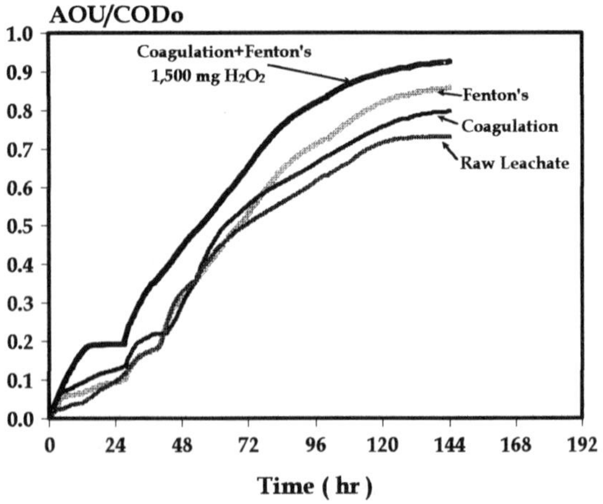

FIGURE 4. Accumulated oxygen uptake change with respect to the chemical pretreatment applied to raw leachate.

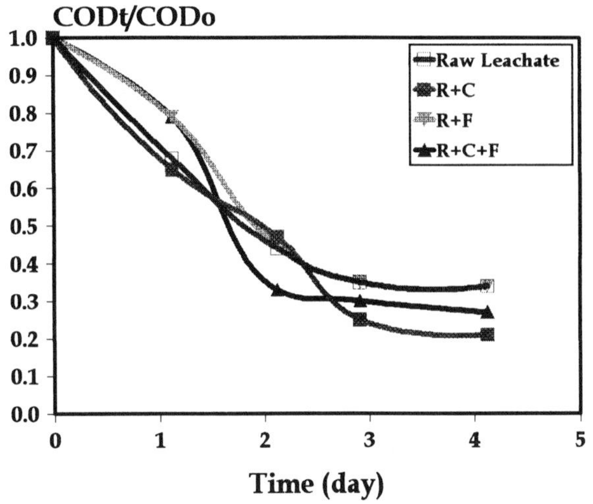

FIGURE 5. Change of COD removal efficiencies by reaction time during bio-oxidation. R: raw leachate, C: coagulation and flocculation, F: Fenton's Oxidation.

Molecular Weight Characterization. The effect of biological process and catalyzed H_2O_2 oxidation process in the leachate treatments on the distribution of molecular weight were displayed on Figure 6 and Figure 7 respectively. K1, K2 and K3 were calculated from TOC of the samples collected before and after each treatment process. Biological process was effective for removing the leachate organics below molecular weight of 500(Figure 6). This trend was consistent with the bioavailability of leachate organics to microorganism in biological process. Organics above molecular weight of 500 was effectively removed by catalyzed H_2O_2 oxidation process(Figure 7). Coagulation process accompanied by catalyzed H_2O_2 oxidation process also showed good performances of removing the organics of above molecular weight of 500.

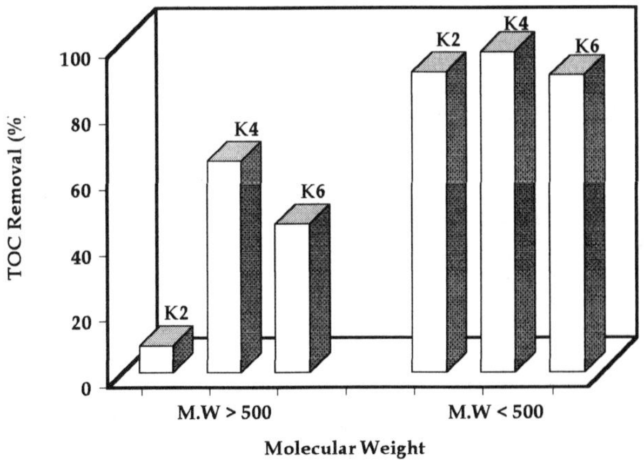

FIGURE 6. Removal efficiencies of the biological treatment with respect to molecular weight.

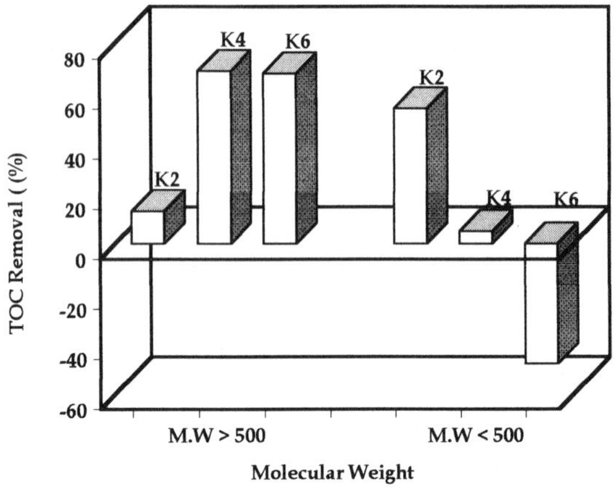

FIGURE 7. Removal efficiencies of the catalyzed H_2O_2 oxidation with respect to molecular weight.

CONCLUSIONS

Chemical pretreatment is necessary treatment process to achieve the effluent guideline of Korea when on-site treatment of leachate alone is considered. Combination or separate application of coagulation with $FeCl_3$ and oxidation with catalyzed H_2O_2 was found to be the proper treatment process to reduce the organic loading to the biological treatment unit and/or to enhance the biodegradability of the non-biodegradable organics.

Biological process was effective for removing the leachate organics below molecular weight of 500, while chemical processes showed opposite results. This fact also suggests chemical treatment process is preferable to treat the leachate containing the organics above molecular weight of 500.

ACKNOWLEDGEMENT

This study was supported by Ministry of Environment of Korea.

REFERENCES

APHA, AWWA and WEF. 1995. *Standard Methods for Examination of Water and Wastewater*. 19th Edition

Chian, E. S. K and Dewalle, F. B. 1976. "Sanitary Landfill Leachates and Their Treatment" *J. of the Enivironmental Engineering Division. ASCE*. 102(EE2).

Christensen, T. H., Cossu, R. and Stegmann, R. 1992. *Landfill of Waste: Leachate* Science Publishers Ltd.

Slater, C.S., Uchrin, C.G., and Ahlert, R.C. 1985. "Ultrafiltration Processes for the Characterization and Separation of Landfill Leachate" *J. Environ. Sci. Health* A20(1).

CATALYTIC STIMULATION OF PERCHLOROETHYLENE REDUCTIVE DECHLORINATION

Kastli D. Schaller, Brady D. Lee, William A. Apel, and Mary E. Watwood

ABSTRACT: Chlorinated solvents, such as perchloroethylene (PCE), are commonly found as contaminants in groundwater and unsaturated subsurface soils. Transition metal cofactors can enhance biological electron transfer, thereby enhancing dechlorination rates and increasing the feasibility of bioremediation strategies at contaminated sites. Sulfate reducing bacteria (SRB) are suitable organisms for this type of catalytic stimulation due to a cytochrome system of relatively low redox potential which exists in the periplasmic space and can serve as a conduit for extracellular electron transfer.

Batch experiments using concentrated microbial cell suspensions were used to investigate the catalytic activity of four transition metal cofactors on the reductive dechlorination of PCE. The four catalysts used were cyanocobalamin, cobalt (II) sulfonatophenylporphine (CSC), iron (II) sulfonatophenylporphine (ISC), and nickel (II) sulfonatophenylporphine (NSC). Subsurface soil samples were enriched under anaerobic conditions for SRB. Dechlorination assays were conducted under 100% hydrogen gas, using a 20X concentrated cell suspension from the SRB enrichment culture. Cell suspensions were inoculated into HEPES buffer, and each reaction flask also contained a single transition metal cofactor, riboflavin as a redox mediator, and 5 μg/mL PCE. Quantitation of PCE and dechlorination products was performed via gas chromatography.

The highest dechlorination rates were exhibited by cyanocobalamin amended cultures. These cultures removed 70% of PCE within 3 days and exhibited 100% PCE removal within 22 days. A corresponding increase of trichloroethylene (TCE) occurred, with a maximum level of 3.9 μg/mL appearing on day 8, followed by an apparent removal of TCE and the appearance of trace amounts of dichloroethylene. The other three catalysts also stimulated reductive dechlorination, albeit to a lesser degree than cyanocobalamin, compared with control cultures which received no catalytic amendment.

APPLICATIONS OF FENTON OXIDATION TECHNOLOGIES TO REMEDIATE TNT- AND RDX-CONTAMINATED WATER AND SOIL

S.D. Comfort, Z. Li, M. Arienzo, E. Bier, and P.J. Shea

ABSTRACT: Industrial wastes produced at the former Nebraska Ordnance Plant have contaminated more than 6,000 m^3 of soil with TNT and RDX. Innovative technologies are urgently needed to ensure environmental and public protection. We determined the potential of the Fenton reagent to remediate water and soil contaminated with TNT (2,4,6-trinitrotoluene) and RDX (hexahydro-1,3,5-trinitro-1,3,5-triazine). Experimental treatments tested the effects of light, clay mineralogy, soluble organic matter (humic and fulvic acid), temperature, Fe^{2+} source, and Fenton reagent application methods on rates of destruction. Treating a ^{14}C-TNT solution (70 mg TNT L^{-1}) by Fenton oxidation in the dark mineralized 40% of the ^{14}C-TNT and produced oxalate as the primary ^{14}C degradation product. The Fe(III)-oxalate complex formed during Fenton oxidation of TNT is resistant to Fenton oxidation in the dark but was easily oxidized in light. Use of ferrous oxalate in place of $FeSO_4$ in the Fenton reagent increased TNT destruction rates with substantially less peroxide and negated the need for pH adjustment. Treatment of TNT- and RDX-contaminated soil slurries was optimum when temperature was increased (45°C) and the Fenton reagent was added in step-wise increments. These results demonstrate the potential of the Fenton reagent to remediate munitions-contaminated water and soil. Results from on-going pilot scale (60 L) tests are being compared to laboratory-based observations to evaluate applications of Fenton oxidation technologies for on-site remediation.

THE ROLE OF FENTON'S REAGENT IN SOIL BIOREMEDIATION

Karen E. Stokley, Evelyn N. Drake and Roger C. Prince
(Exxon Research and Engineering Co., Annandale, New Jersey)
Gregory S. Douglas (A.D. Little Inc., Cambridge, Massachusetts)

ABSTRACT: We have evaluated the addition of Fenton's Reagent either before or after bioremediation as part of an overall remediation strategy for enhanced treatment of aged hydrocarbon contaminated soils. The integration of biological and chemical processes may enhance the degradation of persistent compounds or those of regulatory concern. Laboratory microcosms containing aged hydrocarbon-contaminated soil were treated with Fenton's Reagent either before or after nutrient-stimulated bioremediation. Gas chromatography/mass spectrometry (GC/MS) analysis of polycyclic aromatic hydrocarbons (PAHs) and Freon/Infrared analysis (IR) of total petroleum hydrocarbons (TPH), were used to measure hydrocarbon contamination. Hopane and oleanane were used as conserved internal markers to determine to what extent specific hydrocarbon species were removed. The application of a mild Fenton's treatment to a field soil prior to bioremediation resulted in an observable lag in the subsequent biodegradation of both total petroleum hydrocarbons and the PAH's on the USEPA priority pollutant list. In contrast, application of a rigorous Fenton's Reagent treatment to both a successfully lab bioremediated soil sample and soil from a successful field bioremediation demonstration resulted in significant decreases (a further 50%) in PAH levels. TPH measurements on all soils were unaffected by the chemical treatment.

INTRODUCTION

Regulatory authorities require that remediation of soils contaminated with petroleum hydrocarbons meet specific clean-up criteria. In situ bioremediation has proven to be an acceptable treatment option, where applicable, to reduce the levels of total extractable hydrocarbons to meet regulatory endpoints. When bioremediation is allowed to proceed, significant degradation of extractable hydrocarbons and PAHs with two to five rings is observed (Drake et al.,1995). Additional treatment may be necessary to reduce the levels of specific compounds such as the polynuclear aromatic hydrocarbons on the USEPA priority pollutant list. The endpoint criteria for several of these compounds are such that excavation of the soil and incineration or thermal destruction may be the only available treatment option. The ability to treat the soil in situ would significantly reduce the overall cost of the remediation project.

The addition of Fenton's Reagent, the reactive species formed when hydrogen peroxide and ferrous iron interact, has been used to chemically oxidize a variety of organic contaminants in water and soil. Early work examined the chemical oxidation of chlorinated organics in sand by addition of hydrogen peroxide and several iron species including naturally occurring iron minerals (Watts et al., 1993). Laboratory soil column studies using clay soils contaminated with trichloroethylene demonstrated the feasibility of injecting and mixing hydrogen peroxide solutions to achieve in situ chemical oxidation (Gates and Siegrist, 1995). Fenton's Reagent has been shown to rapidly oxidize 2-methylnaphthalene in a soil/slurry system (Chen et al., 1995). A great deal of work has been conducted on the application of Fenton's Reagent in the treatment of PAHs in manufactured gas plant wastes (Kelley et al., 1990).

Objective. The objective of this research was to determine the feasibility of applying hydrogen peroxide or hydrogen peroxide plus ferrous iron (Fenton's Reagent) to aged hydrocarbon contaminated soils as an adjunct to bioremediation. The goal was to promote oxidation and enhance the removal of PAHs in situ, by direct application of solid and/or aqueous phase reagents to untreated soils or to soils which have been bioremediated. If successful, this treatment would preclude excavation of surface soil, disposal as hazardous waste or treatment by thermal destruction.

MATERIALS AND METHODS

All of the refinery soil samples described in this work were originally obtained from a potential remediation site and the soil characteristics have been described (Drake et al.,1995). The metals content of the soil was determined by Inductively Coupled Plasma/Atomic Emission Spectroscopy (ICP/AES). Estimates of the numbers of heterotrophic and oil-degrading bacteria were determined with the most-probable-number (MPN) technique (Brown and Braddock, 1990). Soil samples for PAH measurements were extracted with methylene chloride and analyzed by GC/MS operated in the selected ion mode (Douglas et al., 1992). TPH was determined by a modified soxhlet extraction method (USEPA 418.1, 1983, Drake et al.,1995).

An original field sample, frozen at -80°C was used for the laboratory microcosm experiment where a mild Fenton's Reagent treatment was applied prior to bioremediation. Following the chemical treatment, the microcosm was amended with soluble nutrients, tilled and maintained at a constant moisture level. A laboratory bioremediated soil obtained from a 52 week microcosm experiment, was subsequently treated with Fenton's Reagent. Finally, soil from a field bioremediation plot at the remediation site was sampled and brought into the lab and treated with Fenton's Reagent at two levels. The study was performed in duplicate. The bioremediated soils received Fenton's Reagent concentrations two to six times higher than that which was applied to the original soil sample which was bioremediated after chemical treatment. All of the experiments were carried out at room temperature in pyrex dishes. $FeSO_4 \cdot 7H_20$, when added, was mixed into the soil as a dry ingredient. Hydrogen Peroxide was added to achieve the appropriate weight percent on the soil (wet weight basis). It was added dropwise by pipette, covering the surface of the soil. The liquid was allowed to penetrate the soil. The soils were stirred and covered. Thermocouples were placed in the soils during treatment with Fenton's Reagent and soil temperature was monitored throughout the experiment. Samples were taken for analysis after 24 hours.

RESULTS AND DISCUSSION

Soil Metals Analysis. The results of the soil metals analysis are listed in Table 1. Several researchers have suggested that iron amendments may not be required to promote Fenton-like reactions in soil (Watts et al., 1992) since iron and manganese oxides present in the native soil matrix may catalyze such reactions.

TABLE 1. Metals analysis of soil from the refinery remediation site.

Metal	Value (mg/kg)	Metal	Value (mg/kg)
Aluminum	46700	Magnesium	6360
Calcium	3710	Manganese	428
Cobalt	15.8	Sodium	7880
Copper	37.1	Nickel	33
Iron	36800	Lead	95.5
Potassium	13900	Zinc	202

Microbial Enumeration. The most probable number determinations of heterotrophic and oil-degrading bacteria in microcosm soils treated with Fenton's Reagent before and after bioremediation are given in Table 2. A mild chemical treatment prior to bioremediation resulted in a slight decrease in bacterial numbers.

TABLE 2. Most Probable Numbers of Heterotrophic and Oil-Degrading Bacteria in Soils exposed to Fenton's Reagent

Soil Sampled	Time	Heterotrophs/g soil	Oil-degraders/g soil
Original Soil		2.95×10^7	3.56×10^6
Fenton's Treat + Nutrients	0	5.98×10^6	4.11×10^4
Control + Nutrients	0	4.20×10^7	5.09×10^5
Fenton's Treat + Nutrients	1wk	4.91×10^7	3.87×10^6
Control + Nutrients	1wk	1.26×10^8	1.00×10^6
Fenton's Treat + Nutrients	8wk	1.87×10^8	3.34×10^6
Control + Nutrients	8wk	5.44×10^8	3.29×10^5
52wk Bioremediated Soil		4.80×10^7	5.76×10^4
Bioremediated + pH3 Fenton's	24hrs	3.17×10^3	<180
Bioremediated + pH6 Fenton's	24hrs	1.02×10^4	<168
Bioremediated + pH3 Fenton's	48hrs	<172	<172
Bioremediated + pH6 Fenton's	48hrs	<157	<157
Bioremediated + pH3 Fenton's	72hrs	<175	<175
Bioremediated + pH6 Fenton's	72hrs	<186	<186

These microcosms recovered within 1 week after nutrient addition. In contrast, when 52 week bioremediated soil from the laboratory microcosm was subjected to the periodic addition of high levels of Fenton's Reagent, bacterial numbers fell below the statistical detection limits of the test. Values provided are the average of quadruplicate samples taken at each time period.

TPH and PAH Analysis. Relative concentrations of TPH and methylene chloride extractable PAHs, including the 16 USEPA priority pollutants and their associated alkylated homologues, were used to evaluate the efficacy of Fenton's Reagent to enhance the rate and/or endpoint of contaminant removal during bioremediation. Figure 1 shows that pre-treatment of soil with a mild Fenton's Reagent treatment inhibited the removal of total petroleum hydrocarbons and PAHs during a subsequent 8 week bioremediation experiment

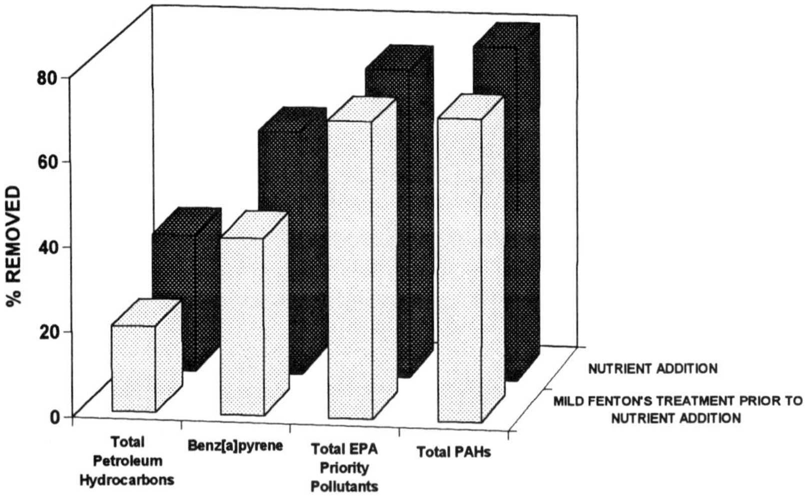

FIGURE 1. Application of a mild Fenton's reagent treatment prior to bioremediation inhibits removal of target analytes at 8 weeks.

In Figure 2, the relative contributions of bioremediation and chemical treatment to the total removal of specific PAHs are given for soil successfully bioremediated in laboratory microcosms. Bioremediation alone was effective in reducing both the total priority pollutants as well as specific PAHs by 80%. Fenton's Reagent, removed up to 50% of what remained. TPH measurements were unaffected by the chemical treatment. Hopane and oleanane levels were not significantly reduced by the Fenton's Reagent treatment. Soil temperatures increased from 22° C to 52°C. These observations were confirmed when the soil from a successful field bioremediation experiment was exposed to the same chemical treatment regime.

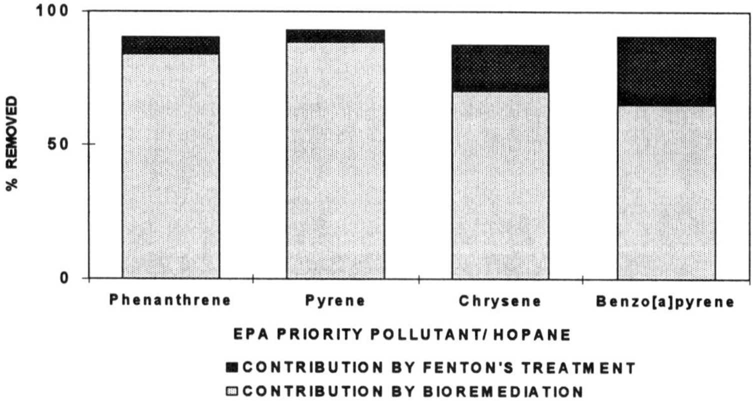

FIGURE 2. Effect of bioremediation followed by chemical treatment on the removal of EPA priority pollutants from soil in a laboratory microcosm.

Fenton's Reagent appears to target the removal of PAH's, particularly the more recalcitrant 4-6 ring compounds. In Figure 3, the behavior of BAP, as an individual PAH is compared to the overall trend observed for the priority pollutants as a total. Once again, hopane was used as an internal biomarker in order to normalize the PAH data.

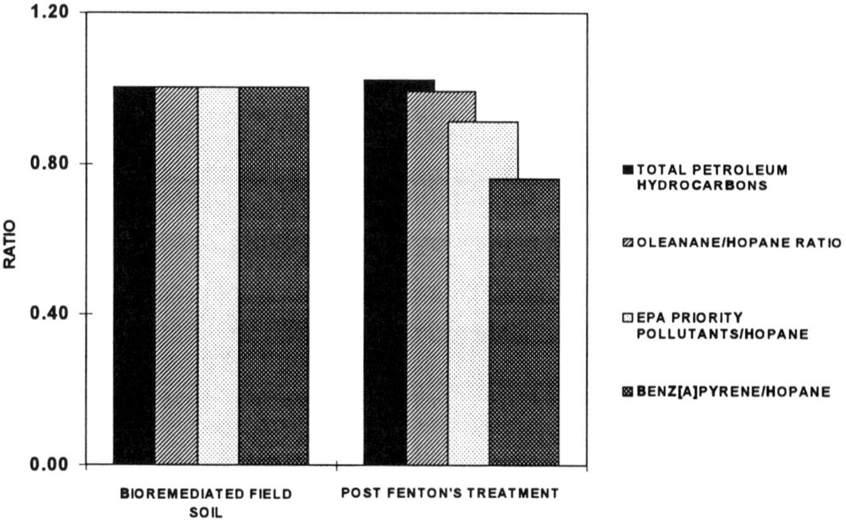

FIGURE 3. Fenton' s Reagent targets PAHs

Thus Fenton's Reagent can further reduce the total amount of priority pollutants in bioremediated soil. Under the conditions studied there was no benefit to chemically pre-treating contaminated soil with Fenton's Reagent prior to undertaking bioremediation.

The in situ, direct application of Fenton's Reagent to a soil which has been successfully bioremediated results in significant further reductions in the levels of regulated polynuclear aromatic hydrocarbons. This may prove to be a cost effective strategy with minimal environmental impact.

REFERENCES

Brown, E.J., and J.F. Braddock. 1990. "Sheen Screen, a Miniaturized Most-Probable-Number Method for the Enumeration of Oil-degrading Microorganisms." *Appl. Environ. Microbiol.* 56: 3895-3896.

Douglas, G.S., J.K. McCarthy, D.T. Dahlen, J.A. Seavey, W.G. Steinhauer, R.C. Prince and D.L. Elmendorf. 1992. "The Use of Hydrocarbon Analyses for Environmental Assessment and Remediation." *J. soil Contam.* 1: 197-216.

Drake, E.N., K.E. Stokley, P. Calcavecchio, R.E. Bare, S.J. Rothenburger, G.S. Douglas and R.C. Prince. 1995. "Nutrient-Stimulated Biodegradation of Aged Refinery Hydrocarbons in Soil." In R.E. Hinchee, G.S. Douglas and S.K. Ong (Eds.), *Monitoring and Verification of Bioremediation*, pp.19-28. Battelle Press, Columbus OH.

Chen, C.T., A. Tafuri, M. Rahman, M.B. Foerst and W. Coates. 1995. "Chemical Oxidation Treatment of Hydrocarbon Contaminated Soil Using Fenton's Reagent." Presented at the I&EC Special Symposium American Chemical Society, Atlanta, GA., September 17-20, 1995.

Gates, D.D. and R.L. Siegrist. 1995. "In Situ Chemical Oxidation of Trichloroethylene using Hydrogen Peroxide." *J. of Env. Engineering* -ASCE 121: 639-644.

Kelley, R.L., W.K. Gauger and V.J. Srivastava. 1990. "Application of Fenton's Reagent as a Pretreatment Step in Biological Degradation of Polyaromatic Hydrocarbons." Presented at the Third International IGT Symposium on Gas, Oil and Environmental Biotechnology, Chicago, Ill.

Watts, R.J., M.K. Foget, M.A. Dippre, A. Kenny and P.C. Stanton. 1992. "Process Considerations in the Promotion of Fenton-Like Reactions for the Treatment of Contaminated Soils." Presented at the I&EC Special Symposium, American Chemical Society, Atlanta, GA., September 21-23, 1992.

Watts, R.J., M.D. Udell and R.M. Monsen. 1993. "Use of Iron Minerals in Optimizing the Peroxide Treatment of Contaminated Soils." *Water Environ. Res.* 65: 839-844.

FIELD PILOT STUDY OF BIOAUGMENTATION FOR REMEDIATION OF TCE CONTAMINATION IN FRACTURED BEDROCK

Matthew Walsh, Robert J. Steffan, and ***Mary F. DeFlaun***

ABSTRACT: A field pilot study was performed that combined the technologies of bioaugmentation with pneumatic fracturing to remediate TCE contamination in fractured bedrock. Site investigation had shown that in the area of a former TCE storage tank, bedrock fractures were a source of residual TCE to the groundwater. Laboratory treatability studies investigated both a biostimulation and a bioaugmentation approach for this source area. Biostimulation of an indigenous TCE-degrading population did not result from methane addition to site microcosms, however, TCE was rapidly degraded in bioaugmented microcosms.

The specialized TCE-degrading organism used in this study, ENV435, was developed with several properties that enhance its applicability for in situ remediation. This organism is adhesion-deficient which allows effective application and transport throughout the contaminated area. In addition, ENV435 constitutively degrades TCE, eliminating the need to add inducers in situ. This strain is also cultured in a manner that allows production of high levels of poly-β-hydroxybutyrate (PHB) within the cells to provide an internal carbon source, minimizing the need to add additional substrates to support degradative activity.

During a field demonstration of the process, pneumatic fracturing, a patented process developed by the New Jersey Institute of Technology (NJIT) and licensed to Accutech Remedial Systems, Inc. was used to increase the permeability of the bedrock. A pneumatic injection process was used to rapidly introduce ENV435 and nutrients into the pneumatically fractured bedrock. Dispersal of bacteria throughout the test area was accomplished with a specialized injector nozzle attached to a high-pressure air supply and fluid pump that atomized and propelled the bacteria into the fractures.

The results of the pilot study indicate that ENV435 was effectively dispersed throughout a 25-foot radius test area resulting in a 50 to 90% reduction in TCE concentrations within the inoculated test zone. These results were used to develop a conceptual full-scale design for the application of bioaugmentation combined with pneumatic fracturing at this site.

INDUCER-FREE MICROBE FOR TCE DEGRADATION AND FEASIBILITY STUDY IN BIOAUGMENTATION

T. Imamura, S. Kozaki, A. Kuriyama, M. Kawaguchi, Y. Touge, T. Yano, E. Sugawa, and ***Y. Kawabata*** (Canon Research Center, Kanagawa, Japan)
H. Iwasa, A. Watanabe, M. Iio, and Y. Senshu (Raito Kogyo, Tokyo, Japan)

ABSTRACT: An inducer-free microbe (strain JM1) for trichloroethylene (TCE) degradation was obtained by chemical mutation of indigenous microbes in soil. Strain JM1 was an aerobic microbe, and could degrade TCE by oxidation without inducers such as phenol and cresol. For bioaugmentation demonstration, several scale-up experiments were achieved using a glass vial (68 ml), and a small and large containers (2.1 and 41 l) filled with a sandy soil. For example, 20 ppm TCE was almost degraded by JM1 in the large container. The TCE degradation efficiency strongly depended both on the injection volume and concentration of JM1 and on the diffusion of TCE in the soil. Further scale-up experiments has been achieved using a lysimeter (48 m^3) filled with the sandy soil contaminated by TCE. Delivery of JM1 by the injection and the growth of JM1 were evaluated. However, significant degradation of TCE was not observed due to a short period of a high degradation activity and a lack of oxygen in the lysimeter. The latter problem was roughly solved by the selection of a nutrient and a degradation temperature.

INTRODUCTION

Environmental pollution is a serious problem because some of the pollutants causes severe disease in human being. TCE is one of the typical pollutants in soil and groundwater. In-situ bioremediation is expected for cost-effective treatment compared with the physicochemical remediation such as soil vapor extraction. Especially, bioaugmentation for TCE-contaminated soil is an attractive approach, and many microbes have been reported as TCE degraders.

Aerobic microbes for TCE degradation are almost classified to either aromatic compounds-oxidizer or methanotroph. These microbes can degrade TCE in the presence of inducers such as phenol and methane since TCE is co-metabolized with these inducers. But these inducers are exogenous chemicals, and additional contamination is arisen as the another problem. Furthermore, such inducers competitively inhibit degradation of TCE since these inducers should be metabolized by the microbes. Thus inducer-free microbes for TCE degradation are necessary to overcome these problems.

On the other hand, the actual contaminated site is large, and laboratory study is sometimes far from the actual site experiments in its scale. There are many differences in laboratory and site experiments since delivery of microbes and material transportation such as TCE diffusion depend on the scale of the contaminated site. To clarify this problem, several bioaugmentation experiments were achieved in different scales from a glass vial to a large lysimeter in this study.

MATERIAL AND METHODS

Inducer-Free Microbe JM1. Indigenous microbes which could degrade TCE in the presence of phenol were obtained from various natural soil. One of the best microbes was incubated in the M9 medium containing 0.1 % of yeast extract, 200 ppm of phenol, and the trace amount of minerals such as magnesium ion at 30 °C. The late-log-phase microbes were harvested after 14 h incubation, and it was resuspended to the M9 medium containing 200 ppm of phenol. A mutagen of nitrosoguanidine was added to the medium, and the concentration of the mutagen was adjusted to 30 μg / ml culture. After shaking for a few hours, this medium was inoculated on an agar plate containing the M9 medium, 0.1 % of yeast extract, and 200 ppm of indole. Indole was well known as an indicator for an oxygenase since it was transformed to blue indigo through indoxyl by the oxygenase. Thus, the color of a colony quickly changed to dark blue when the microbes produced oxygenase constitutively. Such colony was obtained as inducer-free microbes (JM1) for TCE degradation.

Constitutive Degradation in Glass Vial. Strain JM1 was inoculated into the medium containing 2.0 % of sodium malate and the M9 medium. After incubation for 3 days at 15 °C, JM1 was grown to 3-5 x 10^8 cfu / ml. This culture was diluted 100 times with a medium containing 1.0 % of sodium citrate and the M9 medium. Appropriate amount of unsterilized sandy soil was put into the glass vial (68 ml). Then, the diluted culture (10 weight % against the amount of soil) was added to the soil. The moisture of the soil was increased from 10 to 20 % after the addition of the diluted culture. The initial concentration of JM1 was approximately 10^5 cfu / g wet soil. It was incubated for 1-2 days in the atmosphere. When the growth of JM1 was reached to the late-log-phase, the concentration of JM1 became 10^8 - 10^9 cfu / g wet soil. Then, the glass vial was sealed by a butyl rubber septum and an aluminum crimp. The aqueous solution of TCE (saturated, ca. 1000 ppm) was prepared in the sealed glass bottle, and the gaseous TCE in the bottle was sampled by a syringe. Then, the gaseous TCE (0.1 ml) was injected through the rubber septum. The headspace gas (0.1 ml) in the glass vial was sampled by a syringe, and the concentration of TCE was determined by gas chromatography. TCE was well known to adsorb to the organic carbon in the soil. But the sandy soil used had little organic carbon, and adsorption of TCE was negligible.

TCE Degradation in Small and Large Containers. Two stainless steel containers with different size were constructed for evaluation of TCE degradation. A small one had a cylindrical shape with an inner diameter of 116 mm and a height of 202 mm. A injection tube for JM1 was installed at the center of a top surface of the container, and was extended near to the bottom of the container. A small gravel (250 ml) was first put into the container, and the sandy soil (1.6 l) was filled on it. Finally, the sandy layer was covered with the small gravel (250 ml), and the container was sealed with a stainless cover.

The aqueous TCE solutions with different concentrations (50 ppm, 100 ppm, and 200 ppm, each volume was 500 ml) was prepared in a glass bottle (1 l), and gaseous TCE in the bottle was circulated overnight in the container through the injection tube for contamination by TCE. JM1 was incubated with nutrients and the M9 medium for 2 days, and the concentration of JM1 was increased to 3 x 10^8 cfu / ml. After the incubation, the organic nutrient was almost consumed by JM1, and the condition of the microbes became the resting state. Then, the JM1 solution was diluted two times with the M9 medium, and it was introduced into the container through the injection tube. When the diluted JM1 solution overflowed from the container, the injection was stopped. The injection volume was approximately 850 ml. The injection solution was circulated in the container for 3 h in order to equilibrate TCE and the microbes in the soil. The soil water in the sandy layer was sampled (0.5 ml), and the TCE concentration was determined by gas chromatography.

The large container had a cylindrical shape with an inner diameter of 380 mm and a height of 360 mm. The injection tube was installed at the center of the top surface, however, the end of the injection tube was located at near the center of the container. The small gravel (7 l), the sandy soil (27 l), and the gravel (7 l) were laid, and the container was closed with a gas-tight cover. Since a lot amount of resting JM1 was necessary to fill the container, JM1 and some nutrients were first injected in this experiment. Then, JM1 was grown up in the soil and TCE degradation was observed. The diluted JM1 solution (3-5 x 10^6 cfu / ml, 252 ml or 600 ml) by the nutrients and M9 was injected into the soil, and fresh air was introduced into the container for the growth of JM1. The concentration of JM1 became 10^8 - 10^9 cfu / g wet soil after 2 days, and gaseous TCE was circulated through the container. The circulation gas (0.1 ml) was sampled by a syringe, and the concentration of TCE was determined.

Lysimeter Experiments. Soil and groundwater contamination usually extends to a large area, and feasibility study for the laboratory experiments is necessary in order to apply bioaugmentation technique to in-situ remediation. Two soil lysimeters were constructed in this study, one of which had 3.0 m wide, 3.5 m long, and 4.6 m depth. An inner surface of the lysimeter was coated with a vinyl ester resin to avoid leaching of TCE from the lysimeter. The small gravel (9 m^3) was laid on the bottom of the lysimeter. Succeedingly, the sandy soil (36 m^3) was put on the gravel, and the small gravel (1 m^3) was laid on the sandy soil. The top of the gravel layer was covered with a concrete, and it was coated with a polyurethane resin. There were four injection tube, four oxygen sensors, three thermometers, and electrodes for the measurement of conductivity in each lysimeter. There were 16 stainless tubes in each lysimeter to sample TCE gas.

A nutrient solution (1.2 m^3) was prepared containing 5 wt % of sodium malate, 1.2 wt % of disodium hydrogen phosphate, 0.6 wt % of potassium dihydrogen phosphate, 0.4 wt % of ammonium chloride, and 0.1 wt % of sodium chloride. JM1 was cultured in a small fermenter (28 l), and it was mixed with the nutrient solution. The initial concentration of JM1 was adjusted to 1 x 10^7 cfu / ml. There were two injection port (lower and upper ports) in each injection tube. The JM1 solution (140 l) was injected through the lower port, and fresh air (600 l) was succeedingly injected at the same port. The same procedure was achieved at other 3 lower ports. Then the same procedure was done at 4 upper ports. The concentration change of oxygen was monitored by the oxygen sensors. The soil sample was obtained by boring, and it was used to measure the concentrations of JM1 and indigenous microbes.

RESULTS AND DISCUSSION

Constitutive Degradation in Glass Vial. TCE degradation was observed after incubation without phenol to clarify constitutive degradation. The results are shown in Figure 1. The concentration of TCE (based on the moisture in the soil) remaining in the vial was immediately decreased in the presence of JM1. Contrary, the glass vial without including JM1 showed roughly the same TCE concentration. Thus it was apparent that JM1 can degrade TCE constitutively. The first measurement of TCE was achieved after 14 min from the addition of TCE. This time period was necessary to prepare the sample for gas chromatography.

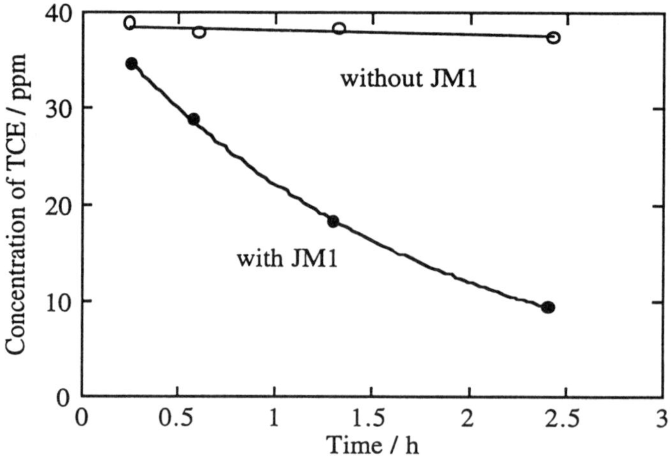

FIGURE 1. Constitutive degradation of TCE by inducer-free microbe JM1

TCE Degradation in Small and Large Containers. For the small containers, the aqueous TCE solutions with different concentrations (50, 100, and 200 ppm) were used in contamination of soil. Assuming no leaching of TCE during the injection of JM1, the concentrations of TCE (based on the moisture in the soil) were calculated to be 8.7, 17.3, and 36.6 ppm. The measured concentrations were 4.5, 10.0, and 19.2 ppm, respectively. Thus about half amount of TCE in the small container was leached out by the injection. The TCE concentrations after injection were shown in Figure 2. When the resting microbes were injected, the TCE concentration rapidly decreased during 5 h from the injection, and gradually decreased after 5 to 22.5 h. The TCE concentration of 19.2 ppm was finally decreased to 1.5 ppm for example. On the other hand, the TCE concentration remained the initial concentration (10 ppm) when the M9 medium without JM1 was injected. For three initial TCE concentrations, the amount of TCE degraded was largest for high initial concentration of TCE (19.2 ppm) since the degradation rate was proportional to the TCE concentration. The final TCE concentration was 0.2 ppm even for the lowest initial concentration of TCE (4.5 ppm). The complete degradation of TCE was not achieved since the degradation activity of JM1 was decreased after 1 day and the encountering efficiency between JM1 and TCE was decreased.

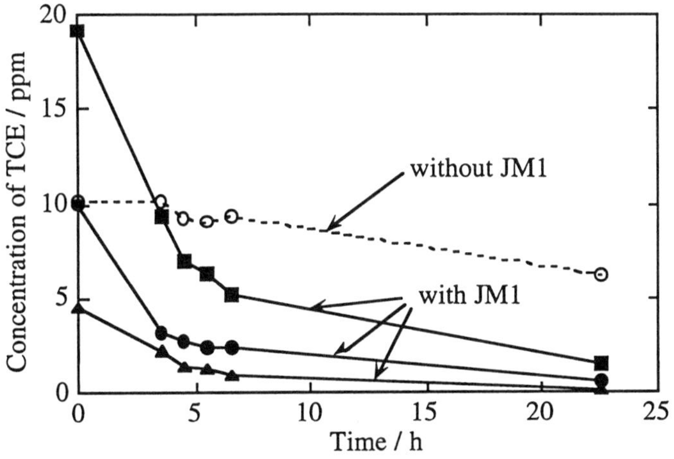

FIGURE 2. Degradation of TCE at various concentrations in small containers

For the large containers, the different volumes of JM1 (252 and 600 ml) were injected and the TCE degradation was observed. When 600 ml of JM1 was injected, 20 ppm of TCE (based on the moisture in the soil) was significantly degraded within 25 h. Contrary, the TCE concentration was slightly decreased due to the leaching of TCE from the container in which only nutrients was injected. Furthermore, the half amount of TCE could be degraded when the volume of injected JM1 was decreased to 252 ml. This implied that the total amount of TCE degraded was proportional to the amount of JM1. Therefore, the volume of the microbes solution injected should be increased for much higher TCE concentration.

Lysimeter Experiments. After the injection of JM1, the electric conductivity in the injection area increased immediately by immersion of the solution. Thus the size of the injection area was measured by the change of the conductivity. The shape of the injection area was spherical with a diameter of 40-50 cm just after the injection, and the injection solution fell down in the soil by its weight gradually. The response of the oxygen sensors was shown in Figure 3. The oxygen concentration at the injection area decreased from 19.5 to 2.6 % after 19.5 h from the injection. This oxygen consumption was due to the rapid growth of JM1. Fresh air was introduced into the lysimeter in order to avoid the lack of oxygen, but oxygen was consumed again for a few hours. This result showed that the amount of oxygen in the soil was not enough for the growth of JM1, and the supply of oxygen was essential for in-situ bioremediation. On the other hand, the consumption of oxygen was slow when the nutrients without containing JM1 was used. The oxygen concentration decreased 19.5 to 12.1 % after 26.8 h from the injection. This slow decrease in the oxygen concentration was due to the gradual increase of the indigenous microbes. The competition in survival between JM1 and the indigenous microbes was occurred after 24 h from the injection, and the selective survival of JM1 should be necessary.

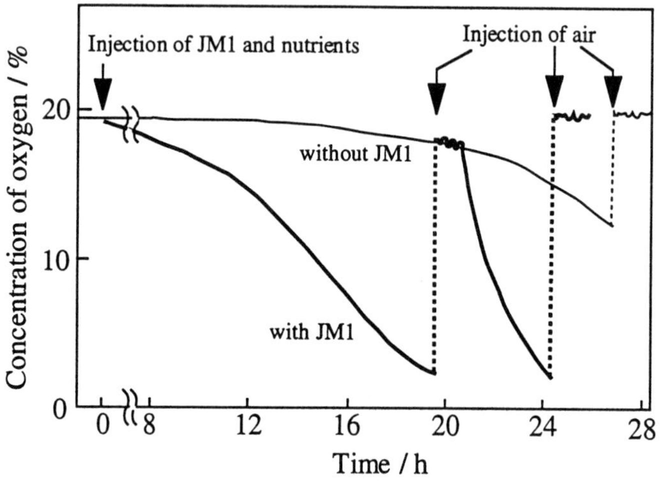

FIGURE 3. Response of oxygen sensors in lysimeters

The soil in the lysimeters was sampled after the injection for the measurement of the concentration of JM1 and the indigenous microbes. The initial concentration of JM1 was 1-2 x 10^6 cfu / g wet soil, and it agreed well with the calculated value from the JM1 concentration in the injection solution and the moisture in the soil. After 1 day from the injection, the concentration of JM1 increased to 2 x 10^8 cfu / g wet soil, and it remained for 2 days. The concentrations of the indigenous microbes were 2 x 10^5 cfu / g wet soil just after the injection, 2 x 10^6 for 1 day, 2 x 10^7 for 2 days after the injection. Thus the concentration of JM1 was one order larger than that of the indigenous microbes, and the selective growth of JM1 could be achieved by the use of the suitable kinds and concentrations of the nutrients.

The experimental temperature was adjusted to 15 °C for biodegradation in the glass vial, and small and large containers since the soil temperature was around 15 °C at the deep from the subsurface. However, the temperature in the lysimeter strongly depended on the environment. The soil temperature was 19-20 °C, the temperature of the injection solution being 24-26 °C. This caused the rapid growth of JM1 and the indigenous microbes, and resulted the lack of oxygen. Therefore, the degradation activity in the lysimeter was much decreased compared with that of the laboratory experiments. Furthermore, the diffusion of TCE in the soil was very slow, and the degradation rate was limited by the low concentration of TCE. Thus the survival of JM1 in the natural soil, the enlargement of the injection area, the diffusion of oxygen and TCE in the soil, and the high degradation activity for TCE were still problems for TCE decontamination in bioaugmentation.

BIOAUGMENTATION WITH *BURKHOLDERIA CEPACIA*: TRICHLOROETHYLENE COMETABOLISM VS. COLONIZATION

Junko Munakata-Marr (Colorado School of Mines, Golden, Colorado)
V. Grace Matheson, Larry J. Forney and James M. Tiedje (Michigan State University, East Lansing, Michigan)
Perry L. McCarty (Stanford University, Stanford, California)

ABSTRACT: Periodic bioaugmentation with *Burkholderia cepacia* strains G4 and $PR1_{301}$ enhanced trichloroethylene (TCE) cometabolism in small-column aquifer microcosms but eventually led to dissolved oxygen limitations and declining TCE transformation. After bacterial addition was stopped but TCE and primary substrate addition continued, dissolved oxygen levels recovered and phenol-fed, previously bioaugmented microcosms maintained significant levels of TCE removal relative to a phenol-fed nonbioaugmented microcosm. A strain-specific genetic probe applied to microcosm effluents identified *B. cepacia* G4 and $PR1_{301}$ regularly throughout the bioaugmentation period, but soon after bioaugmentation was halted, the two strains were no longer detected in the effluents. The probe also did not detect *B. cepacia* in DNA extracted from the solids of previously bioaugmented microcosms. In addition, the samples did not hybridize to a catabolic genetic probe for the toluene ortho-monooxygenase contained in *B. cepacia*, nor to probes for genes encoding enzymes in three other characterized toluene degradation pathways. Bioaugmentation did not result in establishment of *B. cepacia* within the indigenous microbial community. Rather, the efficient TCE-transforming microorganisms are likely to have been indigenous organisms selected by operating conditions, microorganisms that possibly used divergent or even unique pathways for the observed TCE transformation.

INTRODUCTION

Bacterial cultures grown aerobically on primary substrates such as methane (*e.g.* Wilson and Wilson 1985; Fogel et al., 1986), aromatics (*e.g.* Nelson et al., 1986; Wackett and Gibson 1988), ammonia (Arciero et al., 1989), isoprene (Ewers et al., 1990), and propylene (Ensign et al., 1992) have been observed to produce oxygenases that fortuitously degrade or cometabolize trichloroethylene (TCE), a common groundwater contaminant. Stimulation of indigenous organisms with a specific primary substrate for *in situ* bioremediation of TCE may enrich for a population that is unable to cometabolize the target compound or else does so slowly. Bioaugmentation of contaminated groundwater systems through addition of bacterial cultures known to transform TCE rapidly may enhance *in situ* biodegradation or even provide the sole means of degradation in systems without indigenous TCE-degrading organisms. Such biological enhancement has been investigated for degradation of many different compounds (*e.g.* Daughton and Hsieh 1977; Chatterjee et al., 1982) , but successful demonstrations of bioaugmentation have tended to be associated with readily metabolizable target compounds and with added organisms distributed throughout readily aerated topsoils. Bioaugmentation for the degradation of cometabolic substrates has shown limited success (*e.g.* Focht and Brunner 1985; Nelson et al., 1990). A comparative laboratory-scale aquifer microcosm study was conducted to evaluate the potential of bioaugmentation with organisms known to cometabolize TCE.

MATERIALS AND METHODS

Small column microcosms (17 mL total volume, 5.0-6.5 mL pore volume) containing aquifer material from Moffett Federal Air Station, Mountain View, CA (Moffett Field) were prepared and operated in a batch-fed mode as described previously (Munakata-Marr et al., in press). The microorganisms used for bioaugmentation were *Burkholderia* (*Pseudomonas*) *cepacia* G4 (G4), a strain isolated from a holding pond at an industrial waste treatment facility in Pensacola FL (Nelson et al., 1986), and *B. cepacia* $PR1_{301}$ (PR1), a chemically induced mutant strain of G4 (Munakata-Marr et al., 1996). G4 cometabolizes TCE with the enzyme toluene ortho-monooxygenase (Tom), which is normally induced by phenol or toluene (Shields et al., 1991), while PR1 constitutively expresses Tom while grown on substrates such as lactate (Munakata-Marr et al., 1996). Discussion will focus on the microcosms to which G4 or PR1 were added along with phenol as a support substrate.

Phenol, dissolved oxygen, and TCE concentrations were measured as described previously (Munakata-Marr et al., in press). The presence of G4 or PR1 in microcosm effluents was tested as described previously (Munakata-Marr et al., in press).

DNA was extracted directly from aquifer solids using the method of Zhou, et al. (Zhou et al., 1996) except that the masses and volumes used were altered slightly as follows. For the first extraction, 4.4 g of aquifer material was mixed with 8.1 mL of DNA extraction buffer, 100 μL of 10 mg/mL proteinase K, and 0.9 mL of 20% sodium dodecyl sulfate (SDS). Two subsequent extractions were performed using 2.7 mL of extraction buffer and 0.3 mL of 20% SDS. Supernatants were combined and extracted with 13 mL of chloroform-isoamyl alcohol (24:1, v/v). Pellets were resuspended in 150 μL of sterile MilliQ water. Polyvinylpolypyrrolidone (PVPP) was not used.

REP-PCR was used as described previously (Munakata-Marr et al., in press) to amplify the extracted DNA. After PCR, 12 μL of each reaction was electrophoresed in 1.5% agarose with 0.5X TAE buffer (Sambrook et al., 1989). Amplified DNA was transferred to a Hybond-N nylon membrane (Amersham) by capillary blotting (Southern 1975). DNA extracted from aquifer material and isolated cultures was tested for toluene degradative pathways on dot blots using Amersham Hybond-N nylon membranes. All DNA samples were first brought to 250 μL total volume with autoclaved MilliQ water. The samples were then mixed with 250 μL of alkaline solution (0.8 M NaOH, 20 mM EDTA) and placed in boiling water for 10 minutes. The dot blot was rinsed by adding 0.5 mL of 1x TE buffer to each well under gentle vacuum, then samples were added, again under gentle vacuum, after pulsing in a microcentrifuge. After denaturing samples with 0.5 mL of 0.4 M NaOH in each well under full vacuum, the membrane was removed from the dot blot apparatus, rinsed in 2x SSC (17.53 g NaCl, 8.82 g sodium citrate, pH 7.0) and allowed to air dry. All DNA was cross-linked to membranes by 5 min exposure to a Fotodyne UV transilluminator.

Prehybridization was performed for 1 hr minimum at 62°C in a rotating glass tube within a hybridization oven (Robbins), using at least 1 mL of solution/5 cm^2 membrane area. High stringency hybridization was similarly conducted for 8 hr minimum, using at least 1 mL of hybridization solution/10 cm^2. Probe concentrations were 10 ng labeled DNA/mL of hybridization solution. The probes used to determine toluene degradation pathways are listed in Table 1. Chemiluminescent detection was performed as described in the Genius System User's Guide (Boehringer Mannheim), using Boehringer Mannheim reagents and X-OMAT Scientific Imaging film (Eastman Kodak).

TABLE 1. Probes for toluene oxygenase genes used.

Genes	Oxygenase	Original Strain	Plasmid	Lab Source
TOM	toluene ortho mono	*B. cepacia* G4	pMS80	M. Shields
tmoA,B,C,D	toluene para mono	*P. mendocina* KR1	pMY421	M. DeFlaun
TOL, xylM xylA	methyl mono	*P. putida* PaW1	pGSH2836	S. Harayama
tod C1C2BA	toluene di	*P. putida* F1	pDTG601	D. Gibson

Probes were stripped after hybridization as outlined in the Genius System User's Guide (Boehringer Mannheim), except that the stripping solution contained 0.4 N NaOH and the membranes were incubated twice for 10 min each at 37°C. After verifying a neutral pH, prehybridization was started as described above.

RESULTS AND DISCUSSION

The averaged results from the phenol-fed bioaugmented microcosms are shown in Figure 1. These microcosms, all fed phenol and bioaugmented with lactate- or phenol-grown G4 or PR1, behaved similarly and degraded the most TCE of the microcosms tested. The results for C8 during the brief period that it was fed lactate (days 26 to 61) were discussed previously (Munakata-Marr et al., 1996) and are omitted from the averages. Before complete oxygen depletion around day 80, about 120 µg/L TCE was removed compared to the control. As DO levels decreased, TCE removal dropped. After bioaugmentation was stopped and DO once again became available, steady TCE removal of approximately 100 µg/L TCE was achieved for 140 days, showing no signs of loss of activity. In comparison, a phenol-fed nonbioaugmented control microcosm achieved a maximum degradation of about 100 µg/L relative to the control within 60 days of exposure to TCE, with declining degradation after that (Munakata-Marr et al., in press).

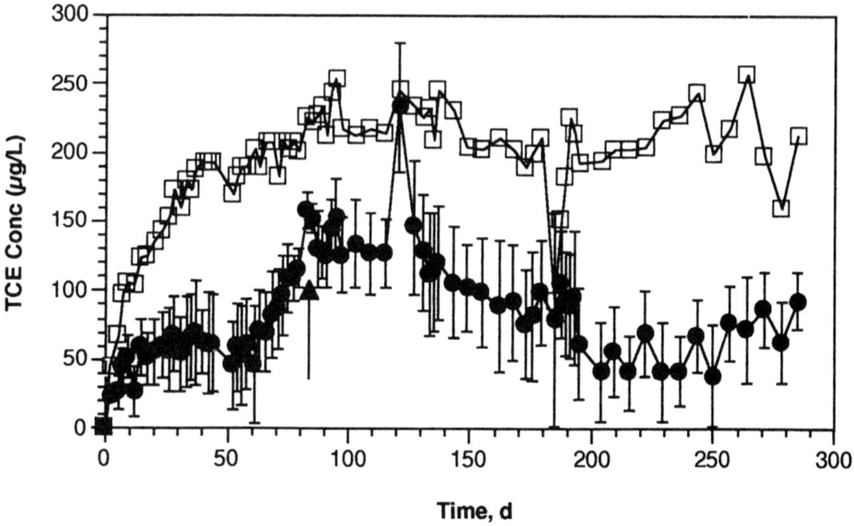

Figure 1. Average effluent TCE concentrations in microcosms augmented with G4 or PR1 and phenol; error bars display 95% confidence intervals. (□) control; (●) average of C4-C8.

G4 and PR1 were never observed in the effluents from nonbioaugmented microcosms C1 and C2. In contrast, G4 and PR1 were regularly detected in the bioaugmented column effluents throughout the bioaugmentation period (days 0 to 83), and, indeed, for several fluid exchanges beyond that (Figure 2). However, by the day 93 exchange, G4 or PR1 was found in only one column effluent and was not detected after that in any of the column effluents (days 95, 97, 105, 113, 137, 191, 236 and 237).

The relatively efficient TCE degradation by microcosms C4-C8 after bioaugmentation was stopped is curious, as there is no evidence that bioaugmented microcosm activity was due to the persistence of the added organisms. In addition to their absence in effluent samples, G4 and PR1 were not found in the aquifer materials from microcosms C6 and C7 after they were dismantled at the end of the experiment (data not shown), at which time good TCE degradation was still being observed.

No cause for the good performance of these bioaugmented microcosms in the latter period of this study could be identified. Transfer of the plasmid encoding for Tom to indigenous organisms either did not occur or the transconjugants were not selected, as DNA samples extracted from previously bioaugmented microcosms did not hybridize to a Tom-specific catabolic gene probe (data not shown). Selection of more effective TCE-degrading organisms from within the indigenous community in these microcosms seems the most plausible explanation. It may be that addition of G4 or PR1 altered the relative dominance of organisms, either through direct competition for substrates or through the development of anaerobic conditions. It is also possible that addition of trace nutrients in the culture media during bioaugmentation may also have improved the effectiveness of the indigenous community or shifted the population toward a more effective organism.

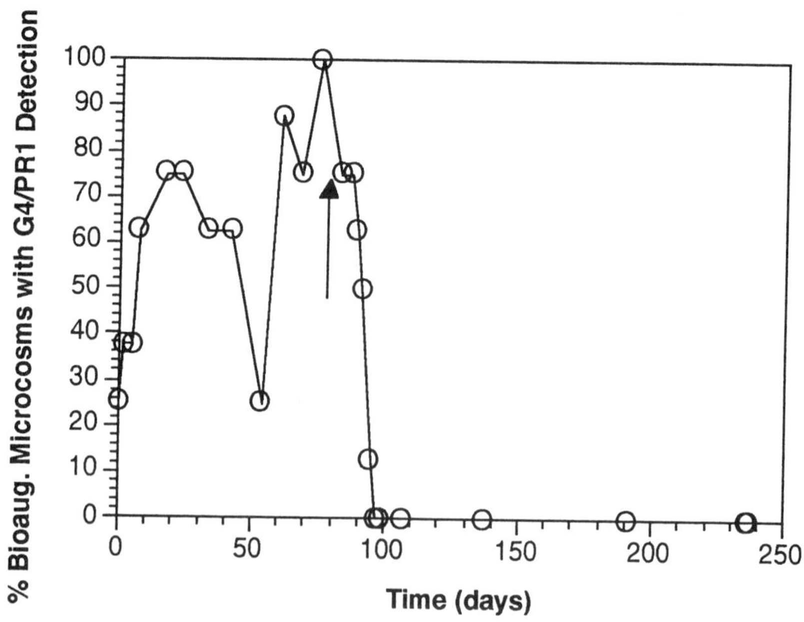

Figure 2. Percentage of the eight bioaugmented microcosms that had detectable levels of G4 or PR1 in the microcosm effluents on different sampling days. The arrow indicates when bioaugmentation was stopped.

Interestingly, DNA extracted from these microcosms did not hybridize to catabolic probes for three additional toluene degradation pathways (data not shown), suggesting that the microorganisms responsible for the efficient TCE degradation may have used divergent or even unique pathways from those previously studied.

CONCLUSIONS

Regular bioaugmentation with *B. cepacia* G4 and $PR1_{301}$, though enhancing TCE removal while the strains were added to the microcosms, did not result in establishment of the strains within the indigenous microbial community. The predominant microorganisms in the previously bioaugmented microcosms are likely to have been efficient TCE-transforming indigenous organisms selected by conditions within the microcosms such as anaerobiosis. Because DNA extracted from these microcosms did not hybridize to catabolic probes for four characterized toluene degradation pathways, the predominant microorganisms may have used divergent or even unique pathways for the observed TCE transformation.

ACKNOWLEDGMENTS

This work was supported through cooperative agreement CR 822029 with the U.S. Environmental Protection Agency, Gulf Breeze Environmental Research Lab, and by a National Science Foundation graduate fellowship. As it has not been subjected to agency review, no official endorsement should be inferred.

REFERENCES

Alvarez-Cohen, L. and P. L. McCarty. 1991. "Effects of Toxicity, Aeration, and Reductant Supply on Trichloroethylene Transformation by a Mixed Methanotrophic Culture." *Appl. Environ. Microbiol. 57*(1): 228-235.

Arciero, D., T. Vannelli, M. Logan and A. B. Hooper. 1989. "Degradation of Trichloroethylene by the Ammonia-Oxidizing Bacterium *Nitrosomonas europaea*." *Biochem. Biophys. Res. Comm. 159*(2): 640-643.

Boehringer, M. "The Genius SystemTM User's Guide for Filter Hybridization, Version 2.0."

Broholm, K., B. K. Jensen, T. H. Christensen and L. Olsen. 1990. "Toxicity of 1,1,1-Trichloroethane and Trichloroethylene on a Mixed Culture of Methane-Oxidizing Bacteria." *Appl. Environ. Microbiol. 56*(8): 2488-2493.

Chatterjee, D. K., J. J. Kilbane and A. M. Chakrabarty. 1982. "Biodegradation of 2,4,5-Trichlorophenoxyacetic Acid in Soil by a Pure Culture of *Pseudomonas cepacia*." *Appl. Environ. Microbiol. 44*(2): 514-516.

Daughton, C. G. and D. P. H. Hsieh. 1977. "Accelerated Parathion Degradation in Soil by Inoculation with Parathion-Utilizing Bacteria." *Bull. Environ. Contamin. Toxicol. 18*(1): 48-56.

Ensign, S. A., M. R. Hyman and D. J. Arp. 1992. "Cometabolic Degradation of Chlorinated Alkenes by Alkene Monooxygenase in a Propylene-Grown *Xanthobacter* Strain." *Appl. Environ. Microbiol. 58*(9): 3038-3046.

Ewers, J., D. Freier-Schroder and H.-J. Knackmuss. 1990. "Selection of Trichloroethylene (TCE) Degrading Bacteria that Resist Inactivation by TCE." *Arch. Microbiol. 154*:410-413.

Focht, D. D. and W. Brunner. 1985. "Kinetics of Biphenyl and Polychlorinated Biphenyl Metabolism in Soil." *Appl. Environ. Microbiol. 50*(4): 1058-1063.

Fogel, M. M., A. R. Taddeo and S. Fogel. 1986. "Biodegradation of Chlorinated Ethenes by a Methane-Utilizing Mixed Culture." *Appl. Environ. Microbiol. 51*(4): 720-724.

Fox, B. G., J. G. Borneman, L. P. Wackett and J. D. Lipscomb. 1990. "Haloalkene Oxidation by the Soluble Methane Monooxygenase from *Methylosinus trichosporium* OB3b: Mechanistic and Environmental Implications." *Biochemistry. 29*(27): 6419-6427.

Harker, A. R. and Y. Kim. 1990. "Trichloroethylene Degradation by Two Independent Aromatic-Degrading Pathways in *Alcaligenes eutrophus* JMP134." *Appl. Environ. Microbiol. 56*(4): 1179-1181.

Hopkins, G. D., J. Munakata, L. Semprini and P. L. McCarty. 1993. "Trichloroethylene Concentration Effects on Pilot Field-Scale In-Situ Groundwater Bioremediation by Phenol-Oxidizing Microorganisms." *Environ. Sci. Technol. 27*(12): 2542-2547.

Munakata-Marr, J., V. G. Matheson, L. J. Forney, J. M. Tiedje and P. L. McCarty. in press. "Long-Term Biodegradation of Trichloroethylene Influenced by Bioaugmentation and Dissolved Oxygen in Aquifer Microcosms." *Environ. Sci. Technol.*

Munakata-Marr, J., P. L. McCarty, M. S. Shields, M. Reagin and S. C. Francesconi. 1996. "Enhancement of Trichloroethylene Degradation in Aquifer Microcosms Bioaugmented with Wild Type and Genetically Altered *Burkholderia* (*Pseudomonas*) *cepacia* G4 and PR1." *Environ. Sci. Technol. 30*:2045-2052.

Nelson, M. J., J. V. Kinsella and T. Montoya. 1990. "*In Situ* Biodegradation of TCE Contaminated Groundwater." *Environ. Prog. 9*(3): 190-196.

Nelson, M. J. K., S. O. Montgomery, W. R. Mahaffey and P. H. Pritchard. 1987. "Biodegradation of Trichloroethylene and Involvement of an Aromatic Biodegradative Pathway." *Appl. Environ. Microbiol. 53*(5): 949-954.

Nelson, M. J. K., S. O. Montgomery, E. J. O'Neill and P. H. Pritchard. 1986. "Aerobic Metabolism of Trichloroethylene by a Bacterial Isolate." *Appl. Environ. Microbiol. 55*(2): 383-384.

Roberts, P. V., L. Semprini, G. D. Hopkins, D. Grbic-Galic, P. L. McCarty and M. Reinhard. 1989. *In-Situ Aquifer Restoration of Chlorinated Aliphatics by Methanotrophic Bacteria.* U.S. Environmental Protection Agency Technical Report, EPA 600/2-89/033, R. S. Kerr Environmental Research Laboratory, Ada, OK.

Sambrook, J., E. F. Fritsch and T. Maniatis. 1989. *Molecular Cloning: a Laboratory Manual.* Cold Spring Harbor Laboratory, Cold Spring Harbor, New York.

Shields, M. S., S. O. Montgomery, S. M. Cuskey, P. J. Chapman and P. H. Pritchard. 1991. "Mutants of *Pseudomonas cepacia* G4 Defective in Catabolism of Aromatic Compounds and Trichloroethylene." *Appl. Environ. Microbiol. 57*(7): 1935-1941.

Southern, E. M. 1975. "Detection of Specific Sequences Among DNA Fragments Separated by Gel Electrophoresis." *J. Mol. Biol. 98*:503-517.

Wackett, L. P. and D. T. Gibson. 1988. "Degradation of Trichloroethylene by Toluene Dioxygenase in Whole-Cell Studies with *Pseudomonas putida* F1." *Appl. Environ. Microbiol. 54*(7): 1703-1708.

Wackett, L. P. and S. R. Householder. 1989. "Toxicity of Trichloroethylene to *Pseudomonas putida* F1 Is Mediated by Toluene Dioxygenase." *Appl. Environ. Microbiol. 55*(10): 2723-2725.

Wilson, J. T. and B. H. Wilson. 1985. "Biotransformation of Trichloroethylene in Soil." *Appl. Environ. Microbiol. 49*(1): 242-243.

Winter, R. B., K.-M. Yen and B. D. Ensley. 1989. "Efficient Degradation of Trichloroethylene by a Recombinant *Escherichia coli*." *Bio/Technol. 7*:282-285.

Zhou, J., M. A. Bruns and J. M. Tiedje. 1996. "DNA Recovery from Soils of Diverse Composition." *Appl. Environ. Microbiol. 62*(2): 316-322.

EVALUATION OF BIOAUGMENTATION TO REMEDIATE AN AQUIFER CONTAMINATED WITH CARBON TETRACHLORIDE

Michael J. Dybas [1], Serguei Bezborodinikov [1], Thomas Voice[2], David C. Wiggert[2], Simon Davies [2], James Tiedje [1], and Craig S. Criddle [2] ([1]NSF Center for Microbial Ecolgy and [2]Department of Civil and Environmental Engineering, Michigan State University, East Lansing, MI)
Orest Kawka and Michael Barcelona (University of Michigan, Ann Arbor, MI)
Timothy Mayotte (Golder Associates, East Lansing, MI)

ABSTRACT: A field experiment was performed in a carbon tetrachloride (CT) impacted aquifer at Schoolcraft, Michigan to evaluate bioaugmentation with *Pseudomonas stutzeri* strain KC for CT remediation. The experiment demonstrated that the aquifer environment could be modified by alkalinity addition to allow colonization by strain KC. Following introduction of strain KC approximately 50-80% of the CT in the groundwater was removed. Final corings indicated that 60-88% of the sorbed CT was removed from those sections of the aquifer that were communicative with the injection well.

INTRODUCTION

Pseudomonas stutzeri KC (DSM deposit no. 7136, ATCC deposit number 55595) is a denitrifying bacterium that converts carbon tetrachloride (CT) to carbon dioxide (40-50%) and non-volatile compounds (45-55%) without production of detectable chloroform under most growth conditions (Criddle et al., 1990; Lewis and Crawford, 1993). Other microorganisms typically convert CT to chloroform under denitrifying conditions. Conditions required for rapid transformation are: 1) an anoxic environment, 2) an electron donor (such as acetate), 3) nitrate, and 4) iron-limiting conditions. Iron-limiting conditions can be achieved by increasing pH (Criddle et al., 1990; Tatara et al., 1993).

In the laboratory, CT transformation has been obtained in groundwater, soil, and aquifer material by raising the pH of the growth environment to around 8.0 (Tatara et al., 1994; Dybas et al., 1995). A one year field study was conducted to determine the feasibility of bioaugmentation using strain KC for CT remediation *in situ.*

Objective. Laboratory studies suggested strain KC would be a suitable candidate for field application, and the objective of this study was to determine: (1) Would it be possible to adjust and maintain the pH of a subsurface environment to the slightly alkaline levels (pH>7.6) needed for growth and CT transformation? (2) Would strain KC survive in the face of competition from indigenous microflora under field conditions? and (3) Would strain KC remain active and perform CT transformation under field conditions?

Site Description. The field experiment was performed in a CT-contaminated aquifer at Schoolcraft, Michigan. The Schoolcraft aquifer contains relatively homogeneous glacial outwash sands, and exhibits high hydraulic conductivity ($\sim 10^{-2}$ cm/s). The CT plume (Plume A as designated by the State of Michigan Department of Environmental Quality) extends approximately 1200 m in length by 400 m in width. The CT is found in the saturated zone only, from 12 to 23 m below ground surface. The water table is approximately 6 mbelow ground surface The test site layout is shown in Figure 1.

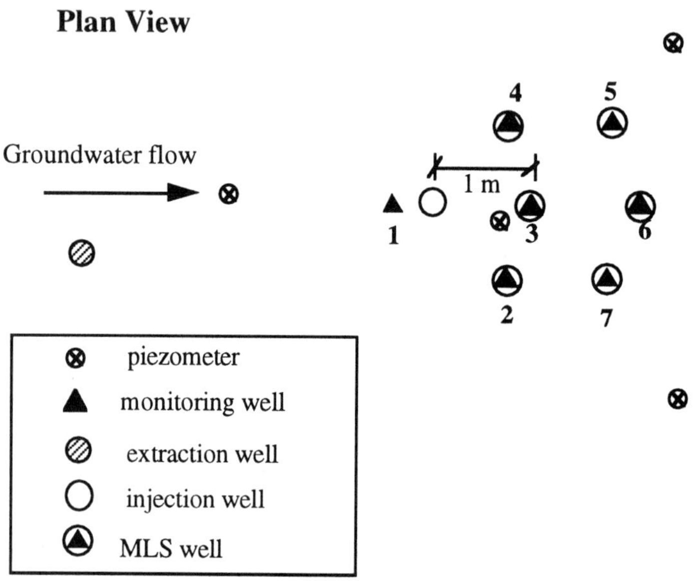

FIGURE 1. Layout of the field test site located in Schoolcraft, MI.

MATERIALS AND METHODS

The field experiment was conducted in six phases: (1) Pre-Operational Phase (including days 0-42): grid construction, initial core sampling, and groundwater sampling to assess baseline conditions in a test grid located in the heart of Schoolcraft Plume A; (2) Operational Phase A (days 43-79): preparation of the grid for inoculation by pulsing with base- and phosphate-amended groundwater to achieve a pH of 7.5-8.0; (3) Operational Phase B (day 80): inoculation of the test grid with *P. stutzeri* KC; (4) Operational Phase C (days 81-107): distribution of strain KC cells throughout the test grid by weekly pulses of groundwater amended with acetate, phosphorus, and base, followed immediately by a chase pulse of acetate-free groundwater; (5) Operational Phase D (days 108-142): maintenance of CT transformation by weekly pulses of acetate, phosphorus, and base, without the acetate-free chase pulse, and (6) Post-Operational Phase (days 143-170): final drilling to determine CT removal from solids within the test grid and groundwater monitoring to document recontamination of the test section. Between each pulse in the Operational Phase, the injected groundwater was allowed to drift through the test grid at the prevailing groundwater velocity (15 cm/d).

Site pH Titration. To titrate the grid and prepare it for introduction of strain KC, CT-contaminated groundwater pumped from an upgradient extraction well (screened at 18.3-19.8 m) was supplemented with base(to pH 8.2, approximately

40 mg/l NaOH), phosphate (10 mg/ml) and bromide (30 mg/l), then injected in discrete pulses of 3000 liters into the grid. Bromide served as a conservative tracer and indicator of hydraulic communication between the monitoring wells and the injection well. Carbon tetrachloride contaminated groundwater extracted from a well located upgradient of the test grid was used for all titration and nutrient delivery injections to prevent dilution of CT within the test grid due to pumping operations. Good hydraulic communication was obtained with wells 2 and 3, intermediate communication with wells 5, 6, and 7, and poor communication with well 4. In the most communicative sections of the grid, pH increased to the required levels.

Inoculation of Test Site. For inoculation, groundwater (1890 liters) was pumped from a shallow upgradient extraction well, filter sterilized, supplemented with acetate (1.6 g/l), phosphate (10 mg/l) and NaOH (to adjust pH to 8.2) and inoculated with a strain KC starter culture (12 liters, 24 hour aerobic nutrient broth grown culture). The resulting culture was grown aerobically for 24 hr to 2 $\times 10^7$ cfu/ml (OD_{660} = 0.1); supplemented with acetate, bromide, and CT; then rapidly pumped into the aquifer. Inoculation was performed one time only (day 80).

Nutrient Delivery. After inoculation, microbial activity was sustained by weekly pulses of CT-contaminated groundwater extracted from well EW-D and supplemented with acetate (100 mg/l), phosphate (10 mg/l), bromide (30 mg/l), and base (NaOH, approximately 40 mg/l to pH 8.2).

Groundwater Sampling and Analysis. Groundwater samples were obtained over 8 depth intervals using passive multilevel sampling devices equipped with 10 micron nylon screens to allow bacterial, volatile and ionic compounds present in the groundwater (Margan MLS L.T.D., Netanya, Israel). Volatiles were analyzed by gas chromatography (GC) using a Perkin Elmer Autosystem equipped with a electron capture detector and a DB-624 column. Solid phase associated volatiles were analyzed by GC and identity verified by GC-mass spectral analysis. Anions were analyzed by EPA Method 300.

Microbial Monitoring. A strain KC-specific DNA probe was used to monitor transport and colonization of strain KC. To generate the specific probe, twenty random clones of approximately 1 kb from the strain KC chromosome were screened against 200 isolates from Schoolcraft groundwater to determine their specificity for KC. Five fragments showing no non-specific hybridization were selected for further evaluation. One of the fragments was selected as superior for specificity and reliable PCR-amplification. Two sets of primers of 20-22 mer (P1F, P2F and P2R-P1R) were selected and tested against aquifer isolates, aquifer groundwater and Genbank and found to be specific for strain KC. Levels of strain KC and aquifer flora were enumerated by plate counts. Populations of denitrifyers were quantified by the most probable number (MPN) method, using the presumptive test to screen for nitrate and nitrite disappearance (Tiedje, 1994).

RESULTS AND DISCUSSION

Carbon Tetrachloride Removal and Microbial Colonization. Rapid initial removal of approximately 60% of CT was observed in the most communicative wells. Wells that were most communicative with the injection well were effectively colonized by strain KC. As shown in Figure 2, well 2 was colonized at all 8 depth intervals for most of the Operational Phase.

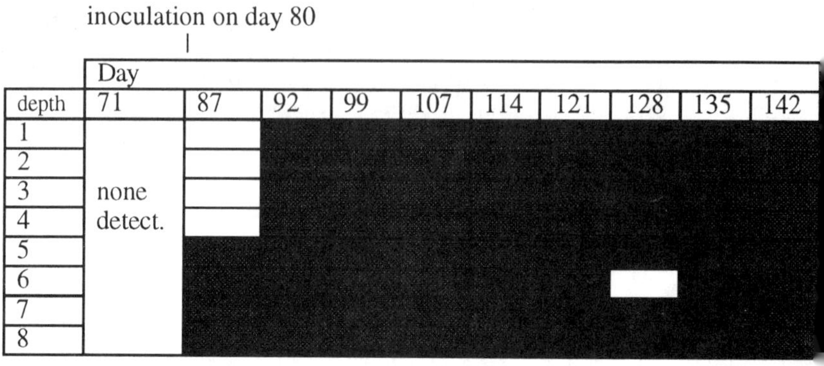

inoculation on day 80
Day
depth
71
87
92
99
107
114
121
128
135
142
1
2
3
4
5
6
7
8
none
detect.

FIGURE 2. Colonization of well 2 as indicated by DNA probe hybridization. A filled cell indicates that strain KC was detected. An empty cell indicates that strain KC was not detected.

Figures 3 and 4 illustrate the response of well 3. Initial rapid CT degradation was observed, however by day 107, use of the acetate-free chase pulse in Operational Phase C had resulted in less efficient removal of CT and nitrate and loss of KC in several wells. When the chase was subsequently eliminated, more efficient removal was observed (Operational Phase D).

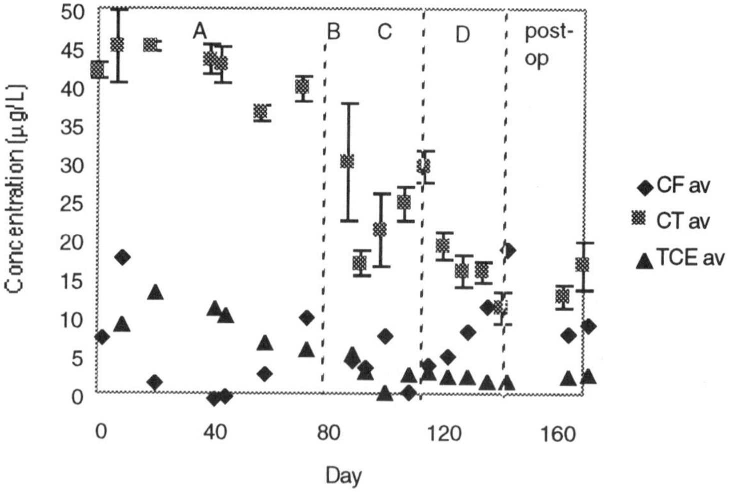

FIGURE 3. Depth averaged volatiles at well 3.

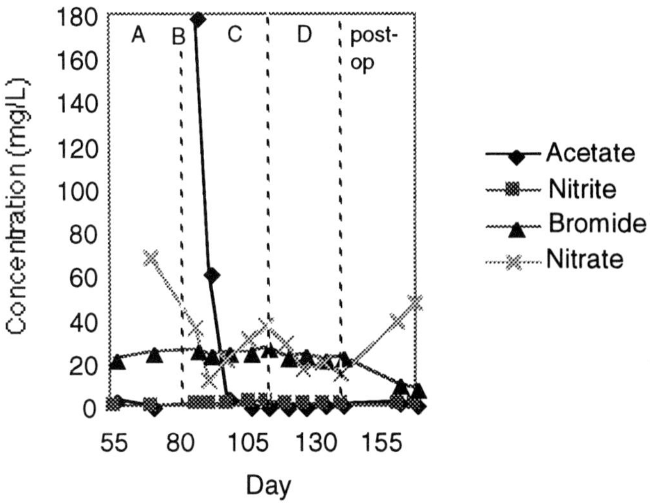

FIGURE 4. Depth averaged anions at well 3.

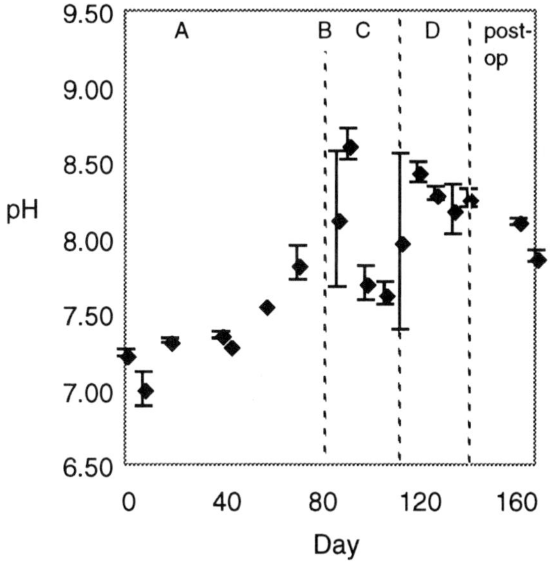

FIGURE 5. Depth averaged pH at well 3.

Overall, nearly 50% of the CT mass that entered or was initially present in the grid was removed, as was approximately 70% of the nitrate. Final core analyses of solids within the grid revealed that 60-88% of the sorbed CT was removed from those portions of the aquifer that were most communicative with the injection well. In the Post-Operational Phase, monitoring wells that were communicative with the injection well behaved as expected: bromide levels dropped, CT levels increased, nitrate levels increased, and pH levels decreased (Figures 3 - 5).

We conclude that the pH adjustment strategy employed in this experiment successfully prepared the aquifer environment for colonization, that strain KC can colonize the subsurface environment after pH adjustment, and that strain KC was activated for CT transformation under in-situ field conditions. Comparison of field and laboratory results suggests that the efficiency of CT removal was primarily limited by the engineering challenges associated with establishing hydraulic control and ensuring effective delivery of substrate, alkalinity, and strain KC throughout the treatment zone.

ACKNOWLEDGMENTS

Funding for this work was provided by the Department of Environmental Quality of the State of Michigan.

REFERENCES

Criddle, C.S., J.T. DeWitt, D. Grbrić -Galić , and P.L. McCarty. 1990. Transformation of carbon tetrachloride by *Pseudomonas* sp. strain KC under denitrifying conditions. *Appl. Environ. Microbiol.* 56:3240-3246.

Dybas, M.J., Tatara, G. M, and C.S. Criddle. 1995. Localization and Characterization of Carbon Tetrachloride Transformation Activity of *Pseudomonas* sp. Strain KC. *Appl. Environ. Microbiol.* **61**:758-762

Dybas, M.J., Tatara, G.M., Knoll, W. H., Mayotte, T.J., and C. S. Criddle. 1995. "Niche Adjustment for Bioaugmentation with *Pseudomonas* sp. strain KC," In R.E. Hinchee, A. Leeson, and L. Semprini (Eds.), *Bioaugmentation for Site Remediation.* Battelle Press, Columbus, OH.pp.77-84.

Lewis, T. A., and R. L. Crawford. 1993. Physiological factors affecting carbon tetrachloride dehalogenation by the denitrifying bacterium *Pseudomonas* sp. strain KC. *Appl. Environ. Microbiol.* 59:1635-1641.

Tatara, G.M., M.J. Dybas, and C.S. Criddle 1995. "Biofactor-Mediated Transformation of Carbon Tetrachloride by Diverse Cell Types" In R.E. Hinchee, A. Leeson, and L. Semprini (Eds.), *Bioremediation of Chlorinated Solvents.* Battelle Press, Columbus, OH.pp.69-76.

Tatara, G.M., M.J. Dybas, and C.S. Criddle. 1993. Effects of medium and trace metals on kinetics of carbon tetrachloride transformation by *Pseudomonas* sp. strain KC. *Appl. Environ. Microbiol.* 59:2126-2131.

Tiedje, J.M. 1994. *Denitrifiers*. ASA-SSSA, Madison, WI. pp. 815-820.

AEROBIC BIOREMEDIATION OF TCE-CONTAMINATED GROUNDWATER: BIOAUGMENTATION WITH *BURKHOLDERIA CEPACIA* PR1$_{301}$

Al W. Bourquin, Douglas C. Mosteller, Roger L. Olsen, and Michael J. Smith
(Camp Dresser & McKee Inc., Denver, Colorado)
Kenneth F. Reardon (Colorado State University, Fort Collins, CO)

ABSTRACT: Groundwater in downtown Wichita, Kansas is contaminated with trichloroethene (TCE) and dichloroethene (DCE). A field pilot test was conducted to evaluate the feasibility of aerobic bioremediation using bioaugmentation with *Burkholderia cepacia* PR1$_{301}$ (PR1), which constitutively expresses toluene ortho-monooxygenase, to degrade the chlorohydrocarbons. The field study included two PR1 injection periods. In the first microbial addition phase, PR1 concentrations in the injection well (IW) were 10^9 cells/mL. The cells degraded to non-detect all contaminants in the aerobic system within 24 h and maintained that level for 4 d (test termination). During this test period, the degradation rate was 94.5 µg/L/h. However, the soil formation plugged at this cell concentration. In the second injection phase, PR1 was added step-wise to the IW to increase cell concentration and then glucose was pulsed in an attempt to maintain cell mass. Groundwater contaminants were removed completely only when cell concentrations reached 10^8/mL. Plugging was evident at this concentration but was not as severe as during the first injection. In neither phase did the injected cells move efficiently downgradient. This was the first field demonstration that bioaugmentation could effectively degrade CAHs using a laboratory developed strain.

INTRODUCTION

Background. A variety of microorganisms have been isolated that aerobically cometabolize chlorinated aliphatic hydrocarbons (CAHs), including those using toluene, phenol, and methane as growth substrates (Fogel et al., 1986; Nelson et al., 1987). Biodegradation of CAHs by these environmental isolates occurs when an oxygenase with relaxed specificity is induced. This requirement for an inducer is problematic for field applications since the required compounds are regulated contaminants (toluene, phenol) and/or sparingly soluble in water (methane, toluene). However, successful field demonstrations have been conducted using these inducers.

Natural attenuation of CAHs is a strategy that has received substantial attention. Sequential reductive dehalogenation of trichloroethene (TCE) and other CAHs has been demonstrated under anaerobic conditions (Semprini et al., 1995). The extent of dehalogenation is known to be influenced by environmental conditions, including the presence of competing electron acceptors (McCarty 1996). At some sites, such as the one described here, biodegradation of TCE does not proceed beyond cis-dichloroethene (cis-DCE) and thus natural attenuation is not an option.

Bioaugmentation, the addition of microorganisms to a site, is an attractive bioremediation strategy when the desired metabolic capability is absent from the site or present at unacceptably low levels. In the case of TCE bioremediation, the desirable capability is constitutive (not requiring an inducer) expression of a nonspecific oxygenase. *Burkholderia cepacia* PR1$_{301}$ (PR1) has this characteristic. This strain, a nonrevertible regulatory mutant of *B. cepacia* G4, constitutively expresses toluene ortho-monooxygenase and can biodegrade TCE when grown on lactate or other noninducing compounds (Munakata-Marr et al., 1996).

Bioaugmentation using PR1 could take several forms, including injection into wells for dispersion across a site and incorporation in biological permeable reactive barriers. Regardless of this engineering choice, PR1 must demonstrate constitutive CAH-degrading activity in the field. Transport of the added cells through an aquifer may also be important. This report describes a field study conducted in Wichita, KS to evaluate the feasibility of aerobic bioremediation of TCE-contaminated ground water using bioaugmentation with PR1.

Objectives. The overall goal of this project was to evaluate the performance of PR1 in the field for CAH biodegradation. Specific objectives were (1) to measure CAH biodegradation in the absence of inducer; (2) to determine the concentration of PR1 required to achieve complete degradation; and (3) to determine the ability of PR1 to move away from the injection point.

Site Description. Ground water throughout the central business district of Wichita, KS is contaminated with TCE and DCE with some areas of perchloroethene (PCE). The depth to ground water is approximately 4.5 m below ground surface (bgs) and bedrock is generally reached at 9 m bgs. The aquifer material is sandy with hydraulic conductivities ranging from 115 to 225 m/d. In the field test zone, typical background CAH levels were 125 μg/L TCE, 95 μg/L cis-DCE, and 4 μg/L trans-DCE.

System and Study Design. The demonstration site was designed with one injection well (IW), one extraction well (EW), and a series of multi-depth, multi-port monitoring wells (MW), as shown in Figure 1. A unique injection well was designed to recirculate groundwater for mixing purposes and to contain tubing for bromide, cell and oxygen addition. The outer well casing of the IW was 15 cm in diameter and had 2.1 m of screening (4.95 to 7.08 m bgs). Five-cm diameter PVC was installed within the 15-cm diameter well to house all sampling, injection and recirculation tubing. The pumping rate at the EW was 13.5 L/min to maintain a positive flow across the site. At each monitoring well location except MW6, samples could be taken at approximately 5, 6, 7.5, and 8.8 m bgs; at MW6, only the upper two depths could be sampled. Because the lower aquifer levels contained PCE (not aerobically biodegradable), the bioaugmentation test was conducted in the upper levels of the aquifer (above 7 m bgs) and the lower levels were monitored occasionally to determine if any effects had occurred.

The field test was conducted in five phases, as presented in Table 1.

MATERIALS AND METHODS

Assays. Groundwater samples from the injection and monitoring wells were assayed for dissolved oxygen and pH using standard electrochemical probes. Bromide tracer concentrations were measured with a bromide probe. Organic contaminant concentrations were measured by GC using a modified Method 8020. (detection limit 1.0 μg/L). Two types of microbial counts were performed using dilution plate counting: total heterotrophs were enumerated on 50% trypticase soy broth agar and phenol degraders were counted on a basal salts+phenol agar (Munakata-Marr et al., 1996).

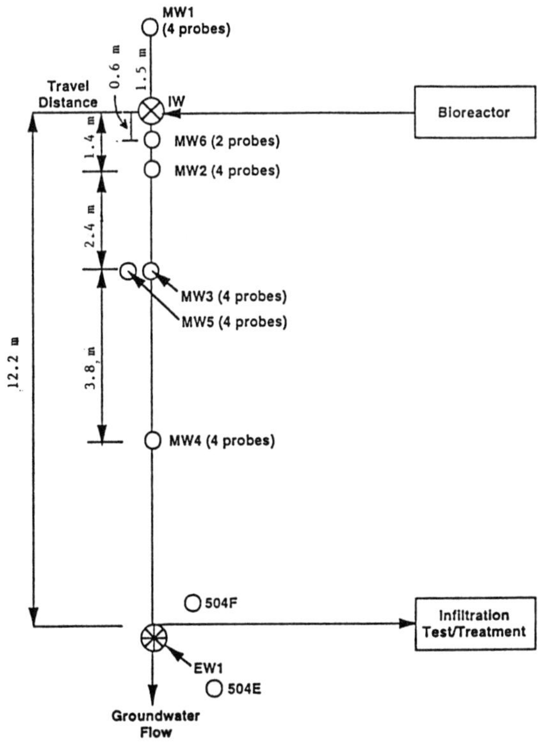

FIGURE 1. Plan view of aerobic bioaugmentation pilot demonstration system. IW: injection well, MW: monitoring well, EW: extraction well.

TABLE 1. Description of bioaugmentation field test.

Phase	Purpose	Duration (d)
Background	To determine baseline chemical and microbiological characteristics under conditions of enhanced groundwater flow	9
Optimization	To determine effects of oxygenation on groundwater, velocity, and tracer transport trends	19
Biotreatment I (followed by a recovery phase)	To evaluate the extent of bioremediation on CAHs following continuous injection of microorganisms	4 (22 for recovery)
Biotreatment II	To evaluate bioremediation following a single injection of cells and subsequent in situ feeding	4
Post-Test	To determine effect of the treatment process following shutdown and evaluate contaminant desorption	15

Cell cultivation. PR1 was grown according to a drain-and-fill batch protocol in a 16-L stirred-tank bioreactor. The growth medium was a basal salts medium that contained 7.2 g/L glucose and small amounts of yeast extract. The bioreactor provided agitation (500 RPM), aeration (95% oxygen), and temperature (30 °C) and pH control. The pH was initially 6.8 and was allowed to fall to 6.0 before being controlled at that value by 5N sodium hydroxide addition.

RESULTS AND DISCUSSION

Background and Optimization Phases. A borehole dilution test using bromide showed that the ground water was moving through the IW at approximately 1.7 L/h. Using the tracer, the groundwater velocity was calculated as 24 cm/d at MW2.

Total CAH concentrations were stable at 200-250 µg/L in the IW and in MW1 (upgradient) during both of these phases, indicating limited volatilization of contaminants during aeration.

Transport of dissolved oxygen (DO) through the aquifer was problematic. During the Optimization phase, sparging with 95% oxygen achieved an average DO in the IW of 15 mg/L . However, DO was not detected in downgradient monitoring wells for many days; a retardation coefficient of 13 was calculated. This retardation could not be explained by microbial utilization (no significant increase in cell concentration) or chemical consumption due to reduced metals in the ground water (e.g., Fe^{2+} concentrations were ca. 0.2 mg/L). The addition of 100 mg/L hydrogen peroxide did not increase the breakthrough of DO.

Since the low downgradient DO levels could not support the activity of PR1 (an aerobe), the biotreatment phases focused on the IW.

Biotreatment Phase I. PR1 was grown to high concentrations (ca. 10^{12} cells/mL) on glucose in the bioreactor and added to the IW to achieve a concentration of 10^9 cells/mL. The ground water in the IW was recirculated from top to bottom at 19 times the groundwater flow-through rate to achieve a good degree of mixing. In less than 24 h following injection, PR1 biodegraded TCE and cis-DCE in the ground water to nondetectable levels (Figure 2). The contaminant concentrations in the IW did not increase until the PR1 injection was terminated. The CAH biodegradation rate was estimated to be 94.5 µg/L·h.

PR1 was detected downgradient of the IW at MW6, approximately 30 cm from the IW, eight days following injection. The arrival of the microorganisms at MW-6 coincided with a decrease in DO at this monitoring point. However, only 0.005% of the injected population of cells were detected downgradient of the IW.

Following termination of PR1 injection, CAH concentrations did not increase as expected and bromide tracer concentrations increased, indicating decreased ground water flow through the IW. The most likely cause for this was microbial plugging of the downgradient side of the IW. This plugging was relieved by mechanically agitating the IW groundwater.

Biotreatment Phase II. Contaminant and microbial concentrations in the IW during this phase are shown in Figure 3. Initially, cells were added to the IW at different concentrations to determine the lowest concentration needed to completely remove contaminants with the goal of avoiding the plugging observed in the previous biotreatment phase. At 10^7 cells/mL, CAH concentrations in the IW were reduced substantially but degradation to nondetect levels was not achieved until PR1 concentrations reached 10^8 cells/mL in the IW. Once cells had been added to reach this concentration, glucose and nutrients were pulsed into the IW at intervals but the cell population could not be maintained at this level.

The desired PR1 concentration of 10^8/mL was eventually maintained by continuously adding cells from the bioreactor. Bromide was continuously added to the IW to monitor groundwater flow.

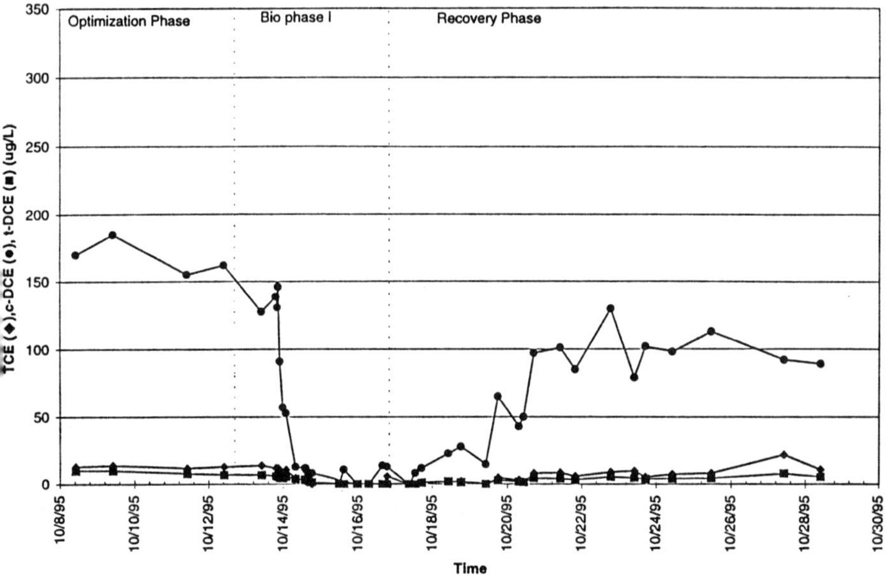

FIGURE 2. CAH concentrations in the IW during the Optimization, Biotreatment I, and Recovery phases.

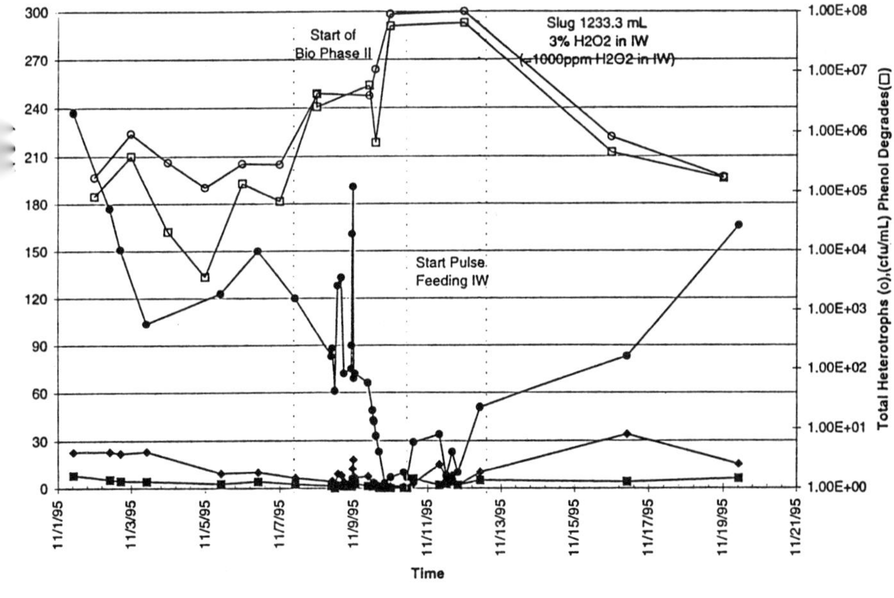

FIGURE 3. CAH and microbial concentrations during the Recovery, Biotreatment II, and Post-Treatment phases.

CONCLUSIONS

This was the first field demonstration that bioaugmentation with PR1 could effectively degrade TCE and cis-DCE. PR1 degraded approximately 250 µg/L CAHs to nondetect levels when present at 10^8 cells/mL. CAHs were degraded at lower cell concentrations but the rates were not sufficient to degrade all of the contaminants in the IW. PR1 did not transport well in the sandy aquifer soils and plugging was evident after both biotreatment phases. This aspect of bioaugmentation was studied further and is reported in a companion paper.

The results of this field test support bioaugmentation strategies such as point injection and biological permeable reactive barriers. Given the oxygen transport problems encountered at this site, the latter approach is of particular interest.

ACKNOWLEDGMENTS

This work was funded by the City of Wichita, Kansas, as part of the Remedial Design for the Gilbert-Mosley site. The authors wish to thank Tracy Squires and Leigh Anne Boyack, Colorado State University, for their technical assistance. Dr. Malcolm Shields, University of West Florida, generously supplied the organism *B. cepacia* $PR1_{301}$.

REFERENCES

Fogel, M. M., A. R. Taddeo, and S. Fogel. 1986. "Aerobic Metabolism of Trichloroethylene by a Methane-Utilizing Mixed Culture." *Appl. Environ. Microbiol. 52*: 720-724.

McCarty, P. 1996. "Biotic and Abiotic Transformations of Chlorinated Solvents in Ground Water." *In Symposium on Natural Attenuation of Chlorinated Organics in Ground Water* (Abstracts), EPA/540/R-96/509, pp. 5-9. US EPA, Washington, DC.

Munakata-Marr, J., P. L. McCarty, M. S. Shields, M. Reagin, and S. C. Francesconi. 1996. "Enhancement of Trichloroethylene Degradation in Aquifer Microcosms Bioaugmented with Wild Type and Genetically Altered *Burkholderia* (*Pseudomonas*) *cepacia* G4 and PR1." *Environ. Sci. Technol. 30*(6):2045-2052.

Nelson, M. J. K., S. O. Montgomery, W. R. Mahaffey, and P. H. Pritchard. 1987. "Biodegradation of Trichloroethylene and Involvement of an Aromatic Pathway." *Appl. Environ. Microbiol. 53*: 949-954.

Semprini, L., P. K. Kitanidis, D. H. Kampbell, and J. T. Wilson. 1995. "Anaerobic Transformation of Chlorinated Aliphatic Hydrocarbons in a Sand Aquifer Based on Spatial Chemical Distributions." *Water Resources Res. 31*: 1051-1062.

BIODEGRADATION OF DIOXINS IN SOIL

R. U. Halden, B. G. Halden, and D. F. Dwyer (University of Minnesota)

ABSTRACT: *Sphingomonas* sp. RW1 belongs to a small group of laboratory-cultured bacteria able to degrade dioxin-like chemicals via initial dioxygenation in the angular position. Microcosm experiments were conducted to determine the potential usefulness of these strains for in situ bioremediation. Unsaturated soils (55% field capacity) were spiked with dibenzofuran (DF), dibenzo-*p*-dioxin (DD), and 2-chlorodibenzo-*p*-dioxin (2CDD; 10 ppm each), equilibrated (24 h), amended with various densities of RW1 (10^3 to 10^9 CFU/g dry weight soil), and incubated (34 days; 21°C). Microcosms (triplicates) were sacrificed periodically to determine densities of bacteria and concentrations of the chemicals. Bacteria were extracted from soil and enumerated on selective plates; colonies of RW1 were identified by catechol 2,3-dioxygenase activity. Dioxins were extracted with acetonitrile by shaking, and extracts were analyzed by high pressure liquid chromatography/photodiode array detection.

Survival of RW1 in soil was better in the presence than in the absence of dioxins. Cell counts decreased by one (DF), three (DD), and four (2CDD) orders of magnitude depending on the chemical present. All three target pollutants were degraded. The rate and extent of degradation was dependent on the nature of the pollutant and densities of RW1. Effective initial densities ranged from 10^6 (DF, 35% loss) over 10^7 (DD, 20% loss) to 10^8 CFU/g dws (2CDD, 10% loss). Maximal loss of model pollutants was observed when larger inocula were used (DF, 10^7, 99%; DD, 10^9, 99.5%; 2CDD, 10^9, 90%). Biodegradative activity was best during the first 7 to 14 days and rapidly decreased thereafter.

This is the first report of in situ degradation of unsubstituted and chlorinated dioxins in soil by introduced bacteria. On the basis of our results we suggest that recalcitrant dioxin-like chemicals are amenable to bioremediation via bioaugmentation. However, it might be necessary to inoculate soils repeatedly in order to achieve pollutant concentrations in the low parts-per-billion range.

ENVIRONMENTAL RESTORATION THROUGH INTRODUCTION OF MICROBIAL CONSORTIA

Jun Yoshitani

ABSTRACT: Bioremediation as an effective environmental cleanup technology has gained much credibility and recognition in recent years. Hundreds of sites under the framework of CERCLA, RCRA, and voluntary initiatives are now being cleaned up using various forms of bioremediation. While further R&D is still required to take the technology to full maturity, a considerable body of knowledge has been developed.

A related area, but receiving much less attention, is the emerging use of microorganisms for general environmental restoration; that is, for application in such areas as treatment of urban and suburban nonpoint water pollution, agricultural waste management, and even environmental degradation caused by chemical management of lawns, parks, and golf courses. This area of environmental degradation, due to its enormous scope and ubiquitous nature, may in fact be more important to the overall health of our environment than the more publicized hazardous waste sites. Microorganisms used for general environmental restoration typically consist of co-existing species which are cultivated ex situ and introduced into the impacted environment. Use of indigenous microorganisms, while theoretically possible, appears to be economically and technically infeasible at this time.

The use of introduced microorganisms embodies the characteristics of sustainability, non-toxicity, and low cost. With proper information dissemination, it holds the potential of becoming a tool, much like backyard composting, for use by the general public. This paper reviews the state of beneficial microorganisms developed for general environmental restoration and provides a case history of one such microorganism consortium that is being used for agricultural and environmental purposes. The review addresses important characteristics of microorganisms, methodology of application, the underlying economics, obstacles and impediments to gaining wider acceptance, factors which contribute to their failure, effect on the environment, and regulations which apply to their use. The author addresses these and other pertinent issues surrounding the use of microorganisms for general environmental restoration purposes.

APPROACHES TO CREATION OF BACTERIAL CONSORTIUM FOR EFFICIENT BIOREMEDIATION OF OIL-CONTAMINATED SOIL

Mikhail U. Arinbasarov, Alexander V. Karpov, Sergei G. Seleznev, Vladimir G. Grishchenkov and Alexander M. Boronin
(Institute of Biochemistry and Physiology of Microorganisms, Russian Academy of Sciences, Pushchino, Russia)

ABSTRACT: The technologies for bioremediation of soil polluted with oil and its products are primarily based on the ability of some microorganisms to degrade the above pollutants. Our approach includes both the activation of local microflora and application of the respective biopreparations. Strains capable to degrade oil and oil products were isolated from contaminated soil from different climatic regions. Their degradative activities toward mazut in model systems were characterized by weight and element analyses. Degraders' activities toward different mazut components were estimated using the mathematical analysis of FTIR spectra. Groups of the cultures with the similar character of mazut degradation and/or transformation were revealed by cluster analysis. Several combinations of degraders together with biogenic additives and an additional nutrients were used as constituents of several biopreparations. The residual content of alkanes on diesel fuel-contaminated sites was reduced twice compared to control within two months after treatment with biopreparation. As for mazut degradation, application of biopreparation during two months reduced the amount of heavy fractions in mazut-contaminated soil by 30-50%.

INTRODUCTION

Oil and oil products are widespread and dangerous pollutants of soil and water. Diverse chemical composition of these pollutants and variability of climate conditions and other environmental factors (humudity, availability of phosphate, nitrogen and potassium sources, pH, etc.) are significant for vital functions of microorganisms-degraders (Leahy and Colwell, 1980). This requires development of quite a number of biopreparations that should contain not one or two strains but be associations of several strains, each degrading certain oil components and together efficiently degrading oil and its products. Creation of an efficient association requires a collection of degrader strains and an express method for testing their activities with respect to the components of oil and oil products. The objective of this study is to characterize degradative activities of strains toward a most complex and hard to degrade oil product - mazut, to construct the bacterial consortiums for efficient bioremediation, and to carry out the field testing of them.

MATERIALS AND METHODS

Isolation and identification of bacterial strains capable of mazut degradation, incubation of them on mazut as the sole carbon and energy source in liquid medium and in model soil systems, weight analysis of residual mazut in liquid and in soil samples were carried out as described earlier (Boronin et al., 1997).

Determination of initial mazut composition. To separate the asphaltene fraction, a mazut sample was poured over with pentane and kept in a dark place for complete precipitation of asphaltenes. The asphaltene precipitate was filtered, washed with pentane and dried out in a vacuum exsiccator over anhydrous calcium chloride and paraffine. Composition of the initial mazut after separation of the asphaltene fraction was determined by the method of liquid-adsorption chromatography with the use of double sorbent.

Element composition of different samples of mazut. The element composition was determined on a CHN-analyzer 1106 (Carlo Erba, Italy). The content of oxygen in the decalcified mazut extract was calculated by the difference of 100%-(C+H+N)%; the content of S was not taken into account.

IR spectra. Research objects were: a) mazut samples subjected to biodegradation by various microbial cultures (subsequently referred to as residual mazut samples); b) standard samples: standard (initial) mazut, mazut fractions, their mixtures, and no-treatment control mazut sample.

All samples were dissolved in carbon tetrachloride, and then the spectra of these solutions were obtained on a PE-1710 FTIR spectrophotometer (Perkin-Elmer, USA). The spectra were obtained under the following conditions: range from 3600 cm^{-1} to 850 cm^{-1} by step 1 cm^{-1} (wavelength) as an average of 10 scans; KBr cell 0.181 cm path length.

Mathematical analysis of IR spectra. For quantitative analysis the spectrum of the solvent was subtracted from the spectrum of a sample using the rule of Lambert-Bouger-Beer and the least squares method. All spectra obtained were centered and normalized for further use.

Changes in fraction composition were determined by a modified method of principal components regression (Fredricks et al., 1985). Changes in fraction composition were evaluated using this method for series of spectra of standard samples with the known content of a given fraction. To construct microbial clusters according to changes in fraction composition of residual mazut, the hierarchical agglomerative procedure of average nonweighing bond was used.

Field testing. Experiments were carried out in natural conditions on soil contaminated with mazut (level of contamination 8 l/m^2) and diesel fuel (level of contamination 4 l/m^2). The duration of experiments was two months. Contaminated soil was treated with biopreparations twice: at the beginning and one month later.

RESULTS AND DISCUSSION

In the course of isolation of strains capable of mazut biotransformation over 20 active strains were selected. Most of them (over 70%) belonged to the *Pseudomonas* genus, 2 strains (M04 and M05) - to the *Rhodococcus* genus, 1 strain (M12) - to the *Xanthomonas* genus, and 2 strains - to the Gram-variable nonsporeforming rods.

The ability of degrader strains to utilize mazut was determined gravimetrically. The most active strains were M05, M08, and M12 utilizing 16-18% of mazut in a liquid medium and M02, M03, M04, M13 utilizing 20-25% of mazut in soil.

To investigate the action of various microbial cultures on the mazut components and to estimate the changes in its chemical composition, the element composition of mazut after extraction from the culture liquid or soil was studied. Based on the results obtained for each strain, the ratios of C/H, C/N, H/O, and C/O were calculated. Comparison of the ratios obtained with the data for the control sample allows us to characterize changes in the samples.

A decrease of the H/C ratio in residual mazut indicates the uptake of a large portion of mazut aliphatic components. This was observed in most microbial cultures and maximally manifested in the liquid medium for M08 and M12 cultures (Figure 1). Predominance of the uptake of aromatic components and/or the break of aromatic rings leads to an increase of H/C, which was observed in the liquid medium for M04, M11, M13 cultures. All cultures tested in soil showed a lower H/C ratio than the control, which indicated the predominant uptake of aliphatic components. Interesting by this was peculiar of the cultures (M04, M11, M13) that in experiments on a liquid medium showed a higher H/C ratio than the control, which indicates the predominant uptake of aromatic components.

The state of residual mazut oxidation gives an idea about the enzymic potential of the cultures. The level of residual mazut oxidation after the action of various cultures in a liquid medium and in soil was evaluated by the changes in C/O and H/O ratios (Figure 1). Most cultures in liquid medium and all cultures in soil showed a lower level of residual mazut oxidation than the control. Based on the values of oxidation, the microorganisms leaving after their action the compounds with a high oxidation level can be supplemented in the course of creating an association by microorganisms able to transform them to compounds with a low oxidation level.

An express method of estimating the residual mazut composition using modified principal components analysis of FTIR spectra has been developed. Estimation of residual mazut composition after its biotrans-

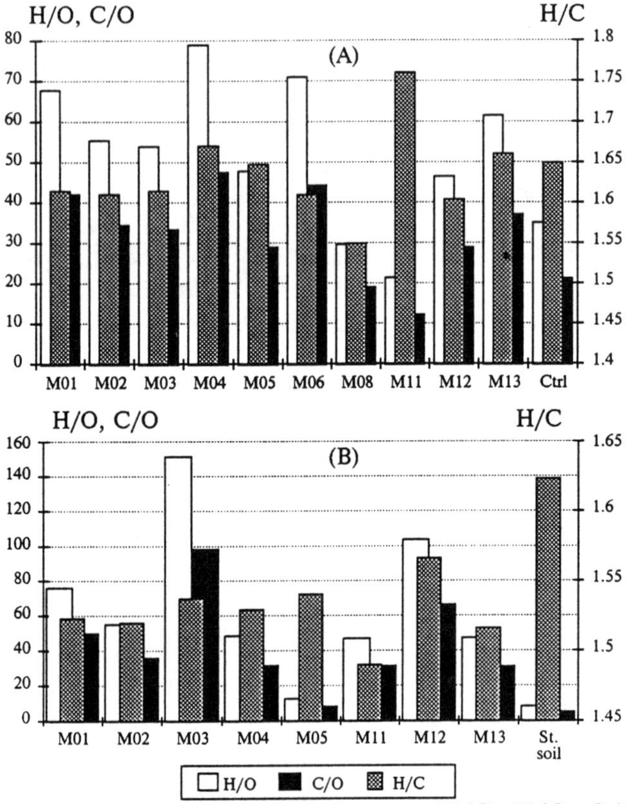

FIGURE 1. A ratio of the chemical elements H/C, H/O, C/O in residual mazut after biotransformation in liquid medium (A) and soil (B).

formation (including also biodegradation) by several microorganisms (Figure 2) shows that activities of strains significantly vary for different components.

Clustering of microorganisms by changes of residual mazut composition is presented as a dendrogram in Figure 3 revealing groups with similar character of biotransformation and/or biodegradation. For instance, strains M02, M04, M06 are included in one group. Hence, these

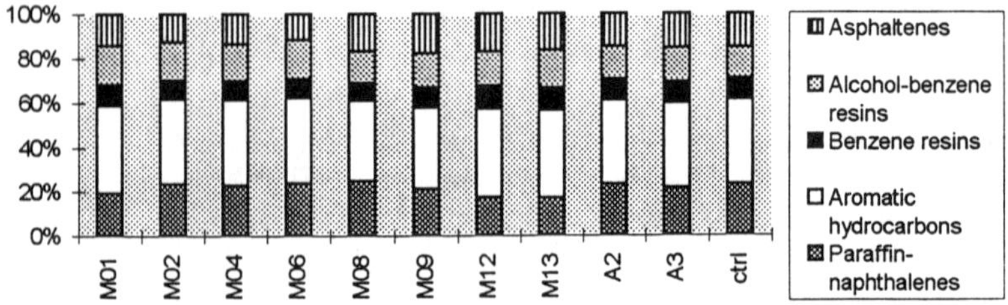

FIGURE 2. Comparative diagram of the fractional composition of residual mazut after its biotransformation by various microorganisms.

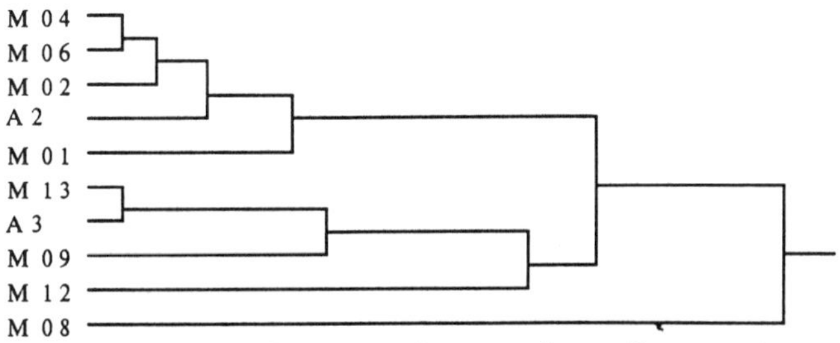

FIGURE 3. Classification of mazut-degrading strains according to composition changes after biodegradation.

strains have similar characteristics of mazut biotransformation and can be interchangeable when choosing bacterial mixtures. On the other hand, strains M01, M04 and M09 refer to different groups. A mixture of these strains is expected to be more efficient than its constituents.

Classification of the cultures into groups of similarity of transforming activities allows one to make a tentative set of combinations of degrader strains that can be specified based on the data of weight and CHN analyses. While making combinations, we also took into account the character of contamination and species relatedness of microorganisms. Thus, to degrade diesel fuel, we applied biopreparation BDNP1 which included bacteria effectively utilizing light fractions of oil products. As for mazut degradation, we mostly employed biopreparations that incorporated bacteria able to consume also heavy fractions.

The data presented in Figure 4 show that treatment of diesel fuel-contaminated sites with preparation BDNP1 allowed to reduce twice the residual content of alkanes during the experiment after two months compared to no-treatment control site. Yet, it should be noted that we also observed the consumption of alkanes in the control contaminated site: at such level of contamination with diesel fuel its degradation proceeded rather effectively both due to the action of local microflora and abiotically.

Figure 5 demonstrates the results obtained when the soil was contaminated with mazut. As was the case with diesel fuel, the consumption of light fractions, paraffins in particular, was observed both in

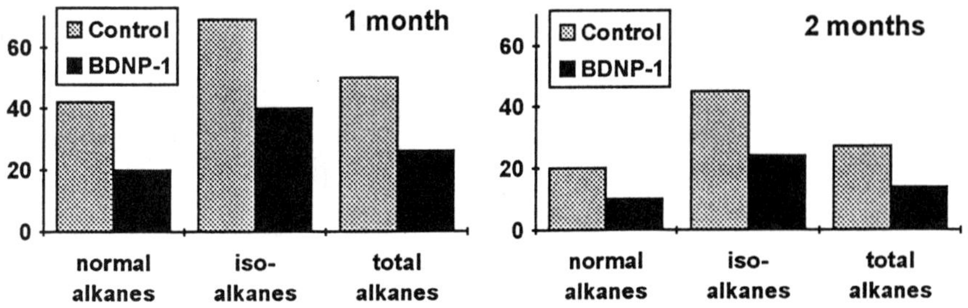

FIGURE 4. Residual percentage of normal and iso-alkanes in soil samples contaminated by diesel fuel and treated by biopreparation.

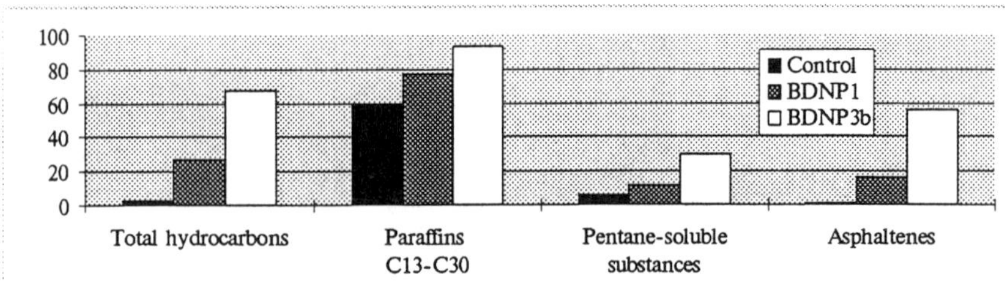

FIGURE 5. Consumption (% of initial level) of mazut components in soil samples contaminated by mazut and treated by different biopreparations.

the tested and control sample. Yet, the level of degradation of total hydrocarbons, pentane-soluble substances or asphaltenes differed drastically in the tested sample compared to control where it did not exceed several percent. So, the application of biopreparations for soil decontamination from mazut is fully justified and efficient.

CONCLUSIONS

Collection of bacterial strains capable to degrade oil and oil products was developed. Data base on their degradative activities toward mazut and its components in model systems was created. Development of an express method based on the mathematical analysis of FTIR spectra allowed estimation of degrader activities against mazut components without direct cultivation of the cultures on each component. Cluster analysis allowed us to reveal groups of the cultures with the similar character of mazut degradation and/or transformation. Several biopreparations for different types of oil contaminations were constructed using combinations of degraders together with biogenic additives and an additional nutrients. Field testing of constructed BDNP biopreparations showed their efficiency for bioremediation of soils contaminated by diesel fuel and mazut.

REFERENCES

Boronin, A. M., V. G. Grishchenkov, A. V. Karpov, S. G. Seleznev, V. G. Tokarev, M. U. Arinbasarov, R. R. Gajazov, and N. P. Kuzmin. 1997. "Degradation of mazut by selected microbial strains in model systems." *Proc. Biochem. 32*(1): 13-19.

Fredricks, P. M., J. B. Lee, P. R. Osborn, and D. A. J. Swinkels. 1985. "Rapid characterization of iron ore by Fourier transform infrared spectrometry." *Anal. Chem. 57*: 1947-1950.

Leahy, J., and R. R. Colwell. 1990. "Microbial degradation of hydrocarbons in the environment." *Microbiol. Rev. 54*: 305-315.

PILOT CLEAN-UP OF CHERNAYA RIVER (MOSCOW REGION) FROM OIL POLLUTION

Valentyna P. Murygina and Mikhail U. Arinbasarov (Institute of Biochemistry and Physiology of Microorganisms, Russian Academy of Sciences, Pushchino, Moscow region, Russia)
Elena V. Korotayeva, Anna V. Stolyarova and Leonid R. Peterson (All-Russian Oil and Gas Research Institute/VNIIneft, Moscow, Russia)

ABSTRACT: Experimental clean-up of oil polluted water surface of the Chernaya river, Lukhovitsy, Moscow Region, with bioremediation solution "Rhoder" was carried out in summer 1995. Bioremediation product "Rhoder" is an association of two Rhodococcus strains - active destructors of hydrocarbons. The above product is a stabil concentrated suspension of living active cells (10^{10} -10^{12} cells/ ml) ready to be used without their preliminary revival or activation. The bioremediation solution "Rhoder" contain bacteria (appr. 10^6-10^7 cells/ml) and nutrients (sources of nitrogen, phosphorus and potassium in a concentration of 0.005-0.02%) for maintenance a high biogeradation activity. Bioremediation solution was applied to polluted surface by sprinkling.

The area of oil polluted surface of the Chernaya river was 100 m^2; the initial pollution was 440.0 mg/l. In two weeks after the first treatment, the level of pollution decreased to 52.0 mg/l. Two-fold use of the bioremediation solution "Rhoder" during June 1995 with a 14-day break between the treatments decreased the pollution to 0.04 mg/l.

In two weeks after the second use of the solution "Rhoder", the quantity of indigenous bacteria returned to the initial level: 10^3 cells/ml for heterotrophs, and 10^2 cells/ml for hydrocarbon-oxidizing bacteria.

INTRODUCTION

The problem of environmental protection from oil pollutions has recently become quite pressing. Clean-up methods sometimes used (plowing, burning out, or raking and removal of polluted layer) are far from always promoting soil and vegetation renewal, and often cause a long-term ecological damage.

Despite patented bioremediation products for environmental clean-up from oil pollutions existing in Russia and abroad, it seems quite necessary to work in this direction. This is due to diversity of polluting objects, climatic and geochemical conditions in polluted places. Besides, microbial activity may be and is affected by humidity and structure of soils, their mineralogical and chemical composition, acidity, presence of toxic compounds, activity of indigenous bacteria and so on. Many research groups in Russia are concerned with development of some environmental clean-up methods. A clean-up method was developed at the Oil and Gas Research Institute (VNII neft) as well. It is based on bioremediation product "Rhoder" comprising two highly active *Rhodococcus* strains - oil degraders. We have a permission of Expert Committee of the State Ecological Expertise, Russian Department of Ecology, of 4 April 1994 to use these hydrocarbon-oxidizing strains on a pilot plant scale.

The goal of this work was experimental testing of the efficiency of the above method for clean-up of water surface from oil pollution under natural conditions in Moscow Region. This testing involved the microbiological and chemical monitoring of oil polluted water surface, chemical composition of the water in the Chernaya river, and natural bacterial cenosis in the cleaned object before, during and after using the clean-up method.

MATERIALS AND METHODS

Site description. Water surface of the Chernaya river in Lukhovitsy, Moscow Region, was polluted in spring 1995. Polluted area was a backwater in the Chernaya river of 100 m^2. Source of pollution was a tributary of the Chernaya river passing through the pipes over the territory of Lukhovitsy oil bulk plant. Due to leakage, the oil product got into the tributary and then into the Chernaya river. The maximal concentration of oil product was found in the place of tributary flowing into the river: the film there was up to 5 mm thick. The major area of polluted surface was covered with 1 mm film, and iridescent film was observed along its edges.

The bioremediation product "Rhoder". It comprises an association of two strains: *Rhodococcus sp.* 1418 (*ruber*) and *Rhodococcus sp.* 1715 (*erythropolis*) that are active oil degraders. The product is a concentrate of living bacterial cells with a quantity of 10^{10} - 10^{12} cells/ml ready to be used without preliminary revival or activation. Before application the bioremediation product is diluted in such a way that the number of active cells does not exceed 10^4 cells/ml in water according to the maximum permissible concentration (MPC) determined for the above bioremediation product by the Committee on Fishing of Russia. To intensify the vital functions of bacteria in "Rhoder", the working solution of the product is supplemented before application with the salt solution as a source of nitrogen, phosphorus and potassium in a concentration of 0.005-0.02%.

The working solution of "Rhoder" was prepared in a separate capacity and then poured into a water-cart tank. The water surface of the Chernaya river was treated with the bioremediation solution sprinkled under pressure of 5 atm from a 5 m^3 water-cart with evenly along the perimeter of the polluted area. The water surface was treated with 1 liter of the working solution per 1 m^2. The treatment was performed twice with a break of 14 days.

Water Sampling and Analysis. The process of clean-up was controlled. The quantity of heterotrophic bacteria was determined by counting the colonies on meat-peptone agar before, during and after using the technology. The quantity of hydrocarbon-oxidizing bacteria was determined by the method of maximum dilutions on the liquid Raimond medium with oil as a sole carbon source.

The levels of oil pollution in water samples were analyzed gravimetrically (Petrov, 1984). The chemical analysis of the water was performed titrametrically and colorimetrically (Lurye and Rybnikova, 1984). pH of the medium was assayed potentiometrically.

RESULTS AND DISCUSSION

Experimental clean-up of oil polluted water surface of the Chernaya river in Lukhovitsy, Moscow Region, with the bioremediation solution "Rhoder" was carried out in June 1995. Water samples were taken before, during and after the application of the clean-up method to analyze the level of oil pollution, quantity of the indigenous bacteria, quantity of the bacteria introduced with the solution "Rhoder", and chemical composition of the water in the Chernaya river.

Changes in the quantity of heterotrophic and hydrocarbon-oxidizing bacteria (HCO) at the application of the clean-up method are presented in Table 1.

TABLE 1. Quantity of microorganisms in the water of the Chernaya river at the application of the clean-up method.

Date	Step	Quantity of cells/ml	
		heterotrophs	HCO
25.05	before treatment	10	10
07.06	after 1st treatment	10	10
20.06	after 2nd treatment	10	10

The data show that the quantity of hydrocarbon-oxidizing bacteria increased 100-fold after each treatment. During application of the clean-up method this quantity was maintained at the same level, which was a determining factor for the success of this method. The quantity of heterotrophic bacteria also increased 10-fold due to introduction of certain amounts of phosphate, nitrate and potassium sources with the bioremediation solution "Rhoder". In 14 days after the last treatment, the quantity of hydrocarbon-oxidizing bacteria returned to the initial level.

The analysis of water samples from the Chernaya river in the course of the clean-up showed that the content of hydrocarbons changed (Table 2).

TABLE 2. Content of hydrocarbons in the water of the Chernaya river at application of the clean-up method.

№	Monitoring,	Content of hydrocarbons	
	days	mg/ml	%
1	0	440.0	100
2	14	52.0	12
3	28	0.04	0.009

The data show that the initial content of hydrocarbons in the water of the Chernaya river was 440.0 mg/ml, which corresponds to the average level of pollution. In 14 days after the first treatment, the content of hydrocarbons decreased 8.4-fold and made 52.0 mg/ml. By day 28 the content of hydrocarbons in the water was 0.04 mg/ml, which is lower than the MPC level (MPC = 0.05 mg/ml).

The chemical analysis performed before, during and after using the bioremediation product for the water clean-up from oil pollution showed that the chemical composition of the water did not change significantly (Table 3).

TABLE 3. Chemical composition of the water in the Chernaya river at application of the clean-up method.

Monitoring, days	pH	HCO_3^-, mg/l	Ca^{2+}, mg/l	Mg^{2+}, mg/l	Cl^-, mg/l	Na^++K^+, mg/l
0	7.0	305	280	24	3053	725
14	6.8	590	250	19	2767	753
28	6.6	610	200	0	1221	791

The content of phosphates, nitrates and sulfates was controlled before and after application of the clean-up method (Table 4).

TABLE 4. Content of phosphates, nitrates and sulfates before and after application of the clean-up method.

Date of sample taking	Step	Phosphates, mg/l	Nitrates, mg/l	Sulfates, mg/l
25 May 1995	before application	3.51	0.99	52.3
27 June1995	after application	0.54	0.58	51.8

Phosphates and nitrates are most intensively utilized in the ecosystem of the Chernaya river. After application of the clean-up method the content of phosphates decreased - 6.5-fold, and that of nitrates - 1.7-fold, although all the above ions were additionally introduced as constituents of the bioremediation solution "Rhoder". The content of sulfates actually did not change.

Thus the application of the bioremediation product "Rhoder" for clean-up of water surfice the Chernaya river from oil allowed us to completely eliminate the pollution in 28 days.

Introduction of certain amounts of easily assimilable sources of nitrogen, phosphorus and potassium resulted in neither phosphate nor nitrate pollution of the Chernaya river. Introduction of hydrocarbon-oxidizing bacteria with the bioremediation product "Rhoder" caused no significant damage to the river ecosystem either. In two weeks after application of the clean-up method, the quantity of indigenous bacteria in the Chernaya river was completely restored.

REFERENCES

Lurye, Yu. Yu., and Rybnikova, A. I. 1984. *Analytical chemistry of industrial waste waters* [in Russian]. Moscow, Khimiya eds..

Petrov, A. P. 1984. *Modern methods for the analysis of oil.* A monograph [in Russian]. Moscow, Nauka eds..

REMEDIATION OF PENTACHLOROPHENOL CONTAMINATED SOIL BY BIOAUGMENTATION USING SOIL BIOMASS ACTIVATED IN A BIOREACTOR

Carole Barbeau[1] , Louise Deschênes[1] , Yves Comeau[2] and ***Réjean Samson[1]***.

[1](NSERC Industrial Chair on Site Bioremediation, BIOPRO Research Center, Department of Chemical Engineering, Ecole Polytechnique of Montreal, Quebec, Canada).
[2](Environmental Engineering Section, Department of Civil Engineering, Ecole Polytechnique of Montreal, Quebec, Canada).

ABSTRACT: The use of an indigenous microbial consortium, acclimated and immobilized on soil particles (activated soil), was studied as a bioaugmentation method for aerobic biodegradation of pentachlorophenol (PCP). A completely mixed bioreactor (CMB) with a 10% soil slurry was used to produce the activated soil biomass. Results showed that the CMB was very effective in producing a PCP acclimated biomass. Within 30 days, PCP degrading bacteria increased up to 10^8 CFU/g of soil. The mineralization of the PCP added to the reactor was demonstrated by chloride release in solution. The immobilized consortium, produced in the CMB reactor, was inhibited at PCP concentrations greater than 300 mg/l. This high level of tolerance can be attributed to the positive effect of the presence of soil particles. The activated soil biomass was used to stimulate bioremediation of PCP impacted sandy soil which proved to have no indigenous microorganisms able to degrade PCP. Results showed that bioaugmentation of this soil by acclimated biomass was necessary, and reduced PCP concentration by 98 % (from 400 to 5 mg/kg). Decontamination level reached the provincial government criteria (5 mg PCP/kg) in 130 days.

INTRODUCTION

Pentachlorophenol (PCP) is one of the most prevalent wood preservatives worldwide. Its widespread use had led to the contamination of considerable volumes of soil. Traditional methods dealing with PCP contaminated soil include storage in secure landfill sites, incineration and other physical and chemical treatment methods. Many bacterial strains are known to degrade PCP but these are not often found in natural soil even after years of contamination with this chlorinated compound. Bioaugmentation, in which an acclimated microbial consortium is added to a contaminated soil, is a promising approach to the restoration of sites contaminated by PCP. In this study, the use of an indigenous microbial consortium, PCP adapted and immobilized on soil particles (activated soil), was studied as a bioaugmentation method for aerobic biodegradation of PCP in a contaminated soil. The objective of this research was to study this new bioaugmentation technique called soil activation at bench scale. This technique is based on the cultivation of biomass from a fraction of contaminated soil for its subsequent use as an inoculum in which the bacteria are immobilized on soil particles.

MATERIALS AND METHODS

Soil. A first PCP contaminated sandy soil sample (soil 1) was obtained from a wood-mill site. Another PCP contaminated soil sample (soil 2), a silty loam soil, was obtained from a pretreated pole storage site. Table 1 presents physical and chemical characteristics of the two soil samples. Soil 2 was used to produce the PCP acclimated consortium in a bioreactor and soil 1 was treated by bioaugmentation using activated soil as inoculum.

TABLE 1: Physical and chemical characterization of Soil 1 and Soil 2

Parameters	Soil 1	Soil 2
PCP (mg/kg)	500	80
pH	6.2	7.5
Organic matter content (%)	1.9	11.5
PCP degrading activity	negative	positive

Biomass Production. The production of the activated soil biomass was performed in a 125 l fed-batch completely mixed soil slurry bioreactor. The bioreactor was provided with an aeration system (15 litters of air per minute) and a mechanical mixer which maintained soil particles in suspension. The growth medium was composed of 10% (w/v) soil (10 kg of Soil 2 in 100 l of mineral salts medium). The bioreactor was fed with PCP dissolved in 0,25 N NaOH, with increasing PCP spikes (50, 100, 150, 200, 250, 300 mg/l) each added only once the PCP concentration in aqueous phase had decreased below the detection limit. Slurry samples were taken from the bioreactor every day to determine the PCP concentration in the aqueous phase by HPLC and the soluble chloride concentration by use of a specific electrode. The pH of the slurry was monitored continuously with a pH-meter. The bioreactor was operated for 31 days.

Bioaugmentation. The macrocosms in which the treatment by bioaugmentation was performed were 4 l glass jars containing 2 kg of Soil 2. The inoculation level of PCP degrading biomass was 10^5 UFC/g of dry soil. Soil humidity was adjusted to 20%. Two forms of PCP-degrading inocula were used for bioaugmentation: slurry activated soil biomass (immobilized bacteria) and centrifuged activated soil (to remove excess of water and chloride ions). After vigorous mixing, all macrocosms were covered and stored at 20°C in darkness. To ensure good oxygen transfer and soil homogenization, all soil macrocosms were manually mixed every 3 days. Adjustment of soil pH during treatment was performed by the addition of 10.00 g of $CaCO_3$ to the soil under treatment. Periodically, soil samples were taken for analysis of soil pH, PCP concentration and total bacterial heterotrophic counts.

Analytical Methods. PCP-degrading bacterial counts were performed using the spread plate method. The total heterotrophic counts were performed using the multiple-tube fermentation method. PCP in soil sample was extracted with a Soxtec Extractor unit HT 1043 (Tecator, Sweden).

RESULTS AND DISCUSSION

Soil Activation. Activated soil biomass was produced with increasing PCP additions. During the subsequent increases in PCP spikes, the PCP degradation rate by the consortium increased by a factor of 24 (from 7 to 167 mg l^{-1} d^{-1}) in 21 days of operation (figure 1). Studies on PCP degradation rates using epilithic consortia (Brown et al., 1986) showed similar degradation rate, whereas PCP-degradation activity by pure culture (*Flavobacterium)* reached 40 to 90 mg l^{-1} d^{-1} (Saber and Crawford, 1985). The maximum degradation rate (200 mg PCP l^{-1} d^{-1}) was observed on day 30, after the temporary inhibition of the consortium. Reaching this high rate suggested that chloride ions accumulation did not reduce the consortium activity since that Cl^- concentration was also at its highest level (800 mg/l) on day 30.

PCP degradation by the consortium was inhibited at concentrations greater than 300 mg PCP/l (figure 1), which compared favorably with tolerance levels of 160 to 200 mg PCP/l reported for *Pseudomonas sp.* (Radehaus and Schimdt, 1992), *Flavobacterium sp.* (Gonzales and Hu, 1991) and *Arthrobacter* strain NC (Stanlake and Finn 1982). As shown in figure 1, approximately 50 % of the liquid phase PCP added to the CMB seemed to disappear in the first five minutes after spiking. It is possible that a portion of the added PCP was first adsorbed onto soil particles and then mineralized by the biomass attached on those same soil particles. Thus, the presence of soil particles seemed to significantly decrease the toxicity of PCP for the biomass by adsorbing the pollutant and allowing cultivation to proceed at higher PCP concentrations. This phenomenon was also reported by Ehrhardt and Rehm (1985) in a study on phenol degradation in bioreactor by *Pseudomonas* sp. immobilized on activated carbon.

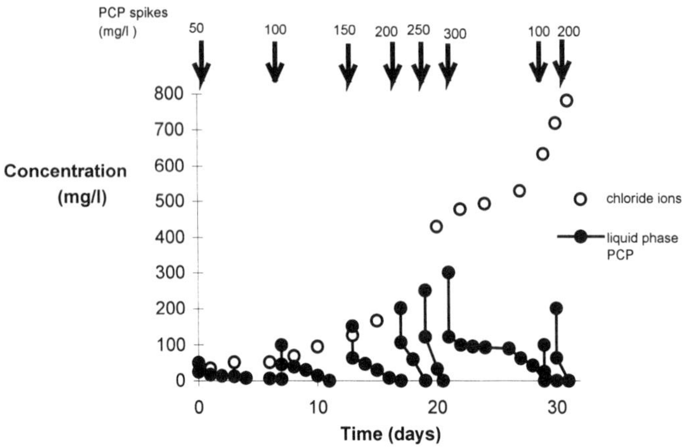

FIGURE 1: Evolution of liquid phase PCP and chloride ions concentration during the production of activated soil biomass in bioreactor

Soil activation in CMB for a period of 31 days resulted in a total heterotrophic bacterial biomass increase from 10^7 to 10^{10} CFU/g of soil, and a PCP-degrading bacterial biomass increase from 10^5 to 10^8 CFU/g of soil.

Bioaugmentation. The efficiency of the activated soil biomass was tested for the treatment of PCP contaminated soil (soil 1) that had no indigenous microorganisms able to degrade PCP. Two forms of inoculum (slurry activated soil and centrifuged activated soil) were introduced into static soil macrocosms to evaluate the effect of chloride concentration in the slurry activated soil and the effect of the biomass immobilization on the overall efficiency of the treatment. As showed in figure 2-A, PCP concentration in inoculated soil decreased from 400 mg/kg to 5 mg/kg in 130 days, independently of the inoculum type used, while PCP concentration remained constant in the non-inoculated soil. Thus, acclimated consortiums were able to degrade 98 % of the PCP initially found in the soil.

A statistical analysis at a 95 % confidence level did not indicate any significant difference between PCP concentration evolution associated with the type of inoculum (slurry activated soil or centrifuged activated soil). Consequently, the chloride concentration associated with the slurry activated soil inoculum did not appear to affect the efficiency of bioaugmentation in the conditions studied. The maximum degradation rates were about the same in all inoculated soil (25 mg PCP kg^{-1} d^{-1}).

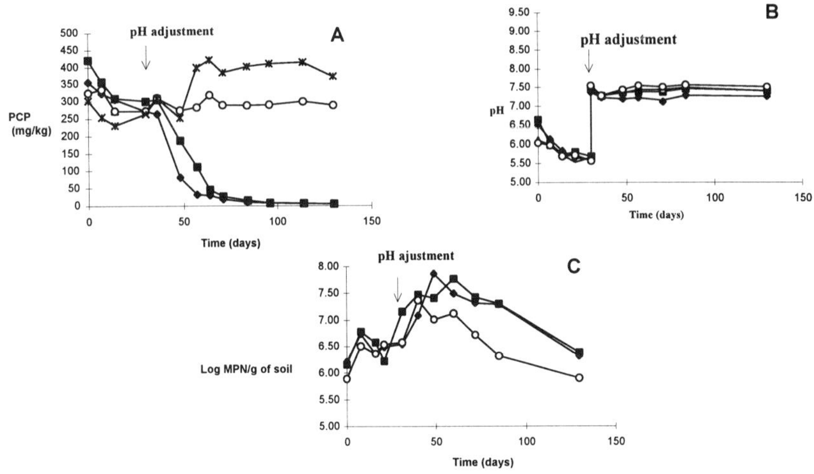

FIGURE 2: Evolution of PCP concentration (A), pH (B) and the biomass density (total heterotroph) (C) in soil inoculated with centrifuged activated soil (□), slurry activated soil (■), non inoculated soil (O) and sterilized non inoculated soil (✶).

After 30 days of treatment, pH of the inoculated soils decreased greatly due to the degradation of PCP (figure 2-A). Initially at a value of 6.6, the pH in the inoculated soil bioreactor decreased to 5.7. This acidification seemed to inhibit the biomass activity since its level decreased by more than one log when the pH level fell under 6.0 (figure 2-C). Additionally, the PCP concentration in the inoculated soil reached a stationary level between 250 and 300 mg PCP/kg in the same period (figure 2-B). After pH adjustment on day 30, however, the PCP degradation rate resumed at the previous levels as the biomass density reached a level of $10^{7.5}$ MPN/g. Similarly, Stanlake and Finn (1982) had observed biomass inhibition when soil pH had fallen below 6.15 in a study on PCP degradation by *Flavobacterium* sp. In that study, the PCP-degrading activity had also recovered once the pH had been adjusted to 7.1. Therefore, pH control during soil biorestoration, especially with halogenated compounds, is crucial to the success of the treatment and must not be overlooked.

The results of this study indicate that the soil activation process is a very effective method for the production of an acclimated biomass. The use of soil particles for indigenous biomass production significantly decreased PCP toxicity to the biomass by adsorbing the pollutant and allowing cultivation to proceed at higher PCP concentrations. Results also indicated that the activated soil biomass represents an excellent solid inoculum for bioaugmentation of contaminated soils from the wood-preserving industry.

represents an excellent solid inoculum for bioaugmentation of contaminated soils from the wood-preserving industry.

ACKNOWLEDGMENTS

This work was supported by Alcan, Analex, Browning-Ferris Industries, Cambior, Hydro-Quebec, Petro-Canada, Premier tech, SNC-Lavalin Environnement inc., the Centre québécois de valorisation des biomasses et des biotechnologies and the Natural Sciences and Engineering Research Council of Canada (NSERC). The authors would also like to acknowledge the technical assistance of Manon Leduc and Denis Bouchard.

REFERENCES

Brown, E.J., J.J. Pignatello, M.M. Martinson, and R.L. Crawford. 1986. ''Pentachlorophenol degradation: a pure culture and an epilithic microbial consortium''. Appl. Environ. Microbiol. 52 (1): 92-97

Ehrhardt H.M., and H.J Rehm. 1985. ''Phenol degradation by microorganisms adsorbed on activated carbon''. Appl. Microbiol. Biotechnol. 21: 32-36.

Gonzalez J.F., and W.S. Hu. 1991. ''Effect of glutamate on the degradation of pentachlorophenol by *Flavobacterium* sp.''. Appl. Microbiol. Biotechnol. 35:100-104.

Radehaus P.M., and S.K. Schmidt. 1992. ''Characterization of a novel *Pseudomonas sp.* that mineralizes high concentrations of pentachlorophenol''. Appl. Environ. Microbiol. 58:2879-2885.

Saber D.L., and L. Crawford. 1985. ''Isolation and characterization of *Flavobacterium* strains that degrade pentachlorophenol''. Appl. Environ. Microbiol. 50 (6):1512-1518.

Stanlake G.J., and R.K. Finn. 1982. ''Isolation and characterization of a pentachlorophenol degrading bacterium''. Appl. Environ. Microbiol. 44:1421-1427.

IN SITU DEEP SOIL BIOREMEDIATION OF PETROLEUM HYDROCARBONS

Vicki H. Bess and ***Richard P. Murray***

ABSTRACT: Separate sites of aged diesel and waste oil contaminated soil, the result of leaking underground storage tanks at an automobile sales and service dealership in the southwestern United States, were contaminated at depths of 10-40 feet and consisted of distinct sand and clay soil horizons. Prior to treatment, diesel concentrations were as high as 17,000 mg/Kg total petroleum hydrocarbons (TPH) and waste oil concentrations were as high as 40,000 mg/Kg TPH. An extensive mass balance treatability study using diesel and waste oil site soils from both clay and sand layers was conducted to optimize bioremediation supplements for the site. The treatability study indicated that a one-time treatment with nutrients, an oxygen supplement, bacteria supplements, and a soil amendment could reduce the diesel and waste oil concentrations 44-73 percent over 12 weeks. The treatability study information was used to develop full-scale in situ treatment, implemented by the use of the Dual Auger® mixing and injection system to deliver the bioremediation supplements.

The Dual Auger® system uses twin 5-foot-diameter augers, powered and moved by a standard backhoe. The hollow-shaft augers drill into contaminated soil, allowing the bacteria, nutrient, and oxygen supplements to be continually injected through a controlled nozzle system. The auger flights break the soil loose, allowing mixing blades to blend the microorganism and nutrient mixture in the soil. Post-treatment soil sample analyses indicated a uniform distribution of the injected reagents throughout the treated area.

Diesel was reduced to below 100 mg/Kg in all soil horizons 120 days after treatment, with the exception of one surface area re-contaminated by others after treatment was completed. Waste oil was reduced to below 6000 mg/Kg 120 days after treatment. The regulated closure concentration for both diesel and waste oil is 7,000 mg/Kg.

LOW INTERVENTION SOIL REMEDIATION APPROACHES FOR NATURAL GAS PIPELINE FACILITIES

Ralph J. Portier (Louisiana State University, Baton Rouge, Louisiana)
S. Reddy Chitla (Envirosystems, Inc., Lafayette, Louisiana)

ABSTRACT: Contamination from natural gas pipeline operations usually extends underneath the structures and pipeline appurtenances, thus posing a greater challenge for remediation efforts. An in situ bioremediation method was developed and evaluated in petroleum hydrocarbon (PHC) contaminated soils at compressor stations for a natural gas pipeline located in Louisiana. The in situ protocol, involving immobilization of adapted microbes to porous media, was developed and implemented with minimal excavation and disruption to the ongoing operations at the facility. The remedial approach was used for remediating soil contaminated with petroleum based lubricants and other petroleum products resulting from chronic leakage of lubricating oil. Initial total petroleum hydrocarbon (TPH) measurements revealed values up to 12,000 mg/kg at one of the remediation areas. The aim of the remediation project was to reduce TPH concentration in the contaminated soils to a level of <100 mg/kg, a level acceptable to state and federal regulators. After operating the system for 122 days, all remediation areas showed greater than 99.5% reduction in TPH concentrations.

INTRODUCTION

Because of man's increasing reliance on fossil fuels, and the substandard handling practices associated with these fuels in the pre-environmental awareness era, wide spread contamination has resulted. Traditional bioremediation approaches have been use to treat petroleum based contamination in readily accessible and relatively permeable soils. Over the last 30 years, leakage and accidental discharge of petroleum based lubricating oil used to maintain the pipelines and related equipment of a gas transmission company have resulted in PHC contamination. An in situ bioremediation technique was developed and implemented at the contaminated areas. The areas with contamination extending to depths greater than 1 m were remediated with minimal excavation using in situ immobilized bioreactors (Bioplugs™). The bioreactors were placed in networks within and adjacent to the contaminated areas to generate sufficient biomass for hydrocarbon mineralization. The contamination resulting from lubricating oil consisted of hydrotreated, heavy paraffinic petroleum distillates, obtained by treating a petroleum fraction with hydrogen in the presence of a catalyst, containing a relatively large proportion of saturated hydrocarbon with carbon numbers predominantly in the range of C20 to C50.

EXPERIMENTAL DESIGN

The bioreactors, constructed of 7.62 cm slotted pipe sealed on top and bottom, were designed to facilitate petroleum hydrocarbon mineralization by placing PHC-degrading bacteria. Each Bioplug™ was filled with inert support media (W.R. Grace™, Type Z carrier) serving as a support matrix for the organisms. Operational flow was maintained by initiating a pressure gradient in the reactor using compressed air. This accentuates water flow (either groundwater or site water) through the immobilized bed. The movement of water through the bed allows mineralization of organics and generation of excess biomass in the form of whole cell bleed-off from the bed. The elevated biomass or microbial front is allowed to escape through the perforations of the Bioplug™ into the surrounding soil. Thus, an enriched, adapted microflora is introduced into the subsurface for effective degradation of organics (Figure 1).

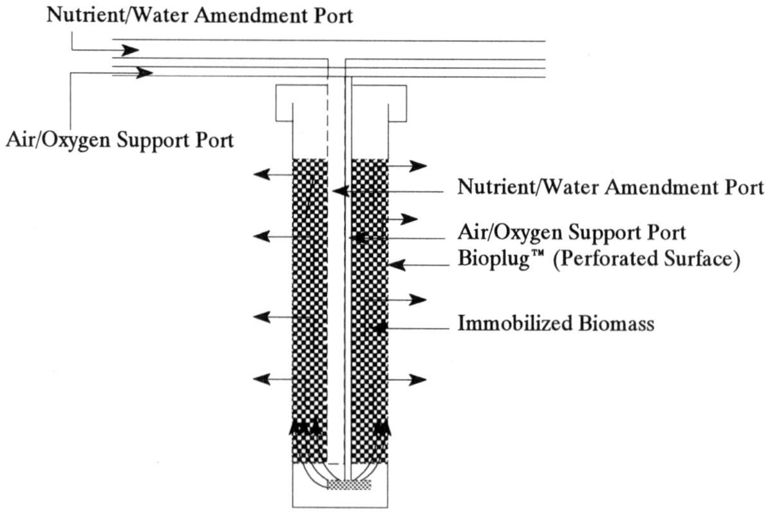

FIGURE 1. Typical Cross-Section of a Bioplug™.

Forty six (46) Bioplugs™ were installed in one of the remediation areas comprising an area of 600 m^2. It was estimated that approximately 800 m^3 of soil was impacted within this area. Bioreactors were designed with ports for the addition of nutrients and air. Selective enrichment was used to obtain the microbial consortia used for immobilization to the support media for placement into the reactors. Microorganisms from site soils were isolated on petroleum-amended media. The microbial strains with the capability to metabolize lubricating oil as a sole carbon source were selected and combined to form a stock culture. The stock culture consists of gram negative organisms including members of the species *Serratia* and *Pseudomonas*.

Sampling and Analysis. Within the remediation area, seven distinct subdivisions were created. In each subdivision, sampling locations were randomly chosen for each sample date so that the exact areas within each subdivision were not sampled twice. Monitoring samples were collected using a standard 8.255 cm diameter, stainless steel auger in 0.3 m intervals. At each interval, excavated soils were thoroughly mixed with triplicate composite samples taken with the corresponding depth, date, and subdivision. Over the course of the 120 day study, a total of 360 samples were taken from the subject area at intervals of 0, 14, 30, 60, 90 and 122 days.

A modified version of U.S. Environmental Protection Agency (USEPA) Test Method 418.1 (spectrophotometric, infrared) was used to analyze the samples for the presence and amount of petroleum hydrocarbons. The total Polynuclear Aromatic Hydrocarbon (PAH) components of soil samples were qualitatively and quantitatively determined by using a gas chromatograph (GC) equipped with a mass spectrometer. These assays were coupled with microbial measurements of total heterotrophic and total petroleum degrading bacteria, and direct measurements with acridine orange fluorescent staining. Changes in the population of petroleum hydrocarbon degrading microorganisms over time in the subject area were determined by averaging the values obtained from all samples collected within the area.

To determine statistical differences in TPH concentrations in reference to time and depth, various tests were run on the statistical analysis program SAS®. A one-way analysis of variance (ANOVA) was performed to determine if the mean TPH concentration for each sample date was statistically different from that of the others. In addition, a Fisher LSD test was performed to determine which means were statistically different from the others. An analysis of covariance was performed on TPH concentration over time and depth to determine if TPH concentrations varied according to depth.

RESULTS AND DISCUSSION

During the course of the experiment TPH levels for four of the seven subdivisions were reduced to less than 100 mg/kg. The remaining subdivisions were found to have TPH levels greater than 100 mg/kg. At the conclusion of the remediation test (Day 122), TPH levels at all subdivisions were less than 100 mg/kg by location and depth. Figure 2 presents data on TPH reductions.

An ANOVA was performed to determine the variation of mean TPH concentration for each sample date from the others. If a significant difference was noted, a Fisher LSD test was used to identify the variation. The results indicated that the mean TPH for Day 30 was different from the mean TPH levels of Day 60, Day 90 and Day 122 ($p>0.001$). The means of Day 60, Day 90 and Day 122 were similar to each other. It appears that degradation occurred at a significant rate from Day 0 to Day 60, but not thereafter.

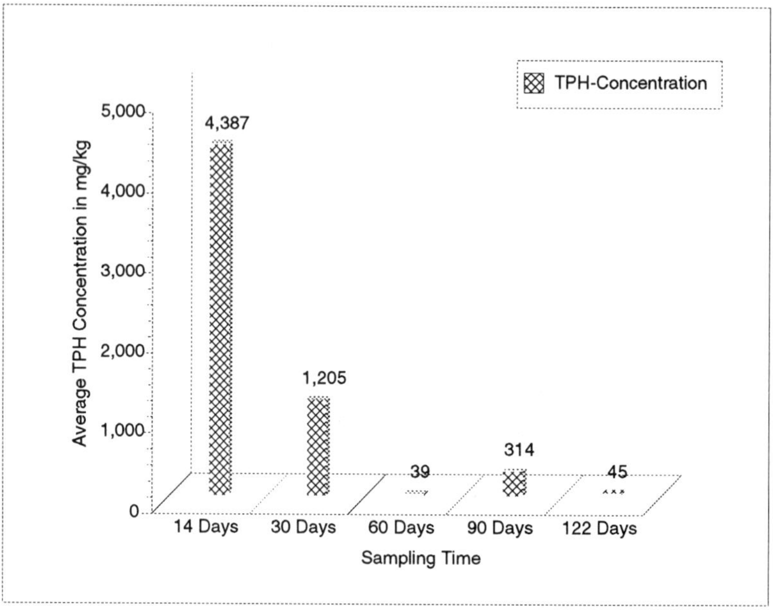

FIGURE 2. Reduction of Average TPH Concentrations Over Time.

Statistical analysis also showed that there was a significant relationship between TPH and time with radial distance from each Bioplug™. Analysis of covariance indicated that TPH degradation at 0.3 m, 0.6 m and 0.9 m radial distances occurred at an equivalent rate throughout the course of the experiment.

Samples from two subdivisions (SD1 and SD5) were analyzed at Day 14 and Day 90 for PAH content and concentration (Table 1). Fourteen (14) PAH compounds were present in varying concentrations. Phenanthrene, chrysene and benzo(a)anthracene were present on Day 14. Phenanthrene was easily degraded but chrysene and benzo(a)anthracene were recalcitrant. A majority of the PAH compounds present after Day 90 were larger molecular 4-, 5-, and 6-ring compounds such as flouranthene, benzo(a)pyrene, benzo(b)flouranthene and indeno(123cd)pyrene.

Samples from Day 14 at all depths in two subdivisions (SD1 and SD5) contained large clusters of high-end saturated alkanes. By Day 90, the composition of saturated alkanes in the samples had been altered. The Day 14 samples from SD1 contained a large number of both high and low end saturated alkanes. The 0.3 m sampling depth contained fewer light-end compounds, while the 0.6 m sampling depth contained low and high molecular weight compounds distributed evenly. The samples from the 0.9 m depth contained mostly high-end compounds.

By Day 90, there were considerably fewer PAH compounds at all depths. At all three sample depths, the high molecular weight alkanes were present, but in relatively reduced amounts. For the 0.3 m depth, the peak dropped from

TABLE 1. PAH Concentration Changes: Subdivisions SD1 and SD5.

Depth	Abundant Compounds	Concentration (D14, D90)
Subdivision SD1		
0.3 m	Phenanthrene	(0.52mg/kg, *)
	Chrysene	(0.04mg/kg, 3.41mg/kg)
	Anthracene	(0.10mg/kg, *)
	Benzo(a)anthracene	(0.37mg/kg, 1.79mg/kg)
	Benzo(b)fluoranthene	(*, 3.17mg/kg)
0.6 m	Benzo(b)fluoranthene	(1.37mg/kg, 1.25 mg/kg)
	Benzo(a)anthracene	(0.87mg/kg, 0.31mg/kg)
	Chrysene	(0.37mg/kg, *)
	Phenanthrene	(*, 0.18mg/kg)
0.9 m	Phenanthrene	(0.98mg/kg, *)
	Benzo(a)anthracene	(0.82mg/kg, 1.01mg/kg)
	Chrysene	(0.48mg/kg, 2.43)
	Benzo(b)fluoranthene	(*, 1.44mg/kg)
Subdivision SD5		
0.3 m	Chrysene	(0.42mg/kg, *)
	Anthracene	(0.14mg/kg, *)
	Benzo(a)anthracene	(0.46mg/kg, *)
	Fluoranthene	(1.12mg/kg, *)
	Benzo(a)pyrene	(*, 0.05mg/kg)
0.6 m	Benzo(k)fluoran	(1.98mg/kg, 1.25mg/kg)
	Chrysene	(0.42mg/kg, *)
	Fluoranthene	(0.44mg/kg, *)
	Benzo(b)fluoranthene	(*, 1.25mg/kg)
	Indeno(123cd)pyrene	(*, 5.10mg/kg)
0.9 m	Benzo(k)fluoran	(1.90mg/kg, 1.50mg/kg)
	Chrysene	(0.09mg/kg, 0.43mg/kg)
	Benzo(b)fluoranthene	(0.07mg/kg, 0.83mg/kg)

* - not present in significant amounts on sample date indicated

3,783,814 to 595,864, indicating a decrease in the total number of compounds present. The number of compounds also decreased significantly, with peak dropping from 3.113 x 10^7 to 338,569 for the 0.6 m sampling depth, and to 476,244 from 1.538 x 10^7 for the 0.9 m sampling depth.

The petroleum degrading microbial population increased 57% from Day 14 to Day 90. The ratio of heterotrophic microorganisms to petroleum degrading microorganisms remained fairly constant for the first 90 days. The microbial growth and the degradation of the contamination occurred exponentially within the first 30 days. Microbial growth was observed to decrease after 30 days when the majority of the easily degraded petroleum hydrocarbons, saturated alkanes and PAHs containing fewer than three rings had been depleted by 72%.

CONCLUSIONS

The TPH concentrations in the subject area were reduced by >99.5% from Day 14 to Day 122. The majority of the degradation, 72%, occurred by Day 30. Prior to Day 30, a very rapid increase in microbial numbers was noticed. This increase can be attributed to the abundance of degradable substrates, such as saturated alkanes and PAH compounds with less than three rings. By Day 30, the more easily degradable substrates were at or near depletion. The adapted microflora appears to have experienced a lag in growth after Day 30. This may be attributed to recalcitrant PHCs remaining as substrates. The remediation of all areas was accomplished by using the in situ approach with a significant cost savings.

ACKNOWLEDGEMENTS

The authors are greatly thankful to Ms. Kimberly K. Barton who performed the experimental work as partial fulfillment for the master degree at Louisiana State University (LSU). We would like to duly acknowledge the effort of Ms. Barton on collection and analysis of data. The work was supported by Envirosystems, Inc. of Lafayette, Louisiana under a research agreement with LSU.

REFERENCES

U.S. Environmental Protection Agency. 1986. *Test Methods for Evaluating Solid Waste. Volume 1A:Laboratory Manual Physical/Chemical Methods*. Office of Solid Waste and Emergency Response, Washington, DC. SW-846.

Barton, K. K. 1994. "In Situ Bioremediation of Petroleum Contaminated Soils." M.S. Thesis, Louisiana State University, Baton Rouge, LA.

EVALUATION OF COMMERCIAL PRODUCTS USED TO BIOREMEDIATE RAILROAD BALLAST

N.R. Chrisman Lazarr, ***L.T. LaPat-Polasko***, J. Heinicke, E. H. Honig, and B.D. Stewart
(Woodward-Clyde Consultants, Omaha, NE and Phoenix, AZ)
(Union Pacific Railroad Company, Houston, TX)

ABSTRACT: A series of treatability studies were conducted to evaluate the ability of various products to stimulate the biodegradation of aged petroleum hydrocarbons in a composited mixture of soil, gravel, and railroad ballast. The products tested consisted of microorganisms, nutrients, surfactants, and unidentified additives. Important considerations associated with product evaluation included: reductions in total petroleum hydrocarbon (TPH) concentration, length of time and level of effort required for product application, and the cost of treatment.

In the laboratory, 15 different products were applied per each vendor's instructions to batch microcosms containing approximately 0.1 cubic yard of contaminated soil composite. Three controls were also tested to evaluate the effect of no treatment, a moisture amendment, and a moisture/oxygen amendment. In the five microcosms which exhibited the highest reductions in TPH, microcosms which initially contained TPH ranging from 34,000 mg/Kg to 45,000 mg/Kg were reduced to concentrations ranging from 10,000 to 19,000 mg/Kg. A comparison of TPH data collected from treatment and control microcosms indicates that at least 7 of the 15 vendor products tested significantly decreased the TPH concentrations compared to the three controls. The highest percent TPH removals were associated with products that contained microorganisms.

At Union Pacific Railroad's Settegast Yard in Houston, Texas, four of the products tested in the laboratory were evaluated in the field. The test plots were 10-foot wide by 40-foot long strips centered on the railroad tracks. The products were applied to surface soils and ballast. The product amendment that resulted in the highest percentage reduction in TPH at the bench-scale level (approximately 68%) also showed the highest percent reduction in TPH at the pilot-scale level (approximately 79%). This same test plot also displayed the highest increase in petroleum degrading microorganisms during the 6-week study. These results suggest that bioaugmentation may be an appropriate strategy for reducing TPH concentrations in surface soil and ballast material.

BIOSTIMULATION AND BIOAUGMENTATION OF ANAEROBIC PENTACHLOROPHENOL DEGRADATION

Siwei Zou, Krista M. Anders, and ***John F. Ferguson***

ABSTRACT: Anaerobic transformation of pentachlorophenol (PCP) is studied with soils from PCP-contaminated sites to evaluate the biostimulation of reductive dehalogenation with mineral nutrients and glucose and the effect of bioaugmentation with a methanogenic consortia that mineralizes PCP. The study, which will be completed in autumn 1996, involves soils from two wood treatment sites and bioaugmentation from an enrichment in a fluidized bed reactor that mineralizes PCP via reductive dehalogenation to phenol. Preliminary studies with soil from a wood treatment site have shown that biostimulation results in slow, but complete PCP removal with negligible accumulation of lesser chlorinated phenols. Rates were much lower and transformation was incomplete when glucose was not added. Bioaugmentation (accompanied by glucose addition) increased the PCP removal rates by a factor of six, again with transient formation and disappearance of lesser chlorinated phenols.

The current study is intended to explore the effect of different levels of glucose amendment and different levels of inoculation with the PCP enrichment, as well as to determine if mineral nutrients enhance degradation. PCP levels of about 1 μM to 40 μM will be used to explore the effects of low and high, possibly inhibitory, concentrations of PCP. Conventional serum bottle bioassay procedures are used; PCP and phenolic metabolites are acetylated and analyzed by gas chromatography, using electron capture and flame ionization detection. Gas composition, including methane, carbon dioxide and hydrogen, is also determined by gas chromatographic methods. Controls are included to assess sorption and abiotic degradation in the soil and possible abiotic effects of the inocula.

The study is expected to develop useful information on in situ bioremediation of PCP-contaminated sites and on the practical utility of bioaugmentation with a complex dehalogenating consortia. Related studies are being conducted to characterize the consortia in terms of the rates of dehalogenation of several chlorophenol congeners, including those that appear to be rate limiting in mineralization. Further enrichment of the consortia is also underway with the objective of attempting isolation of novel dehalogenators. These studies will be used in interpreting the results with the soil microcosms.

RATES AND MECHANISMS FOR MODELING MICROBIAL TRANSPORT FOR BIOREMEDIATION SYSTEMS

Dr. H. Scott Fogler (University of Michigan, Ann Arbor, MI)
David Maurer (University of Michigan, Ann Arbor, MI)
Dong-Shik Kim (University of Michigan, Ann Arbor, MI)

ABSTRACT: The attachment and detachment mechanisms of microorganisms to soil surfaces is dependent upon the properties of the microbe and the surrounding environment. Consequently, the flowrate and composition of the feed solution will have a profound effect on microbial transport. The effect of a stepwise increase in flowrate, varied ionic strength, and carbon source depletion (starvation) were studied in a glass micromodel, a porous media flow matrix. Additional experiments were performed in batch and rotating disk experiments to isolate key effects. The data below show a critical shear rate which causes the detachment of a significant amount of attached biomass. This average critical shear rate increased nearly 100-fold with an increase in the ionic strength of the nutrient solution through the addition of 0.05M $CaCl_2$ within a flow system. An increase in microbial detachment with increasing shear stress was also observed in rotating disk experiments. During starvation, cell diameter decreased from 2.1 μm to 0.9 μm over a period of 25 days. This result suggests that *Leuconostoc mesenteroides* decayed into small non-vegetative cells known as ultramicrobacteria.

INTRODUCTION

Effective design of *in-situ* bioremediation systems requires an understanding of the mechanisms and rates of microbial attachment and detachment. Obviously, if microbes are not present in the proper location of the contaminant, the contaminant will not be eliminated. Therefore, the movement of microbes in the subsurface is extremely important to the control and capability of bioremediation as a treatment technique.

Current research directed at determining factors controlling attachment, detachment, and aggregation rates have identified exopolymer production and biofilm interactions as important influences on transport characteristics. These influences, in turn, are affected by environmental conditions surrounding the microbes such as feed composition and flow rate. Unfortunately, a clear understanding of these characteristics and the subsequent correlation of transport rates with parameters controlling effects of adsorption that would allow selective placement has not yet been achieved.

In the case of *Leuconostoc mesentroides*, a change in the nutrient composition from a mixture of glucose and fructose to sucrose causes the production of an insoluble exopolysaccharide. Exopolysaccharides can strongly influence cell transport through porous media by adsorption, adhesion, retention, and sloughing of cells to and from soil surfaces. Another significant nutrient condition is no nutrient at all. If bioremediation is effective and the contaminant of interest is degraded then the microbes remaining will starve. How this starvation influences subsurface water flow must be understood.

Greater insight into both phenomenological and analytical properties that contribute to microbial transport is needed for the expanded implementation of bioremediation systems. To determine these mechanisms, methods that visualize the physical situation are necessary. Previous work has shown that in-depth understanding of microbial transport mechanisms can be obtained from micromodels.

MATERIALS AND METHODS

Microorganisms: Three different microbes have been selected for this study. The microbes include: 1) a toluene degrader isolated from a Superfund site soil sample, *Adriaens Strain A (ASA)*, isolated by Dr. Peter Adriaens at the University of Michigan; 2) *Pseudomonas stutzeri KC (Ps KC)* - ATCC 55595, a carbon tetrachloride degrader which is currently being introduced into a contaminated aquifer by Dr. Craig Criddle at Michigan State University; and 3) *Leuconcostoc mesenteroides (L. mesenteroides)* - ATCC 14935, a model microbe which produces an insoluble exopolysaccharide under certain nutrient conditions.

Micromodel: A micromodel (Figure 1) is a network of pores and pore throats etched between two pieces of glass which are fused to enclose the network. The dimensions of the system are approximately ~300 microns in diameter for the pores and ~50 microns in diameter for the pore throats. The network is pre-programmed by the researcher allowing for a variety of different network structures. The end result in a transparent network of channels that simulate porous media, such as soil.

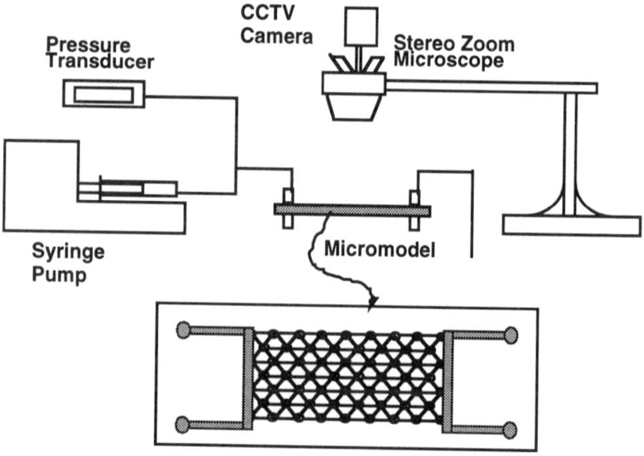

FIGURE 1. Micromodel Apparatus.

The micromodel provides a two-dimensional visualization of the process of cell aggregation and biofilm development under dynamic flow conditions. This tool allows one to magnify, watch, record, and analyze via video camera and VCR the physical mechanisms of deposition.

Flow Detachment System: Feed conditions in both batch and flow experiments were varied to obtain an understanding of the attachment, detachment, and growth characteristics of *ASA* and *Ps KC*. In addition to different feed nutrient compositions, the flow rate was increased step-wise in each experiment. The effluent microbial concentration was measured versus time at a specified flowrate.

Rotating Disk Experiments: In addition to flow experiments, rotating disk experiments have been performed. A biofilm was grown on a thin glass disk and this disk was then rotated in growth media solution at a specified radial velocity value for 1 min. These experiments were performed for 5 different rotation rates as the concentration of detached microbes was measured.

Deterioration of Biozone: In all micromodel experiments, cells were first grown to stationary phase on glucose and fructose plus non-carbon growth medium. Initially, nutrient with sucrose was supplied to allow *L. mesenteroides* to produce exopolysaccharide for two days. The carbon nutrient feeding was stopped, and starvation began, as buffer solution (pH 7) was injected for 7 days. The buffer solution contains trace amount of sodium, phosphate, potassium, and chloride ions, and ascorbic acid and acetate. Video images, pressure drop across the micromodel and effluent cell concentration were collected over time.

RESULTS AND DISCUSSION

Varied Flow Conditions: A micromodel experiment for *ASA* with nutrient feed at 15 g/l sucrose in phosphate buffer was performed over a step-wise increase in flowrate. The outlet bacteria concentration is plotted versus time with the increase in flowrate also indicated on the right abscissa (Figure 2a). The peak observed is attributed to the detachment of a large number of microorganisms. Therefore, the corresponding flowrate of 0.05 ml/min could be considered to impart a critical shear rate on the biofilm to cause most of the microbes to detach.

For comparison, another micromodel experiment was performed using an increased ionic strength nutrient by the addition of 0.05M $CaCl_2$. According to DLVO (Derjaguin-Landau-Verwey-Overbeek) theory, increasing ionic strength will decrease the electric double layer between two particles and therefore cause the microbes to attach more strongly to each other and the surface. This theory is supported by comparing the results of Figure 2a with Figure 2b. In Figure 2b, the peak effluent concentration occurs at a higher flowrate (5 ml/min, if the ordinate was expanded), indicating that a greater shear rate was necessary to remove the microbes from the system.

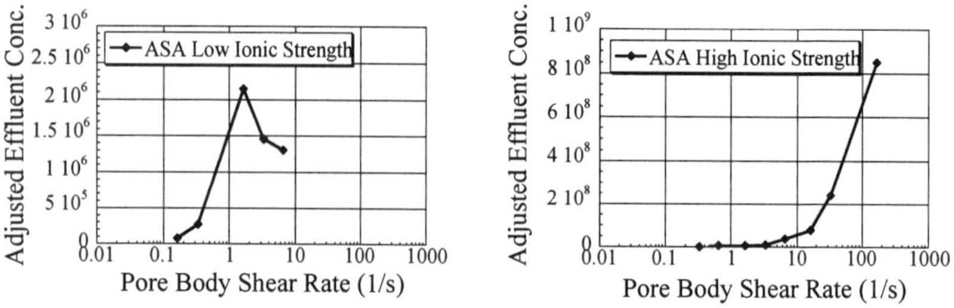

FIGURE 2. Comparison of *ASA* Effluent Concentration at Varied Feed Ionic Strengths. (a) Low Ionic Strength (b) High Ionic Strength

Similar experiments were performed for *Ps KC*. In the first experiment, *Ps KC* was grown on 15 g/l sucrose in phosphate buffer while in the second *Ps KC* was grown on 1.6 g/l acetate in artificial groundwater. Again, curves similar to Figure 2b were observed, illustrating the probable existence a critical shear rate that may be used to characterize bacterial sloughing. These critical values are:

Experimental System	Critical Pore Body Shear Rate (s^{-1})
ASA Low Ionic Strength	2
ASA High Ionic Strength	164
Ps KC on Sucrose	165
Ps KC on Acetate	33

Micromodel images on the evolution of the biofilm are shown in Figure 3. The effluent microbial concentration was measured as the flow rate was increased step-wise. Sudden growth was observed as the flowrate was quickly stepped from 0.001ml/min to 0.050 ml/min (Fig 6a). Biofilm sloughing was observed without increased flow (Fig 6b to 6c). Increasing the flow caused more biomass detachment, but significant sloughing was not observed until an extremely high flowrate (5 ml/min) was used. This flowrate washed out almost all of the bacteria in the micromodel (Fig 6d).

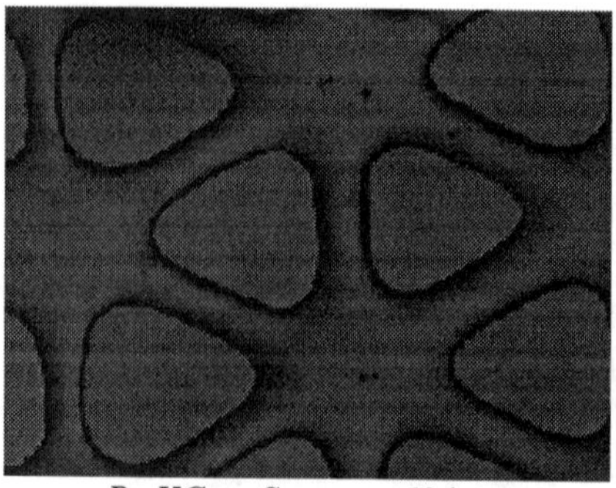

a. *Ps KC* on Sucrose, 69 hours
Flowrate = 0.001 ml/min

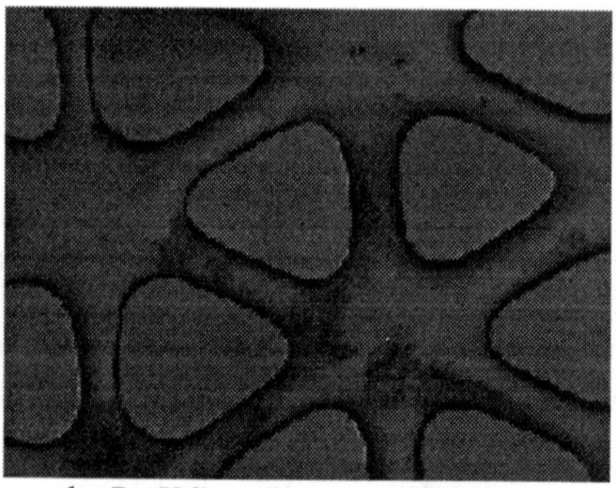

b. *Ps KC* on Sucrose, 75 hours
Flowrate = 0.05 ml/min

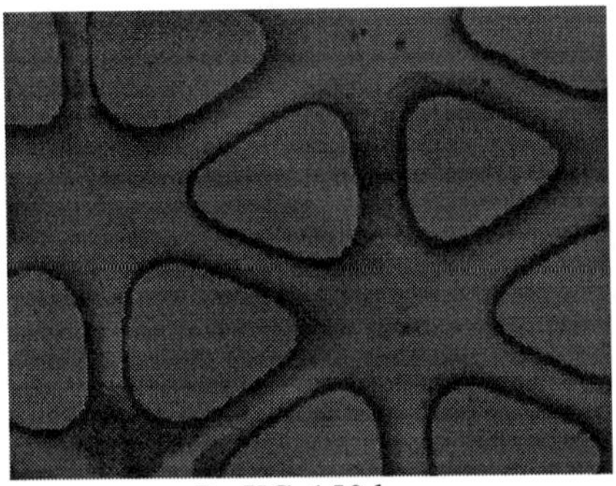

c. *Ps KC*, 150 hours
Flowrate = 0.05 ml/min

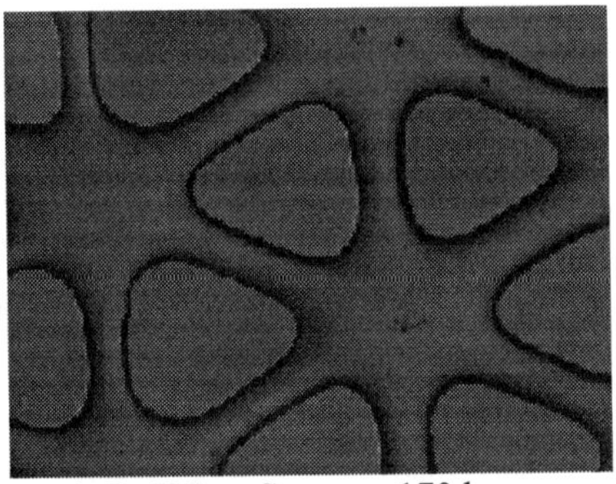

d. *Ps KC* on Sucrose, 170 hours
Flowrate = 5.0 ml/min

FIGURE 3. *Ps KC* Growth and Transport in Micromodel with Increasing Flow Rates

Rotating Disk: The rotating disk apparatus was used to produce a gradient of shear stresses on the biofilm as a function of radius. The faster the disk is rotated the greater the maximum stress present at the edge of the disk and the greater the average shear stress imparted on the entire biofilm. Experiments have illustrated an increase in the amount of biofilm detached from the rotating disk at high radial velocities. These results were confirmed by a physical inspection of the biofilm remaining on some of the disks after rotation. Studies are continuing to see if a critical shear stress is apparent.

Biozone Deterioration: Since exopolysaccharide production has been shown to cause drastic flow diversion, *L. mesenteroides* was studied in both flow and batch starvation experiments. In micromodel experiments, starvation caused detachment and re-orientation of the biofilm. This result was determined by the physical movement of the biofilm downstream and more open space upstream as indicated by the use of a non-toxic fluorescent dye in the feed stream.

Cell size distribution was measured during the period of starvation in a batch reactor using a Coulter Counter. The average diameter of normal cells is about 2.1 μm: after four days of starvation the average diameter about 1.8 μm (Figure 4) and after 25 days the cells shrank to a diameter of about 0.9 μm while the overall number of cells remained essentially constant. This physiological response of *L. mesenteroides* to starvation is a form of ultramicrobacteria (UMB).

Furthermore, more biomass sloughed off, on average, after the switch to deionized water as the feed solution. The pressure oscillations were thought to occur primarily due to biomass sloughing. Dead bacteria and detached biomass trapped in the micromodel pore throats could cause a pressure increase. However, these blockages are eventually eluted, which could result in pressure decreases. Cell size reduction, soluble exopolymer diffusion or exopolymer consumption by starving bacteria may be important in biomass evolution during starvation.

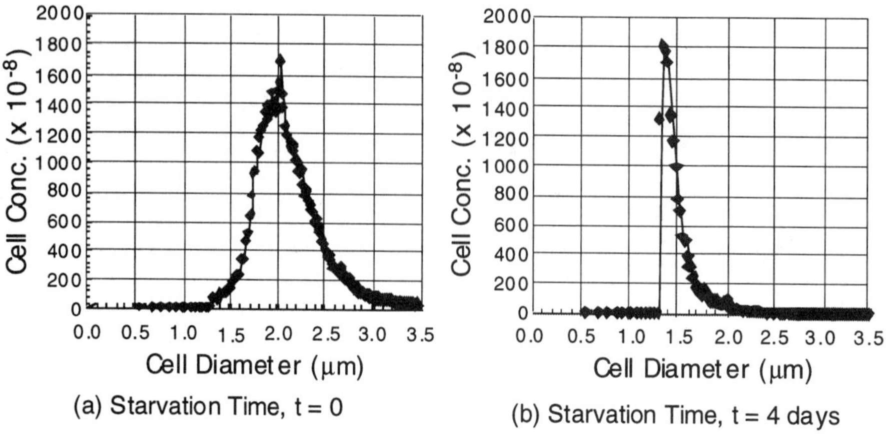

FIGURE 4. Cell size distribution change during starvation.

CONCLUSIONS

Microbial hydrodynamic parameters need to be quantified so that descriptive mathematical models can be used to enhance the management of *in-situ* bioremediation sites. Micromodel experiments have shown there exists a critical shear rate that may be used to characterize microbial sloughing.

A biozone can deteriorate from predation, bactericide exposure, or nutrient stress. Carbon source depletion caused significant biomass sloughing and associated pressure oscillations in porous media. *L. mesenteroides* decayed into small non-vegetative cells known as ultramicrobacteria (UMB).

REFERENCES

Fletcher, M., M.J. Latham, J.M. Lynch, and P.R. Rutter. 1980. "The Characteristics of Interfaces and Their Role in Microbial Attachment." In R.C.W. Berkeley, (Ed.) *Microbial Adhesion to Surfaces*, Ellis Horwood Ltd., Chichester.

Lappan, R. E. and H. S. Fogler. 1994. "*Leuconostoc mesenteroides* Growth Kinetics with Application to Bacteria Profile Modification." *Biotechnology and Bioengineering.* 43: 865-873.

Lappin-Scott, H. M. and J. W. Costerton. 1990. "Starvation and Penetration of Bacteria in Soils and Rocks." *Experientia.* 46: 807-812.

Rutter, P.R. and B. Vincent. 1984. "Physiochemical Interactions of the Substratum, Microorganisms, and the Fluid Phase." In *Microbial Adhesion and Aggregation.* Springer-Verlag. New York.

MICROBIAL TRANSPORT IN A PILOT-SCALE BIOLOGICAL TREATMENT ZONE

Michael A. Malusis, Daniel J. Adams, ***Kenneth F. Reardon***, and Charles D. Shackelford (Colorado State University, Fort Collins, Colorado)
Douglas C. Mosteller and Al W. Bourquin (Camp Dresser & McKee Inc., Denver, Colorado)

ABSTRACT: A critical question for the success of most bioaugmentation strategies is whether the added cells will move readily through an aquifer or plug the formation downgradient of the treatment zone. To investigate this question for two constitutive TCE-degrading strains, a 50% scale model of a biological permeable reactive barrier was constructed and tested under approximate field conditions. Ground water flow in the system was characterized and microbial transport was measured by collecting cells from different sampling points. Although some cells of *Burkholderia cepacia* PR1 were transported, the majority of the cells added to the system became entrapped or attached to the soil within a few inches of the zone of addition. *B. cepacia* ENV-435, an adhesion-deficient variant, was also tested. The majority of these cells were transported rapidly through the soil used in this test although some were retained. Plugging of an aquifer following addition of ENV-435 cells is thus unlikely to be a problem, although it would be a concern with PR1.

INTRODUCTION

Background. Bioaugmentation is the addition of microorganisms to promote bioremediation. Such inoculation is needed only when the desired metabolic capability is absent or at unacceptably low levels. Added cells must be able to survive in their new environment, be transported to the contaminated zone, and retain the capacity to degrade the pollutant(s).

Although the indigenous microorganisms at most sites are capable of biodegrading the contaminants found there, bioremediation practitioners and researchers are finding that bioaugmentation is an important tool for recently contaminated sites, for sites with inhibitory co-contaminants (e.g., heavy metals), and for sites impacted by certain difficult-to-degrade pollutants. In recent years, bacterial (Weber and Corseuil, 1994, Mayotte et al., 1996) and fungal (Lestan et al., 1996) cultures have been used to test the effectiveness of bioaugmentation.

A compound that has proven difficult to biodegrade is trichloroethylene (TCE). Anaerobic reductive dehalogenation can be an excellent approach for some sites, but at others, the dechlorination process stops at cis-dichloroethylene due to the presence of competing electron acceptors or other factors. Aerobic cometabolism takes advantage of nonspecific, inducible oxygenases to degrade TCE. However, it can be difficult to conduct aerobic cometabolism in the field, in part because the supply of the inducer compound is problematic due to low aqueous solubilities (methane) or their classification as a Priority Pollutant (phenol, toluene).

Some of these problems may be overcome by the use of a mutant of *Burkholderia cepacia* G4 identified as strain $PR1_{301}$ (Munakata-Marr et al. 1996). Strain PR1, developed at the EPA Gulf Breeze Laboratory, constitutively expresses toluene ortho monooxygenase (TOM), a nonspecific enzyme that catalyzes the oxidation of TCE. With this strain, only an energy source needs to be added to the site (this energy source also provides carbon for growth). Many inexpensive, nontoxic, water soluble compounds can serve this purpose, including glucose and lactate.

The results of the bioremediation pilot demonstration at the Gilbert-Mosley site (Wichita, KS) showed that biodegradation of trichloroethene (TCE), dichloroethene (DCE), and vinyl chloride (VC) was possible using bioaugmentation with the microorganism *Burkholderia cepacia* $PR1_{301}$ (PR1). In that test, PR1 was added to an injection well at various concentrations and the removal of contaminants was monitored. The minimum concentration of cells required to achieve removal of contaminants in the well was 1×10^8 cells/mL. At a higher cell concentration (1×10^9 cells/mL), the hydraulic conductivity of the soil formation immediately downgradient of the treatment area decreased substantially, presumably because the cells adhered to the soil particles, blocking groundwater flow paths.

Recently, Envirogen, Inc. developed ENV-435, an adhesion-deficient variant of PR1. Laboratory tests conducted by Envirogen suggest that ENV-435 moves rapidly through soils. However, these tests were performed in small soil columns and for short time periods. Furthermore, no data are available on the effect of cell concentration on cell transport.

Whether the bioaugmentation is to be carried out in an *in situ* bioremediation trench (IBT) or in the form of Geoprobe injections of cells, flow blockage is undesirable. This study was performed in a pilot-scale IBT to investigate the "plugging" phenomenon and, in particular, to evaluate the performance of ENV-435 for possible use at the Gilbert-Mosley site.

Objectives. The overall goal of this study was to evaluate aquifer plugging that might occur in full-scale bioaugmentation systems at the Gilbert-Mosley site. The specific aims of this study were (1) to determine if aquifer plugging results from the addition of high PR1 cell concentrations to the IBT; (2) to measure groundwater flow and microbe transport within and downgradient of the IBT; and (3) to determine if an adhesion-resistant strain of PR1 will decrease the effects on hydraulic conductivity;

MATERIALS AND METHODS

Pilot-Scale IBT. A scale model of a proposed IBT system was constructed from Plexiglas and stainless steel. The pilot-scale IBT was 122 cm long, 30 cm wide, and 70 cm tall. The treatment zone (the center 19 cm) was filled with pea gravel and the upgradient and downgradient zones contained Wichita soil (repacked after each test). This soil is sandy; when packed to a bulk density of 1.6 kg/L, the hydraulic conductivity was 75-150 m/d. Groundwater from the site was pumped into the inflow end of the system and lengthwise through the unit. A groundwater recirculation loop within the treatment zone induced flow in the vertical direction to provide mixing. The recirculation loop included an oxygenation unit, a peristaltic pump, and a T connection through which cells were added periodically. Cells were injected at the top of the trench and removed at the bottom. Sample ports with small-bore tubes were installed in all three sections at three different levels. The labeling system used was to number the ports along the direction of flow and to identify them as being in the top, middle, or bottom of the box. Thus, port 5M was in the middle of the trench and port 9T was near the outlet in the top level.

Microbial Cultivation. Strains PR1 and ENV-435 were cultivated in a 2-L New Brunswick BioFlo I bioreactor. The bioreactor provided agitation (500 RPM), aeration, and temperature (30 °C) and pH control. The pH was initially 6.8 and was allowed to fall to 6.0 before being controlled at that value by 5N sodium hydroxide addition. The growth medium was a basal salts medium supplemented with 7.2 g/L glucose. The growth protocol consisted of repeated batch growth phases followed by partial drainage of the bioreactor contents to the IBT and a waste jar. Each batch phase lasted 8 h; at this point, the glucose was depleted and the cell concentration was nearly 10^{11} cells/mL.

Liquid Sample Analysis. Liquid samples (ca. 10 mL) were removed from the sampling ports and analyzed for (a) total cell concentration (optical density); (b) viable cell concentration (nonspecific plate counts); (c) dissolved oxygen (oxygen probe); and (d) bromide, using a colorimeteric assay (Hach).

RESULTS AND DISCUSSION

PR1 Bioaugmentation and Transport Test. Following 12 d of groundwater flow (to equilibrate the system), a period in which additions of 10^8 cells/mL were added to the trench every 8 h was initiated. After 4 d, the volume of cell culture added was increased to achieve > 5 × 10^8 cells/mL in the trench zone.

Cell concentration vs. time data for the PR1 continuous injection experiment (second cell addition phase) are presented in Figure 1. Data are shown only for the downgradient section of the box for the mid-level sampling ports. Although very high concentrations of cells were present in the trench, the measurements in the downgradient section were relatively low, indicating that the transport of most cells was retarded.

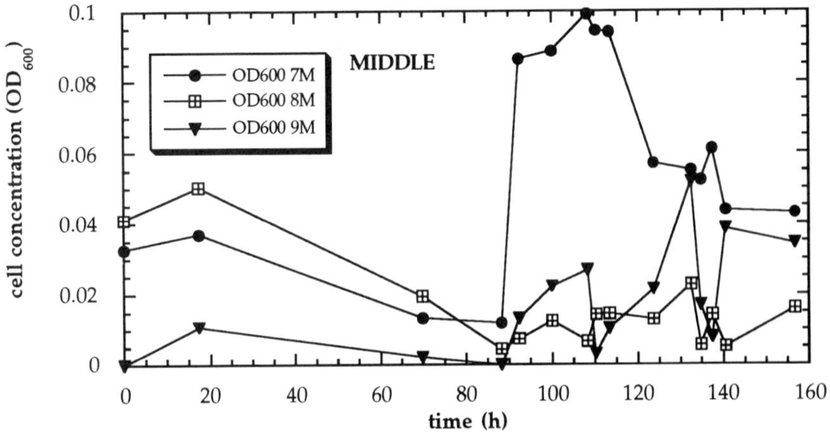

Figure 1. Cell concentration (turbidity) vs. time for the PR1 test.

ENV-435 Bioaugmentation and Transport Test. After a 2-d equilibration period, ENV-435 was added every 8 h to achieve 10^8 cells/mL in the trench zone. After 6 d of this routine, groundwater was pumped through the soils for 6 d to prepare for a single, high concentration (10^9/mL) pulse of cells. This pulse test was followed by a bromide tracer pulse test.

Data from the continuous addition phase of the ENV-435 test are shown in Figure 2. Although the concentration of ENV-435 cells added in this phase was much less than in the PR1 experiment, higher optical density values were obtained downgradient. Furthermore, the ENV-435 concentrations along the length of the downgradient section were nearly the same toward the end of this phase, evidence that most cells were moving through the soil. Dissolved oxygen concentrations were adequate for activity and survival of these bacteria during this test.

Results from the pulse addition of a high concentration of ENV-435 cells are shown in Figure 3. The slug of cells can clearly be seen moving downgradient. However, when these profiles are integrated, it became clear that fewer cells passed ports 8 and 9 that were detected at port 7. Bromide tracer data for port 7B are shown in Figure 4. These data are compared with analogous cell concentration data. It is readily apparent that the ENV-435 cells were transported at essentially the rate of the tracer (and thus the groundwater).

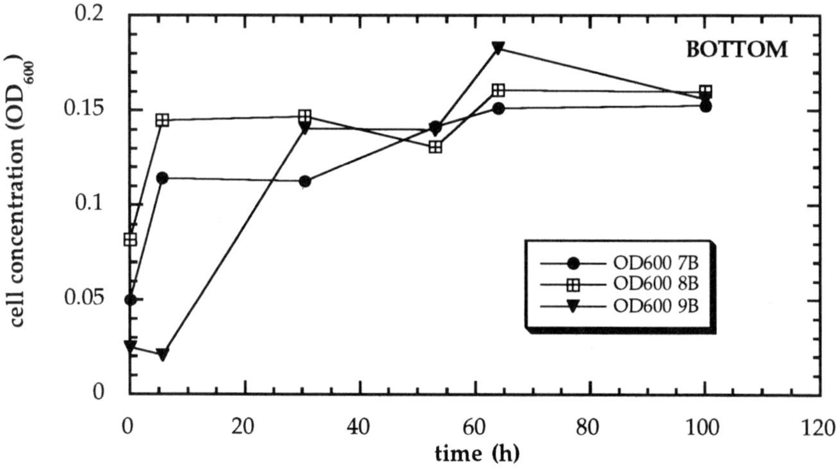

Figure 2. Cell concentration (turbidity) vs. time for the continuous injection phase of the ENV-435 test.

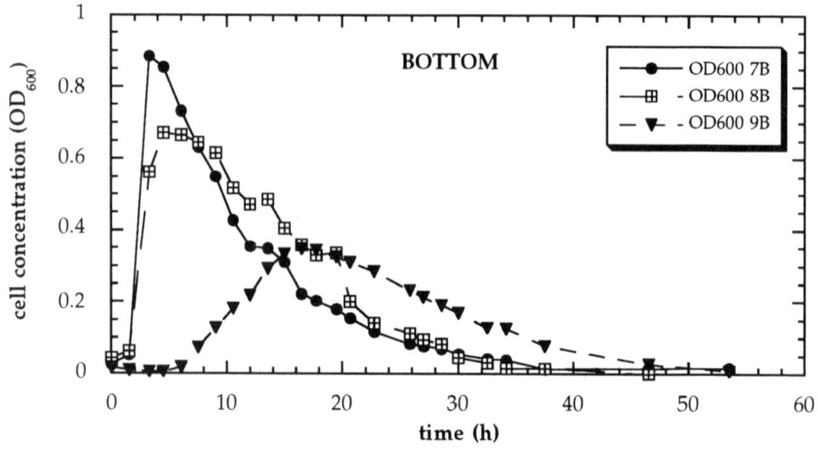

Figure 3. Cell concentration (turbidity) vs. time for the pulse injection phase of the ENV-435 test.

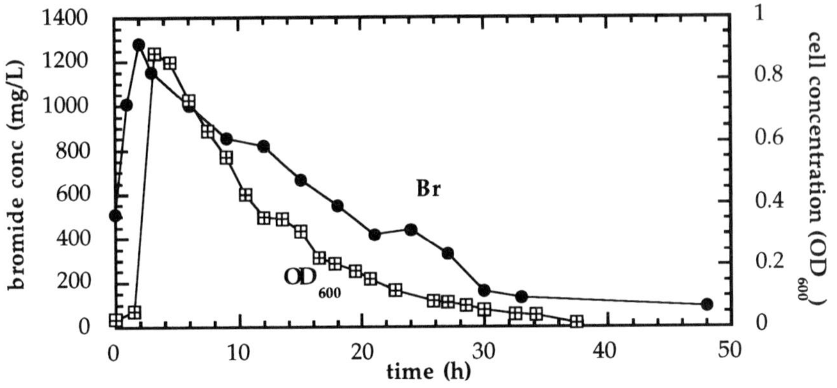

Figure 4. Comparison of bromide and cell (turbidity) pulse injection data at sampling port 7B.

CONCLUSIONS

Under the conditions of these experiments, ENV-435 cells moved rapidly through the Wichita soils and relatively few became attached or entrapped in the soil. In contrast, a high proportion of the PR1 cells were found to attach to the soil and fewer cells moved downgradient.

These results suggest that bioaugmentation has a greater chance for success when performed with an adhesion-deficient microorganism such as ENV-435.

ACKNOWLEDGMENTS

The researchers gratefully acknowledge Dr. M. Shields for donation of PR1 and Envirogen, Inc. for providing ENV-435. R. Callahan assisted with the PR1 experiments. Partial funding for this study was obtained through Grant MSS-9257305 to C. Shackelford from the National Science Foundation. Additional funding was obtained from the City of Wichita.

REFERENCES

Lestan, D., M. Lestan, J. A. Chappelle, and R. T. Lamar. 1996. "Biological Potential of Fungal Inocula for Bioaugmentation of Contaminated Soils." *J. Ind. Microbiol.* **16**(5):286.

Mayotte, T. J., M. J. Dybas, and C. S. Criddle. 1996. "Bench-Scale Evaluation of Bioaugmentation to Remediate Carbon Tetrachloride-Contaminated Aquifer Materials." *Ground Water* **34**(2):358.

McCarty, P. 1996. "Biotic and Abiotic Transformations of Chlorinated Solvents in Ground Water." *In Symposium on Natural Attenuation of Chlorinated Organics in Ground Water* (Abstracts), EPA/540/R-96/509, pp. 5-9. US EPA, Washington, DC.

Munakata-Marr, J., P. L. McCarty, M. S. Shields, M. Reagin, and S. C. Francesconi. 1996. "Enhancement of Trichloroethylene Degradation in Aquifer

Microcosms Bioaugmented with Wild Type and Genetically Altered *Burkholderia* (*Pseudomonas*) *cepacia* G4 and PR1." *Environ. Sci. Technol.* **30**(6):2045.

Weber, W. J. and H. X. Corseuil. 1994. "Inoculation of Contaminated Subsurface Soils with Enriched Indigenous Microbes to Enhance Bioremediation Rates." *Water Res.* **28**(6):1407.

BACTERIAL ATTACHMENT AND TRANSPORT THROUGH POROUS MEDIA: THE EFFECTS OF BACTERIAL CELL CHARACTERISTICS

K. L. Duston, M. R. Wiesner, and C. H. Ward

ABSTRACT: The effects of several microbial, chemical and physical parameters on bacterial attachment were studied in column transport studies. Specifically, the role of bacterial characteristics, ionic strength of solution, roughness of solid surfaces, and hydraulic conductivity were tested. Using particle removal theory, the collision efficiency factor (a) was calculated from the breakthrough of the cells (C/Co) in the columns. The values varied over three orders of magnitude. The distance that the cells were predicted to travel through the porous medium was calculated assuming C/Co=0.05 (95% retention). Predicted travel ranged from less than 5 cm to almost 10 km.

Bacterial surface characteristics that apparently affected attachment were electrophoretic mobility, hydrophobicity, and biopolymer production. Electrostatic interactions, described by DLVO theory, governed attachment of hydrophilic cells to hydrophilic surfaces. Hydrophobic interactions dominated attachment of hydrophobic cells. Steric hindrance may have resulted from the adsorption of extracellular biosurfactant onto cell and solid surfaces.

Size also influenced retention of the cells in the porous medium. Cells less than 0.1 μm and greater than 3.0 μm in diameter were retained to a greater extent than cells with diameters within this range.

Increased ionic strength resulted in increased attachment of cells. Attachment to sand grains was greater than attachment to glass beads, possibly because of the rougher surfaces. Lower hydraulic conductivity resulted in greater retention of cells.

ENHANCING BIOCOLLOID TRANSPORT TO IMPROVE SUBSURFACE REMEDIATION

Bruce E. Logan, Terri Camesano, Brock Rogers and Yan Fang
(Department of Chemical and Environmental Engineering,
University of Arizona, Tucson, Arizona)

ABSTRACT:
Bioaugmentation can be an effective method of subsurface remediation but well clogging must be minimized by reducing bacterial adhesion to soil particles in the vicinity of the well. The retention of radiolabeled bacteria in soil columns was studied in order to determine the most effective methods for increasing bacterial transport distances in groundwater aquifers. The presence of a non-aqueous phase liquid (NAPL) increased bacterial transport. Air sparging increased bacterial transport distances compared to those obtained when bacteria were suspended in an artificial groundwater, but air sparging was not as effective as low IS water in decreasing cell attachment. It is recommended that for soil remediation by bioaugmentation that the bacteria be suspended in low ionic strength (<0.01 mM) water and that low pumping velocities (~1 m/d) be used to minimize bacterial attachment.

INTRODUCTION

Subsurface remediation can be enhanced by injecting bacteria into the aquifer via a procedure known as bioaugmentation, but bacteria-sized (~1 μm) colloidal particles readily adhere to soil particles and may not be transported more than one meter in soils leading to well clogging. In order to efficiently treat large volumes of soils by injecting bacteria from wells, bacterial transport distances of 10's of meters must be obtained (Gross and Logan 1995). Measuring factors that could sufficiently reduce bacterial attachment to achieve these large transport distances would require laboratory columns 1 to 10 m long using a conventional approach of measuring cell concentrations in column breakthrough tests (Jewett *et al.* 1993).

Using the MARK method (Gross *et al.* 1995) we have examined the effects of solution ionic strength (IS), dissolved and sorbed organic matter, and various chemical additives on bacterial transport and attachment (Gross and Logan 1995). Attachment is quantified using filtration theory in terms of α, defined as the ratio of the rate that bacteria stick to a soil grain to the rate that they strike it. Of the chemicals examined to date, surfactants such as Tween-20 have been found to be the most effective, reducing α by 2.5 orders of magnitude on glass surfaces (Gross and Logan 1995). Dissolved organic matter has little effect on α (Johnson and Logan 1995). The most consistent method of increasing bacterial transport in a variety of soils and porous media is decreased IS (Jewett *et al.* 1995). Decreasing the IS from that of growth media (10^{-1} M) to that of ultrapure water (10^{-5} M) decreased α of *Pseudomonas fluorescens P17* from 0.18 to 0.026 (Gross and Logan 1995). The attachment of *Alcaligenes paradoxus* to glass beads in ultrapure water produced the lowest measured α=0.0016 (Li 1997).

We summarize here the results of ongoing research in our laboratory to determine the most effective method of promoting bacterial transport in natural soils. Since our previous work suggests that low IS water is a key factor in promoting bacterial transport in porous media, we have concentrated on comparing effective methods of maximizing transport in relation to low IS solutions. In column studies we examined the effect of water velocity and air sparging on bacterial transport. Since non-aqueous phase liquids (NAPLs) may be present at many sites we also examined the effect of NAPLs on bacterial transport.

METHODS

Column experiments were performed by adapting the MARK procedure, which consists of measuring the retention of radiolabeled cells in short columns (Gross *et al.* 1995), to transport experiments using longer columns (7 to 15 cm). The bacterium used for all experiments was *Pseudomonas fluorescens* strain P17, a Gram negative motile rod. Cells were radiolabeled by incubating a cell suspension with ^3H leucine. In MARK tests cells are pulled by vacuum through an open-ended short column. For the longer column experiments reported here we pumped bacterial suspensions through capped glass or stainless steel columns packed with a soil collected from the North Fallow Field at the University of Arizona farm from a depth of 3 to 6 feet below the surface. This soil is on average 90% sand, 7% silt, 3% clay and 0.1% organic carbon, and it was passed through a 500 μm mesh producing an average soil grain diameter of 127 μm.

After the cells were passed through a column, the column was rinsed and sliced into sections. The fraction of bacteria retained in each slice, R_i, was

$$R_i = N_i / \left(N_o - \sum N_{i-1}\right) \qquad (1)$$

where N_0 is the concentration of bacteria added to the sample (dpm), N_i is the concentration of bacteria in the slice, and N_{i-1} is the total number of bacteria retained in previous slices. The extent of bacterial attachment was evaluated using a steady state clean bed filtration equation

$$C/C_o = \exp\left(-\alpha \lambda L\right) \qquad (2)$$

where C_0 and C are the concentrations of bacteria entering and leaving the column of length L, $\lambda = 3(1-\theta)\eta/2d_c$ is the filter coefficient, θ=0.40 the bed porosity, $\eta \approx 0.005$ the collector efficiency calculated using the RT model (Logan *et al.* 1996), d_c=127 μm the soil grain diameter, and α the sticking coefficient defined as the fraction of particle collisions that result in successful attachment.

The fraction of bacteria removed in the column, R, is related to the concentration of bacteria in the column effluent, C/C_0, by $R=1-C/C_0$. Therefore, the sticking coefficient for bacteria in each slice, α_i, can be calculated as a function of the thickness of each slice, L_i using $\alpha_i=[\ln(1-R_i)]/(\lambda L_i)$. The overall sticking coefficient for the whole column was calculated from the total removal of cells in

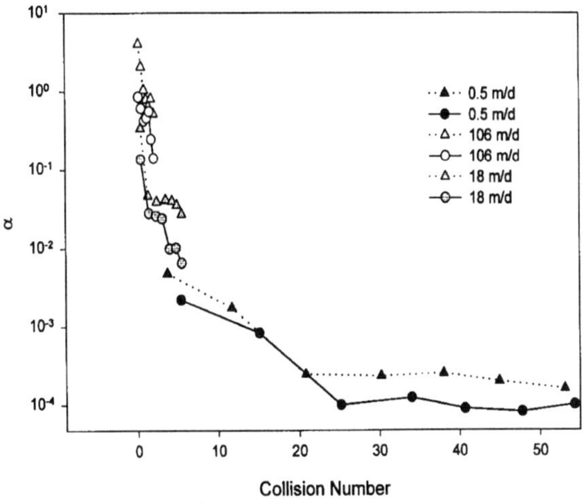

Figure 1. Bacterial transport is increased using water with a low ionic strength, although the attachment is also a function of flow velocity (Triangles; IS=4.1 mM; circles, IS=0.0011 mM).

the column calculated by adding up the removal from all slices.

RESULTS AND DISCUSSION

Ionic Strength (IS). Low IS water (I = 0.0011 mM) reduced bacterial attachment by approximately one-third compared to that in a higher IS artificial groundwater (AGW; IS =4.14 mM) (Figure 1). The reduction in attachment under low IS conditions is consistent with the results obtained in previous studies (Li, 1997) using natural soils at high flow velocities in MARK tests. Cell attachment to a surface is a function of both surface charge (electrostatics) and hydrophobicity. P17 has a zeta potential (ζ) of -40.45 mV and is relatively hydrophobic. Due to its very negative surface charge, however, the attachment of P17 is probably governed by electrostatics and not hydrophobic effects. Because the soil and cells are both negatively charged they repel each other at neutral pHs used in our experiments. This repulsive force is enhanced using low IS water and decreases attachment.

Average Flow Velocity. According to filtration theory, the fraction retained should decrease as velocity increases in order to maintain a constant α. Our experiments found the opposite to be true (Figure 2). The fraction retained increased as velocity increased for velocities ranging from 0.5 to 106 m/d. At a velocity of 0.5 m/d, the overall α was 0.003 over the 7-cm column. Under typical groundwater conditions, this α would result in a one-log reduction in cell concentration after the bacteria had traveled 7 m. At 106 m/d, α was high enough (0.575) to cause > 70 % of the bacteria

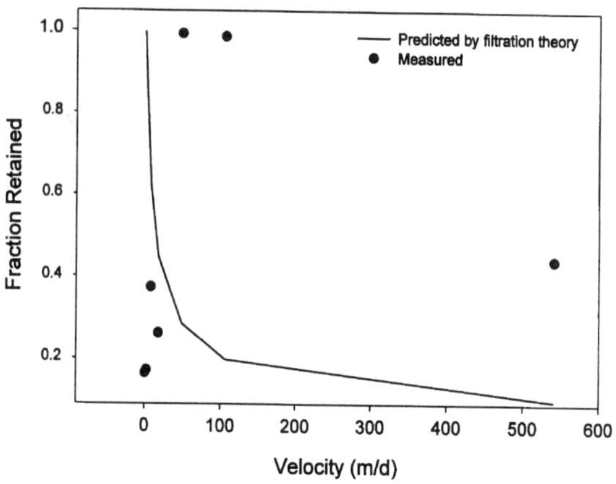

Figure 2. Fraction of cells retained in a soil column in artificial groundwater (IS=4.1 mM) as a function of flow velocity although filtration theory (α=0.1) predicts retention in the column should decrease.

to be retained in the 7-cm column. At groundwater velocities, this α would cause 90% of the bacteria to be retained in the soil after 3 cm. Bioaugmentation would not be feasible at this high cell retention since cells would probably clog the injection well. As the flow velocity increased from 106 m/d to 500 m/d the fraction of bacteria retained decreased from 0.99 to 0.542, resulting in a nearly constant α = 1.2. Thus, filtration theory appeared to be obeyed at the higher flow velocities.

Cell motility may be a factor in explaining the observed changes in cell retention with fluid velocity. Bacteria may swim at rates up to 3.5 m/d (Mercer *et al.*, 1993). Bacterial motility would have little impact on the collision frequency between cells and surfaces at high velocities, but at low velocities, the bacteria may be able to swim at rates faster than the advective flow; swimming may result in their more rapid forward advection in the column than for passive tracers.

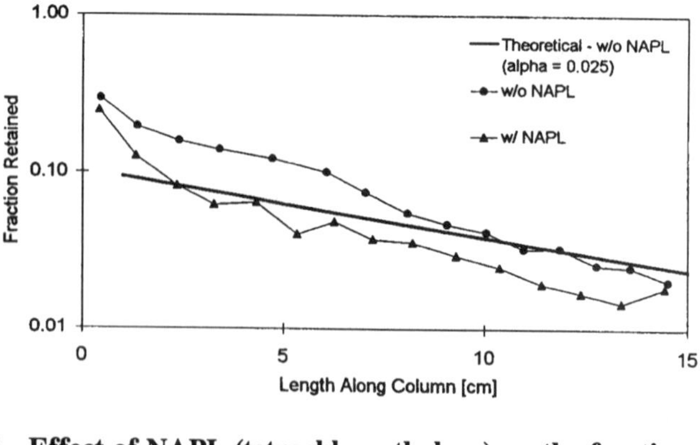

Figure 3. Effect of NAPL (tetrachloroethylene) on the fraction of bacteria retained in a soil column versus that predicted by filtration theory.

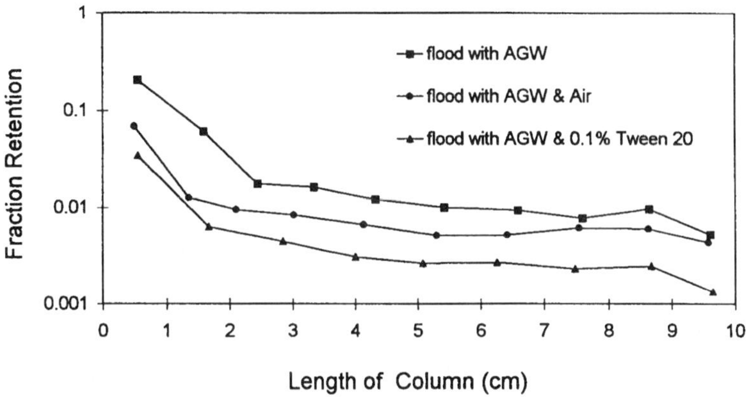

Figure 4. Air sparging during bacterial transport decreases the fraction of bacteria retained in the quartz-media column.

Effect of NAPLs on Transport. The presence of a NAPL residual (21.3 % of void spaces filled) increased the overall transport of bacteria by a factor of 1.6 (Figure 3). The cumulative fraction retained along the length of the column decreased from R=0.78 (α=0.025) to R=0.61 (α=0.016). The theoretical fraction retained was determined using a constant collision efficiency (α=0.025 and a collector efficiency η=0.00585.

The reduction in bacterial attachment in the presence of a NAPL phase may be a result of the occupation of stagnant flow regions, or the micropore spaces, by the NAPL. This would have the effect of blocking a portion of the flow paths in the column, thus increasing the pore velocity of the bacterial suspension. Filtration theory predicts that an increase in velocity will decrease the collector efficiency (η), which would also have the effect of reducing the fraction retained. Another possible explanation for the enhanced transport could be that the NAPL is blocking favorable sites on the collector surface. If these sites are inaccessible to the bacteria, the probability of their attachment will decrease thereby enhancing their transportability. A decreased attachment to the soil surfaces, due to hydrophobic repulsion between P17 and the NAPL was also considered as a possible explanation. However, hydrophobicity tests indicated that P17 favorably partitions into PCE. Therefore, the presence of a hydrophobic phase should have increased, not decreased bacterial retention in the column as observed in our experiments.

Air Sparging. Bacterial transport is enhanced by air sparging, but the enhancement due to the air flow is less than that obtained using a low IS water or with surfactants (Figure 4). The fraction of bacteria retained in a 10-cm long column was reduced from R=0.35 to 0.13 by air sparging while water was simultaneously being pumped through the column. Addition of a non-ionic surfactant (0.1% v/v Tween 20) solution reduced the fraction of bacteria retained to R=0.06.

The enhancement of bacteria transport by air sparging is likely due to the mobile air-water interface. Because P17 is a relatively hydrophobic bacterium we expect that cells will preferentially sorb to the air-water interface. The sorption of bacteria on the gas-water interface is essentially irreversible since capillary energy provides a large attractive force to hold the bacteria on the air-water interface. Thus, the movement of the air-water interface during air sparging results in the increased forward transport of the cells in comparison to saturated water conditions. However, the increase in bacterial transport due to sparging is not as large as that possible using surfactants or low IS water.

CONCLUSIONS

Bacterial transport can be increased in natural soils by using low IS water (˜0.01 mM). For the motile species examined here cell attachment is minimized by pumping bacteria at low water velocities (˜1 m/d) even though filtration models predict that attachment will be decreased with increased water velocity. Air sparging will not be as effective as low IS water in increasing bacterial penetration in groundwater aquifers.

REFERENCES

Gross, M.J., O. Albinger, D.G. Jewett, B.E. Logan, R.C. Bales, and R.G. Arnold. 1995. "Measurement of bacterial collision efficiencies in porous media." *Wat. Res.*, 29:1151-1158.

Gross, M.J. and B.E. Logan. 1995. "Influence of different chemical treatments on transport of *Alcaligenes paradoxus* in porous media." *Appl. Environ. Microbiol.*, 61(5):1750-1756.

Johnson, W., P. and B.E. Logan. 1996. "Enhanced transport in porous media by sediment-phase and aqueous-phase natural organic matter." *Wat. Res.*, 30(4):923-931.

Jewett, D.G., T.A. Hilbert, B.E. Logan, R.G. Arnold, R.C. Bales. 1995. "Bacterial transport in columns and filters: influence of ionic strength and pH on collision efficiency." *Wat. Res.*, 29(7):1673-1680.

Li, Q. 1997. M.S. Thesis, University of Arizona, Tucson.

Logan, B.E., D.G. Jewett, R.G. Arnold, E. Bouwer and C.R. O'Melia. 1995. "Clarification of clean-bed filtration models." *J. Environ. Eng.* 121(12): 869-873.

Mercer, J.R., R.M. Ford, J.L. Stitz, and C. Bradbeer. 1993. "Growth rate and effects on fundamental transport properties of bacterial populations." *Biotechnol. Bioengin.* 42:1277-1286.

QUANTITATIVE CHARACTERIZATION OF BACTERIAL MIGRATION THROUGH POROUS MEDIA

Roseanne M. Ford and Minghui Jin (University of Virginia, Charlottesville, VA)
Peter T. Cummings (University of Tennessee, Knoxville, TN and Oak Ridge National Laboratory, Oak Ridge, TN)
Kevin C. Chen (University of Virginia, Charlottesville, VA)

ABSTRACT: We are interested in characterizing the movement of motile and chemotactic bacteria through saturated subsurface environments. We use experiments, mathematical models and computer simulations to study the behavior of individual cells and bacterial populations in porous media. The objective of these studies is to characterize the behavior of bacteria in terms of intrinsic parameters which can be incorporated into macroscopic balance equations to predict bacterial and contaminant distributions for field-scale applications. Individual *E. coli* and *P. putida* cells were tracked in bulk fluid and their trajectories were analyzed in terms of swimming speed, run time and turn angle distribution. These parameters were used as input to Monte Carlo simulations of bacterial swimming behavior within a bed of uniform diameter spheres. The simulation results compared well with experimental measurements of bacterial migration in saturated sand columns. Monte Carlo simulations were also used to explore the impact of grain diameter and chemoattractant gradients on bacterial distributions within porous media. In a complementary approach, the porous medium is conceptually regarded as a bundle of capillary tubes with an effective pore diameter and tortuosity. Bacterial collisions with the walls of the capillary reduce their run times and consequently, reduce their effective transport coefficients. Microscopic balance equations were used to derive relationships between the individual cell properties and macroscopic transport coefficients.

Using these approaches we have calculated effective values of the transport coefficients in porous media as a function of their values in bulk liquid. These coefficients have been incorporated into transport equations for mathematically modeling contaminant and bacterial distributions within a field-scale system. The model is used to assess the impact of motility and chemotaxis relative to other factors such as growth, advection and adsorption under various field conditions. We present results from these parametric studies which are useful in the design and optimization of bioremediation.

BIOAUGMENTATION AND NUMERICAL SIMULATION OF CARBON TETRACHLORIDE TRANSFORMATION IN GROUNDWATER

Michael E. Witt (Michigan State University, East Lansing, Michigan)
David C. Wiggert (Michigan State University, East Lansing, Michigan)
Michael J. Dybas (Michigan State University, East Lansing, Michigan)
K. Colleen Kelly (Michigan State University, East Lansing, Michigan)
Craig S. Criddle (Michigan State University, East Lansing, Michigan)

ABSTRACT: *Pseudomonas stutzeri* strain KC is a natural aquifer isolate that rapidly transforms carbon tetrachloride to carbon dioxide and nonvolatile end products without the production of chloroform. Laboratory experiments were performed in a model aquifer column to assist in the evaluation of chemical delivery strategies and placement of a carbon tetrachloride-transforming zone for implementation at a field site. After inoculation with strain KC, and addition of nutrients with appropriate pH adjustment to 8.2, a carbon tetrachloride-transforming zone was established near the region of inoculation. A carbon tetrachloride removal efficiency of up to 97% was achieved without the production of detectable concentrations of chloroform. Pilot-scale test results suggest that transformation of carbon tetrachloride can be achieved in situ using similar chemical delivery strategies coupled with bioaugmentation using strain KC.

INTRODUCTION

Extensive use of carbon tetrachloride (CT) in recent decades has resulted in significant contamination of groundwater and drinking water supplies. Once widely used as a solvent and degreasing agent, CT has been found in groundwater throughout the United States and has been targeted as a priority pollutant by the United States Environmental Protection Agency. Due to the large volumes of CT-contaminated groundwater that exist in the United States, significant amounts of money have been invested in remediating these plumes by conventional means.

Remediating groundwater contaminated with CT and other volatile organics has been achieved by pumping the contaminated groundwater to the surface and transferring the contaminant from one medium (water) to another (air). This method of remediation can be costly and labor-intensive because multiple plume volumes need to be removed in order to fully acquire the desired removal efficiency from both the solids and the groundwater.

An alternative method of remediation involves the in-situ transformation of CT to non-hazardous endproducts, bypassing the need to pump the contaminated groundwater to the surface for treatment. This method of bioremediation has the potential to reduce cleanup costs (National Research Council, 1993). Bioremediation involves stimulating indigenous microbes to transform a target compound (biostimulation) or adding non-indigenous microbes to the subsurface for the purposes of transforming a target compound (bioaugmentation). In the case of carbon tetrachloride, chloroform is a common endproduct of transformation in both laboratory and field environments (Criddle et al., 1990; Egli et al. 1987 and 1988; and Semprini et al., 1992). Chloroform is also a suspected carcinogen that is more persistent in many environments.

A laboratory-scale model aquifer column was constructed to examine the transformation of CT by strain KC in aquifer material. Results from this bioaugmentation study are presented.

MATERIALS AND METHODS

One model aquifer column was constructed and packed with material from a Schoolcraft, Michigan, aquifer. This model aquifer column consisted of one polyvinyl chloride tube (183 cm long and 5.1 cm inside diameter) capped at both ends. Sampling ports were installed along the length of the column at three-inch (7.6-cm) intervals. A slug injection zone was built on the column to uniformly deliver nutrients and KC cells into the model column at a specified location (see Figure 1). The model aquifer column, along with feed syringes, were placed inside of a climate control room where the temperature was held constant at 10°C. Schoolcraft groundwater spiked with 100 µg/L CT was delivered to the column at a rate yielding an average linear flow velocity of 15 cm/day, which is identical to the flow velocity of groundwater through the Schoolcraft aquifer.

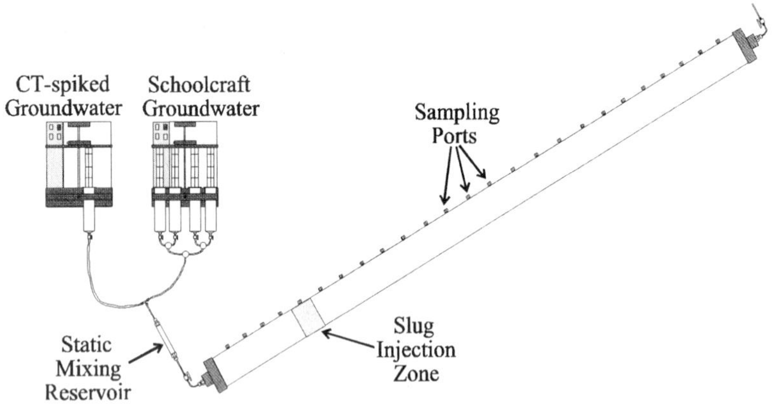

FIGURE 1. Diagram of model aquifer column design.

After achieving complete CT saturation in the column, alkalinity adjustment was performed in the slug injection region of the column. Each adjustment consisted of a slug of Schoolcraft groundwater (containing 100 µg/L CT) augmented with 10 mg/L phosphate and pH adjusted to 8.2 using a sodium hydroxide solution. These alkalinity adjustments were carried out to provide strain KC with a competitive advantage over the indigenous Schoolcraft microflora (Knoll, 1994).

Inoculation of the column involved injecting 320 mL of aqueous strain KC culture after 24 hours of growth. The inoculation was performed by injecting th liquid culture at two locations on the column while simultaneously extracting liquid from two different locations. This method of injection aimed to achieve a cylindrical slug of strain KC culture in the slug injection zone on the column.

Transformation of CT by strain KC was evaluated by analyzing the CT concentration in the headspace of a 1.5-mL sample vial containing a 200 μL aqueous sample. Analytical samples were obtained from each port on alternate days and analyzed by gas chromatography.

Exactly one week after inoculation, a slug of Schoolcraft groundwater containing 100 mg/L acetate, 10 mg/L phosphate, and pH-adjusted to 8.2 was injected into the slug injection zone. This injection of nutrients was repeated each week for a total of eleven weeks.

Enumeration of strain KC was achieved by serial dilution and spread plate techniques. Samples were serially diluted, then portions of each dilution were aseptically spread on sterile petri dishes containing R2A agar. The plates were incubated at room temperature for five days before counting the colonies grown on each plate.

RESULTS AND DISCUSSION

Aqueous-phase CT Analysis. The influent concentration of CT entering the model aquifer column was 100 μg/L. Figure 2 shows the concentration profile of CT in the model aquifer column throughout the study. Note that day 0 was the date of inoculation using strain KC. After 63 days the effluent CT concentration was 3 μg/L, yielding an overall CT removal efficiency of 97%.

Nutrient slug injections were stopped at day 77 and no additional acetate, phosphate, or base were added. CT concentrations began to rise gradually after this date, indicating a loss in the ability of the reaction curtain to transform CT in the groundwater. However, residual strain KC activity was evidenced by continued CT transformation in the region downgradient from the slug injection zone. In addition, there was no evidence of chloroform production.

Strain KC Enumeration. Bacterial enumeration experiments were performed on days 0, 85, and 155 during this study. On day 0, the only detectable aqueous-phase strain KC in the column was near the slug injection zone at port 5. Eight days after the final nutrient addition to this column (day 77) the only detectable strain KC was again in the region of port 5. This was expected since the stationary reaction curtain of CT transformation was located in this region. On day 155, however, detectable concentrations of strain KC were found in samples obtained from ports 5 and 15. The strain KC detected on day 155 explains why evidence of CT transformation was still occurring in the model aquifer column downgradient from the slug injection zone.

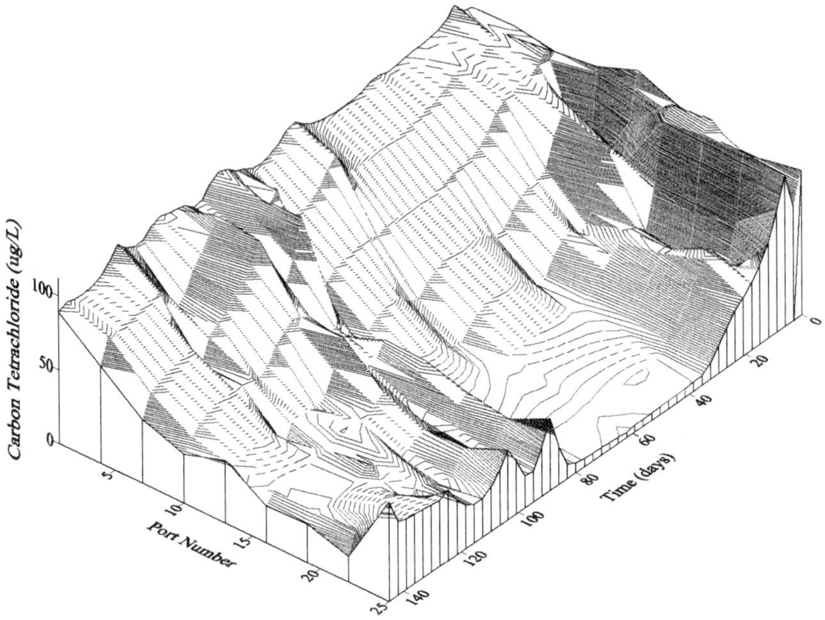

FIGURE 2. Carbon tetrachloride concentration profile in samples obtained from model aquifer column.

Figure 3 shows the results of strain KC enumeration assays performed on solid samples obtained from all odd-numbered ports on days 85 and 155. The concentration of strain KC per gram of soil did not change significantly after feeding was stopped on day 77. High strain KC concentrations were detected on solid samples taken from ports 5 through 13. This, again, explains why some CT transformation continued in the column after nutrient addition was stopped.

Numerical Simulation. Following earlier work (Witt et al., 1995) a numerical model was implemented to predict CT degradation in the laboratory columns. Simulations were made to predict the concentrations of CT, strain KC, acetate, and nitrate as functions of time and position along the column. Propagation of uncertainty in the numerical model was performed using the vector-state-space perturbation method (Hondzo and Stefan, 1992 and 1994). The estimated uncertainty in predicted concentrations was based on the perturbation in the input parameters, or measurands, in the mass balance equations. The coefficient of variation for each measurand was less than 0.3.

The numerical model accurately predicted CT degradation. It also precisely detailed the formation of a CT transformation zone in the laboratory scale model aquifer column. The uncertainty analysis showed that the prediction of CT concentration is strongly affected by its retardation coefficient (see Figure 4). The variation of experimental data, as compared to the predicted CT concentration profile, was attributed to the desorption of CT from the solids into the liquid phase. The analysis provided an accurate initial estimate of processes occurring in the column, and it also demonstrated the effectiveness of strain KC in degrading CT.

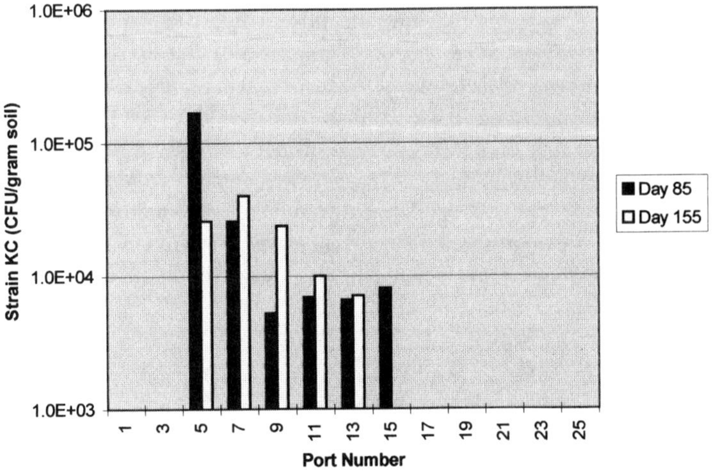

FIGURE 3. Strain KC enumeration results from solid samples obtained from the model aquifer column.

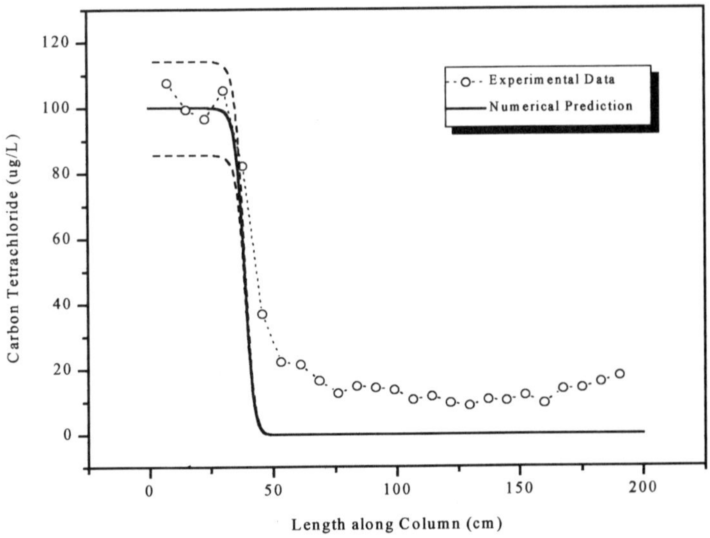

FIGURE 4. Carbon tetrachloride concentrations along the length of model aquifer column at day 45.

SUMMARY AND CONCLUSIONS

The biotransformation of CT by *Pseudomonas stutzeri* strain KC has been closely monitored using a prototype model aquifer column. A model aquifer column was impacted with CT-spiked Schoolcraft groundwater using syringe pumps for delivery. The column was inoculated with 4 x 10^7 KC cells per mL and weekly addition of nutrients (acetate and phosphate) was performed. Aqueous-phase CT concentrations were used as a measure of the level of bioremediation achieved using this type of technology.

Addition of nutrients to the column on a weekly basis provided strain KC cells with sufficient nutrients to reduce the effluent CT concentrations up to 97%. The results of this study indicate that under acceptable conditions in an aquifer setting, CT can be biologically transformed to non-harmful byproducts by *Pseudomonas stutzeri* strain KC. This remediation approach shows promise for application at many contaminated sites in the United States and around the world.

ACKNOWLEDGMENTS

Funding for this research was provided by the State of Michigan Department of Environmental Quality.

REFERENCES

Criddle, C. S., J. T. DeWitt, D. Grbic-Galic, and P. L. McCarty. 1990. "Transformation of Carbon Tetrachloride by *Pseudomonas* sp. Strain KC under Denitrification Conditions." *Appl. Environ. Microbiol.* 56: 3240-3246.

Egli, C. R. Scholtz, A. M. Cook, and T. Leisinger. 1987. "Anaerobic Dechlorination of Tetrachloromethane and 1,2-Dichloromethane to Degradable Products by Pure Cultures of *Desulfobacterium* sp. and *Methanobacterium* sp." *FEMS Microbiol. Lett.* 43: 257-261.

Egli, C. T. Tschan, R. Scholtz, A. M. Cook, and T. Leisinger. 1988. "Transformation of Tetrachloromethane to Dichloromethane and Carbon Dioxide by *Acetobacterium woodii.*" *Appl. Environ. Microbiol.* 54: 2819-2823.

Hondzo, M. and H. G. Stefan. 1992. "Propagation of Uncertainty Due to Variable Meteorological Forcing in Lake Temperature Models." *Water Resources Research.* 28: 2629-2638.

Hondzo, M. and H. G. Stefan. 1994. "Riverbed Heat Conduction Prediction." *Water Resources Research.* 30: 1503-1515.

Knoll, F. H. 1994. "Factors Influencing the Competitive Advantage of *Pseudomonas* sp. Strain KC fo Subsequent Remediation of a Carbon Tetrachloride Impacted Aquifer." M.S. Thesis, Michigan State University, East Lansing, MI.

National Research Council. 1993. *In Situ Bioremediation.* National Academy Press, Washington, D.C.

Semprini, L., G. D. Hopkins, P. L. McCarty, and P. V. Roberts. 1992. "In-situ Transformation of Carbo Tetrachloride and Other Halogenated Compounds Resulting From Biostimulation Under Anoxi Conditions." *Environ. Sci. Technol.* 26: 2454-2461.

Witt, M. E., M. J. Dybas, R. L. Heine, S. Nair, C. S. Criddle, and D. C. Wiggert. 1995. "Bioaugmentatio and Transformation of Carbon Tetrachloride in a Model Aquifer." In R. E. Hinchee, J. Fredrickson, and E C. Alleman (Eds.), *Bioaugmentation for Site Remediation*, pp. 221-227. Battelle Press, Columbus, OH.

1997 AUTHOR INDEX

This index contains the names and affiliations of all authors represented in the technical notes and abstracts included in the five volumes published as a result of the Fourth International In Situ and On-Site Bioremediation Symposium, held in New Orleans in 1997. The back cover of this volume provides ordering information. The number in parentheses corresponds to the volume number in the 1997 set. For example, Brent Adams is an author of two, one beginning on page 395 of Volume 1 and another on page 9 of Volume 4.

Aagaard, Per (University of Oslo/Norway) 4(5):267
Abel, J. A. (Chevron Research & Technology Co./USA) 4(2):69
Abou-Rizk, Jacqueline A.M. (Union Carbide Corp./USA) 4(4):169
Abrams, Scott (University of Minnesota/USA) 4(5):283
Abriola, Linda M. (University of Michigan/USA) 4(1):55; 4(1):323
Acheson, Carolyn M. (U.S. EPA/USA) 4(3):391
Acomb, Lawrence (Geosphere, Inc./USA) 4(1):309
Acree, Steve D. (U.S. EPA/USA) 4(4):261
Acuña, M. (Universidad A. Metropolitant-Iztapalapa/Mexico) 4(5):223
Adams, Brent (Utah State University/USA) 4(1):395; 4(4):9
Adams, Daniel J. (Colorado State University/USA) 4(4):559
Adams, Jeffrey A. (University of Illinois at Chicago/USA) 4(1):165
Adamson, David T. (University of Iowa/USA) 4(3):15
Adriaens, Peter (The University of Michigan/USA) 4(3):173
Aelion, C. Marjorie (University of South Carolina/USA) 4(2):315
Ahrens, Bruce W. (Fluor Daniel GTI, Inc./USA) 4(3):463
Aislabie, Jackie (Landcare Research/New Zealand) 4(2):219
Aitken, Michael D. (University of North Carolina/USA) 4(2):289; 4(3):469
Ajwa, H. (U.S. Dept of Agriculture/USA) 4(3):303
Al-Abed, Souhail R. (University of Cincinnati/USA) 4(3):101
Albiter, Veronica (Universidad Nacional Autonoma de Mexico) 4(2):447
Aldrett, Salvador (Texas A&M University/USA) 4(2):213
Alesi, Eduard J. (IEG Technologie GmbH/Germany) 4(2):353
Allan, Jason P, (Stanford University/USA) 4(3):71
Allan, Katherine A. (Texas A&M University/USA) 4(5):629
Alleman, Bruce (Battelle/USA) 4(1):331; 4(1):347; 4(1):403
Almeida, Jonas S. (Orange County Water District/USA) 4(5):563
Almond, Richard A. (Tennessee Valley Authority/USA) 4(3):315
Alonso, Cristina (University of Cincinnati/USA) 4(5):205
Alphenaar, P. Arne (TAUW Milieu/The Netherlands) 4(5):431; 4(5):591
Al-Tabbaa, Abir (University of Birmingham/UK) 4(4):233
Altman, Dennis J. (Westinghouse Savannah River Company/USA) 4(5):275; 4(5):307; 4(5):315
Alvarez, Gilberto (U.S. EPA) 4(1):103
Alvarez, Pedro J.J. (University of Iowa/USA) 4(1):53; 4(3):367
Alvarez-Cohen, Lisa (University of California Berkeley/USA) 4(2):299; 4(4):15; 4(4):83

Ambert, Jack (Battelle Europe/Switzerland) 4(1):467
Ampleman, Guy (Defense Research Establishment Valcartier/Canada) 4(2):3; 4(2):567; 4(5):171; 4(5):393
Anazawa, Tania A. (Campinas State University/Brazil) 4(2):551
Anders, Krista M. (University of Washington/USA) 4(4):551
Anderson, Barbara L. (Black & Veatch Special Projects Corp./USA) 4(2):131
Anderson, Daniel B. (Battelle-Pacific Northwest National Laboratory/USA) 4(3):9; 4(5):411
Anderson, David W. (Roy F. Weston, Inc./USA) 4(5):425
Anderson, Rachel (Westinghouse Savannah River Co./USA) 4(5):355
Anderson, William (S&ME/USA) 4(4):299
Andersson, Erik (Lund University/Sweden) 4(2):535
Andreotti, Giorgio (AGIP S.p.A./Italy) 4(1):453; 4(1):505; 4(4):393
Antonetti, Michael D. (Dames & Moore, Inc./USA) 4(5):363
Apel, William A. (Idaho National Engineering Laboratory./USA) 4(3):399; 4(3):435; 4(4):483; 4(5):327
Arias, R. (CORPOVEN/Venezuela) 4(4):409
Arias, Y. Meriah (Scripps Inst. of Oceanography/USA) 4(3):411
Arienzo, M. (University of Nebraska-Lincoln/USA) 4(4):485
Arinbasarov, Mikhail U. (Russian Academy of Sciences/Russia) 4(4):523; 4(4):529
Arlotti, Danielle (Foster Wheeler Environmental Italia S.R.L./Italy) 4(1):453; 4(4):393
Armstrong, James E. (Komex International/Canada) 4(1):443
Armstrong, John (The Traverse Group Inc./USA) 4(3):219
Arp, Daniel J. (Oregon State University/USA) 4(3):93; 4(3):107
Arroyo, A. (Colorado State University/USA) 4(5):181
Arvin, Erik (Technical University of Denmark/Denmark) 4(2):119; 4(4):7
Auria, Richard (ORSTOM/Mexico) 4(5):223
Aust, Steven D. (Utah State University/USA) 4(2):481
Autenrieth, Robin L. (Texas A&M University/USA) 4(2):17; 4(2):213; 4(2):569; 4(3):355; 4(5):49
Axelrod, Russell B. (PTI Environmental Services/USA) 4(1):111
Axtel, C. A. (University of Northern Iowa/USA) 4(2):537
Aziz, Carol (University of Texas at Austin/USA) 4(3):79
Bader, Darlene (U.S. Army Environmental Center/USA) 4(3):315
Badgley, Anne (The Dow Chemical Company/USA) 4(3):293
Bahr, J. M. (University of Wisconsin-Madison/USA) 4(5):13
Bailey, John (Mission Research Corporation/USA) 4(1):253
Bailey, Scott (University of Tulsa/USA) 4(4):383
Baker, Ralph S. (ENSR Consulting & Engineering/USA) 4(1):235
Baker, Robert R. (Radian Corporation/USA) 4(5):173
Baldwin, Connely K. (Utah State University/USA) 4(2):329
Baldwin, Jon J. (Texaco Inc./USA) 4(5):419
Ball, H. A. (Stanford University/USA) 4(5):9
Ballapragada, Bhasker S. (University of Washington/USA) 4(5):95
Ballard, Sandy (Hydrotechniques/USA) 4(1):253
Ballerini, Daniel (Institut Français du Pétrole/France) 4(4):395; 4(4):411; 4(5):213
Banerjee, Dwijen K. (University of Alberta/Canada) 4(5):163
Banks, M. Katherine (Kansas State University/USA) 4(3):305
Bantle, Jack A. (Oklahoma State University/USA) 4(4):19; 4(5):11

Bañuelos, Gary (U.S. Department of Agriculture/USA) 4(3):303
Barbaro, Jeffrey (University of Waterloo/Canada) 4(5):21
Barbeau, C. (École Polytechnique de Montréal/Canada) 4(4):535
Barber, Christopher (CSIRO/Australia) 4(1):339
Barbieri, Sioni Maluf (Universidade Federal de São Carlos/Brazil) 4(2):247
Barbosa, Ricardo A.M. (Universidade Catôlica Portuguesa/Portugal) 4(5):1
Barcelona, Michael J. (University of Michigan/USA) 4(1):21; 4(1):55; 4(4):507
Barclay, Clayton (Jacques Whitford Environment Limited/Canada) 4(1):277
Barclay, Michelle (University of Kent at Canterbury/UK) 4(2):531; 4(2):533
Barden, Michael (Wisconsin Department of Natural Resources/USA) 4(1):103
Bargi, Rupinder (Simon Fraser University/Canada) 4(2):499
Barker, James F. (University of Waterloo/Canada) 4(2):127; 4(4):181; 4(4):249; 4(4):255; 4(5):21
Barnes, Paul (Waste Management, Inc./USA) 4(3):339; 4(5):295
Barr, Kelton D. (Delta Environmental Consultants, Inc./USA) 4(1):1; 4(5):283
Bartha, Richard (Cook College, Rutgers/USA) 4(2):191
Bartholomae, Philip G. (BP Oil Company/USA) 4(2):371; 4(2):373
Bartlett, Craig L. (DuPont Co./USA) 4(3):289
Barton, Larry L. (University of New Mexico/USA) 4(1):435; 4(4):195
Basinet, Rick (U.S. Navy/USA) 4(1):205
Bass, B. D. (University of Montana/USA) 4(2):537
Bass, David H. (Fluor Daniel GTI/USA) 4(1):117; 4(1):153; 4(1):243; 4(2):371; 4(2):373
Bastiaens, L. (VITO/Belgium) 4(4):159
Baten, H. H. (Bioclear Environmental Biotechnology b.v./The Netherlands) 4(5):29
Battermann, Gerhard (TGU-GmbH/Germany) 4(5):7; 4(5):27
Baubron, Jean-Claude (BRGM/France) 4(5):465
Baun, Anders (Technical University of Denmark/Denmark) 4(4):1
Bavière, Marc (Institut Français du Pétrole/France) 4(2):559
Bayard, Remy P. (Institut National des Sciences Appliquées/France) 4(5):577
Beabes, Lorie A. (Parsons Engineering Science, Inc./USA) 4(1):227
Becerra, M.A. (Mycotech Corp./USA) 4(2):537
Beck, Frank P. (U.S. EPA/USA) 4(2):309; 4(5):11
Beckman, Scott W. (Science Applications Intl Corp (SAIC)/USA) 4(3):323
Becvar, Erica S.K. (ARA, Inc./USA) 4(2):603; 4(3):39; 4(4):141
Bedard, Adam H. (Applied Hydrology Associates Inc./USA) 4(3):233
Bedient, Philip B. (Rice University/USA) 4(1):43; 4(1):75; 4(5):11
Beeman, Ralph E. (DuPont Co./USA) 4(1):393
Beenackers, A. A. C. M. (University of Groningen/The Netherlands) 4(3):73
Behrends, Leslie (Tennessee Valley Authority/USA) 4(3):315
Bekins, Barbara (U.S. Geological Survey/USA) 4(5):547
Belcher, David M. (ABB Environmental Services, Inc./USA) 4(3):139; 4(3):231
Beller, Harry (Stanford University) 4(5):9
Bender, Judith (Clark Atlanta University/USA) 4(3):373
Benediktsson, Catherine (Federal Aviation Administration/USA) 4(1):309
Bennett, Marlene L. (BHP Research/Australia) 4(1):339
Benoit, Jacques (AGRA Earth & Environmental, Inc./Canada) 4(4):307
Benson, Leigh A. (Parsons Engineering Science, Inc./USA) 4(1):95; 4(1):303; 4(3):191

Bereded-Samuel, Yared (Washington State University/USA) 4(3):399
Berg, Matthew (University of Texas at Austin/USA) 4(5):603
Berglund, Scott (Federal Aviation Administration/USA) 4(1):309
Berry, Christopher J. (Westinghouse Savannah River Co./USA) 4(3):345; 4(5):85; 4(5):275; 4(5):307
Bess, Vicki (BBC Laboratories, Inc./USA) 4(4):541
Bettermann, Alan D. (BioRenewal Technologies, Inc./USA) 4(3):267; 4(5):589
Betts, W. Bernard (University of York/UK) 4(4):167
Bezborodnikov, Sergei (Michigan State University/USA) 4(4):507
Bhattacharya, Sanjoy K. (Tulane University/USA) 4(4):439; 4(5):233; 4(5):239; 4(5):277
Bhunia, Prasanta (Weston & Samson Engineers, Inc./USA) 4(1):297
Bhupathiraju, Vishvesh K. (University of California at Berkeley/USA) 4(2):299
Bianco, Vito (Foster Wheeler Italiana/Italy) 4(1):453
Bielefeldt, Angela R. (University of Colorado/USA) 4(5):37
Bienkowski, Paul R. (University of Tennessee at Knoxville/USA) 4(5):103
Bier, E. (University of Nebraska-Lincoln/USA) 4(4):485
Bilbao, E. (CORPOVEN/Venezuela) 4(4):409
Bishop, Arlen (DuPont Co./USA) 4(2):137
Bishop, Dolloff (U.S. EPA/USA) 4(5):195
Bison, Pier Luigi (Aquater S.p.A./Italy) 4(1):423
Bjerg, Poul L (Technical University of Denmark/Denmark) 4(2):15; 4(5):313
Blagoev, V. V. (Institute of Bioorg Chem & Petrochem/Ukraine) 4(3):407
Blaszkiewicz, Hanna (Connolly Environmental/Australia) 4(2):151
Blazicek, Tracy L. (New York State Electric & Gas Corp./USA) 4(3):463
Bledsoe, S. A. (DuPont/USA) 4(3):21
Blomker, Kimberly R. (Leggette, Brashears & Graham, Inc./USA) 4(2):103
Blotevogel, Karl-Heinz (Universität Oldenburg/Germany) 4(2):9; 4(2):39; 4(2):49
Bocard, Christian F. (Institut Français du Pétrole/France) 4(2):559
Böckle, Karin (DVGW Technology Center Karlsrühe/Germany) 4(5):97
Boersma, Paul (CH2M Hill Inc./USA) 4(1):123
Boggs, Mark (Tennessee Valley Authority/USA) 4(1):23
Boggs, Michael L. (SES Environmental Services/USA) 4(1):213
Bohan, Dave (Envirologic Technology/USA) 4(5):475
Boles, Jeff L. (Tennessee Valley Authority/USA) 4(5):189
Bolton, Harvey (Battelle-Pacific Northwest National Laboratory/USA) 4(3):437
Bonner, James S. (Texas A&M University/USA) 4(2):213; 4(2):569; 4(3):355; 4(4):119; 4(5):49
Bono, John (J.R. Simplot Company/USA) 4(2):61
Boonstra, B. (Axis Genetics Ltd/UK) 4(3):337
Boopathy, Ramaraj (Argonne National Laboratory/USA) 4(2):59
Borazjani, Abdolhamid (Mississippi State University/USA) 4(2):97
Borchert, Susanne (SBP Technologies, Inc./USA) 4(2):353; 4(4):149
Borden, Robert C. (North Carolina State University/USA) 4(1):29
Boronin, Alexander M. (Russian Academy of Sciences/Russia) 4(4):523
Bos, Martinus (University of Twente/The Netherlands) 4(5):565
Bosma, Tom N.P. (TNO Inst. of Environmental Sci./The Netherlands) 4(3):155
Boone, Jennifer (Battelle/USA) 4(1):403
Bossert, Ingeborg D. (Rutgers University/USA) 4(5):495
Bosshard, Bruce (EnviroPacifica, Inc./USA) 4(4):433; 4(5):445

Boufadel, Michel C. (University of Cincinnati/USA) 4(4):267
Bourgon, Greg (Northwestern University/USA) 4(5):379
Bourquin, Al W. (Camp Dresser & McKee Inc./USA) 4(1):311; 4(4):513; 4(4):559; 4(5):89; 4(5):315
Bouwer, Edward J. (Johns Hopkins University/USA) 4(2):125; 4(5):387
Bovendeur, J. (Heidemij Realisatie/The Netherlands) 4(3):155
Bowling, Leon (U.S. Marine Corps/USA) 4(1):7; 4(1):331
Boyd, Thomas J. (Naval Research Laboratory/USA) 4(2):399; 4(4):125
Boyle, Mike F. (University of Washington/USA) 4(5):95
Bracco, Angelo A. (GE Corporate R&D Center/USA) 4(3):297
Brackett, Kim A. (U.S. EPA) 4(3):385
Braddock, Joan F. (University of Alaska/USA) 4(4):283
Braida, Washington J. (Iowa State University/USA) 4(1):187
Bramwell, Debbie-Ann (Florida International University/USA) 4(2):595
Brasaemle, Joan E. (EMG Inc./USA) 4(5):289
Braun, R. (Institute for Agrobiotechnology/AUSTRIA) 4(5):617
Brauner, J. Steven (Virginia Polytechnic Institute & State University/USA) 4(1):35
Breedveld, Gijs D. (Norwegian Geotechnical Institute/Norway) 4(5):267
Breitung, Jürgen (Institute of Immunology & Environmental Hygiene/Germany) 4(2):9
Brennan, Jr., Michael J. (GE Corporate R&D Center/USA) 4(3):45
Bresette, D. (Woodward-Clyde Consultants/USA) 4(3):125
Breure, Anton M. (RIVM/LAE/The Netherlands) 4(5):643
Breyfogle, Bob (Argonne National Laboratory/USA) 4(5):471
Bricelj, Mihael (University of Ljubljana/Slovenia) 4(2):231
Brigmon, Robin L. (Westinghouse Savannah River Co./USA) 4(3):345; 4(5):275
Brinkmann, Madonna (Remediation Technologies, Inc./USA) 4(2):301
Brockman, Fred (Battelle-Pacific Northwest National Laboratory/USA) 4(5):317; 4(5):325; 4(5):345
Broetzman, Gary (Colorado Center for Environmental Mgt/USA) 4(4):323
Broholm, Mette M. (Technical University of Denmark/Denmark) 4(2):119
Brough, Matthew J. (University of Birmingham/UK) 4(4):233
Brourman, Mitchell D. (Hanson North America/USA) 4(2):301; 4(4):149
Brown, Chris L. (University of Texas at Austin/USA) 4(2):565
Brown, David A. (Parsons Engineering Science, Inc./USA) 4(1):429
Brown, Derick G. (Princeton University/USA) 4(2):581
Brown, Kandi L. (IT Corp./USA) 4(1):83; 4(3):227
Brown, Michael J. (Komex International, Ltd./Canada) 4(4):181; 4(4):249; 4(4):255
Brown, Richard A. (Fluor Daniel GTI/USA) 4(1):117; 4(2):245; 4(4):457
Brownell, Kurt A. (U.S. Army/USA) 4(5):13
Bruce, Cristin Lee (Arizona State University/USA) 4(4):99
Brunia, A. (TNO Environmental Sciences/The Netherlands) 4(4):203
Bruns-Nagel, Dirk (University of Marburg/Germany) 4(2):9; 4(2):49
Brusseau, Mark L. (University of Arizona/USA) 4(2):597
Buchanan, Lisa R. (Camp Dresser & McKee/USA) 4(1):311
Buchanan, Ronald J. (DuPont Co./USA) 4(3):21; 4(3):289; 4(4):315
Buchholz, Kurt (Battelle/USA) 4(4):445
Buck, Paul (University of Tulsa/USA) 4(4):383
Buers, Katy L.M. (University of Kent/UK) 4(3):413

Bulko, O. V. (Institute of Bioorg Chem & Petrochem/Ukraine) 4(3):325
Bullock, J. Michael (Matrix Environmental Technologies, Inc./USA) 4(4):147
Bumgarner, B. E. (Westinghouse Savannah River Company/USA) 4(5):307
Buongarzone, E. (Aquater S.p.A./Italy) 4(1):423
Burbage, Greg (Westinghouse Savannah River Company/USA) 4(1):385
Burgess, Kristine S. (SECOR International, Inc./USA) 4(1):67
Burlage, Robert S. (Oak Ridge National Laboratory/USA) 4(4):77
Burns, Richard (University of Kent at Canterbury/UK) 4(4):155
Burton, David L. (University of Manitoba/Canada) 4(4):21
Buschbom, Robert (MidAmerican Energy/USA) 4(3):457
Buscheck, Timothy E. (Chevron Research & Technology Co./USA) 4(3):149
Butcher, Mark G. (Battelle-Pacific Northwest National Lab/USA) 4(4):273
Butler, Barbara J. (University of Waterloo/Canada) 4(2):129; 4(5):21
Buxton, Bruce (Battelle/USA) 4(5):349
Buzzelli, Maurizio (Ambiente SpA/Italy) 4(1):423
Byerley, Brian T. (University of Waterloo/Canada) 4(4):209
Caballer, Reyes (Simon Fraser University/Canada) 4(2):499
Camacho, JoAnn (USEPA/USA) 4(2):251
Camesano, Terri (The University of Arizona/USA) 4(4):567
Campos-Velarde, Maria D. (Center for Advanced Studies & Research (CINVESTAV)/Mexico) 4(3):273
Capps, Mark (EMCON/USA) 4(1):89
Cargnel, Daniel C. (Rust Environment & Infrastructure, Inc./USA) 4(5):43
Carlson, Keith A. (Integrated Science & Technology Inc./USA) 4(1):321
Carman, Eric P. (Geraghty & Miller/USA) 4(3):347
Carmichael, Lisa M. (North Carolina A&T University/USA) 4(3):279; 4(3):283; 4(5):641
Carnahan, John B. (AGRA Earth & Environmental, Inc./USA) 4(1):297
Carreira, L.H. (L.C. Enterprises/USA) 4(3):353
Carrera, Paolo (Eniricerche S.p.A/Italy) 4(4):385; 4(4):131
Carroll, S.L.. (Oak Ridge National Laboratory/USA) 4(5):103
Carter, Kim J. (Willamette Industries/USA) 4(2):413
Carter, Sean R. (Matrix Environmental Technologies/USA) 4(4):147
Cartwright, Colin Daniel (University of Kent/UK) 4(4):155
Carvalho, D. F. (Campinas State University/Brazil) 4(4):91
Caso, Osvaldo (Centro de Ecofisiologia Vegetal/Argentina) 4(2):157
Catbas, K. Hayati (University of Cincinnati/USA) 4(3):101
Cattaneo, Maurice (National Research Council of Canada/Canada) 4(2):3
Cazier, Fabrice (CREID/France) 4(2):79
Chacón, Michael (New Mexico Environment Department/USA) 4(4):323
Chaconas, James D.J. (ENSR Consulting & Engineering/USA) 4(1):497
Chandler, Darrell P. (Battelle-Pacific Northwest National Laboratory/USA) 4(5):345
Chapelle, F.H. (U.S. Geological Survey/USA) 4(3):287
Chapman, Steven W. (University of Waterloo/Canada) 4(4):209
Charbonnier, Patrick (BRGM/France) 4(5):405
Chaudhry, Tanwir (Anteon Corp./USA) 4(1):497
Chen, Kevin C. (University of Virginia/USA) 4(4):573
Chen, Yung-Ming (University of Michigan/USA) 4(1):55
Cheng, Jiayang (University of Cincinnati/USA) 4(2):47
Cherry, Robert S. (Idaho National Engineering Laboratory/USA) 4(2):319
Chevron, Florence (CNRSSP/France) 4(5):405

Chiang, Chen Yu (Shell Oil Product Company/USA) 4(5):413
Chin, Arthur (Exxon Biomedical Sciences, Inc./USA) 4(3):161
Chitla, Santhosh Reddy (Envirosystems, Inc./USA) 4(5):535; 4(4):543
Chiu, Ying-Chih (National I-Lan Institute of Agriculture & Technology/R.O.C.-Taiwan) 4(3):51
Cho, Jae-Chang (Seoul National University/Republic of Korea) 4(2):239
Cho, Jong Soo (U.S. EPA/NRMRL/USA) 4(1):97; 4(2):309
Cho, Soon Haing (Ajou University/Republic of Korea) 4(4):477
Choi, Sang-Il (Kwangwoon University/Republic of Korea) 4(2):589
Choi, Young Soo (Ajou University/Republic of Korea) 4(4):477
Chollak, Darrel (Canadian Occidental Petroleum Ltd./Canada) 4(5):269
Chrisman-Lazarr, Natalie (Woodward-Clyde Consultants/USA) 4(3):125; 4(4):275; 4(4):549
Christen, P. (ORSTOM/Mexico) 4(5):223
Christensen, Thomas H. (Technical University of Denmark/Denmark) 4(2):15; 4(5):313
Christodoulotos, C. (Stevens Institute/USA) 4(5):107
Christopher, Michael (Roy F. Weston, Inc./USA) 4(5):507
Chu, Kung-Hui (University of California Berkeley/USA) 4(4):15
Churchill, Sharon (Washington State University/USA) 4(3):435
Cifuentes, Luis A. (Texas A&M University/USA) 4(2):317; 4(2):353; 4(4):297
Clark, K.W. (Maxim Technologies, Inc./USA) 4(1):503
Clayton, Wilson S. (Fluor Daniel GTI/USA) 4(1):159; 4(1):233
Cleland, Dale D. (University of Kansas/USA) 4(4):105
Clément, Bernard (École Polytechnique de Montréal/Canada) 4(4):227
Clement, T. Prabhakar (Battelle-Pacific Northwest National Lab/USA) 4(1):37
Clemente, Andrea R. (Campinas State University/Brazil) 4(2):551
Cline, Patricia (CH2M Hill/USA) 4(3):213
Coffin, Richard B. (Naval Research Laboratory/USA) 4(4):297; 4(2):317; 4(4):125
Colberg, Patricia J.S. (University of Wyoming/USA) 4(2):125
Cole, Jason D. (Oregon State University/USA) 4(4):239
Collantes, Marta (CEVEG/Argentina) 4(2):157
Collins, Richard (Indiana University/USA) 4(4):41
Colwell, Frederick S. (Idaho National Engineering Laboratory/USA) 4(4):111
Comeau, Yves (École Polytechnique de Montréal/Canada) 4(2):335; 4(2):457; 4(4):535
Comfort, Steven D. (University of Nebraska-Lincoln/USA) 4(1):421; 4(2):287; 4(4):485
Connolly, Mark D. (Connolly Environmental/Australia) 4(2):151
Connor, James M. (DuPont/USA) 4(2):137
Conrad, Mark (Lawrence Berkeley Laboratories/USA) 4(2):299
Cook, James A. (Beazer East Inc./USA) 4(1):413
Coonrod, Steven H. (Tennessee Valley Authority/USA) 4(3):315
Coover, Merv P. (Remediation Technologies Inc./USA) 4(2):55
Corseuil, Henry X. (Federal University of Santa Catarina/Brazil) 4(1):53
Cortsen, J. (Technical University of Denmark/Denmark) 4(5):313
Covell, James R. (EG&G TSWV Inc./USA) 4(1):173
Cowan, Robert M. (Cook College/Rutgers University/USA) 4(1):17; 4(3):187
Cox, Evan E. (Beak Consultants/Canada) 4(3):203; 4(3):205; 4(3):261
Cox, William (Regenesis Bioremediation Products/USA) 4(4):247

Craig, A. Morrie (Oregon State University/USA) 4(2):41
Crawford, Donald L. (University of Idaho/USA) 4(2):23; 4(2):505
Crawford, Ronald L. (University of Idaho/USA) 4(2):505; 4(2):525
Criddle, Craig S. (Michigan State University/USA) 4(3):33; 4(3):59; 4(4):507; 4(4):575
Croft, Barry C. (AEA Technology plc/UK) 4(1):441; 4(5):83
Crossman, Tom L. (Geraghty & Miller/USA) 4(3):347
Cummings, Peter T. (University of Tennessee/USA) 4(4):573
Cunningham, Al B. (Montana State University/USA) 4(5):385; 4(5):387
Cunningham, Scott D. (DuPont/USA) 4(3):353; 4(3):319; 4(4):17
Cusick, John P. (W. L. Gore & Associates, Inc./USA) 4(2):341; 4(1):495
Cyr, Gerald G. (Laidlaw Environmental Services, Inc./USA) 4(1):473
Dablow, Jay (Fluor Daniel GTI, Inc./USA) 4(5):439
D'Addona, John J. (The Traverse Group Inc./USA) 4(3):219
Dale, Bruce E. (Michigan State University/USA) 4(3):361
Daley, Paul F. (Lawrence Livermore National Laboratory/USA) 4(2):299
Dalmazzone, Christine (Institut Français du Pétrole/France) 4(4):395; 4(4):411
Damera, Ravi (General Physics Corp./USA) 4(1):199
Daniel, Robert A. (North Carolina Department of Environment, Health, & Natural Resources/USA) 4(1):29
Dao, H. (University of Nebraska-Lincoln/USA) 4(3):421
Dao, H. T. (University of Nebraska-Lincoln/USA) 4(3):421
Das, Amorjyoti (Arizona State University/USA) 4(1):135
Dasgupta, Anindya (Delphinus Engineering/USA) 4(5):225
Daun, Gregor (Fraunhofer Institut für Grenzflachen & Bioverfahrenstech/Germany) 4(2):1
Davidson, Mark (Naval Facilities Engineering Service Center/USA) 4(3):231
Davies, Shaun (AEA Technology plc/UK) 4(5):83
Davies, Simon (Michigan State University/USA) 4(4):507
Davis, G. (Microbial Insights Inc./USA) 4(5):319
Davis, Greg B. (CSIRO/Australia) 4(1):19; 4(1):339; 4(2):53; 4(4):241
Davis, John W. (The Dow Chemical Company/USA) 4(3):281; 4(3):287; 4(3):289; 4(3):293
Davis, Katherine L. (DuPont/USA) 4(2):137
Davis, Kimberly L. (University of Tennessee/USA) 4(4):317
Davis, Tommy M. (Southern Wood Piedmont Co./USA) 4(5):115
Davis-Hoover, Wendy J. (U.S. EPA/USA) 4(3):101; 4(3):385
Dawson, Dave (Naval Facilities Engineering Service Center/USA) 4(5):445
Day, Michael (Applied Hydrology Assoc. Inc./USA) 4(3):233
Dean, Sean (The University of Michigan/USA) 4(3):173
Dean, Warren T. (Environmental Systems & Technologies, Inc./USA) 4(3):161
Deardorff, Therese M. (U.S. Army/USA) 4(1):295
de Best, Jappe H. (TNO Inst of Environmental Sciences/Netherlands) 4(3):3
de Blanc, Phillip C. (University of Texas at Austin/USA) 4(2):565; 4(5):597
Deeb, Rula Anselmo (Univ. of California at Berkeley/USA) 4(4):83
Defives, Claude (USTL de Lille/France) 4(5):405
DeFlaun, Mary F. (Envirogen, Inc./USA) 4(4):493
Degher, Alexandra B. (Oregon State University/USA) 4(2):279
Delshad, Mojdeh (University of Texas at Austin/USA) 4(5):597
Delwiche, Jim (Wisconsin Department of Natural Resources/USA) 4(1):103
Deming, Jody W. (University of Washington/USA) 4(4):427

Denham, Miles (Westinghouse Savannah River Co./USA) 4(5):355
Dennis, Darrhyl (Georgia Power Co./USA) 4(1):461
de Noyelles, Frand (University of Kansas/USA) 4(2):117
de Ridder, E. J. (TNO-MEP/The Netherlands) 4(2):487
Desai, Sandeep (University of Cincinnati/USA) 4(5):195
Deschênes, Louise (École Polytechnique de Montréal/Canada) 4(2):453; 4(4):221; 4(4):227; 4(4):535
Deshusses, Marc A. (Univ. of California Riverside/USA) 4(5):175; 4(5):219
Desilets, B.K. (Colorado State University/USA) 4(5):89
de Vals, Bruno (ELF Aquitaine Production/France) 4(2):559
Devine, Kate (Biotreatment News/USA) 4(4):329
Devlin, J F (University of Waterloo/Canada) 4(4):181; 4(4):255
DeWeerd, Kim (GE Corporate R&D Center/USA) 4(3):45; 4(3):281; 4(3):289; 4(3):297
de Wit, Johannes C.M. (Tauw Milieu b.v./The Netherlands) 4(5):431; 4(5):591
Dey, Jeffrey (Resource Control Corp./USA) 4(1):445
De Zoysa, D. Stanley (CSIRO/Australia) 4(4):241
Dial, Craig E. (Clayton Environmental Consultants, Inc./USA) 4(5):527
Diaz, Juan Anthony (CH2M Hill/USA) 4(1):455
Diehl, Susan V. (Mississippi State University/USA) 4(2):97
Diels, Ludo (VITO/Belgium) 4(4):159
Di Luise, Giancarlo (Agip S.p.A./Italy) 4(1):109; 4(1):453; 4(1):467; 4(4):393
Dirkse, E. H. M. (Dirkse Milieutechniek bv/The Netherlands) 4(3):65
Dittmar, Charles W. (The Traverse Group, Inc./USA) 4(3):219
Doddema, Hans J. (TNO Inst of Environmental Sciences/Netherlands) 4(3):3
Dolan, Mark E. (Stanford University/USA) 4(3):71
Dolliver, Elson (Waste Management Inc./USA) 4(1):493
Domroes, David (Occidental Chemical Corporation/USA) 4(3):321
Dooley, Maureen A. (ABB Environmental Services, Inc./USA) 4(3):253; 4(2):271; 4(3):231
Dosani, Majid (IT Corp./USA) 4(5):131; 4(2):85
dos Santos Ferreira, Ruy (Universidade Federal de Santa Catarina/Brazil) 4(1):53
Dott, Wolfgang (Inst. of Hygiene & Env. Health/Germany) 4(2):273; 4(5):459
Doucette, William (Utah State University/USA) 4(1):395; 4(4):9
Dougherty, David E. (University of Vermont/USA) 4(4):135
Douglas, Gregory S. (Arthur D. Little, Inc./USA) 4(1):271; 4(2):175; 4(2):205; 4(4):487
Downer, Roswell (Texas A&M University/USA) 4(2):317
Downey, Douglas C. (Parsons Engineering Science, Inc./USA) 4(1):95; 4(1):303; 4(1):341
Downey, S. (U.S. Department of Agriculture/ USA) 4(3):303
Draaisma, René B. (TNO-MEP/The Netherlands) 4(2):487
Drake, Evelyn N. (Exxon Research & Engineering/USA) 4(2):175; 4(2):205; 4(4):487
Drijber, Rhae A. (University of Nebraska Lincoln/USA) 4(1):421; 4(2):287; 4(4):101
Drzyzga, Oliver (Universität Oldenburg/Germany) 4(2):39; 4(2):49
Duaime, Terence E. (Montana Bureau of Mines and Geology/USA) 4(1):311
Dubarry, Jean Louis (ELF Aquitaine Production/France) 4(2):559
Dubois, Charles (Defense Research Establishment Valcartier/Canada) 4(2):3; 4(2):567; 4(5):171; 4(5):393
Dubourguier, Henri-Charles (Institut Superieur d'Agriculture/France) 4(2):79; 4(5):405

Dubrock, Duane D. (Massar Environmental Technologies, Inc./USA) 4(4):335
Ducreux, Jean (Institut Français du Pétrole/France) 4(2):559
Duda, Peter (Chevron Products Co./USA) 4(5):145
Dudas, Marvin J. (University of Alberta/ Canada) 4(5):163
Dudziak, Suzanne (Port of Tacoma/USA) 4(1):283
Duffy, James (Occidental Chemical Corp./ USA) 4(3):321
Dullaghan, Edward M. (URS-Greiner, Inc./USA) 4(1):179
Dumont, D. D. (Camp Dresser McKee Inc./USA) 4(5):89
Duncan, Kathleen (University of Tulsa/ USA) 4(4):383
Dunn, David P. (Massar Environmental Technologies, Inc./USA) 4(4):335
Dupont, R. Ryan (Utah State University/ USA) 4(1):395; 4(1):443; 4(2):329; 4(3):189; 4(4):9
Durrant, Lucia Regina (Campinas State University/Brazil) 4(4):91; 4(2):551
Duston, Karen Lansford (Rice University/ USA) 4(4):565
DuTeau, N. M. (Colorado State University/USA) 4(4):81
Dutta, Sisir K. (Howard University/USA) 4(4):65
Duval, Marie (Institut Superieur d'Agriculture/France) 4(2):79
Dvorak, Bruce I. (University of Nebraska Lincoln/USA) 4(3):127; 4(5):55
Dybas, Michael J. (Michigan State University/USA) 4(3):59; 4(3):187; 4(4):507; 4(4):575
Dyreborg, Soren (Technical University of Denmark/Denmark) 4(4):7
Eaton, Scott (Citgo Petroleum/USA) 4(1):193
Eberhard, Wayne (Federal Aviation Administration/USA) 4(1):309
Eberle, Michael F. (Dames & Moore/ USA) 4(2):303; 4(5):301
Eccles, Harry (British Nuclear Fuels plc/ UK) 4(3):393; 4(3):429
Edwards, Elizabeth A. (McMaster University/Canada) 4(3):203; 4(3):261
Ei, T.A. (DuPont Company/USA) 4(3):287
Eisenbeis, Martina (University of Stuttgart/Germany) 4(3):13
Ekanemesang, Udoudo Moses (Clark Atlanta University/USA) 4(2):57
Ekuan, Gordon (Washington State University/USA) 4(3):321
Elliott, Robert (U.S. Air Force/USA) 4(3):141
Ellis, David E. (DuPont Specialty Chemicals/USA) 4(1):393; 4(3):21; 4(3):287; 4(3):289; 4(4):315
Elthon, Thomas E. (University of Nebraska-Lincoln/USA) 4(4):101
Englande, Andrew J. (Tulane School of Public Health/USA) 4(3):25
Epstein, Eliot (E&A Environmental Consultants, Inc./USA) 4(2):71
Ermolli, F. (Aquater S.p.A./Italy) 4(1):423
Evans, E. (BHP Research Melbourne Labs/Australia) 4(1):339
Evans, Jill M. (Associated Western Universities, Inc./USA) 4(2):291
Everett, Lorne (Geraghty & Miller Inc./USA) 4(5):373
Fadullon, Frances Steinacker (SAIC/ USA) 4(2):263
Fam, Sami (Innovative Engineering Solutions, Inc./USA) 4(4):193
Fan, Jingzhao (Princeton University/USA) 4(2):193
Fang, Jiasong (University of Michigan/ USA) 4(1):21
Fang, Yan (The University of Arizona/ USA) 4(4):567
Fayolle, Françoise (Institut Français du Pétrole/France) 4(5):213
Fedder, Richard P. (RAM Environmental, LLC/USA) 4(1):285
Félix, Crisandel (Clark Atlanta University/ USA) 4(3):373
Feng, Yongsheng (University of Alberta/ Canada) 4(5):599
Fennell, Donna E. (Cornell University/ USA) 4(3):11

Ferguson, John F. (Univ of Washington/ USA) 4(2):421; 4(4):551; 4(5):95
Fernandes, Fermiano M. (Adesol Produtos Quimicos Ltda/Brazil) 4(5):71
Fernández-Villagómez, G. (CENAPRED & UNAM/Mexico) 4(3):273
Ferro, Ari (Phytokinetics/USA) 4(3):309
Field, Jim A. (Wageningen Agricultural University/The Netherlands) 4(2):31; 4(2):487; 4(5):87
Finci, Aka G. (Mission Research Corporation/USA) 4(1):259
Findlay, Margaret (Bioremediation Consulting, Inc./USA) 4(1):265; 4(1):267; 4(1):269; 4(1):271; 4(2):69; 4(2):191
Finnesgaard, Dale W. (Barr Engineering Co./USA) 4(2):173
Finton, Christopher D. (Battelle/USA) 4(1):109
Fiorentine, Anthony M. (Fluor Daniel GTI, Inc./USA) 4(3):445; 4(3):479
Fiorenza, Stephanie (Rice University/ USA) 4(1):347
Fischer, Jeffrey (U.S. Geological Survey/USA) 4(3):115
Fisher, J. Berton (Gardere and Wynne/USA) 4(1):73; 4(4):383
Fitch, Mark W. (University of Missouri-Rolla/USA) 4(3):113
Flanagan, William P. (GE Corporate R&D Center/USA) 4(3):45
Flanders, C. M. (DuPont/USA) 4(3):353
Flathman, Paul E. (OHM Remediation Services Corporation/USA) 4(2):365
Flax, Jodi (Northwestern University/USA) 4(5):379
Flemming, C. (University of Tennessee/USA) 4(5):319
Fletcher, John S. (University of Oklahoma/USA) 4(2):115
Fliermans, Carl B. (Westinghouse Savannah River Co./USA) 4(5):275
Foeller, J. R. (University of Missouri Columbia/USA) 4(3):85
Fogel, Samuel (Bioremediation Consulting, Inc./USA) 4(1):265; 4(1):267; 4(1):269; 4(1):271; 4(2):191
Fogler, H. Scott (University of Michigan/USA) 4(4):553
Follett, Ronald F. (USDA - ARS/USA) 4(3):415
Foote, Eric A. (Battelle/USA) 4(1):403
Forbes, Peter (U.S. Air Force/USA) 4(1):361
Ford, Roseanne M. (University of Virginia/USA) 4(4):573
Forget, Dominique (École Polytechnique de Montréal/Canada) 4(4):221
Forney, Larry J. (Michigan State University/USA) 4(4):501
Forsman, Mats (FOA/National Defence Research Establishment/Sweden) 4(5):333
Forsyth, III, John V. (All Phase Environmental Srvcs, Inc./USA) 4(5):107
Fosbrook, Cristol (U.S. Army/USA) 4(1):61
Fotiades, Stephen M. (E.I. du Pont de Nemours & Company/USA) 4(1):393
Fraley, Rose H. (University of North Dakota/USA) 4(5):113
Francis, M. McD. (Nova Chemicals/ Canada) 4(5):529
Franck, Marilyn M. (Westinghouse Savannah River Co./USA) 4(3):345; 4(5):275
Frankenberger, Jr., William T. (University of California/USA) 4(4):277
Frankenfeld, Katrin (FZMB e.V./ Germany) 4(5):259
Franzmann, P. (CSIRO/Australia) 4(1):339
Fredrickson, Herb (U.S. Army Corps of Engineers/USA) 4(5):157
Freedman, David L. (Clemson University/USA) 4(3):255; 4(4):465; 4(5):1
Fricke, James (Advanced Geoservices Corp./USA) 4(3):401
Frind, Emil (University of Waterloo/ Canada) 4(2):129
Frisbie, Andrew (Purdue University/USA) 4(5):623
Fritsche, Wolfgang (Friedrich-Schiller-Universität Jena/Germany) 4(2):493

Frontera-Suau, R. (Medical Univ of South Carolina/USA) 4(4):79
Froud, Susan (University of Waterloo/ Canada) 4(4):181; 4(4):249
Fu, Chunsheng (University of Cincinnati/ USA) 4(2):195
Gagnon, Josée (National Research Council of Canada/Canada) 4(2):335
Galkin, A. P. (Institute of Bioorganic Chemisty & Petrochemistry/Ukraine) 4(3):325; 4(3):407
Gallagher, John R. (University of North Dakota/USA) 4(5):269
Gallagher, Mark N. (Mobil Oil Corp./ USA) 4(1):181
Gannon, David John (Zeneca Corp./ Canada) 4(3):285; 4(3):299
Ganzeveld, Ineke (University of Groningen/The Netherlands) 4(3):73
Gao, Jianwei (Battelle-Pacific Northwest National Laboratory/USA) 4(3):49; 4(3):57; 4(5):561
Garrett, William (Southern Co. Services Inc./USA) 4(1):461
Gartner, M. (Institute for Agrobiotechnology/AUSTRIA) 4(5):617
Garusov, A.V. (Kazan State University/Russia) 4(2):33
Gatliff, Edward G. (Applied Natural Sciences/USA) 4(3):347
Gaudette, Carol (U.S. Air Force/USA) 4(5):649
Geer, Michael A. (U.S. Air Force/USA) 4(1):23
Gemoets, Johan (VITO/Belgium) 4(4):159
Gemsa, Diethard (Institute of Immunology & Environ Hygiene/Germany) 4(2):9
Georgiou, George (University of Texas-Austin/USA) 4(3):79
Germond, Bart (Focus Environmental/ USA) 4(3):161
Gerritse, Jan (University of Groningen/ The Netherlands) 4(5):431; 4(5):591
Ghetti, Marc (H_2O Options, Inc./USA) 4(4):299
Ghosh, Mriganka M. (University of Tennessee/USA) 4(2):575
Gibbs, James T. (Battelle/USA) 4(1):287
Gidarakos, Evangelos (Battelle Ingenieurtechnik GmbH/Germany) 4(4):445; 4(4):451
Gilbert, Richard O. (Battelle-Pacific Northwest National Laboratory/USA) 4(5):357; 4(5):367
Gilcrease, Patrick C. (Colorado State University/USA) 4(5):169
Gillespie, Rick (Battelle/USA) 4(2):67
Gillette, W.K. (Beltsville Agricultural Research Center/USA) 4(4):65
Gillham, Robert (University of Waterloo/ Canada) 4(4):249
Girvin, Don C. (Battelle-Pacific Northwest National Laboratory/USA) 4(3):437
Glaser, John A. (U.S. EPA/USA) 4(2):85; 4(5):123; 4(5):131
Glass, David J. (D. Glass Associates Inc./USA) 4(4):51; 4(4):307
Glasser, Howard (Amoco Corporation/ USA) 4(1):103
Glucksman, Andrew (Westinghouse Savannah River Company/USA) 4(5):275
Godbout, Judith G. (École Polytechnique de Montréal/Canada) 4(5):393
Godfrey, J. M. (Integrated Science & Technology Inc./USA) 4(1):453; 4(4):393
Godsy, E. Michael (U.S. Geological Survey/USA) 4(5):547
Goetz, James E. (Fluor Daniel GTI, Inc./USA) 4(4):289
Gökçay, Celal Ferdi (Middle East Technical University/Turkey) 4(2):519
Golchin, J. (Iowa Department of Natural Resources/USA) 4(3):487
Goldflam, Rudy (Mission Research Corporation/USA) 4(1):259
Goltz, Mark N. (Air Force Institute of Technology/USA) 4(3):71
Gonzalez, Adrian M. (IT Corporation/USA) 4(3):393
Goodroad, Lewis (Waste Management, Inc./USA) 4(5):295
Gorby, Yuri A. (Battelle-Pacific Northwest National Laboratory/USA) 4(3):47

Gorder, Kyle (Utah State University /USA) 4(3):189
Gordon, Milton (University of Washington/USA) 4(3):321
Gorontzy, Thomas (Universität Oldenburg/Germany) 4(2):9; 4(2):39; 4(2):49
Gossard, R. (Maxim Technologies, Inc./ USA) 4(1):503
Gossett, James M. (Cornell University/ USA) 4(3):11; 4(3):23; 4(3):39; 4(3):61
Gostomski, Peter (Idaho National Engineering Laboratory/USA) 4(5):105
Goszczynski, Stefan (University of Idaho/USA) 4(2):505
Gottschal, Jan C. (University of Groningen/The Netherlands) 4(5):591
Gould, Timothy (Harding Lawson Associates/USA) 4(1):295
Gourdon, Remy P. (Institut National des Sciences Appliquées/France) 4(5):577
Govind, Rakesh (University of Cincinnati/USA) 4(2):195; 4(5):195
Grabinski, Carl (Joslyn Manufacturing Co./USA) 4(2):173
Graham, David W. (University of Kansas/USA) 4(2):117; 4(4):105
Graham, Lori (BioRenewal Technologies Inc./USA) 4(3):267
Granzow, Silke (University of Stuttgart/ Germany) 4(3):13
Grasmick, Dan (Fluor Daniel GTI/USA) 4(1):497
Gravel, M.J. (École Polytechnique de Montréal/Canada) 4(5):197
Graves, Anita E. (Parsons Engineering Science, Inc./USA) 4(3):141
Graves, Duane (IT Corp./USA) 4(1):461; 4(3):393; 4(3):221
Graves, Melody (Battelle/USA) 4(2):67
Graves, Robert (Parsons Engineering Science, Inc./USA) 4(3):141
Gray, Julian (Integrated Science & Technology, Inc./USA) 4(5):365
Gray, Murray R. (University of Alberta/Canada) 4(5):163
Green, Roger B. (Waste Management, Inc./USA) 4(1):487; 4(1):493
Greer, Charles W. (National Research Council of Canada/Canada) 4(2):3; 4(2):335; 4(2):457; 4(2):567; 4(5):393
Grimberg, Stefan J. (Clarkson University/USA) 4(2):587
Grishchenkov, Vladimir G. (Russian Academy of Sciences/Russia) 4(4):523
Gromicko, Gregory J. (Fluor Daniel GTI, Inc./USA) 4(3):439; 4(3):471
Grøn, Christian (Risø National Laboratory/Denmark) 4(4):85
Grotenhuis, J.T.C. (Wageningen Agricultural University/The Netherlands) 4(1):367; 4(5):643
Groudev, Stoyan N. (University of Sofia/Bulgaria) 4(3):409
Grow, Ann E. (Biopraxis/USA) 4(3):411
Grunenwald, Susan A. (Ciba-Geigy Corporation/USA) 4(2):289; 4(3):283
Grutters, Mark (TAUW Milieu/The Netherlands) 4(5):591
Guha, Saumyen (Princeton University/ USA) 4(2):557; 4(2):581
Guillerme, Michel (Elf Aquitaine/France) 4(4):411; 4(4):451
Guiot, Serge R. (National Research Council of Canada/Canada) 4(2):567; 4(4):197; 4(5):171
Guo, J. (University of Cincinnati/USA) 4(2):195
Gupta, Hari S. (Woodward-Clyde Consultants/USA) 4(3):229
Gupta, Munish (Parsons Engineering Science/USA) 4(3):391
Gupta, Neeraj (Battelle/USA) 4(1):7
Gusek, James (Knight Piesold, LLC/USA) 4(3):401
Guy, C. (École Polytechnique de Montréal/Canada) 4(5):197
Haas, Patrick E. (U.S. Air Force/USA) 4(1):273; 4(3):147
Haderlein, Stefan (Swiss Federal Inst for Env Sci & Tech/Switzerland) 4(2):15
Hadley, Paul (CA Dept of Toxic Substance Control/USA) 4(4):347; 4(4):323
Hage, André (TNO Inst of Environmental Sciences/The Netherlands) 4(3):3
Häggblom, Max M. (Rutgers The State Univ of New Jersey/USA) 4(2):293

Hägglund, Lars (FOA/National Defence Research Est/Sweden) 4(5):333
Haglund, Peter (Umeå University/Sweden) 4(5):333
Haikola, Bruce M. (Remediation Technologies Inc./USA) 4(4):391
Haines, John (U.S. EPA/USA) 4(4):359; 4(4):419
Halasz, Annamaria (National Research Council of Canada/Canada) 4(2):567
Halden, B. G. (University of Minnesota/USA) 4(4):519
Halden, Rolf U. (University of Minnesota/USA) 4(4):519
Hall, Barbara L. (Utah State University/USA) 4(2):329
Hallgarth, Michael R. (Texas A&M University/USA) 4(2):17
Hamblin, Gerard M. (WMF-Midwest/USA) 4(1):481
Hamed, Maged M. (Rice University/USA) 4(1):43; 4(1):75
Hamers, Paulus Henricus Johannes (University of Twente/The Netherlands) 4(5):565
Hamilton, W. (Axis Genetics Ltd/UK) 4(3):337
Hampton, Mark L. (U.S. Army Environmental Center/USA) 4(5):151
Hanashima, Masataka (Fukuoka University/Japan) 4(3):63
Hansen, Hannah Houmann (Technical University of Denmark/Denmark) 4(4):7
Hansen, Jerry (U.S. Air Force/USA) 4(3):147; 4(1):303; 4(3):191
Hanson, Greg (DuPont/USA) 4(3):289
Harayama, Shigeaki (Marine Biotechnology Institute, Co. Ltd./Japan) 4(4):371
Harder, Wim (TNO Inst of Environmental Sciences/The Netherlands) 4(3):3
Hardisty, Paul E. (Komex Clark Bond/UK) 4(1):443
Harkness, Mark R. (GE Corporate R&D Center/USA) 4(3):283; 4(3):289; 4(3):297
Harmsen, Joop (DLO Winand Staring Centre/The Netherlands) 4(2):143; 4(2):153
Harris, Ken (Stone Container Corp./USA) 4(1):209
Harris, William J. (University of Aberdeen/UK) 4(3):337
Harshman, Lawrence G. (University of Nebraska - Lincoln/USA) 4(3):421
Hart, A. (British Gas/UK) 4(2):531; 4(2):533
Hartten, Andrew S. (DuPont/USA) 4(2):137
Harvey, Steve (U.S. Army Corps of Engineers/USA) 4(5):157
Hasegawa, Takeshi (Ebara Research Co. Ltd./Japan) 4(5):31
Hashsham, Syed (Michigan State University/USA) 4(4):465
Haston, Zachary C. (Stanford University/USA) 4(3):31
Hater, Gary R. (Waste Management, Inc./USA) 4(1):481; 4(1):487; 4(1):493; 4(5):295
Hautzenberger, I. (Institute for Agrobiotechnology/AUSTRIA) 4(5):617
Havighorst, Mark B. (Oregon State University/USA) 4(2):455
Hawari, Jalal (National Research Council of Canada/Canada) 4(2):3; 4(2):567; 4(5):171; 4(5):393
Hayakawa, Toshio (Railway Technical Research Institute/Japan) 4(2):445
Hayes, Thomas (Gas Research Institute/USA) 4(3):451; 4(3):477; 4(3):487
Hazen, Terry C. (Westinghouse Savannah River Company/USA) 4(1):385; 4(5):85; 4(5):307; 4(5):315; 4(5):467
Hecox, Gary (IT Corp./USA) 4(3):221
Heersche, J.A.N.M. (BION Overijssel/The Netherlands) 4(5):565
Heilman, Paul (Washington State University/USA) 4(3):321
Heine, Robert L. (EFX Systems Inc./USA) 4(5):251
Heinicke, J. (Woodward-Clyde Consultants/USA) 4(4):549
Heislein, D. (ABB Environmental Services Inc./USA) 4(3):139
Heitkamp, Michael A. (Monsanto Company/USA) 4(3):281; 4(3):283; 4(3):289; 4(3):291

Henley, Mark Q. (Texaco Exploration & Production Inc./USA) 4(4):391
Henrysson, Tomas (Lund University/ Sweden) 4(2):535
Henssen, Maurice J.C. (Bioclear Environmental Biotechnology/Netherlands) 4(3):65
Herbert, Bruce (Texas A&M University/ USA) 4(5):629
Herbst, Julie (Navy Public Works Center/ USA) 4(1):455
Herlihy, Fran (Battelle/USA) 4(2):67
Hermann, P. (Frisby Technologies, Inc./USA) 4(4):133
Hernandez, Mark (University of Colorado/USA) 4(2):299
Herre, A. (Friedrich-Schiller-Universität Jena/Germany) 4(2):493
Herrington, R. Todd (Parsons Engineering Science/USA) 4(1):303
Herwig, Russell P. (University of Washington/USA) 4(4):427
Hess, Thomas F. (University of Idaho/ USA) 4(2):525
Hettenbach, Susan (University of Tulsa/ USA) 4(4):383
Heuckeroth, Deborah M. (Dames & Moore/USA) 4(2):303; 4(5):301
Hickey, Robert (EFX Systems, Inc./USA) 4(5):251; 4(3):451
Hickey, W. J. (University of Wisconsin-Madison/USA) 4(5):13
Hickman, Gary (CH2M Hill/USA) 4(3):149; 4(1):455
Hicks, John (Parsons Engineering Science, Inc./USA) 4(3):191
Hicks, Patrick (Regenesis/USA) 4(1):209
Hightower, Mike (Sandia National Laboratories/USA) 4(4):261
Hill, B.M. (U.S. EPA/USA) 4(5):19
Hill, Donald O. (Mississippi State University/USA) 4(5):157
Hill, Michael R. (General Physics Corporation/USA) 4(1):199
Hill, Steve R. (Coleman Research Corporation/USA) 4(4):347
Hillyer, M. (Chevron Research & Technology Co./USA) 4(4):277
Hinchee, Robert E. (Parsons Engineering Science/USA) 4(1):135; 4(1):453; 4(3):141
Hines, Robert D. (U.S. Filter/Envirex, Inc./USA) 4(5):43; 4(5):89
Hirl, Patrick J. (University of Illinois at Chicago/USA) 4(3):87
Ho, YiFong (U.S. EPA/USA) 4(2):249
Hoeppel, Ronald E. (U.S. Navy/USA) 4(1):7; 4(1):331; 4(1):403
Hoffman, Dennis J. (Battelle/USA) 4(2):321
Hofmann, Holger (UFZ Centre for Environmental Research/Germany) 4(2):185
Hogan, Eppie (University of Alaska-Anchorage/USA) 4(1):397
Holbrook, Matthew F. (Apex Environmental, Inc./USA) 4(5):483
Holder, Anthony W. (Rice University/ USA) 4(1):75
Holder, Edith L. (University of Cincinnati/USA) 4(4):359; 4(4):419
Holderness, Brian (Roy F. Weston, Inc./USA) 4(2):251
Hollander, David J. (Northwestern University/USA) 4(5):379
Hollowell, G.P. (Howard University/USA) 4(4):65
Holman, Hoi-Ying (Lawrence Berkeley National Lab/USA) 4(2):299; 4(1):81
Holmes, Marty (DuPont Co./USA) 4(3):353
Holroyd, Chris (British Nuclear Fuels, plc./UK) 4(3):393; 4(3):429
Honig, E. H. (Union Pacific Railroad/ USA) 4(4):549
Hooker, Brian S. (Battelle-Pacific Northwest National Laboratory/USA) 4(1):37; 4(3):9; 4(3):57; 4(4):273; 4(5):411; 4(5):561
Hoover, Darryl G. (Louisana State University/USA) 4(5):115
Hooyberghs, Litiane (Vlaamse Instelling voor Tech Onderzoek (VITO)/ Belgium) 4(4):159
Hopkins, Gary D. (Stanford University/ USA) 4(3):71; 4(5):9

Horst, Garald L. (University of Nebraska-Lincoln/USA) 4(4):101; 4(1):421
Hou, L.-H. (Howard University/USA) 4(4):65
Houtman, Simone (Novem b.v./The Netherlands) 4(5):431
Hsu, Francis (DuPont Co./USA) 4(4):17; 4(3):353
Huang, Su-Ying (National Chung Hsing University/R.O.C.-Taiwan) 4(3):51
Huang, Weilin (The University of Michigan/USA) 4(5):611
Hubbard, Perry (Integrated Science & Technology/USA) 4(1):321; 4(5):365
Hughes, Ed (Environmental Resources Management, Inc./USA) 4(1):275
Hughes, Joseph (Rice University/USA) 4(3):179
Hugler, Walter (Universidad A. Metropolitant-Iztapalapa/Mexico) 4(5):223
Hullman, Aaron (Integrated Science & Technology/USA) 4(1):453
Hundal, L. S. (University of Nebraska-Lincoln/USA) 4(2):287
Hunt, Craig S. (The University of Iowa/USA) 4(1):53
Hunt, James R. (University of California at Berkeley/USA) 4(2):299
Hunter, David W. (Manaaki Whenua Landcare Research New Zealand Limited/New Zealand) 4(2):219
Hunter, William J. (USDA/USA) 4(3):415
Hupe, Karsten (Technische Universität Hamburg-Harburg/Germany) 4(5):137
Hurme, Taina Susanna (Tampere University of Technology/Finland) 4(2):427
Hutchins, Stephen R. (U.S. EPA/USA) 4(4):19; 4(5):11; 4(5):19
Hutson, Karen (IT Corporation/USA) 4(3):221
Hyman, Michael R. (Oregon State University/USA) 4(2):387
Iio, Masatoshi (Raito Kogyo Co., Ltd./Japan) 4(4):495
Imamura, Shigeyuki (Asahi Chemical Industry Co./Japan) 4(4):59
Imamura, Takeshi (Canon, Inc./Japan) 4(4):495
Infante, Carmen (INTEVEP/Venezuela) 4(4):409
Irvine, Robert L. (University of Notre Dame/USA) 4(3):87
Irwin, Walter G. (South Carolina Electric & Gas Company/USA) 4(3):439
Ishihara, Masami (Marine Biotechnology Institute Co. Ltd./Japan) 4(4):371
Islam, Mesbah (Environmental Systems & Technologies, Inc./USA) 4(4):341
Istok, Jonathan D. (Oregon State University/USA) 4(2):387
Iwasa, Hiromu (Raito Kogyo Co., Ltd./Japan) 4(4):495
Jack, Thomas (Nova Husky Research Corp./Canada) 4(5):529
Jacob, Paulo R. (Quiminas/Brazil) 4(5):71
Jaffé, Peter R. (Princeton University/USA) 4(2):557; 4(2):581
Jagosz, Bogdan (Czechowice Oil Refinery/Poland) 4(5):467
Jakobsen, Rasmus (Technical University of Denmark/Denmark) 4(4):85
James, Mark R. (SES Environmental Services/USA) 4(1):213
Jankowski, Michael D. (Texas A&M University/USA) 4(2):17
Janssen, Dick B. (University of Groningen/The Netherlands) 4(2):597; 4(3):3; 4(3):73
Janzen, Paul (Manitoba Hydro/Canada) 4(1):319
Järvinen, Kimmo T. (Technical University of Denmark/Denmark) 4(2):465
Javanmardian, Minoo (Amoco Research Center/USA) 4(1):73
Jennings, Eleanor (University of Tulsa/USA) 4(4):383
Jennings, Olin (The Jennings Group, Inc./USA) 4(4):305
Jensen, Thomas M. (Parsons Engineering Science, Inc./USA) 4(3):141
Jerger, Douglas E. (OHM Remediation Services Corporation/USA) 4(1):245; 4(2):365; 4(2):603; 4(4):141; 4(5):411
Ji, C. (Rice University/USA) 4(3):179
Ji, Wei (University of Arizona/USA) 4(2):597

Jiménez, Blanca (Universidad Nacional Autonoma de Mexico) 4(2):447
Jin, Guang (Tulane University/USA) 4(3):25
Jin, Minghui (University of Virginia/USA) 4(4):573
Jin, Peikang (Global Remediation, Inc./USA) 4(5):233
Johnsen, Steen K. (S. Dyrup & Co. A/S/Denmark) 4(5):57
Johnson, Camdon T. (University of California Riverside/USA) 4(5):175
Johnson, Christian D. (Battelle-Pacific Northwest National Laboratory/USA) 4(3):9
Johnson, E.M. (Colorado State University/ USA) 4(5):89
Johnson, Jeffrey A. (Environmental Systems & Technologies/USA) 4(3):161
Johnson, Jeffrey G. (Gram, Inc./USA) 4(4):215
Johnson, Paul C. (Arizona State Univ./USA) 4(1):129; 4(1):135; 4(1):153
Johnson, Raymond H. (Applied Hydrology Associates Inc./USA) 4(3):233
Johnson, Richard (Oregon Graduate Institute/USA) 4(1):129; 4(1):135
Johnson, Sherri (U.S. Department of Energy/USA) 4(4):299; 4(5):225
Johnston, Carl G. (Mycotech Corp./USA) 4(2):537
Johnston, Colin (CSIRO/Australia) 4(1):339; 4(4):241
Johnstone, Donald L. (Washington State University/USA) 4(2):343; 4(3):399; 4(3):435; 4(4):97
Jones, Demetrius (Westinghouse Savannah River Company/USA) 4(3):345
Jones, D. Martin (University of Newcastle/UK) 4(4):401
Jongema, Hetty (TNO Inst of Environmental Sciences/The Netherlands) 4(3):3
Jordan, Daniel R. (Compliance Corporation/USA) 4(1):275
Joyner, William (U.S. Environmental Protection Agency/USA) 4(2):251
Julik, Joseph K. (Minnesota Pollution Control Agency/USA) 4(5):283
Jung, Carina (California State University/USA) 4(5):649
Kaake, Russell (J.R. Simplot Company/ USA) 4(2):61; 4(2):265; 4(2):291
Kaback, Dawn (Colorado Ctr for Environmental Management/USA) 4(4):323
Kabrick, Randolph M. (Remediation Technologies, Inc./USA) 4(4):391
Kalia, P. (University of Missouri-Columbia/USA) 4(3):127
Kampbell, Don H. (U.S. EPA/USA) 4(1):15; 4(1):347; 4(3):147; 4(3):191; 4(3):285; 4(3):299
Kanally, Robert (Cook College, Rutgers/ USA) 4(2):191
Kane, Allen C. (U.S. Geological Survey/ USA) 4(3):115
Karamanev, Dimitar (École Polytechnique de Montréal/Canada) 4(2):453; 4(4):221
Karkalik, Edward J. (Parsons Engineering Science, Inc./USA) 4(1):227
Karns, Jeffrey (U.S. Department of Agriculture/USA) 4(2):263
Karpov, Alexander V. (Russian Academy of Sciences/Russia) 4(4):523
Karscig, George (Occidental Chemical Corp./USA) 4(3):321
Kaslik, Peter J. (Oregon State University/ USA) 4(4):239
Kastner, James R. (Westinghouse Savannah River Company/USA) 4(1):385
Katamneni, S. (University of Idaho/USA) 4(2):525
Kathinokkula, Kumar (University of Missouri-Rolla/USA) 4(3):113
Katic, Dennis J. (University of Waterloo/ Canada) 4(4):255; 4(4):181
Kawabata, Yuji (Canon Research Center/ Japan) 4(4):495
Kawaguchi, Masahiro (Canon, Inc./Japan) 4(4):495
Kawka, Orest (University of Michigan/ USA) 4(4):507

Kayser, Kevin J. (Institute of Gas Technology/USA) 4(3):487
Kearney, Theresa (British Nuclear Fuels plc/UK) 4(3):429; 4(3):393
Keegan, James (Terra Vac Corp./USA) 4(5):505
Keener, William K. (Lockheed Martin Idaho Technologies Co./USA) 4(5):327
Kelley, Cheryl A. (Naval Research Laboratory/USA) 4(2):317; 4(4):297
Kelley, Robert L. (Institute of Gas Technology/USA) 4(2):439; 4(3):477
Kelly, K. Colleen (Michigan State University/USA) 4(4):575
Kemblowski, Marian W. (Utah State University/USA) 4(3):189
Kennedy, Jean (Phytokinetics/USA) 4(3):309
Kennett, Roger (Arizona Department of Environmental Quality/USA) 4(4):323
Kerch, Paul E. (BDM/USA) 4(2):407
Kershner, Mark (U.S. Air Force/USA) 4(1):215
Keuning, Sytze (Bioclear Environmental Biotechnology b.v./The Netherlands) 4(3):65; 4(5):29
Khaliullina, L. (Kazan State University/Russia) 4(2):33
Khanna, Sunil (Tata Energy & Resources Institute/USA) 4(4):407
Kiehlmann, Eberhard (Simon Fraser University/Canada) 4(2):499
Kiernan, Christian (University of Tennessee/USA) 4(4):317
Kilbane, John J. (Institute of Gas Technology/USA) 4(3):487
Kilkenny, S.T. (Santa Fe Pacific Pipeline Partners/USA) 4(5):89
Kim, Byung J. (U.S. Army Corps of Engineers/USA) 4(5):205
Kim, Byung R. (Ford Motor Co./USA) 4(5):205
Kim, Byungtae (Cook College/Rutgers University/USA) 4(3):187
Kim, Dong-Shik (University of Michigan/USA) 4(4):553
Kim, Jongo (University of Missouri-Rolla/USA) 4(5):211
Kim, Kwang-Soo (Jeonju University/Republic of Korea) 4(2):589
Kim, Sanggoo (Indiana University/USA) 4(4):41
Kim, Sang-Jong (Seoul National University/Republic of Korea) 4(2):239
Kim, Young (Oregon State University/USA) 4(3):107
Kimball, Darek O. (Utah State University/USA) 4(2):181
Kimbara, Kazuhide (Tokyo Institute of Technology/Japan) 4(2):445
Kimura, Toshiaki (Toyota Motor Corporation/Japan) 4(3):133
Kimura, Yosuke (University of Texas at Austin/USA) 4(5):605
King, Mark W.G. (University of Waterloo/Canada) 4(2):127
Kirtland, Brian (University of South Carolina/USA) 4(2):315
Kitamori, Shigeji (Fukuoka Institute of Health & Env Sciences/Japan) 4(3):63
Kittel, Jeffrey A. (Midwest Research Institute/USA) 4(1):273; 4(5):345
Klecka, Gary M. (Dow Chemical/USA) 4(3):281; 4(3):287; 4(3):289; 4(3):293
Kleijwegt, J. (Delft Geotechnics/The Netherlands) 4(1):329
Klemm, David Erik (S&ME, Inc./USA) 4(1):193
Klier, Nancy (Dow Chemical Co./USA) 4(3):287
Klingel, Eric (IEG Technologies Corp./USA) 4(4):149
Knackmuss, Hans-Joachim (Fraunhofer IGB/Germany) 4(2):1
Knecht, S. (Technische Universität Hamburg-Harburg/Germany) 4(5):137
Knight, Derek (Phytokinetics/USA) 4(3):309
Knowles, C. J. (University of Kent Canterbury/UK) 4(2):531; 4(2):533; 4(3):413
Knudsen, S. (Technical University of Denmark/Denmark) 4(2):15
Knutson, Ryan (University of North Dakota/USA) 4(5):269
Koenen, Brent A. (U.S. Army Corps of Engineers/USA) 4(1):297

Koenigsberg, Stephen (Regenesis Bioremediation Products/USA) 4(2):469; 4(4):247
Kolhatkar, Ravindra V. (University of Tulsa/USA) 4(1):73
Kollé, Jack J. (Hydropulse/USA) 4(5):541
Komisar, Simeon J. (Rensselaer Polytechnic Institute/USA) 4(3):331
Kondo, Ryuichiro (Kyushu University/Japan) 4(2):545
Koning, Michael (Technische Universität Hamburg-Harburg/Germany) 4(5):137
Korda, Athanassia (Oiko Technics Institute/Greece) 4(4):445; 4(4):451
Korotayeva, Elena V. (All-Russian Oil & Gas Research Institute (VNIIneft)/Russia) 4(4):529
Kosegi, Jeremy (University of Illinois Urbana-Champaign/USA) 4(4):135
Kosson, David S. (Rutgers University/USA) 4(5):495; 4(5):519
Kotun, Ronald J. (Fluor Daniel GTI, Inc./USA) 4(3):471
Kozaki, Shinya (Canon, Inc./Japan) 4(4):495
Kraatz, Matthias (Carnegie Mellon University/USA) 4(5):669
Kratzke, Robert J. (U.S. Navy/USA) 4(1):479; 4(1):511
Kraus, Jason (CH2M Hill Inc./USA) 4(1):123
Krauter, Paula (Lawrence Livermore National Lab/USA) 4(2):299
Kreye, William C. (Environmental Resources Management, Inc./USA) 4(1):275
Krieger-Brockett, Barbara (University of Washington/USA) 4(4):427
Krishna, Anand (Louisiana Technical University/USA) 4(1):141
Krishnan, E. Radha (IT Corporation/USA) 4(2):85; 4(5):131
Krishnan, Gopal (University of Nebraska-Lincoln/USA) 4(4):101
Krishnan, Srinivas (IT Corp./USA) 4(2):85
Krog, Marianne (Aarhus University/Denmark) 4(4):85
Kuch, David (U.S. Air Force/USA) 4(1):287; 4(5):481
Kuiper, John (AGRA Earth & Environmental Inc./USA) 4(3):241
Kulpa, Charles C. (University of Notre Dame/USA) 4(5):471
Kuntz, Charles S. (Dames & Moore/USA) 4(5):363
Kuriyama, Akira (Canon, Inc./Japan) 4(4):495
Kurz, Marc D. (University of North Dakota/USA) 4(5):113
Kusk, K. Ole (Technical University of Denmark/Denmark) 4(4):1
Kuykendall, L. David (U.S. Department of Agriculture/USA) 4(4):65
Kvetensky, Joee (University of Missouri-Rolla/USA) 4(3):113
Kwon, Gi-Seok (Korea Research Inst of Bioscience & Biotech/Republic of Korea) 4(2):467
Labelle, Suzanne (National Research Council of Canada/Canada) 4(5):393
Lackey, Laura (Tennessee Valley Authority/USA) 4(5):189
Laha, Shonali (Florida International University/USA) 4(2):595
Lal, Banwari (Tata Energy & Research Institute/INDIA) 4(4):407
Lamb, Steven R. (GZA GeoEnvironmental Inc./USA) 4(2):393
Lambertz, D. A. (Chevron Research & Technology Co./USA) 4(2):69
Lampe, David G. (University of Nebraska - Lincoln/USA) 4(3):423
Lang, John R. (University of Michigan/USA) 4(1):323
Lapat-Polasko, Laurie T. (Woodward-Clyde Consultants/USA) 4(2):469; 4(3):125; 4(4):275; 4(4):549
Larsen, M. (Remediation Technologies, Inc. (RETEC)/USA) 4(2):113
Larson, John R. (Dames & Moore/USA) 4(3):167
Larson, Steven L. (U.S. Army Corps of Engineers/USA) 4(3):301
Laubacher, Richard C. (BP Oil Co./USA) 4(5):43

Launen, Loren (Simon Fraser University/ Canada) 4(2):499
Lawlor, Kevin (University of Tulsa/USA) 4(4):383
Lawrence, Christopher A. (CH2M Hill/ USA) 4(3):213
Lawrence, William J. (DuPont/USA) 4(2):137
Lawson, Peter W. (CH2M Hill Inc./USA) 4(1):61
Leahy, Maureen C. (Fluor Daniel GTI, Inc./USA) 4(2):245; 4(3):445; 4(3):463; 4(3):479; 4(4):457
Leandri, Daniel C. (Rust Environment & Infrastructure Inc./WMX Biopit/USA) 4(1):379
Lear, Paul (OHM Corporation/USA) 4(2):365
Learmonth, D. (University of Aberdeen/ UK) 4(3):337
Leavitt, Maureen (Scientific Applications International Corp. (SAIC)/USA) 4(4):169
Lebron, Carmen (Naval Facilities Engineering Service Center/USA) 4(5):9
Lecomte, Paul (BRGM/CNRSSP/France) 4(5):405; 4(5):465
Lee, Brady D. (Idaho National Engineering Laboratory/USA) 4(4):483
Lee, Chi-Mei (Natl Chung Hsing University/R.O.C.-Taiwan) 4(2):433
Lee, Eui Sin (Daewoo Corporation/USA) 4(4):477
Lee, Kenneth (Fisheries & Oceans Canada/Canada) 4(4):365; 4(4):401
Lee, Kwang P. (Samsung Corporation/ Republic of Korea) 4(2):233
Lee, Michael D. (DuPont Co./USA) 4(2):137; 4(3):21; 4(3):281; 4(3):283; 4(3):289; 4(3):295; 4(3):299
Lee, Ming-Kuo (Auburn University/USA) 4(3):379; 4(5):583
Lee, Pak-Hing (Iowa State University/USA) 4(2):281
Lee, Sung-Gie (Korea Research Inst of Bioscience & Biotech/Republic of Korea) 4(2):467
Leeson, Andrea (Battelle/USA) 4(1):129; 4(1):135; 4(1):287; 4(5):481
Leethem, John T. (ERM-Southwest, Inc,/USA) 4(3):167
Le Floch, Stéphane (Centre de Documentation de Recherches/France) 4(4):411
Lehman, R.M. (INEL/USA) 4(4):111
Lehman, Roy L. (Texas A&M University Corpus Christi/USA) 4(4):119
Lehmicke, Leo L. (Beak Consultants/ USA) 4(3):203; 4(3):205
Leibel, Stan (Waste Management Inc./USA) 4(1):493
Leigh, Daniel (OHM Remediation Services Corp./USA) 4(2):603; 4(4):141
Leigh, Mary Beth (University of Oklahoma/USA) 4(2):115
Leins, Christoph (IEG Technologie GmbH/Germany) 4(2):353
Leland, Tom (Utah State University/USA) 4(3):317
Lendvay, John (University of Michigan/ USA) 4(3):173
Lenke, Hiltrud (Fraunhofer IGB/Germany) 4(2):1
Leoshina, L. G. (Institute of Bioorg Chem & Petrochem/Ukraine) 4(3):325
Lepo, Joe E. (University of West Florida/ USA) 4(4):401
Le Roux, Françoise (Institut Français du Pétrole/France) 4(5):213
Lesage, Suzanne (National Water Research Institute/Canada) 4(2):185; 4(4):465
Leskovšek, Hermina (University of Ljubljana/Slovenia) 4(2):231
Le Thiez, Pierre (Institut Français du Pétrole/France) 4(2):559
Lettinga, Gatze (Agricultural University Wageningen/The Netherlands) 4(2):31
Leung, K. (University of Tennessee/USA) 4(5):319
Leung, S.Y. (University of Missouri-Columbia/USA) 4(2):375; 4(3):85
Leuschner, A. (Remediation Technologies, Inc. (RETEC)/USA) 4(2):301
Leuteritz, D. (Remediation Technologies Inc./USA) 4(2):301
Levetin, Estelle (University of Tulsa/USA) 4(4):383

Lewandowski, Gordon A. (New Jersey Institute of Technology/USA) 4(5):577
Lewis, Ronald F. (U.S. EPA/USA) 4(2):603; 4(4):141
Lewis, Thomas A. (University of Idaho/ USA) 4(2):23; 4(2):525
Li, Dong X. (Unocal Corporation/USA) 4(1):373
Li, Shu-mei W. (Battelle-Pacific Northwest National Laboratory/USA) 4(5):345
Li, Wan-Chi (Environment Canada/ Canada) 4(2):185
Li, Xiaomei (Alberta Environmental Center/Canada) 4(4):35; 4(5):599
Li, Zhengming (University of Nebraska-Lincoln/USA) 4(4):485
Libelo, E. L. (U.S. Air Force/USA) 4(1):23
Ligé, Joy E. (EA Engineering Science & Technology, Inc./USA) 4(2):125
Lima, Luiz Mário Queiroz (LM Tratamento de Residuos LTDA/Brazil) 4(2):419
Lin, Y. (EMG Inc./USA) 4(5):289
Lipp, Brian A. (Battelle/USA) 4(2):321
Litherland, Susan Tighe (Roy F. Weston, Inc./USA) 4(5):425; 4(5):507
Liu, Dickson (Environment Canada/ Canada) 4(2):185
Lloyd-Jones, Gareth (Manaaki Whenua Landcare Research New Zealand Limited/New Zealand) 4(2):219
Loehr, Raymond C. (The University of Texas at Austin/USA) 4(2):159; 4(5):603; 4(5):605
Logan, Bruce E. (The University of Arizona/USA) 4(4):567
Loibner, Andreas P. (University for Agricultural Sciences/AUSTRIA) 4(5):617
Lombard, Kenneth H. (Westinghouse Savannah River Co./USA) 4(1):385; 4(5):275; 4(5):315; 4(5):467
Longstaff, M. (Axis Genetics Ltd/UK) 4(3):337
López-Mercado, V. (Ctr for Advanced Studies & Rsrch/Mexico) 4(3):273
Lorah, Michelle M. (U.S. Geological Survey/USA) 4(3):207
Lotrario, Joseph B. (Oregon State University/USA) 4(2):41
Lu, Chih-Jen (National Chung Hsing University/R.O.C.-Taiwan) 4(2):433; 4(3):51
Lubbers, Renate (TAUW Milieu/The Netherlands) 4(5):431
Ludwig, Maria (FZMB e.V./Germany) 4(5):259
Lummus, Stanford (S&ME, Inc./USA) 4(1):193
Lunardini, Robert (ABB Environmental Services, Inc./USA) 4(3):231
Lundgren, Tommy S. (SAKAB/Sweden) 4(1):487
Lundh, Tore (NAT/Norway) 4(4):411
Lunel, Tim (AEA Technology plc/UK) 4(4):365
Lunt, Anne (Safety Kleen Corporation/ USA) 4(4):193
Lüth, Joachim C. (Technical University of Hamburg-Harburg/Germany) 4(5):137
Luthhardt, Walter (ReFIT e.V./Germany) 4(5):259
Luthy, Richard G. (Carnegie Mellon University/USA) 4(5):669
Lutz, Edward (DuPont Environmental/ USA) 4(3):279; 4(3):287; 4(3):289
Lutz, R.L. (Mycotech Corp./USA) 4(2):537
Lutze, Werner (University of New Mexico/USA) 4(1):435
Lybrand, Mike S. (Eco-Systems/USA) 4(2):97
MacFarlane, Ian D. (EA Engineering Science & Technology/USA) 4(2):125; 4(3):485
Maciey, Anthony (CH2M Hill Inc./USA) 4(1):123
MacIntyre, W. G. (College of William and Mary (VIMS)/USA) 4(1):23
Mackay, Douglas (University of Waterloo/ USA) 4(4):209; 4(4):187
Macnaughton, Sarah J. (Microbial Insights Inc./USA) 4(5):319
Madden, Patrick C. (Exxon Research & Engineering Co./USA) 4(2):205

Madura, Richard (Tulane University/USA) 4(5):277
Magar, Victor S. (Battelle/USA) 4(3):39
Mage, Roland (Battelle Europe/Switzerland) 4(1):109
Maheux, Pierre J. (Jacques Whitford Environment Limited/Canada) 4(1):221
Mahieu, Benoit (Institut Français du Pétrole/France) 4(5):213
Mailloux, Michael P. (Chevron U.S.A. Products Company/USA) 4(1):429
Major, David W. (Beak Consultants/Canada) 4(3):203; 4(3):205; 4(3):287
Major, William R. (U.S. Navy/USA) 4(1):479; 4(1):497; 4(1):511
Maki, Hideaki (Marine Biotechnology Institute Co., Ltd./Japan) 4(4):371
Malachowska-Jutsz, Anna (Silesian Technical University/Poland) 4(4):43
Malina, Grzegorz (Technical University of Czestochowa/Poland) 4(1):367
Malloy, Jim (IT Corporation/USA) 4(3):227
Maloney, Stephen W. (U.S. Army Corps of Engineers/USA) 4(5):251
Malusis, Michael A. (Colorado State University/USA) 4(4):559
Man, Alex (Morrow Environmental Consultants, Inc./Canada) 4(1):319
Mandel, Dilip (New Jersey Institute of Technology/USA) 4(5):577
Manning, John F. (Argonne National Laboratory/USA) 4(2):59; 4(5):471
Männistö, Minna K. (Tampere University of Technology/Finland) 4(2):421
Manuel, Michelle F. (National Research Council of Canada/Canada) 4(4):197
Maraqa, Munjed (Michigan State University/USA) 4(5):657
Marchand, Ed (U.S. Air Force/USA) 4(1):341
Marchetti, Gabrio (Ambiente S.p.A./Italy) 4(1):423
Marchi, Daniela D. (Campinas State University /Brazil) 4(4):91
Marcott, Keith (Safety Kleen Corporation/USA) 4(4):193
Marczely, D. W. (EMG Inc./USA) 4(5):289
Marlow, Hal (Hart Crowser Inc./USA) 4(5):473
Marshall, Gary E. (Valero Refining Company/USA) 4(5):173
Marshall, Timothy R. (Woodward-Clyde Consultants/USA) 4(3):229
Martin, Robert (University of Birmingham/UK) 4(4):233
Martinovich, Betty (Polytechnic University/USA) 4(5):469
Martins dos Santos, Vitor A.P. (CSIC, Estacion Experimental del Zaidin/Spain) 4(4):57
Masson, C. (National Research Council of Canada/Canada) 4(2):3
Massry, Ihab W. (Rutgers University/USA) 4(5):495
Matamala, Antonio (Wageningen Agricultural University/The Netherlands) 4(2):31
Matherne, Carla (IT Corporation/USA) 4(3):221
Matheson, V. Grace (Michigan State University/USA) 4(4):501
Matrubutham, Uday (University of Tennessee/USA) 4(4):77
Matsueda, Takahiko (Fukuoka Institute of Health & Env Sciences/Japan) 4(2):545
Matsufuji, Yasushi (Fukuoka University/Japan) 4(3):63
Matteau, Yanick (École Polytechnique de Montréal/Canada) 4(5):199
Maurer, David (University of Michigan/USA) 4(4):553
Maurice, Robert D. (University of Manitoba/Canada) 4(4):21
Maybach, Gerry B. (Electric Power Research Institute/USA) 4(3):463
Maymó-Gatell, Xavier (Cornell University/USA) 4(3):23
Mayotte, Timothy (Golder Associates/USA) 4(4):507
Mazur, Margaret (Connolly Environmental/Australia) 4(2):151
McCarthy, Kevin (Battelle Ocean Sciences/USA) 4(1):505

McCartney, Daryl (The University of Manitoba/Canada) 4(1):319
McCarty, Perry L. (Stanford Univ./USA) 4(3):1; 4(3):31; 4(3):71; 4(4):501
McCauley, Paul T. (U.S. EPA/USA) 4(5):131
McCleary, Gloria (EA Engineering, Science, & Technology, Inc./USA) 4(1):251; 4(3):485
McCormick, M. L. (University of Michigan/USA) 4(3):173
McCutcheon, Steven (U.S. EPA/USA) 4(3):301
McDermott, Ray (U.S. Army/USA) 4(1):199
McDonald, Thomas J. (Texas A&M University/USA) 4(2):213; 4(2):569; 4(3):355; 4(5):629
McFarland, Beverly L. (Chevron Research & Technology Company/USA) 4(2):69; 4(4):277
McIntyre, Terry (Environment Canada/Canada) 4(4):329
McKay, Daniel (U.S. Army Corps of Engineers/USA) 4(1):235; 4(1):309
McKee, Ronald C.E. (O'Connor Associates/Canada) 4(1):221
Mckenzie, David E. (Monsanto Company/USA) 4(3):291
McKinney, Daene C. (University of Texas at Austin/USA) 4(2):565; 4(5):597
McLean, J. E. (Utah State University/USA) 4(3):189
McMaster, Michaye (Beak International/Canada) 4(4):181; 4(4):249; 4(4):255
McMillan, Nancy (Battelle/USA) 4(5):349
McMillen, Sara J. (Chevron Research & Technology Company/USA) 4(2):69
McNally, Tom (ChemFree Corp./USA) 4(3):373
McNeil, Thomas K. (EMS/USA) 4(2):245
McPhee, Wayne (Fluor Daniel GTI/Canada) 4(1):153; 4(1):243
McWhorter, David (Colorado State University/USA) 4(1):135
Mechaber, Richard A. (GEI Consultants, Inc./USA) 4(3):203
Medearis, Sarah (Ogden Environmental & Energy Srvcs, Inc./USA) 4(1):497
Medina, Victor F. (U.S. EPA/USA) 4(3):301
Medvedeva, T. V. (Inst. Bioorganic Chem. & Petrochem./Ukraine) 4(3):325
Meganatha, Shankar G. (University of Nebraska - Lincoln/USA) 4(5):55
Meier, H.C. (Applied Hydrology Associates, Inc./USA) 4(1):413
Meier, Karl (Beling Consultants, Inc./USA) 4(5):457
Meier-Löhr, Matthias (TGU-GmbH/Germany) 4(5):7; 4(5):27
Melchor, A. (INTEVEP/Venezuela) 4(4):409
Melin, Esa S. (Tampere University of Technology/Finland) 4(2):465
Mendoza, Rodolfo E. (CEVEG/Argentina) 4(2):157
Meo, III, Dominic (CH2M Hill/USA) 4(4):277
Messmer, Mindi F. (Innovative Engineering Solutions Inc./USA) 4(4):193
Metzger, Lowell W. (OHM Remediation Services Corp./USA) 4(1):447
Metzinger, Charles S. (EMCON, Inc./USA) 4(1):89
Meulenberg, Rogier (TNO-MEP/The Netherlands) 4(2):487
Meyers, J. D. (Montana State University/USA) 4(5):385
Michels, Jochen (Friedrich-Schiller-Universität Jena/Germany) 4(2):493
Mickelson, George (WI Dept of Natural Resources/USA) 4(1):103
Miesner, Elizabeth (ENVIRON Corp./USA) 4(3):213
Miksch, Korneliusz (Silesian Technical University/Poland) 4(4):43
Miles, David (University of Kansas/USA) 4(2):117
Millar, Kelly (National Water Research Institute/Canada) 4(2):185; 4(4):465
Miller, Aaron (ASARCO, Inc./USA) 4(3):401
Miller, Dennis E. (U.S. EPA/USA) 4(5):11
Miller, Jerry C. (Arco Chemical Co./USA) 4(1):413
Miller, Michael E. (Camp Dresser & McKee/USA) 4(5):89

Miller, Paul J. (EA Engineering, Science, & Technology Inc./USA) 4(3):485
Miller, R. Scott (AGRA Earth & Environmental, Inc./USA) 4(1):283
Miller, Rod A. (OHM Remediation Services Corp./USA) 4(1):447
Miller, Suzanne (Clarkson University/USA) 4(2):587
Millette, Denis (École Polytechnique de Montréal/Canada) 4(2):129
Milligan, Peter W. (Rutgers The State Univ of New Jersey/USA) 4(2):293
Mills, John (Oklahoma City Air Logistics Center/USA) 4(1):287; 4(2):321
Minsker, Barbara Spang (University of Illinois/USA) 4(4):135; 4(4):353
Miralles-Wilhelm, Fernando (Northeastern University/USA) 4(5):601
Mitchell, David J. (AEA Technology plc/UK) 4(4):401
Mixter, Phillip (Washington State University/USA) 4(2):343
Miyahara, Takashi (Tohoku University/Japan) 4(3):391
Mocoroa, J. (University of Groningen/The Netherlands) 4(3):73
Moison, Edwin (TNO-MEP/The Netherlands) 4(2):487
Molnaa, Barry (Fluor Daniel GTI/USA) 4(1):495
Monin, Nicole (Institut Français du Pétrole/France) 4(2):559
Montgomery, Michael T. (Geo-Centers, Inc./USA) 4(2):399; 4(4):125; 4(4):297
Montney, Paul A. (Georgia-Pacific Corporation/USA) 4(1):321
Moore, Brent J. (Komex International Ltd./Canada) 4(1):443
Moore, Margo Marie (Simon Fraser University/Canada) 4(2):499
Morales, M. (Universidad A. Metropolitant-Iztapalapa/Mexico) 4(5):223
Moreno, Terry (CH2M Hill/USA) 4(2):73
Mormile, Melanie R. (Associated Western Universities/USA) 4(5):265
Morris, David (Southern Co. Services Inc./USA) 4(1):461
Morris, Henry (Foster Wheeler Environmental Corp./USA) 4(5):439
Morris, Pamela J. (Medical Univ of South Carolina/USA) 4(2):213; 4(4):79; 4(5):629
Morrison, Jeffrey M. (CH2M Hill/USA) 4(1):455
Morse, William R. (Sun Company, Inc./USA) 4(4):147
Moryama, A.S. (Universidade Federal de São Carlos/Brazil) 4(2):247
Mosbæk, Hans (Technical University of Denmark/Denmark) 4(2):15
Moser, Lori E. (Zeneca Corp SPEL/Canada) 4(3):285; 4(3):299
Mosley, S.M. (Ciba-Geigy Corporation/USA) 4(3):283
Mosteller, Douglas C. (Camp Dresser McKee Inc./USA) 4(4):513; 4(4):559; 4(5):89; 4(5):315
Mott-Smith, Ernest (Fluor Daniel GTI/USA) 4(1):243
Mountain, Stewart A. (Integrated Science & Technology Inc./USA) 4(1):321; 4(1):453; 4(4):393
Moutoux, Dave (Parsons Engineering Science/USA) 4(3):147; 4(3):191
Muehlberger, Eric W. (Geomatrix Consultants, Inc./USA) 4(1):209
Muehleck, Joe (EA Engineering, Science, & Technology, Inc./USA) 4(1):251
Mueller, C. (Colorado State University/USA) 4(2):319
Mueller, James G. (SBP Technologies Inc./USA) 4(2):249; 4(2):353; 4(4):125; 4(4):149
Mueller, John (U.S. Air Force/USA) 4(1):361
Muiznieks, Indulis A. (Washington State University/USA) 4(3):321
Mukerjee Dhar, G. (Railway Technical Research Institute/Japan) 4(2):445
Mukherji, Suparna (University of Michigan/USA) 4(5):635
Mulder, Hendrikus (Wageningen Agricultural University/Netherlands) 4(5):643
Munakata-Marr, Junko (Colorado School of Mines/USA) 4(4):501

Murali, Dev (General Physics Corporation/USA) 4(1):199
Murphy, Vincent (Colorado State University/USA) 4(5):169
Murray, Richard P. (In-Situ Fixation Inc./USA) 4(4):541
Murray, Willard (ABB Environmental Srvcs, Inc./USA) 4(3):253
Murygina, Valentina P. (Russian Academy of Sciences/Russia) 4(4):529
Musser, Dannie (Alliance Technology/USA) 4(5):251
Naber, Steve J. (Battelle/USA) 4(5):349
Nagafuchi, Yoshitaka (Fukuoka Institute of Health & Env Sciences/Japan) 4(3):63
Naumova, Rimma P. (Kazan State University/Russia) 4(2):33; 4(5):231
Ndon, Udeme James (San Jose State University/USA) 4(5):77
Nelson, Christopher H. (Fluor Daniel GTI/USA) 4(1):233; 4(3):457; 4(3):479; 4(4):289; 4(4):457
Nelson, G.L. (Sam) (MidAmerica Energy/USA) 4(3):457; 4(3):487
Nelson, James D. (Louisiana Technical University/USA) 4(1):141
Nelson, Sheldon (Chevron Research & Technology Co./USA) 4(1):123
Newberg, Scott (Stanford University/USA) 4(3):1
Newell, C. (Axis Genetics Ltd/UK) 4(3):337
Newman, Lee A. (University of Washington/USA) 4(3):321
Newman, William A. (Delta Environmental Consultants, Inc./USA) 4(1):1; 4(5):283
Nielsen, Peter B. (Kruger A/S/Denmark) 4(5):57
Nieman, J. Karl C. (Utah State University/USA) 4(2):181
Nies, Loring (Purdue University/USA) 4(5):623
Nitz, David C. (University of North Carolina at Chapel Hill/USA) 4(3):469
Nobles-Harris, Ellen (Mobil Oil Corp./USA) 4(1):181
Nojiri, Hideaki (University of Tokyo/Japan) 4(4):59
Nolen, C. Hunter (Camp Dresser & McKee/USA) 4(1):311
Noll, Mark (Applied Research Associates/USA) 4(2):407
Noordman, Wouter H. (University of Groningen/The Netherlands) 4(2):597
Nordrum, S. B. (Chevron Research & Technology Co./USA) 4(2):69
Norris, Robert D. (Eckenfelder, Inc./USA) 4(1):147; 4(1):445; 4(4):175
Novak, Paige J. (University of Iowa/USA) 4(4):463
Nowak, Joseph (New Jersey Dept of Environ Protection/USA) 4(5):419
Numata, Koichi (Toyota Motor Corporation/Japan) 4(3):133
Nuttall, Eric (University of New Mexico/USA) 4(1):435
Nyholm, N. (Technical University of Denmark/Denmark) 4(4):1
Obraztsova, A. Ya. (University of California/USA) 4(3):411
O'Brien, Robert (Battelle-Pacific Northwest National Laboratory/USA) 4(5):357; 4(5):367
O'Cleirigh, Declan (Roy F. Weston, Inc./USA) 4(5):507
O'Connell, Sean P. (Idaho State University/USA) 4(4):111
O'Connor, Laurel E. (University of Tennessee/USA) 4(5):103
Oda, Yasushi (Toyota Motor Corporation/Japan) 4(3):133
Odell, Karen (ABB Environmental Services/USA) 4(3):139; 4(3):253
Odencrantz, Joseph (Tri-S Environmental/USA) 4(4):215
Odom, J Martin (DuPont Co./USA) 4(3):281; 4(3):283; 4(3):287
O'Flanagan, Barry (Delta Environmental Consultants Inc./USA) 4(5):283
Oh, Hee-Mock (Korea Research Inst of Bioscience & Biotech/Republic of Korea) 4(2):467
Oh, Keun-Chan (Rutgers University/Cook College/USA) 4(3):187

Okamura, Yukio (Toyota Motor Corporation/Japan) 4(3):133
Olie, J. J. (Delft Geotechnics/The Netherlands) 4(1):241; 4(3):239
Olsen, Lisa D. (U.S. Deparment of the Interior/USA) 4(3):207
Olsen, Roger (Camp Dresser & McKee Inc./USA) 4(4):513
Olson, Gail (Idaho National Engineering Laboratory/USA) 4(5):105
Olson, Paul (University of Oklahoma/USA) 4(2):115
Olstad, Gunnar (University of Oslo/Norway) 4(5):267
O'Mara, Mary Katherine (U.S. Army Corps of Engineers/USA) 4(4):289
Omori, Toshio (The University of Tokyo/Japan) 4(4):59
Ong, Say Kee (Iowa State University/USA) 4(1):187; 4(2):281
Oolman, Timothy (Radian Corporation/USA) 4(5):173
Oram, Douglas (EA Engineering Science & Technology Inc./USA) 4(1):251
O'Reilly, John V. (Monsanto/USA) 4(3):283; 4(3):291
O'Reilly, Kirk T. (Chevron Research & Technology Co./USA) 4(2):73; 4(2):387; 4(3):149; 4(5):145
Orr, Michael (J.R. Simplot Company/ USA) 4(2):291
Orth, Robert G. (Monsanto Company/USA) 4(3):283; 4(3):291
Ortiz, Enrique (Carnegie Mellon University/USA) 4(5):669
Otte, Marie-Paule (National Research Council/Canada) 4(2):457
Ouyang, Ying (Computer Data Services, Inc./USA) 4(5):11; 4(5):19
Owens, Victor (Foster Wheeler Environmental/USA) 4(4):393
Oya, Shunji (University of Illinois/USA) 4(5):571
Pacha, Jerzy (Silesian Technical University/Poland) 4(4):43
Page, Cheryl A. (Texas A&M University/USA) 4(2):569
Pal, Nirupam (California Polytechnic State University/USA) 4(2):511
Palumbo, Anthony V. (Oak Ridge National Laboratory/USA) 4(3):281; 4(3):295; 4(4):77; 4(4):117; 4(5):347
Pardieck, Daniel L. (Ciba Geigy/USA) 4(3):279; 4(3):283; 4(3):289; 4(5):641
Park, Jayne (Rensselaer Polytechnic Institute/USA) 4(3):331
Park, Jong Moon (Pohang University of Science & Technology/Republic of Korea) 4(2):233
Park, Keeyong (Cook College/Rutgers University/USA) 4(1):17
Parker, Jack C. (Environmental Systems & Technologies, Inc./USA) 4(4):341
Parker, Joel (The Traverse Group Inc./ USA) 4(3):219
Parkin, Gene F. (University of Iowa/USA) 4(3):15; 4(4):463
Parsons, Eric (Jacques Whitford Environment Limited/Canada) 4(1):277
Paszczynski, Andrzej (University of Idaho/USA) 4(2):505
Paterek, James R. (Institute of Gas Technology/USA) 4(3):477
Patterson, Bradley M. (CSIRO/Australia) 4(1):339; 4(2):53
Paul, M. (Technische Universität Hamburg-Harburg/Germany) 4(5):137
Payne, C. (University of Tennessee/USA) 4(3):295
Payne, Craig A. (Battelle/USA) 4(1):7; 4(1):331
Payne, Edward (Mobil Oil Corp./USA) 4(1):181
Peccia, Jordan L. (Montana State University/USA) 4(5):387
Pekarek, Susan (Kansas State University/USA) 4(3):305
Penmetsa, Ravi K. (Eco-Systems/USA) 4(2):97
Peramaki, Matthew P. (Leggette, Brashears & Graham, Inc./USA) 4(2):103
Percy, Bonnie S. (Maxim Technologies/ USA) 4(1):503
Perez, Alain (ALF Aquitaine Production/France) 4(2):559

Pérez, F. (Universidad A. Metropolitant-Iztapalapa/Mexico) 4(5):223
Pergrin, David E. (EA Engineering, Science, & Technology/USA) 4(3):485
Perina, T. (IT Corporation/USA) 4(1):83
Perkins, Richard E. (DuPont Co./USA) 4(5):399
Persiani, Luciano (Ambiente S.p.A./Italy) 4(1):423
Peters, Catherine A. (Princeton University/USA) 4(2):193; 4(2):557
Petersen, James N. (Washington State University/USA) 4(1):37; 4(2):343; 4(3):399; 4(3):435; 4(4):97; 4(4):157
Peterson, David M. (Fluor Daniel GTI/USA) 4(3):457
Peterson, K. J. (University of Nebraska - Lincoln/USA) 4(3):421
Peterson, Leonid R. (All-Russian Oil & Gas Research Institute (VNIIneft)/Russia) 4(4):529
Petkovsky, P. D. (Shell Oil Product Company/USA) 4(5):413
Petrovskis, Erik A. (McNamee, Porter & Seeley, Inc./USA) 4(2):149
Peyton, Brent M. (Battelle-Pacific Northwest National Laboratory/USA) 4(1):287; 4(3):9; 4(3):49; 4(4):273; 4(5):265; 4(5):325
Pfiffner, Susan M. (Oak Ridge Institute for Science & Education/USA) 4(3):281; 4(3):295; 4(4):117
Phelps, Tommy J. (Oak Ridge National Laboratory/USA) 4(3):295; 4(5):103
Philbrick, Steve (University of North Dakota/USA) 4(5):269
Phillips, Peter (Clark Atlanta University/USA) 4(3):373
Phipps, Donald W. (Orange County Water District/USA) 4(5):563
Picard, M.P. (CDM Engineers & Constructors/USA) 4(5):89
Picardal, Flynn W. (Indiana University/USA) 4(4):41
Pickard, Michael A. (University of Alberta/Canada) 4(5):163
Pickering, Edward W. (GZA GeoEnvironmental Inc./USA) 4(2):393
Pickering, Ingrid J. (Stanford Synchrotron Radiation Laboratory/USA) 4(2):381
Pier, Paul A. (Tennessee Valley Authority/USA) 4(3):315
Pierdinock, Michael J. (RAM Environmental, LLC/USA) 4(1):285
Pijls, C.G.J.M. (TAUW Milieu/The Netherlands) 4(1):241
Pinizzotto, Sam (Mobil Oil Corp./USA) 4(1):181
Pinto, Linda (Simon Fraser University/Canada) 4(2):499
Pittman, Suzie (AEA Technology plc/UK) 4(5):83
Place, Matthew C. (Battelle/USA) 4(1):511
Plaehn, William A. (Michigan State University/USA) 4(3):361
Poeton, Tom (University of Washington/USA) 4(4):425
Poggi-Varaldo, Héctor M. (P3 Consulting Engineers/Mexico) 4(3):273
Pollack, Albert J. (Battelle/USA) 4(1):403; 4(2):321
Pon, George W. (Oregon State University/USA) 4(3):247
Pope, Gary (University of Texas at Austin/USA) 4(2):565
Pope, Jeffery L. (Clayton Environmental Consultants, Inc./USA) 4(5):527
Porta, Augusto (Battelle Europe/Switzerland) 4(1):467; 4(1):505
Portal, René (Total Austral S.A./Argentina) 4(2):157
Porter, Andrew J. (University of Aberdeen/UK) 4(3):337
Porter, Regina S. (Southeastern Technology Center/USA) 4(4):299; 4(5):225
Portier, Ralph J. (Louisiana State Univ./USA) 4(4):543; 4(5):115; 4(5):535
Potter, Carl L. (U.S. EPA/USA) 4(2):85
Power, Terry R. (CSIRO Division of Water Resources/Australia) 4(2):53
Pradhan, Salil (Institute of Gas Technology/USA) 4(3):477
Prince, Roger (Exxon Research & Engineering/USA) 4(2):175; 4(2):205; 4(2):381; 4(4):487

Principe, Jan Marie (GE Corporate R&D Center/USA) 4(3):45
Pritchard, P. H. (U.S. Naval Research Laboratory/USA) 4(2):249
Protzman, Roger (Iowa State University/USA) 4(2):281
Puhakka, Jaakko A. (Tampere University of Technology/Finland) 4(2):421; 4(2):427; 4(2):465
Putscher, Andrea (Camp Dresser & McKee, Inc./USA) 4(5):469
Qiu, Xiujin (Utah State University/USA) 4(3):317
Qu, Mingbo (Tulane University/USA) 4(5):239
Quinton, Gary (DuPont Corporate Remediation Group/USA) 4(4):315
Raber, Tim (Universität Oldenburg/Germany) 4(2):49
Radecki, Mike (U.S. Navy/USA) 4(4):433
Radway, Joann C. (Savannah River Technology Center/USA) 4(1):385; 4(4):133; 4(5):85
Raetz, Richard M. (Global Remediation Technologies, Inc./USA) 4(5):63
Rafalovich, Alex (Metcalf & Eddy Inc./USA) 4(5):649
Ragusa, Santo R. (CSIRO/Australia) 4(4):241
Rahming, Rory (Florida International University/USA) 4(2):595
Rajan, Raj V. (EFX Systems, Inc./USA) 4(3):451
Ramay, Michael J. (Delphinus Engineering/USA) 4(5):225
Ramos, Juan Luis (CSIC, Estacion Experimental del Zaidin/Spain) 4(4):57
Ramsay, Bruce A. (École Polytechnique de Montréal/Canada) 4(5):199
Randall, Andrew A. (University of Central Florida/USA) 4(5):77
Ranganathan, Gauri (Cook College/Rutgers University/USA) 4(3):187
Raphael, Thomas (Umweltberatung Dr. Raphael/Germany) 4(4):307
Raterman, Kevin (Amoco Production Co./USA) 4(1):73
Rathbone, Karrie (Kansas State University/USA) 4(3):305
Rathfelder, Klaus M. (University of Michigan/USA) 4(1):323
Ratz, John (Parsons Engineering Science Inc./USA) 4(1):341
Raumin, Jeff (Kleinfelder Inc./USA) 4(4):433; 4(5):445
Rawlin, S. (Medical University of South Carolina/USA) 4(4):79
Raymond, H. A. (Remediation Technologies, Inc. (RETEC)/USA) 4(2):113
Rayner, John L. (CSIRO/Australia) 4(4):241
Razo-Flores, Elías (Instituto Méxicano del Petroléo/Mexico) 4(2):31
Reardon, Kenneth F. (Colorado State University/USA) 4(2):319; 4(4):81; 4(4):513; 4(4):559; 4(5):89; 4(5):169; 4(5):181
Reddy, Krishna R. (University of Illinois at Chicago/USA) 4(1):165
Reed, Derek (Dudek & Associates, Inc./USA) 4(3):197
Reed, Gregory D. (University of Tennessee/USA) 4(4):317
Rege, Mahesh (Washington State University/USA) 4(3):399
Rehm, Bernd W. (RMT, Inc./USA) 4(3):267
Reinhard, Martin (Stanford University/USA) 4(5):9
Reis, Inês M.L. (Universidade Católica Portuguesa/Portugal) 4(5):1
Reisinger, H. James (Integrated Science & Technology Inc./USA) 4(1):321; 4(1):453; 4(4):393; 4(5):365
Renfro, Norman (Valero Refining Company/USA) 4(5):173
Reynolds, Charles M. (U.S. Army/USA) 4(1):297; 4(4):283
Rice, James A. (South Dakota State University/USA) 4(2):181
Richard, Ellen K. (Barr Engineering Co./USA) 4(2):173
Rickerson, Glenn (Georgia Power Co./USA) 4(1):461
Ricotta, Angela (REMTECH Environmental Services/USA) 4(5):489

Ridgway, Harry (Orange County Water District/USA) 4(5):563
Riedstra, Durk (Bion Overijssel B.V./The Netherlands) 4(5):565
Rifai, Hanadi S. (Rice University/USA) 4(3):179
Riffel, Allison (Northwestern University/ USA) 4(5):379
Rijnaarts, Huub H.M. (TNO-MEP/ Netherlands) 4(2):487; 4(3):155; 4(4):203
Ringelberg, David B. (University of Tennessee/USA) 4(4):117
Ríos-Leal, E. (Ctr for Advanced Studies & Research/Mexico) 4(3):273
Rishindran, T. (Polytechnic University/USA) 4(5):513
Ritter, Kevin J. (Applied Hydrology Associates Inc./USA) 4(3):233
Rittmann, Bruce E. (Northwestern University/USA) 4(5):379
Robbins, Jennifer (North Augusta/USA) 4(5):225
Roberge, François (École Polytechnique de Montréal/Canada) 4(2):453; 4(5):197
Robertiello, Andrea (Eniricerche S.p.A./ Italy) 4(4):131; 4(4):385
Roberts, David B. (Oregon State University) 4(4):239
Robinson, James D.F. (Rust Environmental/UK) 4(3):339
Rock, Steve (U.S. EPA/USA) 4(3):323
Rockne, Karl L. (University of Washington/USA) 4(4):377
Rodgers, David (U.S. Navy/USA) 4(5):439
Rodriguez, J. (DuPont/USA) 4(1):393
Rodríguez, J. (INTEVEP/Venezuela) 4(4):409
Rodríguez-Eaton, Susana (Clark Atlanta University /USA) 4(3):373
Roelke, Lynn A. (Texas A&M University/USA) 4(2):317
Roemmel, Janet S. (Secor International Inc./USA) 4(1):67
Rogers, Brock (The University of Arizona/USA) 4(4):567
Rogers, H. (Medical Univ of South Carolina/USA) 4(4):79
Rogers, J. D. (Colorado State University/USA) 4(4):81
Rogers, Louis C. (Texas Natural Resource Conservation Commission/USA) 4(4):347
Romich, Mark S. (Ayres Associates/USA) 4(5):589
Rooney, Daniel J. (Applied Research Associates, Inc./USA) 4(2):405
Roote, Diane S. (Fluor Daniel GTI, Inc./ USA) 4(3):439
Rosenthal, Nettie (Colorado Center for Environmental Mgt/USA) 4(4):323
Rosenwinkel, Paul A. (Resource Control/USA) 4(1):445
Ross, Nathalie (École Polytechnique de Montréal/Canada) 4(4):227
Ross, Randall R. (U.S. EPA/USA) 4(4):261
Roudier, Pascal (Antipollution Techniques Entreprise/France) 4(1):411
Rouse, Steve (Shell Western E&P/USA) 4(5):413
Roy, Denise V. (University of North Carolina at Chapel Hill/USA) 4(3):469
Rügge, Kirsten (Technical University of Denmark/Denmark) 4(2):15; 4(5):313
Rulkens, Wim H. (Wageningen Agricultural University/The Netherlands) 4(1):367; 4(5):643
Russell, S. (Oregon State University/USA) 4(3):93
Ruszaj, Martin (Occidental Chemical Corporation/USA) 4(3):321
Rutherford, Kyle W. (Fluor Daniel GTI/USA) 4(1):153; 4(1):211
Rykaczewski, Michael J. (Dames & Moore/USA) 4(2):303; 4(5):301
Ryoo, Doohyun (Jeonju University/ Republic of Korea) 4(2):589
Saberiyan, Amireh G. (NEEK Engineering/USA) 4(3):241; 4(1):283
Sacceddu, Pascale (Eniricerche S.p.A./ Italy) 4(4):385; 4(4):131
Sakai, Kokki (Kyushu University/Japan) 4(2):545
Salata, Gregory G. (Texas A&M University/USA) 4(2):317

Salcedo, M. (INTEVEP/Venezuela) 4(4):409
Salvo, Joseph J. (General Electric Company/USA) 4(3):281; 4(3):287
Samson, Réjean (École Polytechnique de Montréal/Canada) 4(2):129; 4(2):335; 4(2):453; 4(2):457; 4(4):221; 4(4):227; 4(4):535; 4(5):197
Sandefur, Craig A. (Regenesis Bioremediation Products/USA) 4(4):247
Santas, Photeinos (Oiko Technics Institute/Greece) 4(4):445; 4(4):451
Santas, Regas (Oiko Technics Institute/Greece) 4(4):445; 4(4):451
Santo Domingo, Jorge W. (Westinghouse Savannah River Co./USA) 4(1):385; 4(4):133; 4(5):85; 4(5):307
Sarouhan, Brian (Bechtel National, Inc./USA) 4(2):359
Sass, Bruce (Battelle/USA) 4(1):7; 4(1):331
Sattler, Donna L. (Louisana State University/USA) 4(5):115
Saunders, James A. (Auburn University/USA) 4(3):379; 4(5):583
Sawatsky, Norman (Alberta Environmental Center/Canada) 4(4):35
Sayler, Gary S. (The University of Tennessee/USA) 4(4):49; 4(4):77
Sayles, Gregory D. (U.S. EPA/USA) 4(1):353; 4(3):285; 4(3):299; 4(3):391
Scalzi, Michael (Sybron Chemicals, Inc./USA) 4(5):419
Schaad, David E. (Parsons Engineering Science, Inc./USA) 4(1):227
Schäfer, Wolfgang (University of Heidelberg/Germany) 4(5):555
Schaffner, I. Richard (GZA GeoEnvironmental, Inc./USA) 4(2):393
Schaller, Kastli D. (Idaho State University/USA) 4(4):483
Schanke, Craig (BioRenewal Technologies, Inc./USA) 4(3):267
Shaver, Don W. (U.S. Navy/USA) 4(1):275
Scheibner, Katrin (Friedrich-Schiller-Universität Jena/Germany) 4(2):493
Schlegl, M. (Institute for Agrobiotechnology/AUSTRIA) 4(5):617
Schlett, Wendy (Envirologic Technology Inc./USA) 4(5):475
Schmauder, Hans-Peter (Forschungszentrum fur Medizintechnik & Biotechnologie e.V./Germany) 4(5):259
Schmetzer, Michael J. (Harding Lawson Associates/USA) 4(1):295
Schmid, Henry (Oregon Department of Transportation/USA) 4(3):241
Schmieman, Eric (Washington State University/USA) 4(3):435
Schmitz, Richard J. (Long Island Lighting Company/USA) 4(3):445; 4(3):479
Schneider, Larry L. (Massar Environmental Technologies, Inc./USA) 4(4):335
Scholz-Muramatsu, Heidrun (University of Stuttgart/Germany) 4(3):13
Schraa, Gosse (Wageningen Agricultural University/The Netherlands) 4(5):87
Schreiber, Madeline E. (University of Wisconsin-Madison/USA) 4(5):13
Schrock, E.J. (Oklahoma State University/USA) 4(4):19
Schroth, Martin H. (Oregon State University/USA) 4(2):387
Schuring, John R. (New Jersey Institute of Technology/USA) 4(5):495; 4(5):519
Schwab, A. Paul (Kansas State University/USA) 4(3):305
Seagren, Eric A. (University of Maryland/USA) 4(5):379
Seaman, Mark (Fluor Daniel GTI/USA) 4(3):457
Segar, Jr., Robert L. (University of Missouri Columbia/USA) 4(2):375; 4(3):85; 4(3):127; 4(5):55
Seidel, Heinz (UFZ Centre for Environmental Research/Germany) 4(2):185
Sekerka, Patrick (IT Corporation/USA) 4(1):83
Seleznev, S. G. (Russian Academy of Sciences/Russia) 4(4):523
Selivanovskaya, Svetlana Yu. (Kazan State University/Russia) 4(5):231
Semprini, Lewis (Oregon State University/USA) 4(3):93; 4(3):107; 4(3):247
Senshu, Yuri (Raito Kogyo Co., Ltd./Japan) 4(4):495

Sepehrnooi, Kamy (University of Texas at Austin/USA) 4(5):597
Šepic, Ester (University of Ljubljana/Slovenia) 4(2):231
Sepúlveda-Torres, Lycely del C. (Michigan State University/USA) 4(3):33
Sera, Nobuyuki (Fukuoka Institute of Health & Env Sciences/Japan) 4(3):63
Setier, Jean Claude (ELF Aquitaine Production /France) 4(2):559
Severn, Shawn (PTI Environmental Services/USA) 4(1):111
Sewell, Guy W. (U.S. EPA/USA) 4(3):39; 4(4):261
Seybold, April L. (EFX Systems Inc./USA) 4(3):451
Shackelford, Charles D. (Colorado State University/USA) 4(4):559
Sharma, Pramod (Stanford University/ USA) 4(3):1
Sharman, Ajay (IBS Viridian Ltd./UK) 4(2):531; 4(2):533; 4(3):413
Sharp, Susan L. (BP Oil Co./USA) 4(5):43
Shea, Patrick J. (University of Nebraska-Lincoln/USA) 4(1):421; 4(2):287; 4(4):485
Shen, Chun-Fang (National Research Council of Canada/Canada) 4(2):567; 4(5):171
Shen, Y. (University of Calgary/Canada) 4(5):339
Sherhart, Thomas (Federal Aviation Administration/USA) 4(1):309
Sherwood, Juli L. (Washington State University/USA) 4(4):157
Shi, Y. (University of Wisconsin-Madison/ USA) 4(5):13
Shields, D. H. (University of Manitoba/ Canada) 4(1):319
Shikaze, S. G. (University of Waterloo/ Canada) 4(4):149
Shimizu, Toshio (Asahi Chemical Industry Co., Ltd./Japan) 4(4):59
Shimomura, Tatsuo (EBARA Research Co Ltd/Japan) 4(5):31
Shimura, Minoru (Railway Technical Research Inst/Japan) 4(2):445
Shin, Chi-Yon (University of Idaho/USA) 4(2):23
Shine, Gene (Westinghouse Savannah River Co./USA) 4(5):355
Shipley, Steven R. (MidAmerican Energy Co./USA) 4(3):487
Shoemaker, Stephen H. (DuPont Co./USA) 4(1):393; 4(4):315
Short, Amy (3M/USA) 4(1):111
Sibbett, Bruce (IT Corp./USA) 4(3):227
Sick, Marc (IEG Technologies Corp./ Germany) 4(4):149
Sieglen, U. (Fraunhofer IGB/Germany) 4(2):1
Siegrist, Robert (Colorado School of Mines/USA) 4(4):117
Sikora, Frank J. (Tennessee Valley Authority/USA) 4(3):315
Simmons, Paul (University College Dublin/Ireland) 4(5):245
Simpkin, Thomas J. (CH2M Hill Inc./USA) 4(2):73
Sims, C. L. (Integrated Science & Technology Inc./USA) 4(1):453
Sims, Ronald C. (Utah State University/ USA) 4(2):181
Sinclair, James L. (Mantech Experimental Research Srvcs Corp./USA) 4(1):15
Singh, J. (University of Nebraska-Lincoln/ USA) 4(2):287
Singh, Manjari (DuPont Co./USA) 4(4):17
Singleton, Ian (University of Adelaide/ Australia) 4(5):245
Sinha, Rajib (IT Corporation/USA) 4(5):251
Sinnenberg, Steven (American Biotech Corporation/USA) 4(5):451
Sipkema, E. Marijn (University of Groningen/The Netherlands) 4(3):73
Sisk, Wayne E. (U.S. Army Environmental Center/USA) 4(5):151
Sisson, James (Idaho National Engineering Laboratory/USA) 4(5):105
Skeen, Rodney S. (Battelle-Pacific Northwest National Laboratory/USA) 4(3):9; 4(3):49; 4(3):57; 4(4):157; 4(5):411; 4(5):561
Smalley, J. Bryan (University of Illinois Urbana-Champaign/USA) 4(4):353
Smallwood, David S. (David S. Smallwood & Assoc., Inc./USA) 4(1):413

Smart, Ross (Chevron Research and Technology/USA) 4(5):145
Smith, Barrett L. (formerly U.S. Geological Survey/USA) 4(3):207
Smith, John R. (Aluminum Co. of America (ALCOA)/USA) 4(2):159
Smith, Michael J. (Camp Dresser & McKee/USA) 4(4):513
Smith, Tom S. (Ciba-Geigy Corporation/ USA) 4(5):641
Smith, Val H. (University of Kansas/USA) 4(2):117; 4(4):105
Smyth, David J.A. (University of Waterloo/Canada) 4(4):209; 4(4):149
Soares, Amilcar (Instituto Superior Téchnico/Portugal) 4(5):349
Sobczak, Kelly (CH2M Hill/USA) 4(2):73
Solek, Susan (DuPont Co./USA) 4(3):21
Sommer, Robert (George Air Force Base/USA) 4(1):83
Song, Chang Soo (Samsung Corporation/ Republic of Korea) 4(2):233
Song, Ki S. (Samsung Corporation/ Republic of Korea) 4(2):233
Song, Kuang-Chung (National Chung Hsing University/R.O.C.-Taiwan) 4(2):433
Song, Qi (University of Cincinnati/USA) 4(2):195
Soni, Bhupendra (Institute of Gas Technology/USA) 4(2):439; 4(3):477
Sonier, Dyane (Washington State University/USA) 4(2):343
Sorensen, Darwin L. (Utah State University/USA) 4(1):395; 4(3):189; 4(3):317; 4(4):9
Sorensen, James University of North Dakota-Grand Forks/USA) 4(5):269
Spadaro, Jack T. (AGRA Earth & Environmental Inc./USA) 4(3):241
Spargo, B. J. (Naval Research Laboratory/ USA) 4(2):399; 4(4):125
Speitel, Gerald E. (University of Texas at Austin/USA) 4(2):565; 4(3):79; 4(5):597
Spence, Stephen G. (Hercules Inc./USA) 4(1):179
Spieles, Patrick (USPCI/Laidlaw Environmental Services, Inc./USA) 4(1):473
Spivack, J. L. (GE Corporate R&D Center/USA) 4(3):45; 4(3):297
Spivak, Jay (General Electric/USA) 4(3):289
Spivey, John (BDM Management Services Co./USA) 4(1):287
Springael, Dirk (VITO/Flemish Inst for Technol Research/Belgium) 4(4):159
Spuij, Frank (TAUW Milieu/The Netherlands) 4(5):431; 4(5):591
Srivastava, Vipul J. (Inst. of Gas Technol./ USA) 4(3):477; 4(2):439; 4(3):487
Stahl, David A. (Northwestern University/USA) 4(5):325; 4(5):379
Stamm, Jurgen (University of Karlsrühe/ Germany) 4(4):165
Stams, Alfons J.M. (Wageningen Agricultural University/Netherlands) 4(5):87
Stancel, Steven G. (Applied Hydrology Associates Inc./USA) 4(3):233
Stang, Keith M. (Fluor Daniel GTI, Inc./USA) 4(3):471
Starreveld, Malcolm (Ministry of Defense/UK) 4(1):441
Staton, M. A. (Mycotech Corp./USA) 4(2):537
Stauffer, Thomas (U.S. Air Force/USA) 4(1):23
Steed, Vicki S. (University of Cincinnati/USA) 4(3):391
Steele, Scott (General Electric CRD/USA) 4(3):281
Stefanoff, James (CH2M Hill Inc./USA) 4(1):455
Steffan, Robert J. (Envirogen, Inc./USA) 4(4):493
Stegmann, Rainer (TU Hamburg-Harburg/ Germany) 4(5):137
Stehmeier, Lester G. (University of Calgary/Canada) 4(5):339; 4(5):529
Stein, Carol (PTI Enviromental Services/ USA) 4(1):111
Steinbach, Klaus (Institute of Immunology & Environ Hygiene/Germany) 4(2):9
Steiof, Martin (Technical University of Berlin/Germany) 4(5):459
Stensel, H. David (University of Washington/USA) 4(4):377; 4(4):425; 4(5):37; 4(5):95

Stepan, Daniel J. (University of North Dakota/USA) 4(5):113
Stephens, M L (GE Corporate R&D Center/USA) 4(3):45
Steward, Charles (University of Tennessee/USA) 4(4):77
Stewart, Brian (Union Pacific Railroad/USA) 4(4):549
Stoelting, Ray (Amoco Corporation/USA) 4(1):103
Stoffers, N. (Oregon State University/USA) 4(3):93
Stokley, Karen E. (Exxon Research & Engineering Co./USA) 4(4):487
Stolpmann, Holger H. R. (W.L. Gore & Associates GmbH/Germany) 4(2):341
Stolte, L. Miller (Imation/USA) 4(1):111
Stoltzfus, Donn (City of Phoenix/USA) 4(1):211
Stolyarova, Anna V. (All-Russian Oil & Gas Research Institute (VNIIneft)/Russia) 4(4):529
Stone, Peter (S.C. Dept of Health & Environmental Control/USA) 4(2):315
Stoyanov, Jivko V. (Technical University of Denmark/Denmark) 4(4):1
Strand, Stuart E. (University of Washington/USA) 4(3):321; 4(4):377; 4(4):425
Strauss, Robin (CH2M Hill/USA) 4(2):279
Stromberg, M. (Remediation Technologies Inc./USA) 4(2):301
Strong-Gunderson, Janet (Oak Ridge National Laboratory/USA) 4(5):347
Stroo, Hans F. (Remediation Technologies, Inc. (RETEC)/USA) 4(2):55; 4(2):113; 4(2):301
Strzempka, Christopher P. (OHM Remediation Services Corp./USA) 4(1):245
Stuart, Sheryl L. (Oregon State University/USA) 4(2):475
Studer, James Edward (INTERA, Inc./USA) 4(2):345
Sturman, Paul J. (Montana State University/USA) 4(5):385
Su, Benjamin Y. (GEI Consultants Inc./USA) 4(3):203
Sublette, Kerry L. (University of Tulsa/USA) 4(1):73; 4(4):383
Subramanian, T. V. (Anna University-Madras/INDIA) 4(5):257
Sudicky, Edward (University of Waterloo/Canada) 4(4):149
Sudirgio, Vivien (Washington State University/USA) 4(4):97
Sugawa, Etsuko (Canon, Inc./Japan) 4(4):495
Suidan, Makram T. (University of Cincinnati/USA) 4(1):353; 4(2):47; 4(3):391; 4(4):267; 4(5):205; 4(5):663
Sukesan, Suma (Idaho State University/USA) 4(5):183
Sun, Yun Wei (Washington State University/USA) 4(1):37
Sunahara, Geoffrey (National Research Council of Canada/Canada) 4(2):3; 4(5):393
Surmacz-Górska, Joanna (Technical University of Silesia/Poland) 4(4):43
Swannell, Richard P.J. (AEA Technology/UK) 4(4):401
Swanson, Matthew A. (Parsons Engineering Science, Inc./USA) 4(3):147; 4(3):191
Sweed, H.G. (Rice University/USA) 4(5):11
Swider, Kenneth (Quest Environmental, Inc./USA) 4(5):419
Swindoll, C. Michael (DuPont/USA) 4(2):137; 4(3):353; 4(5):399
Szafranski, Michael J. (Michigan State University/USA) 4(5):657
Taat, Jan (Delft Geotechnics/Netherlands) 4(1):329; 4(3):155; 4(3):239
Tabak, Henry H. (U.S. EPA/USA) 4(2):195
Tabe, Margaret Egbe (Aluminum Co. of America (ALCOA)/USA) 4(2):159
Taffinder, Sam (U.S. Air Force/USA) 4(1):95
Tagami, Shiro (Fukuoka Institute of Health & Env Sciences/Japan) 4(3):63
Takada, Satoshi (Fukuoka Institute of Health & Env Sciences/Japan) 4(2):545

Takami, Wako (Asahi Chemical Industry Co., Ltd./Japan) 4(4):59
Tamburini, Davide (Battelle Europe/ Switzerland) 4(1):467
Taseli, Basak (Kilic) (Middle East Technical University/Turkey) 4(2):519
Tatara, Gregory M. (The Traverse Group Inc./USA) 4(3):59; 4(3):219
Tatem, Henry E. (U.S. Army Corps of Engineers/USA) 4(4):65
Tebo, Bradley M. (University of California/USA) 4(3):411
Tedaldi, Dante (Bechtel National Inc./USA) 4(2):359
Teets, David B. (Parsons Engineering Science Inc./USA) 4(1):341
Templeton, Alexis (Lawrence Berkeley Laboratory/USA) 4(2):299
Tenente, Angela (Oiko Technics Institute/Greece) 4(4):445; 4(4):451
TerKonda, Purush (University of Missouri - Rolla/USA) 4(5):211
Terry, N. (U.S. Dept of Agriculture/USA) 4(3):303
Tester, Al E. (Global Remediation Technologies, Inc./USA) 4(5):63
Tett, Vanessa (University of Kent at Canterbury/UK) 4(2):531; 4(2):533
Theoret, Dennis R. (O'Brien & Gere Engineers Inc./USA) 4(1):215
Thevanayagam, S. (State University of New York at Buffalo/USA) 4(5):513
Thevanayagam, V. (Polytechnic University/USA) 4(5):513
Thiboutot, Sonia (Defense Research Establishment Valcartier/Canada) 4(2):3; 4(2):567; 4(5):171; 4(5):393
Thirumirthi, Dan (Technical University of Nova Scotia/Canada) 4(1):277
Thomas, Alison (U.S. Air Force/USA) 4(2):407; 4(5):11
Thomas, J. M. (Rice University/USA) 4(5):11
Thomas, Mark (IT Corporation/USA) 4(1):83
Thomas, Mark H. (EG&G, TSWV Inc./USA) 4(1):173
Thomas, R.C. (University of Georgia/ USA) 4(3):379
Thompson, I.P. (Institute of Virology & Environmental Microbiology/UK) 4(3):413; 4(4):155
Thompson, Keith S. (Maxim Technologies, Inc./USA) 4(1):503
Thomson, Bruce M. (University of New Mexico/USA) 4(4):195
Thomson, James A. M. (Applied Hydrology Associates, Inc./USA) 4(3):233
Thorn, Patti (BioTechnical Services/USA) 4(2):271
Thullner, Martin (University of Heidelberg/Germany) 4(5):555
Tiedje, James M. (Michigan State University/USA) 4(4):501; 4(4):507
Till, Brian A. (The University of Iowa/USA) 4(3):367
Timian, Steve (Applied Research Associates Inc./USA) 4(2):405
Timmermann, D. (Technische Universität Hamburg-Harburg/Germany) 4(5):137
Tinholt, Mark (Morrow Environmental Consultants, Inc./Canada) 4(4):29
Tinoco, R. (Universidad Nacional Autonoma de Mexico) 4(2):225
Tischuk, Michael D. (Beazer East, Inc./USA) 4(4):149
Todd, Timothy (Kansas State University/USA) 4(4):383
Tokunaga, Takashi (Fukuoka Institute of Health & Env Sciences/Japan) 4(3):63
Tonnaer, Haimo (Tauw Milieu b.v./The Netherlands) 4(5):591
Torrents, Alba (University of Maryland/ USA) 4(2):263
Torres, Eduardo (Universidad Nacional Autonoma de Mexico) 4(2):225
Torres, Luis Gilberto (Universidad Nacional Autonoma de Mexico) 4(2):447
Touge, Yoshiyuki (Canon, Inc./Japan) 4(4):495
Tovanabootr, Adisorn (Oregon State University/USA) 4(3):93
Toze, Simon G. (CSIRO Div of Water Resources/Australia) 4(2):53
Tozzolino, Pierre (Elf Petroleum Norge/ Elf Akvamilj/Norway) 4(4):411

Trefry, Michael G. (CSIRO/Australia) 4(4):241
Trent, Gary L. (Amoco Production Co./USA) 4(1):73
Trinidad, R. (Universidad A. Metropolitant-Iztapalapa/Mexico) 4(5):223
Trobaugh, Darin J. (Oregon State University/USA) 4(2):413
Trovato, Antonino (Battelle Europe/ Switzerland) 4(1):505
Trowbridge, Bretton E. (Terra Vac, Inc./USA) 4(5):505
Troy, Marleen A. (DuPont Environmental Remediation Services/USA) 4(5):399
Truex, Michael J. (Battelle-Pacific Northwest National Laboratory/USA) 4(3):9; 4(4):273; 4(5):325
Trust, Beth A. (Finnigan Inc./USA) 4(2):317
Tsang, Yvonne (Lawrence Berkeley Laboratory/USA) 4(1):81
Tucker, Mark David (Sandia National Laboratories/USA) 4(4):195
Tuckfield, R. Cary (Westinghouse Savannah River Technology Ctr/USA) 4(5):355
Tuomi, Elona (Remediation Technologies, Inc./USA) 4(2):55
Turick, Charles E. (Idaho National Engineering Lab/USA) 4(3):435
Tyler, Tony (NSW Department of Agriculture/Australia) 4(2):257
Tyner, Larry (IT Corp./USA) 4(1):83
Uberoi, Vikas (Rutgers - The State University of New Jersey/USA) 4(5):277
Uchiyama, Hiroo (National Institute for Env Studies/Japan) 4(5):31
Underhill, Scott A. (Bechtel Environmental, Inc./USA) 4(1):361
Ünlü, Karaman (Middle East Technical University/Turkey) 4(2):463
Uraizee, Farooq (U.S. EPA NRMRL/ USA) 4(5):663
Valo, Risto (Soil and Water, Ltd./ Finland) 4(2):91
Valocchi, Albert J. (University of Illinois at Urbana-Champaign/USA) 4(5):571
Valtere, Sarma (Riga Technical University/Latvia) 4(2):149
van Aalst, Martine (TNO-MEP/The Netherlands) 4(3):155; 4(4):203
van Andel, Johan G. (RIVM/LAE/The Netherlands) 4(5):643
Van Benthem, Mark (U.S. Air Force/USA) 4(1):253; 4(1):259
van Dam, Theo (University of Twente/The Netherlands) 4(5):565
van de Akker, J.J.H. (DLO-Winand Staring Centre/The Netherlands) 4(2):153
Van Deinse, Harold (ENSR Consulting & Engineering/USA) 4(1):497
van den Beld, Henk (University of Twente/The Netherlands) 4(5):565
van den Boogaart, J. (Delft Geotechnics/ The Netherlands) 4(3):239
van den Brink, Karin (TAUW Milieu/The Netherlands) 4(5):431
van den Toorn, A. (DLO Winand Staring Centre/The Netherlands) 4(2):143; 4(2):153
Vanderglas, Brian R. (Parsons Engineering Science, Inc./USA) 4(1):95
van der Marel, N. (Bioclear Environmental Biotechnology b.v./Netherlands) 4(3):65
van der Waall, R. (Dirkse Milieutechnik b.v./The Netherlands) 4(3):65
van der Waarde, Jaap J. (Bioclear Environmental Biotechnology b.v./The Netherlands) 4(3):65; 4(5):29
van Dijk-Hooyer, O.M. (DLO Winand Staring Centre/The Netherlands) 4(2):143; 4(2):153
van Eekert, Miriam H. A. (Wageningen Agricultural University/The Netherlands) 4(5):87
van Eyk, Jack (Van Eyk Environmental Consultants/The Netherlands) 4(1):329
Van Houtven, D. (Vlaamse Instelling voor Tech Onderzoek (VITO)/Belgium) 4(4):159
van Hylckama Vlieg, J. E. T. (University of Groningen/Netherlands) 4(3):73
Vanneck, Peter (Seghers Engineering Water/Belgium) 4(2):211

van Ree, C.C.D.F. (Derk) (Delft Geotechnics/Netherlands) 4(3):239
Van Rijn, Karen (City of Tucson/USA) 4(4):289
Van Zwieten, Lukas (NSW Department of Agriculture/Australia) 4(2):257
Vardy, James A. (U.S. Coast Guard/USA) 4(3):181; 4(2):309
Vargas, Gary (Dudek & Associates, Inc./USA) 4(3):197
Vasiliev, Aleksej N. (Inst. of Bioorg Chem & Petrochem/Ukraine) 4(3):325; 4(3):407
Vassar, Tyler M. (OHM Remediation Services Corp./USA) 4(1):245
Vaughn, Curtis (Roy F. Weston, Inc./USA) 4(5):507
Vazquez-Duhalt, Rafael (Universidad Nacional Autonoma de Mexico) 4(2):225
Veenis, Yvo M. M. (Fluor Daniel GTI/The Netherlands) 4(2):371; 4(2):373
Vemuri, Ramu (Roy F. Weston, Inc./USA) 4(2):251
Venkatraman, Sankar N. (McLaren/Hart Environmental Engrg Corp./USA) 4(5):495; 4(5):519
Venosa, Albert (U.S. EPA/USA) 4(2):47; 4(4):267; 4(4):359; 4(4):419; 4(5):663
Verce, Matthew F. (Umversity of Illinois/ USA) 4(3):255; 4(5):1
Vernalia, Jane (Woodward-Clyde Consultants/USA) 4(2):469
Verstraete, Willy (University of Ghent/ Belgium) 4(2):211
Verzaro, Francis (Institut Français du Pétrole/France) 4(2):559
Vesper, Stephen J. (University of Cincinnati/USA) 4(3):101; 4(3):385
Vetter, Karl R. (Dames & Moore/USA) 4(5):107
VialeRigo, M. (INTEVEP/Venezuela) 4(4):409
Vivek, S. A. (University of Missouri Columbia/USA) 4(3):85
Vogel, Catherine (U.S. Air Force/USA) 4(1):129; 4(1):347; 4(3):39; 4(3):285; 4(3):299; 4(5):481
Vogel, Timothy M. (Rhône-Poulenc Industrialisation/France) 4(1):411
Voice, Thomas C. (Michigan State Univ/ USA) 4(3):361; 4(4):507; 4(5):657
Volpi, Richard W. (Parsons Engineering Science/USA) 4(1):227
von Fahnestock, F. Michael (Battelle/ USA) 4(1):511; 4(2):67; 4(1):479
von Löw, Eberhard (Institute of Immunology & Environmental Hygiene/ Germany) 4(2):9
Voordouw, G. (University of Calgary/Canada) 4(5):339
Waddill, Dan W. (Virginia Polytechnic Inst & State Univ/USA) 4(1):35; 4(5):553
Wagner, Dan (EFX Systems Inc./USA) 4(5):251
Walden, Ronald L. (CNG Transmission Corporation/USA) 4(5):363
Walker, Amy (Naval Facilities Engineering Service Center/USA) 4(1):511
Wallace, Mark N. (U.S. Army Corps of Engineers/USA) 4(1):61; 4(1):295
Walsh, Matthew (Envirogen Inc./USA) 4(4):493
Walter, U. (UMWELTSCHUTZ NORD GmbH & Co./Germany) 4(2):1
Walters, Glenn W. (University of North Carolina/USA) 4(2):289
Waltz, Michael (REMTECH Services Group/USA) 4(5):489
Walworth, James L. (University of Alaska Fairbanks/USA) 4(4):283; 4(1):397; 4(1):297
Wang, H. (Rice University/USA) 4(3):179
Wang, K. G. (Oregon State University/ USA) 4(2):279
Ward, C.H. (Herb) (Rice University/USA) 4(4):565; 4(5):11
Wardwell, David (Mission Research Corporation/USA) 4(1):253
Warikoo, Veena (Stanford University/ USA) 4(3):1
Warrelmann, Juergen (Umweltschutz Nord GmbH & Co./Germany) 4(2):1
Warren, Ean (U.S. Geological Survey/ USA) 4(5):547

Washburn, Fatina A. (Westinghouse Savannah River Company/USA) 4(5):275
Watanabe, Akira (Raito Kogyo Co., Ltd./Japan) 4(4):495
Watkins, Todd (Southern Co. Services Inc./USA) 4(1):461
Watts, Richard J. (Washington State University/USA) 4(2):525
Watwood, Mary E. (Idaho State University/USA) 4(2):291; 4(4):111; 4(4):483; 4(5):183; 4(5):327;
Weathers, Lenly J. (University of Maine/USA) 4(3):367
Weaver, James (U.S. EPA/USA) 4(1):97
Weber, Walter J. (The University of Michigan/USA) 4(5):611; 4(5):635
Webster, Matthew (University of Texas at Austin/USA) 4(5):603
Webster, Todd S. (University of California-Riverside/USA) 4(5):219
Weesner, Brent (Lockheed Martin Specialty Components/USA) 4(4):261
Weigand, M. Alexandra (IT Corporation/USA) 4(3):221
Weijling, Annemieke (TNO Inst. of Environ. Sci./Netherlands) 4(3):3
Weise, Andrea (Fisheries and Oceans Canada/Canada) 4(4):365
Wells, Harrington (University of Tulsa/ USA) 4(4):383
Wells, John H. (University of Idaho/USA) 4(2):525
Wen, Lian-Kai (University of Petroleum/ China-P.R.C.) 4(4):337
Wenaas, Chris (Texas A&M University/ USA) 4(5):49
Wendling, Gilles (Morrow Environmental Consultants, Inc./Canada) 4(4):29
Wenzell, Lisbeth (Kruger A/S/Denmark) 4(5):57
Werner, Peter (Technische Universität Dresden/Germany) 4(5):7; 4(5):97
West, Robert J. (Dow Chemical Co./USA) 4(3):283; 4(3):287
Westerterp, Klaas Roel (University of Twente/The Netherlands) 4(5):565
Westervelt, W. Winslow (CH2M Hill/USA) 4(1):61
Weststrate, F. A. (Delft Geotechnics/The Netherlands) 4(3):239
Wetzstein, Doug W. (Minnesota Pollution Control Agency/USA) 4(5):283
Wheeler, Mark (U.S. Air Force/USA) 4(3):141
Whitaker, M.J. (Westinghouse Savannah River Co./USA) 4(4):133
White, David C. (University of Tennessee/USA) 4(5):319
White, Roger (Westinghouse Savannah River Company/USA) 4(3):345
White, Thomas (Ciba/USA) 4(2):289
Whiteside, Steven E. (Shell Offshore Inc./USA) 4(4):439
Whittaker, Harry (Environment Canada/Canada) 4(2):149
Wickramanayake, Godage B. (Battelle/USA) 4(1):479; 4(1):511
Widdowson, Mark A. (Virginia Polytechnic Inst & State University/USA) 4(1):35; 4(5):553
Wieck, James M. (GZA GeoEnvironmental Inc./USA) 4(2):393
Wiedemeier, Todd H. (Parsons Engineering Science, Inc./USA) 4(3):141; 4(3):147; 4(3):191
Wieggers, H. J. J. (DLO Winand Staring Centre/The Netherlands) 4(2):153
Wiesner, Mark (Rice University/USA) 4(4):565; 4(5):11
Wiggert, David C. (Michigan State University/USA) 4(4):507; 4(4):575
Wijjfels, Piet (Apinor/France) 4(2):79
Wikström, Per B. (FOA/National Defence Research Establishment/Sweden) 4(5):333
Wilde, Edward W. (Westinghouse Savannah River Co./USA) 4(4):133; 4(5):85
Wildeman, Thomas (Colorado School of Mines/USA) 4(3):401
Williams, Steve E. (U.S. Air Force /USA) 4(5):11
Williamson, Derek (University of Texas Austin/USA) 4(5):605
Williamson, Kenneth J. (Oregon State University/USA) 4(2):41; 4(2):279; 4(2):455; 4(4):239

Willis, Alyson (University of Sunderland/ UK) 4(4):401
Willis, Guy D. (EA Engineering, Science & Technology/USA) 4(5):11
Willson, Elizabeth H. (University of Alaska Fairbanks/USA 4(1):397
Willumsen, Pia Arentsen (National Environmental Research Institute/ Denmark) 4(2):249
Wilson, Ashley J. (University of York/UK) 4(4):167
Wilson, Barbara H. (U.S. EPA/USA) 4(3):181
Wilson, David J. (Eckenfelder, Inc./USA) 4(1):147; 4(4):175
Wilson, John T. (U.S. EPA/USA) 4(1):15; 4(1):97; 4(1):347; 4(2):309; 4(3):147; 4(3):181; 4(3):191
Wilson, Ryan D. (University of Waterloo/Canada) 4(4):187; 4(4):209
Wilson, Timothy P. (U.S. Geological Survey/USA) 4(3):115
Wimpee, M. (Microbial Insights Inc./USA) 4(5):319
Windfuhr, Claudia (University of Stuttgart/Germany) 4(3):13
Winters, Alec T. (EA Engineering Science & Tech Inc./USA) 4(1):251
Witt, Michael E. (Michigan State University/USA) 4(3):59; 4(4):575
Woertz, Jennifer (Clemson University/ USA) 4(5):1
Wolf, Jerry (Terra Vac, Inc./USA) 4(5):505
Wolf, Lorraine W. (Auburn University/ USA) 4(3):379; 4(5):583
Wolfe, N. Lee (U.S. EPA/USA) 4(3):301
Wolfram, James H. (Idaho National Engineering Laboratory/USA) 4(1):435
Wollenberg, John (U.S. Navy/USA) 4(1):205
Woo, Seung Han (Pohang University of Science & Technology/Republic of Korea) 4(2):233
Wood, Kenneth N. (DuPont Co./USA) 4(2):137
Wood, Terri M. (Texas A&M University/ USA) 4(4):119
Woodhull, Patrick M. (OHM Remediation Services Corp./USA) 4(1):245; 4(1):447; 4(2):365; 4(2):603; 4(4):141
Woods, Sandra L. (Oregon State University/USA) 4(2):41; 4(2):279; 4(2):413; 4(2):455; 4(2):475; 4(3):47; 4(4):239
Woodward, David S. (Rust Environment & Infrastructure, Inc./USA) 4(5):43; 4(1):379
Woolard, Craig (University of Alaska/ USA) 4(1):397; 4(4):283
Workman, Darla J. (Oregon State University/USA) 4(2):413; 4(3):47
Worsztynowicz, Adam (Inst. for Ecology of Industrial Areas/ Poland) 4(5):467
Wrenn, Brian A. (Environmental Technologies & Solutions, Inc./USA) 4(4):267
Wright, Chris F. (GZA GeoEnvironmental Inc./USA) 4(2):393
Wright, Curtis S. (Fluor Daniel GTI/USA) 4(1):211; 4(4):289
Wu, J. L. (DuPont/USA) 4(3):353
Wu, M. H. (University of Nebraska - Lincoln/USA) 4(3):421
Wu, Nerissa T. (E&A Environmental Consultants, Inc./USA) 4(2):71
Xing, Jian (Global Remediation Technologies, Inc./USA) 4(5):63
Xu, Bing (Clark Atlanta University/USA) 4(2):57
Xun, Luying (Washington State University/USA) 4(3):437
Yagi, Osami (National Institute for Env Studies/Japan) 4(5):31
Yakovela, G. Yu. (Kazan State University/ Russia) 4(2):33
Yakushijin, Fumiko (ABB Environmental Services Inc./USA) 4(2):271
Yamane, Hisakazu (University of Tokyo/Japan) 4(4):59
Yamashita, Masami (Toyota Motor Corporation/Japan) 4(3):133
Yang, Jen-Rong (Jackson State University/USA) 4(4):71
Yang, Wen-Hsun (Jackson State University/USA) 4(4):71
Yang, Xiaoping (Amoco Corporation/ USA) 4(1):103
Yang, Yujing (U.S. EPA/USA) 4(2):249

Yano, Tetsuya (Canon, Inc./Japan) 4(4):495
Yeom, Ick Tae (Korean Institute of Sci & Tech/Republic of Korea) 4(2):575
Yergovich, Tom (J.R. Simplot Co./USA) 4(2):61; 4(2):265
Yerushalmi, Laleh (National Research Council of Canada/Canada) 4(4):197
Yonamine, E.K. (Universidade Federal de São Carlos/Brazil) 4(2):247
Yonge, David R. (Washington State University/USA) 4(2):343; 4(3):399; 4(3):435; 4(4):97
Yoo, Hee Chan (Daewoo Institute of Construction Technology/Republic of Korea) 4(4):477
Yoon, Byung-Dae (Korea Research Inst of Bioscience & Biotech/Republic of Korea) 4(2):467
Yoon, Je Yong (Ajou University/Republic of Korea) 4(4):477
Yoon, Woong-Sang (Battelle/USA) 4(1):511
Yoshitani, Jun (EME Environmental, Inc./USA) 4(4):521
Yost, Eric C. (Barr Engineering Co./USA) 4(2):173
Young, Jill D. (Westinghouse Savannah River Technology Center/USA) 4(5):315
Young, Riki G. (Cornell University/USA) 4(3):61
Yu, George (Battelle/USA) 4(2):321
Zachary, Scott (Metcalf & Eddy Inc./USA) 4(5):373
Zappi, Mark (Mississippi State University/USA) 4(5):157
Zappia, Luke (CSIRO Div of Water Resources/Australia) 4(2):53
Zarina, Dzidra (University of Latvia/ Latvia) 4(2):149
Zaripova, Sania K. (Kazan State University/Russia) 4(2):33
Zettler, Berthold (TU Berlin/Germany) 4(2):273
Zhang, Tian C. (University of Nebraska - Lincoln/USA) 4(3):423; 4(1):421
Zhang, Wei-xian (Lehigh University/USA) 4(5):387
Zhang, Yanfang (Texas A&M University/ USA) 4(5):49
Zhao, Xianda (Michigan State University/USA) 4(3):361; 4(5):657
Zheng, Yuan-Yang (University of Petroleum/China-P.R.C.) 4(4):337
Zhou, Jizhong (Oak Ridge National Laboratory/USA) 4(5):347
Zhu, Xueqing (University of Cincinnati/ USA) 4(5):205
Zilber, Barbara (National Research Council of Canada/Canada) 4(5):393
Zimbron, J. (Colorado State University/ USA) 4(5):181
Zimmerman, Christian T. (Battelle/USA) 4(1):331
Zinder, Stephen H. (Cornell University/ USA) 4(3):11; 4(3):23; 4(3):39
Zingmark, R. (University of South Carolina/USA) 4(4):133
Zotzky, L. K. (Montana State University/ USA) 4(5):385
Zou, Siwei (University of Washington/ USA) 4(4):551
Zumwalt, Gary (Louisiana Technical University/USA) 4(1):141
Zweers, A. J. (DLO Winand Staring Centre/The Netherlands) 4(2):153
Zwick, Thomas C. (Battelle/USA) 4(1):7; 4(1):331; 4(1):403
Zwolinsky, M. (University of Wisconsin-Madison/USA) 4(5):13